9th International Symposium on Display Holography

(ISDH 2012)

Journal of Physics: Conference Series Volume 415

Cambridge, Massachusetts, USA
25-29 June 2012

ISBN: 978-1-62276-993-3
ISSN: 1742-6588

Printed from e-media with permission by:

Curran Associates, Inc.
57 Morehouse Lane
Red Hook, NY 12571

Some format issues inherent in the e-media version may also appear in this print version.

Printed by Curran Associates, Inc. (2013)

Institute of Physics
Dirac House, Temple Back
Bristol BS1 6BE UK

Phone: 44 1 17 929 7481
Fax: 44 1 17 920 0979

techtracking@iop.org

Additional copies of this publication are available from:

Curran Associates, Inc.
57 Morehouse Lane
Red Hook, NY 12571 USA
Phone: 845-758-0400
Fax: 845-758-2634
Email: curran@proceedings.com
Web: www.proceedings.com

TABLE OF CONTENTS

ART AND HOLOGRAPHY

TECHNIQUES AND MATERIALS

DIGITAL TECHNIQUES

Introduction to the Proceedings of the 9th ISDH

V. Michael Bove, Jr. and Seth Riskin

MIT Media Laboratory, 20 Ames St., Room E15-448, Cambridge MA 02139 USA;
MIT Museum, 265 Massachusetts Avenue, Building N51, Cambridge MA 02139 USA

vmb@media.mit.edu, riskin@media.mit.edu

1. The Proceedings

As co-chairs of the 9th International Symposium on Display Holography, we welcome readers of this collection of papers and posters presented at the event. We hope that both attendees of the event and others pursuing the art, science, and business of holography and 3D imaging will find the authors' contributions of lasting interest and importance.

2. The Event

Since its creation at Lake Forest College in 1982 by Prof. Tung H. Jeong, ISDH has followed a model that differentiates it from other scientific conferences. The 9th ISDH continued this history, fully occupying a floor of the MIT Media Lab for five days. The single-track conference opened with reports on the state of holography in the various nations represented by the attendees, followed by a series of presentations on Education and Holography. One and one-half days of papers on Art and Holography followed, then sessions on Techniques and Materials, Digital Techniques, and Commercial and Applied Holography. A poster session permitted more in-depth discussion between authors and the audience. Two exhibitions of holographic works opened at ISDH: an informal display area at the symposium, and a 15-month-long MIT Museum exhibition, *The Jeweled Net: Views of Contemporary Holography*.

The success of an event of this sort requires the help of many people and organizations. We wish especially to recognize our Honorary Conference Chairs: Tung H. Jeong and Joseph W. Goodman; our Technical Program Committee: Hans I. Bjelkhagen, Frank Fan, Nasser Peyghambarian, and Hiroshi Yoshikawa; and our Arts and Exhibition Committee: Betsy Connors-Chen, Melissa Crenshaw, John Durant, Dieter Jung, Linda Law, Martin Richardson, Jonathan Ross, and Sally Weber. Betsy also coordinated the on-site exhibition. Kristin Hall at the MIT Media Lab made local arrangements, while registration was handled by MIT Conference Services.

We also gratefully acknowledge support from Lake Forest College, holographer.org, and authentibrand.com, and endorsement by the Optical Society of America.

Finally, and most importantly, we thank the authors for contributing the papers that follow.

Figure 1: Attendees of 9th ISDH at MIT Media Lab.

Figure 2: Presentation session at ISDH.

Figure 3: Poster session at ISDH.

Figure 4: Informal exhibition space at ISDH.

Figure 5: Exhibition at MIT Museum.

Australia

P Dawson,

The College of Fine Arts, University of New South Wales, Paddington, NSW, 2021, Australia

I am going to report briefly in 3 areas - artists, galleries and courses

1. Artists

Margaret Benyon's work continues to be an important influence. An early laser transmission work has been shown recently at a group exhibition on the theme of the environment at Macquarie University Art Gallery. Margaret has donated *White Rainbow*, *Tiresias* and *Counting the Beats* to the Newcastle Gallery and five more of Margaret's works have joined the other two pieces previously acquired by the National Gallery of Australia. Congratulations to Martina Mongrovious on the completion of her PhD on holography though the Spatial Information Architecture Laboratory within RMIT University's School of Architecture + design and her exhibition at Postx in Gent, Belgium. Melinda Menning who has recently graduated with a Master's degree in education is also nearing the completion of her PhD on art holography at Macquarie University. Dave Warren has produced a new major one-kilometer x 3.5cm white light transmission achromatic hologram titled *Hololujah* on the theme of twitter. Amy Rush has been continuing with the *Love Motel* project and has received funding for new skills in holography from the Australia Council for the Arts. Linnea Rundgren has been working collaboratively with Nils Abramson on displacement and distortion, and ways to display medium to large-scale holographic images of small objects as part of her Bachelor in Scientific Photography at RMIT.

2. Galleries, curators and conservator

The National Gallery of Australia is continuing the pioneering interest and support of the medium. The senior curator of photography, Gael Newton has listed 14 works in the collection - 8 by Margaret Benyon, 1 by Paula Dawson, 3 by Robyn Beeche/Edwina Orr and 2 by Shu Min Lin and has indicated that she anticipates that more 3D work in various formats will be acquired in the future. Gallery M in South Australia hosted an exhibition of Dave Warren's new works. The Newcastle Gallery commissioned a portrait of the much loved art patron Ann Lewis AO and Director, Ron Ramsey co-curated a major retrospective of my holographic works together with Director, Rhonda Davis and her collaborative team at the Macquarie University Art Gallery. Andrea Wise, the senior Paper Conservator of the National Gallery of Australia has written a paper outlining conservation issues relevant to reflection and transmission holograms in the collection. Andrea will present this paper at the Australian Institute for the Conservation of Cultural Materials conference in Brisbane and will continue seeking analytical methods to determine types of deterioration with a post-doctoral physics student at the Australian National University Canberra.

3. Courses and research

Philip Wilksch, school, of Applied Physics, RMIT has supervised projects in the design of holographic optical elements for the reduction of lens aberrations, and for optical trapping of colloidal particles in multiple locations, for study of crystallisation processes. At RMIT, as part of a building refurbishment program, the holography lab has moved into new premises. The elective course on holography for science students has not run for the past few years, because of resource restrictions and policies to do with minimum class sizes. Patrick McMillan continues to teach the elective in optical holography to Physics students at the University of New South Wales and at the College of Fine Arts, the fully online digital hologram course, has been taught twice by Leo Martin, despite low students enrolment numbers. An online optical hologram course has also been prepared however it has not run yet due to changes in Laser Safety standards in Australia. A five-year research project to develop software for haptic tools which enable a fine art method of hand drawing content for holograms, in collaboration between me, Masa Takatsuka, Hiroshi Yoshikawa and Brian Rogers is funded by the Australian Research Council. Finally one of the highlights was the extended visit by Hiroshi Yoshikawa to Sydney to collaborate with researchers at the Australian National Fabrication Facility, New South Wales Node.

Journal of Physics: Conference Series **415** (2013) 011002 doi:10.1088/1742-6596/415/1/011002

Report from China

Frank C. Fan AFC Technology Co., Ltd., China

1. Government Policies Inspire Holography for 3D display

In the "Twelfth Five Year Plan of National Science and Technology", Chinese government encourages holography researches for real 3D display as a new type of display technology in future. It is the first time in recent ten years that holography is definitely supported by the government project guideline in both areas of material science and information technology. The holoTV system invented by AFC Technology Co., Ltd. has got 2.5m RMB financial support for further innovation. This work has been published on OPTICS LETTERS (December 15, 2009/V01.34, No. 24, P. 3803-3805) after its first demonstration at the 8th International Symposium on Display Holography in Shenzhen, China. It was also reported and commented by 《Scientific American》, 《Optics & Photonics News》 and " www.optics.org ". (Please see: http://optics.org/article/41424& http://www.osa-opn.org/Content/ViewFile.aspx?Id=12928). Related works and further explanations are also published on APPLIED OPTICS (Vol.49, No.14/10 May 2010, P. 2676-2685) and OPTICS EXPRESS (20 December 2010/Vol. 18, No.26, P.27821-27826).

2. Traditional Hologram Industries Encounter Big Challenges

Embossing hologram industries have developed so rapidly in China that holograms are immerged into various merchandises more or less, from small logos to the integral packaging, aimed at security and promotion. It is a trend for this market that holographic features could finally be imbedded in to the packaging industry more properly as a special color for colorful printing. But the unexpected deformation caused by traditional thermo embossing technology makes it impossible to register the holographic pattern with the printing colors in high precision. That's why the price of present hologram film decreases to the edge profit for non-registered usage. For example, the metalized shimless "big rainbow" BOPP film decreases to 0.5 RMB/m^2, less than the half of its price one years ago.

3. UV Casting Techniques Bring a New Chance for Holographic Packaging

UV casting technology can overcome the deformation effectively because no temperature is needed. It can also precisely produce three dimensional nanometer structures with much huger dynamic ranges than the traditional thermo embossing, therefore, to produce microstructure such as Fresnel Lens. Now, we can arrange different type holographic element in the same master by UV recombining to make each specified element precisely registered to the afterward printing colors. This master can be used for UV embossing to produce the no deformed film for the final product of Registered Holographic Paper. Such paper is now widely used in the package of Chinese wine. See the following samples:

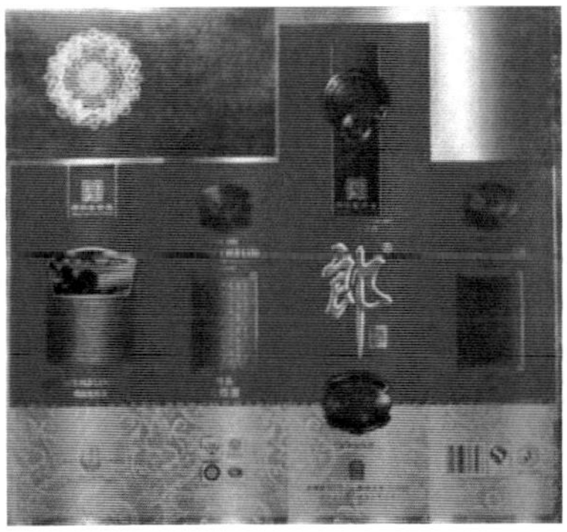

Journal of Physics: Conference Series **415** (2013) 011002 doi:10.1088/1742-6596/415/1/011002

Germany

Martina Mrongovius and Julius Schmiedel

Academy of Media Arts (KHM) Peter-Welter-Platz 2, 50676 Köln, Germany

mlm@khm.de

Holography in Germany is an active part of a number intuitions and companies, with the development of commercial products as well being engagement in reach and artistic projects.

1. Holography Companies

In 2010 Bayer MaterialScience launched a new class of photopolymer materials under the trademark Bayfol® HX. The standard grade is a panchromatic material which can record volume holograms and requires no chemical processing after exposure. Bayer MaterialScience is working with a number of industry and research partners to develop industrial and consumer applications.

Pronova continues to make HoloPro holographic screens for video projection. The screens are printed as an array of holographic optical elements, which direct the light from a video projector into viewing area. The individual holographic optical elements that make up these screens are now much smaller than the pixel resolution of video projectors. Besides many other screens three HoloPro™ were installed in 2011 as part of an extensive renovation of Moscow's Planetarium. Pronova also continues to develop dot-matrix HoloSigns. Michael Bleyenberg created a new series of artwork using multichannel HoloSigns, these works were presented at Gallery A22 in Budapest as part of a light event by the International Kepes Society in 2011.

2. Holography education and research

The Fraunhofer Institute in Berlin developed a auto-stereoscopic display which is now available through the company Real Eyes. Holographic displays are also being investigated however none of this research has yet been published. The Institute also uses holography as an analysis tool. At the Cologne University of Applied Sciences, Physical Optics and Holography is one of the researches and learning laboratories with most of their research focused on optical measurement techniques. The Electron Microscopy and Electron Holography workgroup at the Technical University Berlin also use holography as tool for material analysis, and moved to a new a new disturbance resilient building in 2011.

The Academy of Media Arts, Cologne (KHM) has a holography laboratory founded by Dieter Jung and we are currently researching how to enable the next generation of artists to work with new holographic technologies.

3. In memory

In December 2009 Matthias Lauk passed away. Lauk was a prolific and passionate collector of holography. He established *Museum for Holography and New Media* in Pulheim in 1979. His memoir *'I saw the light'* is available in German.

Italy

Dora Tassinari
Artist, Rome

Generally speaking, the contemporary Italian context in holography is not very well developed. Concerning holography and art, there are no holography courses currently being taught at any of the Italian Academies of Art, or even within university departments concentrating on scientific studies. Currently, the display of holography is disconnected, with no institutional links between the arts and sciences. The only professional lab exploring this connection is the HOLOLAB run by Mauro Melotti in Formigine (MO) Italy, and his focus is specialized only in Denysiuk recording holograms. I have been working with Mr. Melotti for the last 3 years to realize my series of holograms assemblages calls "Documents" (of which I show one piece at the MIT Museum). For the reasons mentioned above, very few artists are involved in holography. There is however an underground movement in Italy of individuals who are interested in the display of holographic art. Most of them work by themself for hobby or personal research, seeking connections on the web, but unable to align themselves with private or public institution of research. There one Italian HoloLaser group on the web that organizes an annual informal meeting .

Holography workshop are rare, and I know of only one in Trieste where a company, "Tutto tondo", sometimes offers private courses on analog and digital holography.

An additional resource is the Leonardo Longo International Academy for Laser Medicine and Surgery Borgo Pinti, in Firenze, Italy. Its work is focused on the medical profession using a variety of lasers of different types. Paula Dawson gave a paper there in 2009.

There are several other companies that offer holographic solutions for the packaging and print industry, and for anti-counterfeiting applications.

For the reasons I mentioned my goal in Italy is improve the Hololab of Mauro Melotti, creating a connection with the Art Academy in Bologna (IT) and with the Department of Physics in the University of Bologna. My connection is the "Marino Golinelli Foundation" in Bologna, www.fondazionemarinogolinelli, they support a lot art and sience, and they are very active in this sense with cultural projects and exhibition. Also I will work in order to organize some exhibition of holography in Bologna and Rome, concomitantly with the exhibition at MIT Museum. The ISDH was very useful for connections with the international holography community, so will be much more easy for me working in this direction in Italy.

Journal of Physics: Conference Series **415** (2013) 011002 doi:10.1088/1742-6596/415/1/011002

Report from Japan

Hiroshi Yoshikawa

Department of Electronics and Computer Science, College of Science and Technology
Nihon University
7-24-1 Narashinodai, Funabashi-shi, Chiba 274-8501, Japan

E-mail: yoshikawa.hiroshi@nihon-u.ac.jp

1. Standardization on holography

Two standards on holography, "Methods of measuring diffraction efficiency and associated optical characteristics of holograms" and "Methods for measurement of hologram recording characteristics" have been registered to Japanese Industrial Standard (JIS) on Jan., 2011. Those standards are also submitted to the International Organization for Standardization (ISO) as working draft 17901 on Feb., 2012 under TC 172/SC 9/WG 7.

2. Publication of "Development of holography in Japan"

The book, written in Japanese, covers technology, business and art from' 60s to 2010. It also includes Prof. Mitamura ' s interview with 26 holographers. The book was edited by some members of HODIC and editor in chief is Dr. Fujio Iwata. Its ISBN is 978-4-915851-41-4.

3. International conferences

International Workshop on Holography and Related Technologies 2011 (IWH 2011) was held in Utsunomiya from Nov. 16-18, 2011.

OSA ' s topical meeting on Digital Holography and Three-Dimensional Imaging (DH 2011) was held in Tokyo from May 9 to 11, 2011.

4. Tribute to Dr. Akira Tonomura

Dr. Akira Tonomura passed away on May 2, 2012 during a treatment on pancreatic cancer. He has developed holography electron microscope and verified the Aharonov-Bohm effect. His recent achievements on development and application of an atomic-resolution holography electron microscope can be found at http://www.first-tonomura-pj.net/e/.

9th International Symposium on Display Holography (ISDH 2012) IOP Publishing
Journal of Physics: Conference Series **415** (2013) 011002 doi:10.1088/1742-6596/415/1/011002

Current research and development activities for holography in Korea

SeungHyun Lee
Dept. of Holography and 3D Contents, Kwangwoon Univ.
26 Kwangwoon-Gil, Nowon-Gu, Seoul, 139-701, Korea

E-mail: shlee@kw.ac.kr

Korea has established the world class level of high speed telecommunication inrastructure. Users' interests in online games and 3D animations has been gradually increasing in Korea. In addition, Korea's display industry is developing remarkably under the dynamic power of the current Korean government and related industries which consider 3D technological fields to be the next generation industry by putting a lot of investments. Holography is a technology that provides ultimate sense of dimension. Korea has been carrying out steady holographic research from optical holography that uses laser and holographic films to digital holography.

1. Research Institutes

ETRI(http://www.etri.re.kr) : Carrying out following researches, "Development of core technologies of holographic 3D video system for acquisition and reconstruction of 3D information," and "Development of basic technology based on digital holography ". **KETI(http://www.keti.re.kr)** : Carrying out technology development of high speed CGH algorithm, holographic interactive system, depth camera 3D application software, and holographic printer. **KIST(http://www.kist.re.kr)** : Carrying out researches and development to adjust 3D image information obtained from microscope and digital hologram and to generate realistic 3D objects combined real images.

2. Universities

Kwangwoon Univ.(http://www.kw.ac.kr) : Carrying out researches on high speed CGH(Computer-Generated-Hologram) algorithm, digital holographic display of real objects, and color video holographic stereogram. **SeJong Univ.(http://www.sejong.ac.kr)** : Carrying out research and development of 3D image acquisition technology based on scanning electro-holography technology.
Seoul National Univ(http://www.snu.ac.kr) : Carrying out research and development of holographic display prototype which can have wide viewing angle with spatial light modulator upgraded horizontal resolution to curve array format. **Korea National University of Arts (http://www.karts.ac.kr)** : In process of produce high resolution and high efficiency optical hologram. Carrying out systematic and efficient research on development of a new exhibition technique and light system

3. Companies

Samsung(http://www.samsung.com/) : Carrying out development of digital hologram generation technology optimization and holographic display ability evaluation technology. Developed "1 lens 1 sensor" way of Full HD RGB/Depth combining sensing 3D camera. **LG(http://www.lge.co.kr)** : Carrying out development of digital hologram spatial light modulation technology, light source, and optics design technology. **Hangyo IC(http://www.hangyo97.co.kr)** :Carrying out development and design of digital hologram recording optical engine and color holography technology

Journal of Physics: Conference Series **415** (2013) 011002 doi:10.1088/1742-6596/415/1/011002

Portugal

I Azevedo

ARCA-EUAC, University School of Arts of Coimbra, Portugal, ID+ (Research Institute for Design, Media and Culture), Post-Doc Researcher from the Fundação da Ciência e Tecnologia in Faculty of Sciences, Department of Physics and Astronomy, University of Porto, Rua do Campo Alegre, 687, 4169-007, Porto, Portugal and in De Montfort University, Leicester, Faculty of Technology, The Imaging & Display Research Group, 2.27 Queens Building, The Gateway, LE1 9BH, UK

E-mail: azevedo.mariaisabel@gmail.com

ABSTRACT: Activities about holography in Portugal are related to the areas of scientific research, teaching, education and art.

The development of holography in Portugal has taken place in different laboratories located at University of Porto, University of Aveiro, research centres and institutes: CLOQ (Centro de Lasers e Óptica Quântica), INETI (Instituto Nacional de Engenharia e Tecnologia Industrial) and in a state printing company INCM (Imprensa Nacional Casa da Moeda).

The research and development activities are supported by National and European programs and projects involving national and international partners. The areas included in these activities are:

- Fundamental principles of holography and their applications to optical processing
- Real time holography in photorefractive crystals
- Pulse holography
- Holographic interferometry with applications to medicine and mechanical engineering
- Identification holographic tags with recorded images or coded information
- Volume holography in dichromate gelatine and polymers
- Colour control on white-light holography

At the Universities of Porto, Aveiro and Lisboa, the holographic principles and technologies are taught in graduate and undergraduate courses. Summer schools, seminars, talks and exhibitions are organized regularly to the specialists and the public in general. The Science Museum of University of Porto, the "Fábrica", connected with University of Aveiro, and programs like "Ciência Viva", are all strongly promoting that kind of actions. One of the strongest is the implementation of a net of basic holographic labs in secondary schools.

Art Holography has been implemented at University of Porto and more recently at University of Aveiro. The teams are composed by physicists and artists and their objective is to obtain transmission and reflection holograms to be reconstructed with laser or white light. Rainbow holograms and multiple image holograms have been more deeply researched. Colour control on reflection holography has been obtained by means of chemical methods, other than the preswelling on silver halide photographic emulsions. Collaboration between University of Porto and De Montfort University, Leicester, UK, carry out developments in digital art holography. University of Porto has recently produced a transmission hologram with a reconstruction of seven meters' depth.

Isabel Azevedo thanks Fundação da Ciência e Tecnologia, Portugal, for the post-doc research grant.

Journal of Physics: Conference Series **415** (2013) 011002 doi:10.1088/1742-6596/415/1/011002

Sweden

Nils H. Abramson

I am sorry to say that in Sweden the interest for Holography has during the last years decreased both as a research tool and as an art form.

Jonny Gustafsson at the Royal Institute of Technology in Stockholm (KTH) is at the moment developing the most important work on holography in Sweden. The *Interaction table* presents a computer-generated, auto stereoscopic, three-dimensional image that can be viewed and interacted with. It is believed that a display fulfilling the requirements for the interaction table will be an important tool in all applications which require a three-dimensional display, combined with good possibilities for communication between people. Examples are military command, medical imaging, city planning, surveying, scientific visualization, chemistry, games, and a number of education- and information applications.

Nils Abramson also at KTH is retired since 15 years but still working on his holodiagram which was already presented at the very first ISDH conference at Lake Forest 1982. It was designed for holographic measurement of large objects, but has later developed into a practical tool for understanding optical information in general, including special relativity. During the years he has published his ideas at a number of international conferences.

Hans Bjelkhagen left KTH in1983 to join CERN, Fermi Lab, Northwestern University etc. In 2011 he published New Recording Materials for Holography.

Mona Forsberg, Holomedia AB, Sweden, is still selling commercial holograms.

Journal of Physics: Conference Series **415** (2013) 011002 doi:10.1088/1742-6596/415/1/011002

Switzerland

Sebastian Mader and Wayne Tompkin

Optical Technology Division, OVD Kinegram AG, 6301 Zug, Switzerland

Email: sebastian.mader@kinegram.com

Abstract. In Switzerland there are a couple of companies working on diffractive optics, while academic activities focus rather on fabrication technologies, which can be used for holography rather than holography itself.

3D AG in Baar (www.3dag.ch) was founded in the year 1989. 3D AG is a supplier of holographic security elements for high security applications, brand protection and decorative patterns. 3D AG is also producing functional surfaces with micro or nano structures used in various applications (tooling for injection moulding, solar cells, LCD displays, anti-reflective surfaces, surfaces for light absorption or direction). Services include design and origination, electroforming, embossing and UV recombination.

CSEM SA with the headquarter in Neuchatel and several branches across other Swiss cities http://www.csem.ch is performing applied research in the field of micro optics. The research includes zero-order color filters, anti-reflection structures and diffractive structures for security applications in injection molding.

Holography Switzerland (http://www.holography.ch) in Urdorf, near Zurich, was formed in 2010. It offers consultancy for holographic projects for brand owners. The company uses the origination technologies including 2D, 3D, 2D/3D, and dot matrix.

Holosolutions/Holodesign-Midolo (http://www.holosolutions.ch) in Zurich is a design agency for holographic projects. The core competence of the design agency is the realisation of optically appealing and counterfeit-resistant designs meeting the specific customer requirements. The design agency develops the designs which are originated by third parties; the origination technologies include 2D, 2D/3D, dot matrix and e-beam.

Institute of Micro-Engineering (IMT) of the Ecole Polytechnique Fédérale de Lausanne in Neuchatel (http://sti.epfl.ch/) is performing applied research in micro-fabrication technologies. The Optics & Photonics Technology Lab (OPT Lab) of Prof. Hans Peter Herzig has longstanding activities in micro-optics. The group carries out research projects concerning optical and photonic nanostructures; the OPT Lab has developed a number of capabilities spanning the design, fabrication, and characterization of nanostructured optical and photonic devices.

OVD Kinegram AG in Zug (http://www.kinegram.com) was established in 1985 and is a member of the KURZ Group. OVD Kinegram AG is the only manufacturer of the KINEGRAM®, which is a proprietary, non-holographic diffractive optically variable device. The KINEGRAM® is reserved for high-security applications and is used to protect banknotes and governmental ID documents of over 105 countries. OVD Kinegram AG currently employs approximately 200 people.

PWO of the Interstaatliches Hochschule für Technik Buchs (NTB) in Buchs (www.ntb.ch) does research in the fields of Speckle Interferometry, Digital Holography and optical 3D measurement. The research is done in the Technical Optics of the NTB led by Prof. Andreas Ettemeyer.

9th International Symposium on Display Holography (ISDH 2012) IOP Publishing
Journal of Physics: Conference Series **415** (2013) 011002 doi:10.1088/1742-6596/415/1/011002

Report of the holography development in Taiwan

Ya-Ling Huang
Associate Professor, Department of Visual Communication Design
Kun Shan University

E-mail: ylhuang @mail.ksu.edu.tw

1. Brief history of hologram development in Taiwan

The development of holography in Taiwan started from Yan Qing Ling Industry Development Centre in National Taiwan University since 1980's. The function was helping the industry to develop high technology and advance science. At that time, Dr. Hun, Jin-Biao was the first one to create the hologram lab system in Taiwan and to develop hologram related technology into industry in the begging of 90's.
Almost the same time, the Artist Yong, Ying-Feng was impressed by the hologram during his trip to Japan in the late 1970's and then he worked very hard to bring the laser hologram into art and specially focusing on the sculpture.
The above is the brief history of the hologram development in Taiwan. Now there are a few of companies focusing on security, recording and displaying, most of the professionals are from photoelectric or physics academy background. However, Kun Shan University is the only one that focusing on the holographic design and 3D image displaying application.

2. Holographic exhibitions to open the public's view

Exhibition is always a good way to share the new technology, new concept, and new media to public. In year of 1980's, the first hologram group exhibition started from Dr. Hu and Yong, Ying-Feng who brought the first vision of hologram application on art. In late of year 1990's, Shu-Ming Lin brought traditional philosophy with hologram and he bring wider vision for young artists about hologram and new Medias.

Many hologram artists bring many great hologram arts to Taiwan and also hold workshops or lectures for students that build up the important base to the development in Taiwan. This connection start from Dieter Jung been invited by Kun Shan University in 2004 to have workshop, he had two solo exhibitions in Taipei and Kaohsiung Fine Arts Museum in 2005, and he hold workshop at Kun Shan University in the same year. Martin Richardson start to have hologram workshop at Kun Shan University in 2007, Setsuko Ishii has workshop and lecture at Normal University in 2008, Juyong Lee starts to hold workshop at Kun Shan University from 2010. Martina Mrongovius starts to come in 2010. All the great connects bring the beauty of hologram into visual design, museum display, and product design in Taiwan academy and move forward to the art and culture industry now.

Report from the Nations: Holography in UK

H I Bjelkhagen

Hansholo Consulting Ltd., The Bungalow Offices, St-Asaph, LL17 0LQ, UK
E-mail: hansholo@aol.com

UK is rather active in the field of art and display holography. A short summery of the main players and activities is provided here (in alphabetic order).

Aberdeen University in Scotland is involved in holographic recording of marine particles by *John Watson* and his team. He has constructed several holographic cameras for underwater imaging of plankton and other marine particles. These devices use both analogue and digital holography to record water volumes up to a cubic metre. Recent advances made in digital holographic recording on electronic sensors coupled with fast numerical reconstruction led to the development of smaller under-water holographic recording equipment, the *eHoloCam*, allowing rapid capture and storage of images including holographic video of moving objects down to one hundred metres depth.

Centre for Modern Optics (CMO) in Wales has been closed after *Hans Bjelkhagen* retired. He is now a consultant working through Hansholo Consulting Ltd in Wales. Most of his time last year, he was writing a book: "Ultra-realistic imaging – advanced techniques in colour holography" together with *David Brotherton-Ratcliffe*, Geola UK, to be on the market later in 2012. *Ardie Osanlou*, former CMO researcher, is at Glyndŵr University in Wales involved in teaching.

Colour Holographic Ltd in London run by *Michael Medora* has become a main player through its manufacturing of holographic silver-halide emulsions which now includes a panchromatic version for recording colour holograms. *Manuel Ulibarrena* has been responsible for this work and high-quality colour holograms are recorded in the company. Colour holograms of products and museum artefacts are now taking place within the company. A new wooden display box with integrated LED light for colour holograms has been introduced. The company is involved in advertising including for example movie promotions. Colour Holographic had the expertise and technology to mass produce the *Moviposter*™, a large format full colour rainbow stereogram.

De Montfort University in Leicester is actively involved in teaching holography at both undergraduate and post-graduate levels under *Martin Richardson*. Steven Smith (former Lasersmith in Chicago) got a PhD from the university last year. Richardson is also active in creating art holograms, his own as well as together with visiting artist in his laboratories. In particular he is working on creating digital art colour holograms which are printed at Geola in Lithuania. He is also making lenticular prints. In addition Richardson is involved in projects in India and Taiwan.

Gallery 286, *Jonathan Ross'* art gallery in London is a place where several hologram exhibitions have been organised, for example:

- *Holograms and lenticulars* by Isabel Azevedo and Paul Scattergood during March 2011.
- *Holographic Portraits and Figure Studies* from the Ross' collection during July/August 2011.
- *Sculpting with Light* - Abstract holograms from the Ross' collection during April 2012.

Jonathan Ross has arranged and toured two hologram exhibitions in the UK with a selection of holograms from his collection:

- *HOLOGRAMS, the first 60 years,* went to six museums and galleries.
- *HOLOGRAMS 2*, the follow-up exhibition, has visited three museums so far and is going to Rugby, where Dennis Gabor invented holography, in August 2012.

In addition he has contributed the Richmond Holographic Studios hologram of Boy George to the Victoria & Albert Museum's travelling exhibition on Postmodernism as well as curated: *Ultimate 3D - The Wonder of Holography*, a travelling exhibition of 175 holograms for United Exhibits Group, which opened at the National Museum of Science & Technology in Kaohsiung, Taiwan in January 2012, and will be at Songshan Cultural and Creative Park in Taipei from July to September 2012.

Harman Technology Ltd (former Ilford) introduced in 2011 new holographic silver-halide materials on the market. HARMAN ***HOLO FX*** fine grain holographic plates are coated with a very fine grain holographic emulsion (crystal sizes typically 30-40 nm) and are available in both red and green sensitive:

- *HOLO FX RED SENSITIVE* holographic plates are sensitive over 600 nm – 694 nm with peak sensitivity at 660 nm.
- *HOLO FX GREEN SENSITIVE* holographic plates are sensitive over 488 nm – 560 nm with peak sensitivity at 532 nm.

Holoxica Ltd is a relative newcomer to the hologram scene, its founder, *Javid Khan*, is a PhD student at Herriot Watt University in Edinburgh, Scotland. He is researching holographic display techniques based on a set of spatially multiplexed interference patterns to make a holographic screen. By changing the pattern the reconstructed image can change. This can be done in any sequence or combination to make simple 3D animations. The company offers digital hologram design services suitable for digital holographic printers. The focus is on technical and scientific imagery with notable designs including biotech images and making a full-length hologram of the human body for anatomy teaching. The recent Mummy hologram will be on display at the conference.

In addition to his work on embossed security holograms at **Light Impressions**, *Inaki Beguiristain* operates an independent studio in Chelmsford, Essex, specialising in pseudocolour reflection work. His private studio is one of the only dedicated multicolour hologram facilities in the world.

Nottingham Trent University is where *Andrew Pepper* is based working at the School of Art and designed. He has been engaged in a number of holography and contemporary art exhibitions at the university, for example:

- *miniMents,* April - May 2011, Pepper installed a holographic roof (an elevated 'lake') onto the top of one of the townscape's flat top buildings. "Tiny architectural holography"
- *Contemporary Sculpture: Transit* in December featuring art work with its transport and packing material. This allowed a classic hologram (*Point Addition* made in the 1980's in association with Light Fantastic) to be displayed resting on its original Agfa packing box – complete with 'red strips'
- *Archipelago*, January - February 2012, Pepper included a floor based reflection hologram, of the shadow of water, illuminated by a sculptural Tizio table lamp

Pepper was represented at an exhibition in London: *Sculpting with Light*, May 2012 at Gallery 286 in London. A selection of abstract holograms, from the Jonathan Ross Collection, by British & American artists from the 1980s and 90s. Two works by Pepper were included. Pepper's work has also been included in international exhibitions such as *Garden of Light* at Gallery 175 in Seoul, Korea, in November 2011 organized by Professor Juyong Lee. Pepper has delivered an invited public lecture at the National Centre for Craft & Design, Sleaford, Lincs, UK, and has acted as an external examiner for a PhD in creative holography at the School of Architecture, Ghent, Belgium.

Reconnaissance International Ltd with *Ian Lancaster* as the director is involved in arranging international conferences, mainly focussed on the document security and brand protection. The company is publishing the monthly newsletter *Holography News* with world-wide distribution. Associated is International Hologram Manufacturers Association (IHMA) which is a not-for-profit membership organisation registered in the UK to represent the interests of hologram manufacturers and the hologram industry worldwide.

Royal Photographic Society (RPS) The former RPS Holography Group in UK has launched the new *RPS 3D Imaging & Holography Group* with a possibility for the *Stereoscopic Society* members to be able to join. RPS Advisory Board representative Alan Cooper, chairman of the Stereoscopic Society, is to take over from Graham Saxby in this position.

Southampton University, where *Pearl John,* artist and educator is based, and works with local school and college students - for the School of Physics and Astronomy at the university - inspiring an interest in science and photonics in particular. She has a holography studio at the university where she continues her work as a holographic artist. In addition she also has an outreach programme - travelling to schools around the country doing holography workshops in classrooms.

Spatial Imaging, with *Rob Munday* has developed several areas of holographic business over the last few years. In particularly the company is now the world leading provider of digital hologram mastering systems to the embossed hologram industry. It has also gained an international reputation for its in-house mastering of security holograms and produces master holograms for some of the world's biggest hologram and security printing companies. Spatial Imaging continues to develop 3D imaging techniques, technologies, products and applications and to create new business opportunities in the 3D Imaging field. The Spatial Imaging Group consists of:

- *SITECH* - a specialist provider of hologram mastering systems *Lightgate* and ancillary equipment to the holographic industry
- *HOLOVISTA* - 3-D and animated holograms, lenticular images and stereographic displays for point of sale and promotional applications
- *FACESTATION* - manufactures and sells a range of automated digital camera and printing systems for the production of lenticular portraits
- *THE HOLOGRAPHIC PORTRAIT ARCHIVE* - an archive of holographic portraits
- *LIVING POSTER* - 3-D lenticular posters and prints
- *SECURITY IMAGING* - hologram and OVD design and mastering services for the security print and embossed hologram industry

Spatial executed in 2004 an important hologram portrait *Equanimity* of *Her Majesty the Queen*. Holographer *Rob Munday* and the artist *Chris Levine* created the stereogram hologram portrait. On 1 June 2012 a £10 holographic stamp with an embossed hologram of the Queen portrait was issued by Jersey Post to celebrate *Queen Elisabeth II*'s Diamond Jubilee. The hologram for the stamp was created especially for Jersey Post by Rob Munday and is a 3-D achromatic digital embossed stereogram.

Jersey *The Queen* hologram stamp

Time Magazine June 4, 2012 Vol. 179, No.22

United States

V. Michael Bove, Jr.

MIT Media Lab, 20 Ames St., Room E15-448, Cambridge MA 02142 USA

vmb@media.mit.edu

1. Events

Besides ISDH, major recent US holographic imaging conferences included SPIE's Photonics West (January 2012 in San Francisco) and the OSA's Digital Holography and Three-Dimensional Imaging conference (April 2012 in Florida). Holography presentations were also featured in other venues such as SID's Display Week, SMPTE's Annual Conference on Stereoscopic 3D for Media and Entertainment, and OSA's 3D Display Technology, Perception and Application Incubator Meeting. In conjunction with ISDH, the MIT Museum in June opened an exhibition titled *The Jeweled Net: Views of Contemporary Holography.*

2. Research

Academic research in holographic imaging – particularly in digital methods – continues in the US, with papers published recently by groups from (at least) Duke University, Harvard, Michigan Technological University, MIT, University of Arizona, University of Colorado, University of Connecticut, and Virginia Polytechnic Institute. IARPA issued a call for proposals for a large full-parallax holographic video display, though as of this writing the funding recipients have not yet been announced.

3. Industry

"Hologram" remains a powerful marketing word, even if applied to things that aren't really holographic. The most recent example is the "hologram" of deceased performer Tupac Shakur, which made an appearance on-stage in April 2012 at the Coachella Valley Music and Arts Festival. Tupac was actually a "Pepper's Ghost" projection setup, with graphical re-creation by Digital Domain.

Over 200 US patents related to holography have issued in the past twelve months.

Peer review statement

All papers published in this volume of *Journal of Physics: Conference Series* have been peer reviewed through processes administered by the proceedings Editors. Reviews were conducted by expert referees to the professional and scientific standards expected of a proceedings journal published by IOP Publishing.

Journal of Physics: Conference Series **415** (2013) 012001 doi:10.1088/1742-6596/415/1/012001

Artistic Representation with Pulsed Holography

S Ishii

Independent Artist
1-2-3-513, Kohinata, Bunkyo-ku Tokyo 112-0006, Japan

E-mail: gp5s-isi@asahi-net.or.jp

Abstract. This thesis describes artistic representation through pulsed holography. One of the prevalent practical problems in making holograms is object movement. Any movement of the object or film, including movement caused by acoustic vibration, has the same fatal results. One way of reducing the chance of movement is by ensuring that the exposure is very quick; using a pulsed laser can fulfill this objective.

The attractiveness of using pulsed laser is based on the variety of materials or objects that can be recorded (e.g., liquid material or instantaneous scene of a moving object). One of the most interesting points about pulsed holograms is that some reconstructed images present us with completely different views of the real world. For example, the holographic image of liquid material does not appear fluid; it looks like a piece of hard glass that would produce a sharp sound upon tapping. In everyday life, we are unfamiliar with such an instantaneous scene.

On the other hand, soft-textured materials such as a feather or wool differ from liquids when observed through holography. Using a pulsed hologram, we can sense the soft touch of the object or material with the help of realistic three-dimensional (3-D) images. The images allow us to realize the sense of touch in a way that resembles touching real objects.

I had the opportunity to use a pulsed ruby laser soon after I started to work in the field of holography in 1979. Since then, I have made pulsed holograms of activities, including pouring water, breaking eggs, blowing soap bubbles, and scattering feathers and popcorn. I have also created holographic art with materials and objects, such as silk fiber, fabric, balloons, glass, flowers, and even the human body. Whenever I create art, I like to present the spectator with a new experience in perception. Therefore, I would like to introduce my experimental artwork through those pulsed holograms.

1. Introduction

One of the salient features of holography is the reconstruction of realistic three-dimensional (3-D) images. There are many restrictions associated with shooting holograms, however, if a real subject is chosen. Vibration during shooting is a serious problem to artists from the viewpoint of artistic representation since it restricts subject choice and limits hologram size.

One of the solutions to this problem is to use the pulsed laser for shooting. The pulsed laser makes it possible to shoot such subjects as the human body, fluid and liquid material, moving objects, and falling and floating objects, which are not appropriate for use with a continuous laser. A pulsed hologram records scenes instantaneously at the rate of a few dozen nanoseconds per scene. Observing the reconstructed images of such scenes made this author aware of some interesting phenomena. For example, some images do not appear as the original subjects. They portray completely different aspects of the real world.

This paper describes the features of reconstructed images of pulsed holograms according to variations in subjects. Additionally, it introduces the application these features for artistic representation through use of the author's practical works of art.

2. Frozen time

2.1. Liquid subject

One of the early holograms for which this author/artist used pulsed laser was *Crystal White* (Fig.1), a rainbow hologram, measuring 30 cm × 40 cm, produced in 1979 and depicting a breaking egg [1]. This master hologram was shot using a ruby laser. At that time, the author of this study also attempted to shoot several different subjects with the ruby laser, including roses, pineapples, and hands. I became aware that the reconstructed images of those subjects were unchanged from the subjects themselves. On the other hand, the reconstructed image of an egg white does not appear fluid; it looks like a piece of glass that would produce a sharp sound upon tapping. Further, an observer can see different optical effects through the egg white with shifts in vantage points, just as if it were a real glass [2].

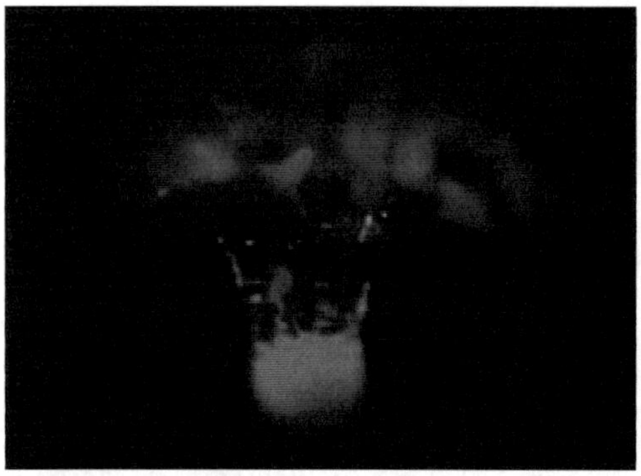

Figure 1. *Crystal White,* 1979
Rainbow hologram, 30 cm × 40 cm

Figure 2. *Self-portrait*, 2000–02
Laser transmission hologram, 50 cm × 60 cm

Figure 3. *Frozen Time,* 1983
Rainbow hologram, Handmade Japanese Paper

We recognize phenomena of the real world in the streaming of time. We understand that time flows, but we are not able to understand that it is liquid until an instantaneous scene suggests that all movement has frozen. Suddenly, we are not able to recognize the differences between liquid and solid material. In everyday life, we are unfamiliar with such scenes. The instantaneous world portrays completely different features, notwithstanding reconstructed images coming from the real world. Thus, our vision is extended with holograms.

Self-portrait (Fig. 2) is a laser transmission hologram (50 cm × 60 cm, 2000–2002); water was poured into a glass bottle containing two different transparent liquids–water and oil. In the course of pouring the water, the boundary plane between the two liquids (with different specific gravities) was broken and appeared as a complicated shock wave. In this hologram, all liquids froze, and the oil and water in the glass bottle looked like solid crystal glass instead of liquids [3].

Frozen Time (Fig.3) [4] is the finished installation work that comprises *Crystal White*. The image of an egg, which has been popular with many artists as an object of artistic representation, is absorbing for me as well because it evokes images of organic/natural/eternal life and fluidity when it is broken. The upper part of this installation, *Frozen Time*, comprises three different holograms (images of eggs); handmade Japanese paper on which the traces of flowing egg fixed with tempera painting represents the basis for the lower part. Albumin in the egg is well known as a strong painting medium; the technique of egg tempera painting has been used for many years by western artists.

The basic concept for creating this installation was the showcasing of contrasts between the new medium (just formed) and the original (historical) medium, as well as contrasts between a real image and a holographic image.

2.2. Reality

Soft-textured materials such as feathers or wool differ from the abovementioned liquids. Reconstructed images from these materials appear just the same as they appear in the real world. We can experience the sense of soft touch with the help of realistic 3-D images, despite the fact that only visual information is available (Fig. 4, Fig. 5). These images provoke an absorbing experience in perception。

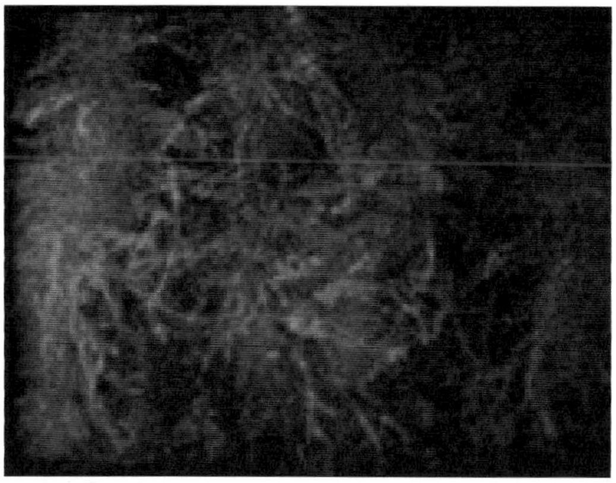

Figure 4. *Visible Temperature α* (part)
Subject: Feather

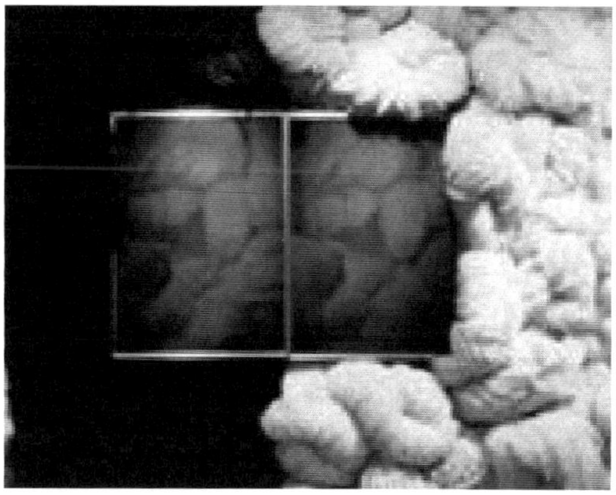

Figure 5. *Visible Temperature α,* (part)
Subject: Wool

Journal of Physics: Conference Series **415** (2013) 012001 doi:10.1088/1742-6596/415/1/012001

Figure 6. *Visible Temperature α* (complete view), 1987–88, 220 cm × 750 cm

Wall hanging art pictured in *Visible Temperature* α (Fig. 6.) comprises reflection-type image holograms (Fig. 4, Fig. 5); dimensions of each are 30 cm × 40 cm. They exhibit various textures, such as those of feathers, wool, cotton, hemp, rope made of chemical fiber, steel wire, and nails [5]. A reflection-type hologram is excellent for reproducing texture, and an observer will surely memorize an illusion from which a soft feel is transmitted. This experience is attained by placing an actual finger on the reconstructed image of feathers or wool, which loom several centimeters away from the glass side of the hologram. Thus, *Visible Temperature α* is a work that affects vision and the tactile sense [6]. It indicates that a realistic holographic image enables one to experience the sense of touch as if he or she is touching an actual object. A contrast is presented between texture and the sense of touch for different materials, as well as for real objects versus reconstructed 3-D images.

2.3. Subject: Scene of floating in the air

The following images (Fig. 7, Fig. 8) are instantaneous scenes of a feather and popcorn scattering into the air.

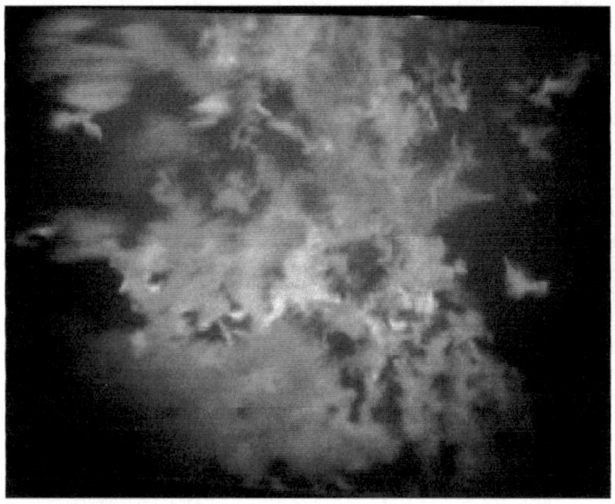

Figure 7. Subject: Feather,
Multi-color, multi-channel hologram
85 cm × 110 cm

Figure 8. Subject: Popcorn,
Multi-color, multi-channel hologram
85 cm × 110 cm

Each work is a pulsed master hologram transferred to a multi-channel, multi-color hologram [7]. Reconstructed images appear on both sides, in front of, and behind the holograms. As mentioned above, the feather has been reconstructed into a realistic 3-D image. It appears to float on the wind, giving a natural impression. On the other hand, the reconstructed image of popcorn portrays an unfamiliar scene.

We also attempted to shoot soap bubbles as subjects of interest. We were able to shoot several, from a few to 10 cm across on a 50 cm × 60 cm hologram. When we carefully observed the reconstructed images with a laser, we found that they differed completely from the familiar ones. In everyday life, we look at soap bubbles under white light, sunlight (outdoors), or artificial white light (inside). A large floating bubble on the breeze changes form continuously and organically. Its unique and beautiful colors come from the effect of interference colors under white light. These colors undergo kaleidoscopic changes. On the other hand, the reconstructed holographic image loses the array of colors and looks like a transparent glass ball. The edge of the sphere is barely recognizable, as are the two reflected bright points, one on the outside surface and the other inside..

If we shoot the real soap bubble as the subject by using three lasers each oscillates different wave length (red, green, blue), is it possible to reconstruct the beautiful realistic soap bubble colors? In the near future, is it possible to create a natural color hologram from real soap bubbles? This is an absorbing theme for prospective research.

3. Body Wrapped with Fabric

The most popular subjects of pulsed holograms are portraits and human bodies. Historically, these subjects have been important to painters and sculptors.

Figure 9. *Body Wrapped with Fabric A,*
Multi-color,multi-channel hologram
2000–08, 110 cm × 134 cm,

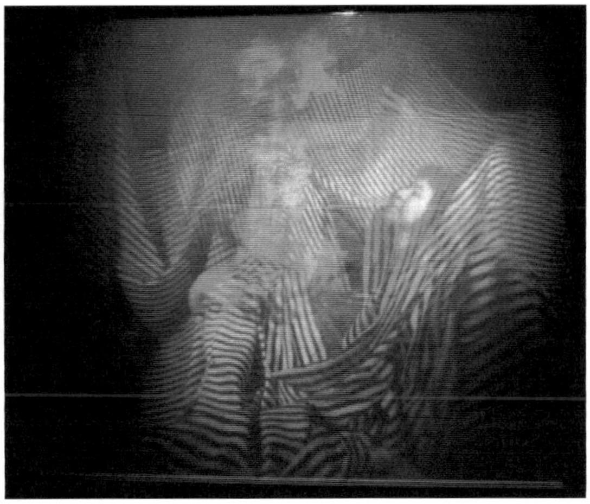

Figure 10. *Body Wrapped with Fabric B*
Multi-color, multi-channel hologram
2000–08, 110 cm × 134 cm

. *Self-portrait —Body Wrapped with Fabric A* (Fig. 9) and *Fabric B* (Fig.10) are multi-color, multi-channel holograms (110 cm × 134 cm); each hologram comprises about 20 different images. Each image has its own limited field of view, and reconstruction images of the hologram appear to occupy their own positions in space. Each scene appears dissimilar when observed from a different viewpoint. When our perspective shifts vertically upward or downward, we observe a change in the hologram's color combinations. When our viewpoints shift horizontally from left to right, we notice different images. Three-dimensional collage can be produced using this method. Since it is not possible to look at all images in the hologram from one vantage point simultaneously, we unconsciously interact with the holograms. If we want to discover the various embedded images, we have to interact voluntarily with physical movement. Holography gives a new understanding of the physical space in which we live; this aspect makes it an attractive art medium [8].

4. Conclusion

The examples described in this paper have conclusively shown that the various characteristics of pulsed holography enable new expressions that cannot be realized through existing mediums of expression. The illustrations indicate that realistic 3-D holographic images affect the relationship between vision and the tactile sense. Some instantaneous scenes reconstructed with the pulsed holograms portray a completely different aspect of the real world. Our vision is extended with holography.

On the other hand, there are presently several difficult problems associated with the creation of the pulsed hologram. First, amateurs will find that the pulsed laser is expensive and difficult to operate. Secondly, recording materials must be improved for appropriate sensitivity and durability. More devoted and astute scientists and engineers are needed to solve these problems.

References

[1] Catalogue 1979–80 New Space–The Holographer's Vision, The Franklin Institute Press, USA.
[2] Encyclopedia of Practical Photography 1982 Kodansha, Tokyo, p. 89.
[3] Catalogue 2001 Technique of Contemporary Art-Light and Its Representation, Nerima Art Museum, Tokyo.
[4] Catalogue 1983 Realism Now, The Museum of Modern Art, Saitama, p. 63.
[5] Catalogue 1988 Artware–Kunst und Elektronik, Deutschen Mess AG und der, Siemens, Germmany.
[6] Ishii S. and Tsujiuchi J. 2008 Where are We Going in Art Holography?, Holography and Speckle, America Scientific Publishers, pp.95–113.
[7] Catalogue 2008 DOMANI, The National Art Center, Tokyo.
[8] Ishii S. 2006 Artistic Representation with Holography, Forma, Vol. 21 (No. 1), pp. 81–92.

LightLeaves: computer controlled kinetic reflection hologram installation and a brief discussion of earlier work

Betsy Connors Chen
ACME Holography, Somerville, Massachusetts USA 02144
E-mail: acmeholo@gmail.com

Abstract. *LightLeaves* is an installation combining leaf shaped, white light reflection holograms of landscape images with a special kinetic lighting device that houses a lamp and moving leaf shaped masks. The masks are controlled by an Arduino microcontroller and servo-motors that position the masks in front of the illumination source of the holograms. The work is the most recent in a long series of landscapes that combine multi-hologram installations with computer controlled devices that play with the motion of the holograms, the light, sound or other elements in the work.

LightLeaves was first exhibited at the Peabody Essex Museum in Salem, Massachusetts in a show titled "Eye Spy: Playing with Perception".

Figure 1 LightLeaves 2012

To explore new ways to create the most realistic images of the natural world, to capture the light falling on a subject, record it full color with motion and sound in three dimensions, and instantly transmit the images around the planet. Holography and computer generated imagery are the latest scientific display technologies with the potential to realize this, and artists have helped shaped the development of all these technologies & mediums.

1. INTRODUCTION

My work in holography is concerned with using nature as a primary subject and using multiple image installations as the form. Adding a kinetic element for hologram motion or lighting effect, usually controlled via a computer, creates an *ambient* feeling of the natural world. In a previous paper [1], I talked about using landscapes as a subject, and explored my interest in juxtaposing technology with nature, as well as bringing attention to the concerns many artists have about environmental issues, and the creation of art works to heighten an awareness of the damaging changes taking place on our planet. In this paper I would like to talk about the creation of LightLeaves and comment on the application of computer controllers to create motion in my work. I have used a variety of control systems in previous works over the years, and now the electronics are much smaller, cheaper, and accessible for everyone to use.

My interest in creating motion, sound, and lighting effects in my work in holography, has roots in my earlier work in film and video, and parallels (but in a different way) the quest to create holographic electronic moving images.

2. BACKGROUND

When I began working on LightLeaves and other recent works, LightRain (I and II), I considered the fact that in creating installations that express a connection to nature and a concern for the environment, with the awareness that these works do themselves have an environmental "footprint'; using precious resources to create the holograms which in turn put chemicals and pollutants, etc. back into the ecosystem. I also became aware of the amount of materials used in holography, when I inventoried, assessed, and catalogued forty years of research, done by Dr. Stephen Benton and his students and colleagues at the former Polaroid Corp. and MIT's Spatial Imaging Group. There was a near four to one ratio in holographic plate use over the years, which was similar to my own lab habits. I made many multiple hologram installations, and produced many extra "test" plates to achieve a single usable one, so I decided to create a new work by using fragments of the useable parts of *already existing* older work. The pioneer video artist Nam June Paik, with whom I had the privilege of working in video, often recycled older videotape imagery into newer sculptural forms in his large TV installation works. This idea influenced my holography installation work. Paik was a personal mentor and master of mechanical invention and form, and a great influence on many artists.

Figure 2 TV Garden Nam June

Figure 3 LightLeaves (side view)

3. LIGHTLEAVES

LightLeaves is a white light reflection hologram installation, made up of leaf-shaped reflection holograms of images of plants, flowers, trees and leaves; from both real and fabricated landscape subjects. Many of the master holograms for the work were produced using a pulsed laser, during an artist-in residency at The Center for Holographic Arts in New York. Other masters and all transfers were produced in my lab, ACME Holography, in Somerville MA. The holograms are mounted at different depths with threaded rods onto six, 9"x9" black lacquered panels. The panels are approximately 1 ½" deep, mounted off the wall and the holograms are mounted one to four inches out from the panel surface.

The installation is illuminated by a special effects lighting device that houses sensors, motors, a lamp, and leaf shaped light "blockers" that move across the work in front of the light, controlled by a small microprocessor. The motion of the light blocking elements responds through a sensor to the temperature in the gallery where the work is displayed; causing the motors to move the masks faster when the temperature is higher.

3.1 Out of the Frame - Off the Wall

The holograms are laminated to plastic of different opacities and then cut from stencils created from actual maple leaves, for a one to one size ratio. They are mounted to the panels with half inch spacing between each panel. Some of the holograms were mounted on clear plastic, some on fifty percent opaque plastic and some on solid black (Figure 4). There is an interesting play of light across the different substrates, at certain angles, and outside the illumination areas. The various imaging depths (in the z-axis) of the holograms, interacts with the material substrate and the other hologram planes, to create an awareness of the 3-D flow across the images. The illumination device with rippling leaf shadows creates a subtle effect of simulated sunlight and wind moving across the three-dimensional leaf layers of three-dimensional imagery.

Figure 4

3.2 Illumination Device 1

Figure 5

When LightLeaves was first installed at The Peabody Essex Museum (Salem, MA), the location of the work in the gallery was under a soffett, so I had to design and construct the illumination device specifically to work in that space (Figure 5). The light was fixed directly above the work, pointed into the space towards the device, and then directed with a mirror about twelve feet away to light the work. The device was built into a box, approximately twelve inches wide, by fourteen inches high, by ten inches deep, with the mirror mounted inside to produce the correct illumination angle. The box also contained an Arduino microcontroller attached to two motors and a sensor. The Museum asked that I camouflage the device into the Museum space, so the box was painted to match the gallery, and was placed and tuned for the correct illumination position, and then glued onto the ceiling above and in front of the hologram. Two lightweight wires with small aluminum leaf shaped blockers were mounted onto the motors controlled by the Arduino, and programmed to simulate jittery wind motion. A temperature sensor was also attached to the Arduino to control the speed of the motors. When the sensors detected a rise in the temperature in the gallery, the Arduino program made the motors move faster and so the leaves "blew" a little more wildly. I don't know if the viewers sensed the difference, it is a very subtle effect.

3.3 Illumination Device 2

I designed a new lighting device for the newer version of LightLeaves for the ISDH exhibit in 2012, which I built out of aluminum framing (Figure 6) since the work would be lit traditionally for reflection holograms—from above and in front of the work. I bought a PAR 36 fixture and lamp for the best illumination distance to the holograms, and built an aluminum box around it with holders for the motors and leaf blockers, now directly in front of the lamp, but at a distance of approximately ten inches for one motor, and sixteen inches for the other (Figure 7). The Arduino microcontroller, the same used in the PEM show, was attached to the back of the housing away from the lamp. I also added plastic baffles to block light from spilling into the gallery space. Although the device

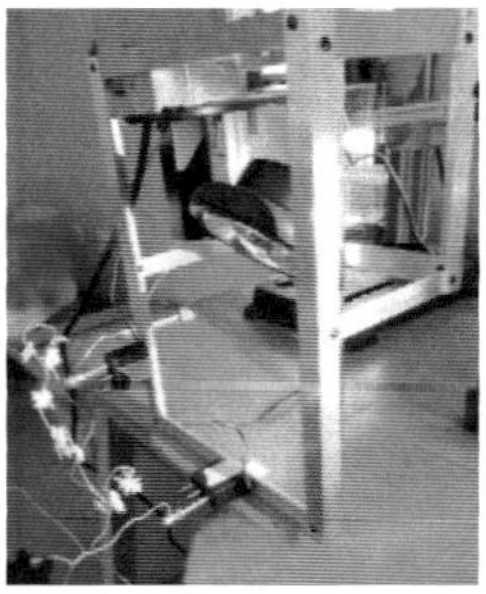

Figure 6

Figure 7

was not built with a milling machine or CAD fabrication equipment, it proved to be a very stable platform for the motor control and the distance from the leaf blockers to light can be controlled pretty precisely. The programing for the Arduino controller was done by Dave and Cody Chen, for the first version of LightLeaves and modified only slightly for version two.

4. LIGHTLEAVES CONTROLLER

The Arduino is an open-source electronics prototyping tool for integrating sensors with motor-control and is a perfect system for creating interactive installations. The baseboard includes a microcontroller, 14 digital input/output pins, 6 analog inputs, USB connector and a power jack. The free Arduino software includes a "Servo" library that uses 2 of the available digital outputs to generate the PWM signals needed to control the in-expensive off-the-shelf servomotors that we used for LightLeaves. The LightLeaves installation also uses a simple temperature sensor through one of the analog inputs to control the speed and direction of motion for the leaves.

The basic Arduino board is very inexpensive (less than $50) while the Vex servomotors we used on each side of the frame were about $20 each. The Arduino software is available as a free download and runs on Windows, Mac and Linux. The development environment is very user-friendly and lets you easily download compiled firmware through the USB port of your computer and see debugging output from the board.

The Arduino programming language is based on *Wiring* and is implemented in C/C++; and is very similar to the *Processing* language invented at MIT. The language includes all the expected control and arithmetic operators, data types and ability to define sub-routines. The program we wrote for LightLeaves is about 150 lines long and includes routines for moving each branch in a specific direction for a set distance. To simulate leaves moving in a changing breeze we tied in the temperature controller with a random number generator to create an environmentally sensitive affect.

Figure 8

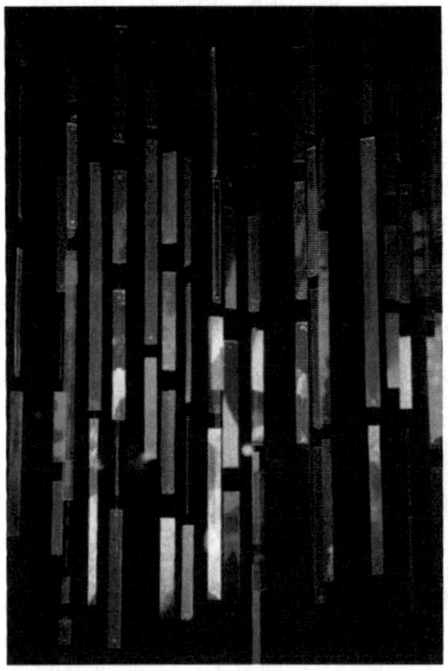

Figure 9 LightRain II

5. LIGHTRAIN CONTROLLER

We used a different motion control system for the holograms in a previous installation, LightRain (figure 9), which is a hanging kinetic sculpture of over eighty narrow 'rainbow' transmission hologram shards of landscape images. The narrow holograms are attached with filament and suspended from a housing device containing both motion-capable and stationary hologram holders. The motion holograms are controlled by a VEX robotic microprocessor, which is programmed to operate two motors and using a variety of sensors to run simple vertical movements of the hologram shards. The vertical movements of the hologram display "jewel like" shimmering effects, viewable with "water like" reflections below and behind the sculpture. Like the Arduino, the Vex microcontroller also provides a simple, easy to program platform for creating a broad range of motor control applications with sensor inputs. But because it is geared to "off the shelf" use in the school education market, the Vex is considerably more expensive—with Starter Kits in the $400 range. The standard kit comes with quite a few components including the microcontroller, motors, servomotors, optical encoders, light sensors, ultra sound distance sensors, gearbox components and "Erector Set" type metal motor mounts and frameworks for assembling small custom structures.

For LightRain (Figure 10) we set up two independently controlled banks of holograms attached to 30-inch wooden dowels, separated by about six inches. Each bank had its own independent motor and spun the dowel with a gear ratio reduction of about four-to-one. Each bank also had its own optical encoder to measure the number of revolutions that the rod would turn. The holograms were then attached with filament onto the rods as well as stationary hooks.

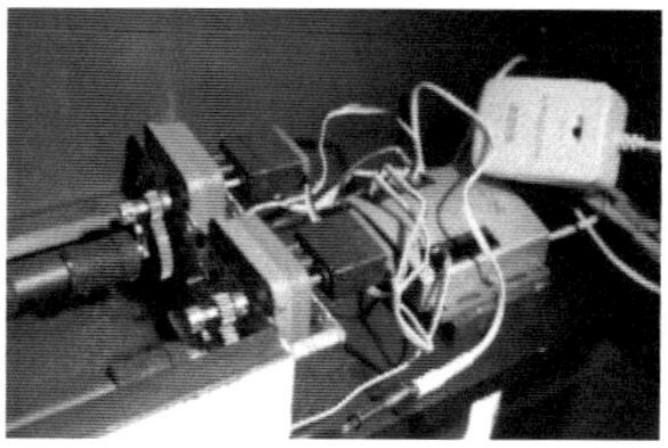

Figure 10

We started with a simple control strategy of cumulating the count from the encoders to measure the total distance travelled, but we found very quickly that this number was not accurate due to mechanical backlash and play in the system. Thus, to avoid wrapping the holograms around the wooden axle, we had to add a separate switch to each bank to tell us when the filaments were close to their extents.

6. FUTURE GARDENS CONTROLLER

The Future Gardens holographic installation [2] (Figure 11) was my first work using computers to control lighting and sound. Working with composer David Atherton, we set up a MIDI music mixing board to program a theatrical lighting grid with the cues triggered by David's music composition. The composition was approximately 12 minutes long and there were individual lighting cues for over forty holograms. The installation was seventeen feet long by ten feet high by seven feet deep. The images are of traditional western landscape imagery: combined with traditional eastern landscape elements inspired by a visit to Japanese gardens in Kyoto.

Figure 11 Future Gardens

The sound mixing board computer program was limited to on/off and any rate of dimming the lights but it could be run by the music cues, so in that way the sound and lighting were in perfect sync. The forty individually controlled lights had to be run through expensive lighting dimmer packs, made more for the 500Watt or 1K to 3K lighting fixtures used in theater productions.

7. LIGHTFOREST AND FUTURE GARDENS CONTROLLER

LightForest: the holographic rain forest, [3] (Figure 12) installation was even more ambitious with the installation of well over two-hundred 12"x16" holograms of rain forest plants, leaves, tree trunks and flowers. Multiple white light rainbow transmission holograms—from fifty to sixty unique masters—were created, laminated, and mounted behind the walls of a specially designed space. It was a room within a room, within a Museum, with the illumination hidden behind the walls, above the ceiling, and from below a false floor. LightForest was a Star Trek 'holodeck 'inspired space, with the idea to hide the illumination sources, as best as possible, from the viewer. The lighting for the holograms was controlled using a MAC MIDI system that extended what was done for Future Gardens. One idea for the water image holograms installed in the floor was to have the computer controlled lighting 'twinkle' across the holograms the way sunlight dances across a lake or pond.

Back in the late 1990's when planning the sensors and lighting control that would be used for LightForest, MIDI was the most widely used protocol by artists working with physical interfaces, computers, lighting and synthesizers. Inspired by early work of Hiroshi Ishi's *Tangible Media Group* at the Media Lab, we programmed the lighting for LightForest in the MAX graphical programming environment developed at IRCAM in France.

Max lets you easily "patch" together lighting and sampled sounds with sensor inputs, firing events from a preset musical track and perform changes on the fly. We used proximity and sound sensors as additional triggers for interactive lighting effects and tried to create a unified experience as groups of people explored the LightForest space.

Figure 12 LightForest

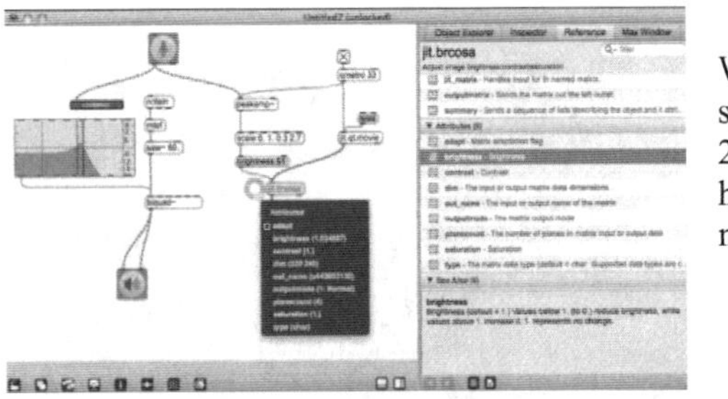

We had to invest a hefty amount into the LightForest control system; $500 for the MAX software, a small MIDI controller, 2 light banks for high-voltage lamps, and a myriad of homemade sensors. When compared to the cost and effort needed for LightLeaves, the difference is *amazing.*

Look deep, deep, deep into nature and then you will understand everything. Albert Einstein

ACKNOWLEDGEMENTS

I would like to acknowledge with thanks the help given by David Chen for support and assistance and the programming of the many of the light devices mentioned in the paper and general support of these projects. I would also like to thank Cody Chen for programming and robotics assistance and Darcy Chen for assistance in the pulse mastering lab. I would also like to acknowledge an artist-in-residence grant from the Center for the Holographic Arts and thank Ana Maria Nicholson, Sam Moree and Martina Mrongovius; and Janie Winchell and the staff of the Art and Nature Center of the Peabody Essex Museum in Salem MA.

REFERENCES

[1] B. Connors, "LightRain: computer controlled kinetic transmission hologram installation" S.P.I.E. *Proceedings of the Eighth International Symposium on Display Holography,* Holography--Culture, Art, and Information Technology, ISBN: 978-7-5023-7262-0, Shenzhen, China 2009
[2] B. Connors, "Future Gardens Holographic Installation", *International Symposium on Display Holography,* Proc. SPIE vol. 1600, p. 423 – 426, Lake Forest, Il 1991
[3] B. Connors, , "LightForest: the holographic rainforest", *Practical Holography XI and Holographic Materials III,* Proc. SPIE vol. 3011, p.146 – 150, San Jose, Ca 1997
[4] B. Connors, "LightForest and the MIT Museum Holography Education Project", *Sixth International Symposium on Display Holography,* Proc. SPIE vol. 3358, p.160 - 167, Lake Forest, Il 1997

Holography without frames: sculptural installations incorporating 'drawn' elements

A Pepper

Senior Lecturer in Fine Art, Nottingham Trent University, Nottingham, UK

E-mail: andrew.pepper@ntu.ac.uk

Abstract. Since its adoption by artists, shortly after the process became practical in the 1960's, holography has struggled to be accepted as a legitimate medium within the visual arts. A number of key milestones, during the early developmental process of creative holography, are acknowledged with reference to sparse critical discussions. The value of 'traditional', framed, wall-based display is explored and a number of sculptural installations by the author, which have attempted to question this method of presentation, are discussed. Floor-based installations, which involve the integration of drawn holographic elements within site-specific locations, are examined.

1. Historical parameters

When any medium, which might be termed 'new', becomes available to creative practitioners, there is often a period of insecurity. This manifests itself within the research activities of the 'early adopters', as practical difficulties for those wishing to exhibit the results, and with curators who are unfamiliar and unsure where, in a historical context, this new manifestation might 'sit'. Holographic images are no exception.

If we consider that holography developed using two distinct methods of display, dictated entirely by their optical production and reconstruction requirements, it is possible to examine distinct influences on creative development.

The very early laser transmission holograms demonstrated by Leith and Upatneiks in the USA [1], presented a dichotomy for display. Viewing their first holograms required laser light to be directed through the holographic plates using the same path taken by the original off-axis recording beam. The display light shone through the plate from behind, out into the space occupied by the viewer - a display configuration diametrically opposed to more 'traditional' methods of presentation. As practitioners in the field, we embrace this off-axis transmission configuration without question. It is, after all, the optical path, which allows this type of holography to make its recordings and was a huge practical development compared to the earlier, in-line, recording and display geometry proposed and demonstrated by Gabor. [2]

For the scientists making the holograms, this method of display is a completely acceptable one. It mimics, with clear logic, the process of recording. It also embedded the display of these early holograms within the laboratory, not just because of the need for coherent light from a laser, but the physical support structures of the optical elements used in the recording. They were often displayed on the optical tables used to record them. Margaret Benyon was one of the first artists to identify issues of display and presentation in relationship to holography. In her 1971 exhibition at the University of Nottingham she presented laser transmission holograms on a large rotatable 'table' so that each framed transparent plate could be moved into the reconstruction light of the laser. [3] This was an unusual display device in a gallery context in that it attempted to accommodate the display light from the side.

In Russia and Europe, where progress was being made using reflection holography, a completely different, and more familiar, display system was being used. As reflection holograms required the display reference beam to be reflected from the surface of the holographic plate (a more 'traditional' approach compared to side/back illuminated laser transmission holograms) and could use non-coherent white light, holograms recorded using this process were easy to frame, hang on a wall, and illuminate from above – a traditional gallery configuration.

2. Critical engagement

As Rebecca Coyle and Philip Hayward rightly identify, "Critical discourse around holography in the 1970's was minimal". [4] This is an accurate reflection of the early gestation period where minimal discourse, and some basic analysis, tended to be either defensive, mainly from makers and promoters in the field, or dismissive, mainly from the few established critics who attempted to engage intelligently with this 'new' medium. During these early stages, there appears to have been an attempt at assimilation of this unfamiliar, phenomenon-based, recording and display process by artists, scientists and critical observers.

Coyle and Hayward go on to pinpoint one of the key issues in this' struggle' for 'acceptance': "Despite the special facilities required for illumination of the holographic image, much holographic art fits far easier within the traditional gallery context of two dimensional display than contemporary technology performance or computer interactivity". [5] The early framing of holography has been either instructional or an attempt to conform to that traditional gallery context. Visitors needed help to 'access' visually, physically and conceptually, the contradictory images they were being presented with; visually solid objects with no physical (real) substance.

3. Frame as boundary

Unsurprisingly, the traditional frame offered an opportunity for familiarity and, possibly, acceptance. For reflection holography, it became the norm, both practically and aesthetically. The glass holographic plates, using traditional frames, allowed work to be securely 'hung' on gallery walls (a clear practical advantage, particularly for curators and installers). The accepted boundaries of the frames provided viewers with a familiar and defined basis on which to engage with the content the frame encased. It is this issue of containment which caused a significant visual and conceptual fissure.

Viewers bring with them an acquired (subliminal) knowledge of how to view 'art' on walls, which includes an understanding of where the work stops (the edge of the frame), orientation (we know where the top and bottom are), the nature of completion (the work is deemed 'finished') and spatial location through the definition of the picture plane (the surface contains the visual illusion of volume). Applying these learnt, and now standard, visual indicators, to holographic imaging causes difficulties of interpretation on both a visual and a conceptual level. Holographic space has a set of parameters which do not appear to conform to our post-Renaissance understanding of the depiction of volume on a flat surface. [6] There is also a dangerous clash of values. We use 9 sources of information to perceive space in a natural environment (occlusion, height in the visual field, relative size, relative density, aerial perspective, binocular disparities, accommodation, convergence, and motion perspective) [7]. When we view a flat 'picture', we understand that an illusion is taking place on the picture plane and compensate perceptually and conceptually. Unfortunately, when holograms are framed in a format which makes them 'appear' like flat pictures, but display stored holographic volume, a visual and perceptual conflict occurs. What appears to be a traditional, framed picture of the illusion of space actually contains information which is three-dimensional but we still apply the 9 ways of assessing them, normally reserved for the natural (real) world. What was an attempt to introduce the unfamiliar observer to an unusual imaging system, through familiar methods of display, actually causes a mismatch of expectations and increased difficulty in processing visual and conceptual aspects of the work on show.

4. Frame as support

For those artists working with transmission holography (laser and white light), the physical requirement of the display (reconstruction) light to pass through the transparent holographic plate from behind, has allowed frames to be used as a support rather than a physical defined edge (although in some examples they perform both functions).

This approach to framing is fundamentally different from that of wall-based examples as it 'offers up' the transparent glass plate, which can be looked through. It is this acknowledgement of the 'window' principal, and the understanding that there is 'space' behind the plate which does not logically connect to the content of the hologram. It can trigger an extra element of investigation on the part of both the artist and the viewer. As Achim Lipp identifies when discussing how glaziers approach and view recently delivered (holographic) glass sheets: "Their eyes don't stop at the surface of the glass. One clearly sees how their pupils continually change focus; over and over again they shift their eyes to a new point at a new distance. Sometimes the look concentrates on a thing quite near to them, maybe just a yard away from the pane of glass, sometimes the pupil relaxes, reaches out into the distance – and the whole time the pane that the 'look' must pass through is leaning on the wall of the art building." [8]

When we 'look' through transparent planes such as these, we appear to be comfortable with the concept of there being 'space' behind. I would, therefore, suggest that this helps us visually disengage from the framing in a way which does not happen in framed reflection holograms. The viewer can accept that any frame involved is there to support the glass sheet. It is peripheral, whereas with reflection holograms it is acknowledged (through historical precedent) as a boundary.

This subtle difference in how we visually and conceptually engage with the frame has allowed artists, working with transmission holography, to begin incorporating the physical aspects of the frame. This is succinctly demonstrated in Rudie Berkhout's '12mW Boogie' [9] and in versions of 'The New Territories' [10] where transparent holographic plates are edged with black frames and mounted on black industrial tripods. The framing device shifts in character to become an element in a recognisable, sculptural, installation. Here the emphasis of the holographic plate shifts to become a demonstrable optical device, particularly when mounted on tripods where the inference of the telescope, and 'looking' into a different world, is difficult to ignore.

4. The hologram as frame

There are examples of artists who have not employed physical framing, using instead either the actual glass substrate of the holographic plate as the main support system for the image area, or lamination onto similarly sized, or larger, sheets of glass and plastic sheeting. 'Cartesian Memory' [11] by Sam Moree 'sits' in the top right of a larger column of glass which extends out of the gallery floor. The addition of luminous krypton tubing contributes to the overall structure of the installation and offers a visual pathway from the gallery floor, up to a section of the complex holographic image, and back down to the floor. The fact that there is also a 'window' effect in being able to look through the transparent sheets, to the gallery space behind, establishes the work within the architectural structure in which it is shown.

Holograms are incorporated into installations where they are integral to the architecture and materials which surround them and therefore become framed through the context in which they are viewed. Vito Orazem's 'Stop' sign, although not conceived as a visual artwork, does highlight many of the aspects of placement and relationships with existing surrounding elements, as well as the position of the observer and the limited viewing zones of holographic display. His study for a holographic stop sign, an optical 'barrier' for subway tracks, uses as its frame the edges of the public subway platform. [12]

5. Placement

Several works produced between 1989 and 2012 have been exhibited by the author of this paper. They attempt to explore the restrictions of traditional framing [13] and interrogate placement and the use of drawn holographic elements in site-specific installations. Three of these will be described in practical and constructional detail here. They range from initial research stages, through site-specific gallery installations, to the most recent piece incorporating existing, functional objects. No attempt to interrogate the conceptual, critical or experiential aspects of these works is included here, partly because, as in all of these pieces, those elements are the triggers which attempt to activate the work in context, and partly because the results of this research are based on a physical and visual engagement.

Often holograms are, quite naturally, clustered together into media-specific exhibitions, which have not only proved very popular (less so critically) but helped establish a ghetto mentality. [14] There is safety in numbers and the visual fidelity of these mammoth survey exhibitions has hidden many aspects of critical engagement. The relevance of works incorporating holography being seen alongside other media, so that they can be 'tested' using the same criteria of critical interrogation, is paramount.

5.1 Positive Attempt

'Positive Attempt' (figure 1) was originally shown in a group sculpture exhibition alongside a broad mix of media and installation techniques. [15] Apart from attempting to engage with an emotional narrative, it was the first opportunity to use projected light as a device for illuminating holograms. Of particular importance was the need to establish a direct link between the content of the holographic image, in this case hand-drawn scribbles, and the light which made them visible.

Figure 1. Positive Attempt, 1998, multiple reflection hologram fragments, glass circle, projected light. First shown at 3-D '89, National Sculpture Open Exhibition, Watermans Arts Centre, Brentford, UK.

Several reflection holograms of gestural marks, by two individuals, were captured using the Denisyuk recording technique. These were displayed pseudoscopically so that the 'drawn' element, with no discernable content, existed in the space between the observer and the holographic plate. This visual device is used in almost all subsequent works.

Each hologram was sealed on the back (emulsion) side and then broken into pieces. There was no attempt to engineer specific shapes from the broken sections. These fragments, characteristics from two people, were mixed together and placed in a circular area on a large circular glass sheet which was raised off the gallery floor by 2 cm. A circular drawing, incorporating a larger 'patch' off-centre was produced and made into a high contrast black and white 35 mm projection slide. This drawn image was then projected vertically down from the gallery ceiling onto the large glass circle and the arrangement of holograms.

Light from the projector (itself a drawing) illuminated the hologram fragments, allowing each to reconstruct its own, stored, set of marks which then appeared just above the surface of the installation. As the glass circle was transparent, any 'unused' light passed through it to appear on the gallery floor, slightly subdued, due to loss of light as it passed through the dark 'tint'.

What resulted was a series of drawn marks in separated spatial locations, which were only visible from certain angles as viewers moved around the circular structure. As the holographic fragments were randomly orientated, each would 'release' its section of marks at differing viewing positions. A kinetic 'pulsing' was produced, which directly related to the speed and direction of the passing viewer. There was, therefore, a tangible connection between the work and the viewer.

Framing here was undertaken by the circular glass 'platform' and the 'drawn' light which illuminated the piece.

5.2 Sight Lines

'Sight Lines' (figure 2) was constructed specifically for installation at Gallery 286, London, UK [16] and provided an opportunity to engage with the existing 'white box' volume of the gallery space, its white painted floor and the traditional wall areas.

Reflection holograms are often displayed on gallery walls, as mentioned previously, to connect with an accepted method of graphic display. In this case they have been moved away and in front of those walls to act as a barrier (line) to the graphic elements, which are shown in the more 'traditional' location. Nine rusted metal plates, each incorporating a small reflection hologram of the shadow of liquid, are hung from the gallery ceiling and illuminated from a position close to the vertical wall behind. This not only allows reconstruction of the holographic images, but casts a rectangular shadow of the metal plate onto the white floor at a location between the observer and the plates.

Photographs of each plate and hologram were edited to select a single, thin, section through the length of the hanging object and these were hung at eye height on the wall behind the piece. This 'traditional' viewing location is separated from the viewer by the construction hanging in front of it.

A series of lines and boundaries are 'drawn' in the gallery space: On the wall, suspended in front of the wall and on the gallery floor.

A final aspect of the installation is in the 9th metal plate which is embedded in the gallery wall, offering a suggestion that the undulating installation continues through the wall and, possibly, much further across London. The implication of a constructed space, more normally associated with holographic images, is reinforced with a physical indicator and another opportunity to suspend belief by engaging with this 'other' possible narrative.

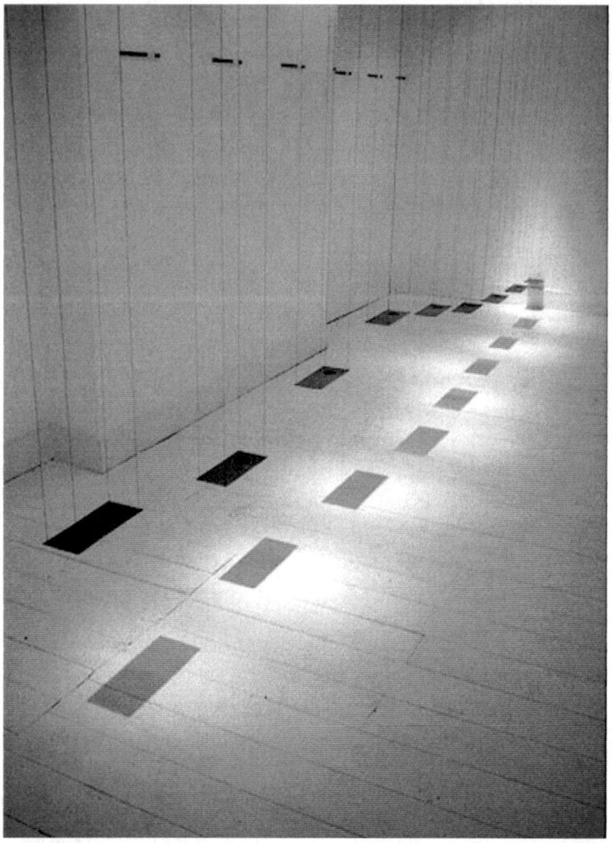

Figure 2, Sight Lines, 2001, shown in Deep Shadows? Solo exhibition, Gallery 286, London, UK, June 14th - July 30th, 2001.

5.3 Tizio Water

'Tizio Water' (figure 3) was installed as part of the Archipelago exhibition [17] which featured experimental practice from a range of art and design disciplines at Nottingham Trent University. Artists were asked to consider themselves and their practice as islands. Those islands could be the work of one practitioner, of an established collaboration, or a group brought together by common concerns.

The installation consisted of two elements; a rectangular reflection hologram of the shadow of liquid and a Tizio table lamp designed by Richard Sapper in 1971. One aim of the installation was to interrogate the symbiosis of the historic 'found' object (recognised, celebrated and archived in major international art and design museums [18]) and the 'historic' holographic recording (unknown, displaced and functionless). The latter is made visible by the presence of the former. The acknowledged supports the unacknowledged.

This small rectangle of 'ground' inverts the normal impression of an archipelago - here the holographic shadow of liquid is the defining terrain, edged by the junction between the glass holographic plate and the surrounding solid architecture (the gallery floor).

Approached from a distance, the black crane-like structure of a Tizio table lamp offers a graphic and defined 'point to view from'. It is, in a way, instructional - guiding the viewer to an observation

Journal of Physics: Conference Series **415** (2013) 012003 doi:10.1088/1742-6596/415/1/012003

point in the same way that traditional frames guide a viewer to the section they should look at. Its light is precisely directed down onto a holographic rectangle, which appears black and flat from a distance.

Common elements of frame, location and illumination are disrupted in an attempt to engage with each component and question their relationship to each other within an historical art and design context.

Figure 3, Tizio Water, 2012, installation with reflection hologram and Tizio table lamp. Bonington Gallery, Nottingham Trent University, UK.

6. Conclusion

The duality of framed holograms, based on their optical recording and reconstruction geometries has, rightly, become blurred as artists engage with the medium as a legitimate process for practical research output.

There will continue to be 'lowest common denominator' displays where framing is a basic first choice. This approach may also be utilized because of issues of practicality, security or architectural display restrictions. However, as artists continue to engage with the critical interrogation of process, placement and contextualisation, the physical frame, and substrate on which holographic emulsions are manufactured, will need to become integral to critical analysis in the same way that other display technologies are absorbed and assessed within the visual arts.

Acknowledgements
I would like to thank Nottingham Trent University School of Art and Design and its research team for supporting the presentation of this paper, my colleagues in fine art, for providing an environment where practice-lead research is considered a viable pedagogic imperative, and to the galleries and curators who have offered considerable freedom to allow me to test out ideas and question issues of 'drawing' through the installations they have supported.

References and Footnotes

[1] Leith E and Upatnieks J June 1965 Photography by laser *Scientific American* vol 212 no 6
[2] Gabor D 1948 A New Microscopic Principle *Nature* vol 161 pp 777-778
[3] Benyon M 1973 Holography as an art medium, *Leonardo Journal of the International Society for the Arts, Sciences and Technology* vol 6 p 3
[4] Coyle R and Hayward P 1995 *Apparition, Holographic Art in Australia* (Sydney: Power Publications) p 19
[5] Coyle R, Hayward P 1995 *Apparition, Holographic Art in Australia* (Sydney: Power Publications) p 20
[6] Pepper A 1989 Holographic space: a generalised graphic definition *Holography as an Art Medium Leonardo Journal of the International Society for the Arts, Sciences and Technology* vol 22 no 3 & 4 ed Brill L (Oxford: Pergamon Press) p 295
[7] Cutting J 1997 How the eye measures reality and virtual reality, High-performance computing and human vision 1 *Behavior Research Methods, Instruments, & Computers* vol 29 pp 27-36
[8] Lipp A 1985 *More Light!, Mehr Licht, Künstlerhologramme und Lichtobjekte* (Hamburg: Ernst Kabel Verlag) pp 60-61
[9] White light transmission holograms (rainbow) 1978. Three 8" x 10" transparent glass holographic plates, framed and mounded on black tri-pod. Collection MIT Museum, Cambridge, USA. Accession No. MOH-1978.06.
[10] White light transmission holograms (rainbow) diptych, 1984. Fielmann Collection, Hamburg Germany.
[11] White light transmission holograms (rainbow), 1981, laminated onto glass sheet incorporating luminous krypton tube, 30.4 x 160 cm. Lauk Collection, Germany.
[12] Orazem V 1994 Holography as an element of the media architecture Proc. Fith International Symposium on Display Holography SPIE vol 2333 ed Jeong T (Bellingham: SPIE) p 171
[13] These include: Positive Attempt 1989, Positive Attempt Redrawn 1990, Ripped Wall 1992, Wall Drawing 1998, Divided Line 1999, Sight Lines 2001, Curved 2001, Sight Lines Wall 2004, Light Liquid 2011, Vertical Liquid 2011, Tizio Water 2012.
[14] Pepper A 1994 Beyond the gallery ghetto *The Creative Holography Index The International Catalogue for Holography* vol 2 issue 2 (Bergisch Gladbach: Monand Press)
[15] Selected for exhibition in the 3-D '89, National Sculpture Open Exhibition, Watermans Arts Centre, Brentford, UK
[16] Deep Shadows? 2001 Solo Exhibition, Gallery 286, London, UK, 14th June – 30th July
[17] Archipelago 2012 Bonington Gallery, School of Art and Design, Nottingham Trent University, Nottingham, UK. 16th January – 10th February
[18] Sapper's Tizio lamp is acknowledged as a design classic through its inclusion in the Twentieth Century Design Collection, Metropolitan Museum of Art, New York, USA, and the Design Collection of the Museum of Modern Art, New York, USA

A Curious Conundrum; The State of Holographic Portraiture in the 21st Century

R Taylor
President, Forth Dimension Holographic Technologies
PO Box 259 Nashville, Indiana 47448 USA

E-mail: rtaylor@forthdimension.net

Abstract. The technology of producing (true) hologram portraits was first introduced in the late 1960's. From this time, a number of individuals and organizations worldwide have specialized in providing holographic portraiture services with varying degrees of achievement. Yet today, some 45 years later, holographic portraiture remains an obscure and niche form of displaying an individual's likeness. Despite all of this technology's promising and unique attributes, and the astonishing fact of holography being the most accurate and realistic form of imaging available today; true holographic portraits continues to be a form of portraiture largely unknown to the general public and has never achieved large-scale commercial success.

This paper will present a brief history of holographic portraiture, designating the different types of 3-D hologram portraits available today, and their uses. Emphasis will be given to true holographic pulsed portraiture in which the subject itself is recorded holographically using high-energy pulsed lasers. Possible cause and effect for explaining the present demise of this type of portrait making will be discussed along with recent advancements and future developments in this fledgling field which could ultimately lead to a "tipping point" in large-scale consumer and commercial awareness and desirability of the medium.

The author will share his experiences in operating pulsed holographic portraiture studios for over the last 15 years including the vision of a new type of holographic portrait studio for the 21st century which he hopes will attain the level of success enabling a next generation of commercially viable holographic portrait studios for the future.

Keywords: Holographic Portraiture, Hologram Portraits, 3-D Portraits, Pulsed Holography, Pulsed Portraits, Portraiture Holography, 3-D Imaging, Life Holography

1. Introduction

The practice of representing the physical and psychological likeness of an individual is as old as art itself. The principle methods of portrait making; e.g. painting, sculpture, and photography, have long been used successfully to immortalize one's "essence of being". The mystical, magical, and spiritual nature of the portrait has long been associated and accepted within each of these methods.

A relatively recent newcomer, holography and it's unique ability to accurately record and reconstruct original light wave-fronts, *particularly* lends itself to this form of artistic endeavor. It is *exactly* this capability of capturing the "wholeness" of an individual's likeness that makes hologram portraiture so alluring and attractive.

Yet, despite some forty plus years on the scene and even the recent mis-placed hype of the "Tupac hologram" among many other more substantial achievements, holographic portraits remain largely undiscovered and out of the mainstream public's eye. Why is this so? How can it be that the most accurate method of reproducing one's likeness (and thus by its very nature, potentially the most emotional) is also by far the least sought after and under-utilized? What forces and factors are involved, and how and when will this new form of portraiture evolve to a wide-spread audience? This paper will attempt to bring to the forefront some of the existing factors and social conditions that have limited the large-scale success of this method of portraiture to date.

2. Classifications and History

Hologram portraits are generally classified into two distinct types.[1] Pulsed portraits are classically referred to as *true* hologram portraits, in which the original image-capture (H1) and subsequent final copy (H2) are produced using wave-front recording. In this type of hologram, high energy, extremely short exposure ruby or frequency-doubled neodymium-based pulsed lasers are used to successfully record the living subject. Multiplex and/or stereogram portraits are holograms in which the original image capture is generated photographically (P1) and are subsequently utilized in the production of a final H1 and/or H2 hologram (usually a rainbow hologram) with standard continuous wave lasers. A recent sub-category of this type worth mentioning is the now commercially available direct-write, digital "dot-matrix" technology in which the initial image-capture can be generated by digital photography and/or computer generated content and then directly "printed" as a full-color reflection hologram using either CW or pulsed RGB lasers (GEOLA Digital uab Vilnius Lithuania, Zebra Imaging Austin Texas USA, Ultimate Holography Bordeaux France). This paper will principally be referring to the completely holographic method of recording, known classically as "pulsed portraits".[2]

Figure 1 Ronald Reagan pulsed hologram portrait. (1991, Bjelkhagen, Marhic) Author's personal collection.

The first-ever pulsed hologram (self)portrait was produced by Lawrence D. Siebert while he was working as a electrical engineer for the Conductron Corporation on (what must have been a scary) Halloween night in 1967. Since this time, many successful holographers and artists have produced hologram portraits using both pulsed; ex., Margaret Benyon, Hans Bjelkhagen, Richard Bruck, Bruce Nauman, Peter and Ana Marie Nicholson, Edwina Orr, John Webster, and stereogram techniques; ex., Lloyd Cross, Mark Diamond, Larry Lieberman, Jason Sapan, among others.

A historic example of a pulsed transmission hologram portrait was done of Dennis Gabor (1971, R. Rinehart) as homage to Dr. Gabor's Nobel Prize that year for the invention of holography. Since then hologram portraits have been produced of US Presidents Ronald Reagan (P), Bill Clinton (S), Barack Obama (D), and Queen Elizabeth II of England (S) and include many other notable heads-of-state figures of our times.[3] Yet if a sampling where taken, most of the public would report not being aware of these holograms and would be hard pressed to

[1] By the term "hologram portrait", it is defined here as any type of portrait that uses the science of wave-front recording and reconstruction known as holography, either partially or wholly in its method of producing a three-dimensional image of an individual's likeness. The author expressly wishes to distinguish the art of true holography from other popular alternative 2-D and 3-D projection and "Pepper's ghost" effects and techniques that inaccurately use this terminology.

[2] It should be noted that the majority of "pulsed" holographers do not prefer this term for their profession, justifiably citing portrait photographers are not sub-termed "flash" and "available light" photographers.

[3] Denotes- (P) pulsed hologram portrait, (S) stereogram hologram portrait, (D) computer-generated digital hologram portrait.

remember an instance of seeing *any* hologram portrait (albeit possibly the mislabeled "CNN" and "Tupac" variety).

Due to the strict requirements initiated in holographic recording, original holographic image-capture (H1)of living and/or moving things can only be accomplished with extremely short exposure, high energy pulsed lasers. Pulse line-widths in the 25-50 nanosecond range and energy levels of between 1-5 Joules are acceptable for currently available holographic silver-halide recording emulsions such as the Slavich VRP-M for green pulsed holography. Lasers can be found (albeit still relatively expensive) in both the classic ruby or the newer neodymium crystal type and are capable of producing high quality, large-format pulsed holographic portraiture. Optical schemes for the recording of living things as true holograms has not changed much. Multiple diffuse object beams allow maximum creative lighting control for the holographer. When using these types of Class IV lasers, extreme care must be maintained to keep maximum permissible exposures of radiation well below the levels set forth to eliminate ocular risks to the subject(s).

The majority of pulsed hologram portraits made to date have used the ruby crystal rod for light amplification simply because this type of laser happens to be the very first laser invented, and also because many of the original pulsed ruby lasers designed for holography (JK Lasers, Lumonics 1970's) are still around and functional today. The principle disadvantages of using the pulsed ruby laser for holographic portraiture is its output wavelength at 694nm. The deep-red color of this wavelength is at best difficult to use for the imaging of subjects as this color does not reflect well and permeates the outer layers of human skin producing a "glowing effect" if suitable make-up is not instituted. Other disadvantages include the difficulty of keeping the laser properly tuned and aligned to the high optical beam characteristics required for display holography and often the additional need for continuous-wave vibration controlled laser recording systems for the final H2 reflection image transfer and the added costs associated with such. This said, outstanding hologram portraits have been made with this laser design. What is considered by many as one of the finest examples of hologram pulsed portraits of all time, "Lucy in a Tin Hat" (1987, Patrick Boyd) was produced with this original laser design.[4]

Figure 2 "Lucy in a Tin Hat" (1988, Patrick Boyd) Author's personal collection.

Green-based pulsed laser holography using the frequency-doubled neodymium family of lasers (YAG/YLF) first appeared in the former Soviet Union in the late 1980's. In the USA, Ron and Bernadette Olson (Laser Reflections, Poulsbo, Washington) were the first to incorporate this laser design into a holography laser using military surplus optics to successfully produce hologram portraits. The advantages of the Nd:YAG/YLF laser is primarily improvements/corrections in the addressed disadvantages of the ruby, i.e., better reflecting green wavelength light, vastly improved reliability and stability, and the fact that final H2 reflection copies can be produced with the same laser system.
Today it is generally accepted that frequency-doubled neodymium-based pulsed holography lasers are the superior choice for monochromatic, high definition pulsed portraiture[5]

[4] It was the initial viewing of this hologram in 1990 that started the Author's life-long love affair with holographic portraiture.

[5] A interesting note to this trend is the complete turn-around of commercially available recording materials for pulsed holography, with red pulsed sensitive emulsions currently in short supply. All of the major photographic film suppliers including Agfa, Ilford, and Fuji no longer manufacture holographic materials, although some of these emulsions are still available in limited supplies on the surplus market.

3. Technical Advancements

Despite the extremely unique true three-dimensional image reproduction qualities, and ultra-high resolution nature of display holograms, commercial pulsed display holography continues to struggle to gain market share and become a commercially successful business model. Companies such as GEOLA Digital uab, and Ultimate Holography have been leaders in the manufacture of specialty pulsed holography lasers and camera systems specifically designed for pulsed portraiture applications. GEOLA Digital uab continues to produce a line of commercially available pulsed neodymium holography lasers in addition to complete turn-key pulsed holography studio and portable systems that are used world-wide within the industry. These two companies have additionally contributed greatly to the advancement of designing and making commercially available the specialty film and glass plate fine-grain AgBr emulsions (VRP-M, Ultimate) required to successfully manufacture high quality pulsed display holograms.

Due to the extremely short exposure times of pulsed lasers and the effect known as latent-image fading, special developers and processing techniques must be utilized for the manufacture of pulsed holograms. Photographic techniques such as hypersensitization and latensification are used to increase the light sensitivity and reduce the reciprocity failure tendencies of classically slow fine-grain holographic emulsions. The holographic developer SM-6 (Bjelkhagen/Wesly) is now considered the defacto standard for pulsed laser holography.[6] It has been observed that slight reductions in the concentration of phenidone in this developer can be fine-tuned to improve the overall signal to noise ratio of finished holograms while obtaining only a slight decrease in diffraction efficiency. Bleaches common for pulsed holography include multiple versions of the rehalogenating type, such as the PBU-Amidol (Bjelkhagen-Phillips) and Ferric nitrate (Phillips) formulations. Post processing emulsion shrinking color-control techniques have been developed for greatly improving the aesthetic qualities of monochromatic green pulsed reflection holograms using varying concentrations of d-Sorbitol or Citric Acid solutions(Olson-Blyth) .

Although currently impractical for daily use, the multi-step processing technique known as silver-halide sensitized gelatin (SHSG) which renders silver-halide emulsions as “grain-less” with high diffraction efficiency and low scatter similar to dichromated gelatin (DCG) has been shown to produce pulsed display holograms of exceptional brightness and clarity. Further advances in the chemical processing/recording materials will continue to improve the over-all quality and archival properties of pulsed holograms in the future.

In hologram reconstruction, low-wattage, narrow-band LED lighting solutions are now available in lumens sufficient to adequately illuminate standard 30x40 cm and larger monochromatic reflection holograms with higher contrast and increased sharpness over depth-of-field. Additional benefits include lower operating costs, greatly increased bulb life, and reduced heat generation compared to standard halogen spot lamp lighting.

Recently research has begun by the GEOLA Group and others towards the development of two (RG), and ultimately, three (RGB) color high-energy pulsed holography lasers based on the second and third harmonic generation of the neodymium family of lasers. The commercialization of a two-color high energy pulsed holography laser would make possible natural color skin tones along with multi-chromaticity scenes. This advancement would compare in significance to the commercialization of the first photographic portraits from black and white to color.

---

[6] Ed Wesly relates the discovery of this developer by accident. A lab assistant at the time (Sal), mis-measured the prescribed amount of phenidone (6 grams), and “Sal’s Mistake-6 wasinvented.

4. Aesthetics

By the view of the author, current limitations of pulsed display holography such as specific lighting playback requirements, limited viewing angle, and monochromaticity of imagery have little to do with the lack of commercial success of the medium. Moreover, it is felt that factors such as the high-cost of imaging, narrow exposure to within society, lack of a successful marketing strategy, and the complex physiological perceptions of the public play a larger role in the limitation of large-scale commercialization of holographic portraiture. The general public's changing perceptions (and ultimately the desirability) of hologram portraiture is a complex subject that warrants further investigation. Historically, one can certainly compare and see similarities within the early days of photography, when initially a substantial percentage of the population related the photographic portrait with the "capturing of one's soul". "Spooky" and "scary" are two adjectives that are used far too often by some upon witnessing a pulsed portrait for the first time.

Figure 3 "Nathan & Carolyn" Author's children, pulsed transmission hologram. (1993, Bjelkhagen, Bruck)

From the author's experience in the direct marketing of holographic portraiture over the last 20 years, it can be noted that this same type of adverse emotional state towards the "intense realism" and "stillness" of pulsed portraits is experienced by a significant portion of the population today. That said, clearly there is also a growing (largely untapped) audience which can envision the benefits of commissioning a personal pulsed portrait of a loved one and upon completion relate this as a deeply positive emotional experience.[7]

Of course just as within the other classical portraiture mediums, successful commercialization is largely achieved by producing *aesthetically pleasing* portraits. Holographic portraiture is no different in this regard and multiple factors have contributed in a less than exemplary record for pulsed portraiture. Part of the reason for this is the extremely high-costs associated with holographic recording materials. In an age of digital photography, where endless shots and poses can be recorded during a portrait session, pulsed holography must balance a rather limited "contact sheet" from which to select versus the steep cost of it's recording materials. Another factor lies within the sterile laboratory and eerie safe-light conditions of the conventional pulsed holography studio. As can be imagined, small children are especially difficult to image in this environment.

Portraiture Desirability Factors

Technique	Cost	Versatility	Aesthetics	Uniqueness	Realism
Sculpture	High	Low	High	High	Low
Painting	High	Low	High	Medium	Low
Photography	Low/Medium	High	Medium	Low	Medium/High
Holography	High	Low	Low/Medium	High	High

Figure 4 Factors determining the marketability of various portrait techniques.

The absolutely "pure" and unretouchable nature of holography would normally be perceived as a disadvantage for other imaging mediums, however with pulsed portraits it can be used to an unique advantage. The "ultra-accurate" aspect of pulsed portraiture allows the client of this service the

7 In 1997, Forth Dimension Holographics completed a marketing study on the perceptions and desirability of holographic portraiture. 244 participants filled out a short questionnaire while analyzing a pulsed hologram portrait. Copies of this study are available upon request.

assurance of knowing their family keepsake's likeness is exact in the greatest detail possible. Literally a moment in time captured forever. That said, correctly designed H1/H2 transfer optical recording schemes *must be strictly* maintained in order to not introduce three-dimensional distortions of the subject which result in ghastly reflection hologram portraits!

To reduce the material costs, some pulsed holography studios have initiated the on-site manufacture of film/plates (Laser Reflections/Forth Dimension Holographics) in which lower cost/higher quality film emulsions are permanently laminated to glass prior to exposure. This results in a substantial reduction in material costs allowing for more H1 transmission proofs to be made during a portrait session while also achieving the blemish-free glass plate quality required particularly of H2 reflection holograms.

The GEOLA Group has been conducting research in the manufacture of high-speed, large-format shutter systems in which a light sensitive holographic plate is "pre-loaded" in a light safe chamber in much the same way as early photographers did with photographic plates. The shutter is triggered to fully open within a few milliseconds around the nanosecond laser wavelength pulse window. This enables the holographer to pose his clients in an ambient light environment, greatly easing both the subjects themselves and the holographer's task at hand.

5. Authors Personal Journey

In 1992 Forth Dimension Holographics (FDH) was started as a small hologram gallery/retail store in the quaint rural tourist village of Nashville Indiana. From the start, the author had the vision of operating a specialty pulsed portraiture studio in this locale. While attending the 6th International Symposium of Display Holography, the author was introduced to Dr. David Ratcliffe and his team of Russian scientists known as "General Optics Laboratory" (GEOLA). During the conference this company introduced their new line of pulsed lasers and holography studio systems. This meeting set in place the author's lifelong commitment to the advancement of pulsed holographic portraiture.

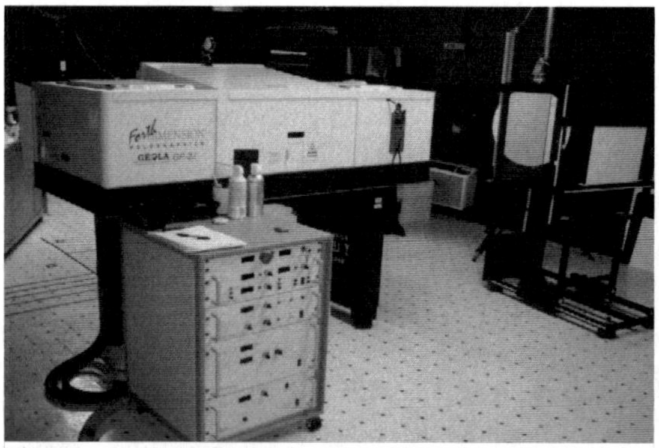

Figure 5 FDH/GEOLA pulsed holography studio, Nashville Indiana USA (2008)

In 1999, FDH imported the first operational GEOLA GP-2J (now designated the HS-2) and opened a pulsed holography studio specializing in portraiture. During the next 10 years, over 200 pulsed portraits were commissioned at the facilities in Indiana. In 2003, a satellite GEOLA pulsed holography studio was installed in St. Charles Missouri. In 2008, FDH was forced to close its Indiana pulsed holography studio operations due to the poor economic climate. That same year, the author was contracted to operate part-time the pulsed holography studio of Ron Olson (Ron Olson Holography) in Las Vegas Nevada. This operation also subsequently closed, and the studio was moved to Poulsbo, Washington (Laser Reflections) where it is operating today. FDH is currently completing it's pulsed portraiture commissions at the St. Charles Missouri satellite location.

Now spanning 15 years, FDH has been the North American distributor and representative of The GEOLA Group. FDH anticipates opening a newly remodeled holography center in southern Indiana with both full-service analog pulsed, and full-color digital hologram production capabilities in the first quarter of 2013.

6. Future Directions

Recently there has been a renewed interest in holography's purest form i.e.; transmission holograms (H1). In this form of display, with the correct laser wavelength characteristics for reconstruction, an *exact* 3-D copy of the original subject is achieved. FDH is currently developing an integrated free-standing commercial display in which low-power CW lasers with simple beam-cleaning/folding optics are used. This will be marketed as an "add-on" sale to existing/new reflection pulsed portrait customers. The unique "magic window" qualities of this type of hologram offer the public a completely different emotional experience than a conventional reflection hologram portrait. This is as *real* as it gets.

Although not comparable to 1:1, high-definition, true 3-D image-capture pulsed portraiture, interesting hologram portraiture avenues have recently opened with the aforementioned new digital hologram technology. Using simple digital devices such as cellphone cameras and web-cams, companies such as GEOLA Digital will produce small, economical, full-color stereogram hologram portraits for anyone who electronically sends them their content.[8] Combining more advanced digital photography equipment such as the GEOLA Holo-Cam in conjunction with a continued direction of smaller and smaller holo-pixel resolutions will enable conventional photographers to offer larger, higher quality custom full-color hologram portraits to their clients with interesting effects such as animation and CGI backgrounds. Indeed it is now possible for the first time to enlist the talents of computer 3-D modeling artists to produce a photo-realistic hologram portrait of the deceased.[9]

7. Conclusions

The future of holographic portraiture is largely an uncertain one. Pulsed holographers continue to remain a very small group (as is reflective of display holography in general). With continued dedication and education, and a little patience, the time may be nearing where the public's awareness, perceptions, and appreciation of a true holographic likeness coincide to produce a watershed event in which holographic portraiture becomes a commercially sustainable industry.

FDH's anticipated new facilities will reflect this approach and be attuned to a next-generation clientele that will bring with them a new level of conditioning to, and acceptance of, new personal technologies such as holographic portraiture. The importance of using pulsed holography as an archival tool *today* cannot be over-stressed. Many current figureheads of our society deserve to be chronicled for historical record.Can you imagine pulsed holographic images of... The Dali Lama..., Nelson Mandela..., Stephen Hawking..., Steve Jobs..., *can you imagine*? We as a society currently have obtainable to us, nothing less than the ability to store and display the most accurate and detailed reproduction of an individual's likeness possible with today's technology. And we are vastly under-utilizing this imaging tool available to ourselves and for the benefit of future generations.

There is a need for more educational institutions, such as the HOLO-Center (Seoul, South Korea) and Ohio State University (Columbus Ohio) with their existing pulsed laser facilities, to continue to instruct and educate the new breed of holographer with the skills needed to evolve to the next level of commercial availability of the medium. Only by continued public exposure and awareness, woven within the complex fabric of an ever-changing social environment will holographic portraiture reach a "tipping point" of commercial success. It is in the view of this author and a growing number of others that this is simply a matter of "when" and not "if" this will occur.

---

[8] http://www.geola.lt/lt/digital_holography/mobile_phone_holography/

[9] http://www.3d-portrait.blogspot.com

8. References

Bjelkhagen H, 1991 "Holographic portraits" Holography *J. SPIE* **8** ed. Greguss P, Jeong T 347-53

Bjelkhagen H, 1992 "Holographic portraits made by pulse lasers" *Leonardo* **25** 443-8

Johnston S, 2006 Holographic Visions *Oxford University Press* 206-9 400 423 426 442 445

Grichine M, Ratcliffe D, Skokov G, 1998 "An integrated pulsed holography system for mastering and transferring onto Agfa or VRP-M emulsions" Sixth International Symposium on Display Holography, *J. SPIE* **3358** ed. Jeong T 203-10

Ratcliffe D, Rodin A, Zacharovas S, 2000 "Evolution of automated turn-key systems for the production of rainbow and reflection hologram runs by pulsed laser" Holography 2000, *J. SPIE* **4149** ed. Jeong T, Sobotka W 359-366

Rodin A, Taylor R, 2003 "Emulsions for pulsed holography: new and improved processing schemes" Practical Holography XVII and Holographic Materials IX *J. SPIE* **5005** ed. Jeong T 75-85

Bjelkhagen H, 1993 Silver-Halide Recording Materials *Springer-Verlag* 215 236-238 274 283-4 291-3

Olson R, 2002 "Holography grows up and gets a job" *unpublished work*

Evstigneeva M, Drozdova O, Mikhailov V, 2002 "Silver-halide sensitized gelatin (SHSG) processing method for pulse holograms recorded on VRP emulsions" Practical Holography XVI and Holographic Materials VIII *J. SPIE* **4659** ed. Benton S, Stevenson S, Trout T 405-12

Johnston S, 2011 "Whatever became of holography?" American Scientist *Sigma Xi, Scientific Research Society* **99/6** 482-9

Zacharovas S, Bakanas R, Pileckas J, Nikolskij A, Brotherton-Ratcliffe D, 2006 "A new device for holographic photography" Advances in Display Holography, Seventh International Symposium on Display Holography *River Valley Press* ed. Bjelkhagen H 222-7

Brotherton-Ratcliffe D, 2006 "Large format digital colour holograms produced using RGB pulsed laser technology" Advances in Display Holography, Seventh International Symposium on Display Holography *River Valley Press* ed. Bjelkhagen H 200-8

Gladwell M, 2000 The Tipping Point *Little, Brown and Company*

Journal of Physics: Conference Series **415** (2013) 012005 doi:10.1088/1742-6596/415/1/012005

Comparing oil paintings and their true color holograms from a painter's perspective.

Mike Finegan

Manhattan, New York USA

mkrjf0@yahoo.com

Abstract. Several researchers have used holography for museography and art conservation. It is often stated that 3 or 4 laser wavelengths are sufficient to create a visual duplicate of any two dimensional work. This claim is tested specifically for the domain of oil painting. Like holograms, oil paintings convey information through a three dimensional light field. In many cases, this is an intentional effect of the painter's process of creating the painting. An overview of related fields is provided since there are dependencies. Classical techniques and palettes are subjected to holographic duplication. The results are classified and may be surprising to both holographer and the gallery owner.

1. Introduction

Using Munsell color space, painter's perspective and leveraging state of the art equipment and media, our goal is to determine if museum quality duplication of oil paintings (over the full visual field of observation) is possible using color reflection holography.

- Claims of holography to create exact color duplicates do not always seem to be accurate
- The full visual field is sometimes ignored in analysis of duplication quality
- Measurement criteria are not always clearly presented and so are not easily reproducible
- Mathematical simulations are not always correlated with actual results
- CIE 1931 is nonlinear across dimensions – but this is not usually mentioned – making error factor appear smaller than visually apparent; object luminance is not explicitly compared
- Munsell color theory can be applied directly at all stages of analysis – to select test cases and subjectively evaluate matches
- Color opponent theory should be leveraged in deciding illumination requirements (special consideration for purple, yellow)
- Illumination requirements are correlated between recording, processing and playback – and should be explicit
- Some paintings are better exemplars than others
- Palette 'primary' pigment spectra should be correlated with the resulting overall spectrum
- Test cases should include representative and reproducible pigments, mediums, techniques, finishing (lead white, lindseed oil, glazing / impasto, varnish, high chroma, etc.)
- The same concepts should be applicable to other forms of painting (watercolor, acrylic, etc.)

2. Relevant Painting Concepts

Reputedly, oil painting was discovered by the stained glass and tempera artisans Jan (and Hubert) van Eyck of Flanders, circa 1400. Between then and now a series of technical improvements have resulted in commercially available paints of consistent high quality.

Each academic school of thought since then has promoted different sets of base pigments/colors to be used as a 'palette.' Different methods of color mixing have been employed ranging from three or four colors to complete grids of premixed colors. Black may be straight from the tube, or may be created with four complementary colors. Sometimes the colors are mixed on the palette, other times on the canvas, some times by layering transparent glazes and yet other times they are simply placed next to each other on the canvas and mixed by the human visual system. Complementary colors may be intentionally placed in proximity of each other to 'vibrate' the HVS (see detail from painting).

Figure 1 – Color Function Painting: ALL THINGS DO LIVE IN THE THREE (Anuszkiewicz)

Studying with Jack Faragasso at the New York Art Students' League introduced me to the premixed grid approach of Reilly and Faragasso. It is distinguished from other premixed grids in that it is based on Munsell Color Theory. Hue selection is performed at constant Value. Value selection is performed at constant Hue. Chroma selection is performed at constant Hue and constant Value. In an advanced variant, the grid can be reduced to a small subset of spectral colors which are sufficient to recreate the full grid. It is primarily aimed at flesh tone reproduction for all ethnicities under all lighting conditions. Additional colors can used independently of the grid primaries to match atypical colors.

Particularly with oil paint, light can be partially absorbed and reflected by the various layers of the painting. This results in a more complex reflection of light and individual wavelengths than highly reflective opaque painting.

The surface of a painting can range from smooth as glass to deep troughs and bare canvas or board. The paint may have additives, such as sand, that create unusual textures. Sometimes this surface is sealed with a varnish and other times not.

So, the light reflected from a painting is not the same as reflected from a photograph. The surface is not uniform in texture or material or reflectance and may even create shadows on its own surface due to deep texture. Finally, the colors may be composed of a wide variety of natural, mineral or synthetic pigments with different absorption spectra.

As a result of the above complexities, some accidental and others intentionally created by the artist, the painting may appear very different from various viewing angles and distances. The illumination of the painting is intimately related to the viewing experience and can sometimes have a negative impact (glare, non-uniform lighting, etc.).

The elements mentioned above have led me to believe that true color holography might be the only way to duplicate the viewing experience of a specific painting and that Munsell Color Theory could be applied to the task.

3. Relevant Lighting Concepts

Lighting is critical at all stages of the creation and enjoyment of an oil painting. Each painter has a specific preference depending on the subject, mood and destination of the painting. Ideally the painting is displayed in the same light as when it was painted. However, even in paintings of the 17th century one can detect instances where part of a painting was finished indoors and part was finished outdoors. Furthermore, many museums used mixed modes of lighting, for a variety of reasons. So it is difficult to point to an ideal historical viewing standard for lighting. Many feel early morning light is perfect, and some museums have tried to adopt this by using slightly warm, broad spectrum, lighting.

Figure 2 Typical Museum Gallery

Lighting is typically muted in the gallery except for spotlights focused on individual artwork. Recently, LED lighting has come to the forefront and some museums are starting to switch over, gallery by gallery, to warm white LED from incandescent Halogen. The museum LED have a spectrum very similar to existing Halogen, but with long life, lower cost of ownership, lower temperatures, etc. In a recent study, the lamps selected by the museum were determined from qualitative assessment by staff and visitors of color samples and artwork. The better LED spotlights were favored over Halogen. Halogens are (subjectively) described as stronger in red hue but weaker in blue and purple hues than LED.

Figure 3 U.S. DOE study at Jordan Schnitzer museum

The technologies used for current LED lighting is not always as would be expected. One of the vendors provides 'white light' using two different technologies. While the result is white in both cases, the color accuracy is likely very different. There has been discussion in some of the lighting studies that color accuracy metrics used to categorize the quality of LED (CRI for example) do not actually correlate well with qualitative assessment.

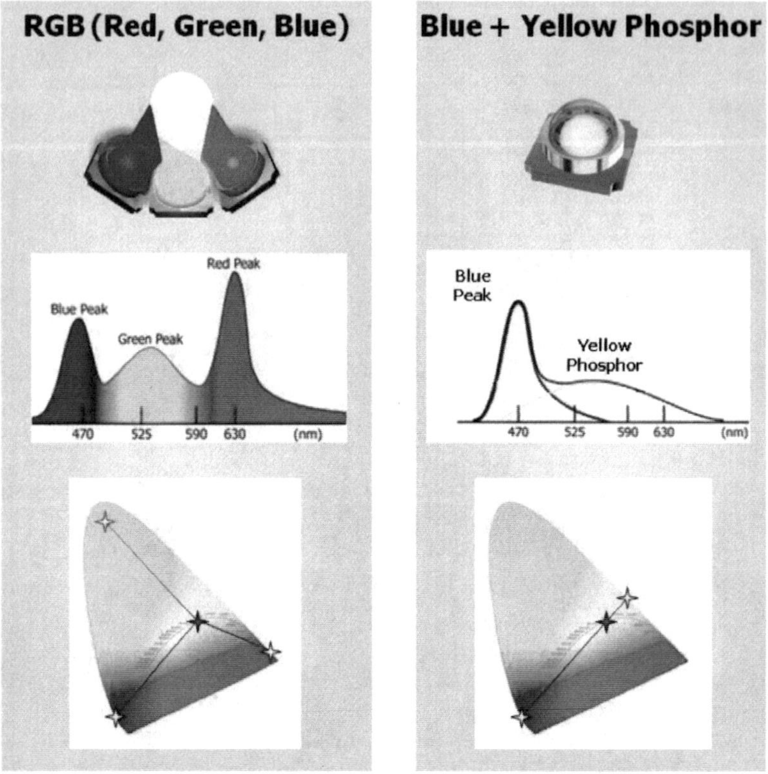

Figure 4 – from CREE: "Two ways to get white"

4. Relevant Holography Concepts

A color reflection hologram (Denisyuk/Lippmann/Bragg) captures the standing waves created between object and emulsion for each laser wavelength used. When processed and illuminated with white light, all colors are filtered out except the wavelengths originally reflected by the object. If the illuminant is the original laser light source, or equivalent wavelengths, with identical geometry then the reproduction quality should be the maximum possible. One criteria for perfect reconstruction is that 50% of observers could not distinguish between hologram and original. Another criteria is color difference in the CIE color space.

Denisyuk has written that the dynamic range of brightness recorded by such a hologram should exceed the dynamic range of photographic film. The 'S' curve typical of film is presented to show that two bright points on film (E2, E3) may actually record the same value (D2), whereas they would not for a hologram since the value for any point on the surface is recorded across the entire hologram and will not over saturate at any point in the holographic film. No mention is made if this will also hold true at the dark end of the brightness range. Intuitively, the answer is no since a very small value would be recorded at each point in the hologram when the object surface is dark. At some point it will be comparable in magnitude to noise.

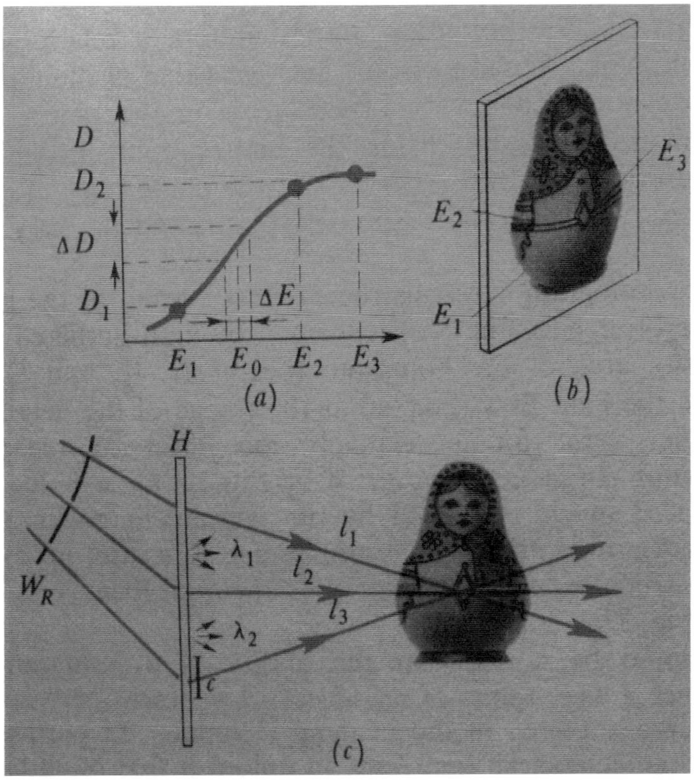

Figure 5 – from Denisyuk: dynamic range comparison

Several authors (Hubel, Bjelkhagen, etc.) have documented issues specific to multi-color reflection holograms, including the reduction in diffraction efficiency as the number of discrete wavelengths are increased. The limit/effect is dependent on the recording material. The same authors have also investigated using the CIE (1931) color space to analyze which, and how many, discrete wavelengths are required to sufficiently cover the space of recordable colors. In many cases, numerical simulations have been used to derive the optimal choices. In others, the evaluation criteria depend on some aspect of the material or its processing. Most of the CIE based analyses ignore the non-linearity of the 1931 CIE color space. None appear to match Luminance explicitly. Many of the emulsion issues are now historical, as current high quality holographic emulsions do exist without the issues mentioned (Gentet emulsion for example).

Flourescent and diffraction effect colors are not recordable with holography, although for some cases they can be illuminated with white laser light.

5. Relevant Neurophysiology Concepts

The Human Visual System (HVS) has two modes – night (scotopic) monochromatic vision using rods and daylight (photopic) color vision using cones. The HVS uses three different 'cone' cells to detect respectively red, green and blue light. The detection is not independent however (they overlap in broadly detected wavelengths). There are more red cones than green cones, and more green than blue. This has led to opponent color theories that describe the further abstraction of color by the HVS. An internal value of yellow (not directly detected as yellow) is synthesized from the red, green and blue inputs. These four values are then used to creed opponent color channels, namely Blue versus Yellow and Red versus Green. There is speculation that this decomposition is the most efficient from an information theory perspective as various reductions in data transfer and processing result.

Yellow is sensed when equal input of Green and Red are present. More interesting than the apparent redundancy of yellow is the existence of purple, which is a unique color only in our mind, resulting from equal Red and Blue inputs. When R-G and Y-B are balanced we sense Gray.

Also, additive color and subtractive color stimulate the HVS slightly differently. For subtractive color the complementary colors must balance. Lasers combine as additive color and light is reflected from a painted surface as subtractive color. For this reason, superposition calculations for color on CIE chart are predictable. Subtractive colors for similarly colored pigments are not predictable. As the spectrum of illuminant changes two pigments that were identical in color may have very different colors as their absorption can change with wavelength. This defines metamerism – colors that are identical only under certain lighting conditions.

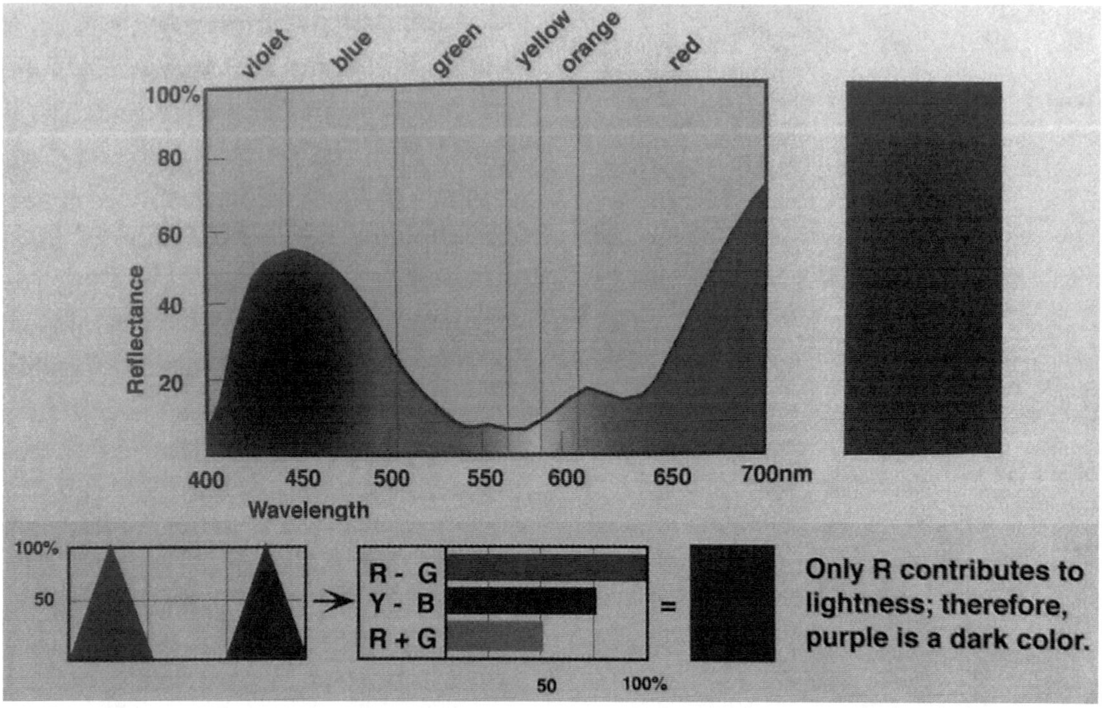

Figure 6 - from Munsell Student Color: Blue containing Chromas are darker

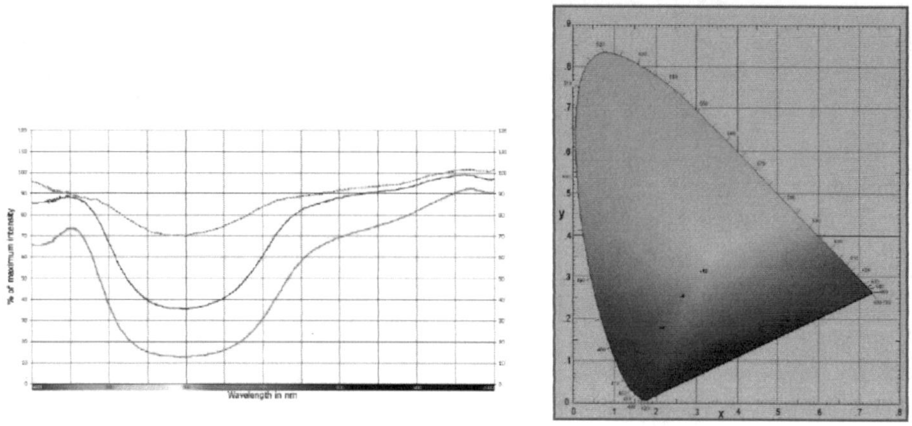

Figure 7 – from Aambø: As Yellow is increased Ultramarine looses Chroma

From these facts one can see that the HVS does not perceive a uniform color space and that chroma can be independent of luminance.

6. Relevant Colorimetry Concepts

6.1. Colorimetry

Colorimetry is the measurement of color specifically as related to HVS perception and typically leveraging CIE color space.

There are several CIE standards. NIST recommend using CIE 1976 L*a*b* since it represents an 'object color space' as opposed to a (source) 'color space' like CIE 1931. The primary differences are the addition of an object with reflectance of the source light, additional dimension of Luminance, and a predefined reference white (X_n, Y_n Z_n). For painting analysis Luminance must be considered along with Hue and Chroma in measuring a difference.

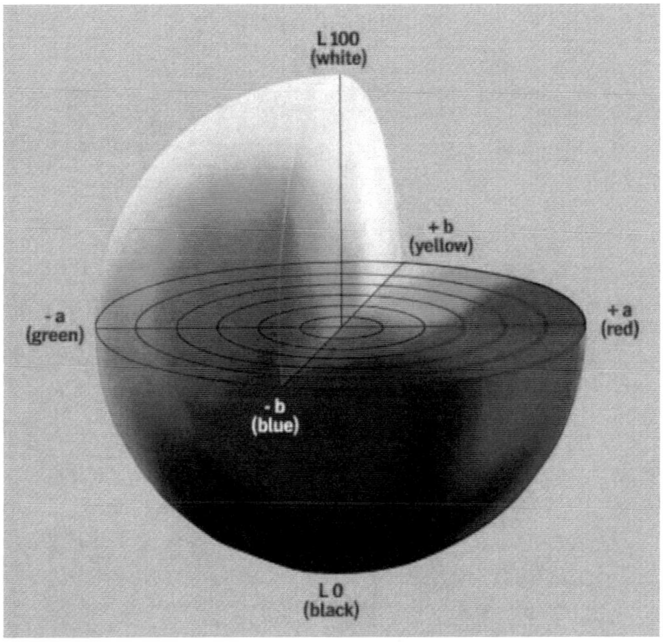

Figure 8 - from Heidelberg: L*a*b* object color space

$$L^* = 116\left(\frac{Y}{Y_n}\right)^{1/3} - 16$$

$$a^* = 500\left[\left(\frac{X}{X_n}\right)^{1/3} - \left(\frac{Y}{Y_n}\right)^{1/3}\right]$$

$$b^* = 200\left[\left(\frac{Y}{Y_n}\right)^{1/3} - \left(\frac{Z}{Z_n}\right)^{1/3}\right]$$

Figure 9 - from NIST: XYZ conversion to L*a*b*

Journal of Physics: Conference Series **415** (2013) 012005 doi:10.1088/1742-6596/415/1/012005

6.2. Spectrophototometry

Spectrophototometry is the physical measurement used as input to Colorimetry. The output is a CIE value for each measured spot (X, Y, Z for example).

Most all Spectrophotometry measurement devices require control over the reference source illuminant. Some provide for a nanometer by nanometer wavelength scan and provide extremely precise measurement.

An example specific to oil painting conservation is shown below; the entire painting has been measured and the data is available to researchers on the internet.

http://www.ifac.cnr.it/webcubview/demo/leofull.php

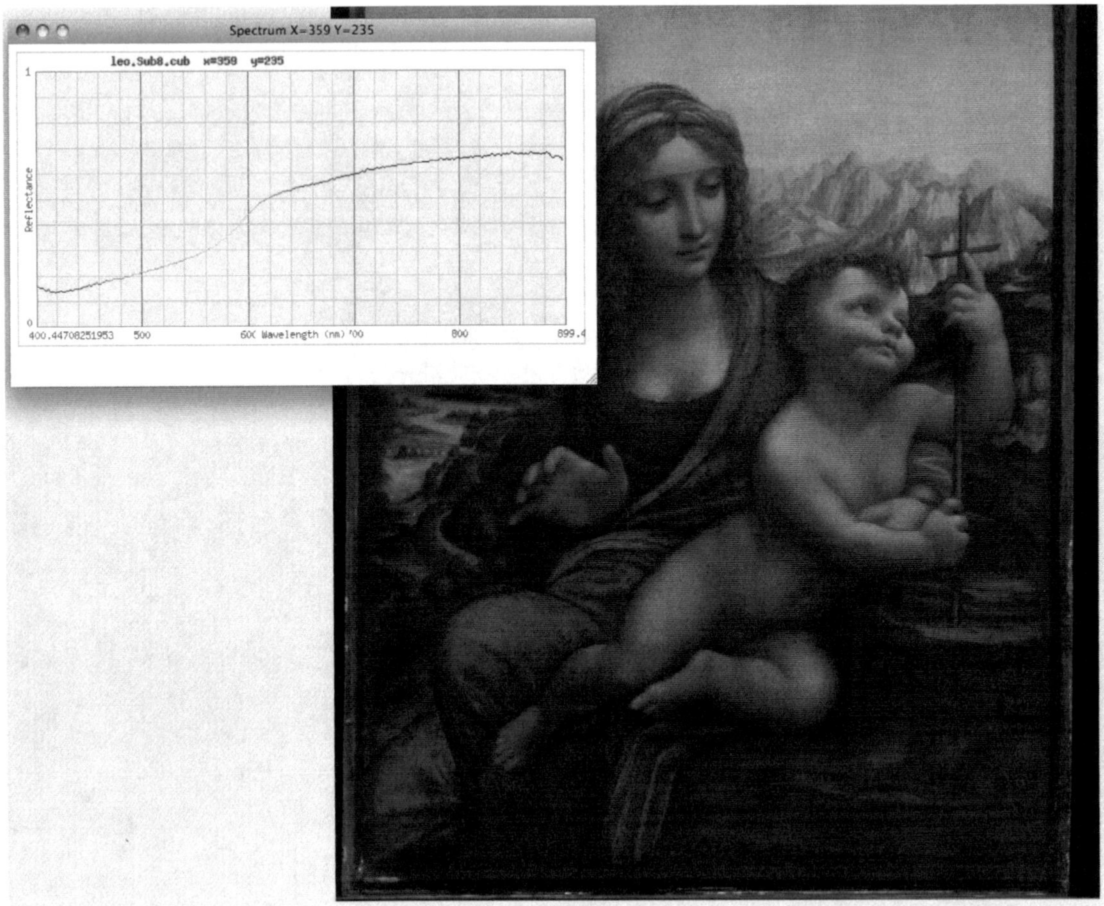

Unfortunately, there does not appear to be such a device that provides for white laser light to image a broad area and then allow for x,y coordinate scanning of the sample. It might be possible to use a very quality digital camera to compare color histograms of two samples under constant laser illumination, and at a next level of detail perform pixel by pixel registration and comparison. Several spectrophotometry manufacturers mention that the spectral response of cameras is not precisely defined enough to allow the same level of quantitative analysis. But one suspects it would provide a reasonable preliminary validation to perform Mean Square Error analysis on the two images once registered for rotation, scale and translation.

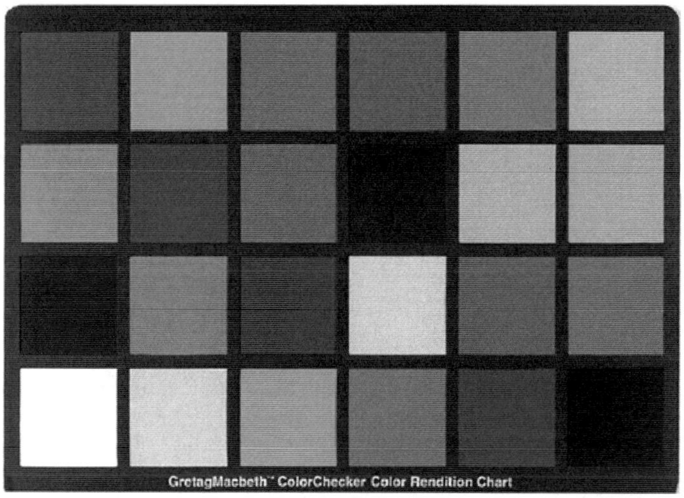

GretagMacbeth ColorChecker Color Rendition Chart.

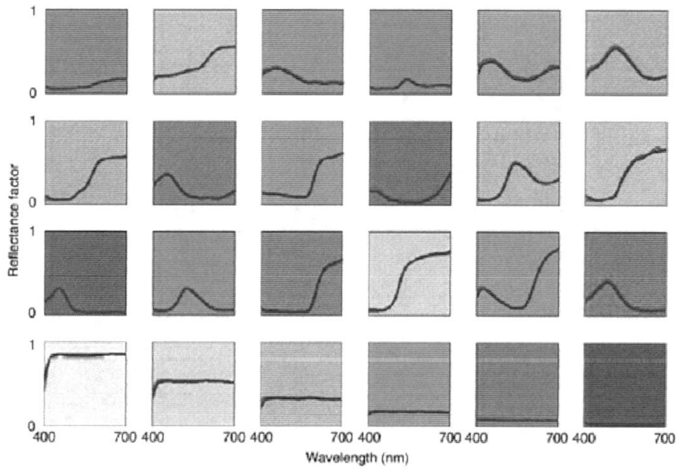

Typical spectral-measurement accuracy

Figure 10 - from Sackler NAS Colloquium (R Berns): Color Sample Spectrum

Some of the papers reviewed leveraged custom equipment to achieve some level of Spectrophotometry, but as mentioned above did not seem to compare Luminance. Above is shown a color test pattern and corresponding measured subranges of spectrum using a hybrid imaging system. Possibly this approach of filtered sampling could be leveraged for reflection holograms.

7. Approach

Compare existing samples visually and if a perfect match occurs, continue analysis with imaging and numerical techniques.

- Select sample art work that would benefit from true color holography; not all work needs
- Apply Munsell Color Theory where feasible
- Correlate cross constraints across holography, museum, artist and testing perspectives
- Prepare testable exemplars that exercise constraints for all parties
- Evaluate existing painting holograms
- Create viewing stage – used for both recording white laser light and playback display lights

- Pre-screen targets under white light
- Compare targets and holograms under white light and display lighting
- If possible create holograms of one or more exemplars
- Film / photograph viewing stage for all media; evaluate subjectively or with software
- At first stage leverage manual white balance on camera for bright white printing paper
- Document wavelengths, media, illumination and setups used; any issues
- Document findings / recommendations

7.1. Viewing Stages

Two viewing stages were used: sample and duplicate with shared illuminant; two different illuminants for one or more samples / duplicates.

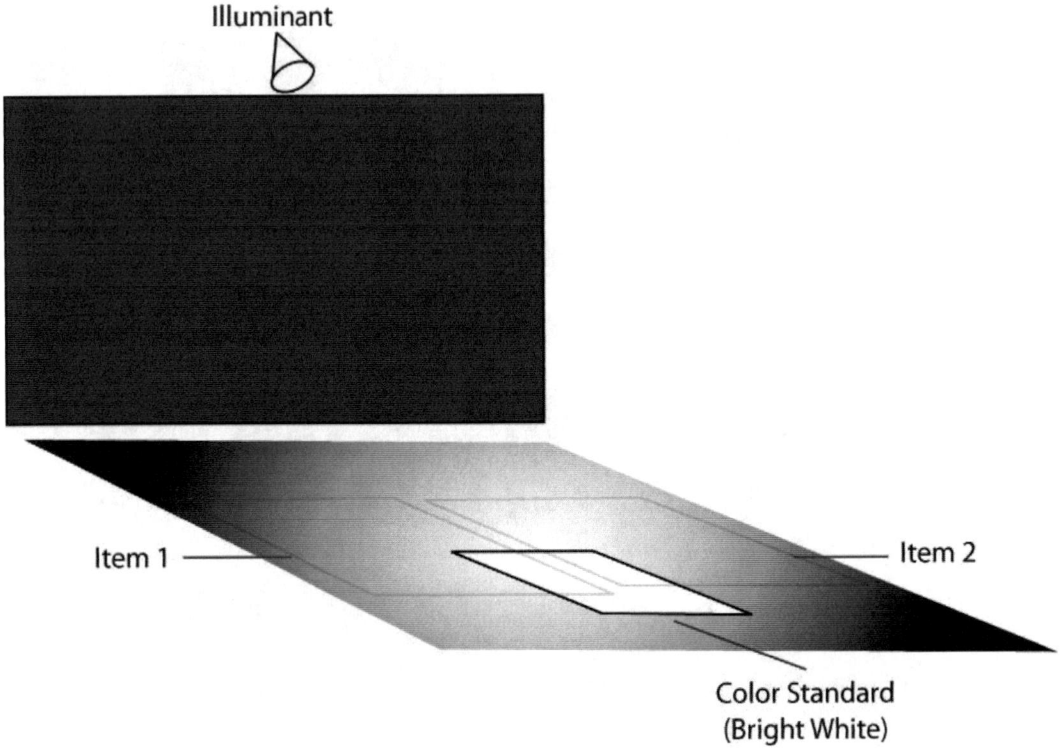

Figure 11 Same Illuminant

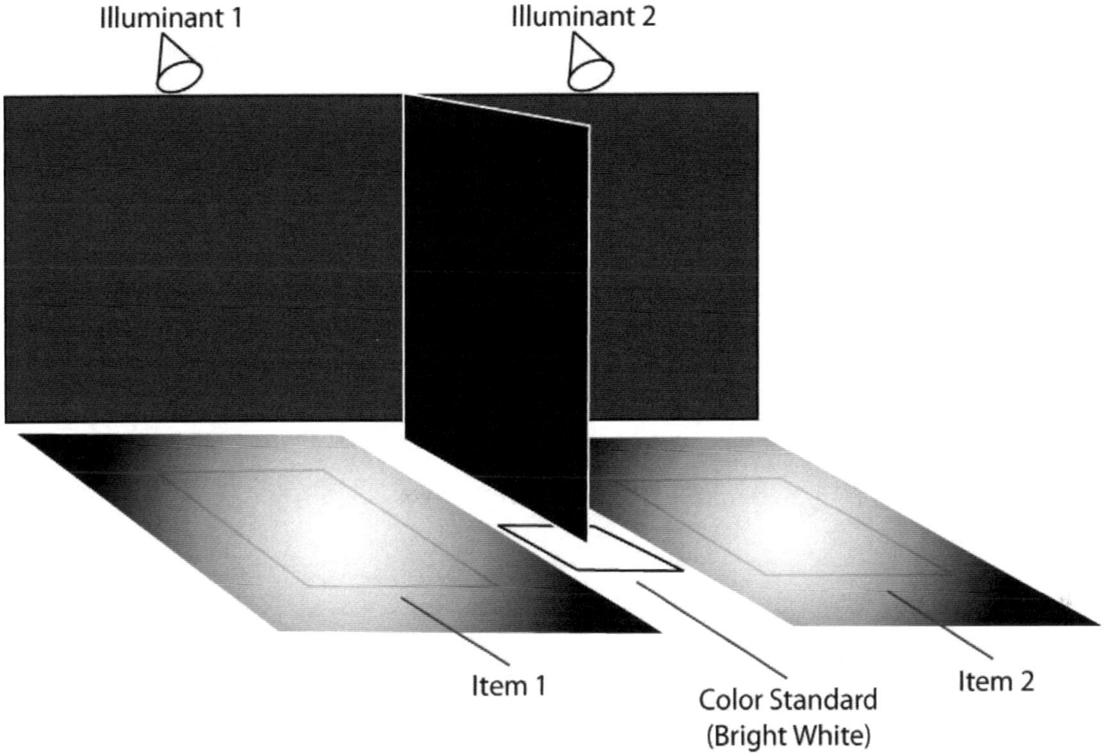

Figure 12 Different Illuminants

8. Issues

The initial expectation was to quickly move from preliminary analysis to detailed computational analysis. This did not occur. However the initial stages of analysis did uncover issues and alternative approaches that will be investigated in the future.

Several independent fields of study require color analysis, each with open issues directly impacting this research.

Media / pigment issues – the color standards are typically highly reflective and of unknown material. It would be useful if the color standards were not proprietary and leveraged several luminance targets. One solution for Oil Painting is to create a set of easily reproducible standards using a consistent manufacturer of oil paint as a reference (for example Windsor Newton Artist Colors).

Holography result is behind glass – most oil paintings are shown without glass. While this impacts display acceptance it does not impact comparison.

Laser issues – glare, reflectance, etc. – Laser illumination for reflection holography involves tangential lighting and is not ideal for viewing of textured surfaces. Initial solution is to tilt the object slightly, but could also include polarization and other modifications if required.

Practical issues – a color holography lab is not always accessible for research

Validation issues

- Simulations and statistical models do not seem to result in equivalent low error results
- Other fields have issues with CIE 1931 and some of the 'standard' color tests, etc.
- Some results are strictly numeric, using CIE metrics
- Many researchers only evaluate hue and seem to ignore luminance and its dynamic range
- Most results are not shown with original
- Failures are not typically discussed – it would simplify testing to share failure modes
- Display and lighting issues – Halogen and LED lighting is targeted to consumer point of sale. This can be contradictory to holography as increased noise, ghosting, and color shift.

- Colorimetry issues – there is not an easily accessible instrument to do high precision colorimetry of holograms, especially for white laser light illumination.

9. Results

The results are preliminary as mentioned above. Lab time was not available to create a sequence of new holograms and some of the existing holograms seem to have color balance issues related to changes in equipment. Based on recent non-painting examples the problem is resolved.

- No existing examples were indistinguishable between original and duplicate.
- For texture and geometry they were visually identical; geometries were identical.
- Color is not a perfect match and the overall reflectance is much lower: particularly for display lighting and in some cases for white light laser illumination (one laser changed as mentioned above).
- Glare is an issue – once locked in it is permanent. It appears polarization and tilt can eliminate.
- Dark values and earth tones have lower value than original; dark hue difficult to judge.
- Display and recording wavelength mismatch; power balance / reflectance mismatch
- Test approach (viewing stages) useful
- Laser color primary combinations visualization / checking useful
- Laser illumination and previsualization helpful

9.1. Images

Test patterns under white laser light are shown first, to validate their use for holographic duplication. These are followed by the two test paintings for which holograms were available in three viewing stages. In future work a subset of the test patterns / paintings will be compared with their holograms.

Figure 13 - Test Color Spectrum: Oil; Aquarelle; Acrylic (repeats)

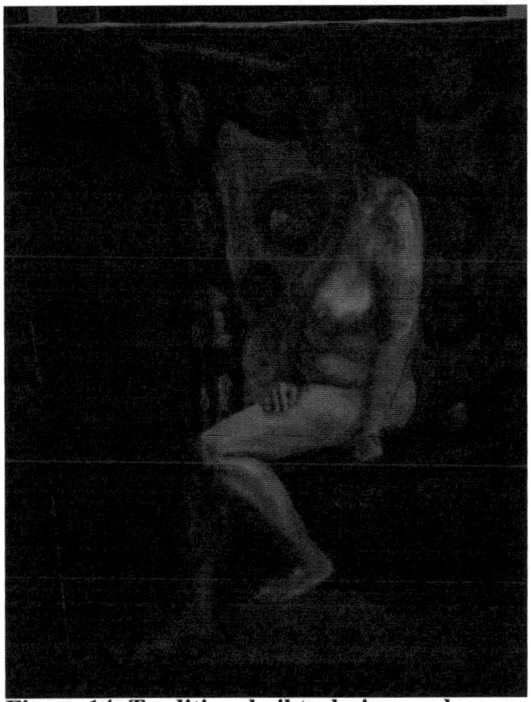

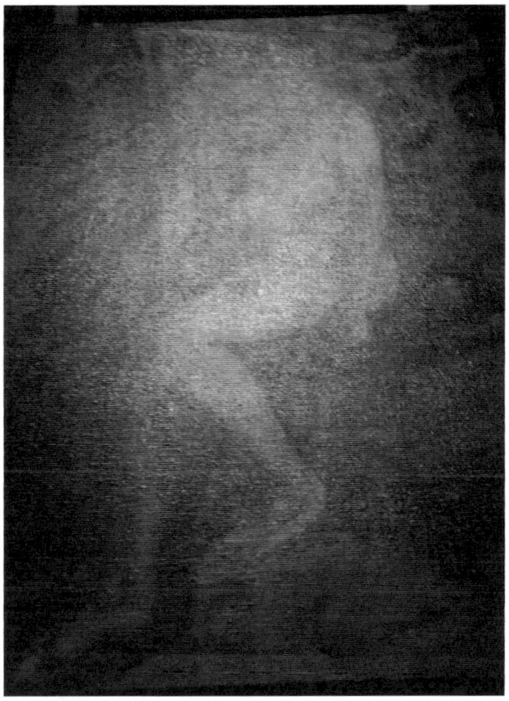

Figure 14: Traditional oil techniques; damar varnish; note glare at some angles

Figure 15 - watercolor sketch

Figure 16 - Still life with RGBYP and B/W

Figure 17 - High Chroma; various skin tones

Figure 18 - Painting 1 and Hologram; White Laser Illumination

Figure 19 - Painting 1 and hologram; warm LED

Figure 20 - Painting 1 and hologram; Halogen

Figure 21 - Painting 2 and hologram; White Laser illumination

Figure 22 - Painting 2 and hologram; warm LED

Figure 23 - Painting 2 and hologram; Halogen

Figure 24 - Kaolin painted Native Indian Kachina doll

10. Conclusions

The conclusions are preliminary based on issues above.

- The claims of greater dynamic range than photography (film) do not seem justified for color holography (at least not yet)
- Analyzing white light reflectance, spectrum and number of wavelengths must be practical and not via simulation.
- Ideal test exemplars should be created and shared; 8 highly reflective color chips are not sufficient; take approach similar to other fields (24 or more)
- An approach that leverages high quality digital camera and ICC profile would simplify process even if a preliminary stage of colorimetry
- Holography will not be identical unless reflectance is 'dimmed' on original
- Dark areas of hologram appear to have less dynamic range than bright areas. Note that the above hologam shows a bright black – but this is because the Kaolin paint used has extremely high reflectance (Kaolin / China Clay is used for porcelain and for glossy print stock).

11. Future Work

The initial need is to resolve the issues mentioned above. Subsequent to that the following is suggested:

- Create (or join) controlled environment focused on true color holography for museography
- Solicit sponsors from Laser, Lighting, Colorimetry and Museum fields
- If possible get access to two or three wavelengths for each primary (RGB); evaluate color accuracy separately from noise minimization; it seems that CIE color space can be covered as a polygon with red to green being a straight line and green through blue requiring additional wavelengths; digital TV is now leveraging 4 laser displays – research what wavelengths Mitsibushi recommends
- Create standard exemplars leveraging Munsell color with CIE equivalent using most uniform and available commercial pigments and with value range and wash/varnish for each pigment
- Get input from museums on standard pigments that correlate with old masters, try different era of paintings (or palette of that era with recreation) to see if they are each reproducible

- Research options to 3 wavelength LED and museum selected Halogen replacement (warm broad spectrum LED)
- Add pastel to the pigments tried? Good usage due to fragility of media and behind glass
- Continue original intent of evaluating subtleties of visual duplicate over full visual field

12. Acknowledgements

The author wishes to acknowledge and thank Atelier d'Art en Holographie for providing access to existing works, material, equipment and assistance in preparing test samples.

The author also wishes to thank Jack Faragasso for providing insight into the use of Munsell color system to match paint colors, Hans Bjelkhagen for sharing research papers and Kremer pigments for providing sample color charts.

References

[1] Denisyuk Y N 1984 *Fundamentals of Holography* (Mir Publishers, Moscow)
[2] Long J T 2011 *The New Munsell Student Color Set* 3rd Ed. (Fairchild Books, New York)
[3] Ratliff F 1996 The Theory of Color and The Practice of Painting Color Function Painting *The Art of Josef Albers, Julian Stanczak and Richard Anuszkiewicz*
[4] Tornari V, Fotakis C, Georgiou S, Zafiropulos V, and Anglos D 2006 *Lasers in the Preservation of Cultural Heritage Principles and Applications* Taylor & Francis pp. 53–94
[5] Hubel P M and Solymar L 1991 Color-reflection holography: theory and experiment (*Applied Optics* vol 30 no 29) pp. 4190-4203
[6] Gentet Y and Gentet P 2000 "Ultimate" emulsion and its applications: a laboratory-made silver halide emulsion of optimized quality for monochromatic pulsed and full color holography *HOLOGRAPHY* Proc. SPIE 4149 pp. 56-62
[7] Gentet Y Shevtsov M 2008 "Ultimate Holography" Technologies for Museum Applications: Denisyuk Colour Mobile Camera, Pulsed Mobile Camera, 3DHoloprint *Holography In The Modern Museum* (Royal Photographic Society Holography Group)
[8] Faragasso J 1979 *The student's guide to painting* (North Light, Westport)
[9] Berns R S 2005 Color-Accurate Image Archives Using Spectral Imaging (Sackler NAS Colloquium) *Scientific Examination of Art: Modern Techniques in Conservation and Analysis Proceedings of the National Academy of Sciences*
[10] Nadal M, Early E A, Bousquet R R2008 *0:45 Surface Color* NIST Special Publication SP250-71
[11] Aambø M 2011 Use of the "Video Spectral Comparator 6000" as a non-destructive method for pigment identification - An experiment (*Program in Conservation of Cultural Property Graduating thesis, BA/Sc*) University of Gothenburg Department of Conservation
[12] Stephenson J 1989 *The materials and techniques of painting* (Thames and Hudson Ltd, London)
[13] Harihan P, 1996 *Optical Holography: principles, techniques, and applications* (Cambridge University Press, Cambridge)
[14] Bjelkhagen H 1993 *Silver-Halide Recording Materials* (Springer-Verlag, Berlin)
[15] Bjelkhagen H 2008 Colour Holography: The Ultimate Imaging Technique for Museums *Holography In The Modern Museum* (Royal Photographic Society Holography Group)
[16] http://www.ifac.cnr.it/webcubview/demo/leofull.php
[17] http://kremerpigments.com/download/KremerPigments_catalog-2009.pdf
[18] http://www.heidelberg.com/h/www/en/binaries/files/prinect/user_guide_colorimetry_en_pdf

Journal of Physics: Conference Series **415** (2013) 012006 doi:10.1088/1742-6596/415/1/012006

For a photon there is only the present

Sally Weber
Resonance Studio,
1406-D Smith Road, Austin, TX 78721 USA
Website: www.sallyweber.com

E-mail: snw@sallyweber.com

Abstract. Two works using holography to explore the unexpected that is always present. One project is an experiment in complexity and the other in light's shadow..

1. Introduction: Wonder exists where the experience does not reflect expectation

My explorations often are nourished by the works of earlier cultures or techniques. The extraordinary 30,000 year old images from Chauvet Cave in France (figure 1) and the Archimedes Palimpsest, a nearly illegible manuscript which revealed lost treatises of Archimedes (figure 2), both show profound insight and technical skill. Looking at these works offers the opportunity to compress the time span between our present and theirs and appreciate the human connection between us and the person who drew these exquisite images on a cave wall or the geometric theorems in Hellenistic Syracuse on Sicily. Displaced in time, but exhibiting a pinnacle of the technologies of their times, these ancient works arc a thread of the human continuum of thought and creativity and yet, they are also the ones most susceptible to the inevitable destruction of the very materials that carry these thoughts to us.

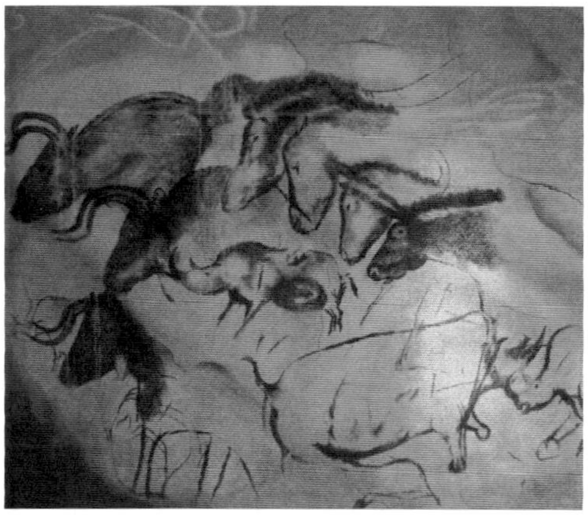

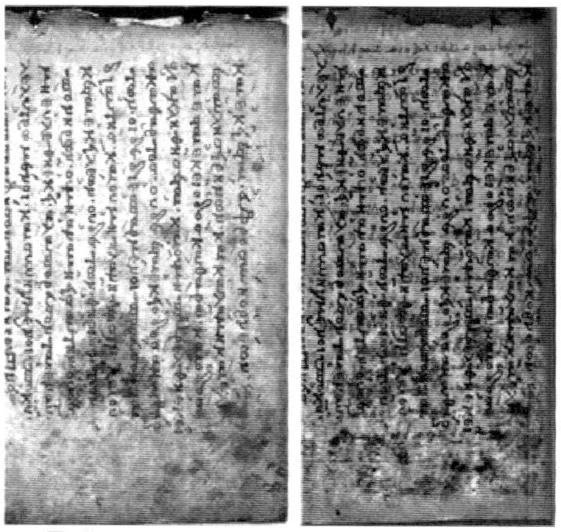

Figure 1: (left) Detail from the horse panel at Chauvet Cave, France.
Figure 2: (right) Archimedes Palimpsest detail of an unrestored page (left) and a pseudocolor page (right) revealing the 10th century Greek translation of Archimedes text in red orthogonal to the 13th century black text of the prayer book.

Intuitive and analytic thinking might inspire insights and innovations differently, but similar human queries re-surface through the ages. The human mind has an affinity for patterns and forms and these re-emerge in different guises. Tiny perturbations in the earliest moments of the universe are thought to be the seeds responsible for the distribution of the stars and galaxies we see now. The COBE image (figure 3) of the cosmic microwave background radiation (CMB) is the remnant, the most ancient of memories, "blown up to astrophysical size by the rapid expansion and (it) becomes the seed for all the structures we see today. The pattern seen on the CMB sky is a giant image of the subatomic world."[1]

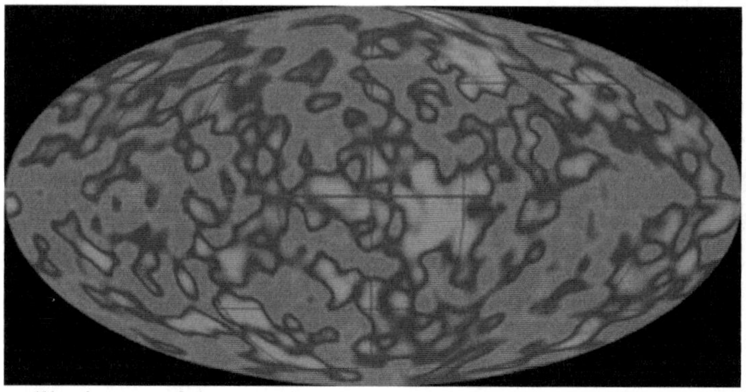

Figure 3: COBE satellite image of the cosmic microwave background (CMB) fluctuations are extremely faint, only one part in 100,000 compared to the 2.73 degree Kelvin average temperature of the radiation field.

The signature left by the creation of the universe still inhabits it. Nothing is lost, just recycled to appear again, potentially to be recognized or uncovered. "According to traditional thermodynamics, there is no such thing as a truly clean slate; every system always retains a memory of its past in the configuration of its atoms."[2]

The experience of looking at the stars compresses our sense of time. When seeing photons from hundreds of millions of light years away, 'now' is neither the origin of the star's light, nor our presence seeing it. It is both simultaneously. Einstein's theory of relativity states and quantum physics confirms that from the point of view of a photon there is no time. At the speed of light time slows down to zero. Since a photon moves at the speed of light, it sees the whole universe at once. A photon is in the perpetual present, everywhere at once. Time and space collapse into 'now'.

"Mathematics would not work at all were there no universal features to be discovered."[3]

2. Infinity in the Complex

Several years ago I produced a pulsed laser hologram at the Center for Holographic Art, New York to explore an idea illustrated by the ancient Hindu story of Indra's Net. The highly specular model created a secondary set of images surrounding the reconstruction source. Upon close examination, the elements of the objects formed a complex Gabor hologram centered in an array of multi-referenced images. (figures 4a & 4b) The inter-related network of images suggests the infinite reflections of the Indra's Net story and parallels both ancient and contemporary concepts of complexity. (For a detailed discussion of this project, please see the 8th ISDH Proceedings 2009, Shenzhen, China.)[4]

fig. 3: COBE satellite image of the cosmic microwave background (CMB) fluctuations are extremely faint, only one part in 100,000 compared to the 2.73 degree Kelvin average temperature of the radiation field.

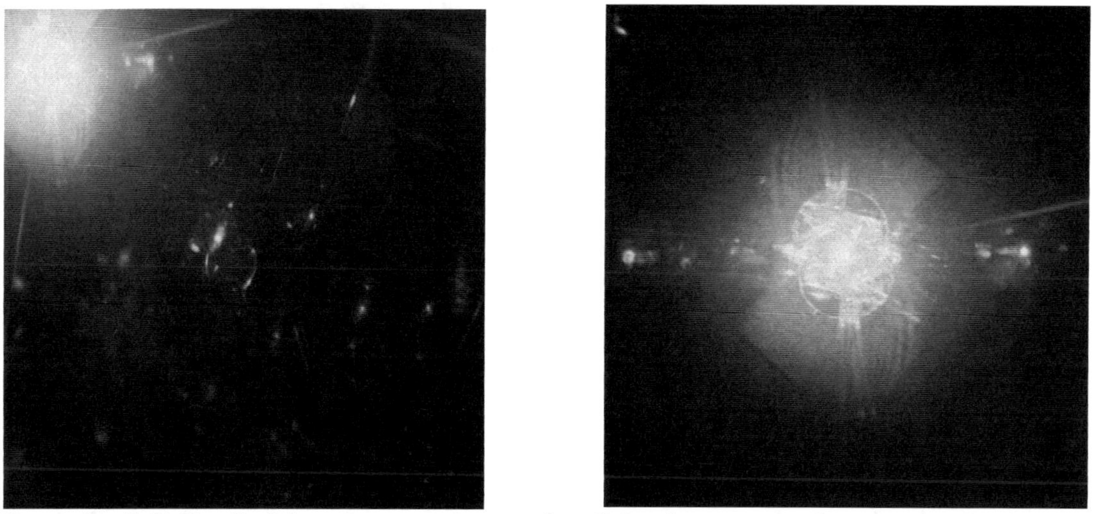

Figure 4a: (left) Detail of the pulse master's secondary Gabor imagery near the reconstruction beam.
Figure 4b: (right) Detail of the *vesica piscis* image surrounding the reconstruction beam.

In the ancient story, Indra, the Vedic Lord of the Gods, had an infinite net decorating his palace. At each juncture of the net there was a faceted jewel that reflected every other jewel in the net. In looking at just one gem one saw the reflection of all of the others in the infinite array. One gem was both a single entity and simultaneously intimately connected to all of the others. As a result affecting or removing one altered them all in some significant way. Francis H. Cook in commenting upon Indra's Net and its implications in Chinese Buddhism of the Tang period, states that:

"..each individual is at once the cause of the whole and is caused by the whole, and what is called existence is a vast body made up of an infinity of individuals all sustaining each other and defining. each other. The cosmos is, in short, a self-creating, self-maintaining, and self-defining organism."[5]

Unintentionally, the complex reflections of light between elements in the holographic model created an image array paralleling an idea of infinite inter-connections intuited by people in 3rd century India. The complexity structure of light in the hologram resulted from the placement of mirrors and other elements in the model which in combination with a very low reference angle of the pulsed laser established the correct configuration to create the internal Gabor hologram. Intention, intuition and recognition played roles. Time compressed for me in appreciation of people who intuited and observed light's inherent principals centuries ago and created an image in a story of vast networks of inter-connections and inter-relationships close to our conception of the complexity structures that underpin fields of study now as well as the network of galaxies. (figure 5)
fig.5: The Infrared Local Universe: this all-sky map shows galaxies in the 2MASS survey color coded by their distance from us with blue showing the nearest sources, through green to the most distant sources shown in red.

Figure 5: The Infrared Local Universe: this all-sky map shows galaxies in the 2MASS survey color coded by their distance from us with blue showing the nearest sources, through green to the most distant sources shown in red.

3. Speed of Shadows

Margaret Wertheim, the Director of the Institute of Figuring in Los Angeles, wrote in *The Shadow Goes* that though light is limited to 186,000 miles a second, a shadow is not. In theory, a shadow can travel faster than the speed of light. Since a shadow has neither mass nor energy it is not constrained by the laws of relativity. The photons emitted by a sweeping source of light, like a rotating hose spewing water, continue straight from the source. However, the rotation of a smaller circle blocking the center of the light source and creating a shadow, (she suggests a giant klieg light in space with the Batman symbol centered on it), can outpace the light as it revolves in a great circle.(figure 6)

"Just as the rim of bicycle wheel moves faster and faster than its hub, so too, away from the source our bat shadow will fly faster and faster, a consequence of the geometry that guarantees the rim of a really big wheel moves faster that a co-rotating small wheel."[6]

This unexpected speed of a shadow, like a disruption of the natural order, or speculation yet to be quantified, suggests that the limit of a theory might sleep within it.

Figure 6: the bat shadow

In *Speed of Shadows*, the installation, the shadow partially obscures a vivid orange array of light and motion beyond. Light, shadow, form, solidity, illusions and reflections juxtapose each other. The holographic real image of a blue cube surrounds the spherical black shadow. A thin, black cable hangs from the ceiling through the holographic shadow while the cable's own shadow bisects the entire holographic image. Below, a solid, black jet sphere hanging from the cable hovers over a spherical mirror. It creates a black mirror when seen from above reflecting off the jet surface. Air turbulence moves the jet sphere slowly altering its shadows and reflection in the mirror and on the sphere itself. These reflections are the active counterpoint to the illusive image above. Jet, a highly compressed wood, was the symbol in Classical lore of the darkest black. Black mirrors or black bowl with water were used for divination in Medieval Europe and for star gazing by the Maya. Lightweight and warm to the touch, jet can burn, drawing together both the shadow and the fiery holographic ring above. (figure 7, figure 8a & 8b)

Journal of Physics: Conference Series **415** (2013) 012006 doi:10.1088/1742-6596/415/1/012006

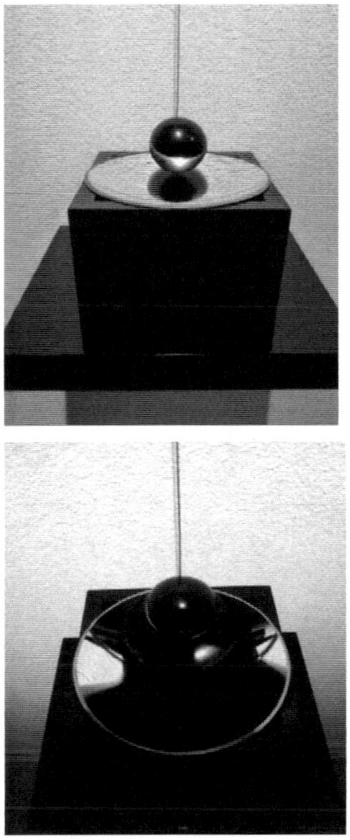

Figure 7 (left) Speed of Shadows (2012)
Figure 8a & 8b (right top and bottom): details of the jet sphere and mirror

Speed of Shadows was not consciously inspired by a solar eclipse, but having seen some it certainly crept in. Experiencing the totality of a solar eclipse is a powerful experience and the lore associated with eclipses seems relevant to the challenges that arose in realizing this work. Throughout the ancient world people feared eclipses. They symbolized death of the sun, the world out of order, illness or the break down of social taboos. Some saw the shadow as either an evil spell or a wild animal attacking and eating the sun. (figure 9) Claude Levi Strauss noted that the orchestrated noise of a 'pandemonium' could scare the creature away or remove the curse.

"In the case of eclipses, it is the disruption of an order causing, by a regular chain of events, the alternation of the sun and the moon, day and night, light and darkness."[7]

Certainly ancient astronomers recognized the recurrent order over time and prediction of an eclipse became a powerful tool of both divination and political strategy. Einstein's 1911 prediction that a star's light would be bent by the Sun's gravity, based on his theory of relativity, was validated by measurements taken during the 1919 solar eclipse.

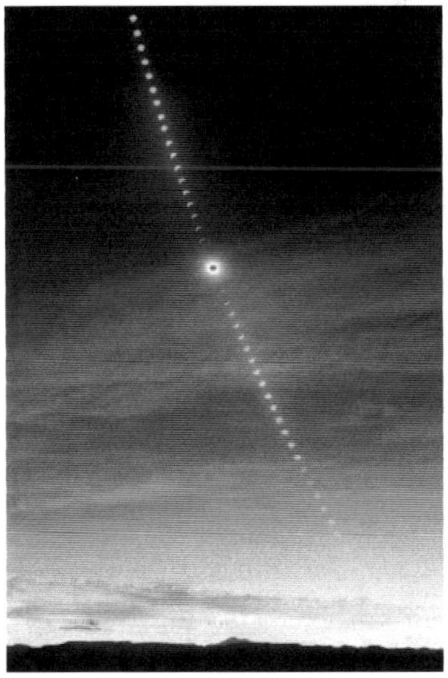

Figure 9: progression of partial solar eclipses and totality

"On the tiny difference, less than one second of arc, rested the choice between two theories of one of the basic forces of nature - gravity."[8] Einstein became instantly world famous reaching "superstar status".

4. Project Production

Speed of Shadows' origin was far more humble. The project began in my studio/lab in Austin, TX in 2007 with a two-part model intended for a holographic master and transfer. (figure 10a & b) Inherent optical effects from glass spheres behind the black sphere in the model burned into the silver halide and obliterated the work's intent in the transfer. Experiments with other means to create a volumetric cube and alternate lighting and optical effects proved unsatisfactory. I set the project aside, but it was always in the background persistently nudging.

Figure 10a: Initial model for Speed of Shadows

Figure 10b: laser illuminated model

In 2009 I began working with Chris Portales, a Master's candidate from Texas A&M. We developed a 3D computer model based on my physical model and specifically worked to capture the speckle of the cube and the optical qualities behind the shadow. (figure 11) The computer generated imagery looked promising but unfortunately, changes in the rendering parameters for Zebra
Imaging holograms at that time could no longer accommodate the raytrace of optical qualities we hoped to accomplish. With the semester over and Chris' thesis pending, we curtailed this attempt as well.

Figure 11: computer model for Speed of Shadows

Two years later while visiting The Light Foundry in Santa Fe, NM the artist and holographer, August Muth, invited me to come and try it again as a dichromate gelatin hologram. I had never worked with dichromates, but was optimistic as our initial test showed that a dichromate hologram was much more forgiving about hot spots and did not burn in as silver halide had. We anticipated two, one beam shots exposures superimposed to create the box and optical effects behind a shadow sphere that I had always visualized.

Figure 12: predistorted new model for cube

Figure 13: new model of spheres

In April of 2012, I went to Santa Fe to work at The Light Foundry. We experimented with my original models but it was soon apparent that they would need to be rebuilt optimized for the dichromate holography setups. A new cube was made of glass, predistorted for a one-beam setup. (figure 12) The optical and shadow spheres model was also rebuilt using a segment of a decorative glass sphere and optic with the black sphere in front. (figure 13) The new model removed the hotspots but decreased the brightness substantially. The lowered the light level necessitated 23 minute exposures for the master. This coupled with Santa Fe's unsettled spring weather conditions, the idiosyncrasies of dichromates, lasers and holography in general converged on the very fickle eclipse-like image and shadow.

Figure 14: Annular solar eclipse

Figure 15: Transit of Venus

There were other unexpected convergences as well. During the holographic production shoots in Santa Fe the shadows of two different planets crossed the sun. The annual solar eclipse of May 20, 2012, (figure 14) passed directly over Santa Fe. Sixteen days later on June 5, 2012 Venus (figure 15) crossed the Sun in its last transit until 2117. We both observed these natural phenomena from Santa Fe, NM and Austin, TX respectively. However, we should have recognized them as signs and with proper foresight organized a pandemonium ourselves. It might have improved our odds.

On June 15, 2012 the hologram for *Speed of Shadows* was finally complete through the tenacity of August Muth's almost alchemistic understanding of dichromate holography. My collaboration with the Light Foundry proved a dance of light, time and space. The paradox of *Speed of Shadows* was its very resistance to be realized. The work defined itself in some ways for only after many attempts was a balance of light, shadow, color and forms achieved. "A shadow is not a thing and has no mass. It is the absence of light, weightless and without energy."[9] But light and shadow, like mass and gravity are tied to and reveal each other.

5. Conclusion
Over time the same questions seek new expressions. These seeds like the structure in the COBE image sleep quietly until they resurface as a thought, an expression, an image, a measurement or an equation. The itch to know is deep. The story of Indra's Net was an attempt to suggest infinity and the interconnectedness of life through imagery and metaphor. The passage of an eclipse is a natural, awe-inspiring occurrence once feared for disrupting the natural order by bringing darkness during the day, but also enabled a measurement that changed the 20th century. For a photon it is all simultaneous, but what does that mean in our experience? Both my projects, the complexity experiment, *Indra,* and the installation, *Speed of Shadows,* developed not by a direct path but by a meander and search through experimenting with optics, materials and means to explore something ephemeral; a simultaneous tug across time that could be realized only in holography.

6. Acknowledgement
With thanks to: Center for the Holographic Arts, New York and The Light Foundry, Santa Fe, New Mexico for their exceptional expertise and assistance in the holographic production of these works.

7. Illustrations
fig.1: Open source: http://en.wikipedia.org/wiki/File:Chauvet_cave,_paintings.JPG
fig. 2: Archimedes Palimpsest Image Bank, http://archimedespalimpsest.net/
fig. 3: NASA's COBE satellite image of CMB fluctuations. http://aether.lbl.gov/www/projects/cobe/
fig. 4a: Author's photograph of secondary holographic image in pulsed holographic master, 2008
fig. 4b: Author's photograph of central *vesica piscis* image in pulsed holographic master, 2008
fig: 5 Jarrett,T., The Scale Structure in the Local Universe: The 2 Mass Galaxy Catalog, 2004, PASA, 21, 396 (Opened source image:2MassRedShiftSurvey https://www.cfa.harvard.edu/~dfabricant/huchra/2mass/)
fig. 6: Opened source: composite Bat-Signal
fig. 7: Author's photograph, Speed of Shadows, holographic installation, 2012
fig. 8a & b: Ibid, details of the hanging jet sphere and spherical mirror below, 2012
fig. 9: Newswanger, C: detail from photograph of solar eclipse, Bolivia, 2003
fig.10a: Author's photograph of initial model for *Speed of Shadows*, 2007
fig.10b:Author's photograph of initial model for *Speed of Shadows* in laser light, 2007
fig.11: Goodman, V., photograph of computer rendering of *Speed of Shadows* model, 2009
fig.12 Author's photograph of one-beam setup for cube at the Light Foundry, Santa Fe, NM, 2012
fig.13 Author's photograph of new model of eclipse element at the Light Foundry, Santa Fe, NM, 2012
fig.14 Opened source, NASA image of May 20, 2012 annular solar eclipse
fig. 15: Opened source, NASA image of detail of the transit of Venus across the Sun

8. References:
[1] Turner, Michael S 2009 Origin of the Universe, *Scientific American*, August
[2] Carroll, Sean M 2008 The Cosmic Origins of Time's Arrow, *Scientific American*, June
[3] Livio, Mario 2011 Why Math Works, *Scientific American*, August, pg 83
[4] Weber, Sally 2009 Indra's Net: A Holographic Approach to Mutual Causality; *8th ISDH Proceedings* Shenzhen, China

[5] Cook, Francis H. 1977 Hua-Yen Buddhism, The Jewel Net of Indra, Pennsylvania State University Press, University Park and London, pp. 3
[6] Wertheim, Margaret 2007 The Shadow Goes, Opinion, *New York Times*, June 20,
[7] Verdet, Jean-Pierre 1992 The Sky, Mysteries, Magic and Myth; Harry N. Abrams, Inc., New York; pp.75
[8] Wilson, Robert 1997 Astronomy Through the Ages; Princeton University Press, Princeton, NJ, pp.135-136
[9] Wertheim; ob sit.

The future of holographic technologies and their use by artists

S Oliveira[1] and M Richardson[2],

[1] Student, De Montfort University, UK, The Imaging & Display Research Group. Leicester LE1 9BH

[2] Professor, De Montfort University, UK, The Imaging & Display Research Group. Leicester LE1 9BH

Abstract. The use of holographic technologies in the past has faced resistance in the artistic field. The most conservative artists and critiques saw the term "holographic" more as a technical subject than an artistic one. Nevertheless, to explore new forms to create art has been a constant challenge for any artist, whatever their field. At the end of the 20-century the concept that art can explore any field or subject, create a vision that is somehow technological, is part of the evolving artistic world. In the last two decades, in the search for new terminologies, scientists and artists have used the expression "Holographic" as a synonym of evolution, but with different meanings. Artists are using it as a new form of art call "Holo Art"; scientists see it as a "new" science technique where light takes an important part in the process, and can be used in various aspects of daily life, such as, security and medicine. This paper will explore artists who take the challenge of combining their art with new technologies and how they are viewed in a world where the question of 'what is and what isn't art' is very debatable. Other questions that will be explored are 'How can these techniques be useful to artists?' and 'How do artists challenge themselves to analyse the pros and cons of the results that can be obtained?'

Key-words: Holographic Technologies, Artistic Field, Holographic, Holo Art, Artists

Introduction

Artists embrace the world that surrounds them in their attempt to explore the different aspects and situations that they are involved in, whether voluntary or involuntary. As a form of expression the use of different techniques became paramount for creativity; taking the artists to a different level of interpretation, of their understanding and view about conscious and unconscious challenges, that appear in the process of communication.

For centuries, the attempt to transmit and represent concepts and belief in all areas of expertise, in original and unique ways, has been systematic. For example, when debating how scientists and artists contribute to knowledge enhancement, the argument inevitably results in the question of how much collaboration is seen has essential or desirable. The scientist seeks clarification and support so the sharing of research into using holographic techniques is necessary so that studies may be replicated and scientific expertise supported and developed. Conversely the artists does not seek such affirmation; they see the use of holography as a way to create pieces that hold a unique presentation of their artistic concepts, and because of their personal way of working, no two pieces of art work will be the same.

However, even though all arguments have been exhaustively debated for decades, there is still no common agreement that the use of holography can be considered as art. Nevertheless, the fact that art and holography can be complementary has a common ground that is still to be explored by both sides, in a way that will push the boundaries as far as possible to bring something new to their field.

For scientists the "abstract" option is not feasible, the facts need to be precisely above any doubt. The point of view for a scientist is that nothing is a mistake; there is an explanation for everything.

Artists are searching for new forms of expression that allow them to explore a visual perception of the world in a different perspective. The "abstract" option is an open door to develop the creativity that artists aspire to achieve with their work. Also, each viewer has their own perception of what is exposed to them, so each interpretation will be different leading to the evolution of newer and more creative methods of art making.
As Krueger said:

> *"Artists today are not technologists. Most are careful to maintain the distinction between the two roles and would be offended if they were considered technicians. Yet, art and technology were once inextricably intertwined" (p.2)*[1] *Myron Krueger (1991)*

The citation by Krueger demonstrates that the line between art and technology is not clear and that artists can contribute to the creation of new technology. However the different point of view between artists and scientists can in some ways create critics from both sides arguing that each interpretation is made to suit their on arguments. Art is an open book; the viewer is part of the interpretation without having specific parameters to adhere to, free of restrictions and perpetuation.

Although the main concern for this paper is connected to holography and art, it would be remiss not to consider holography as a technological technique that has found an acceptable place within the scientific world. Its acceptance by science has perhaps had an easier route than that of holography and the art world. However Holography is part of a group of technological techniques that can be used in both fields.

Artists and Holography

For over 30 years artists have been using holography as a technique to explore space, time, movement and light in artwork. Nevertheless, holographic art is still seen as more of a scientific subject than an artistic subject to the public eye, due mainly to the existence of old conceptions about what art is. The definition that Art is an open book means that "everything" can be art and "nothing" can be art. The artist can be a psychologist, philosopher, doctor, mathematician, politician, teacher, and scientist, whatever needs to be depicted in the subject within their art. Artists have a "motivation" that is close to them which makes them unique, special and an expert in their field. The use of Art as a vehicle to explore and illustrate the variety of a subject is a form that allows artists to express themselves giving the freedom for individual interpretation. Equally Holography has its own characteristics, which provide new oportunities for artists and artwork creating the possibillity of presenting a subject in 3D space on a flat surface Hans Wilhelmsson[2] (1968).

Wilhelmsson (1968) continues:

> *"the question of how holography may be used by artists is left open"*

Wilhelmsson, was discussing the scientific side of holography when at that time holo art was almost unknown by the general public. However at that point he could see the opportunity of this technique being used to create art work as the different methods used, for example photography or paint were unable to produce the combination of space, depth and movement that holography could provide.

The debate about holography being a creative method is continually debated because of the technical requirements needed. However, it is a technique that has gained more visibility with artists in their attempts to explore the visual display of the subjects. All techniques that allow the representation of light in art and creates 3D vision, which is close to that perceived naturally by the human eye, is something that attracts the attention of artists.
The phenomenon of light has been the subject of different interpretation for centuries. *Capturing light in the most natural way is a continual challenge for artists in all fields.*

Artists such as Paula Dowson, Michael Bleyenberg, Betsy Connors, Ikuo Nakamura, Sally Weber, and Pearl John are well known for their artwork with holography. These artists embrace holographic techniques to explore the representation of different subjects. However not all of them embrace the scientific process. Some take time to understand the scientific process and engage with new developments in technology that will help to expand their art works. For example Dowson, who has spent significant time in understanding how holography works and how it can be applied to depict any subject, has presented her with the possibility of creating exceptional works.

Nevertheless, to utilize these techniques in art, requires the artist to understand and develop their scientific knowledge in some way, that will enable the association between art and science to come ever closer.

The fascinating possibility to combine different aspects such as light, time, movement and space into the process of artwork, gives us self-determination for a new level of creativity. This will allow us to expand, away from the restrictions that in some way have been limiting to art, in breaking through the ceiling of what can and cannot be achieved.
A good example of the close association between art and science is the artwork "Changing Thoughts"[3] by Isabel Azevedo. Her interpretation of space, time and movement using digital holography enables the use of the magic of light, to take this art form to another dimension. For all artists and art forms if they use digital holography as an ingredient in their creative process and the presentation stage, the limits of what can be achieved are limitless.

"Women", Digital Reflection Hologram, 60x50cm.[4]

When we look at the history and development of holography it is amazing to see how long artists have been trying to bring the best of this technique to the public eye. To help art and science embrace and develop even more, we need to help artists understand the great benefits and unique possibilities holographic techniques provide. The magic of light brings to life all artists unique characteristics in something amazing that would seduce any follower who is trying to understand the best way to capture and display their subject. Holography is a technique that captures and 'plays' with

Journal of Physics: Conference Series **415** (2013) 012007 doi:10.1088/1742-6596/415/1/012007

light; it plays with light to create symbolism and expression of feeling, yet it captures light to enhance the finished piece when presented to the viewer.

Holography has crept into our daily lives without most people noticing due to it mainly being developed within scientific and technological applications. It is when it is being associated with art that people become more sceptical. Science has seen the opposite reaction, in that the technique is seen to be a tool that can enable artist's scientists to open new windows of creativity within all artistic scientific formats. In many ways, the scientist's view and the artist's view are both the same, it is just that the majority do not realise they are looking at the beginnings of a new era and how this will be influenced by individual and different perspectives. The artist's aim is to project the visual impact of their subject to all humankind in a way that viewers can see the piece in a unique and personal way. This leaves the door open for viewers to associate their experiences, and for the artist's vision to help form a new perspective.

The development of new techniques under the umbrella term holography, such as digital holography, gives more perspectives and freedom for artists to express themselves using light.

The hologram on its own only shows what was recorded by the artist at that precise moment; there is no interaction with viewers. Therefore there is a need for artists to use the beauty of light in their work to create a visual perception for the viewer, creating a space away from the real space – a new universe of art.

Light is the one ingredient that art has used since the start of time, and when we look at the very beginning of time when cavemen painted images using raw basic material, through to Da-Vinci expanding artistic boundaries, light and the positioning of light is the one thing that brings these to life for us. Now when we really embrace holographic science and what this does to the basic ingredient of light we see that what could be achieved is boundless.

CONCLUSION

Whilst some artists have used holography and holographic techniques for a period of time now, others continue to use it at base level. The time is right to call for a new way to view the technology and techniques for all art mediums and give the artist the freedom to collaborate with *current* and future advances found in the scientific domain and use the "light ingredient" to dramatic effect. This can help to transform the cultural perception of artists and art institutions to embrace holographic science and accept that it this is vehicle for taking art to new frontiers. The opportunities are endless and each new advance creates new dimensions and the possibility for the display of a subject in 3D and beyond.

The transition of art to meet the next generations need for more technical artists who can then reach out and help the "pure arts" generations to embrace the opportunity of expansion of light into mainstream art, can be achieved by producing robust art pieces that showcase the endless possibilities.

References

[1] KRUEGER, M W (1991) Artificial Reality II. 2nd ed. USA: Addison-Wesley Publishing Company, Inc.

[2] Wilhelmsson, H (1968) Holography: A New Scientific Technique of Possible use to Artists. *Leonardo, Vol.* ***1****, pp.161-169.*

[3, 4] Azevedo, I and Richardson, M 2011 Changing Thoughts: A Series Of Digital Art Holograms. *In Proceedings of SPIE, Sao Francisco, February 2011. SPIE* Volume: **7957**. ISBN: 9780819484949

Content metamorphosis in synthetic holography.

Jacques Desbiens
Artist, researcher and art historian, Quebec, Canada.

www.i-jacques.com

Abstract. A synthetic hologram is an optical system made of hundreds of images amalgamated in a structure of holographic cells. Each of these images represents a point of view on a three-dimensional space which makes us consider synthetic holography as a multiple points of view perspective system. In the composition of a computer graphics scene for a synthetic hologram, the field of view of the holographic image can be divided into several viewing zones. We can attribute these divisions to any object or image feature independently and operate different transformations on image content. In computer generated holography, we tend to consider content variations as a continuous animation much like a short movie. However, by composing sequential variations of image features in relation with spatial divisions, we can build new narrative forms distinct from linear cinematographic narration. When observers move freely and change their viewing positions, they travel from one field of view division to another. In synthetic holography, metamorphoses of image content are within the observer's path. In all imaging Medias, the transformation of image features in synchronisation with the observer's position is a rare occurrence. However, this is a predominant characteristic of synthetic holography. This paper describes some of my experimental works in the development of metamorphic holographic images.

1. Introduction

When Étienne-Jules Marey[1] uses chronophotography[2], he wants to study body movements by dividing it into a series of static images from a fix predetermined point of view. When the Lumière brothers[3] start experimenting with their cinematograph, they use the same sort of series of images to reproduce movements also from a static position. When Georges Méliès[4] compose his movies, there is no visible camera movement. For him, a camera movement is a trick used to produce special effects[5]. The movement is inside the image. The first travelling is initiated by Alexandre Promio[6] who embarks his camera on a gondola in Venice in 1896 to film the *Panorama of a canal from a boat*. In the beginning of cinema, a camera movement is a rarity. The cinematographic language borrows from photography and theatre. Yet, it took years to develop the cinematographic language in its full capabilities. After more than a century, it is still in development.

Each element composing a visual language has to be experimented, tested in its variations, confronted to the viewer's perceptions and interpretations. Developing the basic elements of a visual language is a process that never ends. It is linked to the graphical and optical characteristics of the imaging technology; it is subjected to the context of production and presentation and it has personal and cultural ramifications.

After a few decencies of experimentation and being at a technological crossroad, we can't expect holography to present itself with a completed and achieved visual language. Working with analog holography is not the same process than working in photography or sculpture. Working in synthetic holography[7] is not the same process than working in cinema, computer graphics or even analog holography. Even though we can establish several similarities between these forms of visual representations, their dissemblance and their particularities makes them distinct Medias with their own codes, their own language. The optical, spatial and temporal characteristics of synthetic holography provide particular formal elements, limitations and effects that imply the development of a proper visual language. I have been experimenting with different characteristics of holographic images produced by a limited set of perspective views to investigate its optical, spatial, interactive and narrative possibilities. One of these characteristics is the metamorphosis of content; the transformation of forms attributes that occurs when the observers moves and changes their angles of view.

2. Fragmentation:

"(...) lumen aliquando diffringi, hoc est partes eius multiplici dissectione separatas (...)" [8].

Francesco Maria Grimaldi, 1665.

To consider the specificities of synthetic holography, and subsequently its possibilities for building narrative forms, we have to realise that this imaging process is based on a series of fragmentations. First of all, as Francesco Maria Grimaldi described diffraction, in 1665 for the first time in history, light itself is divided and fragmented[9]. Synthetic holography goes even further in this fragmentation process. To produce the necessary set of image for a synthetic hologram, the computer generated 3D scene is rendered as a series of several hundred perspective images, each one representing an angle of view on three-dimensional space. This is a division of the field of view. In a direct-write holographic imaging system, these images are fragmented, processed and recombined to fit the specific optical needs of the holographic cells structure. Before imaging the hologram, there is also a chromatic fragmentation in three channels, red, green and blue. And finally, the holographic emulsion itself is divided in thousands of holographic cells. These successive fragmentations allow us to setup the necessary requirements for reconstruction of the multiple viewpoints perspective. Then, the holographic display will show the images in all their graphical, chromatic and spatial dimensions. All in all, synthetic holography is fragmentations to metamorphose virtual space into optical space.

The concept of fragmentation as a mean for representation of space isn't new. You can find it in pre-Renaissance perspective systems in which the pictorial space is divided in horizontal zones, each step showing represented objects in smaller sizes from the bottom of the picture to the top. Such spatial representation system can be seen in many paintings by Hieronymus Bosch[10]. Fragmentation of pictorial space is also visible in Oriental paintings that apply Kuo Hsi's theory of three distances[11] where spatial divisions are used to represents relative depth[12]. Also, oriental horizontal scroll landscape paintings often represent wide spaces by dividing the scroll into multiple viewpoints and accumulating several views to compose a landscape. Each portion of the pictorial space has transition elements that connect the accumulated views as if it was a single panoramic view. To appreciate these compositions, the observer has to move and vary his own points of view. More than spatial representations, these paintings are representations of a path in space[13].

These fragmentations and this observational movement is an opportunity for the artist to introduce gradual modification of forms, content or spatial appearance itself. An example of such a metamorphosis is found in three etchings by M. C. Escher[14] called *Metamorphosis I, II* and *III*. These complex patterns gradually transform into different forms and even between the flatness of bi-dimensional geometrical compositions to the illusory three-dimensionality of perspective. Escher's Metamorphosis transforms the identity of represented space itself[15].

By fragmenting pictorial space in such ways, artists spread the metamorphosis of content, not only over space, but also over time since the observer has to accumulate **what is seen here** and **what is seen there**. In synthetic holography, we would rather say that the observer accumulates **what is seen from here** and **what is seen from there**. This distinction is fundamental in the composition of a synthetic hologram. Instead of producing fragmentation of the pictorial space, we apply fragmentation to the optical space by dividing the field of view of the hologram. In my *Tractatus Holographis* hologram (2005) (figure 1), I divided the field of view in three zones. When the observer is in the left portion of the field of view, he can see two pages of the book. When he is in the right portion of the viewing zone, he sees two other pages. The center part of the field of view is used for the page to turn. This hologram of a fictitious treatise on holography presents a text written in 16th century French that describes the basic principles of holography. By fragmenting the field of view, I could not only augment the content with a metamorphosis of forms, but also demonstrate visually a spatial and interactive characteristic of synthetic holography.

Figure 1. Jacques Desbiens, Tractatus Holographis, 2005.
Three views of the synthetic hologram, 60cm X 4cm.

To go beyond this configuration of two zones with a transition in between, I produced another experimental hologram in which numbers change from 0 to 9 when the observer moves from left to right. The *Numbers* (2009) hologram demonstrates that several transformations can occur in a small hologram. In such a simple configuration, the observers understand easily the purpose of his movement and the necessary orientation.[16]

It is the observer's movement that is the main feature of these holograms. It is what generates the modification of content. To produce this effect I had to work spatially and foresee the optical circumstances of the holographic display. The computer graphic animation wasn't the finality. It was just a tool to introduce content variations in a fragmented space. These transformations are possible due to the anisotropic[17] properties of holograms. Content metamorphosis, in synthetic holography, is produce by the angle variations between the light source, the hologram and the viewer.

It would be tempting to associate this kind of metamorphosis of the holographic image to some cinematographic special effects such as morphing. However, in these holograms, the image movement is different than the cinematographic movement. The metamorphosis rhythm and direction is controlled by the observer. His movements can pace or reverse the transformations. He can be selective; stop here to look, move there to compare. Content metamorphosis in synthetic holography is a fragmentation of movement. In that sense, synthetic holography may be closer to chronophotography than cinema.

3. Spatial synchronisation:

" (...) the change of shape with every step one takes" [18].
Kuo Hsi, 1117.

Anisotropy is a directional property of the transformation of visual appearances. We can imagine several compositions using this property to create content metamorphosis as an aesthetical effect where colors, forms and space itself would transform in synchronization with the observer's movement. Given the fact that synthetic holograms are three-dimensional multiple points of view perspectives and that the observer's two eyes sees different points of view at the same time, these transformations will appear different than what a 2D image would show. This is a field where there has been some experimentation in analog holography. However, synthetic holography offers different effects and controls over the transformations that are particular to the synthetic hologram structure and the computer graphics tools used to compose the images. Beyond the visual effects, the directionality, this link and dependence to the observer's angle of view, can be an asset in the development of narrative tools. To experiment this characteristic, I produced a large hologram in which texts, images, objects and backgrounds transforms in a synchronized composition to build a complex set of relations that the observers can reconstruct at their own rhythm.

This experimental hologram is titled *Graphis*[19] (2009) (figure 2, 3 and 4). It is a panoramic 3D scene produced from 2537 computer generated images showing books, drawings, calligraphy and objects that transforms when the observer walks and change his angle of view on the horizontal axis[20]. All the elements composing this holographic image are references to the history of optics, perspective and research methodology. Superposed to the 3D scene are several historical quotations of Zangzhi[21], Zhang Huaiguan[22] (8th century), Ibn Al-Haytham[23] (965-1039), Kuo Hsi[24] (11th century), Su Shih[25] (1036-1101), Roger Bacon[26] (1214-1294), Leonardo da Vinci[27] (1452-1519), Jean Pélerin Viator[28] (1445-1524) and Francesco Maria Grimaldi[29] (1618-1663). The field of view of this holographic image is divided in several portions but three main zones prevail. From the left portion of the field of view, the observer sees a blurry landscape with a few calligraphic texts floating above (figure 2). The center of the field of view shows a pack of books and papers, five polyhedrons and a mirror. Calligraphic texts are also visible from this center part (figure 3). From the right side of the field of view, the space is filled by a horizontal plane of water and a few images and calligraphy appears (figure 4). The texts are in Latin, Chinese, Renaissance French, Greek, Italian and Arabic.

Therefore, with such a mix of sources and languages, I don't expect that the observer will understand and reconstruct the whole references and significations brought in this hologram without some extensive research. Besides, I didn't include any linear story or discourse in this holographic scene. There is no progression and no predominant point of view. *Graphis* is a chaotic composition from which the observer can take the elements of information according to his wishes or abilities.

Figure 2. Jacques Desbiens, *Graphis*, 2009,
Synthetic hologram, 1.5m X 30cm.
View from the left.

Figure 3. *Graphis*, 2009. View from the center.

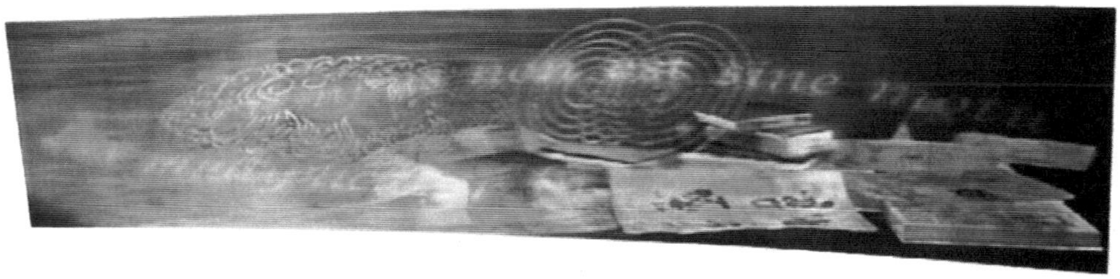

Figure 4. *Graphis*, 2009. View from the right.

Even though the field of view of this hologram is divided in three zones, objects appear and disappear independently and at their own rhythm. This procedure creates more alignments and associations between objects and texts. It also allows transitions that include links between the content of each zones. For example, this kind of transition is visible when the mountain in the left zones dissolves into a Chinese text shaped like the mountain's topography (figure 5). This text is a quotation from Kuo Hsi's painting treatise that can be translated as follow:

"A mountain viewed at a close range has one appearance; a mountain viewed at a distance of several miles has another. When viewed from a distance of scores of miles, it has still another. The change of appearance caused by the varying degree of distance from the object is figuratively known as "the change of shape with every step one takes." The front view of a mountain has one aspect; the side view another; the back view still another. The ever changing view from whatever side one looks is described as "different shapes of a mountain as seen from every side." Thus a single mountain combines in itself several thousand appearances. Should we not realize this fact?"[30]

Figure 5. Jacques Desbiens, *Graphis*, 2009.
Kuo Hsi's calligraphic mountain, detail of a source image.

Obviously, such a transformation is charged with symbolism. Content metamorphosis and spatial synchronisation are tools that can be used in different ways to produce meanings and relationships between the elements occupying a composition. In synthetic holography, animation isn't just movements, it is transformations, alignments, superpositions, juxtapositions, contrasts, transparencies, occlusions and all sorts of variations and attributes that we can apply to elements in momentarily synchronisation with each other and with the observers position. *Graphis* is filled with such configurations to introduce meaningful associations. For another example, the Chinese text visible from the left part of the field of view (figure 6) is a quotation from Zhang Huaiguan's description of the mythological creation of writing by Canjie, a mythical character who develops ideograms from his observation of animal tracks. When the observer starts to move toward the right of the field of view, he will see bird tracks gradually appearing on the plane of the hologram, just behind the calligraphy. In the same way, a graph made by Ibn Al-Haytham is superposed on his text commenting on research methodology[31] (figure 7). This drawing is the first correct depiction of stereopsis in history. These two examples use superposition to create meaning. It is one of the most obvious narrative tools that we can use in the composition of a hologram. Besides, this positioning enhances the three-dimensional effect of the hologram.

Figure 6. *Graphis*, 2009. Detail. Calligraphy : Yunjeung Yang.

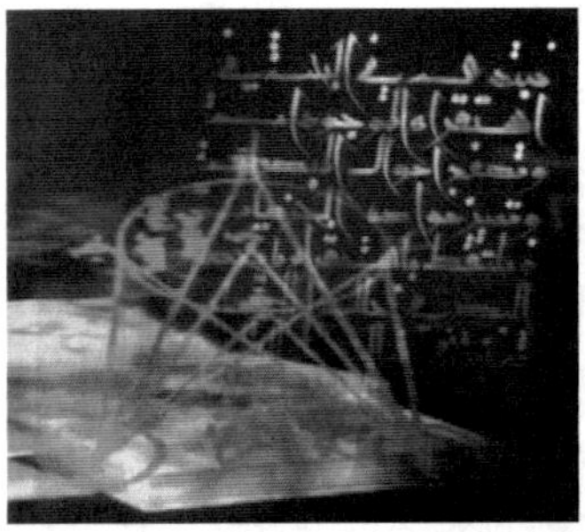

Figure 7. *Graphis*, 2009. Detail. Calligraphy : Joseph El Hourani.

Some quotations, words and images in *Graphis* refer to the optical properties of synthetic holograms. For example, when viewing the hologram while moving horizontally, the observers may notice that most transformations appear as a wavy movement. This is not an effect produce by computer graphics but rather by the holographic cells structure itself. These waves are in fact a form of time-smear[32] in which shapes appear to be distorted. The introduction of water in the third zone is obviously an analogy to this effect. To push further this theme, I added Leonardo da Vinci's drawing of interferences (figure 4). Francesco Maria Grimaldi's description of diffraction[33], from his 1665 treatise on light, can also be seen floating on the water[34].

With the attribution of specific effects to particular elements, in connection with the optical properties of the hologram, a hologram such as *Graphis* becomes a set of spatial relationships between words and objects, space and kinetics, between what is visible here now, what was seen there before and what will be seen later elsewhere. The metamorphic composition of forms, zones and content establish the multiple views in an observation time defined by the observer's free movements. *"Tempus non est sine motu"*[35] as it is quoted in *Graphis*: *Time isn't without movement.*

To build this kind of metamorphosis, where each form, each volume, each object, each graphic element, appear and disappear independently from each other, linked to the observer's

Journal of Physics: Conference Series **415** (2013) 012008 doi:10.1088/1742-6596/415/1/012008

displacements, the rendering process and the optical structure of the hologram needs to provide continuous variations over the whole field of view. The anisotropic properties of the synthetic hologram optical system is giving us a clue for developing narrative forms that uses spatial synchronisation with the observers angle of view. *Tractatus Holographis* presented a narrative composition that was oriented by the subject, an open book. It suggested a chronology. *Graphis* is an attempt in an adaptation of the narrative structure to the disorder of the observation movements. The narration is fragmented but reconstructed in the synchronisation of presences and in random successions of spatial relationships. We may try to order and structure images and optical systems, but these very structures deprive us from controlling the sequence of observation. In synthetic holography, the movement of the gaze is the transformation of forms.

Jacques Desbiens
June 2012

---

[1] Marey, Étienne-Jules (1830-1904), French scientist and physiologist.
[2] Chronophotography is a technique using sequential photography to register a movement decomposed in a serie of successive images. See the work of Étienne-Jules Marey (1830-1904), Eadweard Muybridge (1830-1904) and Georges Demenÿ (1850-1917) among others.
[3] Lumière, Auguste (1862-1954) and Louis (1864-1948), French scientists and inventors.
[4] Méliès, Georges (1861-1938), french illusionist and filmaker.
[5] In *The man with the rubber head* (1902), Méliès use a camera on wheels to move closer or farther to the man's head to produce an illusion of stretching when the head gets bigger or smaller.
[6] Promio, Alexandre (1868-1926), cameraman for the Lumière brothers.
[7] Synthetic holography is a process that register on a photosensible emulsion a large number of images (usually computer generated) representing a succession of points of view on a 3D scene.
[8] *"(...) at times light is divided, to be partitioned, multiplied, severed and separated (...)".* Francesco Maria Grimaldi, 1665, *Physico-mathesis de lumine, coloribus et iridi,* proposition I, num. 5.
[9] *Ibid.*
[10] Bosch, Hieronymus (c. 1450-1516), Dutch Painter.
[11] High distance (*gaoyuan*), deep distance (*shenyuan*) and level distance (*pinyuan*).
[12] See: Hsi, Kuo (1117), *Lofty Messages of Forests and Springs* (*Lin Ch'üan Kao Chih*), in : *An Essay on Landscape Painting,* translated and commented by Shio Sakanishi, London, John Murray, 1935, 70 p.
[13] See my articles: "Nomadic Perspectives: Spatial representation in oriental scroll painting and holographic panoramagrams." In: *Proceedings of the 12th International Conference on Virtual Systems and Multimedia (VSMM)*, Xi'an, China, October 2006, Springer-Verlag Berlin 2006, p. 504-513. And: "The perspectives of synthetic holography", in: *Proceedings of the 9th International Symposium on Display Holography (ISDH),* Shenzhen, China, July 2009.
[14] Escher, Maurits Cornelis (1898-1972), Dutch graphic artist.
[15] Another example of this kind of graphical transformations can be seen in Saul Steinberg drawing *The Line* (1954). This drawing shows a horizontal line where different elements modify the identity of the line by refereeing to different angles of view. Steinberg made this very long drawing on the wall of a maze in Milan. It was an image of spatial metamorphosis that you observe by walking.
[16] On these experiments see my article in the ISDH 2009 proceedings: *Experiments in image composition for synthetic holography*, Jacques Desbiens, ISDH 2009.
[17] Anisotropy describes "a substance that exhibits different properties along different axes of propagation" (Photonics dictionary, www.photonics.com). I use this term in the context of a whole hologram to describe the properties of a hologram to show different aspects and content depending on the angle of view or the angle of lighting, in opposition to "isotropic" which would be the property of 2D graphical images to remain unchanged even with variations of angle of view.
[18] *Op. Cit.* p. 37.

---

[19] *Graphis* : latin word meaning "pencil" or other instruments used for drawing and writing. The hologram exhibited in the MIT Museum (June 2012 to September 2013) is a smaller version (1.5m X 30cm) of a large panoramic hologram (3m X 60cm).

[20] This synthetic hologram has horizontal parallax only.

[21] Zangzhi, (4th century B.C.), Chinese philosopher.

[22] Huaiguan, Zhang (8th century), calligraphy theoretician and historian. Author of several books on painting and calligraphy.

[23] Ibn Al-Haytham (Alhazen) (965-1040), Persian scientist. Author of the treatise *Kitâb fi'l Manazir* (*Book of Optics*) 1021.

[24] Hsi, Kuo (c.1020 - c.1090), Chinese painter and theoretician. Author of a treatise on landscape painting: *Lofty Messages of Forests and Springs (Lin Ch'üan Kao Chih)*, 1117.

[25] Su Shih (Su Dongpo) (1037-1101), Chinese poet, calligrapher and scientist.

[26] Bacon, Roger (1214-1294), English philosopher and scientist. Author of the optics treatise: *Opus Magus*, 1267.

[27] Vinci, Leonardo da (1452-1519), Italian drafter, painter and scientist.

[28] Viator, Jean Pélerin (1445-1524), French artist and diplomat. Author of a perspective treatise: *De Artificiali Perspectiva*, 1505.

[29] Grimaldi, Francesco Maria (1618-1663), Italian astronomer and physicist. Author of the optics treatise: *Physico-mathesis de lumine, coloribus et iridi*, 1665, in which he describe diffraction.

[30] Hsi, Kuo (1117), *Op. Cit.* p. 89.

[31] *" Research begins with observation of present things, examination of optical circumstances and discernment of the specificities of their components. Thus, we observe what concern vision and we progress in research by measures and classification while criticising hypothesis and tackle results with circumspection."* Ibn Al-Haytham, *Kitâb fi'l Manazir*, 1021.

[32] On time-smear see my article in ISDH2009 proceedings: *Experiments in image composition for synthetic holography*, and Teitel, Michael A. (1989), "Animation in holographic stereograms: the time depth paradox", *Practical Holography III*, SPIE Vol. 1951, 205-215.

[33] *"Nobis alius Quartus modus illuxit, quem nunc proponimus, vocamusq ; Diffractionem, quia advertimus lumen aliquando diffringi, hoc est partes eius multiplici dissectione separatas per idem tamen medium in divrersa ulterius procedere, eo modo, quem mox declarabimus."* I translate: *"A fourth lighting method is known, here now how I propose to name it : diffraction. I bring attention that at certain moments light is divided, being partitioned, multiplied, sectioned and separated, this however by the same medium in diverse ulterior procedures. Because of this, this is how I conclude." Francesco Maria Grimaldi, "Physico-Mathesis de Lumine Coloribus et Iride ", (1665).*

[34] Unfortunately, when the light source isn't precisely positioned and strong, this text is quite faint.

[35] Bacon, Roger, (1267), Opus Magus.

Journal of Physics: Conference Series **415** (2013) 012009 doi:10.1088/1742-6596/415/1/012009

"Kunstwerk" in the age of holography

Authors: Dietmar Öhlmann, MA(RCA), Holographer & 3D artist, Syn4D, and Odile Meulien Ohlmann, Collector and Sociologist

Kehl, Germany and Strasbourg, France

DO@syn4D.com, www.syn4D.com , om@artbridge.info , www.artbridge.info

Abstract. In 2007 one auction shocked collectors and artists: Two art works of Rudie Berkhout were offered for one thousand dollars on eBay. The closing down of museums for holography in Cologne, New York and Washington DC paid a big price of respect for true art. Benjamin's vision about "Art in the Age of Mechanical Reproduction" found a new expression in abuse and ignorance. Beautiful sculptures of Art are reduced to "Holograms" like Rudie's "Cairo Aspect" placed in a rummage sale, covered by fingerprints of ignorance.

1. Introduction

The artists of holography failed quite often to use the same professionalism as photographers, possibly not warned by Walter Benjamin uplifted finger. Single artworks or limited editions on glass have been unexpectedly reproduced on open edition on film. Quite a few collectors stopped investing in holographic art because of this practice. To separate technical beauty and true art is with the elaborate possibilities of photonics media challenging. However, difficult color control in holography might have been one (among other) criterion for art critics and visual artists to consider holographic works without paying attention to their artistic content. But history of art shows that colors have been rendered progressively. To study "Painting Material and its use in the painting" from Max Doerner (1992 München-Berlin-Leipzig) you learn that malachite makes a beautiful green in a Fresco.(1) To create a Fresco like Benozzo Gozzoli, e.g. the Procession of the Magic in the Palazzo (1459) Merici Riccardi, you need to have the whole painting already finished before you take the brush in your hand. To appreciate the variety of green in the breathtaking Fresco becomes even greater, if we know that Chromoxid was first described by Vauquelin in 1809, and emerald green patented in 1850 by the French paint producer Pannetier. Then, we can appreciate the pseudo color display holograms of artists like Inaki Begusristain and Margaret Benyon. Still both works cannot be analyzed in the same context. The paper will identify some of the hindrances of the holographic art in regards with these past 40 years in Europe and America, based on the insider experience of the authors. Crossing their perspectives, Odile Meulien Ohlmann, a French/American curator, collectors and culture analyst, and Dietmar Öhlmann, a German artist and 3D designer will show that the education in art history and humanities need to make some evolution or revolution!

2. Art and Reproduction

When Walter Benjamin wrote before 1940 his essay about art in the time of mechanical reproduction, he described the rise of photo and film, not the manual reproduction of Graphic and painting.(2) A permanent exhibition of the live of Frida Karlo in Baden Baden, Germany, used mainly licensed replicas from © Banco de México Diego Rivera & Frida Kahlo Museums Trust and fast no originals.(3) If you have a walk around Montmartre in Paris, there are mainly galleries employing painters who paint the whole day the Eiffel tower. None of them speaks French, a good sample of successful immigration and a good prove of Benjamins thesis about iconic perception about art. For him, just a photo of a portrait has the aura of art. What is the original, replica, licensed replica, limited edition or open edition? Does it make sense to ask to deliver the real photo, the original? Or the true digital images in the internet or on youtube? Benjamin wrote in a small history of Photography: "...Daguerre's images of light were iodinated and he lightened them in camera obscure, forming light-sensitive silver plates which were moved back and forth, till we could see in the correct lightning a recognizable soft gray picture" . They were unique piece; in 1839 the price for one Photo was 25 Gold Frank. Often they have been kept like jewels in caskets..."(4) We could use the same description for a single beam Denisyuk hologram, which was said impossible to copy. The original, iconic work was clear for artists and collectors. The technical evolution of holographic techniques brought the distinction between Master and Copy, Reflection and Transmission, limited edition on glass and open edition on film. People involved in Holography mentioned on the market, that just expert could answer the question: "Do I have the artwork or the Pin Up? Do I have the original first copy or the mass produced hologram?" Already Benjamin said in 1934, it is senseless to ask for the first copy of a photo.(2) Today, digital holography makes the copying of the hologram even easier, and opens a new question: "Is this a copy from the artist or from the manufacture?" Before the little plant of holographic art could grow, the capitalization of the market swapped it together with all junk of simple replicas, gimmicks and "funny staff", no more question about use, emotion, content.

Pic01: Hans Vredeman de Vries, Perspectiva (1604). Some master pieces have been covered by white painted and not preserved. Certainly, there were not much considered by their contemporaries.

3. What is Art?

Vittorio, a figure in the roman "Vittorio the Vampire" from the American author Anne Rice (1941), filled his eyes with bloody tears as he looked at the painting from Fra Filippo Lippi in Florence (1406-1469). As Vittorio saw the master surrounded by true angels raised his thirst to see more creations.(5) Rice's visit in the Italy of the 1400 awakes emotion of ecstasy between fiction and true visual art. If we continue the history of art, we see the need for beauty did not decreased; at the reverse it accelerated the creativity with the spatial paintings of Jacopo Tintoretto (1518-1594) in Venetia, or the flight into space with Giovanni Battista Tiepolo (1696-1770), and Michelangelo (1475 –1564) in Rom.

The world already loved 3D. It was the time of glasses free spatial theatre, the Fan Club of Michelangelo celebrated the stars, till someone else came and made his self portrait, in a way we would say: expressionism. El Greco (1541-1614) was too modern, his style did not fit next to Michelangelo, and people asked: " Is this art?" Lucky that he moved to the next town and created paintings like the Laokoon of the Washington National Gallery. (6)
So the question "Is this Art?" exists in all epochs. Taste and social representation of reality are constantly changing, from the Greek Mimesis to Marinetti and his Manifesto. In matter of fact, it is not the proper of art historians, to anticipate and reveal what their contemporaries will understand years later! Art has always been a process of experimentation, and it is also true with new media. Experimentation conducts to better understand and transform technique and give access to new perceptions. Especially in our time where new visual techniques allows us to discover and see images of the nano and macro universe, a fantastic exploratory trip now available to artists interrogation. In the "Micro Hologram"(13) I fully used this new possibility of exploration, and access to the atomic force microscopy to analyze and compare patterns of the hologram, showing a trip of a spatial landscape, and the holographic dots in which I place a macro image of a woman looking at herself in water. Finally I was showing different realities of a dot, a universe hidden to our visual organs, unfolding another nano dimension and possibly others ones we do not even consider at our present time.
Just a fool cans belief that art and technology are separated. If we read books about history of arts, like those ones of Herbert Read or Rene Huyghes, their books are filled with a kaleidoscope of senses, intuition, rationality, research, techniques. They described artists' quest for conceiving the world by all matter, emotion, accident, misunderstanding, reinterpretation of science, philosophy, and alteration of perception in a flux which offers us new creative works. At school, we do learn to place art in categories, like impressionism and expressionism, so that collectors and spectators are able to adjust, recognize, and identify a form to represent the world.
We find in holographic art a similar evolution of subjects than in painting, like copy of subjects, portrait, nude, and iconic presentation. Margaret Benyons' cosmetic series shows different images than the reflection in a mirror like a Rembrandt 1606-1669, or Félix Vallotton 1865-1925. It did not bring the spectator to the same controversial reaction than works like those ones of Egon Schiele, 1890-1918, rather more indifference and misunderstanding, because unclassified. The process of painting gives the artist more freedom, to observe, to reflect, and to change the concept on the canvas. The creation of a photo or a hologram pushes the artist to plan, create one chosen technical setting and then finally, "Click!" Nevertheless, the creation is not just in the final process of visualization it is before in the wish, the motivation to create an image. This is where we can see the difference between an artwork who reflects a personal, individual expression of his/her time, and the wish to copy an object or subject and make it technically perfect. The first so called artworks were trying to copy and reproduce the nature or be the visual element which would help to reconstruct a story. This, was the function of these images we now call art pieces. The goal, the search makes the difference, like a work of Margaret Benyon and the work of Inaki Begusristain. One intends to express an inner feeling, the other one

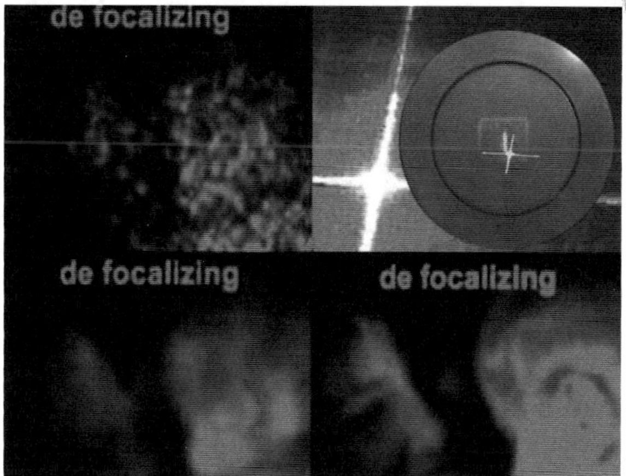

Pic02: "Micro Hologram", Installation from Dietmar Öhlmann. Looking at the red dot, we see speckles, zooming further; appear the full image of a pulsed hologram. We de- focalize, we go beyond surface, which is like looking through the window of a car on a rainy day.

wants to copy an object like a vase. Still in exhibition, and because, spectators are still not able to capture the meaning of the content, an Begusristain 's vase and its wonderful immaterial color might talk more directly to the public than the intrusion of the multidimensional consciousness represented by animals portrayed in Benyon' face. This explains, without excuses for artists, why, in most of exhibitions of holography, some holograms are still placed next to holographic artworks.
Among artist holographers, we have the constructivist. Rudie Berkhout was all of them, his Hologram Cairo Aspect moved me as much as Fra Filippo moved Vittorio in the roman of Anne Rice. John Kaufmann used the notion of time, his series of Rocks replacing time, and substituted it with the matter earth. As I explains in the paper "Artists and Collectors"(7) , 20 years ago, and it is still true, the institutions of the art world are always late in the presentation of the art also some institutionalized fine arts museums already courageously presented holograms to the public. But the press, the critics, and collectors and in general the art communities were not supporting these audacious attempts. And holography remains for the time being a fascinated field of techniques and science, before, in some time from now, entering the pantheon of art.

4. Value?

Benjamin defined art and its function is its secularized ritual, iconic magic or a kind of theology of art:"l'art pour l'art." (2) Photography shacked the tower of profound service of beauty as an "evolutionary means of reproduction" (in the dawn of socialism). Benjamin reduced for further analysis the perception about art to two major parts: the social cultural value and its exhibit value.
One of the first and largest event of Photography was in 1951 the work of Edward Steichen with the exhibition of "Family of Man." (8) It started in Museum of Modern Art New York, traveled until the border of communism and still found its place in Luxemburg. If we are reading the thesis of Silke Walther from the Staatliche Hochschule für Gestaltung Karlsruhe, we find that photography is extremely political.(9) The simplest images can be composed to any kind of social messages, like the idea of the good democrat.

Pic02: Holographic illustration of Zaha Hadid Phaeno in Phaeno

Holography never appeared to be very political. Fascinations of Light in London, Art, Science and Technology in Washington DC, Museum of Holography in New York and Cologne have been the forum of holographic art. However, most of exhibitions have been a strange mixture of holographic art, industrial holography and some nice pictograms. The images in the exhibition screamed at our mind, it transgressed all our visual aesthetic we have been confronted before, still it did not reveal any echo known as art forum. Even when Bruce Nauman and Dali encountered the media, the reaction has been quite and without too much emotion. After more than 800 years history of painting, changing the clerical subject, changing form and color, 100 years of photography and only 60 years of holography, the public is still in expectation and suspicious about the immateriality of these objects of light. I always remember this sentence repeated by several collectors; "Finally with a hologram, what do I buy, a frame, a lass plate, what?

5. No Border?

James Elkins from the School of the Art Institute of Chicago complains, that the history of image making has too much been focused on fine art, otherwise the images of the Hubble Deep Field would be one central work of art of our epoch. His philosophical essay about "Kants's Reflections on the Sublime and the Infinite," questions the meaning of emptiness and infinity of the Universe.(10) While the first photo from the NASA has been illustrated by artist into "pretty pictures," the infinite eye is transgressing all our previous aesthetic about image making.
Andreas Ruby presents Zaha Hadid Architecture, her sketch and her transformation of form as excessive. Nicole Stöcklmayer describes the production of the hologram and its presentation into the building of the Phaeno Museum itself, in Wolfsburg Germany, the tip of her excessive presence. (11) This doubling of this presence, the digitalized sketch illustrated into Andrea Rubys Space ship and transformed by the artist Dietmar Öhlmann into a digital Hologram, makes it an iconic piece of work, as Dr Börner, Curator of the Exhibition at the Phaeno Museum, stated too. Digital Holography makes it possible, to transform CGI (computer generated images) into an art creation using expression and techniques, of the present time, and combining real and virtual reality. However, if you exhibit this Holographic Phaeno out of the context, it becomes just a hologram; the art stays in the paradox.
The art history gets a new form in writing, like the publication from Ingeborg Reichle and Steffen Siegel with their international research of new aesthetic of images making. Analog photography and holography vanished like HTML5 might remove Flash on the market. But the issue of visual creation is the same, just like in language; the grammar and vocabulary are changing.

6. Conclusion

This June 2012, Madonas' live show in Tel-Aviv has used big LED's panels. Everybody called it 3D and holographic projection. In Forum, like the holography group in Linked-in, people deliver a fight about the definition of a hologram. Some people claim that they do art, but talk just about technique. Some holographic artists do not want to be mixed with other media, so for an art historian it is difficult to find out what kind of art they are talking about. I believe, in holography we are back in the quatrocentro of the renaissance, in the experimental stage of Leonardo mixing visual and technique. In the past, the artist has been a hard worker. He mixed paint and placed it in the correct context so that, the authority of clergy and nobility would accept the work and its artist. Some artists need pats of recognition like a cat. Benjamin stated; Artists, like actors are very often cut, separated from their public, e.g. the movie stars and the spectators in cinema. This is the reason why we invented the stars, to give them some pats. If Fluxus and Dadaism is the reaction on photography, holography might wait to find its proper place, and this might take some time. Indeed if the form created by art always anticipated the structure of the societies like demonstrated René Huyghes in "Le pouvoir de l'image" (12), we can wonder what kind of structures holographic art is shaping for our upcoming societies.

References

1. Max Dörner 1994 *Malmaterial* (Thomas Hoppe Leipzig 1921)
2. Walter Benjamin 1963 *Das Kunstwerk im Zeitalter seiner technischen Reproduzierbarkeit* (Edition Suhrkamp)
3. Kunstmuseum Gehrke-Remund, Baden Baden http://www.kunstmuseum-gehrke-remund.de/
4. Walter Benjamin 1963 *Kleine Geschichte der Photographie* (Edition Suhrkamp)
5. Anne Rice 1999*Vittorio* (Knopf NY)
6. Zweiburgen 1982 *Das große Buch der Malerei* Westermann
7. Odile Meulien 1995 *Artists and Collectors* (Proc. SPIE Vol 2333)
8. Edward Steichen exhibition *The family of man* ,Luxemburg, (Schloss von Clerf Clervaux) http://www.family-of-man.public.lu

9. Silke Walther, Ingeborg Reichle / Steffen Siegel *Fotografie als Weltsprache zum Theater der Realität* p379-398 (Maßlose Bilder: Visuelle Ästhetik der Transgression, Munich Fink)
10. James Elkins, *Maßlose Bilder* Immoderate Images. Visual Aesthetics of Transgression*)* ;p 303,.
11. Nicole E. Stöcklmayr *Architektur ohne Maßstab*: *Digitale Visualisierungen im Entwurfsprozess, in* p. 279-294.
12. René Huyghe 1967 *Sens et destin de l'Art, De l'art gothique au XXème sciècle* (Edition Flammarion)
13. Dietmar; Ohlmann, Odile Meulien Hans Danzebrink 2005 Photo The *Micro Hologram* artwork first shown in Montreal and San Jose The philosophy of this ongoing project has been presented in: *The matrix of inspiration* (Practical Holography XIX: Materials and Applications Proc. SPIE 5742 p 71-78)

Autres references

Pic 1: Hans Vredeman de Vries, (1604) *Perspectiva in Monasterie Bad Gandersheim © Germany*, copy with permission of „Portal zur Geschichte"; Zur Abbildung: Bildfeld aus dem "Saal der Perspektive" (Nordwand, Sockelzone) im Fürstlichen Haus der Äbtissin Elisabeth Ernestine Antonie von Sachsen-Meiningen (1716-1726), Kloster Brunshausen, Bad Gandersheim

Pic 2 : Dietmar Öhlmann, 2005 *Micro hologram* © Germany

Pic 3: Dietmar Öhlmann 2007 *Phaeno hologram* © Phaeno Museum with Friends of the Museum's Association, illustration of the architectural model of Zaha Hadid .

Hololujah; A One Kilometre Art Hologram

David Warren

Not affiliated, artistdw@supernerd.com.au

Abstract. This paper will outline the production of the white light transmission achromatic image art hologram titled Hololujah displaying forward projecting real imagery of text and measuring 100,000 x 3.5 cm. **Materials and methods:** The paper will cover the use of Slavich VRP-M film 33.3 x 1.05 metres that was exposed and processed as thirty-three and a third 100 x 105 cm frames. These thirty-three and a third frames were subsequently cut into one thousand 100 x 3.5 cm strips with their ends cold laminated together to form the kilometre length hologram. This paper will expand on the use of a Coherent Compass 315M, 532 nm, 150 mW DPSS laser diode in a lensless setup, using a single beam through diffuse glass, no isolation systems and a two minute exposure time with the film lying flat on the floor. Lastly, this paper will illustrate how the hologram was produced in a 220 x 200 x 300 cm confined area of a suburban bedroom. **Theme:** This artwork is a comment on the social networking site Twitter.

Keywords: White light transmission achromatic image art hologram, dpss laser, VRP-M holographic film, single beam lensless optical setup

INTRODUCTION

This paper describes the production of a one kilometre art hologram titled Hololujah. It is a white light transmission achromatic hologram measuring 100,000 x 3.5cm with front projecting imagery comprising of text and commenting on the social networking site Twitter.

TECHNICAL

It was recorded on VRP-M film, 33.3 x1.05 metres using a Coherent Compass 315M, 150mW, 532nm DPSS laser using a lensless single beam optical geometry and utilising two frosted glass sheets to expand the beam. It was made in my ground floor suburban bedroom within a space measuring 220 cm high x 200 cm wide x 300 cm deep. The laser was mounted 150 cm off the ground and projecting its beam at an angle of 45 degrees for a distance of 235 cm to the centre of the film on the floor. A 120 x120 x1.2 cm MDF wood platform rested on a carpeted floor with a concrete floor beneath. Each holographic film frame measuring 105 x100 cm rested on this wooden platform and was kept stable and flat beneath two panels of 110 x100 x 0.6 cm clear glass. The top glass panel was divided into thirty horizontal strips each 100 x 3.5 cm with a line of text written in each strip with a white opaque marker pen (see Fig 1). As there were two such glass panels, each 0.6cm thick, then this text was 1.2cm from the film during exposure. In the final hologram the text projects approximately this distance in front of the film and appears achromatic when illuminated with a white spot light.

Fig 1: Hand written text on glass lying flat on the ground, 105x100cm.

The single beam optical setup used two sheets of frosted glass to expand the beam instead of a lens. The first, measuring 4x4cm, expanded the beam while the second, measuring 75x40cm and placed 50cm 'down stream', blurred the laser speckle created by the first and provided further beam expansion to cover the 105 x100cm frame of film with a final beam diameter of 125cm (see Fig 2). The use of frosted glass negated the need to continually realign a spatial filter pinhole over the three-month period of the hologram's production. A further benefit was that the frosted glass provided a uniformly smooth beam without any projected dust or dirt aberrations. Finally, there was no traditional isolation system used, although it is felt that the carpet assisted in dampening vibrations caused by the author's close presence during the exposure stage. Extensive initial tests proved that any traditional isolation system was not required.

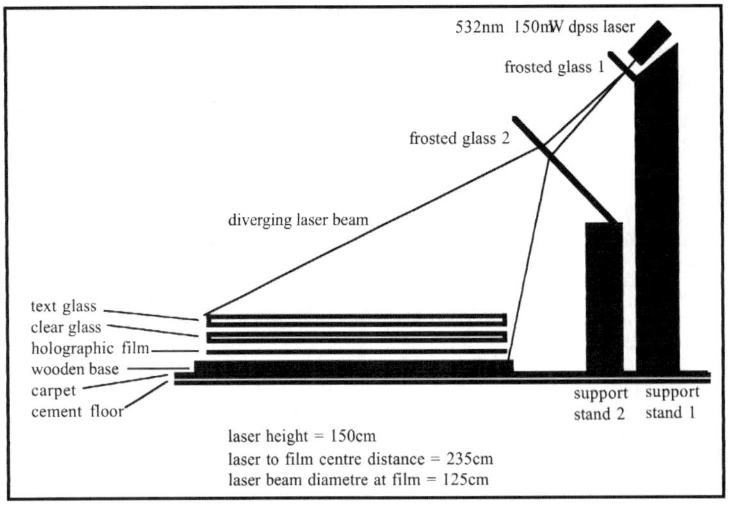

Fig 2: Holojuah optical setup

After a two hour settlement period, the exposure time of each film frame was two minutes followed by one minute development using my Zip 1 developer (see Fig 3), a two minute wash, bleached in EDTA until clear and then washed thoroughly in water before being hung up to dry.

Developer		Bleach	
Zip1		**EDTA**	
Metol	1.0gr	EDTA	30gr
Hydroquinone	1.0gr	Potassium Bromide	30gr
Phenidone	0.5gr	Sodium Hydrogen Sulphate	30gr
Potassium Hydroxide	10.0gr	Ferric Sulphate	30gr
Ascorbic Acid	10.0gr	Water to make	1000ml
Sodium Sulphite	30.0gr		
Water to make	1000ml		

Fig 3: Processing Chemistry

After processing and drying (see Fig 4), each 105x100cm film frame (33.3 in total) was cut into thirty individual 100 x 3.5 cm strips (1000 in total) and cold laminated together to make the one kilometre length hologram.

Fig 4: Section of processed 100x105cm film frame showing reconstructed achromatic holographic text with shadow image of text at the film's plane. From this full frame each strip of text was cut into 3.5cm x 100cm lengths.

AESTHETICS

Hololujah is an art hologram that has the appearance of an extended ribbon of film. While it specifically comments on Twitter, it could be easily associated with similar social networking sites. The reason for its length reflects the magnitude of endless tweets by an organisation that, if it were a country, would be one of the most populated on the planet. The dual image of projecting holographic text and shadow text at the plane of the film reminds us that communication, such as tweets, is a double linguistic event; there is the communication as a thought then the realisation of that thought as typed text on a screen. But to understand the attraction of Twitter, and similar social networking sites, you need to realise that the addiction is not in the posting of the tweet but in the anticipation of the reply. Ignoring this will lead to the dismissal of social networking as an excuse by many to communicate mindless chatter. There is also the bonus of people appearing to belong to a group, a fundamental human need (1).

Three halogen spotlights illuminate Hololujah that is displayed unwound in a triangular clear acrylic cuplike frame (see Fig 5). Viewers walk up to the artwork and are encouraged to pick up strands of the hologram and then, after aligning it approximately 45 degree to one of the three lights, read the selected tweet that will be seen to project in front of the film plane (see Fig 6). It presents an opportunity for viewers to interact or comment on the hologram tweets and, in turn, reflect on the broader issues and mechanics of Twitter. There is also the deliberate presentation of the hologram that encourages it to be discarded onto the floor after being read, echoing the ephemeral nature of tweets as junk messages with limited shelf lives. I also have a dislike for presenting holograms as precious artefacts and so Hololujah provided me with another opportunity to more fully engage on a personal level with viewers. In effect, this method of presentation provided a tactile experience and, in doing so, made the tweets seem more real.

The holographic text is a combination of actual tweets and fictional tweets. Thirty-four topics are covered which equates to the number of metre frames, each frame containing thirty tweets. Tweeters represented range from Lady Gaga (maybe), NASA updates, a girl next door, while topics included B grade movies, diets, how to rob banks, terrorism and the making of Hololujah as well as fictional tweets that discuss UFO's, capitalism, time travel and blondes know best.

Fig 5: Hololujah exhibited

Fig 6: holographic text

Twitter lends itself to scrutiny as it has gone through numerous stages of development. At first, people used it for locating and communicating with family, friends and acquaintances. Then the potential for data collection saw its growth with the opportunity to target commercial sponsors: the era of manipulating tweeters had begun. Now we see the rise of such companies as AdLy and Twittad who match brands and celebrities for promoted tweets, also known as integrated endorsements and in-stream ads. Often these companies write the celebrities tweets. '...we deliver the message to your fans and followers on your behalf.' (2) Or as Twittad puts it, "Do you have 53 followers on Twitter? Shouldn't you be making money off of them somehow, selling their attention to the highest bidder?" (3) The important step is integration; the days of separating the advertiser from the content are fading fast. Statistics suggest (4) that we are becoming savy onliners in terms of visually ignoring what might

be termed ad barriers that pop up prior to going to the prime website, or articles that surround the peripheral area of a site; better to have it integrated into what people *have* to read and look at.
But these changes in social networking from communication to commerce are not going unnoticed as read in the following anonymous tweet: "How naive was I: I thought twitter was all about "real " people communicating?"

In summary, Hololujah provides an excuse to examine the life and times of contemporary social networking.

ACKNOWLEDGMENTS

Sponsored by

ACT Government

Visual Arts grant received from artsACT, Australia, 2009

REFERENCES

[1] Baumeister, R. & Leary, M. (1995). The need to belong: Desire for interpersonal attachments as a fundamental human motivation. Psychological Bulletin, 117, 497-529.
[2] Adly.com, http://adly.com/celebrities/
[3] Schonfeld, Erick, September, 2nd, 2008, http://techcrunch.com/tag/twitad/
[4] Suster, Mark, April, 29th, 2011, http://www.bothsidesofthetable.com/2011/04/29/the-future-of-advertising-will-be-integrated/

Artist Projects at Holographics North

Dr. John Perry

Holographics North Inc.

University of Vermont

jp@holonorth.com

Abstract. The New York Times has declared the concept of holography in art as "laughably dated". And yet fine art remains one of the most durable applications of the medium. Holographics North Inc. has produced work for over 50 artists in 28 years. In many cases, new techniques and systems were required in order to implement the client's vision. The technical and conceptual challenges involved in several of these projects will be discussed, including photos of the work and the systems built to produce it. Among the artists addressed will be James Turrell, Michael Snow, Frank Stella, Michael Hayden, Harriet Casdin-Silver and Chris Levine.

Holographics North Inc (HNI) owns the basement of an old school building in Burlington, Vermont (Figure 1). It is 6,000 square feet with 10 foot ceilings and 3 foot stone foundation walls. The building is now a commercial condominium and the regulations require that no one can produce substantial noise or vibration. We are also located on a quiet, residential street.

Figure 1. HNI building in Burlington, VT.

Figure 2. The old boiler rooom, now our showroom and main office.

Nevertheless it has taken considerable work to render the space into its current configuration. Here, for instance (figure 2), is the asbestos-clad boiler that was removed to make room for our showroom. And we're small. Our staff have varied over the years between 9 people and 1.

But in that modest space, with that tiny staff, we have generated over a thousand large holograms for a variety of clients, in both commercial and fine are realms (Figure 3). Most of those projects required new procedures and equipment, and it is some of those that I shall address, in particular the artist projects. If you want to learn your limitations, work for an artist.

Commercial Production

Accolade	Fluor Daniel	Omni Laboratories
Adidas	General Instruments	Pacificvision
The Aerospace Corp.	GE-RCA	Pennzoil
Air Canada	GTE	Planning Research Corp.
Air France	Hard Rock Cafe Montreal	Pontiac Motor Division
Alabama Space & Rocket Ctr.	Holocom Gmbh	Positive Productions
Alberta Government Telephone	Holomatrix	Pratt & Whitney
Amagic	Holophile	Princess Cruise Lines
Anheuser-Busch	Hue Legwear	Razz!
Association for the Preservation Of Virginia Antiquities	H.W. Exhibits	RC Cola
AT&T	Iacono Studio	Rittenhouse Communications
Avery Dennison	IBM	Roc Company
Bell Laboratories	ICI Bank Security Systems	Rochester Visitor's Association
BF Goodrich Aerospace	Interpore	Rubbermaid
BHA Design	Jet Propulsion Laboratories	Sewoo System Co.
Boeing Services Co.	Kelly Springfield	Spacecraft
Boreal	Korean Broadcast System	Special Projects Group
British Petroleum	Korean Electric Power Co.	Sprint
Canadian National Defense	Lancome	Squibb Pharmaceutical
Case Knives	Laser Express	Stryker Endoscopy
Casino Magic	Laser Image	SVG
CD Medical	Leica USA	Taylor Technologies
Chicago Fish House	Lightrix	Teleglobe Canada
Chrysler	The Limited Companies	Third Millennium Productions
Ciba-Geigy	Lockheed Martin	Trace Design
Compaq	London and Edinburgh Trust	Trim International
Creative Solutions	Lucarelli Design	Tonka Toys
CXC Software	Lucent Technologies	U.C. Irvine
Danavox	Lynch Industries	United States Army
Digital Equipment Corp	M & T Chemical	UNITEL
Disney World	Matra Marconi	University of Indiana
Dupont	McGard	WCAX-TV
Dynabec	Merck&Co.	Weidig Exhibits
Esso	Micro-Tek	The Weinberger Group
Electrochemical Technology	Miles Labs	Wyeth Pharmaceuticals
Exhibit Design Group	Motorola	Xaloy
Faraday LLC	MTU America/Mercedes	Xerox
Farmer Studios	NASA	Zale Corp.

Museum Collections

- American Museum of Natural History, New York, NY
- California Academy of Sciences, San Francisco, CA
- Camera Obscura, Edinburgh, Scotland
- Canadian Museum of Civilization, Ottawa
- Children's Technology and Science Museum, Terre Haute, Indiana
- Center of Science and Industry, Columbus, Ohio
- Cleveland Center for Contemporary Arts, Cleveland, Ohio
- Discovery Museum, Essex Junction, Vermont
- Discovery Place/Nature Museum, Charlotte, NC
- Explora, Montevideo, Uraguay
- Feria Internacional del Pacifico, Lima, Peru
- Franklin Institute, Philadelphia, Pennsylvania
- Fresno Metropolitan Museum, Fresno, CA
- Hologram Varga Miklos, Budapest
- Holography Museum, Chicago, Illinois
- Holo-Service, Basel, Switzerland
- Hong Kong Space Museum
- Interference Gallery, Toronto, Ontario
- International Holography Gallery, Tokyo
- Jamestown Discovery Museum, VA
- Lida Company, Beijing
- Maryland Science Center, Baltimore, Maryland
- MIT Museum, Cambridge, Massachusetts
- Museo Interactivo Infantil, Mexico City
- Museu de Ciencias e Tecnologia, Porto Alegre, Brasil
- Museum Fur Holographie & Neue Visuelle Medien, Cologne, Germany
- Museum of Science & Industry, Tampa, Florida
- National Museum of Natural Science, Taichung, Taiwan
- National Museum of the American Indian, Washington, D.C.
- National Museum of Health & Science, Washington, D.C.
- National Postal Museum, Washington, D.C.
- National Science Center, Fort Gordon, Georgia
- National Science Center of Malaysia
- New Zealand Science Center
- Niagara Falls Visitor Center
- North Museum, Lancaster, Pennsylvania
- Omniplex Science Museum, Oklahoma City
- Ontario Science Center, Toronto
- Pacific Science Center, Seattle, Washington
- Perkins Museum of Geology, University of Vermont
- Placer County Museum, CA
- St. Louis Zoo, St. Louis, MO
- Saskatchewan Science Center, Saskatchewan, Canada
- Sci-Tech Discovery Center, West Perth, Australia
- Smithsonian Institution, Washington, D.C.
- Stoneage, Trieste, Italy
- Southern Vermont Nat. Hist. Museum, Vermont
- Sunrise Museum, Charleston, West Virginia
- Tivoli Gardens Museum, Copenhagen
- The Whitaker Center, Pennsylvania

Artist Collaborations

- Claudette Abrams (2)
- Mary Alton (3)
- Alexander (8)
- Phillippe Boissonet (9)
- Jamie Carpenter (2)
- Harriet Casdin-Silver (16)
- Lori Clarke (1)
- Doug Coffin (1)
- Betsy Connors (3)
- Marie-Andre Cossette (1)
- Paula Dawson (1)
- Ivan Dutrisac (1)
- Georges Dyens (8)
- Evergon (5)
- Hrein Friddfinsson (2)
- Mary Harmon (3)
- Michael Hayden(7)
- Setsuko Ishii (16)
- Dieter Jung (3)
- Juyong Lee (8)
- Chris Levine (1)
- Tulla Lightfoot (1)
- Annikki Luukela (4)
- Sharon MacCormack (3)
- Vincent Mariani (1)
- Marie Christiane Mathieu (8)
- James McIntyre (8)
- Marjorie Minkin (4)
- Momoko Naruke (1)
- Rubin Ninio (1)
- John Perry (12)
- Antonio Peticov (1)
- Christine Remy (6)
- Ariel Rubin (1)
- Sam Saunders (3)
- Matt Schreiber (5)
- Dan Schweitzer (1)
- Tulay Shakir (1)
- Michael Snow (12)
- Michael Sowdon (6)
- Frank Stella (1)
- Alexander Tsiaras (2)
- James Turrell (42)
- Doris Vila (1)
- Elena Verpeta (1)
- David Warren (2)
- Sally Weber (1)
- Ann Weiner (1)
- Hae-Won Won (1)

Figure 3. HNI's client list in December, 2009.

Michael Sowdon of Fringe Research approached me in 1985 to produce a series of large holograms for Canadian artist **Michael Snow**, to be shown at **Expo '86** in Vancouver. Building on the success of his *Walking Woman* series at the Montreal Expo '67, Snow had been given Vancouver's original railroad engine house to fill with art. He chose to show exclusively holograms, hopefully making the work accessible for all viewers - parents, children, art critics etc.

We quickly determined that our 5x10 foot table was too small. And we abandoned the staggered object beams required by our 25mw HeNe, which had no etalon. As Steve Benton put it, we "took the argon plunge," and also poured the 12x22 foot, floor level isolation table that we still use today. It is a 14 inch thick concrete slab, reinforced with fibreglass strands. The slab is supported by a "tic-tac-toe" pattern beneath, rigidly connected to four sono tube legs, each surrounded by a larger, open PVC sleeve (Figures 4, 5 & 6). The slab does not float, but makes no contact with the ground until about 12 feet down, where the legs hit a layer of glacial till. There is also a space between the slab and the surrounding floor. So we are isolated from the building, and Burlington is a quiet city.

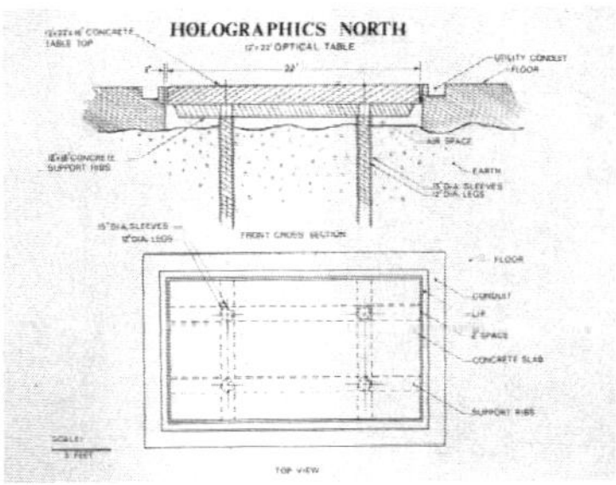

Figure 4. Isolation table design.

Figure 5. Table leg sleeves.

Figure 6. Leg support.

Figure 7. The ice pack to support the concrete slab while it set.

Figure 8. The finished 12x22 foot table, with (l. to r.) Michael Sowdon, Michael Snow and John Perry.

Ice supported the 25 ton slab while it set, producing an air space beneath (Figures 7 & 8).

The Expo 86 exhibition was a spectacular success, representing all modes and sizes of holography, and a wide spectrum of subjects, all displayed in a dazzling environment. The series entitled *Still Life in 8 Calls* (Figure 9), later to be shown in several galleries, including at MIT, garnered particular attention. Eight 24x32 inch holograms showed the same plywood tabletop with a lamp, telephone, coffee cup, car keys and pencil on top. The expo theme was *transportation and communication*. "It's for you" was written on the corner of each tabletop in pencil. And each hologram showed the items in different styles of abstraction - melted down, exploded, cubist etc. Displayed in a straight line, the holograms were surrounded by four actual table legs with a chair in front of each.

Figure 9. *Still Life in 8 Calls*; the table legs and chairs; *Call #8* - exploded (not pulsed).

Over 50 holograms for the exhibition were produced at 3 labs: HNI, Fringe Research and Laboratoire de L'holographie de Paris. A few are shown in Figures 10a and b.

Figure 10a. 1 of 3 HNI 3x5 foot landscapes; reflection holograms by Fringe Research;

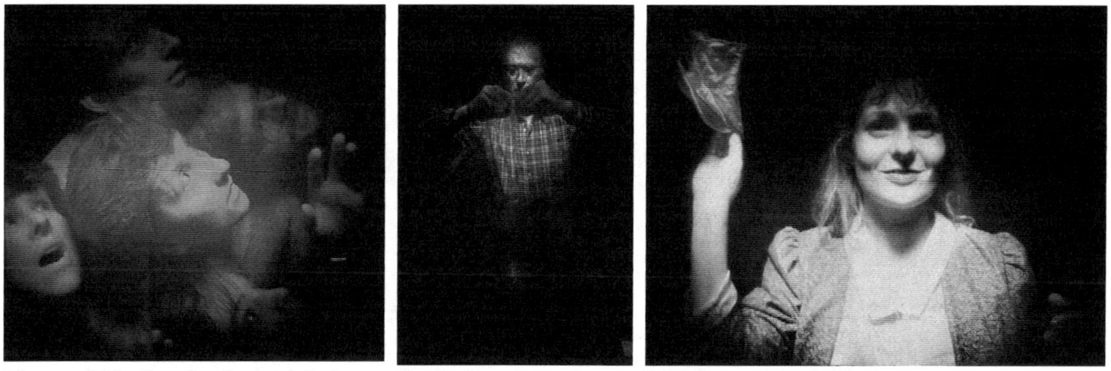

Figure 10b. 3 pulsed portraits by the Laboratoire de L'Holographie de Paris.

The next year David Katzive asked us to render a **Frank Stella** maquette as a hologram for an exhibition in Cleveland, marking the 100th anniversary of the Michelson-Morley experiment.

Production of the 3x3 foot piece was straightforward. But color seemed critical in the original, and was of course not recorded in the single slit rainbow hologram (Figures 11 & 12). We were not in a position back then to produce true color, and I still discourage it today. There are many small holograms with strong, vibrant true color, but I have yet to see a large one. Nevertheless, there are those occasions when it should be attempted, and this was perhaps one of them.

Figure 11. The Frank Stella maquette with John Perry.

Figure 12. The 3x3 foot Stella hologram.

The topic of *Colorful* brings us to the late **Harriet Casdin-Silver**, who was awarded a grant in 1999 to produce an exhibition for **Boston's millennium first night**. Her brilliant concept involved large holographic portraits of people who had lived in Boston for most of the century. Each would be accompanied by a soundtrack of the subject telling anecdotes from his or her life. Sound domes in front of the holograms focused the monologues to individual viewer locations.The exhibition was entitled *A Celebration of Aging* (Figure 13), and production again went smoothly.

Although the holograms were shown for only 2 days at Boston's Hynes Center, they were a major success. The Boston Globe loved them. Even the New York Times liked them, but mistakenly suggested that the holograms were made in Harriet's Boston studio. And that was our only serious problem with this project, as well as others. Like Harriet, many artists employ fabricators. I urge the artists to take credit for their artwork, negotiate with their fabricators whether to grant them credit, but

Journal of Physics: Conference Series **415** (2013) 012011 doi:10.1088/1742-6596/415/1/012011

not pretend that they fabricated the work themselves. This belittles the medium and everyone involved, including their audience. It is unfortunately all too common in holography. I asked Harriet, and then the *Arts* editor for a printed correction, but was denied.

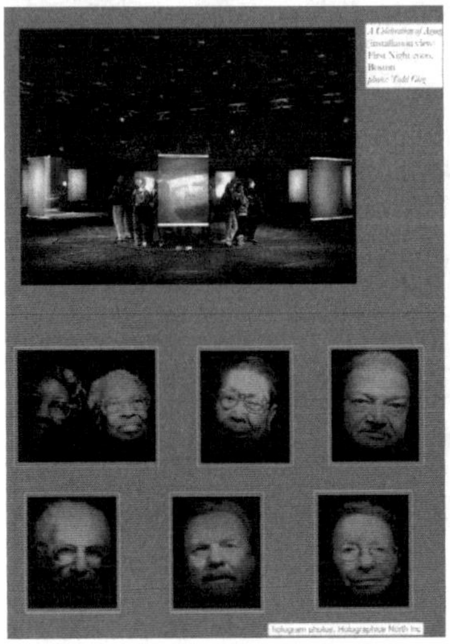

Figure 13. *A Celebration of Aging:* installation and 6 of the 9 images.

Figure 14. *Face to Face,* 16x20 inches.

Figure 15. The 3x5 foot *Fort Point Artists* hologram on *Art New England's* cover.

Celebration was shown several more times in various New England venues, with well deserved positive reviews. We made many more portraits with Harriet in the following years (Figures 14 and 15). She had a particular gift for working with her subjects, and her talents will be missed.

Passing through O'Hare airport, you may have encountered the *Neon Tunnel*, designed by artist **Michael Hayden**. He asked us in 2002 to produce a series of outdoor holograms for the **Charles M. Schulz Museum** in Santa Rosa, California. They were to show Snoopy and Woodstock playing hockey on a 3-foot diameter birdbath (Figures 16 and 17). As you walked around it, you would discover five different images, sticking straight up about 8 inches toward you, like flying over a miniature New York City. We decided on first generation, split-beam, real-image reflection pieces, lit from directly above, for a wide, unshadowed viewing angle.

Figure 16. 1 of 5 holograms for the birdbath.

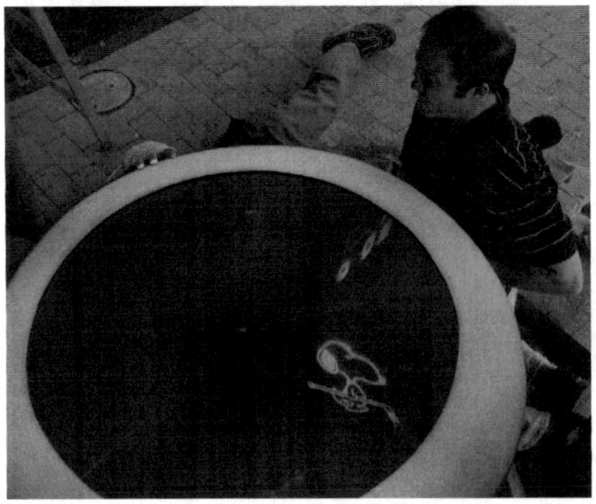

Figure 17. Installation of the holograms.

Production was not difficult, but the operative word of course is "outdoor". We usually discourage this, for a host of reasons: rain, sun, bright ambient light etc. Our first plan, luckily suggested by Hayden himself, was to encase the wedge-shaped holograms in a kind of circular acrylic coffin. We packed silica bags inside under dry conditions and prepared to seal it up tight, but then realized that it would probably explode in cross-country shipment. We drilled tiny breathing holes with instructions to seal them immediately upon arrival. After four months in a humid San Francisco warehouse, our crate was finally opened and the instructions promptly lost. After another four months baking on-site in the California summer sun, our solar oven was warped beyond recognition. We had apparently achieved 180 F, the annealing temperature of acrylic.

The next attempt involved overcoating each holo-pie slice with waterproof, uv-blocking polyester film, and then loosely covering the entire birdbath surface with glass. Hayden also introduced a roof in the form of Lucy's baseball cap (Figure 18), not only to block much of the sun and rain, but also to house an overhead lamp (Figure 19). The first images were reconstructed with sunlight, but variations in angle made Snoopy and Woodstock lose their balance on the ice. The overcoating worked for several years, but eventually these images also faded (Figure 20).

Figure 18. Lucy's 18 foot baseball cap.

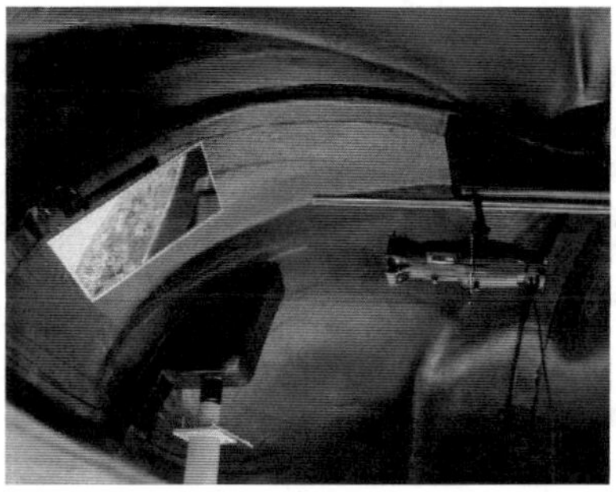

Figure 19. Lighting hardware under the top of the cap.

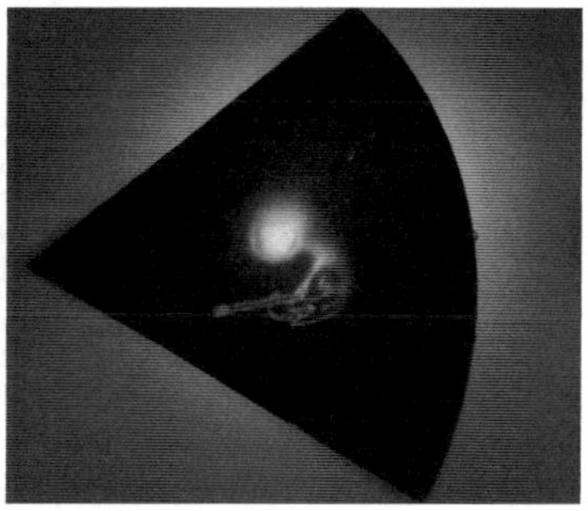

Figure 20. A sun-damaged hologram section.

I know of no successful attempts to embed a holographic display this size in solid resin. Finally we replaced some of them again, with the individual pie-sliced holograms cemented to glass top and bottom. As far as I know, these are holding up well. The film was laminated on the top side with our usual clear, double-sided, cold roll adhesive. Then we cemented the lower side to glass with uv-curing liquid cement. The glass edges were cut to a bevel, creating a pocket in which we could edge-seal the pieces with silicon, probably the weakest link.

Fortunately, Hayden had ordered early on that we make at least 2 backup copies of each hologram along with the original, a precaution I recommend to most clients. In our medium more than most, it is far cheaper to make extra copies when the setup is in place, than to go back later.

In another high profile project, Jeffrey Robb called in 2003, asking if we would be interested in making a holographic portrait of "a member of the royal family" for "a British artist". "And don't tell anyone," he cautioned. The artist with the commission turned out to be **Chris Levine**, the family member was **Queen Elizabeth II** (Figure 21), and the stereogram footage was shot at Buckingham Palace with a rail and digital camera system built by Rob Munday. Being a Colonial, I was not invited, but did later get to hang the queen hologram in Buckingham Palace.

The 1x1.3 meter transfers had to be flawless. But the major challenge was that Chris wanted the image well in front and only in blue - royal blue - a stroke of artistic genius. The size and depth of the image required a white light transmission hologram. But at that time, we required a vertical array of lamps to provide a tall enough viewing field from a "rainbow" hologram lit in single color. An 18 inch tall array at about a 10 feet distance was found to be sufficient (Figure22).

Figure 21. 1x1.3 meter hologram of Queen Elizabeth II.

Figure 22. Installation in Buckingham Palace, with the vertical blue LED array.

Her majesty's eyes projected about 32 inches, and the transfer was shot in red. And so the wavelength shift from HeNe red to LED blue caused her majesty to look too wide. The mismatch of reference to illumination beam divergences made it even worse. After several batch processes in Photoshop, we found just the right predistortion of the photo frames. I used the ratio of eye-to-eye distance to hairline-to-chin distance as our guide. Several copies of the hologram have now traveled around England, with a permanent installation on the Island of Jersey. The same photo footage has been used in the years since to produce other portraits, including small embossed versions and large lenticular renditions. These are mostly true color, but personally I far prefer the original LED blue.

Single-color LED illumination was in fact so stunning that it seemed a valuable technique to pursue. And a call from Matt Schreiber in 2004 provided the opportunity. Matt installs **James Turrell**'s exhibitions around the world, and knew that James was increasingly interested in holography. But he required single color, not rainbows, and not the somewhat muted colors of most reflection work. The transfer sizes would vary from 24x36 to 36x72 inches.

Matt suggested the same concept I was considering at that time, rainbow transfers made from open aperture masters. This provides a tall viewing field when lit in single color, or of course an achromatic image in white light. Furthermore, lamps in different colors can be separated horizontally, to produce several colors of the same image, crossing at the film plane. And these decisions are made upon installation, offering a greater degree of spontaneity.

In stereogram work, this seemed simple enough - lengthen the stereogram exposure slit from the usual few centimeters to at least 16 inches. But the mastering screen must be focused accurately onto the transfer plane, to keep horizontal lines sharp from that tall, open aperture master.

This presented problems. First, we had built our stereogram mastering camera in the abandoned, 5x30 foot heat exchange space of the original 19th century, coal-fired furnace of our building. It wasn't wide enough to accommodate the longer exposure slit and the expanded reference beam. So the whole rig had to be moved out into the open, and reconfigured. This took a few months, and then production began, along with more problems.

For several years we had used a 4x60 inch liquid cylindrical lens to converge the illumination beam across the length of the master for transfer exposures. This also had to be fabricated again, in a 16 inch width, requiring 34 liters of mineral oil (Figure 23). But it was mistakenly made with birefringent polycarbonate surfaces. A polarizing filter quickly confirmed that it had to be remade with acrylic, which we did.

Unfortunately, the transfer images were still not sharp, and we eventually identified astigmatism as the culprit. We needed to converge the H1 illumination beam in both dimensions, to fully focus the H1 screen onto the H2 plane. This is of course not necessary for a slit master, since there is no vertical parallax. So we built another 16x60 inch liquid lens curved in the other direction, to be used in series with the first. Images were now indeed reasonably sharp in both dimensions.

The Turrell transfers were mostly 6 feet tall, made from 16hx44w inch masters. And this posed an even more difficult problem. The master illumination beam struck the top of the transfer film directly (Figure 24). Masking it between the master and transfer did not work well. We eventually learned to insert a knife-edge into the unspread reference beam, before the spatial filter. This produced a soft shadow in the reference beam, preventing the direct object beam from recording.

Next, internal reflections within the liquid lenses caused another, secondary object beam to fall across the transfer just above the center line of the image (Figure 25). After several weeks, we eventually found a way to baffle that beam after the second surface of the last big liquid lens. Each of the 40 transfers required careful re-adjustment of all these baffles.

Figure 23. 16x60 inch liquid cylindrical lens.

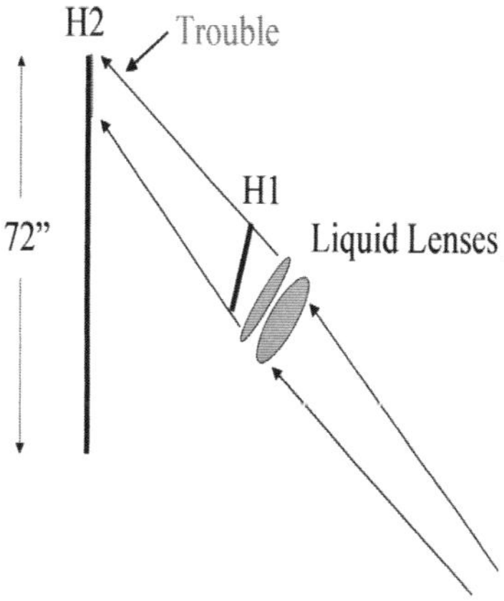

Figure 24. Geometry of the open aperture master illumination beam.

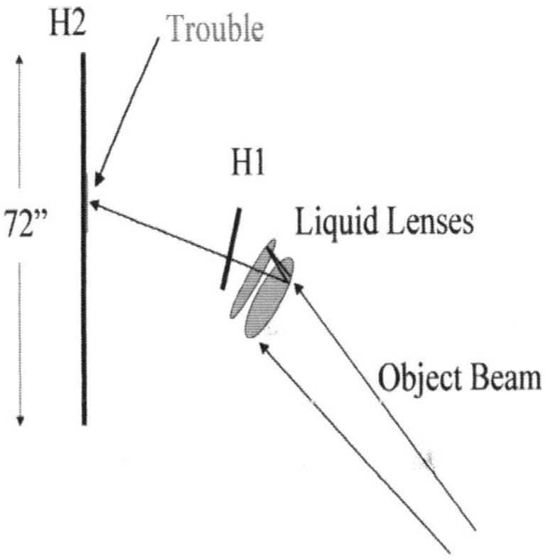

Figure 25. The liquid lens internally reflected beam.

Our biggest challenge, however, was none of the above. It was simply to produce perfect, unblemished 38x72 inch transfers on film, something that I usually claim is simply not possible. This required a massive cleanup of the labs and refiguring of almost every step in our processes. It also reduced our success rate drastically. With each shot taking most of a full day, that was costly.

Eventually the 40 large holograms were successfully completed, laminated to 1/2" acrylic, framed at Schreiber's studio in Florida and finally delivered to Pace Gallery in New York. They have now been shown in several Pace Gallery exhibitions (Figures 26 and 27).

Predictably, the New York Times panned the holograms on Sept. 18, 2009, stating that "The idea of using holograms in art seems laughably dated or, at the very least, kitschy." And yet the public seemed to love them, judging by the response at the openings, and by the multitude of amateur photos to be found online. Perhaps we should yet again question the value of journalistic criticism.

Figure 26. 3x6 foot Turrell hologram in double, and 3x5 in single color lighting.

Figure 27. Opening of the 3rd Pace Gallery show, September, 2009.

But the Turrell work suggests a more important topic for consideration: is there perfection, not only in holograms, but in anything? Should clients, art or commercial, expect it? I would argue no, there is no perfection, and there in fact should be no perfection. As the Navajo weavers would say, it traps our souls in our work. But then what degree of imperfection is acceptable? Is hand-made work in a different category from machine-made, large different from small, simple different from complex? Surely the answers depend upon the application. But they are poorly understood.

HNI's large hologram on display in the ISDH exhibition, for instance, has flaws. Does anyone care? Can we liken them to Navajo *Spirit Lines*? Should they have been "fixed"? I can't say, but for better or worse, we keep trying to chase them away. And in our medium, that is difficult.

Photo credits: John Perry, Kevin Eaton, Pace Gallery, Todd Gieg , unknown from web.

9th International Symposium on Display Holography (ISDH 2012) IOP Publishing
Journal of Physics: Conference Series **415** (2013) 012012 doi:10.1088/1742-6596/415/1/012012

INVESTIGATING THE SPHERICAL NATURE OF LIGHT THROUGH LARGE FORMAT DCG HOLOGRAMS

August Muth

2911 San Isidro ct.

Santa fe, New Mexico 87507

august@augustmuth.com

Resolution of reality

You might think your fifty inch 1080p screen has a pretty high resolution, but reality is a quadrillion times better - a hundred trillion dots per inch. In a collaboration between Fermilab scientists and about six hundred meters of laser tube the GEO600 team determined the resolution of reality. In essence the very pixels of our universe, grains of spacetime are one tenth of a femtometer across, accurate enough to detect even the tiniest ripples in space time.

The interferometer staff constantly battled against unwanted aberration, and were struggling against a particularly persistent signal when Fermilab Professor Craig Hogan suggested the problem wasn't with their equipment but with reality itself. The quantum limit of reality, the Planck length, occurs at a far smaller length scale than their signal - but according to Hogan, this literal ultimate limit of tininess might be scaled up because we're all holograms.

Edited Post by Luke McKinney http://www.dailygalaxy.com

To speak about the spherical quality of light I first need to talk about three aspects of the holographic technique I utilize in the making of holograms. The completed holographic artworks that I prefer to manifest are analog sourced, denisyuk holograms utilizing dicromated gelatin emulsion. In the vast majority of my work no master holograms are used. One reason for this is a loss of dimensional data that occurs with each additional transfer step. The denysuk set-up with a reference angle of 30 to 45 degrees allows for the realization of holograms with more than 180 degrees of horizontal parallax and around 140 degrees of vertical parallax. This totally open aperture configuration allows for the recording of reality in a way that no other holographic set-up can accomplish.

The use of dichromated gelatin as my holographic emulsion of choice is integral in this story. Dichromated gelatin emulsion record holograms on the molecular level. Electron microscope images of DCG show something that has the apperarence of steel wool. Molecular strands of gelatin create a web of gelatin fibers. This structure enables DCG to record holograms with higher diffraction efficiency and greater resolution than any other holographic emulsion. It is my understanding that holograms produced in this manner are the highest resolution manifestations created by man in the recording of space-time. If any of you have any better ideas let me know about it as that is where I want to go.

It all began with this crystal clear memory I treasure. I am seated in the conference center at Lakeforest college. Nils Abramson in showing a short film about light in flight. The film documented a femtosecond parcel of light on its journey through the atmosphere and onward into the universe. As this parcel of light embodying a wavefront progressed it produced a beautiful expanding arch across the sky. I began to contemplate the information that was contained in that femtosecond parcel of light as it progressed outward carrying its record of time intertwined with it. This was that moment, that epiphany when I began to see what truly interested me about holography.

When I retuned to my lab I found myself not looking at the image of the object in the holograms but how the light reconstructed by the holograms was interacting in space and time. Then after making tens of thousands of holograms I made my first hologram true to my sprit. It was a Denisyuk dichromated gelatin hologram of a flat white painted surface. The distance from the emulsion to the white surface was about 1/8 inch. I presumed that the resulting hologram wouldn't have a great deal of depth. So much for presumptions.

As I began to horizontally move around this broadband frequency response white reflection hologram I began to see a dimensional quality at the edges of its parallax. I needed to see further into this edge of parallax. I proceeded to make a narrow band response hologram of the white flat surface I was using as my model. This hologram was a clear blue frequency response hologram. When looking into the outside parallax edges of this narrow band blue hologram I was able to perceive a rudimentary spherical form receding into a volume of space that existed behind the image plane. I thought that this spherical form within the hologram might be caused by the diverging beam I had used in the making this hologram. Next I used collimated light in the set-up and got in essence the same results. The spherical form within the hologram remained intact, and unchanged in its character. Was this spherical form the edge of holographic reality turning in on itself forming a spherical light memory? I began to rotate the hologram off its reference angle and the spherical form began to expand and realize itself within the hologram. It all began to make sense. That femtosecond-arching parcel of light dimensionally expanding in all directions on its journey enveloped in time. Wrapping in upon itself, spherical in character, as it should be.

Traversing Time and Space from the Blessing Window

Ya-Ling Huang
Kun Shan University, Taiwan
Associate Professor of Visual Communication Design Department
E-mail: yaling@mail.ksu.edu.tw

Abstract. The visual graphics for the holographic artwork "Blessing Window" were created from observations of Tainan city, with a focus on the beauty of Chinese characters, their typographic. The concept of movement in the artwork is from a traditional Chinese philosophy, "When the mountain does not move, the road extends, when the road does not extend to the destination, the heart will extend". One multiplex-hologram and an interactive installation were used to combine the visual concepts of typography and the philosophy.

Key words: Chinese typography, totem, historical symbol, holographic, interactive

1. Introduction

From the 18th Century to now, the city of Tainan has been influenced by Dutch, Portuguese, Chinese, Japanese and naive cultures. As the first developed city in early Taiwan, Tainan has distinctive characteristics from its rich history. The people who inhabit this historical city can explore elements aspects of the past from the streets, temples, and many small old corners.

Chinese characters are an old system of writing, which are still used in the modern world. The concept of "Blessing Window" combines elements observed from the city of Tainan. In the spring of 2012 I photographed Chinese characters and traditional totems from temples, buildings, monumental, and famous scrolls. These were then divided into five symbolic subjects – wisdom, love, fortune, health and wealth – and composed into a holographic work.

There is a famous traditional Chinese philosophy:" When the mountain does not move, the road extends, when the road does not extend to the destination, the heart will extend"(山不轉，路轉；路不轉，心轉). According to the philosophy, the work "Blessing Window" was designed to convey blessings from five vases that each depicted one subject – wisdom, love, fortune, health and wealth. The holographic image was digitally composed by placing the five vases within an old window (Figure 1).

Figure 1: The visual design of "Blessing Window"

Each vase is animated rotating one after each other as the viewer moves past the hologram. The movement of the viewer through different viewing angles, animating each symbolic subject symbolizes the transfer of a blessing to the viewer.

2. The historical observation in Tainan

There are at least 55 historical buildings and structures in the city of Tainan. Most of the historical architecture is related to religions, there are also a few forts, the city hall and many monuments. Windows and doors are an important elements of a building, and in a historical context, most windows and doors are only not functional but are also metaphors. The materials and techniques of windows used reveal different the periods of building, and using the materials of wood, brick, ceramics being the most common materials of Tainan's historical architecture. The form and shape of a window represents different symbolic aspects of life. Windows were the first element observed for "Blessing Window", the three popular historical windows in Tainan are as shown below (Figures 2–4).

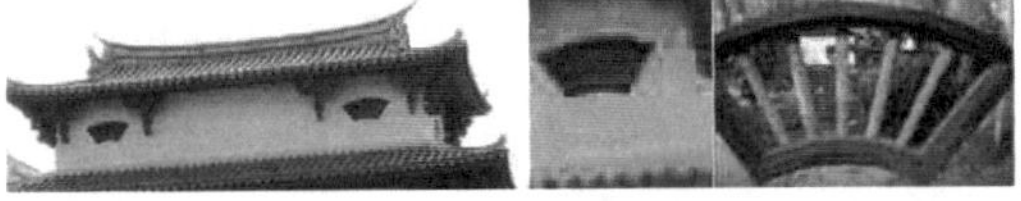

Figure 2: Fan window – the **fan** character (扇＝善) means nice relationship and is pronounced similar too **good serendipity** (善緣)

Figure 3: Book and bamboo window – the **book** shape of the window means **smart and knowledge**. Combined with **bamboo**, which means **rising,** the window symbolizes career advancement. (節節高升)

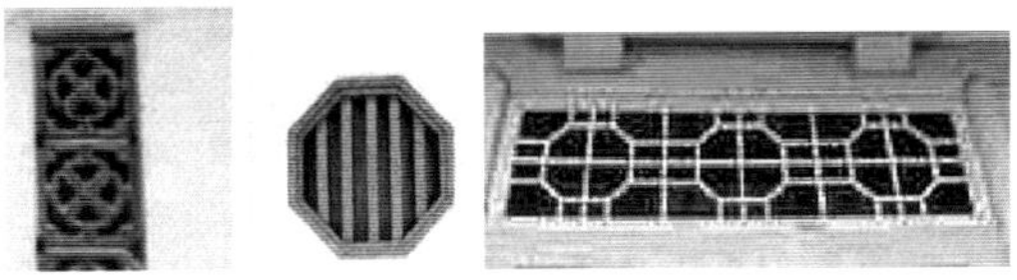

Figure 4: Octagonal shaped window – represents the **eight trigrams** (八卦，a set of symbolic signs created in ancient China to show the elements of life)

The symbol of a book with bamboo window also means knowledge, and knowledge is the invisible heirlooms for generations, and bamboo means good character of people. As Tainan is known as city of culture, a culture that is an heirloom of the people, I choose the book with bamboo window as the main element for "Blessing Window".

Chinese characters and typography have evolved through history modern Chinese characters have a different writing style but are based on ancient hieroglyphs (Figure 5). This form of writing has allowed for the passing of knowledge and symbolic representation from one generation to the next.

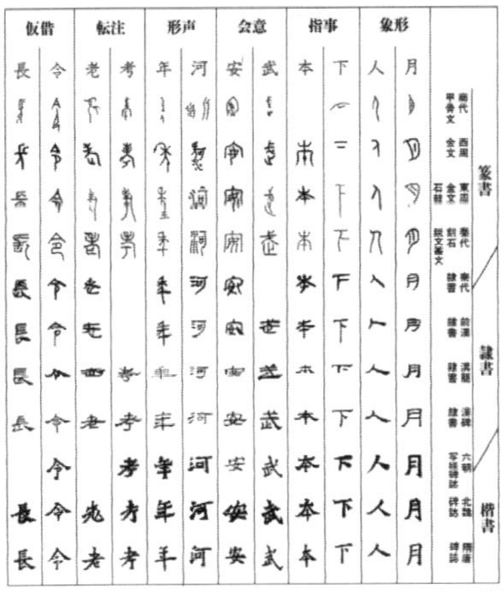

(Figure 5: Chinese characters in different style, from the top to the button are the style development from ca. 500 to present.)

According to the concept of the "Blessing Window", my observation focused on how to convey the five topics – love, fortune, health, wealth and wisdom. The temple is always a good place to invoke different wishes, therefore, the typographic were collected from different temples. Five Chinese characters that relate to the concepts were found (Figure 6–10) and combine with different shaped five vases. The five traditional styles of vases each have a shape that resonates the concept of the character. (Figure11-15).

Fig. 6, Love
"Every Jack has his Jill"
Location: (月老廟) The Temple of Matchmaker (the Master of Moon)

Fig. 7, Wealth
"the wealth start from connotation, from self behavior"
Location: (文昌殿) The Temple of Literature

Fig. 8, Wisdom
"joy, fortune, longevity"
Location: Temple of Tiantan (天壇)

Fig.9, Fortune
"Prosperity"
Location: Temple of three Mountain Tutelary Gods (三山國王廟)

Fig. 10, Health
"**Ills away!**"
Location:Ching-Ji-Gong (興濟宮)

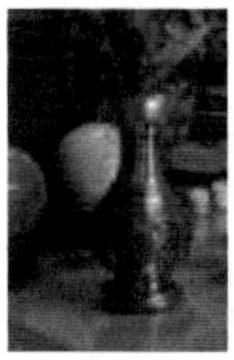
Fig.11, vase for Love

Fig.12, vase for Wealth

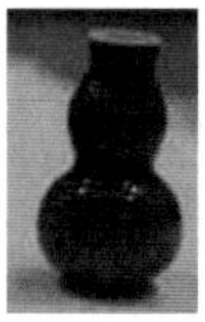
Fig. 13, vase for Wisdom

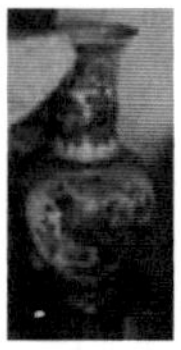
Fig. 14, vase for Fortune

Fig. 15, vase for Health

3. Mediums for display

The "Blessing Window" is exhibited in using two mediums–a reflection multiplex hologram (Figure 16) and an interactive installation with rear projection video and an infrared sensor (Figure17). Both installations allowed for viewers to engage with a holographic composition of the concept.

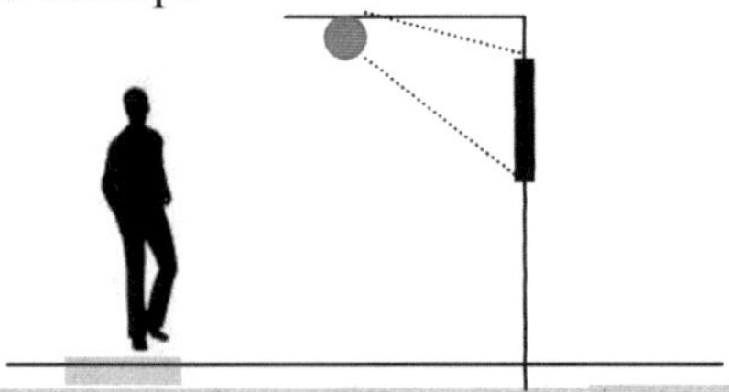
Figure 16: Display design for reflection hologram "Blessing Window"

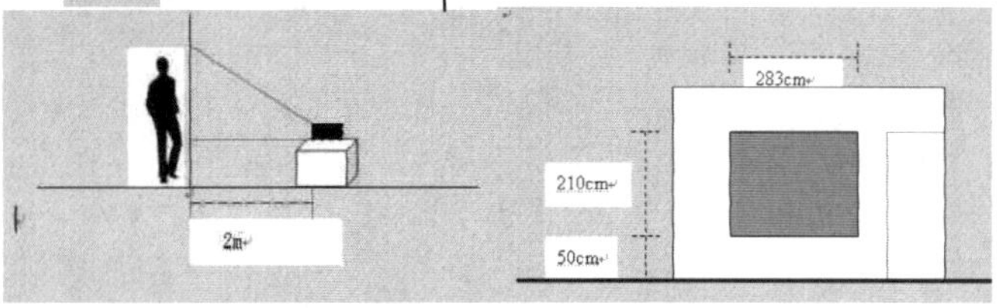

Figure 17: Display idea for interactive installation of "Blessing Window"

In both cases the vases are animated, rotating one after each other, to symbolize the transfer of blessings to the viewer – wisdom, love, fortune, health and wealth. The animation design for each medium, however, was different.

For the hologram, the movement of the viewer through different viewing angles animates each symbolic subject. The spatial animation of the vases has two speeds, either 16 frames, or 30 frames across the viewing angle to consider human eye's movement and the hologram's angle changing. Finally, sending two fields of hologram digital fields which one is with 16 frames and the other is with 30 frames to print by Geola. The installation of this hologram is hung on the wall and the responses of viewers were observed. At the end of the viewing experiment of two holograms, the hologram with 16 frames is easier for viewers to see.

For the installation from the animation design of the interactive installation each vase rotates separately triggered by the viewer's hand touching the vase. From observation of the interactive installation, viewers always want to touch the transparent project from the back. When the video camera catch user's hand shadow and the vase rotating at same time. The interactive installation Blessing window is becomes a famous work during the exhibition at Chinese Culture Centre in Tainan.

4. The final works and conclusion

These works consider to design and to use the historical elements into the high-tech technology in interactive projection and stereo hologram. Each interactive Blessing Window (Figure 18) or stereo hologram (Figure 19) to symbolize the tunnel of time and space journey of the city of Tainan, from modern display to combine with the historical element. These two new materials and techniques to present these Blessing Window is also to echo with Tainan historical windows' materials and techniques which always reveal the different periods.

Figure 18: ... Photographs of interactive installation at the Chinese Culture Center, Jan. 30 to March 30, 2012, size 283cm*210cm

Figure 19, Photographs of stereo hologram installation at 3D Image Lab., Kun Shan University, May 23, 2012. size is 39cm*29cm

Reference

Chang, C., (2009), The betuy of Chinese callgraphy-*dancing of cursive* ， Yuan Liou Publishing Co., Ltd. ， Taipei.

Hsu, H. (2011) ， The word and it antecedents! AZOTHBOOKS', Taipei.

Ho, P.F. (2010) ， Guiding for the historic spot in Tainan ， Tainan Cultural Affairs Bureau, Tainan.

The artistic use of parallax and lenses revealing the invisible in holography

R. Garcia-Robles

Lecturer, University of Seville, Spain
Avda. Reina Mercedes s/n, 41012

Email: rociogarcia@us.es

R. M. Oliveira

Assistant Professor, University of Aveiro, Portugal

ID+ Research Institute in Design, Media and Culture
www.idmais.org

Email: rosaoliv@ua.pt

Abstract. There are many artistic resources offered by holography: third-dimension registration and reconstruction, immateriality, color interpretation, holographic space, realism, etc. But there are a few of them which are very characteristic and singular of that media such as the inversion of parallax, and the possibility of making invisible to turn into visible. Current paper aims to discuss key issues concerning with the aesthetic use of those special features. It is based on theoretical as well as critical analysis of the production by some of the most outstanding holographic artists who have made use of such interesting resources.

1. Making invisible to turn into visible

Although the sentence "making invisible to turn into visible" may suggest magic tricks or deception, there are many ways to get that aim. Good examples are micro- and macro-scopes which allow human to access spatial as well as temporal domains which would remain unreachable otherwise.

But, in relation to making invisible to turn into visible by means of holography, it refers in a poetic manner to other possibilities offered by such media, as the ones listed below:

1. To register optic properties of translucent objects such as lenses. A mirror or a lens transforms the incoming light wavefront into a new one, which is then registered into the holographic plate or film. In fact, holography is mainly a wavefront registration technique. Afterwards the holographic plate or film allows the reconstruction of the registered objects, including the translucent and transparent ones. Figure 1 ("Breakfast", by Patrick Boyd, 1988) and figure 2 ("Self-portrait", by Edwina Orr, 1982) show holographic artworks which have used lenses as resources for the production of the final image.

Journal of Physics: Conference Series **415** (2013) 012014 doi:10.1088/1742-6596/415/1/012014

Figure 1. "Breakfast", Patrick Boyd (1988) **Figure 2.** "Self-portrait", Edwina Orr (1982)

2. It is also possible to use lenses and translucent objects when creating rainbow holograms, supporting the creation of abstract multi-colored compositions. Many artists have explored that resource; some of them even did it a distinctive feature of his/her visual language or style. One outstanding example is Marie Andreé Cossette's holographic art production (see figure 3 "Mémoire", and figure 4 "Luminescence" as paradigmatic examples).

Figure 3. "Mémoire", Marie Andreé Cossette

Figure 4. "Luminescence", Marie Andreé Cossette

HOE ("Holographic Optical Element") is a type of technique that supports the registration of wavefronts themselves. There aren't usually other objects in the scene, but physical or virtual lenses (the last one when they are computer generated). That technique has been used by several artists as aesthetic resources in their corresponding creations. Mainly in two formats: (a) Textured surfaces producing a huge range of spectral colors.

(b) As elements integrated in abstract spatial compositions. Outstanding artists on the field are: Rudie Berkhout, Melissa Crenshaw, Setsuko Ishii, Paul Newman and Orazem et Luck.

Melissa Crenshaw [1] and Setsuko Ishii [2] [3] made interesting land art pieces such as "Water Falls II" (fig. 5) and "Apolinean Gifts series" (fig. 6) respectively. First of them was composed of an holographic film inlayed into two glass surfaces, used to support a waterfall while being backward illuminated creating an spectral spectacle. Second aforementioned work by Ishii was undertaken with the same type of holographic material used as texture of cylindrical sculptures placed in a water pool. The sunlight reflected on the water surface randomly moved by the wind transformed such installation in an evanescent color orchestration.

Figure 5. "Water Falls II", Melissa Crenshaw, (1992)

Figure 6. "Apollonian Gift Series", Setsuko Ishii, (2000)

On the other hand, Paul Newman brilliantly embodied light into holographic sculptures such as "Light Form" (see figure 7). As stated by Eduardo Kac concerning his holographic production "in his holograms it is impossible to identify specific shapes or compositions, because the artist works with the evanescent behavior of the diffracted white light to produce ephemeral appearances" [4].

Figure 7. "Light Form", Paul Newman

Figure 8. "TV HOE", Orazem & Luck

Orazem & Luck are two associated artists who were working together in light installations, integrating different media such as holography, video and computer graphics. In "TV HOE" (figure 8) a digital black & white TV animation was projected through a big HOE placed in front of the TV monitor. Therefore, the monitor changing images work as lighting sources to illuminate the HOE (further information: [5]).

The artistic outcome became "a cinematic spectacle combining the spatial linearity of the animation with the spatial and chromatic dimension of holography" [4].

3. Alternative sources such as acoustic holography or X-rays allow making holograms of the inner part of opaque objects or the human body. It has not been documented any artistic experience in this direction, but there exists that chance to make invisible to turn into visible.

4. Holographic interferometry is a very useful scientific tool to study material deformation. It has been mastery used by some holographic artists such as Margaret Benyon in her works "Counting the Beats" (fig. 9) and "Tiresias", as well as Sally Weber [6] in "Fossil" (fig. 10).

   Making visually explicit human heart beating as well as human skin movements turn subtle life signals into poetic visual patterns, revealing the mystery and beauty of the living nature.

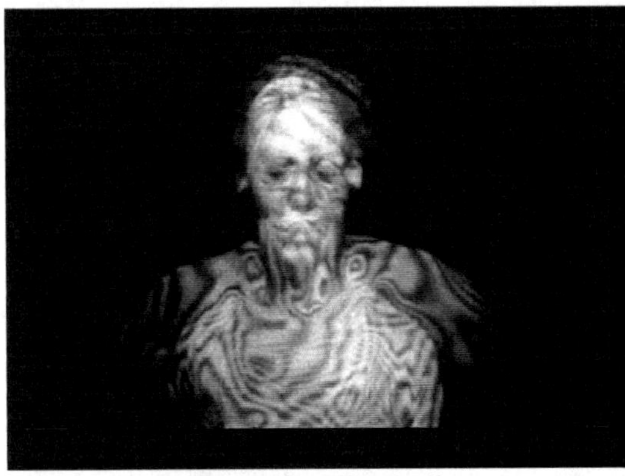

Figure 9. "Tiresias", Margaret Benyon, (1981)

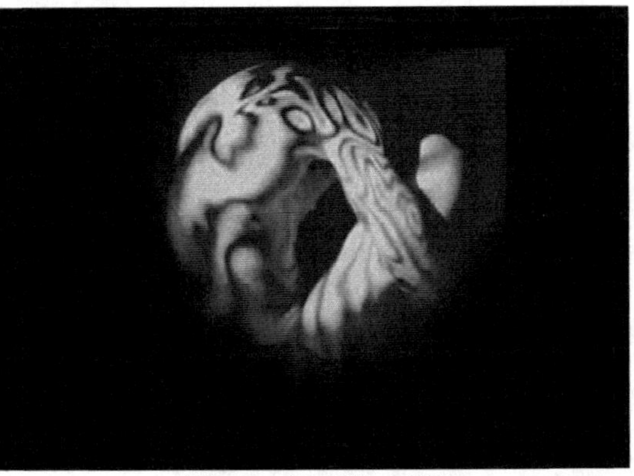

Figure 10. "Fossil", Sally Weber, (2006)

The aforementioned resources illustrate how holography supports the access of human senses to invisible realms in a very characteristic way; and how artists have experienced them according to different original aesthetic approaches.

To finish the survey on the use of that holographic aesthetic resource, it is remarkable how light acts as genesis and reproduction mechanism of the resulting holographic image. In fact the hologram is essentially light itself, i.e. light represented by light, and such self-referentiality is absolute. That circumstance is especially evident in relation to some outstanding holographic artists whose production may be considered color and light abstraction. That's Rudie Berkhout's case [7],[8],[9]. His may be considered connected to the legacy of other previous light art practitioners such as Thomas Wilfred or Laszlo Moholy-Nagy. Other outstanding artists making great contribution to the use of holography as an abstract lighting media are Dieter Jung, Michael Bleyenberg, and Karsten Hobihorst.

2. Parallax effects: Inverting the space concavity

First question to be answered is how it is possible to invert the space concavity. Pseudoscopy is the physical property explaining that issue. It supports changing the proximity-distance spatial relationship, and the parallax effects of the holographed scene. So, objects close to the holographic plate/film during the registration phase appear to be far away when the holographic image is reconstructed.

When, instead of several objects, there is only one, the result is the volume inversion, i.e. making convexity to turn into concavity and vice verse. Such effect has been widely used in traditional sculpture and reliefs throughout the art history from ancient to contemporary practices. For example, the Egyptian ones in the walls of the pyramids (figure 11); or Subirach's sculptures placed in the facade of the Sagrada Familia Cathedral designed by Gaudi and built in Barcelona (figure 12). Positive (convex) or negative (concave) volumes can be achieved and showed in a very different way depending on the light source position illuminating them.

Figura 11. Egyptian Temple at Dendera

Figura 12. Sagrada Familia Cathedral at Barcelona, Josep. M. Subirach

But, what aesthetic possibilities does this resource offer in relation to holography? The artist can find an illusionistic resource to mislead spectators. The human intellect is culturally conditioned in such a way that it understands as concave what actually a convex volume is. That is due to be accustomed to the sun light shining up in the sky, which makes people to observe outstanding volumes rather than negative ones.

The same happens with holography, and artists have used that resource to deceive spectators, but also to transform the perception of the real world. Although there are not so many examples, a few good ones are "Grater" by Harriet Casdin-Silver (figure 13), "Horizon" by Ken Harris (figure 14), and "Vortex II" by Dan Schweitzer (figure 15).

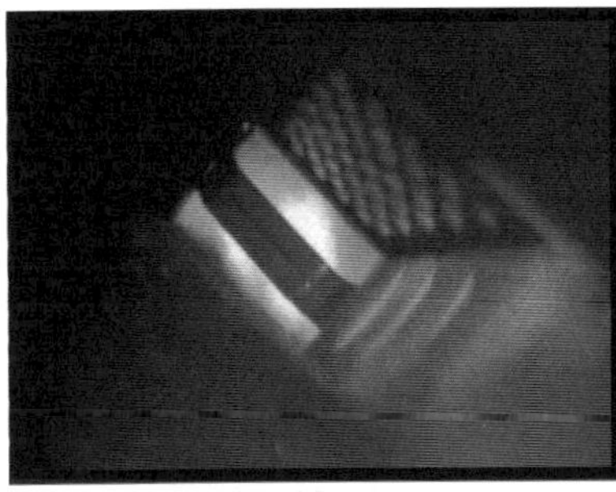

Figure 13 "Grater", Harriet Casdin-Silver, (1991)

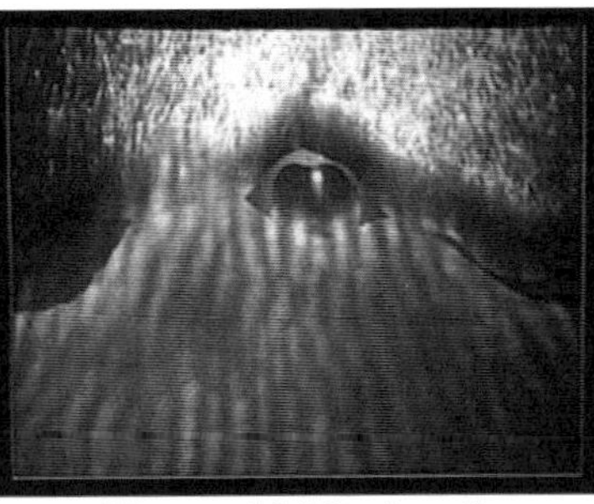

Figure 14 "Horizon", Ken Harris, (1980)

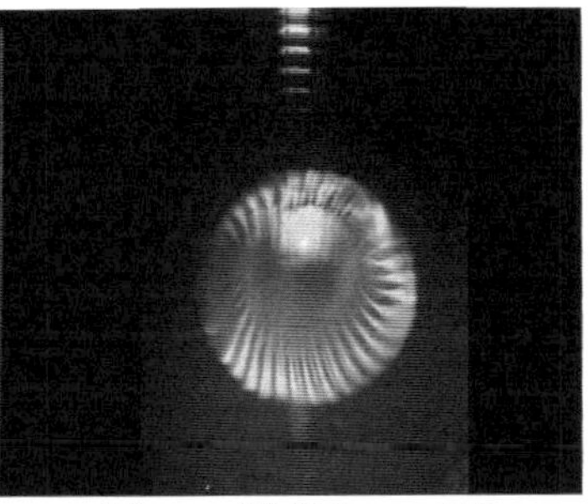

Figure 15 "Vortex II", Dan Schweitzer, (1978)

Casdin Silver transforms a banal kitchen object, inverting the inner volume to outer; transforming her pseudoscopic 'grater' into an abstract light sculpture, as stated by Jonathan Ross (Ross, 2003:Still Life). Similarly to 'ready-mades' by Duchamp, 'Brillo-box' by Warhol, or Claes Oldenburg sculptures, Casdin-Silver decontextualise an ordinary instrument to become an artistic object.
As it can be appreciated, much more abstract are Dan Schweitzer' and Ken Harris' works, with reminiscences of previous artists styles such as Man Ray's.

3. Conclusion
Pseudoscopy as well as the capacity to register invisible light wavefronts are very characteristic features of the holographic media. Authors have been researching on how to get good artistic outcomes based on both resources. Some of them have even based their own visual languages on shaping light, capturing its ethereal and unique nature by means of both features.

4. References
[1] Crenshaw M. 1987 *Psedo-color reflection holography – An artists perspectiva*, Practical Holography II, ed. T. H. Jeong, Proc. SPIE Vol. 747, pp.104-107.
[2] Ishii S. 2006 *Artistic Representation with Holography*, Forma, Vol.21, pp.81-92.
[3] Ishii S. 2006 *Architectural application and maintenance for art holography*, Proc. 7th International Symposium on Display Holography, pp.123-128.
[4] Kac E. 1998 *Beyond the Spatial Paradigm: time and cinematic form in holographic art*, Proc. 6th International Symposium on Display Holography, SPIE, Vol.3358.
[5] Orazem V. 1995 *Holography as an element of the media architecture*, Proc. 5th International Symposium on Display Holography, SPIE Vol. 2333, pp.168-177.
[6] Weber S. 2006 *Beneath the surface: New works form the Strata Series*, 7th International Symposium on Display Holography, St. Asaph, Gales.
[7] Berkhout R. 1989 *Holography: Exploring a new art realm – Shaping empty space with light*, 1989, Leonardo, Vol. 22, 313-316
[8] Berkhout R. 1996 *Investigating the use of HOE's in the holographic image process*, 1996, Practical Holography X, ed. S. A. Benton. Proc. SPIE Vol. 2652, pp. 204-212
[9] Berkhout R. 1998 *A Compact Hologram Display, and the Design for a Full-Color Version, Using Diode Lasers*, Proc. SPIE Vol. 3293, Practical Holography XII.

Acknowledgments
Thanks to artists exploring the holography media, and specially those cited on this paper, for discovering the beauty of light using the discussed resources. Also to collectors such as Jonathan Ross, because his passion for holography supports research such as the one presented in the current paper.

Across Light: *Through Colour II*

I Azevedo [1], M Richardson [2] and L M Bernardo [3]

[1] ARCA-EUAC, University School of Arts of Coimbra, Portugal, ID+ (Research Institute for Design, Media and Culture), Post-Doc Researcher from the Fundação da Ciência e Tecnologia in Faculty of Sciences, Department of Physics and Astronomy, University of Porto, Portugal and in De Montfort University, Leicester, Faculty of Technology, The Imaging & Display Research Group, UK
[2] De Montfort University, Leicester, Faculty of Technology, The Imaging & Display Research Group, 2.27 Queens Building, The Gateway, LE1 9BH, UK
[3] Faculty of Sciences, Department of Physics and Astronomy, University of Porto, Rua do Campo Alegre, 687, 4169-007, Porto, Portugal

E-mail: azevedo.mariaisabel@gmail.com

Abstract. This paper intends to be a reflection on the end product of the process of preparing movies to digital art holograms, comparing the kind of space and movement between those images and the ones in a previous paper. This paper explores too questions about the act of seeing through those images and it is concerned with the surrounding debate of ideas for new experimental methodologies applied to holographic images.

1. Introduction

In a recent paper titled "Across Light: *Through Colour*" [1], we have explained the expressive possibilities of the holographic medium in the creation of digital art holograms and lenticulars. The paper's subject was essentially concerned with time as expressed through change and transformation. Light and colour played an important role in the construction of a special space and special time. Exploring holography as a time-based medium some scenes can be seen left and right or back and forward deliberating a visual dialogue inside holographic plate with the viewer. That dynamic experience of seeing a digital hologram where the image is not static, but in a non-stop transformation, questions us about the act of seeing. In that paper, our method for image capturing was explored using the HoloCam Portable Light System whereby a camera is running parallel to the floor and in front of the scene. In the present paper, we have used another geometry, that explores the same system using the

Canon camera angled at 45 degrees (toward the floor), and produced another series of digital art holograms and lenticulars. In this case the scenes were captured from top to bottom carrying on another kind of set-up and compositions where the floor is the place where things happen. We built another kind of space, projecting the blue, red and green light, angled toward the floor 45 degrees too, collocating objects on the floor, and suspending and moving another, mixing and projecting another colours, on the other objects and on the floor.

2. Concepts in art about representation and illusion

In the act of opening our eyes we receive a mass of visual information that can often become anaesthetic. In everyday life, vision is very limited, because we only see what our instincts require to solve practical problems. We only pay attention to what we need and is likely that we are blind to what we don't "need " to see. In fact perception is connected with symbolic conventions, and the visual stimuli depends upon habits and conditions of observation. When we are reading a book, or listening to music, we enter into a larger space than the room of the activity. Our imagination helps us enter this space, especially in visual arts, and the imaginative space of the artist. *Why you don't enter? If you don't like it you can go out again* [2].

What are images? For example, to Damásio [3], the word images means mental patterns with each sensorial way. Not only related at "visual" images nor static objects, but also sonorous images, or body inside images like those described by Einstein when he was trying to solve problems. In the passing of time, historic generations from each culture and society, art was always expressed to establish different rules of representation and production, because it was always using the technical devices from each epoch. A technological innovation is only important to an innovation in the art, if it implies new relationships, new ideas, new utilizations leading to a new consciousness. When the public viewed Muybridge's [4] pictures depicting the moment of a horse galloping with the four legs in the air, they never looked to a horse galloping in the same way. This is an example of mechanisms that change the way we see. Today representation seems natural to photography, but not natural for holography, because we have learned how to look at a two-dimensional photographic image through a long cultural process that may be traced in history back to Brunelleschi and Alberti. So, some holographic images only seem to be "realistic", if people are changing their notion of the real. Changing their notion of the real, they will be able to establish a symbolic relationship between those images that challenge their senses and their notion of what images are and how they work, and the world [5].

"*Realism is a matter not of any constant or absolute relationship between a picture and its object but of a relationship between the system of representation employed in the picture and the standard system*", as wrote Goodman [6].

Representation is a matter of context and illusion a matter of how understanding the relationship between objects and images that represent objects. The aim of illusion could be to create in the audience the same response that the viewer would have to the object in similar conditions of observation. Perhaps we can say that art has never been a pure representation, "*that was figurative without been representative*", as said Deleuze [7]. An artist depicting an image is never doing that without imprinting himself in each work of art. Even to a photographer, a pure representation is an impossibility. Perhaps we could define representation as an interpretation of another representation in infinitive way: each one interpreting in different and unique way, reinventing the work of art. "*The works of art are done by the viewers*", to quote Marcel Duchamp [8].

It does not mean that traditional ways of making art has been exhausted. Today we live in a time of synchronization where we have a mix of universal language and trans-disciplined multimedia offering new ways of appreciation join together different cultures. The act of taking a hologram might be described as a more accurate way of collecting data about our world than taking a photography, but Denisyuk advised "it is natural to imagine that the property of a three-dimensional hologram is brought about by the fact that its structure somehow aspires to copy a given object's form, but it is very difficult to prove this since no method for predetermining an arbitrary three-dimensional object is known to optics" [9]. Holography is one of many new "Hyper-Medias" that allows a three-dimensional form to superimpose another three-dimensional form, as one may envisage a double exposure in photography, but in real space, in real time where one can walk around the projected images and view them as one would a sculpture [10]. In "reality" this is not something we would see in everyday life. On one hand we need to learn some rules to see some kind of representation as natural, in the other hand we can see new forms of images that may help us to see our "reality" in a different kind of way. Similar to the way photography made profound changes in the way we see ourselves and the world around us, as said Susan Sontag [11]. Holography could do the same, even more if we are understanding holography not to "reproducing" the world but opening new formal play and unexpected imagery, where can coexist photography, film, video and electronic image manipulation into the digital holographic image, perhaps could be a land to develop new ways of seeing ourselves and the world.

Journal of Physics: Conference Series **415** (2013) 012015 doi:10.1088/1742-6596/415/1/012015

3. Digital holographic images

The fusion of digital holographic images is derived from different artistic fields such as photography, film and video. In the previous paper we have explained the expressive possibilities of the holographic medium in the creation of digital art holograms and lenticulars. Having in mind holography as a time-based medium, acting through light and colour, the final images can be seen in transformation, left and right or back and forward, carrying on a visual dialogue inside holographic plate with the viewer, and questioning himself about the act of seeing. In that paper, we used the HoloCam Portable Light System, that is a Syn4D imaging device system that produces a sequence of digital photographic shots of a real scene in a format designed for use with HPO, horizontal parallax only, digital holographic Syn4D printers. The device has been designed for use with holographic Syn4D printer operated by Geola [12] group in Vilnius, Lithuania. The final Syn4D print contains a realistic colour four-dimensional image of the filmed subject.

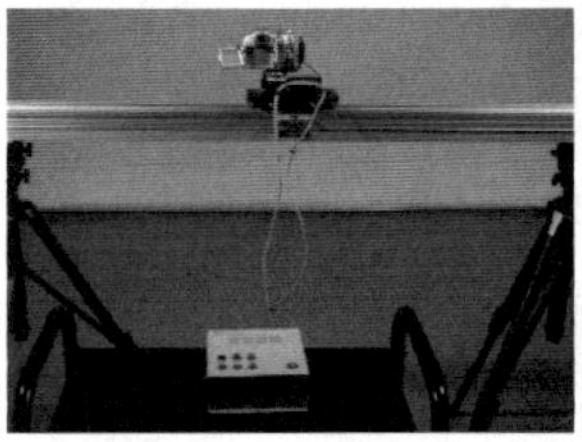
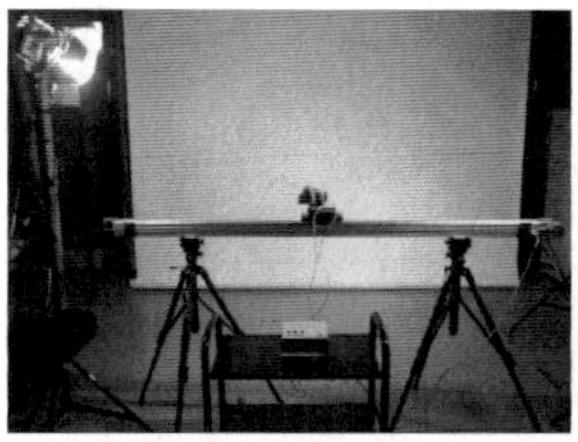

Figure 1 - The HoloCam Portable Light

Figure 2 - The HoloCam in the space where we will built a scenario

When the movies were finished they were sent to Geola in Lithuania to be printed as reflection holograms. The title of this series is "Across Light: *Through Colour*" and are shown in the illustrations. In "Pop Dog" (Fig. 3) there is a singer that suddenly transforms into a dog and the dog transforms again into a singer. There are few moments during the transforming that both occupy the same space. The idea of this work was related at the movement from artists to protect animals, like People for the Ethical Treatment of Animals (PETA). Here the artist transforms into a dog she wants to protect and the dog is acting like an artist too, using the same clothes and wig. As the viewer moves in front of the holographic image, he can see that non-stop transformation, from the left to the right and from the right to the left, forward and backward.

Figure 3 - "Pop Dog", Digital Reflection Hologram, 18x13cm. (Recorded with Sony)

"Velasquez" (Fig. 4) is homage to the painter, and to the painting, Las Niñas. A girl does gumball and the rail camera projects in the girl glasses, showing the moment of the movie record. As the viewer moves left to right or right to left the ball inflates or deflates, and we can see the mouth movement through the transparency of the ball. The space can be seen in the glasses by reflection. The viewer occupies the camera position and he reconstructs his personal image through the one recorded by the camera.

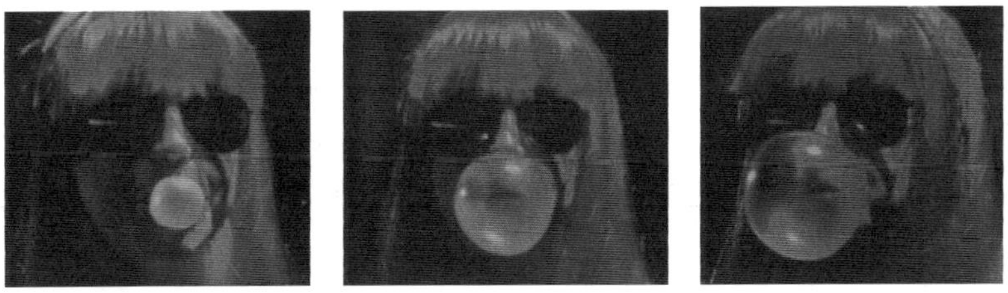

Figure 4 - "Velasquez", Digital Reflection Hologram, 18x13cm. (Recorded with Sony)

In "Talk-Talk" (Fig. 5), there are two figures, one small wearing female glasses, (glasses with a woman's eyes looking out), the second with male glasses, (glasses with a man's eyes looking out). They represent the both sides of human being, the male and the female. A space inside other space, a shape inside another shape. The little figure is speaking from the left to the right and from the right to the left, each time the viewer is moving in front of that holographic image, the figure is speaking and changing position, in a colour movie acting all the time.

Figure 5 – "Talk-Talk", Digital Reflection Hologram, 18x13cm. (Recorded with Canon)

Figure 6 – "Talk-Talk", Digital Reflection Hologram, 18x13cm. (Recorded with Sony)

These images are recorded using Canon IS3 camera (Fig. 5) and Sony HDR CX105 camera (Fig. 6), on the rail, using the HoloCam, to compare different effects.

Approximately thirty movies were recorded, but only four digital holograms were selected for printing. We observed that the movies made with the Sony camera have a tendency to be red, and the red is without alternate transparencies and opacities. Utilizing the Canon camera we found a higher colour content. One may notice the white dotted pattern in the costume of the sitter appears red within the Sony recordings.

Pictures used from the Canon present an image identifiable with a kind of timeless quality, that specificity seems interesting to explore in next digital art holograms.

In the present paper, we have used another geometry for image capturing that explores the same system using the Canon camera angled at 45 degrees (toward the floor) and carrying on a different kind of space and movement as compared to the movies in the previous paper.

In capturing the scenes top to bottom, we used the floor to build another kind of set-ups and compositions and we produced another series of digital art holograms and lenticulars titled "*Across Light: Through Colour II*".

We recorded near fifty movies but we only selected four to printing.

In "Peter Pan" (Fig. 7) we can see feet and legs depicted behind a balloon, as if about to kick it. Three colours were projected to overlap three shadows in the middle of balloons; these balloons are moving slowly - mixing the colour of shadows on the floor. The white leggings also have colours projected onto them which change as the observer moves to view the holographic plate.

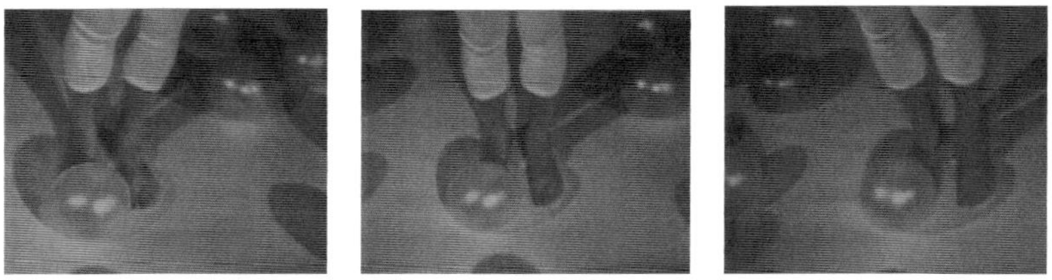

Figure 7 - "Peter Pan", Digital Reflection Hologram, 18x13cm. (Recorded with Canon)

"A Being from Space" (Fig. 8) is a masked being without expression, blowing bubbles in a non-defined space. Like planets in orbit, the bubbles hover inside the hologram as if creating another universe yet to be finished.

Figure 8 - "A Being from Space", Digital Reflection Hologram, 18x13cm. (Recorded with Canon)

In "Earth Lighting" (Fig. 9), a Moon balloon is jumping in direction at the Earth balloon near at a hammer and a shinning stair. It is jumping, up and down depending viewer movements, the hammer is presented like something dangerous about to happen to the planet.

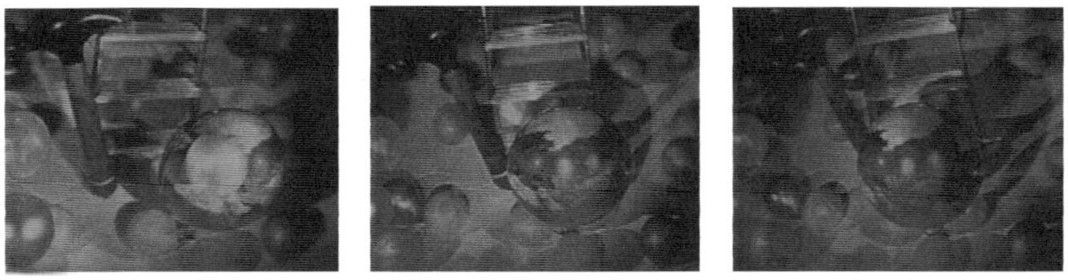

Figure 9 - "Earth Lighting", Digital Reflection Hologram, 18x13cm. (Recorded with Canon)

In "The Gardener" (Fig. 10) a human figure, an entity, is partially showing through the feet and a white dress, inside a celestial garden. A frog is recognizing the entity, and all the elements in the garden are celebrating his presence, shining, moving and projecting colours between them, for all and forever, depending the viewer movements.

Figure 10, "The Gardener". Digital Reflection Hologram, 18x13cm. (Recorded with Canon)

Using different cameras on the HOLOSHOT rail, we discovered several problems occurred with the digital printing process due to non-calibration by GEOLA.

1. Because the cameras were not synchronized, both the Sony HDR CX105 and Panasonic Lumix G2 did not focus to the centre of the required scene because of the predictive focus modes. Whereas the Canon IS3 camera, with changes made by GEOLA, formed the correct focus.
2. Both the Sony and Panasonic cameras are not focusing to the centre of the scene, and in the end the frames have a different sizes, which was not suitable for the digital holographic Syn4D printers in GEOLA.
3. Colour saturation was observed to change in the Sony camera, as it warmed in the studio environment. This was especially noticeable in the digital hologram "Talk-Talk" (recorded with Sony) and after looking the movies in the camera and observing in computers too, the blue colour projected on the black curtain, in the back of the figures, was turned green after two or three movies. We conclude that the camera when was warm has changed colours. Then we decide to use a white support having in mind, if doing that, the colours will not change.
4. Experiments with the cameras angled at 45 degrees toward the floor and with the three light colours projectors, red, blue, green, projecting at the floor near 45 degrees recorded blue, green, red – red, green, blue; blue, red, green – green, red, blue; green, blue, red – red, blue, green. This test was repeated with Canon IS3 camera and the same with Sony HDR CX105 camera and Panasonic Lumix G2 camera. Each of these factors changed the aesthetic of the digital hologram -

different environments with that change and a kind of pallet that we can use depending the kind of space and ambiance we want to create in another holographic images.

4. Conclusion

The quality of these digital holographic images is dependent on true colour observations that play an important role in the construction of space and time. The audience will interact with events that have no beginning, or end, that can be perceived in any direction, left and right or forward and backward, in a symmetrical time, and depending on the relative position of the viewer. The relationship with the scene inside holographic plate and the viewer when looking at the plate is fundamental, the image is activated from a multitude of positions in a similar way that we interact with the "real world".

The authors would like to thank Mr. Jeremy Collingwood and Mr. Stuart Wade for technical support and collaboration. Isabel Azevedo also thanks Fundação da Ciência e Tecnologia, Portugal, for making this work possible through a post-doc research grant.

QUADRO DE REFERÊNCIA ESTRATÉGICO NACIONAL
GOVERNO DA REPÚBLICA PORTUGUESA
UNIÃO EUROPEIA Fundo Social Europeu

References:

[1] Azevedo I, Richardson M and Bernardo L M 2012 Across Light: *Through Colour,* Proc. SPIE 8281, 82810B

[2] Turrel J 1993 in Catálogo da Sala de Exposiciones de la Fundación "La Caixa", 12 de Novembro de 1992 - 10 de Janeiro

[3] Damásio A 2000 *The Feeling of What Happens, Body: Emotion and the Making of Consciousness* (London: Vintage)

[4] Muybridge E 1902 *Animals in motion: an electro-photographic investigation of consecutive phases of animal progressive movements* (London: Chapman & Hall)

[5] www.ekac.org/photonic.webswww.ekac.org/photonic.webs

[6] Goodman N 1976 *Languages of Art* (Indianapolis: Hackett)

[7] Deleuze G 1981 *Francis Bacon, Logique de la Sensation* (Paris: Différence)

[8] Cabanne P 1971 *Dialogues with Marcel Duchamp* (London: Thames and Hudson)

[9] Richardson M 2006 *The Prime Illusion, Modern Holography In The New Age Of Digital Media* (London: The Holographic Image Studio)

[10] Richardson M 2006 *The Prime Illusion, Modern Holography In The New Age Of Digital Media* (London: The Holographic Image Studio)

[11] Sontag S 2008 *On photography* (London: Penguin)

[12] www.geola.lt

Holography at the Butler Institute of American Art

Thomas Cvetkovich
Artist, consultant, USA.
Rayvel Holographics, CN USA
Odhner Holographics, NH USA

Abstract. The Butler Institute of American Art is the first museum in the country to focus on American art. Since 2003 there has been a continuious showing of international holographic artists at the Butler Art Institute, perhaps more then any other single museum. This paper will catalog these exhibits with ccommentary from the artists and staff of the Butler.

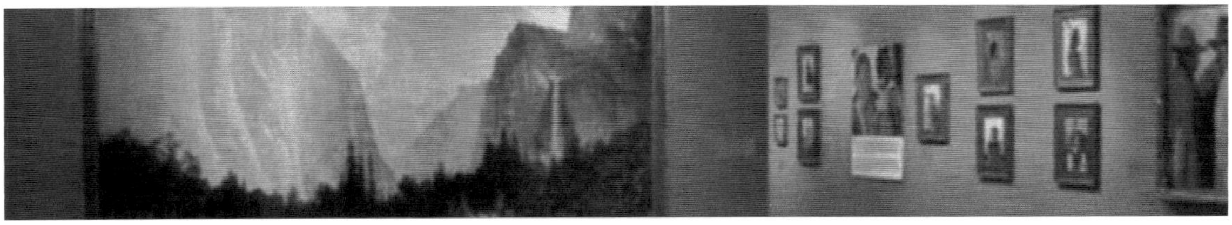

The Butler Institute, A Home for Holography.

The Butler Institute of American Art[1] was born of fire. The home of industrialist Joseph G. Butler suffered a fire destroying his nascent art collection. Vowing that this should not happen again Mr. Butler built a museum of marble. His new collection specialized in American art at a time when only European art was considered worth collecting. It thus became the first museum of American art in 1919.

The museum has grown in stages to encompass many galleries (72,000 square feet) and three locations. It houses acknowledged masters through the decades from The Hudson River School through Winslow Homer, Mary Cassatt and onto contemporary artist like Rauschenberg and Paul Jenkins. Director Dr. Lou Zona presided with great vision over much of the museum's growth.

Dr. Zona remarked: "I fell in love with holograms the first time I saw one...not the early images but rather the technology. The earliest images were of course typical of what scientists would choose such as a plastic model of the (Star Trek) Enterprise. Before I became director of the Butler, I assisted Joe Butler who was then director (around 1980). I had visited the Holographic Museum in Soho and saw that a traveling exhibition was available. I talked Mr. Butler into booking it with the idea that it would have to pay for itself by charging an admission fee. The exhibition was crowded every single day and the exhibit did indeed pay for itself. I recall Mr. Butler asking me about my "little light show". The thing about the show was that while it dazzled everybody with the technology, the imagery was not very creative. I thought back then...just wait until serious artists and not engineers were in control of this medium. So when the Beecher Center was built in 2000, there was no question that I would focus on holography."

That first show, *Through the Looking Glass* opened in 1981 at the Butler. This exhibit included many of the works that defined the early stages of art holography; Loyd Cross- *Kiss ll*, Sam Moree-*Loose End,* Steve Benton- *Rind ll*, William Moltini- *Crayons*, Anait Stephens- *Space Graffiti,* Daniel K Schweitzer- *Movie Theater.* Some 26 pieces were exhibited in all, and it did in deed break attendance records at the Museum.

Loyd Cross & Pam Brazier, *"Kiss ll" (1976)*

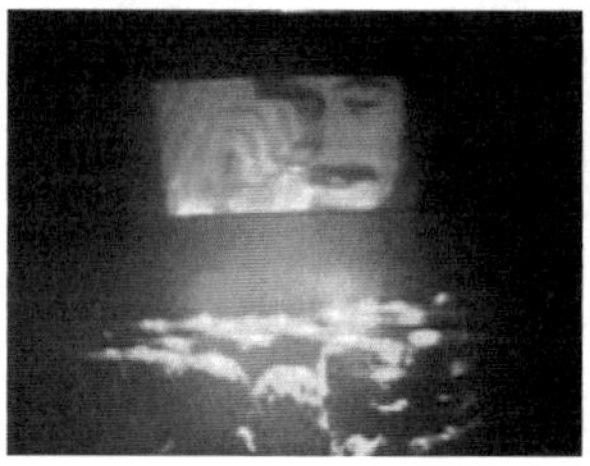

Dan Schweitzer's *"Movie Theater"* (1978) 5 x 7 inch

It was not until 10 years later that the second show, "*Future Sight, Innovations in Art Holography"* appeared. Works for this exhibit included: Yaacav *Aram-Holograph ll*, Paul D Barefoot-*Canarias Magic*, Melissa Crenshaw-*Levels with Light Blocks*.

Both exhibits originated from the now defunct Museum of Holography, New York and were produced and circulated by Paul Barefoot.
"*Looking Glass*, showed it's audiences the artistic capabilities and tools available at that time through holography. Most of the "tricks" of the trade were demonstrated. The true artistic breakthrough did not come until Rudie Burkhout's *"12 Mille Watt Bogie"* that was featured in "Future Sight." That was the beginning of color control and sophisticated imagery that characterized later shows at the Butler. Artistic holography has come a long way since the 70s and The Butler Institute has chronicled the development of the medium through its shows. I am delighted to have participated in that." Paul Barefoot.
……………………………..
Tom Cvetkovich's [2]"*Coruscation,"* 1992, was a large installation of a double spiral walkway into increasingly dense sheets of rainbow.

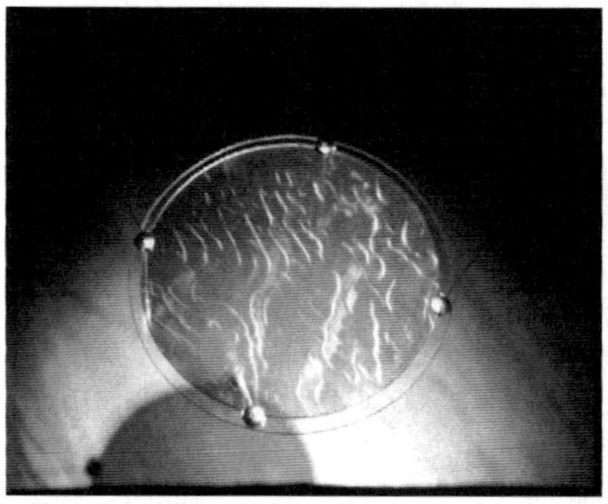

Tom Cvetkovich "*Coruscation"* (1992) 60x60 feet, installation with diffraction gratings and reflection hologram. Left: school children and centerpiece. Right: centerpiece, 12" diameter hanging reflection hologram of waves.

[3]

The newest wing of the Butler, the Beecher Center for Art and Technology is dedicated to new art forms; digital art, large installations, and technical art of all kinds.

Fred Unterseher and **Rebecca Deem**[4] showed a small group of some of their finer work in in the late nineties.

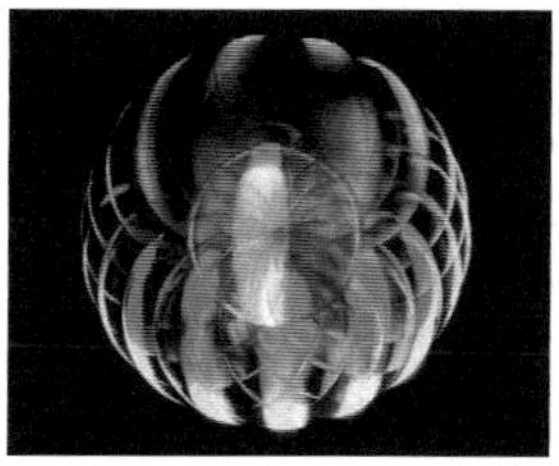

Left **Fred Unterseher** *"Kinetic Mandala 18R"* (1994) 12"x16" Fourier transform lens matrix hologram, silver halide.
Right **Rebecca Deem** *"Three Modes of Knowing"* (1988-90) 22"x22" each custom frames, reflections hologram from pulse master.

"Working with the medium of holography can offer an unusual opportunity to explore the relationships between science and technology as well as art and visual perception. ... The viewer can direct the experience of encountering a hologram and that experience can be different and unique for each of us. The interaction serves as an opportunity for awareness of our participation in the creation and perceptions of our own world and how we share that experience with others." Fred Unterseher

"Much of my work addresses individual human growth and development to that of human development through space/time (Ontogeny / Phylogeny). Holography has proven to be an excellent tool for this exploration since we appear to be living at a time that continually challenges our concepts of both space and time with nearly instant communications world-wide and breaking physical barriers on earth and in outer space or simply space. The hologram can bring to our attention the perceptions we form of the relationship between immaterial and material." Rebecca Deem

....

In 2000 British collector Jonathan Ross[5], brought a small portion of his collection to Ohio for it's first US showing. Artists included; Margaret Benyon, Patrick Boyd, Sue Cowles, Pearl John, Ken Harris, John Kaufman, Ruben Ninez, David Pizanelli, Caroline Palmer, John Milton, Martin Richardson and Jeffrey Robb.

Regarding holography Mr. Ross commented that, "It's the main thing I collect, having been involved for over 30 years. Holography reaches places other art does not...I really admire the way Lou Zona has run his holography program for over 10 years. A voice in the wilderness! Everyone was so friendly and helpful, I enjoyed seeing the permanent collection and meeting other visiting artists. An enjoyable opening reception too, as I recall. American drinks are so much more generous than European ones."

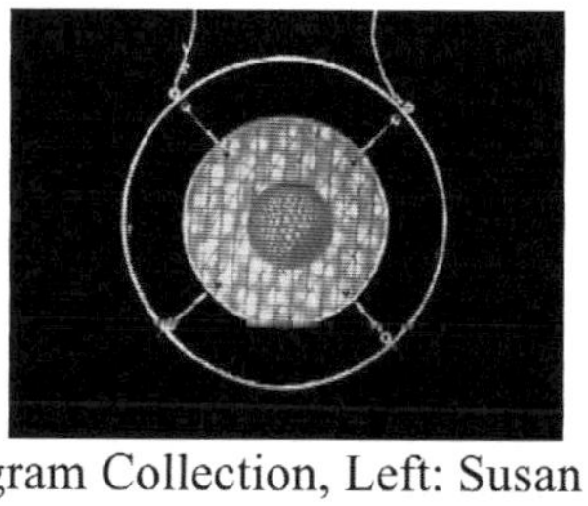

From the Jonathan Ross Hologram Collection, Left: Susan Cowles *"The House of Moons (A stage for the chyncal Theatre)"* (1988) Refelction hologram 2 plates 30x40cm with ink drawing on paper. Center: Jon Milton *"Orgone Accelerator"* (1991). Right: Margaret Benyon *"Tiger Girl"* (1985) 30x33cm reflection hologram and reproduction.

From 2003 forward the Butler has had a continuous showing of holographic art. That year saw five holographers come to the Butler in rapid succession.

………………………………..

The first of which was Shu Min Lin's *"Glass Ceiling and other Works"* beautifully exhibited in the Flad Gallery with high ceilings, walls painted black and ample track lighting.

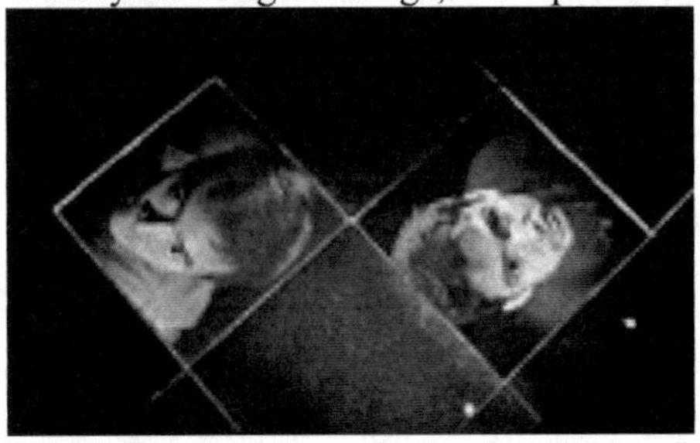

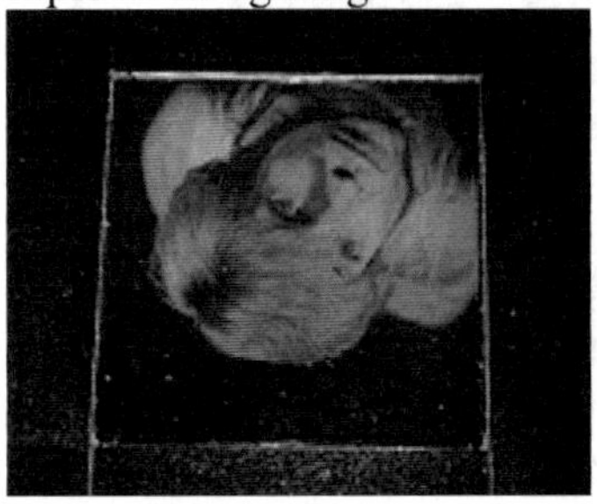

Shu Min Lin "*Glass Ceiling*" (1997) reflection holograms from pulse masters, variable installation size on floor.

"My hologram shows images of people from more than 40 different countries, both men and women, and of all ages (between seven and eighty-one). In a way, the piece creates the malicious feeling of walking on someone's head or face. It therefore confronts the audience with a kind of negation of the other, which occurs in human behavior on a physical, as well as on a psychological level."[6] Shu Min Lin

………………………………..

Andrew Pepper[7] showed his, *"UK Spaces, Holographic Drawings, Volumes and Site Specific Installations"* in March of 2003.

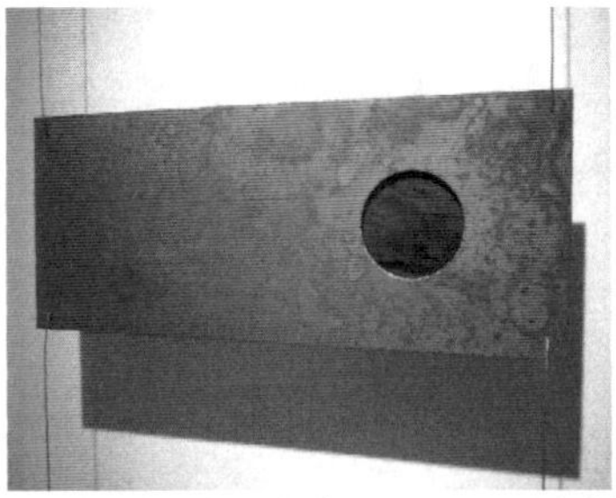

Andrew Pepper *"Sight Lines Wall"* (2001) and detail, refection holograms, variable dimension installation.

"This was an incredibly important venue to show in, not only because of its continuing commitment to creative holography but also being able to exhibit alongside its permanent, academic, collection…By offering the opportunity to release marks from the surface, on which they are traditionally located, the process of holography opens up a series of optical, conceptual and physical investigations which can help to 'put pressure' on our engagement with the 'drawn' mark." Andrew Pepper.

………………………………..

Philippe Boisonnet's[8] *"Efflorescence and other works"* opened in the September of 2003.

"I was happy to present one of my best interactive holographic installation «In-Between» (from 1997), and a few other smaller pieces, which were all related to the idea of encounter. The encounter between the eye and the fleeting images of holograms, of course, but also the encounter between the different viewers through the mean of a small ultrasonic motion device that was switching on and off the lights of the 3 pieces within the installation. A kind of conceptual reference to the physical encounter between the two beams of laser lights that are needed to built an interference pattern that registers the 3D light information."

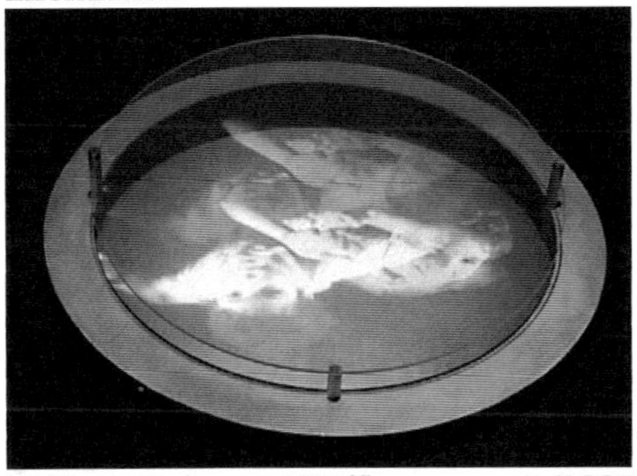

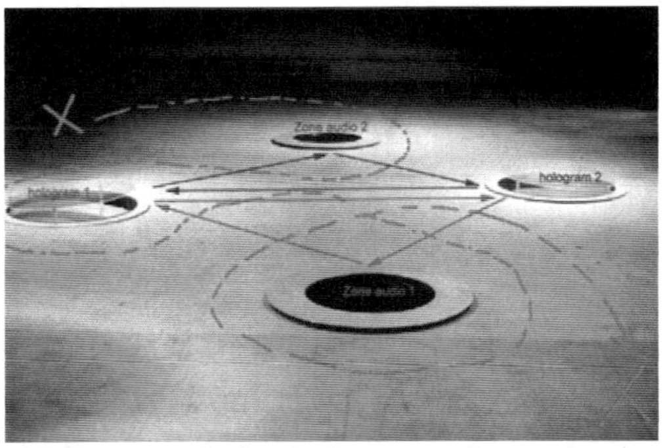

Philippe Boisonnet "*Efflorescence*" (1997-2003) white light transmission hologram; mirror; aluminum; Plexiglas; electronic sensing device for audio system. Left; detail. Right viewer path.

....

Patrick Boyd[9] exhibited his unique photo-to-stereo grams and pulse work in October of 2003. These often-whimsical works revealed the trained eye of a professional photographer.
"Holography is so much more than just a 3-D image. It can display time, motion, fantastic colours; it can be interactive and is capable of capturing the qualities of light itself.... My works of meticulously created installations & captured sequences blur the boundaries between two & three dimensions. I try to present a colorful world where real life, narrative, light & shadow connect & collide." Patrick Boyd

Patrick Boyd Right: "*Breakfast*" reflection hologram form pulse master Left: "*Charles Bridge*" animated stereogram with lenticular background.

....

Brazilian born **Eduardo Kac**[10] used the Bermat Gallery to show "*Amalgam*" and other works form the *Holopoem* series in November of 2003. Of this series he wrote, "holopoetry: fluidity of the verbal sign and semantic interpolation, i.e., mutability of the actual topology of words in space leading to changes in meaning."

Eduardo Kac, *AMALGAM,* 10 X 7.5 cm, Reflection hologram, 1990

Eduardo Kac, *Andromeda Souvenir,* 1990, 30 X 40 cm, Digital transmission hologram)

………………………………..

In 2004 the author, **Tom Cvetkovich**, collected a group of hitherto un-shown works as well as creating a number of new ones for the show, *"Presence."* It was a much-appreciated opportunity to whole heartedly engage in holography as art as opposed to commercial holography.

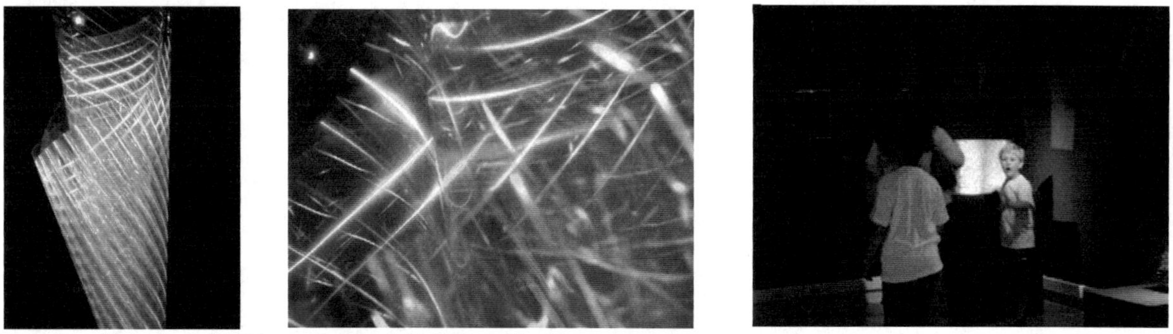

Tom Cvetkovich Left: *"Convolutions"* (2004) 2.5' x 5' UV curable ink on plastic and detail. Right: *(Light Fountain)* (2004) 18"x 22" UV curable ink on plastic, mirror, silver nickel dot matrix hologram.

………………………………..

Pulse master holograms created by **Sally Weber** during her Artist in Residence at the Center for the Holographic Arts, New York, 2001, came to fruition with *"Strata Series"* which opened in the Butler's Bermant Gallery in October 2006. The pulse masters were transferred in Sally's holography studio in Austin, TX. Also exhibited were her early water installation, *"Threshold of a Singularity- A Memorial"* and figurative photographs that were the genesis of the *Strata Series.*

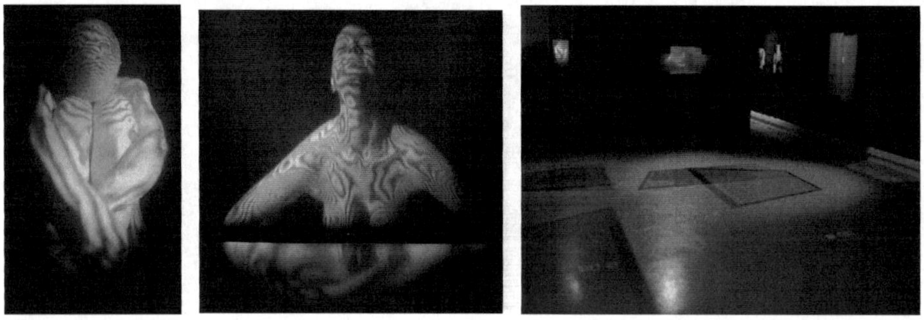

Sally Weber Left: *"Crevass"* (2006) 29"x 18" third of six images from the installation, *"Descent."* Center: *"Laccolith"* (2006) 27"x 25", fifth of six images from *"Descent."* Both images are open-aperture, pulsed light image, illuminated with LED's. Right: Gallery view.

"I was honored to exhibit in the David Bermant Gallery. I had known David Bermant and his deep interest in light, art and technology. The Butler Institute's Executive Director, Lou Zona, was also a long time friend of David's. Both were very early supporters of the field. Their advocacy for light-based art and technology continues to be evident in the Institute's collection and its on-going interest in exhibitions dedicated to the field. ...My experience at the Butler Institute was exceptional as is their staff and Executive Director, Lou Zona. It was certainly a privilege to exhibit there."- Sally Weber

....
In August 2008 **Doris Vila**[11] informed us by her title that, *"Every Time I open my Eyes I See Things."*

"Chart: Space-Time-Sex-Money Continuum," 1988, is a good example of the effective use of projected imagery to illuminate a hologram. (One of several in the show.) Imagery of falling money illuminates the hologram. The money wafts down ward and then appears to get stuck to the hologram, though it is actually an image of dollar bills hitting a floor. A digital projector replaced for the original 8mm projector.
Doris wryly remarked on the work, "Clearly Einstein forgot a few important dimensions."
Other works included the projection feed back of the viewer's feet standing in a holographic fire, in *"Device for the Lifting of Gravity and Other Serious Situations."*

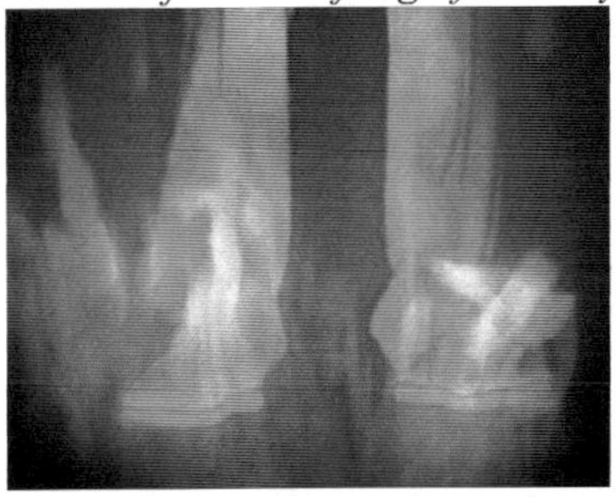

Doris Vila Right: *"Device for the Lifting of Gravity and Other Serious Situations"* (originally 1992, 2008 version shown here) refection hologram with live video feedback.
Left gallery view.

"I make holograms of stories instead of things."
"The Butler suggested a small retrospective of my work. At first, I said "But I'm not that old" to which they answered "yes, you are." The exhibition brought together a sampling of hallmark works, including holographic work with video. It was a pleasure to see threads of thought connecting works that spanned many years." Doris Vila

"Body Electric" was an exhibition of pulse work by the artists, **Ana Maria Nicholson[12] and Ikuo Nakamura.[13]**

....
As director of the Center for the Holographic Arts, New York, Ann Maria Nicholson is a well know figure in the holographic art world. Some 60 artists from around the world have worked there.

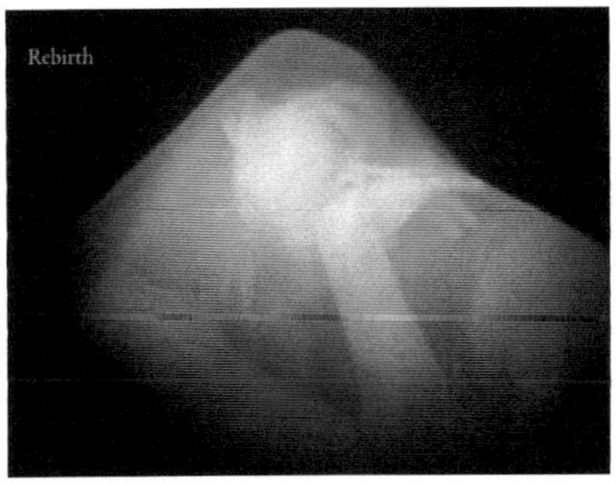

Ana Maria Nicholson Left: *"Rebirth"* (2003) 20"x24" reflection hologram from pulse master. Right:

Journal of Physics: Conference Series **415** (2013) 012016 doi:10.1088/1742-6596/415/1/012016

"Union" (circa 2003) pulse transmission hologram and "Water Dreams" (2000) installation, size variable, hologram 11"x14"

"When I started working in holography I became captivated by the way that the medium transformed the human body. The central theme of my work as an artist has always been the human figure: its beauty, its complexity and its ability to express in gesture and movement the hidden realities that lie at the core of the human spirit and heart. When carefully observed the body is but a thin shield, a permeable membrane that can obscure but not obliterate the psychological and spiritual dimensions." Ana Maria Nicholson

………………………………..

Ikuo Nakamura included *"Fossil,"* and the intriguing *"Thera, "w*hich explores the painting "Madam X" by John Singer Sargent in hologram and digital projection, as if the model could undo the social harm done from the fallen dress strap.

"I shot *"Fossiles"* or *"Thera"* as installation work, not as portrait hologram. By presenting the co-existence with holographic space and real object in location, the audience experience Real-Virtual space." Ikuo Nakamura.

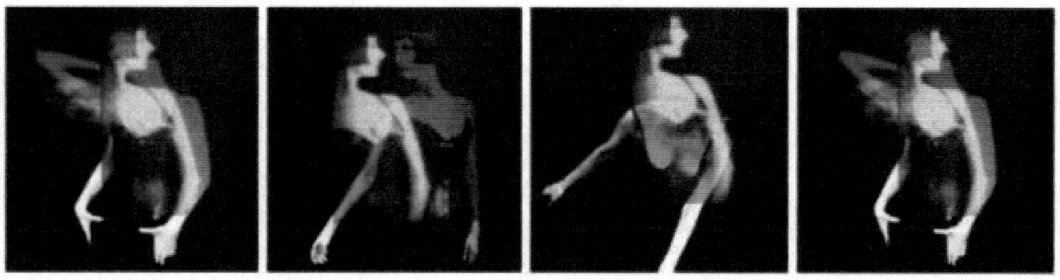

Ikuo Nakamura. Left: *"Fossiles"* (1999) 50cm x 60cm laser transmission hologram, illuminated by DPSS laser with original beam expander. Location: front of Temple 1, Tikal, Mayan ruin complex, Guatemala. Date: March 2006
Right: *"Thera"* (2004) 50cm x 60cm laser transmission hologram with Digital Projection Installation and diode laser illuminated as above.

………………………………..

Ana Maria Nicholson displayed a large selection of her pulse laser work in *"PORTRAITS IN LASER LIGHT"* December of 2008.

Ana Maria Nicholson: Reflection holograms form pulse master, Left: George Plimpton (1990) 11"x14"
Right: Don Alejandro, (2006) 11"x14"

"The power and impact of a holographic portrait never cease to amaze me. The medium has the ability to capture on film a much deeper sense of the person than a photograph and can reveal the reality that lies

behind the façade of personality."
Ana Maria Nicholson
………………………………..
In 2006 The Butler created a small set of galleries devoted to art works requiring a darkened environment.
"Krishna on the Banyan Leaf" depicts the Hindu mythology which can be thought of as the last moment of the Universe before it starts anew."
"*Coruscation ll"* is a celebration of all things bright and colorful, reminiscent of the near death experience." Tom Cvetkovich

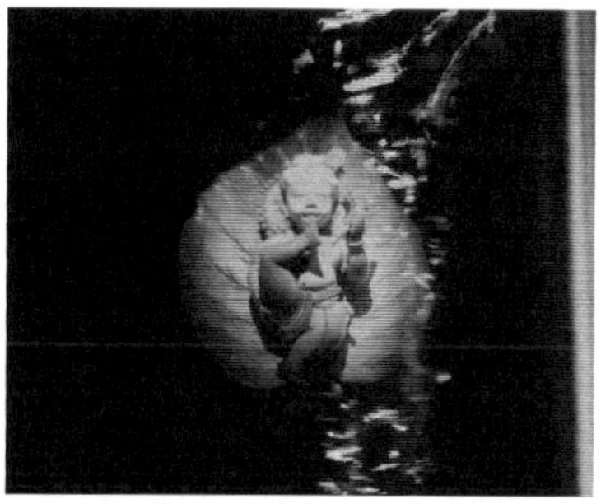

Tom Cvetkovich Left *Coruscation ll*; (2009) 10 wide x 15 high x 40 feet deep, digital projection, custom color balanced LEDs and shaped diffraction gratings. Right; *"Baby Krishna on the Banyan Leaf",* (2005) 14"x 20" inch laser viewable hologram on silver nickel shim.
………………………………..
Mary Harman's[14] sculpture painting and holograms filled the gallery for her show *"Object and Illusion,"* September 2009.
"The philosophy behind all my art seems to be that of a quest to make some sense of the world I live in. That involves filtering information, truth from lie, reality from illusion…My art is about what is real and what is illusion, in both the political and the personal" Mary Harman.

Mary Harman Homage to Magritte, 21 ¼" x 14" x 5 ½", reflection hologram and sculpture. Left: Gallery view with paintings.
………………………………..

***"About Light: Works from the Butler Collection*"** opened in Mar 2011.
This exhibition features light works that utilize a variety of technologies including **August Muth's** "*Acceleration*." Included in the show are works by Ray Howlett, Clyde Lynds, Stephen Knapp, Ted Victoria and Alan Rath, as well as recent gifts from the Bermant Foundation.

Journal of Physics: Conference Series **415** (2013) 012016 doi:10.1088/1742-6596/415/1/012016

August Muth; *"Acceleration" #2* (2010) Holographic emulsion, glass and steel.

"These pieces are part of a body of work that has been evolving over many years. They are recordings of pure light produced by interfering coherent light waves upon themselves. Each piece presents perceptually challenging environments in which neither color nor shape is static. Within each piece is contained a much vaster space as one discovers by physically moving around each one. It is as if someone captured space itself and harnessed it in it's own container." August Muth

…………………………….

Georges Dynes[15] staged *"Genesis;Lux ex Chaos"* with Carole Brissson in Dec 2011. This sweeping exhibit with a multitude of works of various sorts but all related to a creation/environmental theme.

"I consider my work to be monumental, immersive (360° around the spectator), multi media, and developing in time and space). In my exhibition at the Butler, I focused rather on the 2d drawings (on which video images are projected) and 3D drawings in boxes and holophotograms (Earth-Moon)…Also, on the dead black-and-white images of the moon are super impressed coloured holographic events, which symbolizes LIFE on earth. I must add that the enormous quantity of scrap lying underneath the holograms of this series symbolizes the lack of care of many human beings about our earth." Georges Dynes

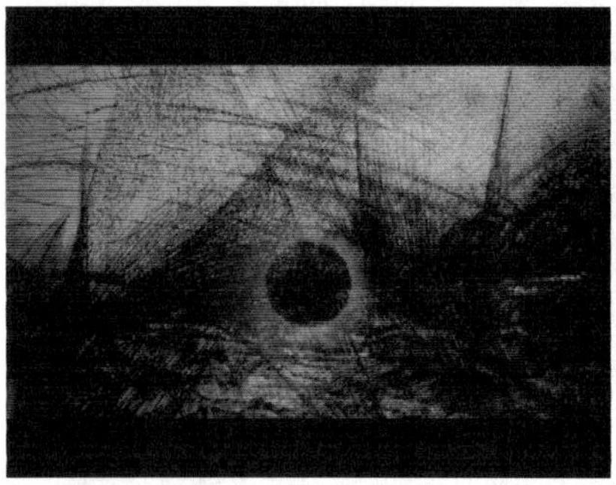

Georges Dynes LEFT: *"Eclipse"* (2000) Graphite and reflection hologram. Right: gallery view.

CONCLUSION.

"Over the years we've been able to show many of the leading holographic artists of the world. It is wonderful to see that the medium is now in the hands of artists although I wish that holography were a little more accessible to an even broader range of artistic talent. The obvious obstacle is that it is so darned technical… But I am not complaining since I believe that many of the holographic artists working today are enormously creative and are showing off the medium to full advantage." Lou Zona

The Butler has also made seventeen purchases of holographic works through the years, often buying an example from an artist's show. Thanks to its outreach program one thousand school children per month visit the Butler. In this way the Museum will continue to champion the still esoteric art of holography.

---

[1] http://www.butlerart.com/

Journal of Physics: Conference Series **415** (2013) 012016 doi:10.1088/1742-6596/415/1/012016

[2] http://www.flickr.com/photos/tomcvet/
[3] Photo from Palo Creative for the Youngstgown Foundation http://www.youngstownfoundation.org/
[4] deem.unterseher@gmail.com
[5] http://www.jrholocollection.com/collection.html
[6] From and interview with Pat Binder of "Universes in Universe."
[7] http://www.apepper.com/butler/info.html
[8] http://www.flickr.com/photos/uqtart
[9] http://boydphoto.co.uk/ and http://www.peekaybee.com/Peekaybee/Patrick_Boyd.html
[10] http://www.ekac.org
[11] http://www.vilamedia.com, photos David S Licht
[12] http://www.anamarianicholson.com/
[13] http://hololab.com/index.php
[14] http://www.maryharman.com/holograms.htm
[15] georgesdyens.com

The shifting appearance/disappearance of holographic images and the dynamic ontology of perceptual and cognitive processes

Philippe Boissonnet[1, 2]

[1] École des arts visuels et médiatiques (Phd Studies), UQÀM, C.P. 8888, succ. Centre-Ville, Montréal, H3C 3P8 (Qc) Canada
[2] Art Professor, Dpt de Philosophie et des arts, UQTR, C.P. 500, Trois-Rivières, G9A 5H7 (Qc) Canada

Abstract. The French philosopher M Merleau-Ponty captured the dynamic of perception with his idea of the intertwining of perceiver and perceived. Light is what links them. In the case of holographic images, not only is spatial and colour perception the pure product of light, but this light information is always in the process of self-construction *with* our eyes, according to our movements and the point of view adopted. According to the aesthetic reception of a work of art, Holographic images vary greatly from those of cinema, photography and even every kind of digital 3D animation. This particular image's status truly makes perceptually apparent the "co-emergence" of light and our gaze. But holography never misleads us with respect to the precarious nature of our perceptions. We have no illusion as to the limits of our empirical understanding of the perceived reality. Holography, like our knowledge of the visible, thus brings to light the phenomenon of reality's "co-constitution" and contributes to a *dynamic ontology of perceptual and cognitive processes*. The cognitivist Francico Varela defines this as the paradigm of *enaction*,[i] which I will adapt and apply to the appearance/disappearance context of holographic images to bring out their affinities on a metaphorical level.

1. A sensation of the *loss of reality*

Despite the half century that has passed since the invention of holography, in the artistic and cultural community a major misunderstanding persists today as to what holograms can contribute on an aesthetic and conceptual level. Too often, holography is seen as a mere extension of photography, an aesthetic determined by a completely different historical context in the arts and sciences, when a fixed point of view and monocular perspective brought with them structuring concepts sufficient to satisfy our need for veracity and for grasping reality. Since then, discoveries in quantum physics and the trans-disciplinary contribution of cognitive science to our representation of the reality principle have made this fixed and distanced vision of the world completely inadequate to an understanding of today's world. Today, reality appears much more complex. Its forever veiled and inaccessible nature, the dissolving of the boundary between objective and subjective and the incorporation of uncertainty, interdependence and chaos have become cornerstones of rethinking our conception of the world. This conception is no longer tied to a single perspective or a centralising point of view. Today many forms of artistic expression are impregnated with this new conception of the world, such as those which explore installation and perceptual questions, but also all those practices which promote interaction, relation or immersion. I thus have no doubt that the aesthetic qualities of the holographic image are among the clearest manifestations of this contemporary epistemological context.

Published under licence by IOP Publishing Ltd

Few people, apart from the artists and scientists directly concerned, have seen holography as an extraordinary means for examining the unstable and transitory nature of our perceptual relationship with reality. It is still too often associated with mere illusion. We must thus conclude that the task of profound reflection amongst art specialists, and that of education amongst the public, remains to be done. And yet already in 1989 the German essayist Peter Zec, in an article published in the journal *Leonardo,*[1] had taken a critical approach, highlighting the extent to which the visual experience provided by a holographic image was based on a sensation of the *loss of reality* (that of our relationship with materiality and the tactile) which should not be confused with the impression of illusion. Illusion has a direct connection to a representational position, whose starting point is the principle of a disjunction between the world and our knowledge of it. The point Zec made in particular concerned instead another consequence of holography's formal immateriality. He spoke of a shift in the image's central aesthetic interest towards that of perception itself, which is to say towards the moment of the cortico-visual formation of the image, whether it is figurative or not. This approach connects holography with an artistic position taken up as early as the 1970s by James Turrell, who paraphrased Marshall McLuhan with the assertion '*perception is the medium*'; this accounts for Turrell having recently embarked on the creation of large-format transmission holograms (Pace Wildenstein Gallery, 2009). This artistic position, although it dates from the entire generation of Op Art and lumino-kinetic artists, still exists today, for example in the recent work of the Danish artist Olafur Eliasson, making it possible now more than ever to highlight the cognitive-perceptual process and not only the representational conventions and technologies underlying them.

The artistic potential of holograms should thus always be presented, not as another 3D photographic technology, but rather as an optically coded memory with which our eyes interact through the use of diffracted light. Information is indeed coded, but without digital values being assigned to the formal elements. The light flux is indeed directly recorded, but there is no analogical resemblance between the interference pattern printed on the surface of the hologram and perceived reality. While this kind of explanation may be more complex than the idea of 3D photography to convey through cultural outreach activities, it is closer to the truth. In addition, long-term benefits would accrue if: a) the conceptual premises of photography and holography were clearly distinguished; b) the optical and physical nature of the hologram was well explained, separately from that of the holographic image; and c) the aesthetic value of the perceptual and cognitive nature of the holographic image was highlighted.

As John Perry from Holographics North laboratory writes, 'a hologram is like a lens with a memory, focusing light into a 3-D image in the space around it'.[2] Clearly, most people would find that more difficult to understand, but the art critic's attention would be less directed towards preconceptions around the mimetic representation of a connection to reality.

2. A temporal dynamic

For the holographic image is neither painting nor photography nor cinema. On the physiological level of vision, it mainly employs the phenomenon of retinal persistence in cases where the source of the images being recorded is an animation or video sequence. It calls natural (human) stereovision into play to a much greater extent. Nor does it project images, or use an apparatus that imposes a speed at which the images pass before our eyes. It is completely different from 3D cinema. The time of the image *is* the time of the observer. There is complete temporal identity between the image seen and the act of seeing. The image is closer to a form of virtual sculpture which evacuates the materiality of the world from objects. Naturally, this form of light sculpture distances us from our commonsensical relationship with reality, but it has the advantage of better expressing the dynamics of our perceptual interaction with light. The holographic image greatly resembles a light simulation of the phenomenon of visual perception itself much more than it resembles a representation of the outside world. The immateriality and artificiality of its appearances distance it from this world. It is, instead, the restitution of an optical event which, in our everyday lives, remains imperceptible to our senses. It is the visual recreation of an undulatory encounter: that of the temporal gap produced in two beams of light at the moment of recording. It is thus not quite right to limit oneself to saying that it represents the three-dimension imprint of a material thing now absent. Rather, the holographic image is the memorised result of an event – a dynamic result which relates the essentially temporal nature of the recording technique as much as its visual reception. For too long it was believed that this point was useful in any fundamental sense only in technical explanations, whereas it is also useful for the entire holographic process, right up to its artistic appreciation. Thus the full worth of the medium's aesthetic and expressive dimension can only be brought out if we start from this base, that of the image's temporal dynamic.

Out of the encounter between the hologram and light (diffraction) is created another kind of encounter, one more temporal and relational: that of human time, the time experienced by each individual who "inhabits" the holographic image through the investigation of their gaze. Beyond the hologram, the holographic image thus brings into play one's entire subjective comprehension, brought to bear by visual perception and the cognitive processes associated with it. Before beginning to seek out the artistic significance of a holographic image it is thus important to think outside the conceptual and predetermined representational framework and to focus on the "appearing" phenomenon.

3. The 'co-emergence' of light and space in display holography and the concept of *enaction*

The holographically experienced image is thus a becoming aware of the human visual process. I refer here to the philosophical concept of '*hervorbringen'*, or making the objects of the world emerge and making us conscious of them. This concept appears in the work of Martin Heidegger and Maurice Merleau-Ponty and is revisited by the neuroscientist Francisco Varela in his research into cognition. Varela remarks that, whether we are in the face of the real world or that of images, it is no longer a matter of working with the simple idea of preconceived things, ready to be grasped "as is" by our vision or our cameras. Varela, like Merleau-Ponty, thus insists on the act of dynamic correlation and the co-constructive effect that is constantly being created between the seer and the seen, between the thing to be known and the person who knows. Using this phenomenological concept of making emerge, Varela defines the paradigm of enaction in order to break loose of the constructivist aims of cognitivists in general and bring out a new approach of cognitive processes.

Briefly, what is meant in neuroscience by the concepts co-emergence and enaction? And what, in particular, can be taken from them with respect to my thoughts here on the holographic visual experience? Enaction, a neologism based on the word "enact", is a concept derived from biology and forged (Varela, 1988, 1993) to rethink the definition of cognition, introducing to it the point of view of human experience and individual temporality. That is to say, by seeing cognition as the act of making emerge both the objective world and the subject, or what Varela calls *embodied cognition*. Cognition, as an enactive principle applied to vision, is a co-emergence of the world and its image "inextricably linked to the history of what is experienced, the same way that a previously inexistent path appears while walking".[3] Or, to employ another metaphor used by Varela,[4] embodied cognition is a middle path between the egg and the chicken which, as we know, are correlative and define each other. We are, indeed, in the world as much as the world is in us.

Understanding the act of seeing through co-emergence is thus also to formulate a certain vision of the world and, as a result, 'of our relationship with our knowledge of the world',[5] in the words of another neuroscientist, Olivier Pénelaud. We can thus migrate the concept enaction from the cognitive sciences to the field of aesthetic ideas by using it to insist on the importance of the processes and of the position of the *viewer-actor*. And this is what takes place quite naturally in any visual (or kinaesthetic) experience of luminous holographic space. There, the gaze is always active, in dynamic correlation and even in complete identity with the optical emergence of the image: the image I see before me, in the real space in which my body is located, emerges in me thanks to both light and vision interactions.

This is not merely an obvious optical and perceptual fact. Nor is it merely a manner of speaking poetically about an aesthetic dimension proper to the reception of works of art. It is a true state of "making emerge" that is manifested both in the act of perceiving a hologram and in theoretical reflection on what perceiving an image is. The "making-emerge" of a holographic image is also a true aesthetic experience which is defined by an enaction (the realm of cognition) and by a position of viewer-actor (the realm of aesthetic reception in the arts). On the level of pure aesthetic value, the holographic visual experience thus conforms optically and aesthetically to what Marcel Duchamp described a century ago with his famous phrase 'the viewer completes the work of art'. This is also in keeping with a general trend in contemporary art in various artistic practices using installation and new media in an effort to make us experience the impression that we are in a world to be shared, to be made to emerge, and not only to represent, as was the case at the time of the invention of perspective and even that of the invention of photography.

4. The processual dynamic of the holographic image: an ontological posture for the artist

In my view, however, the appearance/disappearance of the holographic image is even more emblematic, because the way in which it is formally manifested is as phenomenal as pure light, and thus closer to the visual cognitive process. At a time when, generally speaking, we tend to forget that every image located at the back of our retina is on the threshold of appearance and disappearance, holography openly reminds us of this. Its evanescent array of light in space tells us quite directly that the visibility of things (forms, colours, textures, etc.) is always vacillating, appearing or disappearing. By means of this way of thinking about holography, the medium is no longer a mere technological extension of photography, with its objective aptitude for recording the appearance of reality. It is thus a medium that reveals the entire importance of our subjectivity and mental processes for constructing the image and making it emerge; which is to say our human dimension as perceptual and knowing beings. Although this dimension exists for everything that we perceive in the world, here it is literally REVEALED (put into light!) through the effects of diffraction, which weaves light by interfering with it. Through the holographic perceptual experience, the physical and psychological qualities of light intermingle to MAKE EMERGE IN US the submerged part of our cognitive relation with reality. Meaning the always unfinished and vacillating processual part, which is constantly being woven between the subjective and the objective, between the world and I. To choose holography as one's artistic medium is also a way to adhere to an ontological position which casts in relief the processual dynamic as a postulate of our relationship with the world.

The holographic image thus becomes an embodied revelation of the precarious status of the image in general and more precisely of its dependence on the human perceptual and cognitive process. Perhaps even Marcel Duchamp would have seen in it the imperceptible dimension of what he called the *infra-thin*, an "in-between" conceptual category because it is always ready to invert something or make it disappear (a prefabricated object into a ready-made, for example). For Duchamp the infra-thin was what is found at the minimal threshold of perception, at the interface of two dimensions or states, such as the boundary between the tangible and the mental. A holographic image's shift from an orthoscopic perspective to a pseudoscopic perspective, for example, would undoubtedly have been for Duchamp a highly edifying visual experience. So we might suppose, at least in re-reading this quotation from such a visionary artist:

> *"I simply thought of an idea of a projection, of an invisible fourth dimension . . . in other words that every three-dimensional object, which we look at coldly, is the projection of some four-dimensional thing which we do not know".*

Acknowledgments

Acknowledgments to the *Décanat aux études supérieures et à la recherche (*Université du Québec à Trois-Rivières, Canada) for its financial support, and to Timothy Barnard for English translation.

Notes

[1] Peter Zec, "The Aesthetic Message of Holography", in Louis M. Brill, ed., *Holography as an Art Medium*, *Leonardo* 22 nos. 3-4 (Oxford: Pergamon Press, 1989), 426.

[2] John Perry on the Holographics North Laboratory web site. See http://www.holonorth.com/faq.htm

[3] Francisco Varela, *Invitation aux sciences cognitives* (Paris: Seuil, 1989), 111. A translation of *Cognitive Sciences: A Cartography of Current Ideas* (1988). The sentence quoted here has been translated into English from the French.

[4] In his book *Cognitive Sciences.*The term "enaction" was proposed by Francisco Varela in order to designate a new paradigm in cognitive science, based not on the metaphor of the computer as in classical cognitivism, but instead on the metaphor of living organisms.

[5] Olivier Penelaud, *Le paradigme de l'énaction aujourd'hui (apports et limites d'une théorie cognitive "révolutionnaire"),* consulted on line on 24 April 2012 at http://plasticites-sciences-arts.org/PLASTIR/Penelaud%20P18.pdf

9th International Symposium on Display Holography (ISDH 2012) IOP Publishing
Journal of Physics: Conference Series **415** (2013) 012018 doi:10.1088/1742-6596/415/1/012018

A Holographic Collaboration

Ana Maria Nicholson

64 Toledo Way, San Francisco, CA 94123

Email: ananicholson@aol.com

Abstract. In the Fall of '87 Rudie Berkhout and myself started a very intense and fruitful collaboration producing a series of holographic art pieces that were experimental but that reflected our different artistic sensibilities. The masters were made in my portrait studio in the Museum of Holography in New York using a pulse laser and later transferred in my Long Island City Studio. These pieces were shown at the Holocenter in 2009 and poignantly, it was the last show that Rudie had while he was alive. My paper details the process of an artistic collaboration, its pitfalls and advantages, its conflicts and compromises. It will illuminate the creative process that from two separate and very different streams melded into beautiful and evocative art.

1. The Beginnings

It began on a hot August afternoon. I was at the Museum of Holography on Mercer Street in New York setting up a portrait studio that was to open in conjunction with the "Holographic Instant " exhibition that Ian Lancaster, the director of the museum was organizing. JK lasers had just bought Lumonics and was anxious to extend their markets. They agreed to lend us a one joule pulse laser under the auspices of Holographics Inc. the company I was involved with, and I was to run it. There were delays in delivering it, and John Webster, an expert on lasers, had to go back to England the same day it arrived. Needless to say his tutoring was minimal to none and there I was trying to understand the laser and design and construct a camera and studio in the lower floor of the museum. It was a very stressful time since a high profile opening had been scheduled.

I was working on the laser when I looked up I saw this very handsome and charming man saying "Can I be of help?" It was Rudie whom I had not met before although of course I knew who he was. We

Fig. 1
Ana Maria and laser

talked and talked for hours and so began our friendship.

Fig. 2
Rudie

Rudie would come by on his bicycle every afternoon and we would spend hours at the local coffee shop on Canal Street discussing love, life and especially holographic art. Was it an oxymoron? What did it mean to be an artist; did the technology, the sheer difficulty of making a hologram affect the

Fig. 3
Bruce
Nauman
Hologram

Fig. 4
Dali
Hologram

outcome? Was a hologram a work of art if the artist had not made the hologram, but had hired a technician or a lab to do it? What is the best way to align the object beam?

Holography, being a young medium at this time lacked a cannon to follow or rebel against. There were hardly any art critics writing articles about it or art galleries showing holographic art.

2. An Art Medium Blossoms

There was a lot of chatter and publicity about holography then, but it was all about the possible technological uses of the medium. The most difficult problem that a holographic artist had to confront, aside from how hard it was to make a piece was the dreaded "Gee-Whiz" factor. On seeing a hologram the main reaction of the public was "Gee, how do you do it?"

Photography was early on embraced by sculptors and painters. In holography only a handful of established artists like Bruce Nauman, Dali, Agam and later Chuck Close tried it. Only James Turrell took it up consistently.

One topic that was close to my heart and kept coming up was what made a portrait a work of art. From the Roman Egyptian Fayum portraits to Chuck Close and Karsh the human face has had a great fascination for artists. Cicero said "a face is a picture of the mind."

Holographic portraits presented both a challenge and opportunity. The challenge is that the high fidelity of the recording shows every hair and wrinkle without the benefit of retouching. The opportunity is that that wealth of information can reveal more psychological depth that a photograph.

Fig. 5
Diane Dixon
Hologram

Fig. 6
Walter
Cronkite

I began doing portraits in Hawaii with my show "The Children of Hawaii" in 1973. I became fascinated by the sense of presence in the portraits, the sense that some essential quality of the person had been captured, almost another dimension of their being. I became quite good at it and made portraits of famous and not so famous people.

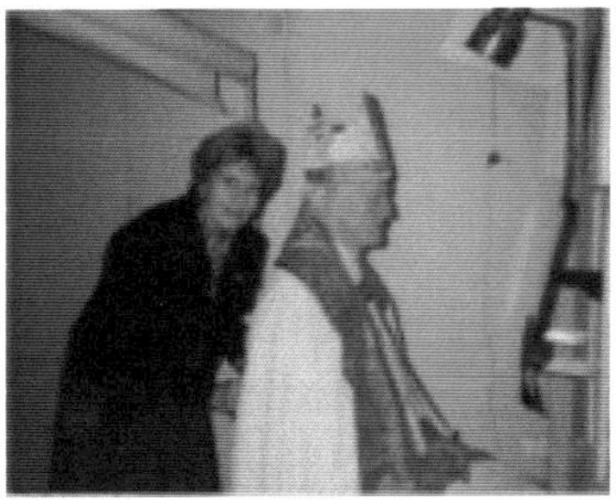

Fig. 7
Ana Maria &
Bishop Moore

Fig. 8
Bishop Moore
Hologram

This was a very exciting time Keith Haring sat for his portrait, as did numerous other notable people.

Fig. 9
Keith Haring
Hologram

Fig. 10
Keith Haring &
Ana Maria

CNN ran a whole segment, hosted by a novice Christiane Amanpour, about the process of making a portrait. There were magazine articles and a sense that portraits would become very popular, the price would drop and there would be portraits in every home.

This momentum gave rise to many, many discussions between Rudie and I. Is a portrait a work of art, are the intentions of the practitioner as important as the presence of the subject. Is the very

Fig. 11
Rudie

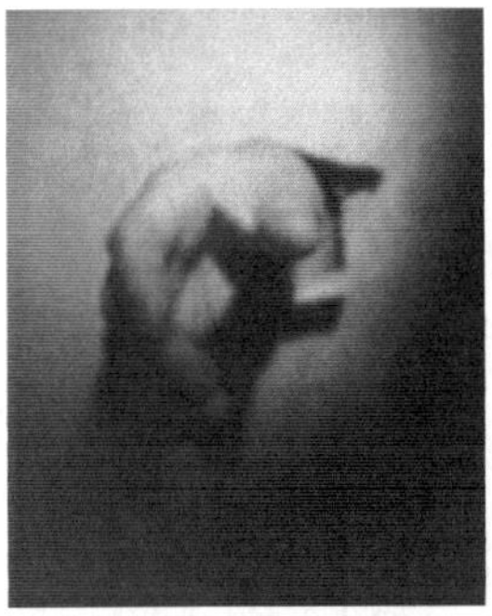

Fig. 12
Rudie/Hand

similitude that made the images so vivid and real a barrier against having an esthetic experience?

By the way, in case you are wondering, Rudie came up with the final edict - a work is art if you say it is art. Those of you who knew Rudie will not be surprised at this. He absolutely believed in the power of the individual to shape his reality and not depend on some expert to tell you what is what.

Fig. 13
Francis Bacon
Self-portrait

This is the great lesson that Rudie taught me and I am-still learning, to play, to be free. He had no respect for the establishment art world with its galleries and articles and rituals. He insisted that the process of creation should be fun, joyous, that we should always be explorers.

I began to experiment with some more elaborate portraits, such as the one of me with ropes and the diptych of Rudie where I wanted to express the conflict between the desire to trust and the necessity of self-protection that was such an ingrained aspect of Rudie's psyche. I placed him behind a wall looking out where he remains weary of the viewer. In the second hologram his hand breaks through the screen in his desire for contact.

I bring this piece up because it was our first collaboration. At that time I had begun to feel a dissatisfaction with the constrains of a holographic portrait. They were, on the whole monochromatic, their depth limited, the space ungenerous. I speak here of reflection holograms.I had gone to see an exhibition of Francis Bacon's portraits and I was overwhelmed by them. Those liquefying faces, tormented by the weight of the flesh haunted me. Rudie and I began our conversations on how best to break the mold. How to have portraits be more abstract, more pliable perhaps more universal.

We were using the pulse laser at the museum's portrait studio. The camera was functional but rather rigid. It had only one object beam illuminating the subject from above and the side.

Fig. 14
David Byrne

Fig. 15
Ana Maria
Self-portrait

We began to use only film and introduced irregularities on the platen behind the film. We got the most wonderful wobbly images that changed look as the viewer moved his head.

Even though the integrity of the image was broken by this method, creating very interesting distortions and effects, the results were haphazard and unpredictable.

To break up the face even more. We cut out different shapes out of pieces of cardboard and removed the diffusing screen from the object beam. This created a small, tight light source so that t only the parts of the face that were uncovered were illuminated. That is why, for safety, our eyes are closed in all the holograms.

Those were the masters, each with a different shape, different size, some recognizable as features some just abstract shapes.

We created the series in a matter of weeks, and then the work of assembling them into transfers began. I took the masters to our lab in Long Island City and some times combined two or more masters, sometimes just used a single image. At this point Rudie's input was greatly diminished. At that time he was not exploring reflection holography as he did later with such beauty. I was doing a lot of color combinations and the whole process of swelling and shrinking the emulsion with

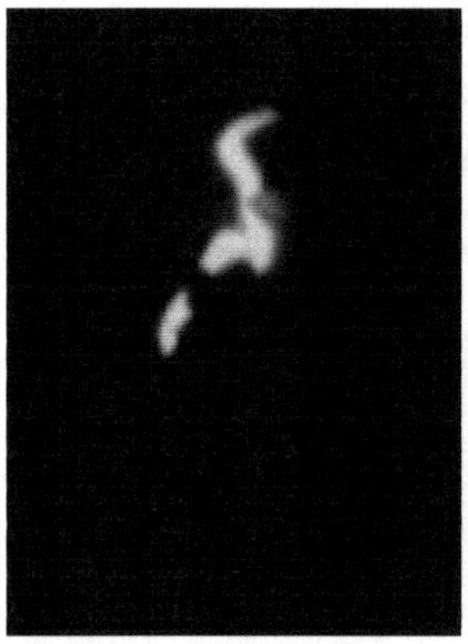

Fig. 16
Rudie
Fragment 1

Fig. 17
Rudie
Fragment 2

triethenolomine was too messy and primitive for him.

I, of course shared the results with him, but by that time our collaboration took on a different direction, more towards learning and practicing different healing modalities. I put the series away in a box and did nothing with them for a number of years. In fact I had forgotten how interesting they were.

One day, Martina Mngovrious and Sam Moree who were working at the Holocenter with me, found the box and were quite enthusiastic about the work. I took a second look, called Rudie and proposed that we have an exhibition.

This was the show HOLOGRAPHIC COLLABORATIONS that was shown at the Center in 2008. We had a splendid opening, the show was a great success, Rudie came down from his mountain for it. It makes me glad but saddens me that it was the last show of his work that our great friend and companion attended.

Journal of Physics: Conference Series **415** (2013) 012018 doi:10.1088/1742-6596/415/1/012018

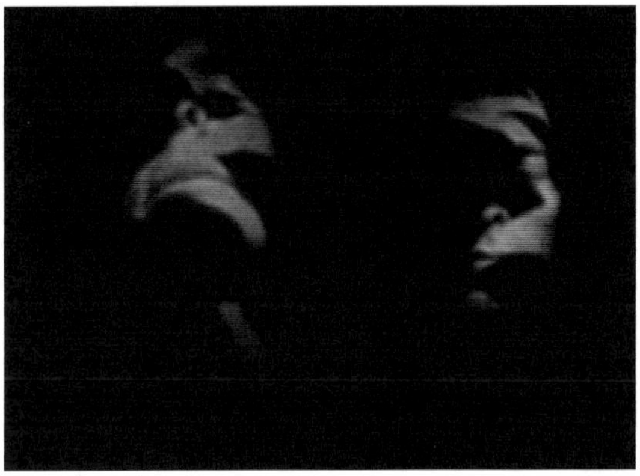

Fig. 18
Rudie & Ana Maria

Rudie's extraordinary work has been an inspiration, not only for myself but also for a generation of art holographers. Hopefully his efforts will be remembered as a key stepping stone in the flourishing of this magical medium.

9th International Symposium on Display Holography (ISDH 2012) IOP Publishing
Journal of Physics: Conference Series **415** (2013) 012019 doi:10.1088/1742-6596/415/1/012019

Coexistence

Ray Park

Korea Electronics Technology Institute, Seoul, South Korea

E-mail: rayhologramart@gmail.com

Abstract. This is a paper on the explanation of coexistence of visible world and invisible world, and the introduction of the world on 'Coexistence' work starting from a Cylindrical Hologram, 'Green Apple' in 2007 as its first work. It's a work improved from the existing way of creating Cylindrical Holograms in optical problems of the unseen lower part of an object by the shadow casting over it.[1] Many Cylindrical Holograms were made in the process of improvement such as Cylindrical Holograms of 360 degrees, of 180 degrees as well as Stereo Holograms and Color Holograms to create and express the existence of visible world and invisible world. Many Cylindrical Holograms were made in the process of improvement such as Cylindrical Holograms of 360degrees, of 180 degrees as well as Stereo Holograms and Color Holograms to create and express the existence of visible world and invisible world. Those created 'Coexistence' Holograms were exhibited 4 times through in group exhibitions and 2 times in private exhibitions at the Gallery Golmok, Itaewon, Seoul on May 2012.

1. Introduction

'Coexistence' means anything or phenomena of more than 2 of them existing together. Our lives are surrounded by Coexistence that it is easy to find in two or more of any things or phenomena existing together. Any visible things such as our possessions, animals, plants, and invisible things such as lights, air, scent, etc are apparent in their coexistence.

'Coexistence' is a word of vary arcane imagination and a word scientifically provable to me thru studies arts and engineering. 'Green Apple', a cylindrical hologram created in 2007 is the starting of my work on 'Coexistence'. I eat an apple every morning and the same kind of apples are always put on the same plate. Eating an apple leaves some seeds and a part of the apple, and yet an apple on the same plate is before me everyday showing the fact that the plate is coexisting with invisible apples. What I found out was the connection of the past and the present of the plate and the invisible apples for my intake of an apple every morning. That is, the existence of the plate is possible only when it coexist with an apple.

Apples Plane Holograms were made as a Coexistence work in 2007, but as the Fig.1, the work made it impossible to grab the image of holograms floating in the air unlike cylindrical holograms putting the meaning of 'Coexistence' in the shade, and so I became to create cylindrical holograms for the first time.

Making of Cylindrical Holograms of 'Green apple' was confronted with a difficulty, which was the impossibility of reappearing a perfect shape of the apple due to the shadow casting over the lower part of the apple.[2] For that reason, I had to improve the making method of existing Cylindrical Holographic Optics into an Apple Hologram perfect in seeing upper and lower parts of an apple. (figure 2) The know-how accumulated by the work on Cylindrical Holograms and research done for

Journal of Physics: Conference Series **415** (2013) 012019 doi:10.1088/1742-6596/415/1/012019

the first work led me to create many Cylindrical Holograms, and then to create the 'Green apple on a small plate' a follow-up work of the 'Green apple' in 2011 (figure 3).

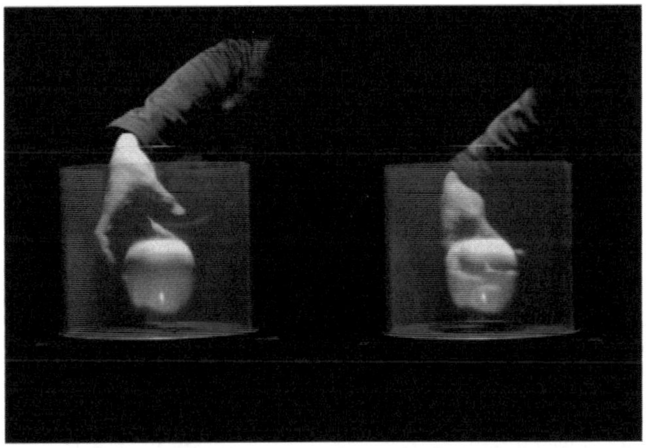

Figure 1. Green Apple (2007)

Figure 2. Green Apple (2007)

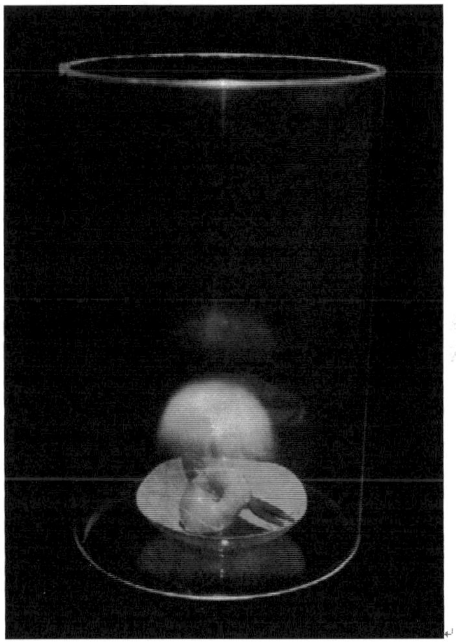

Figure 3. Green Apple on small plate (2011)

2. The Coexistence of Visible World and Invisible World

Everyone has a mobile phone in their hands to talk and send text messages with it. The phones in their hands before being used for contacts are the visible things. But the mobile phones are interacted by the help of invisible radio waves consisting of also invisible waves and particles. We can say this a 'Coexistence' of a visible things and an invisible thing.

It's our Coexistence that enables those mobile phones to work. Like this, there are the visible nature and artifacts together with invisible things on the earth. The existences of some of them are based on interaction. Only Coexistence makes their existences possible thus.

The existence of our 'bodies' ensures the existence of our 'invisible bodies'. The similar meaning of invisible body is 'a spirit'. The existence is possible by the 'Coexistence' of a visible body and an invisible body.

We are living in the world filled with invisible waves. Waves are unseen. The sun and the lights from the bulbs hanging from the ceiling -in other word, waves are invisible, but we can see things by their being.

Waves connecting a thing to another thing through mobile phone, radio and TV, and waves linking a life to another life are ponderable. We are able to see the holograms of the figures of things and lives recorded and restored by waves in the past, present and even in the future.

'Schrodinger's Cat' in the quantum theory discussing a scientific tangible world is a story of a cat whether it would die or survive in the sealed box with a poison in it. The cat's survival depends on the act of opening the box by an observer. It explains that the observing act affects the consequence. Then, if a sequence ahead could come out by the thinking and imagination done before the act, we will be able to see secondly the sequence reflected on visible world firstly by the influence of invisible world.

My work on Holography was the making of the outcome of another 'Coexistence' by way of turning the familiar routine things and pictures around us into Holograms. Those Holograms are showing the Coexistence of the past and the future through life and death, visible thins and invisible things, things to use and to throw away. Working on and exhibiting bulb Hologram and Cylindrical Hologram of 180 and 360 degrees making visible things and invisible things coexist, formed the consensus of 'Coexistence' of the things unseen and the things seen.

3. Fruits Cylindrical Holograms

The Fruits Cylindrical Holograms were made for the 'Picasso & Einstein 3.0' Group Exhibition, which dealt with Science in Art that held Sep 2011 at Seoul Arts Center. Some of natural fruits were turned into 360 degree of Cylindrical Holograms hung in the air to make them look as if they were the fruits on the trees in the nature. Uni-color LED lighting was used instead of laser lighting to restore the existing Cylindrical Holograms. The restoration of the Fruits Cylindrical Holograms into the pastel feel by using Uni-color LED lighting developed a feeling of spirits out of those objects. There were interesting performances by the spectators trying to pick the fruits floating in the air during the exhibition of the Fruits Cylindrical Holograms (figure 4).

4. Green Apple on small plate

The 'Apple on a Small Plate' is a follow-up work of 'Green Apple' which clearly shows the world of my work on 'Coexistence'. It's the work completed by constant supplementation and improvement in optical method of the Cylindrical Holograms that made firstly. This Cylindrical Holograms showed the expressions of the things seen and unseen at a time by disclosing them twice upper and lower parts. Shining the laser lighting from beneath made the broken apple and forks seen clearly, and the shining the LED lighting from above made the hallucination of the apple on a plate in the cylinder appear.

My will to express that started from the beginning of the work on 'Coexistence' was achieved through the application method of optics-based creation. My first 'Green Apple' was a work of a holographic apple on a small white plate aiming to express Holographic Coexistence of visible plate and invisible apple with the existing optics-based creation method, which ended up imperfect. The apple and forks shapes in the Cylindrical Hologram of the work of the 'Green Apple on a small plate' were seen clearly with the help of laser lighting after a research, and as the Figures 5 and 6 shows, they were seen from all directions, therefore 360 degrees looking as if they were actual broken apple on a plate in a cylinder. I could express the Coexistence of the past and the future by the help of LED lighting through the expression of the hallucination of the apple with visible apple and invisible apple on a plate and forks.

Figure 4. The scene in the exhibition hall (2011)

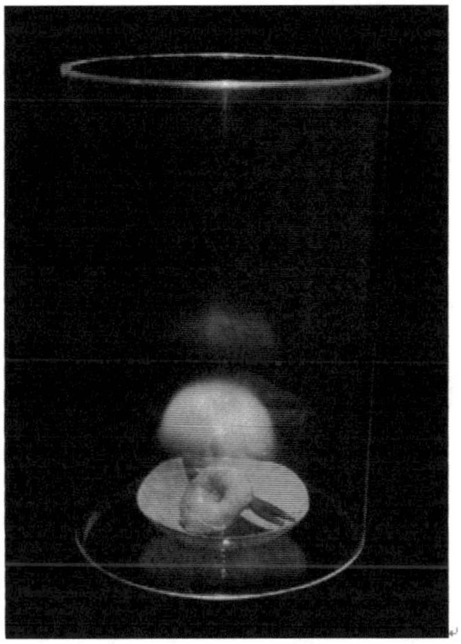

Figure 5. Green Apple on a small plate (2011)

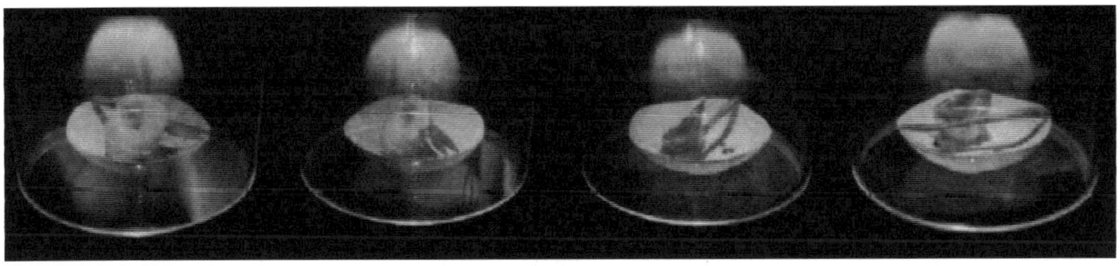

Figure 6. Green Apple on a small plates observed in 360-degree (2011)

Journal of Physics: Conference Series **415** (2013) 012019 doi:10.1088/1742-6596/415/1/012019

5. Exhibition of the Work on Coexistence

I held the 2nd private exhibition on May 2012 at Golmok Gallery at Itaewon, Seoul to show my 5 years works of 'Coexistence'. I put out 14 works of 'Coexistence' together with diverse Holograms including Stereo, Color Holograms beside Cylindrical Holograms.

Figure 7 is a 180- degree Hologram of 'Inspiration' expressed invisible inspiration into a hologram. The inspiration received from God is the greater one than any other ones people get. The writings in the Bible are the products of the most perfect inspiration from God. Visible men wrote those, but the truth is an invisible inspiration that made the physical hands write them. This work is the expression of invisible inspiration with the hologram of a hand, and the expression of writings on the visible paper with the hologram of a hand.

Figure 8 is a work of 180-degree Hologram of 'Coexistence with Death' which warns the people of today leading busy life against their insensibility to death. A hologram of a skull is staring at a man standing before it without knowing death is approaching him.

Figure 9 is a work of 180-degree Hologram of 'Coexistence of a visible banana and an invisible banana'. I kept the peel of a banana I ate on the opening day of the exhibition under my work of the banana hologram for 2 weeks. As time goes by, the banana feel was rotting down, but the hologram of a banana remains intact. It can be used as a comparison of human spirit and human body. It holds a message that human spirit remains intact because of its permanence while human physical body is rotting away.

Figure 7. Inspiration (2012)

Figure 8. Coexistence with Death (2012)

Figure 9. Coexistence – Banana (2011)

Figure 10. Coexistence - Bulbs (2012)

Figure 10 is a work of 180-degree Hologram of 'Bulbs' that expresses the Existence of a visible bulb and an invisible bulb. An exhausted bulb is disposed in the condition of broken and lost in its glow while the Hologram of a bulb remains intact emitting light. A visible bulb emits light by the electric power, but not forever, for it is a mortal thing. On the other hand, an invisible bulb can emits light by other lights. Like this, our physical bodies shine by fame and substance, which never last long after all. Unlike the foregoing, spirits would shine forever by the lights in the world of God.

Figures 11 and 12 are 3 works of Stereo Holograms of hectic routines of city dwellers. People crossing the street at a crosswalk were taken in continuous shooting mode with a camera and then used for the creation of the Stereo Holograms. If an observer moves horizontally to see them, the people crossing the street at a crosswalk would be seen. Green, yellow and red LED lightings were used respectively to express and then restore the people moving busily following the traffic lights.

Figure 11. Coexistence – City dwellers (Parts)

Figure 12. Coexistence – City dwellers (2009)

Figure 13. Coexistence – Perrier Bottles (Parts)

Figure 14. Coexistence – City dwellers (2012)

Figures 13 and 14 are the Color Holograms of Perrier Bottles, the contents of which I usually enjoy. The visual and invisible bottles moving from side to side behind the Hologram film being overlapped into one and create an illusion that actual Perrier bottles are behind the films. We can see the Coexistence of a visible thing and an invisible thing at the same moment when the Perrier bottles move from side to side are not overlapped with films.

6. Conclusions

The work on theme Coexistence was conducted in various ways with the view of Coexistence of 'visible world' and 'invisible world'. Holograms were fine tools that helped me in widening my vision on Coexistence of visible and invisible things, and they also brought me an opportunity to look into the invisible spiritual world. I will continuously make efforts for my next work on Digital Holograms to widen my vision on 'Painting Holograms' and 'Coexistence'.

References

[1] JS Park(Ray Park), JY Park, SS. Woo, YS. Moon, HJ. Kang, Cylindrical hologram using Silver-Halide Holographic film, *Workshop on the Optical Information Processing and Three-Dimensional Display Technology*, Gangneung, Korea, September, 2011.Another reference

[2] G Saxby, Practical holography, *Taylor & Franci*, 2004

The Haptic Lines of Homeland

P Dawson[1], M Takatsuka[2]

1. The College of Fine Arts, University of New South Wales, Paddington, NSW, 2021, Australia

2. ViSLAB, School of IT, University of Sydney, Sydney, NSW, 2006, Australia

Abstract.
This paper discusses the conceptual underpinnings, working processes and the tools used for preparing the scene files of a holographic art work which offers a subjective view point on the idea of homeland. The art work, *Homeland*, an optically formed fringe digital hologram, which is contextualized by the holographic maps used in situational awareness, indicates its subjectivity by strongly referencing the human body, particularly the lines of the palm of the hand.

...because the body belongs to the order of things as the world is universal flesh.

Maurice Merleau-Ponty

1. Background

The military applications of technologies have long been a part of the discourse of art practice engaged with these technologies, but until recently, such issues of military agency were not applicable to the field of holography. Now it is widely known that holograms are playing a role in homeland security. A description of the use of holograms from the Army Times article 12/3/2012, *Intel in 3-D Imaging technology provides new edge in preparing for battle* states:

> This three-dimensional representation of complex urban, nonurban terrain, subterranean bunkers, and other combat and non-combat environments aids the war fighter through providing high-resolution 3-D battlefield infrastructure intelligence ... this is critical for planning and executing combat and noncombat operations (1)

According to Lynn Schnurr the Army intelligence chief information officer and director of the intelligence community's information management directorate:

> ...so far, about 12,000 holographic images have been sent to soldiers in Iraq and Afghanistan...any unit preparing to deploy can ask for one to be custom-made for its area of operations. Run by the Army deputy chief of staff for intelligence (G-2), the Tactical Battlefield Visualization program provides soldiers with these holograms, made by Zebra Imaging in Austin, Texas. (1)

2. Conceptual Underpinning

Dora Apel's article, *Technologies of War, Media, and Dissent in the Post 9/11 Work of Krzysztof Wodiczko*, 2008 provides a view of some of Krzysztof Wodicko's artistic works which engage with the idea of homeland.

> In addition to its formal intertwining with media technology, the multiple environments of Wodiczko's installation demonstrate the infiltration of war technology through the trope of surveillance. Conceptually, the project addresses the effects of contemporary war culture and the incursion of the heightened power of the state into every kind of domestic or homeland space, creating a perceptual state of hyper-vigilance, in which the 'homeland' is always already mobilized for war. If military technology can be domesticated, the domestic also becomes militarized. (2)

These ideas about homeland provide a framing/context for the overarching experience of viewing Krzysztof Wodiczko's works, which is one of deep empathy and compassion for all people. Such an experience of empathy and care is also the objective of the *Homeland* art project. The *Homeland* art project puts on display in a museum /art gallery the way of looking associated with holographic Tactical Battle Visualization in order to propose homeland as a personal subjective domain. This is accomplished by shifting the 'area of operation' to the space and time of human life. The *Homeland* art project image was made using the same technology and is displayed horizontally in the gallery with the bunker purpose-designed rugged Grasshopper lighting stand with rotating turntable used by the military to closely examine holographic images. As the Army times article states:

> Using these holograms, soldiers can improve their understanding, retention and situational awareness of the area of operations... leaders reporting they used the holograms to better analyze, assess and determine different courses of action.(1)

Inviting this strategy of keen awareness in the art gallery viewer, the project visualises homeland as the terrain of traces of the hands of many humans. In a phenomenological sense this visualisation suggests homeland to be the traces of life, registered in the fabric of the body. In the words of the philosopher Maurice Merleau-Ponty:

> The thickness of the body, far from rivalling that of the world, is on the contrary the sole means I have to go unto the heart of things, by making myself a world and by making them flesh. (3)

3. Processes and Pictorial Agency

The ground of the *Homeland* art project image is green and looks similar to the military terrain holograms, which are monochrome green. However there are some significant differences. Describing the military scene preparation process Rick Black, a retired chief warrant officer 4 who is director of defense/intelligence programs for Zebra Imaging states:

> To make the holograms for the army Zebra imaging uses data from the army- classified and unclassified- along with open-source data, photographs, and light deception and ranging, or LIDAR imagery (1)

The 'terrain' of the *Homeland* art project is made in an entirely different way from an aggregate of 'lands', which were formed by directly casting silicone into the palms of several people's hands who were asked to imagine they were holding a piece of light. These positive shapes with their complex organic ridges and valleys are the materialised form of an invisible place, a personal and subjective 'homeland' which people carry with them everyday. In contrast to the significant reduction in scale and resolution of terrain elements in the military holograms, sufficient palm casts were made to cover the 60 x 60cm area to enable each of them to be life size. Each silicone cast was photographed and made into a 3D form using Agisoft Photoscan software and then the forms were assembled in Maya.

The prominent, spatially positive, path of the lifeline is obvious on each of these casts. The lifeline is unique in the repertory of lines encountered in the world in as much as it does not denote a physical edge or boundary of a three-dimensional form, rather it is a body marking which is integral to the body. Its contours and length have a questionable but long established association with life prediction, potential(4) and mortality.

5. Lines

Apart from the terrain itself, sometimes for clearer military pre visualization, 2D lines drawn onto acetate and overlaid on the hologram, are moved around e.g. a printed straight line used to indicate an approach path. When drawing is presented in a two-dimensional format, as the beholder moves through space no new visual information becomes available. For this reason a two-dimensional drawing enables specific types of representations of space such as the use of line to enclose discrete spaces by marking their boundary and illusions of depth from projection systems such as perspective.

In the *Homeland* art project the quality and placement of lines made by hand and placed in three-dimensional space extends the expressive potential of hand made marks. Once the viewer or the hologram is moving and the drawing is distributed throughout a spatial volume, neither of these familiar ways of representing space with line on a two dimensional plane are appropriate. This is because the mobility of the beholders line of sight makes available numerous possibilities of composition from the 3D line relationships. As Emma Dexter writes in her introductory essay to *Vitamin D: New Perspectives in Drawing*:

> Drawing exists at another level within the human psyche it is a locus for signs by which we map the physical world, but it is itself the pre-eminent sign of being. Therefore drawing is not a window on the world but a device for understanding our place in the universe.(5)

Hand drawn line is used in the *Homeland* art project to trace the potent subject of the lifeline. Using the haptic Phantom interface and Holoshop software, lines were drawn by feeling the paths of the lifelines on the three-dimensional templates of the palm casts.

6. Tool Development

A major pictorial objective of Holoshop is the assignment of appropriate visual characteristics to gestures. The Phantom device is used to capture the gestures made by the artists. It is a haptic device, which allows a user to input three-dimensional (3D) spatial movement through a pen-like grip. The grip is attached to a mechanical arm with 6 degrees of freedom, hence it is capable of recording all 3D movements of the user. The device is also capable of generating a synthetic force field based on the computer generated 3D graphics. Various types of synthetic force fields (such as a magnetic field) can be programmablly generated based on the 3D geometries.

For artists to fully express their 3D spatial movement using this device the Holoshop software needs to be able to capture various physical characteristics of those movements through adjusting the Phantom device's mechanical parameters; they are angular and linear tolerance, force, stiffness, and damping parameters. These parameters will determine the physical response of the Phantom device when it interacts with the 3D virtual objects. Those parameters also influence how much detail of artist's movements is captured. Hence, these parameters serve two purposes: 1) adjust the levels of detail of the artist's movement, and 2) adjust the tactile feedback. While these parameters, which are directly associated with the Phantom device, play a certain role in determining the range of expression lines can convey, there are other factors, which have significant impact on how artist's movement is expressed. They are 1) an encoding method of the speed of the artist's hand and 2) computational constraints applied to the movement of the Phantom device.

6.1 Velocity Mode. In order to truthfully capture the characteristics expressed in the form of speed of the artist's hand, Holoshop software has implemented methods to control the thickness (width) of the drawn line depending on the speed of the hand movement. This method transforms a velocity value of the device into a width of a line using a sigmoid function, whose minimum and maximum values correspond to the minimum and maximum width of lines. The normal velocity mode will apply a normal sigmoid function to the velocity so that the faster the speed of the movement, the thicker the line becomes. The reverse velocity mode allows a user to produce a thicker line when the speed is slow and a thinner line when the speed is fast.

6.2 Magnetism Mode. Another factor, which significantly influences how the artist can efficiently work in the 3D environment, is the provision of spatial constraints on the haptic device. Without any spatial constraints, an artist will be able to freely move the device in the 3D space to draw lines. However, if the lines need to be drawn with respect to other geometry in the same space, having the 6DOF device often provides too much free movement. This usually results in the user not being able to have total control in determining the location of the haptic device in the 3D space. In order to reduce this inconvenience, Holoshop software exploited the virtual magnetic field to constrain the movement of the haptic device. When this virtual magnetic field is turned on, the space where the tip of the haptic grip can move around is restricted to the nearby surfaces. This function is particularly useful if the movement of the artist's hand needs to be guided by the underlying geometry template.

With the provision of these parameter controls and different line drawing modes to capture the spatial movement of the haptic device, the changes of momentum caused by the topography of the contours and the speed of the gesture were registered in the varying width and orientation of the lines. The lines are flat, red and ribbon like, referring to approach path planning lines yet they strongly evidence the inflections from changes in direction and speed familiar from direct hand made marks.

Journal of Physics: Conference Series **415** (2013) 012020 doi:10.1088/1742-6596/415/1/012020

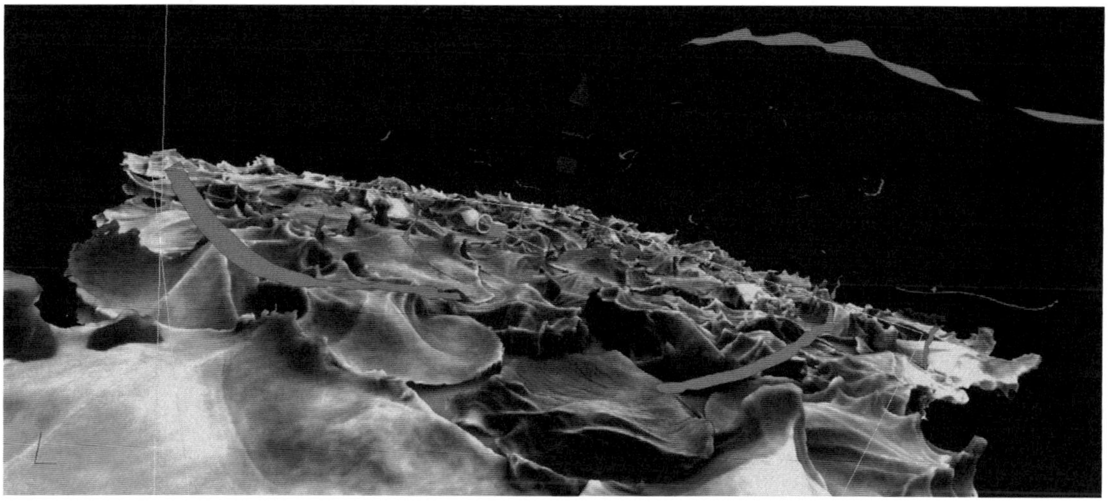

7. Composition of Lines

The red colour of the lifelines is symbolic of the luminous presence of life. (6)The lines were translated at varying heights, above the lifelines templates from which they were formed. When viewed directly from above it is possible to superimpose only one line at a time with its lifeline. As the viewer's line of sight changes, the amount and types of occlusion between lines at different levels obscure or open sections of the terrain.

The depth of Tactical Battle Visualization holograms is intentionally restricted to enable the subject matter to appear sharp. Generally lines superimposed are 2D and on the hologram picture plane. The red 3D contoured lines of the *Homeland* art project are translated at various distances from the holographic plate so they range from sharp to blurred. This is contrary to the perception of real space, in which sharpness and colour fade with distance. Similar to the use of low resolution in the works of Jim Campbell(7), the sliding scale of blur in the Homeland art project image acts as a metaphor for the distant in its objectification to be gradually dissolved into intangible softness with closeness. As the red *Homeland* lines become less opaque the green background shows through reminiscent of green under-painting lending vibrancy to painted flesh.

These dissolving intertwined lines conflate the rational purpose of line as intension with the predictive association of the lifeline and destiny and invite the viewer to reach out and touch. As Dianna Petherbridge notes in the Primacy of Drawing: Histories and Theories of Practice:

> The linear paths that the spectator/interpreter directly perceives or infers in the drawing constitute cognitive mapping. And such readings are inseparable from the affective response to the gestural traces of the hand and the echo of the body, as well as the expressivity of the topic or subject matter of the drawing – or its absence. The difference between these readings constitutes the surplus of the drawing within which meaning is constituted.(8)

8. Conclusion

The expressive potential of drawing is broadened by the availability of drawing tools, which enable the free-hand inscription of space with line. Holoshop software which in conjunction with a haptic interface (Phantom Omni) to feel virtual contoured surfaces provides a means of using forces to modulate line quality through velocity, damping and friction. Future work in Holoshop will focus on expanding the scope and understanding of haptic, hand made, three-dimensional drawing for holograms and other three-dimensional displays. The use of Holoshop software and ZScape™

technology developed by Zebra Imaging enabled a subjective and personal visualisation of homeland, which unites people in an awareness of the mysterious and precious nature of life.

The views expressed in this paper and the art work *Homeland* are those of Paula Dawson and are not necessarily shared by the friends, colleagues, loved ones and institutions involved.

Acknowledgements

This project was made possible with the generous contribution of collaborators Graham Olsen, Masahiro Takatsuka, Daniel Barry, Benjamin Sutas, Hiroshi Yoshikawa, Brian Rogers and Zebra Imaging.

Thanks and appreciation go to each of the people who kindly gave their time and the cast of their palm: Virginia Aghan, Mike Allen, Daniel Barry, Maria Baveas, Jill Bennett, Margaret Benyon, Sue Bignell, Georgie Carnegie, George Dawson, Heatley Dawson, Clancy Dawson, Georgie Dawson, Andrew Dzurak, John Gage, Fiona Hall, Ross Harley, Melanie Harris, Julie Hamann, Ian Howard, Fay Hudson, Zoe Little, Kevin Murray, Craig Newswanger, Claire Parker, Eugenio Papino, Maureen Pitts, Edi Pucci, Ron Ramsey, Sylvia Ross, Sue Rowley, Lindsay Shurey, Ben Sutas, Joanna Szymanska, David Tai, Masa Takatsuka ,Sally Weber, Hiroshi Yoshikawa

This research was supported under Australian Research Council's *Discovery Projects* funding scheme (DP1094613).

Australian Government

Australian Research Council

[1] Tan M. 3-D maps provide new edge in battle preparation. 2012; Available from: http://www.armytimes.com/news/2012/03/army-3d-map-provides-edge-preparing-for-battle-030612/.

[2] Apel D. Technologies of War, Media, and Dissent in the Post 9/11 Work of Krzysztof Wodiczko. Oxford Art Journal. 2008;31(2):261-80.

[3] Fañanás L, Rosa A, Gutiérrez B, Guerra A, Arias B, Fañanás L. Dermatoglyphics and abnormal palmar flexion creases as markers of early prenatal stress in children with idiopathic intellectual disability. Journal of Intellectual Disability Research. 2001;45(5):416-23.

[4] Gage J. Colour and Meaning Art, Science and Symbolism. London: Thames and Hudson; 1999.

[5] Dexter E. Vitamin D: New Perspectives in Drawing. New York: Phaidon; 2005.

[6] Merleau-Ponty M. The Visible and the Invisible. Evanston: Nothwestern University Press; 1968.

[7] Plohman A. Explorations of Meaning in Quantized Information. 2002 [cited 2012 11 June 2012] http://www.fondation-langlois.org/html/e/page.php?NumPage=107

[8] Petherbridge D. The Primacy of Drawing: Histories and Theories of Practice. New Haven and London: Yale University Press; 2010.

Drawing Lines with Light in Holographic Space

Yin-Ren Chang
Imaging & Displays Research Group, Faculty of Technology, De Montfort University
1.32d Room, Queens Building, Gateway, Leicester LE1 9BH, UK
E-mail: yinrenchang@yahoo.com.tw

Martin Richardson
2.27 Room, Queens Building, Gateway, Leicester LE1 9BH, UK
E-mail: mrichardson@dmu.ac.uk

Abstract. This paper explores the dynamic and expressive possibilities of holographic art through a comparison of art history and technical media such as photography, film and holographic technologies. Examples of modern art and creative expression of time and motions are examined using the early 20th century art movement, Cubism, where subjects are portrayed to be seen simultaneously from different angles. Folding space is represented as subject matter as it can depict space from multiple points of time. The paper also investigates the way holographic art has explored time and space. The lenticular lens-based media reveal a more subjective poetic art in the form of the lyrical images and messages as spectators pass through time, or walk along with the piece of work through an interactive process. It is argued that photographic practice is another example of artistic representation in the form of aesthetic medium of time movement and as such shares a common ground with other dynamic expression that require time based interaction.

Keywords: Holographic Art, Cubism, Motions.

1. Introduction

The theme of time has dominated aesthetic consideration in fine art. This is of course invisible but it sets boundaries and the ultimate challenge for artists is in developing, depicting and exploring methods to visualise quantifying rates of its change, not only as a visual abstract concept, but to also as the narrative of human condition.

The invention of photography made visual statements became more accessible for ordinary people; tourists can take pictures to illustrate vivid experiences from their point of view. Light is a compulsory element to create an image exposure on a frame of film, the developing process could be regarded as a novel technique that reveals very personal artistic perspectives, reproducing flakes of reality as the image creator experiences the world and time.

Stereographic based practices such as anaglyph, lenticular and holography offer a new dimension of visual art. That expedition into new forms of expression and aesthetic effects in time and space constitutes one of the most fascinating adventures in the history of art. This paper will trace the links between cubism and photographic artwork and then connect these to the theme of time and movement. This will involve an examination of depicting methodologies and representations.

This Research aims to explore the interaction between space, time, movement and visual perception by using three-dimensional representation tools such as anaglyph, lenticular and holographic technologies. It identifies potential associations between spatial imaging, perception, artistic interpretation and interaction. These practices will be a key for gaining an in-depth understanding to develop knowledge within the context of art.

2. Depicting Time

Following the advent of new technologies since the Industrial Revolution, people found that they could reach and connect the world quicker and broader than before. The invention of photography offered opportunities to make visual recordings more available and accessible for ordinary people. The concept of emerging techniques has had a tendency to lead artists into looking for new ways of communication in visual languages, similar to other new technologies that were crossing the geographical restrictions and bring people to explore diversity through world.

In the early of 20th-century, cubism launched its distinguished abstract style of modern art and introduced new ways of seeing. Cubists questioned the belief that art should be reproduced from natural and also challenged the traditional techniques such as perspective, light shadow, proportion, modelling and foreshortening.

This new way of seeing required a fresh concept of artistic representation. Cubist painters reduced the detailed description of objects and fractured the natural forms into semi-geometric shapes. They depict the simplified object from a multiple or contrasting viewpoints or fold space in this way they can portray the object at different angles and motions at the same time, according to the Heilbrunn Timeline of Art History [1]. The satisfactory results showed on their 2-D canvas, a single object can be seen from multiple perspectives in time.

Cubist painting responded to the innovative technology development; also captured this era's unique characteristics and transform it into a novel visual language. The artistic expression in painting and photography showed unexpected similarities between the manual visual statement and the mechanical visual statement. In near-monochromatic colour tone (e.g. browns, greys or blacks), objects be transformed to a series of overlapping plans, and the iconic artistic expression not only to represent cubist's aesthetic concept, but also could be found the analogous features of photographic works such as Muybridge's photographic works.

Eadweard Muybridge (1830-1904), made pioneering experiments, photographing motion which used multiple cameras and improved the development of the shutter to capture motion in stop-action photographs. Previous failed work was attributed to the lack of a fast shutter. "Woman walking downstairs", one of his remarkable works, viewers could distinguish the whole activity as separate movements through each frame. After this cooperated research programme with University of Pennsylvania, a portfolio "Animal Locomotion" has been published in 1887.

Muybridge's experimental photographic studies are recognised as an influential development on artistic creative practices and captured body movements by representing consecutive still images. These black and white photographs influenced Cubist's, who tried to simplify the colour description, in order to emphasise the fracturing geometrical objects in motions or seen from multitude viewpoints [2].

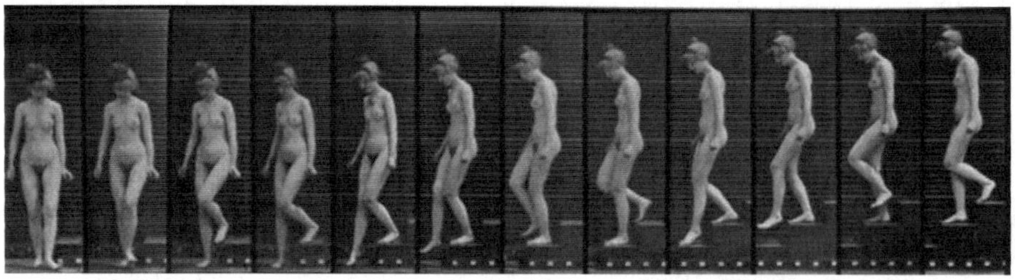

Figure 1-"Woman walking downstairs" (part), 1887, Eadweard Muybridge.

Let's compare it to 1912, Marcel Duchamp presented his distinguished oil painting of perpetual motion, "Nude Descending a Staircase (No. 2)". He responded Muybridge's photographic works, depicted fracturing body forms through successive images of movement in near-monochromatic gold and beige tones to represent a nude female in motions. This mechanistic style has successful signified the idea of visualised time movement, also showing the inspirational links with Muybridge's photographs and films on the analogy of visual narration, according to the website of Metropolitan Museum of Art [3].

Figure 2-"Nude Descending a Staircase (No. 2)" (left), 1912, Marcel Duchamp.
Figure 3-"A nude descends a staircase" (right), 1942, Gjon Mili.

A communication has been created between oil painting and photography, each medium has its unique artistic representation, but also expanding the potential possibility after blending their visual languages. Through Duchamp and Muybridge's work, Gjon Mili has developed a series of investigational works to successfully capture a sequence of action in one photograph by using stroboscopic instruments [4].

Following this emerging unique technology, holography made a strong dynamic visual statement to illustrate the movement. Jacques Desbiens's computer-generated hologram, "The Broken Window" (2006), the motion of a scroll painting from unrolling to expanding; a branch breaks through the window from behind, which brings allows the viewer to experience a new visual effect when they move from left to right, depicting a new narrative possibility connecting time and space [5].

Figure 4 -"The Broken Window", 2006, Jacques Desbiens.

3. Holographic artistic expression

Dieter Jung (2006) has described holographic art as being based on a technical medium, but its diversity and attractive vitality catch the attention of artists, leading them to transform vivid experimental imaginations into the "real virtual space", "shimmering with reflections of truth, fiction and fantasy" [6].

"A new sense of the notion of information has been constructed around the photographic image. The photograph is a thin slice of space as well as time", stated by Susan Sontag (1977) [7]. Compared with painting and photography the natural characteristics of medium, holographic technologies, these three-dimensional image recording mediums could offer multiple dimensions in 2-D materials for artistic representation as well as bring profound experience to audiences through its interactive display.

Setsuko Ishii (2006) has suggested that holographic art demands new forms of artistic expression and also has classified the characteristics of visual effects into three categories: "realistic three-dimensional image reconstruction", "viewing area constraints" and "colour of light". Based on these potential competences and relating the findings of artistic interpretation in cubism and photography, a series of creative artworks, which explore the dynamic, expressive, possibility of holographic art, as well as demonstrating the potential of spatial imaging as applied to the pictorial image [8].

3.1 Lenticular in art practice with still life

Figure 6- a) "Fly away".

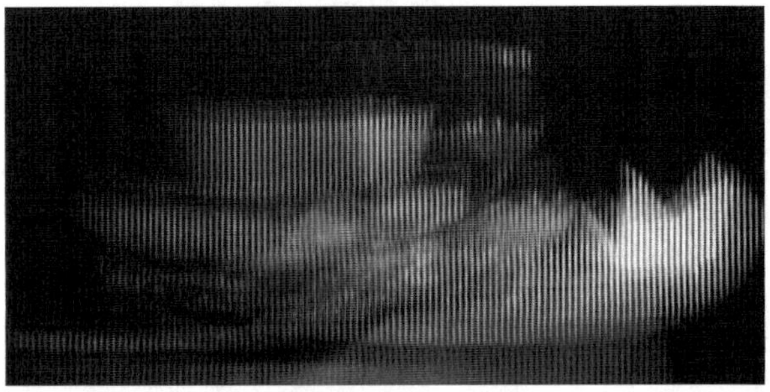

b) Enlarged details view.

Figure 6-"Fly away", 2011. a) Main part, which gives an idea about the working environment in which it was created; b) Enlarged details from the original source, a very strong sense of the semi-fracturing movement continues with the construct.

One of the outstanding characters of lenticulars is the representation of the three-dimensional image. This studio shoot was set-up with a digital camera, artificial lighting, tripod and rail. A sequence of digital photographic shots of the sill life has been made utilizing the movement of the rail system, capturing a multitude of viewpoints. Twenty-five frames have been collected, after the process of interlacing and lamination a view through the lens represents the image of three-dimensional reconstruction.

It is interesting to observe that the interlaced image has delivered a strong graphic statement as it displays without lens. A numbers of semi-fracturing objects form from different perspective when seen through consecutive recordings, these familiar narrative methods could relate to cubism artworks. Moreover, this visual effect has not come from artists' fantasy, that visual effect is the resultant form of computer graphic processing.

3.2 Lenticular in art practice with human motions

Original recorded (Fig. 7) by digital camera filming and editing before processing in Photoshop to create multiple layers. These consecutive frames show a man in motion. This process before interlacing is a vital creative step to check motion and flow. An excited finding shows the connection between photographic techniques and computer software, delivering movement within the still image.

Figure 7-"Passing" displayed in sequential format.

Passing (Fig. 8) was generated from 25 frames; a man just walking past the camera. However, his figure has vanished after the interlacing process because of fast movement. This unexpected result strongly points out that "time" is still is a key factor of image recording. The relation of motion speed and camera shutter could be defined as a key factor as the interplay between light and shadow.

Figure 8-"Passing" displayed in interlaced format.

3.3 Anaglyph in art practice

Anaglyph offers stereoscopic (three-dimensional) effect by focusing each eye's image through applying filters of different colours, usually chromatically opposite (such as red and cyan.) It is explained that Red-Cyan glasses became very popular as 3-D movies became more and more accessible. However, what could be seen in the image as the viewer wears the glasses in a wrong way (right eye in red; left eye in cyan)? Can people visit an exhibition, but perceive two versions at the same time? To imagine it, people wear a pair of red glasses to visit an exhibition, and turning back re-visit the same images but only wearing cyan glasses that is fascinating! In "Good evening" (Fig. 9) there are two versions, viewers need to try on the glasses in both ways, the background may confuse viewers by its three-dimensional array paradox (pseudoscopic view).

Figure 9-"Good evening".

Flip canvas (Fig. 10) refers to the concept of a multitude of viewpoints and traditional perspective law, using a mirror effect to recreate a composition; vanishing points locate the lower centre of this picture. The first glance with naked eyes, a convincing space was rebuilt. However, the reality only existed under the red-cyan world; people can see truth behind the sense through the specialised glasses when closing each eye alternatively.

Figure 10-"Flip canvas".

4. Summary

Moholy-Nagy identified photography is a form of artistic expression, which shared a similar tendency with other creative forms. The main influential factors are dependent on "present technical, scientific sociological trends and their relationships. As these relationships are not obvious to everyone it will be necessary to make an analysis of this statement and show by examples what its meaning is" [9]. It shows the importance of understanding the potentialities of materials, in order to distinguish their unique characteristics of being and how artists reflect their perspective of contemporary tendency by using such medium of artistic expression.

"A space inside other space, a shape inside other shape" says Isabel Azevedo (2012) [10]. In holographic arts practitioner possess an extraordinary instrument for reproduction/recreation. Art can be perceived by exploring multiple layers of meanings. Experiencing and interacting dynamic three-dimensional artwork will help the audience to experience the full concept of an artwork.

Reference

[1] REWALD, S (2004) *Cubism* [WWW] The Metropolitan Museum of Art. Available from: http://www.metmuseum.org/toah/hd/cube/hd_cube.htm [Accessed 15/06/2012].

[2] MUYBRIDGE, EADWEAR (1955) *The human figure in motion*. New York: Dover Publications Inc.

[3] REWALD, S (2004) *Marcel Duchamp (1887–1968)* [WWW] The Metropolitan Museum of Art. Available from: http://www.metmuseum.org/toah/hd/duch/hd_duch.htm [Accessed 15/06/2012].

[4] LIFE (2012) *It's About Time: Gjon Mili's Stroboscopic Portraits* [WWW] Life. Available from: http://life.time.com/photographers/its-about-time-gjon-milis-stroboscopic-portraits/ [Accessed 15/06/2012].

[5] DESBIENS, J. (2009) *The perspectives of synthetic holography* [WWW] I. Jacques. Available from: http://www.i-jacques.com/Text/Perspectives%20of%20synthetic%20holography-JacquesDesbiens-ISDH09.pdf [Accessed 15/06/2012].

[6] JUNG, D. (2006) As I see it-The absence of darkness. In: *Proceedings of the 7th International Symposium on Display Holography*, Wales, July 2006. London: River Valley Technologies, pp. 120-122.

[7] SONTAG, S. (1979) *On photography*. London: erquin Book.

[8] ISHII. S. (2006) *Artistic Representation with Holography* [WWW] Scipress. Available from: http://www.scipress.org/journals/forma/pdf/2101/21010081.pdf [Accessed 15/06/2012].

[9] KOSTELANETZ, R. and MOHOLY-NAGY. L. (eds.) (1970) *Moholy-Nagy*. London: Allen Lane.

[10] AZEVEDO, I.,RICHARDSON, M. and BERNARDO, L.M. (2012) Across Light: Through Colour. In: *Proceedings of the SPIE Photonics West* , San Francisco, January 2012.

Image

Figure 1 MUYBRIDGE, E. (1887) *Woman walking down stairs.* [Photogaphy]. In: MUYBRIDGE, E. (1955). *The human figure in motion*. New York: Dover Publications Inc.

Figure 2 THE METROPOLITAN MUSEUM OF ART (2004) *Nude Descending a Staircase (No. 2)* [Online image]. Available from: http://www.metmuseum.org/toah/hd/duch/hd_duch.htm

Figure 3 LIFE (1942) *A nude descends a staircase* [Online image]. Available from: http://life.time.com/photographers/its-about-time-gjon-milis-stroboscopic-portraits/#7

Figure 4 I.JACQUES (2006) *The Broken Window* [Online image]. Available from: http://www.i-jacques.com/Text/Perspectives%20of%20synthetic%20holography-JacquesDesbiens-ISDH09.pdf

Moving through a multiplex holographic scene

Martina Mrongovius

Center for the Holographic Arts, 39-31 29th Street, Long Island City, NY 11101, USA
Academy of Media Arts (KHM), Peter-Welter-Platz 2, Cologne 50676, Germany

martina@holographics.com.au

Abstract. This paper explores how movement can be used as a compositional element in installations of multiplex holograms. My holographic images are created from montages of hand-held video and photo-sequences. These spatially dynamic compositions are visually complex but anchored to landmarks and hints of the capturing process – such as the appearance of the photographer's shadow – to establish a sense of connection to the holographic scene. Moving around in front of the hologram, the viewer animates the holographic scene. A perception of motion then results from the viewer's bodily awareness of physical motion and the visual reading of dynamics within the scene or movement of perspective through a virtual suggestion of space. By linking and transforming the physical motion of the viewer with the visual animation, the viewer's bodily awareness – including proprioception, balance and orientation – play into the holographic composition. How multiplex holography can be a tool for exploring coupled, cross-referenced and transformed perceptions of movement is demonstrated with a number of holographic image installations. Through this process I expanded my creative composition practice to consider how dynamic and spatial scenes can be conveyed through the fragmented view of a multiplex hologram. This body of work was developed through an installation art practice and was the basis of my recently completed doctoral thesis: *'The Emergent Holographic Scene — compositions of movement and affect using multiplex holographic images'*.

1. Introduction

A culture of viewing holograms is not well developed. Viewing a hologram is a different activity from that of looking at sculptures, screen images and surface pictures. With a hologram what is seen depends on from where the viewer looks, the hologram being a kind of virtual window through which a scene is recorded. As the viewer's movement is coupled to the recorded visual perspective the viewer's sense of space and motion extends into the perception of the holographic scene, giving the scene spatial and/or animate qualities

With a multiplex hologram the virtual window through which the viewer can look around is fragmented, allowing for different perspectives to be mapped into a single scene. I began creating multiplex holograms by hand, using stencils to record stop-motion animations of physical scenes and objects. This work developed into a practice of using sequences of digital images to produce holographic scenes from 200 to 1800 related views. The multiplex holographic recordings were used to explore perceptual dynamics within urban experiences, by mapping the relationships that shaped my sense of place. Each hologram was a composition of re-arrangements and superimpositions of different related perspectives to create a spatially dynamic holographic scene.

With the installations of these holograms I experiment with how the imagery could be unfolded into physical spaces and how viewers tended to move through that space. The main design question revolved around how to establish and utilise relationships between the viewing experience and the spatially-dynamic visual perspective mapped into the scene. To produce an affective relationship or augmentation of motion the holograms would have ideally been larger than the viewer's body – so that the viewer 'looks around' the scene by moving with their whole body. As the cost of producing large-scale[1] holograms was outside my research budget, I explored ways to amplify the viewers' bodily awareness with my installations. The techniques I developed induced a bodily awareness by limiting some aspects of movement while provoking others – a kind of choreography aimed at intensifying the virtual movement in the scene. These techniques involved:

- Confinement to guide the viewer into the viewing field of the hologram and amplify their sense of movement when exploring the scene.
- Heightening an awareness of gravity, to intensify a sense of balance and proprioception.
- Employing armatures to elicit a particular action.
- Playing on the inherent movement through architectural spaces.

These techniques also informed the type of places I chose to capture as well as the way I moved and directed the camera while making the recordings.

2. The holographic scene

The continuity of visual perception experienced when looking at a *trompe l'oeil* painting from its encoded perspective point or when looking around a classic holographic scene of a singular virtual volume allows a sense of space to extend through the image surface. The holographic image however has an additional sense of presence that is physically felt as the visual scene is intrinsically linked to the viewer's movement. Artist Paula Dawson describes laser-viewable transmission holograms[2] as 'concrete'[1] holographic images because of the way they appear to have a physical presence. An example of this kind of holographic image is Dawson's triptych installation of large-format holograms, *'To Absent Friends'* (1989), which captures a bar at three times during a New Year's Eve party. The holograms form a crystallized memory, three frozen moments of a scene that the viewer can peer around.

Such a visually realistic virtual volume can also be created with multiplex and digital holograms by capturing the scene with a matching camera perspective to that adopted by the viewer. Jacques Desbiens describes such holograms as synthetic holograms as they produce an "illusion of volumes and presence"[2]. As David Pizzanelli outlines in his PhD thesis *'Aspects of Spatial and Temporal Parallax in Multiplex Holograms'*[3] the multiplex hologram allows for the recording and perception of motion as well as volume. The holograms that accompanied Pizzanelli's thesis follow the tradition of chronophotography, using the hologram to bring together a sequence of frames to represent continuous movement and space. By contrast I draw together different perspectives to produce non-linear structures and virtual movements that never existed physically.

I am interested in how virtual motions can be established by the holographic scene – how a dynamic perspective can be captured with a holographic image. To do this, my compositions often trace the perspective from a mobile body – such as the urban photographer; or are assembled from multiple different yet related perspectives – such as a collective view through multiple cameras.

3. Crossing into the holographic scene

The hologram allows a 'peering into and around' another space. This is like looking through a keyhole or from an aeroplane because the viewing window is much smaller than both the space in

---

[1] A number of installations were produced with multiple holograms or by tiling multiple pieces of film together. However for the most part the compositions were single 30 x 40 cm panels.

[2] Such holograms are also know as 'masters' as their image can be reconstructed by laser-light and transferred into a 'holographic print'.

which the body is situated and the space beyond the window. The extension of spatial perception through the holographic window is particularly apparent when looking into a laser-viewable transmission hologram as the holographic image can be very deep. As such, these images encourage a sense of an 'over there' that is distinct from 'here', where the viewer's body is.

Holographer Dan Schweitzer, whose background was in theatre and film, made an analogy between the hologram and stage, describing both as spaces where there is a border – a 'space barrier' – between the observer and scene. This 'space barrier' delimits the accessible and the inaccessible, perceived 'virtual' space. Working in a small scale, Schweitzer and his studio partner Sam Moree incorporated miniature figurines into their compositions. In the catalogue of *'Mehr Licht'* Achim Lipp describes Schweitzer's figurines:

> "At the same time they draw attention to themselves and prepare the observer for the appearance of the holographic image"[4].

The figures are a kind of visual protagonist, establishing the scene, and an avatar that suggests a perspective within the holographic scene. By adopting the suggested act of viewing within the scene the viewer imaginatively crosses the 'space barrier'.

For the exhibition *'Explorations of the holographic gaze'* at Gallery 175 in Seoul (2010), I brought together 16 holograms that visually depict the photographic act to emphasise the construction of the multiple-perspective scene. The photographer who appears in these holographic images is a protagonist and an avatar who simultaneously acts as the subject and director of the image. The shared activity of looking – between the construction and viewing – shapes and animates the scene.

In several of the holographic images, including *Shadow-waves at Safety Beach*, the photographer is implied by their shadow. The shadow's movement and view is adopted by the viewer.

Shadow-waves at Safety Beach, 2008/9 200-farme multiplex hologram laminated to laser-etched mirror 40 x 50 cm
Exhibited *'Explorations of the holographic gaze'* (Left) and *'The Emergent Holographic Scene* (Right)

The sequence of views in *Museum Reflection* (2006/9)[3] are also anchored to a shadow of the photographer, who in this case is looking into the Melbourne Museum through its glass façade. There are two references that establish the holographic scene *Museum Reflection*: the animated act of photographing as shown by the photographer's silhouette on the reflective glass surface; and a spatial depth resulting from the parallax between the regularly spaced sequence of seated positions along a concrete block in front of the Museum. Each photographic view optically combines the silhouetted reflection of a building with the view through the glass into a courtyard and reflected again. With the second reflection the building's spatial form shifts about because of the differences in orientation between the glass panels to produce a gentle cubism. Various readings of space and dynamics emerge as the viewer moves past the hologram. The photographer's shadow is anchored to the image frame, which causes it to puncture the virtual image-volume established by the parallax – yet the figure is linked to the rhythm of this animate spatiality. The viewer's movement then creates a connection into the animate scene.

[3] The image sequence was photographed and compiled in 2006 but not printed into a hologram until 2009.

Museum Reflection, 2006/9 Three-colour transmission 200-frame multiplex hologram 48 x 32 cm (laminated)
Source photograph (Left). *'Explorations of the holographic gaze'* (Centre and Right)

4. Movement as a material

The viewer's agency of movement around my holograms is key to establishing an active engagement with perception. Artist David Hockney considers a sharing of time as crucial for linking the experiences of capturing and viewing:

> "And the reason you can't look at a photograph for a long time is because there's virtually no time *in it*—the imbalance between the two experiences, the first and second lookings, is too extreme"[5].

While initially this lead Hockney to disregard photography, he then used the temporal quality of photography to explore an assembled perception through collages of photographs, which he called *'joiners'*. In making the *'joiners'*, Hockney had a rule to never crop the photographic prints, observing that "the evenness of time seems to be tied up with a regularity in the print size"[6]. The visual rhythm of the prints is used to establish a dimension that is temporally reconstituted by the viewer. In making compositions that both encoded movement and require movement to be seen I attempt to educe a sensational reading of the scene, an experience constructed by cross-referencing the felt and the observed.

The concept that holograms encode a movement through space was compared to a projected diagram of dance steps by Bob de Marrais in 1981:

> "A hologram properly viewed is an excuse for TaiChi, requiring you to bob up and down, weave left and right, each artwork having its own implicit set (or sets) of 'Arthur Murray footprints'[4]"[7].

With my compositions and their installation the implicit choreography of the viewer's movement was developed into a poetic aspect of experiencing the holographic scenes. However getting viewers to actively move when engaging with a hologram is not always easy.

The installation *Jumping Jellyfish* comprised five jellyfish holograms hung around a trampoline. When the viewer jumped on the trampoline they moved through the holograms' vertical animation. Not only did one movement more-or-less correspond to the other, the jumping movement was an integral part of 'playing' the holographic animation. Participants felt their own 'squishy' parts as they watched a similar 'blobbing' in the jellyfish. The proprioception of elongating and squishing with each jump was projected onto the jellyfish by the viewer. While the spatial animation is recorded vertically, it is the animate horizontal expansion and contraction that creates a 'shared' sense of squishiness. The aim was for the visual to feed the physical (and vice versa) – to evoke an animated vitality[5] that is ascribed to the jellyfish, whose action is mimetically associated to the viewer's proprioception. Thus a physical-visual feedback loop is shaped by a sense of blobbing.

[4] The Arthur Murray School taught dance steps through mail-order diagrams of footprints.

[5] Vitality affects are described by Susan Langer in *'Feeling and Form'* [8] and Daniel Stern in *'Forms of Vitality'* [9].

Jumping Jellyfish installations. Bushwick Boat, New York Harbour, 2009 (Left) ZKM| Center for Art and Media Karlsruhe, 2011 (Left of centre). *'The Emergent Holographic Scene'* (Right sequence)

5. Disjunctions, amplifications and transformations

The holographic compositions and installations attempt to extend, dislocate and reveal the viewer's perspective so that they can obtain a sense of both the overall structure of multiple viewpoints and a sense of the virtual dynamics and shaping of moving between them. The structural relations generated by using the photographer as a protagonist of looking allowed for scenes that mimetically referenced the viewer's own looking at the holographic scene.

The viewer knows that they are looking into an image (as the photographer knows they are looking into a camera) yet this external perspective is incorporated into an expressive spatial experience. During my image capture for *On the Roof* the visual distortion of looking into the camera created a vertigo and sense of my feet stretching into a visual void in search of solid ground. I layered multiple copies of the video footage with a vanishing zoom to visually exaggerate these sensations.

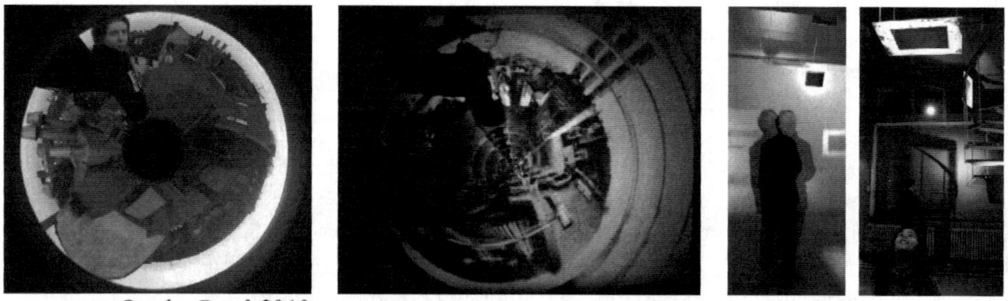

On the Roof, 2010 Two-colour transmission multiplex hologram laminated to laser etched plexiglass, 38 x 49 cm Video still from recording (Right). Photograph of hologram (Centre). Viewers looking up into the image (Left)

The installation of these holograms on the ceiling was designed to evoke a similar vertiginous feeling in the viewing to that experienced during the image capture. When looking up and leaning around to animate a holographic image there is a heightened awareness of proprioception and balance. As one viewer described, "you lean backwards and realise there is gravity"[6] an awareness that I am using to show how activity of looking can shape a sense of place.

6. A multiplex of movement

The multiplex process of producing the holograms is a hidden operation of recording[7], but one which I have used to reveal how the activity of photography shapes the perceived scene by structuring camera views into movement. By setting up a viewing structure tied to a recording of that view, there is both a looking at and looking with the holographic image. This situates the viewing act within separate but

[6] A Korean National University of Arts student visiting the exhibition *'Explorations of the holographic gaze'* described her vertiginous sense of looking up at the holograms.

[7] Prior to this work, my projects including *'Hover…'* (2004) were aimed at revealing technical aspects of holographic recording.

linked processes – the act of recording is accessed through its relationships to the activity of reconstruction. While the differences between these processes highlights the activity of each.

While experimenting with the process of multiplexing I came to think of the pattern of captured perspectives as a kind of reference structure that enabled an encoding of spatial dynamics. As the scene is assembled through multiple related perspectives, the viewer becomes dislocated from a single position to inhabit a structure or action of looking.

For the series *'we're all looking'* the holographic view is composed from the combined photographic act of a group of photographers positioned in formations that related to the surrounding architecture. The group photographic capture through which the images are constructed is reinforced by the appearance of the photographers in the holograms.

'we're all looking' *Exhibition Circle*, 1800-frame multiplex (Left). *Figure8 on Drummond Street, reflection print* (Centre) Photograph of *Figure8 on Drummond Street* by Wouter Cox at *'The Emergent Holographic Scene'*(Right)

In *Figure8 on Drummond Street* the act of photographing – a shared action – hinges the viewer to the scene. As the viewer of the hologram moves past the image their perspective seems to hop through each of the photographers, seeing them, seeing the next photographer through their camera and then seeing through this photographer's camera. The jumps of visual perspective encode a change of position that fuses into a virtual movement. As the viewer moves from side to side the visual perspective moves through a figure '∞' loop, while the effect of seeing through the sequence of perspectives is that of zooming into and out of the image.

At the centre of the image, the visual movement changes direction while the viewer continues to move. This flip is like a momentary feeling of 'weightlessness' or a micro sensation of realising you are lost – in that direction loses its vectorial structuring. The viewer feels forces shift within and around them as the expected result of action inverts. The sense of zooming in and out and the inversion of direction in this virtual perspective creates a complex experience of motion.

7. Resonating an experience of traversing stairs

My experiments to accentuate and amplify bodily awareness led me to realise that provoking vertical movement produces a more acute proprioception than horizontal motion due to a vectorial structuring of space by the experience of gravity. I made a number of installations in stairwells, utilising the vertical delimitation of physical space and the rhythm of viewing in moving between steps. The act of stepping is encoded in a staircase, an action that is drawn into the reading the holographic scene.

Stairwell installations. *'Ex tempore'*, Melbourne, 2006 (Left and Centre). *'Housebroken'*, New York, 2010 (Right)

For a stairwell installation as part of *'Sommerloch'* in Wuppertal (2010), I decided to record my traversing of different staircases and create a vertically animated series of holograms *'Up the Stairs'*. The initial recordings were made using a Bloggie HD video camera with 360-degree lens attachment that gave a 'donut-panorama' view. Each situation was different but all were captured as a single piece of video footage, which was then cut and layered. The layers of video fuse together through the shared rhythm of action – a stepping – activated by the viewer's own stepping. In the first hologram the pace of the various layers was quite different, while in the later compositions each layer had the same pace.

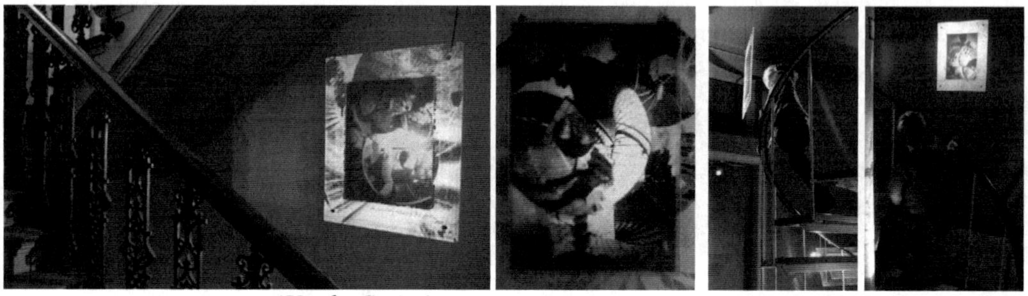

'Up the Stairs' Installation in Wuppertal (Left) *up and down the spiral stairs* (Centre)
Installation at *'The Emergent Holographic Scene'* (Right images)

When installed parallel to the stairs the animate action of the holographic scene unfolds into a suspended act of stepping. While the visual-protagonist climbs through a virtual staircase, the viewer is poised in the action of stepping. A potential act that becomes fulfilled through the image.

The recording for *up and down the spiral stairs* was made going down and then up an outdoor spiral staircase, my movement guided by both the physical staircase and my view through the lens. In one direction of movement, the footage was reversed for the hologram so the virtual movement always matches the perspective of a moving viewer. Using the Bloggie camera to record the action of traversing the spiral staircase there was both a physical circulation and optical circling due to the lens distortion. In the holographic scene the photographer's shadow anchors the view so with each step the stairs seem to rotate underfoot. While only a few steps are captured, there is a distinct sense of spiral circulation from the combined moving perspectives, which one viewer described as an Escher-like infinity that seemed to fold in on itself.

8. Conclusion

My holographic scenes are inspired by the activity of moving through and the conceptualisation of urban spaces. The recorded movements along paths and through patterns become entangled with the physical sense of moving around the hologram. This extends embodied cognition to produce an awareness of virtual movement while the dynamics of the holographic scene fold back into the viewer's sense of a located body.

The appearance of the photographer and camera links a point within the scene to the place from which the scene is viewed – hinging these locations into a structure of looking. As such, there is a simultaneous sense of being at the core of a dynamic scene and standing outside it, reading its diagram of structural shaping. The activities of exploring and recording multiple perspectives coalescence into a dynamic structure of viewing relationships allowing for a complex impression of place and action. The installations enhance the body's role in this dynamic structuring of a visual impression – an experiential mapping of the scene occurs.

As the viewer moves around the animate scene, dynamics are coupled to their proprioception. This produces an awareness of different bodily sensations that shape the holographic scene. With the *Jumping Jellyfish* installation there is an internal sense of organs shifting, and the jellyness of what is under the skin. This is quite different from the sense of balance and vertigo that is an important part of

On the Roof and *This morning of the balcony reminded me of a dream*, different again is the 'lost' sensation within a directional inversion of the relationship between physical and virtual movement.

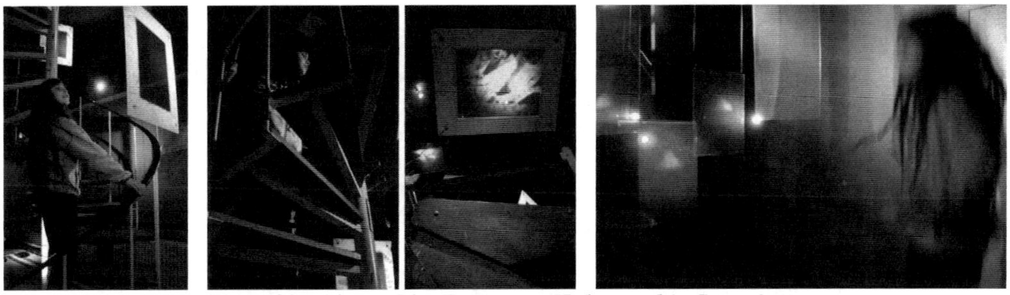

Various activities of looking, *'The Emergent Holographic Scene'*. Photographs by Anna Baróthy

The physical negotiation of the installation space by amplifying, mirroring or contrasting the virtual movements and forces of the image becomes part of how the scene is experienced. The encoded movement between the visual perspectives amplifies and transforms the viewer's action, establishing virtual forces that are drawn from and affect a bodily sensibility. In using a visually-depicted protagonist who is at the core of spatial warping, I aim to assist the viewer of the hologram to navigate through the structure of folds and forces, to sense a different spatiality. The distortions of the image are virtual, yet are felt as an extension of movement – pulling, pushing, causing little jumps and folds in the perception of continuity. What emerges from the activity of viewing is a system of relationships, a kind of perceptual physics that is inscribed by the arrangement of relations.

In physically negotiating the virtual movement, an embodied sensibility is brought forward into conscious awareness and contributes to perceptual activity. I am interested in the way that a 'shaping' by perception can be carried through the image composition – from the act of capturing to the viewer's experience of looking. My holographic image installations attempt to draw-out and amplify relationships and forces to express a perceptual construction of experience.

References

[1] DAWSON, P. 2000. *The Concrete Holographic Image: an Examination of Spatial and Temporal Properties and their Application in a Religious Art Work.* PhD, College of Fine Arts, University of New South Wales, p335.

[2] DESBIENS, J. 2006. 'Nomadic Perspectives: Spatial representation in oriental scroll painting and holographic panoramagrams'. *The International Conference on Virtual Systems and Multimedia (VSMM),* October 2006, Xi'an, China. [source: http://www.i-jacques.com/brokenwindow_en.html], p7.

[3] PIZZANELLI, D. 1994. *Aspects of Spatial and Temporal Parallax in Multiplex Holograms: a study based on appropriated images*. PhD, Royal College of Art, London.

[4] LIPP, A. & ZEC, P. 1985. *Mehr Licht : Künstlerhologramme und Lichtobjekte = More light.* Hamburger Kunsthalle. (Translated by KABEL, E). Hamburg : Fielmann, p144.

[5] WESCHLER, L. & HOCKNEY, D. 2008. *True to life: twenty-five years of conversations with David Hockney*, University of Californina Press, p7

[6] ibid, p29

[7] Bob de Marrais. 1981. Quoted in the catalogue of Sam Moree's exhibition '*Flux*', 12 February – 9 May 1982, Museum of Holography, New York, p7.

[8] LANGER, S. K. 1953 (Fourth Impression 1967). *Feeling and Form : A theory of art developed from phllpsophy in a new key,* Routledge & Kegan Paul Limited.

[9] STERN, D. 2010. *Forms of Vitality: Exploring Dynamic Experience in Psychology, the Arts, Psychotherapy and Development*, Oxford University Press.

Further documentation of the holographic image installations is at www.holographics.com.au

Ultra-realistic 3-D imaging based on colour holography

H I Bjelkhagen

Hansholo Consulting Ltd., The Bungalow Offices, St-Asaph, LL17 0LQ, UK
E-mail: hansholo@aol.com

Abstract. A review of recent progress in colour holography is provided with new applications. Colour holography recording techniques in silver-halide emulsions are discussed. Both analogue, mainly Denisyuk colour holograms, and digitally-printed colour holograms are described and their recent improvements. An alternative to silver-halide materials are the panchromatic photopolymer materials such as the DuPont and Bayer photopolymers which are covered. The light sources used to illuminate the recorded holograms are very important to obtain ultra-realistic 3-D images. In particular the new light sources based on RGB LEDs are described. They show improved image quality over today's commonly used halogen lights. Recent work in colour holography by holographers and companies in different countries around the world are included. To record and display ultra-realistic 3-D images with perfect colour rendering are highly dependent on the correct recording technique using the optimal recording laser wavelengths, the availability of improved panchromatic recording materials and combined with new display light sources.

1. Introduction

This paper is concerned about *colour holograms* which, after the recording process (by analogue or digital methods) and upon processing, display the colours of the recorded object as accurately as possible. There are many techniques of creating holograms displaying different colours, e.g. *pseudo-colour holograms*, which have been popular among artists over many years. However they are not creating holograms of objects in their correct colours and not included here.

Over the last few years it has become popular to produce computer-generated colour holograms. In this case there is a possibility to first record a scene or an object photographically, e.g. using a camera moving along a horizontal rail. Large quantities of 2-D colour images of the object are captured during the recording. Then the images are transferred into a computer-generated pixilated printed hologram. If the hologram is of the reflection type the colour of the image remain fixed and is independent of viewer position. By using the rainbow technique it is possible to produce transmission colour holograms, however, the correct image is only possible to view at a certain position. Any deviation from the viewing position changes the colours of the holographic image. Therefore, to obtain the most realistic-looking holographic images, reflection holograms are the most suitable. The digitally printed holograms can be created with selected colours to fit a particular image. For example completely computer-generated holograms of non-exciting objects have of course no "correct" colours; the image can have any desirable colours. However for museum and archival applications it is important to record 3-D images with extremely accurate colour rendering. To some extend this can be achieved in both analogue and digitally-printed holograms, however improvements are needed in both recording techniques as well as recording materials.

The focus here is on the direct holographic recording of objects to create colour-accurate hologram images. To be able to record or print high-quality colour reflection holograms it is necessary to use extremely low-light-scattering recording materials, which means the use of ultra-fine-grain silver-halide emulsions.[1-6] This type of material has the advantage of higher sensitivity compared to photopolymer and dichromated gelatin (DCG) which are used as alternative materials for colour holography.

Published under licence by IOP Publishing Ltd

Journal of Physics: Conference Series **415** (2013) 012023 doi:10.1088/1742-6596/415/1/012023

Table 1: Total average error for three to seven optimal wavelengths

Number of wavelengths	**Optimal laser wavelengths (nm)**	**Average colour rendering error***
3	466, 545, 610	0.0137
4	459, 518, 571, 620	0.0064
5	452, 504, 549, 595, 643	0.0059
6	451, 496, 544, 590, 645, 655	0.0040
7	445, 482, 522, 560, 599, 645, 655	0.0026

* Computer simulation based on the *Macbeth ColorChecker*® target used to illustrate improved colour rendering with increased amount of laser wavelengths, assuming a perfect recording material.

2. The progress of colour reflection holography

In several publications details and the history of colour holography have been provided.[7-12] Analogue colour holograms of the Denisyuk single-beam reflection type, are the ones which really creates the illusion of viewing a real object behind the plate rather than an image of it.

The possibility to record super-realistic-looking holographic images depends on four factors:

- The recording laser wavelengths;
- The recording setup;
- The recording material;
- The light source used to illuminate the recorded hologram.

In regard to required laser wavelengths at least three laser wavelengths (RGB) are needed, However for perfect colour rendering more than three wavelengths are required.[10,13,14] In **Table 1** three to seven optimal laser wavelengths are listed based on computer simulation to minimize the colour rendering error. The *Macbeth ColorChecker*® target was used to illustrate improved colour rendering with increased amount of laser wavelengths, assuming a perfect recording material which means that the selection of wavelengths is actually independent of the holographic recording process. Bazargan [15] is addressing the laser wavelength selection issue at this conference.

The lasers must be powerful, with long coherence and very stable output. If one wants the most realistic-looking images, parallax in both vertical and horizontal directional are desirable. This combined with a viewing field of 180-degrees in both dimensions means that the most suitable direct-recording setup is the single-beam Denisyuk technique. Digitally printed colour holograms can also be created with both vertical and horizontal parallax to make them more realistic looking. In the following both types of colour reflection holograms are described with recent progress and applications. The quality of the recording material is extremely important to obtain realistic-looking holographic colour images. The current problem is very limited supply of suitable commercial materials for colour holography.

2.1. Commercial recording materials for colour holography

The choice of recording material is of equal importance as the choice of laser wavelengths for the recording of a colour hologram. To be able to record high-quality colour reflection holograms it is necessary to use extremely low light-scattering panchromatic recording materials. Currently the main materials in use are special emulsions based on silver-halide and photopolymer materials.

2.1.1 Silver-halide materials. The market offers a very limited choice of commercial silver-halide recording materials for colour holography. In the past Agfa-Gevaert, Ilford and Kodak manufactured materials for monochrome hologram recording. None of these companies have holographic materials on the market now with the exception of *Harman Technology Ltd*[16] (former Ilford) which started a limited production of a new holographic emulsion however with a silver halide grain size not suitable for colour holography. In general, sensitivity of an emulsion depends on many factors, for example the laser wavelength, the exposure time (reciprocity failure), the development, (developer type, processing time, temperature, agitation, etc.) and storage conditions. Holographic sensitivity often varies from batch to batch, which is rare with conventional photographic materials. It is consequently recommended to make exposure and processing tests each time an important holographic recording is to be performed. It needs to be emphasized that for recording materials for colour holograms, high sensitivity is nice to have, but if it comes with an increased grain-size, it is not acceptable. Unfortunately it is much better to accept a longer exposure time (or to use a higher energy laser) to obtain a scatter-free recording.

In Russia today the manufacturing of holographic emulsions suitable for colour holography takes place in two companies, the *SLAVICH Joint Stock Company* [17] and *Sfera-S* [18] Both companies are located in Pereslavl, outside Moscow. Slavich manufactures different types of silver-halide materials for holographic purposes. Panchromatic materials for colour holograms are manufactured by Slavich under the name of PFG-03C. Sfera-S was formed by Yury Sazanov in 2004. Sazanov was earlier responsible for holographic plate manufacturing as director of the Micron plant at Slavich. In his new company, Sfera-S, located in an ex-Slavich building, he is now responsible for manufacturing a high-quality emulsion for colour holography which is available coated both onto glass plates as well as onto film. The main customer is *Geola uab*[19] in Vilnius, Lithuania where the film is used for their production of digitally printed colour holograms, marketed as *i-Lumograms*.[20] The Sfera-S emulsion [5] is currently the highest quality commercial panchromatic colour holography material available. It is sensitive to the short laser-pulse recording primarily used in the pulsed RGB digital holographic printers used by Geola. Both the Slavich and Sfera-S materials are available from Geola or from their international network of distributors.

Colour Holographic Ltd [21] is the primary UK manufacturer of ultra-fine-grain holographic emulsions. Their emulsion is based on the material that Birenheide manufactured in Germany. This type of emulsion (BB emulsion) was launched in 1996. Mike Medora of Colour Holographic acquired the rights to the BB products in 2001. Initially these emulsions were monochromatic and later a panchromatic emulsion was introduced, the BBVPan. Ulibarrena *et al.*[3,4] published reports concerning this new panchromatic ultra-fine-grain emulsion which has a mean grain size of 20 nm. The shrinkage or swelling of the emulsion after the plate is processed is one concern since the emulsion requires TEA pre-treatment to increase its sensitivity. In colour reflection holography a change in emulsion thickness is directly related to the wavelength of reconstruction and so affects the final replay spectrum and the colour reproduction of the image. Ulibarrena is now working at Colour Holographic and is responsible for emulsion manufacturing including the panchromatic emulsion which is now called RBG BBPAN. The company's production facility is based in Maldon, Essex, UK.

The other European manufacturer of colour emulsions *Ultimate – Holography* in France.[22] This emulsion is based on the work by the Gentet brothers.[2] Today Malasy Gentet (Yves Gentet's wife) produces and sells these products. Her expertise in the field of silver halide emulsions introduced a new achievement in quality control and reproducibility of Ultimate materials, including major improvements in their storage lifetime. The plates are manually produced with extreme patience and care. The company produces a panchromatic colour emulsion with a grain size of only 8 nm. However the production capacity of such plates is limited. Ultra-fine-grain monochromatic red and green sensitive emulsions are also manufactured. The Ultimate-08 plates are ready to use (no bath or adjustment needed before recording).

Some details about the Ultimate panchromatic material:

- Ultra-fine grain (8 nm), for unequalled resolution;
- Super high contrast material, with high diffraction efficiency holograms even at high Reference/Object ratio (1/10);
- Materials sensitive to all visible wavelengths;
- Recording of transmission or reflection holograms;
- Best diffraction efficiency on the market (up to 95%);
- Best sensitivity/resolution on the market;
- Substrates available: glass 3 mm thick or triacetate film of 190 μm;
- Storage recommendation at 4°C - lifetime more than 2 years; more than 6 months at room temperature.

At the time of writing it seems that *Colour Holographic* could not guarantee commercial supply of panchromatic silver-halide holographic plates to all customers. *Ultimate Holography* which, in the beginning, did not offer its panchromatic emulsion on the market but is now able to provide colour holographers with its excellent product. Commercial silver-halide holographic recording materials for colour holograms are listed in **Table 2**.

Table 2: Commercial silver-halide holographic recording materials for colour holograms

Material	**Spectral sensitivity** [nm]	**Resolving power** [line pairs/mm]	**Grain size** [nm]	**Substrate**
Colour Holographic				
RGB: BB-PAN	440-650	4000	20-25	Glass
Slavich				
Pan: PFG-03C	450-700	5000	10	Glass/Film
Sfera-S				
Pan: PFG-03CN	457-633	6000	9	Glass/Film
Ultimate				
Pan: 08-COLOR	460-650	10000	8	Glass/Film

2.1.2. Photopolymer materials. Photopolymer materials can be used for recording phase holograms, where applications in the mass production of small (usually monochromatic) security and display holograms as well as optical elements constitute today the main commercial interests. The sensitivity is not as high as the silver-halide materials but the advantages are a low light-scattering noise level as well as an innate suitability for the application of dry processing techniques. *E. I. du Pont de Nemours & Co.* [23] has been the main manufacturer of commercial photopolymer materials and has for a long time marketed these under the name of OmniDex.® The DuPont material requires only a dry processing technique (exposure to UV light and a heat treatment) to obtain a hologram. Since 1995 DuPont has manufactured a panchromatic material for colour holograms. [24,25]

Since early 2000 *DAI Nippon Printing Co., Ltd.* (DNP) [26] used the DuPont panchromatic material for producing photopolymer colour holograms. [27,28] Today DAI Nippon has access to a domestic (Nippon Paint) photopolymer product which is used exclusively for the in-house security colour hologram production.

A new German photopolymer material was introduced in 2011. The Bayfol®HX [29-31] by *Bayer MaterialScience AG* [32] intended for colour holography requires less post-processing as compared with the DuPont materials. The material has many advantages such as long life time, stability, almost no shrinkage and no post-processing (thermal or wet). It comes in rolls of up to 1.2 m wide. Besides creating fascinating optical effects this material can also be used to make ID cards and other documents which are forgery-proof. The current holographic performance of the material is as follows:

- Refractive index modulation $\Delta n = 0.03$ in reflection holograms;
- Diffraction efficiency = 98%;
- Colour sensitivity (450 - 650 nm);
- Suitable for both reflection and transmission holograms;
- Sensitivity for reflection holograms 100 $\mu W/cm^2$ - 50 mW/cm^2;
- 10-25 μm photopolymer thickness;
- Substrates PET or PC in roll format;
- Environmentally stable (UV, heat, humidity).

Since the Bayer photopolymer materials require no subsequent chemical or thermal treatment they are exceptionally suitable for cost-effective mass-production of volume holograms. The Bayer photopolymer has the real potential to become a leading material for light management within a variety of new technologies - for example in improved 3-D digital and analogue holographic displays or for diffusers required in energy efficient lighting technologies such as LEDs. However, like DuPont which restrict their materials for their in-house hologram production and document security producers, Bayer has decided, at least for the moment, to restrict sales of its holographic materials in order to protect the value chain to the security industry. A few selected customers, for example *Zebra Imaging Inc.*[33] employing it for its large digital holograms are able to order the DuPont polymer materials.

Although today's panchromatic photopolymer materials constitute an extremely interesting solution for colour display holography, from a purely practical point of view, there therefore remains a sizeable question mark over whether these materials will be commercially applicable to display holography applications in the near-term.

3. Recording of colour holograms

Most colour holograms today are recorded using three laser wavelengths. Analogue reflection holograms of stable objects can be recorded using cw lasers. Panchromatic silver-halide emulsions or photopolymer materials are used.

3.1. Setup for recording colour holograms

A typical laboratory Denisyuk recording setup for analogue reflection three-laser colour holograms is illustrated in figure 1. The three laser beams are combined using two dichroic mirrors and pass through the same beam expander and spatial filter. The resulting "white" laser beam illuminates both the holographic plate and the object itself through the plate. Often the following three primary laser wavelengths are employed: 476 nm provided by an argon ion laser; 532 nm provided by a CW frequency-doubled Nd:YAG laser; and 647 nm provided by a krypton ion laser. It is possible to control independently the RGB ratio by individually changing the output power of the lasers, while the overall exposure energy is controlled solely by the exposure time.

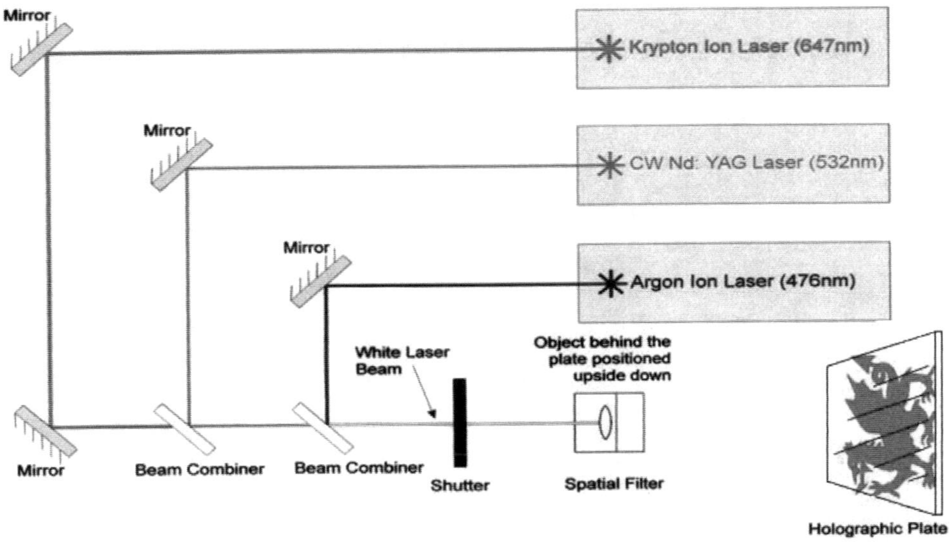

Figure 1. Setup for recording analogue colour reflection holograms

For most display purposes, the very large field of view obtainable in the Denisyuk hologram is very attractive. There are only very few colour hologram recording facilities in the world. For example, the problem with having to bring the objects and artefacts to a recording laboratory limits to wide-spread use of display holography in museums. Therefore there is a need to develop mobile colour holography recording equipment.

3.2 Mobile recording equipment

Gentet and Shevtsov [34] reported that they had fabricated a camera for recording colour holograms on Ultimate glass plates up to the 30 cm by 40 cm format. The camera is equipped with a red (λ= 639 nm) semiconductor laser and a green (λ = 532 nm) and blue (λ = 473 nm) solid-state lasers.

In 2011 the *Hellenic Institute of Holography* in Greece built a small portable three-colour analogue holographic camera, the Z3 RGB Holography YSB1 prototype camera to be used to record holograms of museum artefacts. The camera, depicted in figure 2, is a computer-controlled opto-mechanical device capable of exposing selected, commercially available or experimental, panchromatic silver halide emulsions to combined red, green and blue laser beams at appropriate energy levels. The device consists of a main camera unit (MCU) built on top of a lightweight aluminium honeycomb and a control electronics unit (CEU). The three lasers are:

- Red laser: 638 nm at output power 80 mW (*CrystaLaser* laser);
- Green laser: 532 nm at output power 100 mW (*Cobolt Samba* laser);
- Blue laser: 457 nm at output power 50 mW (*Colbolt Twist* laser).

The lasers produce TEM_{00} emissions with coherence lengths of more than five meters each and the MCU contains suitable optics to generate a clean colinear mixed RGB beam. The Z3 has been successfully tested under wide ambient temperature and humidity ranges. The Z3 camera is accompanied by auxiliary equipment for beam orientation and a flexible vibration-absorbent setup for the object positioning. One example of a hologram recorded with the camera is shown in figure 3. The Hellenic Institute of Holography has also produced a portable darkroom for on-site processing (figure 4). At this conference Sarakinos *et al.*[35] report on this system and Jith[36] describes another mobile colour hologram camera.

Figure 2. The Z3 RGB Holography YSB1 prototype mobile camera

Figure 3. Test colour hologram recorded with the mobile camera

Figure 4. Sarakinos and Lembesis in front of the darkroom tent

4. Lights for illumination display holograms

The light source to display the recorded holograms is also very important. In regard to colour holograms, the selection of a spotlight is actually more critical than for monochrome holograms. The colour temperature of the source has an influence on the colour rendering of the holographic image. The source size determines the image resolution in the parts of the image which appears in front of or behind the holographic plate. Arranging the correct illuminating angle (i.e. the same reference angle as was used for recording the hologram) between the holographic plate and the spotlight determines the correct colour of the displayed image. The distance from the source should be the same as the one used for the recording of the hologram to avoid any image aberrations.

The rapid progress in solid-state LED lighting has opened up new possibilities to illuminate colour reflection holograms. A significant advantage of LEDs is that they possess a much smaller bandwidth than broadband white-light sources. Although typical bandwidths are much larger than those commonly associated with lasers and laser diodes, LED light sources should nevertheless be matched to the recording laser wavelengths (or vice versa). This guarantees that only the white light from the LED source (which is a mixture of the primary LED wavelengths) contributes to creating the holographic image. Using a halogen spotlight, a large part of the light spectrum emitted illuminates the surface of the plate without having any impact on the intensity of the image. Instead this light is scattered, lowering the image contrast. The lack of this scattered light in LED illumination can lead to significantly higher image fidelity. In addition LED light sources have considerable advantages over halogen and other traditional lighting sources, such as

- long life (20,000 to 100,000+ hours);
- small size;
- small étendue;
- high durability and robustness to thermal and vibration shocks;
- low energy usage / high energy efficiency;
- no IR or UV in beam output;
- directional light output;
- digital dynamic colour control – white point tunable;
- relatively low cost.

4.1 Preferred LEDs for colour holography

An alternative white LED technology to the most common phosphor-pumped LEDs is offered by RGB or RGBA LEDs. These combine red, green, and blue or red, green, blue, and amber chips onto one discrete package allowing the generation of white light or any of 256 colours by utilizing circuitry which drives the three diodes independently. In applications requiring a full spectrum of colours from a single point source, this type of RGB diode format is the preferred technique and more suitable for illuminating colour holograms. Currently one of the best methods for the generation of white light is the "RGB LEDs Combined" technology - i.e. the generation of white light using a combination of red, green and blue. Note however that his form of white light relies on the electrical control of three LED chips. Typical FWHM spectral widths are 24-27 nm.

When monochrome colour LED lights were first marketed in 2003, the *OptiLED* chip - a commercial LED spotlamp was introduced to illuminate monochrome reflection holograms. The 4° red, (627 nm, spectral width 20 nm), green (530 nm, spectral width 35nm) and amber (590 nm, spectral width 14 nm) versions provided suitable illumination for such holograms. The 2.5W LED spotlights had electric circuits allowing them to be operated at 90V to 240VAC. The monochrome LED lights were promoted and distributed by *Laser Reflections*.[37]

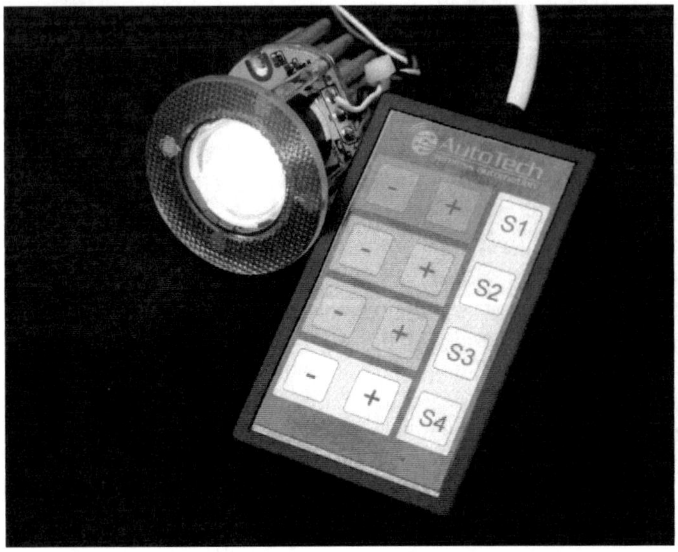

Figure 5. The *HoLoFoS* LED spotlight

4.2 Special LED spotlights for colour holograms

The *Hellenic Institute of Holography* has developed a special LED spotlight to illuminate colour holograms. The *HoLoFoS* LED spotlight, based on Cree LEDs, is manufactured at *AutoTech* [38] and is commercialised by *TAURUS SecureSolutions Ltd.*[39] Through proper choice of the component LEDs in terms of bandwidth and wavelength, the *HoLoFoS* LED spotlight is capable of achieving high quality reproduction of deep colour holograms. The illuminating head contains the RGB LEDs, mixing optics, lenses and heat sinks. The system has an embedded microcontroller for intensity control of each LED. The LED spotlight with its control unit is shown in figure 5.

The optics incorporated in the unit provide for an axial mixing of the LED beams resulting in a homogeneous colour mixing over the full extent of the projected beam. The small footprint of the LED die (approx. 2 mm) is small enough to produce clear and deep holograms even at small illuminating distances. The illuminating head can be fitted with a variety of LEDs at selected wavelengths and more than three different LEDs can be fitted to match various recording wavelengths. For example, an RRGGBB configuration can be achieved. This is important for colour holograms which will be recorded with four or five laser wavelengths to obtain more or less perfect colour rendering. The current prototype unit uses three LEDs with the following spectral characteristics for the red, green and blue LEDs which correspond to the lasers of the *Z3 Holographic Camera* (mentioned in Section 3.2) The dominant wavelengths for the LEDs are:

- Red 620-630 nm;
- Green 520-535 nm;
- Blue 450-465 nm.

To demonstrate the advantage of using this LED light for displaying colour holograms, the same hologram was illuminated with a conventional halogen spotlight (figure 6) and the new *HoLoFoS* LED light (figure 7). The hologram in figure 6 was recorded on a slightly noisy Slavich emulsion to demonstrate the advantage of LED illumination. Note the increased contrast and sharpness obtainable with the LED light. This is also a good illustration of the vital importance of the illumination source in holography.

Figure 6. Halogen spotlight illumination

Figure 7. LED spotlight illumination

5. Colour holography around the world

There are very few companies or institutions that are able to record and produce analogue or digital colour holograms today. The main reasons are the costs, the expensive recording setup, and the demand on the recording material. Currently, the main producers are:

- Colour Holographic Ltd., England, UK;
- DAI Nippon, Japan;
- Geola Digital uab, Lithuania;
- Hellenic Institute of Holography, Greece;
- Holorad LLC, Utah, USA;
- Liti Holographics, Virginia, USA;
- Strasbourg University, France;
- Ultimate Holography - 3D Holoprint, France;
- Zebra Imaging Inc., Texas, USA.

It should be mentioned that RabbitHoles Media Inc. (former XYZ Imaging Co.) in Canada has ceased operation. The company produced art and promotional digital colour holograms, in particular for the movie industry, of which the 2010 *Avatar* hologram is an example. Some companies/universities are using the printing service provided by Geola Digital in Lithuania for printing the digital holograms marketed as *i-Lumograms*. They often have a digital camera system which can move along a horizontal rail to record people and any indoor or outside scenes and objects. Sometimes digital holograms are made from computer-generated images for which the institutions provide the 3-D data.

- 4-D Studio R, Dubai, United Arab Emirates;
- De Montfort University, England, UK;
- Forth Dimension, Indiana, USA;
- Holoxica Ltd., Scotland, UK;
- Tutto-Tondo Srl, Italy.

A recent example of a *Holoxica* hologram with Geola was to reveal the concealed detail of an Egyptian mummy unseen for more than two millennia. Thanks to the digital hologram imaging process the *Rhind Mummy* of the National Museum of Scotland's collection, has been revealed in true 3-D. A series of CT scans at Edinburgh Royal Infirmary carried out by the University of Edinburgh's Clinical Research Imaging Centre (CRIC) were used to create the hologram.[40] The Geola printing service is also used by artists, such as the Australian artist Paula Dawson who has created large-format digital holograms.[41] Martin Richardson in UK is frequently using the Geola printing service for his art holograms.

6. Applications of analogue colour holographic imaging

The most obvious applications of colour holography are for archival and exhibition purposes of rare and highly-valuable artefacts. Colour holograms of the analogue or digital type are also used in advertising and product promotions. Provided that the illumination of the colour holograms at exhibitions can be correctly arranged this application of display holograms may take off in the near future. In this section we focus on current applications of the technologies of full-colour analogue hologram recording and digital holographic printing.

6.1 Holographic copies of museum artefacts

Holography now offers the possibility to essentially duplicate artefacts - and to a point where an observer practically cannot tell whether he is looking at the real exhibit or at a holographic copy. Although such holographic reproduction can never match the value of actually possessing the real artefact, it can allow the museum to fulfil one of its most important functions - to maintain display of the exhibits. Analogue holography does offer a means to preserving a faithful visual recording of unprecedented microscopic detail.

It should be mentioned that digital holographic printing can also be usefully applied to museum recordings. By using a two-dimensional tracking camera system, high resolution digital image data from a museum exhibit can be recorded from over many different angles. This data may then be written onto a very high resolution digital hologram producing a digital holographic copy. Although the resolution of this type of hologram is less than that of an analogue hologram, at the smallest hogel sizes now available (around 250 μm), it can be very difficult for an observer to tell the difference between an analogue and digital hologram with the unaided eye. And the digital hologram offers several sizeable advantages. For instance, new holograms can be generated from the digital data whenever required. This means that as long as the original data is stored securely, there are no image life-time problems. In addition the same image data can be used to produce holograms of small and very large size. So museum exhibits can be exhibited in any format. Finally, such digital images are not constrained to be behind a glass plate.

Holography can also help museums with travelling exhibitions as we shall see below. It is difficult for some people to travel to museums and as a result there is pressure on museums to take exhibits to the people. But transporting priceless artefacts is both hazardous and expensive. Transporting holographic copies on the other hand is not.

A final reason why ultra-realistic full-colour holograms are useful to museums is related to insurance costs of exhibiting within a museum. Most museums have large collections "downstairs" which they do not exhibit. The reason is that it generally costs more to exhibit something than to securely store it - as the risk of damage or loss is greater when an exhibit is on display. Again holography can help solve this problem. If the hologram is indistinguishable from the real exhibit, why not just securely store the real item and display the hologram?

6.1.1 Museum holography in UK. In 2010 the *Colour Holographic Ltd* [21] in London started to record colour holograms in their own ultra-fine-grain panchromatic material. An example of a 20 cm by 25 cm colour museum hologram is shown in figure 8. It displays the front and back of a statuette. The 15 cm blue statuette is an Ushabti Figure (mummiform funerary figure) of the *Princess Nesitanebashru*, (c. 1000 BC) found in the cache of Royal Mummies, Deir el Bahari, Thebes, in 1881.

A new product which the company has introduced is a nice wooden box for the display of colour holograms with an integrated LED light source built into the lid. When the box is opened, correct illumination of the hologram is provided. It is possible to switch holograms and the company offers the box with different holograms for sale. The box and a colour hologram are shown in figures 9 and 10.

 doi:10.1088/1742-6596/415/1/012023

Figure 8. Front and back recordings of an Ushabti Figure from private collection, recorded by *Colour Holographic Ltd* in UK

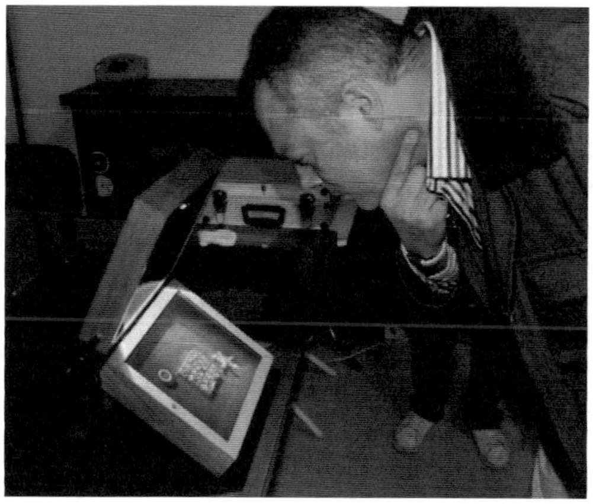

Figure 9. *Colour Holographic* display box

Figure 10. *Colour Holographic* hologram

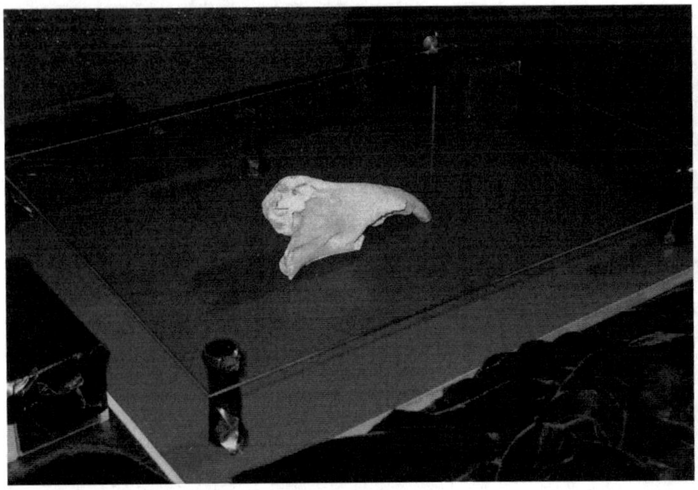

Figure 11. Horse jaw illuminated with "white" (RGB) laser light positioned under the holographic glass plate

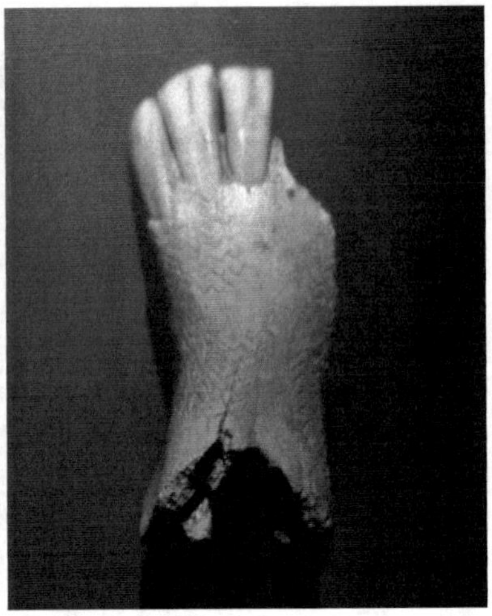

Figure 12. Horse jaw hologram [British Museum 1959, 1203.1]

6.1.2 Virtual museum exhibition. One interesting colour hologram project[42] which was carried out in the Centre for Modern Optics' (CMO) laboratory by the present author in North Wales. The project was funded by the Esmee Fairburn Foundation entitled 'Bringing the Artefacts Back to the People'. The project involved collaboration with a number of major museums including the National Museum of Wales, the British Museum, the Maritime Museum in Liverpool, as well as with the Royal Commission for Ancient and Historical Monuments in Wales. David Crane of Llangollen Museum was responsible for the holography museum project and for arranging the touring exhibition. Colour holograms of various artefacts were recorded using the Denisyuk analogue techniques described in Section 3.1. Each artefact was positioned on a horizontal aluminium plate placed on the recording table. The recording holographic plate was positioned above and very close to the object. During the exposure, the divergent 'white' RGB laser light illuminated the object through the holographic plate at about a 45-degree angle from above. Several original holograms were recorded of each artefact to assure that a perfect hologram could be selected for the exhibition. After processing the exposed hologram plate it was sealed to a black glass plate using an index matching optical cement. The holograms were completed by the end of 2009, after which they were displayed as a travelling exhibition which toured North Wales and its borders. The exhibition first opened at Llangollen Museum in June 2010 and later at the museums of Grosvenor (Chester), Wrexham, Llandudno, Bangor and many others.

One of the recorded artefacts, supplied by the British Museum, was a 14,000-year-old decorated horse jaw bone (British Museum 1959, 1203.1) from the ice age, or late glacial period.[43] The jaw bone was positioned on a red-painted aluminium plate placed on the recording table. The recording setup is shown in figure 11 and the recorded colour hologram in figure 12.

Other recorded artefacts were the *Tudor Owl Jug* and *Sergeant at Arms Ring*. The hologram of the two objects is shown in figure 13. The artefacts were from the Grosvenor Museum in Chester, UK. In total, ten 30 cm by 40 cm colour holograms of different artefacts were included in the touring museum exhibition. The first place where the exhibition appeared was at the upper floor of *Llangollen Museum* in Wales, as shown in figure 14.

Figure 13. *Tudor Owl Jug* and *Sergeant at Arms Ring* hologram [Grosvenor Museum (Chester)]

Figure 14. The travelling colour hologram exhibition at Llangollen Museum in Wales

Figure 15. Hologram of a Ceramic Vase recorded by *Yves Gentet*, *Ultimate,* in France

6.1.3 Museum holography in Greece. The *Hellenic Institute of Holography* in Athens has invested in colour recording equipment and is currently promoting colour holography. The primary goal of the Institute is to record Greek cultural artefacts through the "*HoloCultura*: *Applied Holography in Cultural Heritage*" project. The project consists of three parts:

- *Phase A*: Study necessary for implementation of the colour holography programme;
- *Phase B*: Recording of experimental colour holograms;
- *Phase C*: Pilot project involving the recording of cultural artefacts.

The Institute is active in a country with a unique cultural tradition of worldwide influence extending from classical ancient Greece to orthodox Byzantium Christianity. The use of display holography in the preservation, recording and public visual dissemination of artefacts from this cultural heritage is at the core of the activities of the Hellenic Institute of Holography. In addition to the possibility to record their own holograms, the Institute has produced museum holograms in collaboration with Yves Gentet in France (see figure 15, showing a hologram of a ceramic vase) as well as in cooperation with *Colour Holographic Ltd* in UK (see figure 8). More information about the holography institute in Greece is provided by Sarakinos *et al.* at this conference.[35]

6.2. Medical and scientific applications of colour holograms
An application of small analogue colour holograms used in endoscopy is reported at this conference by Osanlou *et al.*[44]

There are several scientific and medical applications of digital colour holograms. Educational 3-D images are used in biochemistry, medicine and life sciences showing images of protein structures, DNA and cells, for example. *Holorad Inc.*[45] in Utah, USA, has been in the field of medical holography during many years (operating as *Voxel Inc.*[46] providing full-parallax 3-D holograms of MRI and CAT data). The new company is producing scientific and advertising holograms including colour holograms. Hart *et al.*[47] is reporting on a large hologram at his conference.

Figure 16. *Morus Head* digital hologram

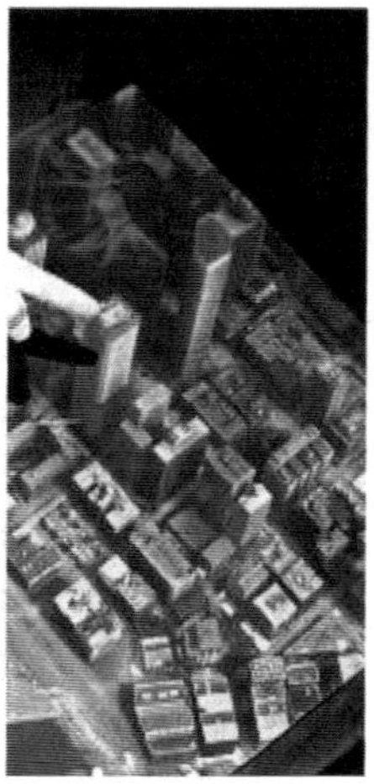

Figure 17. *Zebra* holographic map

7. Computer-generated colour holograms

Today it is not possible to obtain a computer-generated hologram (CGH) with the same huge information content as the laser-recorded analog ones of real objects. What may become possible in the future is a technique to compute and record the interference pattern that is stored in a reflection colour hologram which upon illumination can produce an image like the laser-recorded ones of real objects. The best compromise today is to use *holographic stereograms* which can provide high-quality CGHs. A holographic stereogram can be created by using a series of 2-D photographic images or 2-D images displayed on a LCD screen from which the hologram is recorded. In order to obtain a high-quality holographic image with a large field of view, many 2-D images are needed. It is also possible to use a laser scanner to obtain 3-D information about the object to create a digital hologram. An example of such a digitally-printed hologram by Geola Digital is the *Morus Head,* shown in figure 16. The *Morus Head* was found at the site of the Valle Crucis Abbey in Llangollen, Wales, and is now at National Museum of Wales in Cardiff.

For a long time *Zebra Imaging Inc.*[33] in Texas has produced large colour reflection holograms.[48] These holograms can be produced having both vertical and horizontal parallax with a 100° field of view. The holograms are recorded on DuPont panchromatic photopolymer film. The main work is for military applications to produce holographic maps used both in headquarters as well as in the field. More information about Zebra and its projects are provided by Klug at this conference.[49]

8. Art colour holograms

Over many years *pseudo-colour holograms* of both the transmission and reflection types have been attractive to artists and there are many examples of beautiful holograms recorded by artists such as Inaki Beguiristain, John Kaufman and Rudie Berkhout. These holograms have been created by emulsion thickness manipulation or, in regard to rainbow holograms, the recording geometry. An example of an analogue Denisyuk reflection colour hologram by an artist is the *Stars and Stripes Flag* hologram (figure 17). Anaït Stephens who created this hologram in 1995 together with the present author used a pseudoscopic cast of the flag to make the image appear in front of the plate in the finished Denisyuk hologram. The digital printed colour holograms on ultra-fine-grain panchromatic emulsions have opened new possibilities of producing creative art holograms. Artists have been attracted to the new digital hologram medium. Paula Dawson[41] in Australia, who has always been interested in large-format holograms, has recently created several large digital colour holograms, printed by Geola. The 2007 *Luminous Presence*, a 0.95 m by 1.5 m hologram is shown in figure 18.

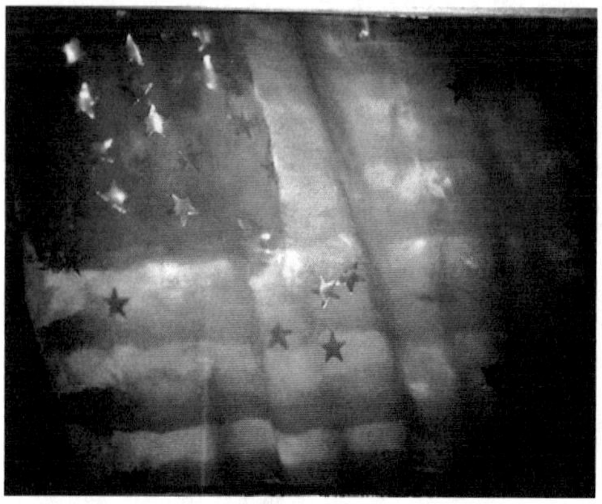

Figure 17. *The Flag* hologram by Anaït

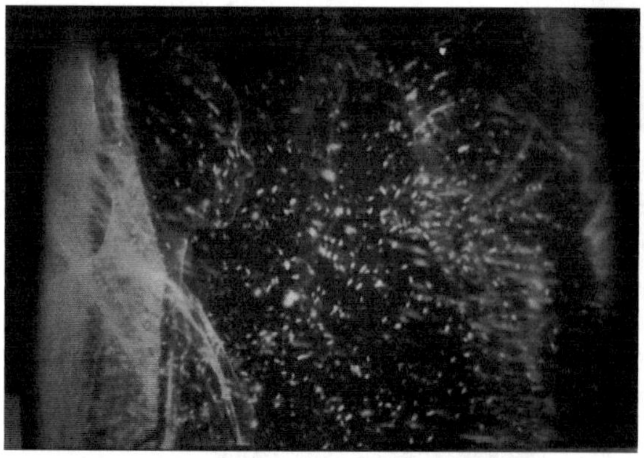

Figure 18. *Luminous Presence* by Paula Dawson

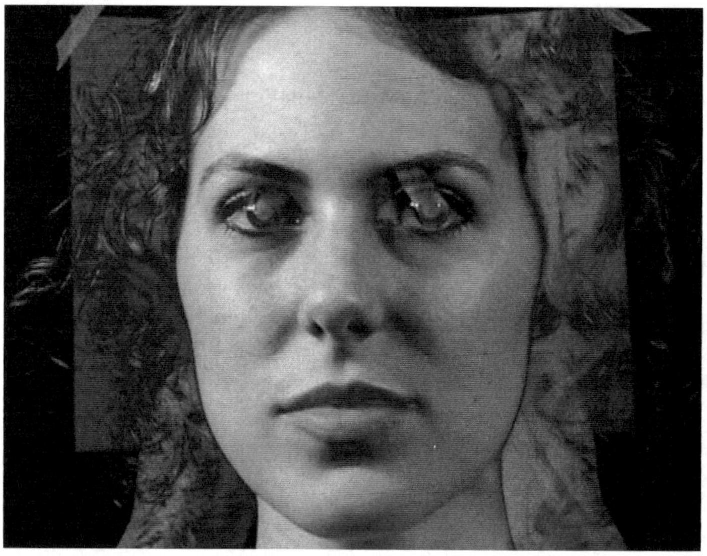

Figure 19. *Psychedelic Amy,* by Martin Richardson, UK

The combination of a photographically recorded 3-D portrait and added computer-generated content is the recent art hologram, *Psychedelic Amy*, size 50 cm by 60 cm, a digital hologram by Martin Richardson (De Montfort University, UK). This 2009 hologram, shown in figure 19, is another example of the collaboration with Geola being responsible for the printing. At first sight, this is a digitally retouched portrait of a young woman, but a ghost-like image can be seen to reflect in the subjects eyes as one moves from left to right, giving the impression of sixties-like psychedelic imagery. Several artists, for example Maria Isabel Azevedo and Sandra Oliveira (in cooperation with Martin Richardson) have started to work with this new and exciting medium.

The digital colour holograms may also become more popular among advertising companies. Currently the limited printing capacity makes it difficult to see a rapid market growth in this field. For both types of colour holograms a copying technique needs to be developed including required hardware. The new Bayer photopolymer would be suitable for commercial mass production of colour holograms provided it will be supplied to display holography companies in the future. Geola has started tests of contact-copying holograms on Bayer photopolymer materials. An example of a copied test hologram is shown in figure 20.

Figure 20. Photopolymer copy test hologram by *Geola*

8. The future of colour holography

The recording of large-format colour reflection holograms in ultra-fine-grain silver-halide and photopolymer materials is going to become more common, in particular for museums and documentation. With an optimal choice of the recording laser wavelengths within the visible spectrum, good colour rendering can be achieved in high-quality analogue colour holograms. The virtual colour image behind a Denisyuk holographic plate represents the most realistic-looking image of an object that can be recorded today. The extensive field of view adds to the illusion of beholding a real object rather than an image of it. The wavefront reconstruction process recreates accurately the laser wavelengths scattered off the object during the recording of the colour hologram.

The large-format digital colour holograms may find increased applications in art and advertising where their possibility to include motion (caused by the viewer moving or by moving the light illuminating the hologram) is an attractive feature not possible in the Denisyuk holograms. New solutions for illuminating the recorded holograms will make it easier to use them in museums and for exhibitions. A new detailed book on ultra-realistic imaging based on analogue and digital colour holography will be published later in 2012 by the present author together with David Brotherton-Ratcliffe.[50]

References

[1] Bjelkhagen H I 1993 *Silver halide recording materials for holography and their processing*, Springer Series in Optical Sciences, vol 66 (Heidelberg, New York: Springer)

[2] Gentet Y and Gentet P 2000 "Ultimate" emulsion and its applications: a laboratory-made silver halide emulsion of optimized quality for monochromatic pulsed and full color holography, in *Proc. SPIE: HOLOGRAPHY 2000* vol 4149 ed T H Jeong and W K Sobotka pp 56-62

[3] Ulibarrena M 2004 A new panchromatic silver halide emulsion for recording color holograms, *Holographer.org* , 1-12

[4] Ulibarrena M, Vale M and Medora M 2006 Colourholographic Ltd BBPan panchromatic emulsions for recording colour holograms, in *Advances in Display Holography, Proc. 7th Int. Symposium on Display Holography* ed H I Bjelkhagen (London: River Valley Press) pp 73-77

[5] Sazanov Y, Gradova O, Zacharovas S, Bakanas R and Gudaitis G 2006 New ultra-fine grain photofilm for pulsed colour holography, in *Advances in Display Holography, Proc. 7th Int. Symposium on Display Holography* ed H I Bjelkhagen (London: River Valley Press) pp 65-68

[6] Bjelkhagen H I, Crosby P G, Green D P M, Mirlis E and Phillips N J 2008 Fabrication of ultra-fine-grain silver halide recording material for color holography, in *Proc SPIE: Practical Holography XXII: Materials and Applications* vol 6912 ed H I Bjelkhagen and R K Kostuk pp 09-1 - 14

[7] Bjelkhagen H I and Vukičević D 1995 Lippmann color holography in a single-layer silver-halide emulsion, in *Proc. SPIE: 5th Int. Symposium on Display Holography* vol 2333 ed T H Jeong pp 34-48

[8] Bjelkhagen H I, Jeong T H and Vukičević D 1996 Color reflection holograms recorded in a ultrahigh-resolution single-layer silver halide emulsion *J. Imaging Sci. Technol.* **40** 134-146
[9] Bjelkhagen H I and Mirlis E 2006 Colour holography – state of the art, in *Advances in Display Holography Proc. 7th Int. Symposium on Display Holography* ed H I Bjelkhagen (River Valley Press) pp 193-199
[10] Bjelkhagen H I and Mirlis E 2008 Color holography to produce highly realistic three-dimensional images *Appl. Opt.* **47** A123-A133
[11] Bjelkhagen H I 2009 Denisyuk color holography for display, documentation and measurement, *J. Holography Speckle* **5** 33-41
[12] Bjelkhagen H I 2011 Progress in colour holography and new recording materials, in *Holography – Culture, Art, and Information Technology, Proc. 8th Int. Symposium on Display Holography* (Shenzhen, China, Waterlands Resort, 13 – 17 July 2009) ed T H Jeong *et al.* (Tsinghua Univ. Press) pp 346-357
[13] Peercy M S and Hesselink L 1994 Wavelength selection for true-color holography *Appl. Opt.* **33** 6811-6817
[14] Kubota T, Takabayashi E, Kashiwagi T, Watanabe M and Ueda K 2001 Color reflection holography using four recording wavelengths, in *Proc. SPIE: Practical Holography XV and Holographic Materials VII* vol 4296 ed S A Benton *et al.* pp 126-133
[15] Bazargan K 2012 What are the ideal wavelengths for full color holography? *ISDH2012* IOP Publishing
[16] Harman Technology Ltd UK; www.harmantechnology.com
[17] SLAVICH Joint Stock Company Russia; www.slavich.com
[18] Sphera–S Ltd, Russia; www.geola.com
[19] Geola Digital uab, Lithuania; www.geola.lt
[20] Geola i-Lumogram; www.i-lumogram.blogspot.co.uk
[21] Colourholographic Ltd UK; www.colourholographic.com
[22] Ultimate – Holography France; www.ultimate-holography.com
[23] DuPont Autentication, USA; www2.dupont.com/Authentication/en_US/
[24] Trout T J, Gambogi W J and Stevenson S H 1995 Photopolymer materials for color holography, in *Proc. SPIE:Applications of Optical Holography* vol 2577 ed T Honda pp 94-105
[25] Steijn K W 1996 Multicolor holographic recording in Dupont holographic recording film: determination of exposure conditions for color balance, in *Proc. SPIE: Holographic Materials II* vol 2688 ed T J Trout pp 123-134
[26] DAI Nippon Printing Co. Japan; www.dnp.co.jp/international/holo
[27] Watanabe M, Matsuyama T, Kodama D and Hotta T 1999 Mass-produced color graphic arts holograms, in *Proc. SPIE: Practical Holography XIII* vol 3637 ed S A Benton pp 204-212
[28] Kodama D, Watanabe M and Ueda K 2001 Mastering process for color graphic arts holograms, in *Proc. SPIE: Practical Holography XV and Holographic Materials VII* vol 4296 ed S A Benton *et al.* pp 196-205
[29] Bruder F-K, Deuber F, Fäcke T, Hagen R, Hönel D, Jurbergs D, Kogure M, Rölle T and Weiser M-S 2009 Full-color self-processing holographic photopolymers with high sensitivity in red – the first class of instant holographic photopolymers *J. Photopolym. Sci. Technol.* **22** 257-260
[30] Bruder F-K, Deuber F, Fäcke T, Hagen R, Hönel D, Jurbergs D, Kogure M, Rölle T and Weiser M-S 2010 Reaction-diffusion model applied to high resolution Bayfol®HX photopolymer, in *Proc. SPIE: Practical Holography XXIIV: Materials and Applications* vol 7919 pp 0I-1-15
[31] H. Berneth H, F.- K. Bruder F-K, T. Fäcke T, R. Hagen R, D. Hönel D, D. Jurbergs D, T. Rölle T and M.-S. Weiser M-S 2011 Holographic recording aspects of high-resolution Bayfol®HX photopolymer, in *Proc. SPIE: Practical Holography XXV: Materials and Applications* vol 7957 ed H I Bjelkhagen pp 0H-1-15
[32] Bayer MaterialScience AG, Germany; www.bayermaterialscience.com
[33] Zebra Imaging Inc. USA; www.zebraimaging.com

[34] Gentet Y and Shevtsov M K 2009 Mobile holographic camera for recording color holograms, *J. Opt. Technol.* **76** 399-401

[35] Sarakinos A, Lembessis and Zervos N 2012 A transportable system for the *in situ* recording of color Denisyuk holograms of Greek cultural heritage artifacts in silver halide panchromatic emulsions and an optimized illuminating device for the finished holograms *ISDH2012* IOP Publishing

[36] Jith A 2012 Holographic color camera for recording artifacts *ISDH2012* IOP Publishing

[37] Laser Reflections, USA; www.laserreflections.com

[38] AutoTech Wireless Automation, Greece; www.autotech.gr

[39] Taurus SecureSolutions Ltd. Greece; www.taurus.com.gr

[40] Hooper R 2012 Hello mummy *New Scientist* No.2806 (26 May) 24-25

[41] Dawson P 2010 *Virtual Encounters; Paula Dawson Holograms* (Sydney, Australia: Macquarie University and Newcastle Region Art Gallery)

[42] Bjelkhagen H I and Osanlou A 2011 Color holography for museums: bringing the artifacts back to the people, in *Proc. SPIE: Practical Holography XXV: Materials and Applications* vol 7957 ed H I Bjelkhagen pp 0B-1-7

[43] Bjelkhagen H I and Cook J 2010 Colour holography of the oldest known work of art from Wales *The British Museum Technical Research Bulletin* vol 4 ed D Saunders (London: Archetype Publications) pp 87-94

[44] Osanlou A, Bjelkhagen H, Mirlis E, Crosby P, Shore A, Henderson P and Napier P 2012 Optimized processes in full color holographic endoscopy *ISDH2012* IOP Publishing

[45] Holorad Inc. USA; www.holorad.com

[46] Voxel Inc. USA; www.voxel.com

[47] Hart S, Brockbrader B, Burman D, Molteni W, Padiyar D, Padiyar J, Parris W, Reetz D, Shara M and Stevens C 2012 A stellar hologram *ISDH2012* IOP Publishing

[48] Klug M A 2002 Display applications of large scale digital holography, in *Proc. SPIE: Holography: A Tribute to Yuri Denisyuk and Emmett Leith* vol 4737 ed H J Caulfield pp 142-149

[49] Klug M 2012 Holograms by/for the masses *ISDH2012* IOP Publishing

[50] Bjelkhagen H I and Brotherton-Ratcliffe D 2012 *Ultra-realistic imaging – advanced techniques in colour holography* (London, New York: CRC Press Taylor & Francis Group) (in the print)

Journal of Physics: Conference Series **415** (2013) 012024 doi:10.1088/1742-6596/415/1/012024

A transportable system for the in situ recording of color Denisyuk holograms of Greek cultural heritage artifacts in silver halide panchromatic emulsions and an optimized illuminating device for the finished holograms.

A Sarakinos[1], A Lembessis[2] and N Zervos[3]

[1] Scientific Director, Hellenic Institute of Holography, 28 Dionysou Street, 15234, Chalandri, Greece.
[2] Chief Administrator, Hellenic Institute of Holography, 28 Dionysou Street, 15234, Chalandri, Greece.
[3]Electronics Eng. RnD, Hellenic Institute of Holography, 28 Dionysou Street, 15234, Chalandri, Greece.

Email: [1]rnd@hih.org.gr ,[2] director@hih.org.gr, [3] eleng@hih.org.gr

Abstract. In this paper we will present the Z-Lab transportable color holography system, the HoLoFoS illuminator and results of actual in situ recording of color Denisyuk holograms of artifacts on panchromatic silver halide emulsions. Z-lab and HoLoFoS were developed to meet identified prerequisites of holographic recording of artifacts: a) in situ recording b) a high degree of detail and color reproduction c) a low degree of image distortions. The Z-Lab consists of the $Z3^{RGB}$ camera, its accessories and a mobile darkroom. HoLoFoS is an RGB LED-based lighting device for the display of color holograms. The device is capable of digitally controlled intensity mixing and provides a beam of uniform color cross section. The small footprint and emission characteristics of the device LEDs result in a narrow band, quasi point source at selected wavelengths. A case study in recording and displaying '*Optical Clones*' of Greek cultural heritage artifacts with the aforementioned systems will also be presented.

1. Introduction

Holography is a form of optical information storage method that can record our three-dimensional world on a two-dimensional recording medium and playback the original object or scene to the unaided eyes as a three dimensional image. Holographic image differs from photographic image as it has perspective and depth and can be viewed from various angles. In 1948, Denis Gabor invented the technique of recording phase information by means of a background wave, which converts phase differences into intensity differences. The general principles are described by Leith et al [1] and holography theory and its applications are well described in the literature [2]. In 1962, Dr. Y.N. Denisyuk from Russia combined holography with 1908 Nobel Laureate Gabriel Lippmann's work in natural color photography. Denisyuk's technique produced a reflection hologram which, for the first time, could be viewed under ordinary incandescent bulb light [3]. By using 3 or more laser lines, one can record color Denisyuk holograms of artifacts [4].

1.1. *HoloCultura: Color holograms of cultural heritage artifacts*
Current technological progress in various disciplines such as solid-state lasers, panchromatic emulsions, LED illumination etc., allow for the first time the use of display holography in cultural heritage in a practical and tangible manner. A renewed interest in applications of display holography for Museums and worldwide activities in this sphere have ensued [4].

In response to a public call for proposals financed under the EU Structural Funds and involving display holography issued by the Greek State General Secretariat for R&D in Sep 2009, the Hellenic Institute of Holography submitted its proposal titled 'HoloCultura' specifying -among other- a transportable recording system and LED illuminating devices for Denisyuk color holograms of objects of Greek cultural heritage [5]. Unfortunately, no project involving holography was financed in spite of the explicit specifications of the original call.

Nevertheless, the Hellenic Institute of Holography and its partners decided to embark into this project on their own capacity. During Phase I of the 'HoloCultura', the necessary transportable equipment, processes and software for the in-situ recording of artifacts and their subsequent interpretation in either analog or digital display holograms have been developed. The project is now in Phase II where experimental work is done by using different materials for producing display holograms of acceptable standard by the prospective end-users. Results of Phase I and Phase II (May 2012) will be discussed in the next sections.

2. The Z3^{RGB} and HoLoFoS : Theoretical and Practical Considerations
In theory, true color Denisyuk holograms of high quality can be recorded in silver halide emulsions and subsequently be displayed reproducing life-like three-dimensional images, if certain prerequisites are fulfilled [6] [7] [8] [10] e.g. :

- Suitable selection of three or more laser wavelengths
- Panchromatic recording plates with mean grain size well below 10 nm
- Optimized processing of the exposed plates
- Suitable recording geometry to eliminate dispersion
- Mechanical and thermal stability
- Optimized illumination of the color hologram to enhance depth reconstruction, color rendition and minimize blur

Moreover the transportability of a holographic camera for the in situ recording of artifacts introduces more challenges to the overall optical and mechanical design of such a system. We will elaborate on the aforementioned prerequisites in the following sub sections.

2.1. *Selection of Laser Wavelengths*
The human eye has photoreceptors (called cone cells) for photopic vision, with sensitivity peaks in short (420–440 nm), middle (530–540 nm), and long (560–580 nm) wavelengths. Thus, in principle, three parameters describe a color sensation. These tristimulus values of a color can be conceptualized as the coordinates of three primary colors in a tri-chromatic additive color space like the CIE XYZ.

Two light sources made up of different mixtures of various wavelengths may appear to be the same color; this effect is called metamerism. Such light sources have the same apparent color to an observer when they produce the same tristimulus values, no matter what the spectral power distributions of the sources are. In this aspect researchers in color holography have aimed at choosing three suitable wavelengths for recording a true color hologram, which would cover a sufficient area of the CIE chromaticity diagram.

Thornton [9] found that the reflectivity of an object at three well-specified wavelength bands (near 450, 540 and 610nm) have a disproportionately high color rendering index. Using these wavelengths, a wide range of highly saturated colors can be produced. Hubel and Ward [11] in 1989 recorded color reflection holograms using 647, 528 and 458nm lasers with a wide color gamut. In 1991 Hubel and Solymar [12] found another good practical choice of wavelengths in using a combination of 442, 532 and 633nm. Bjelkhagen and Mirlis [6], Peercy and Hesselink [14] and Kubota et all [13] claim that more than three wavelengths are needed for true color rendition. In our transportable Z3^{RGB} holography camera we use 457, 532 and 638nm solid state lasers with space provision to add more.

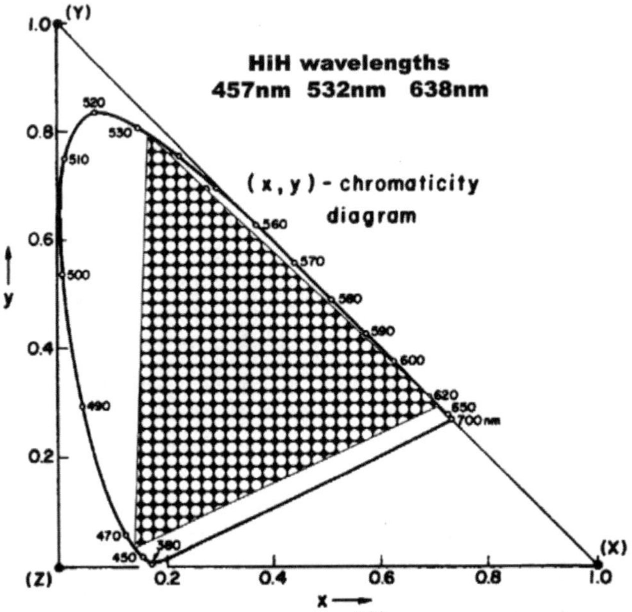

Figure 1. Color gamut of the Z3^{RGB} laser wavelengths.

This combination of wavelengths gives a gamut (figure 1) that includes yellow, purple, dark blue and violet as the Hubel and Ward one, and overlaps the Wintringham data. An important reason for this specific selection of wavelengths was the close matching to the emission characteristics of available power LEDs for the HoLoFoS hologram illuminating device.

2.2. *Recording Media and Chemical Processing*

To be able to record high quality color Denisyuk holograms, it is necessary to use panchromatic emulsions with mean grain sizes less than 10nm. Grain size affects the maximum spatial frequencies that can be recorded, the signal-to-noise ratio and the exposure energy required. Currently two commercially available emulsions, the Slavich PFG-03c and 'Ultimate' 08 by Yves Gentet [15], have suitable characteristics for color reflection holography. 'Colour Holographic' have recorded on their BBpan emulsion high quality holograms but the emulsion is not commercially available yet. Bjelkhagen [4] has recorded museum artifacts on the experimental Silver Cross emulsion

There are many formulae for suitable developers and bleaches described in the literature [16]. The chemical processing of the exposed plates has to be carefully adjusted to an emulsion in order to achieve high diffraction efficiencies and low level of scattering, especially towards the blue end of the spectrum.

At the current phase of the 'HoloCultura' project, we decided to follow the developing processes proposed by the manufacturers of the test plates. Practically all of our experimental recordings, while testing the Z3^{RGB}, have been carried out, so far, on the 'Ultimate' 08 emulsion and processed with the 'Ultimate' developer and bleach. It is our aim, at this phase of the project, to test in comparison all commercially available emulsions and publish our results in the near future. Some characteristics of the PFG-03c and 'Ultimate' 08 as published by their manufacturers are summarized in Table 1.

In our short experience with the 'Ultimate' plates we have confirmed the exposure energy levels quoted, high diffraction efficiencies and tolerable scattering in the blue if exposed properly.

Table 1. Characteristics of Slavich and 'Ultimate' emulsions.

Silver halide material	PFG-03c	'Ultimate' 08
Grain size	10nm	8nm
Resolution	10000 lines/mm	10000 lines/mm
Blue sensitivity	2mJ/cm2	120-150 μJ/ cm2
Red sensitivity	3mJ/cm2	120-150 μJ/ cm2
Green sensitivity	3mJ/cm2	120-150 μJ/ cm2

2.3. *Recording geometry*
Color reflection holograms exhibit less dispersion than transmission holograms due to their inherent sampling over the illuminating source frequencies. It is possible to minimize dispersion by using a reference angle of 30 degrees while keeping the plate inclined at 15 degrees [7]. This arrangement results in 45 degrees illumination angle during playback, a situation where the reflection of the illuminating source just misses the line of sight of an observer. However, not all objects can be displayed in this configuration, as the object appears inclined relative to the plate's surface.

2.4. *Mechanical and Thermal Stability*
Although the Denisyuk setup is less susceptible to mechanical disturbances, we prefer to heavily isolate the object and the plate using a Thorlabs 60 by 60 cm passive isolation platform on which we place a thin aluminium optical breadboard sitting on Sorbothane™ legs. We use a specially designed mobile tent chamber that encloses the object/plate space in order to minimize air currents and ensure thermal stability during the exposures. Both the isolation system and the chamber are standard parts of the Z-Lab holography system.

2.5. *Optimized illumination*
Though reflection holograms do filter the illuminating light, variable bandwidths of the recording frequencies may occur. Image blur due to dispersion is heightened in a broadband plate under a broadband illumination. On the other hand, image blur due to the finite dimensions of the illuminating source forces the placement of the light source inconveniently far from a hologram. The HoLoFoS LED RGB, a narrow band and quasi-point illuminating source, was designed as a device which addresses these problems.

3. The Z3^{RGB} Transportable Color Holography Camera
The Z3^{RGB} is actually a light mixing device built with transportability in mind and using high quality optical and mechanical parts (figure 2, figure 3). At the moment it incorporates three solid state lasers emitting at 457, 532 and 638nm with corresponding powers of 50,100 and 80 mW of single mode TEM^{oo} radiation. The coherence lengths of the lasers are more than 5 meters each. The camera is PC-driven by software developed in house (figure 4). The lasers and optics are mounted on a 60 by 60 cm honeycomb optical breadboard which in turn is enclosed in a fibber glass protective housing which contains the DAC boards and power supplies. The housing (figure 5) stands on three removable pneumatic legs allowing height and tilt adjustments and serves the thermal stability of the camera.

At early camera development stages we encountered laser jitter and mode hoping due to thermal instability. To overcome this behavior we had to build optimized TEC driving electronics for one of the lasers and fit the layout in a housing to ensure thermal stability. At the same time we added a Fabry–Pérot scanning interferometer to continuously monitor the laser beams and coupled the lasers monitoring and control lines to the Z3 controlling software. A thermistor was also fitted inside the camera housing and coupled to the software in order to read the temperature inside the housing compartment. In this way we now have continuous monitoring and control of the lasers which ensure mode-hop-free exposures except under extreme humidity and temperature gradients (as in one particular case of a windstorm when our red DL substituted our barometer!).

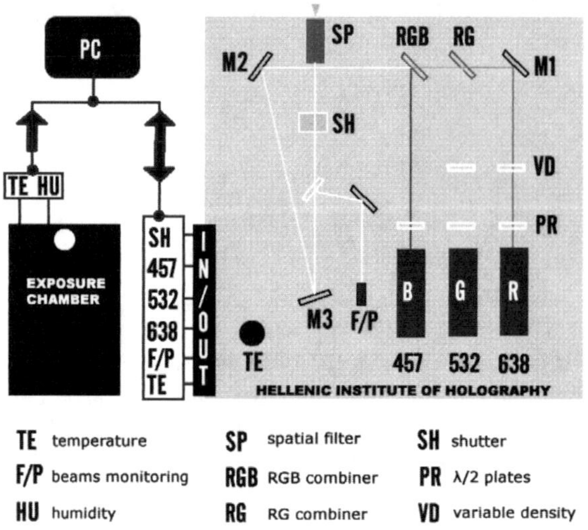

Figure 2. Z3 ^{RGB} Schematic.

Testing the camera under actual or simulated varying environmental conditions, with the camera being thermally stable to a high degree, we realized that we had to enhance vibration and thermal isolation in the plate-object space. So we assembled our two layered isolation platform where we place the object and the plate and manufactured our light-proof exposing chamber to enclose the plate/object space. The chamber is fitted with temperature and humidity sensors which feed their data to the Z3 controlling software. The Z3 software creates a record of each exposure including emulsion type, exposure energy, temperature inside the camera and the exposing chamber as well as humidity in the exposing chamber.

Figure 3. Inner works & beam.

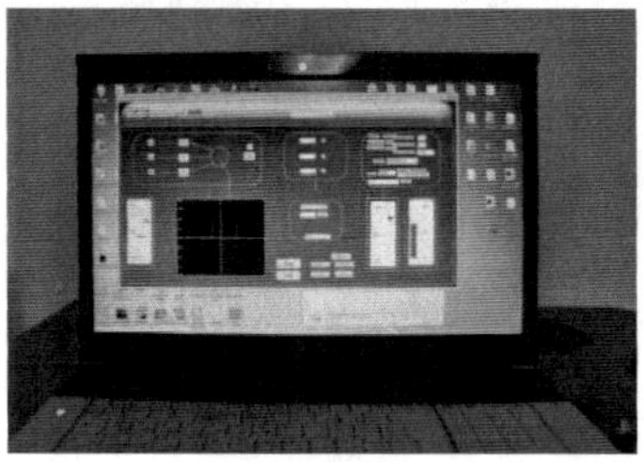

Figure 4. PC control software.

Figure 5. The Z3^{RGB}.

The camera's 'white light' output beam can be directed to the plate either by adjusting the housing's legs or by using a relay mirror, which is our preferred method. Typical times to reach thermal stability inside the camera are 30 to 60 min while typical exposure times for a 30X40 cm 'Ultimate' 08 plate are 6 to 12 seconds.

An autonomous and easily assembled darkroom completes the Z-Lab setup for the in situ recording of artifacts. The darkroom is equipped with vertical development tanks to save space and chemicals, can be connected to an external water supply and sink or -alternatively- use the fitted water and waste collection tanks.

4. HoLoFoS: An Optimized Light Source for Color Holograms Display

For decades display holograms have been illuminated either by cumbersome theatrical lights or simple halogen spots. This kind of illumination has many drawbacks including -among other- image blurring, high power consumption and thermal degradation of the illuminated hologram. On the other hand, an RGB-LED illuminant is characterized by narrow bandwidth per color to a high degree and small size of source, which minimize blur [8], while power consumption and thermal effects are reduced. Consumer available RGB LED fixtures suffer, from holography's view, of non-axial placement of the RGB LEDs and poor spatial mixing of the discrete beams.

'HoLoFoS' is our intelligent LED illuminating device that can supply to a hologram the necessary narrow bandwidth in red, green and blue light with peaks near the recording wavelengths . At the same time the small footprint of the HoLoFoS LED dies (2mm by 2mm per color) allows placing the HoLoFoS at a closer distance to a hologram while retaining image sharpness and enhanced depth. Electronic remote control of the LEDs driving current permits the accurate intensity mixing of the emitted red, green and blue bands in order to fine tune the colors of a hologram. The HoLoFoS provides coaxial mixing of the LEDs beams with the use of dichroic combiners (figure 6).

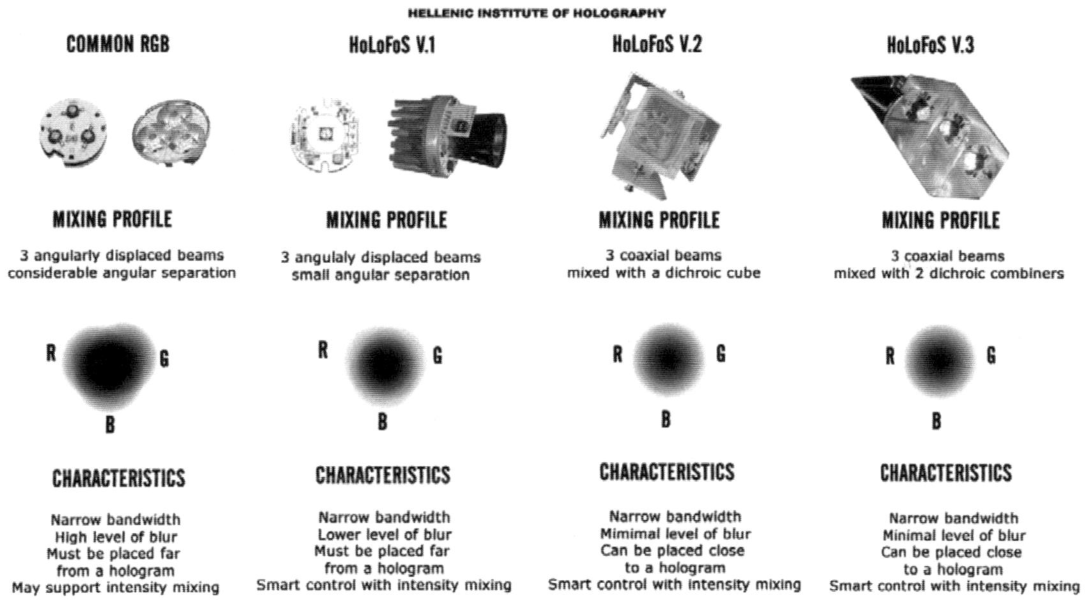

Figure 6. Evolution of HoLoFoS smart LED lighting for holograms.

The 'HoLoFoS' can be fitted with any combination of 3 or 4 LEDs in various wavelengths. Qualitative inspection of color reflection holograms under HoLoFoS illumination show a dramatic increase in image crispness while RGB intensities mixing help to 'correct' many faults in color rendition.

5. The 'Chios Epitaph' : An Actual Recording with the Z3^{RGB}

The 'Chios Epitaph' proved to combine all the requirements for testing our equipment : original artifact, natural multi-level depth, bright natural colours, wooden 'breathing' substrate, lots of fabric cloth, free-standing 'vibrating' metal flowers, ideal size and a fragile nature.

Several alternative layout setups were tested for the final hologram and various slightly different holographic images were thus obtained. What has been selected to go on public show has been recorded with the object artifact contained within a wooden box featuring a yellow-painted background in order to enhance colour contrast. The recording setup is depicted in figure 7.

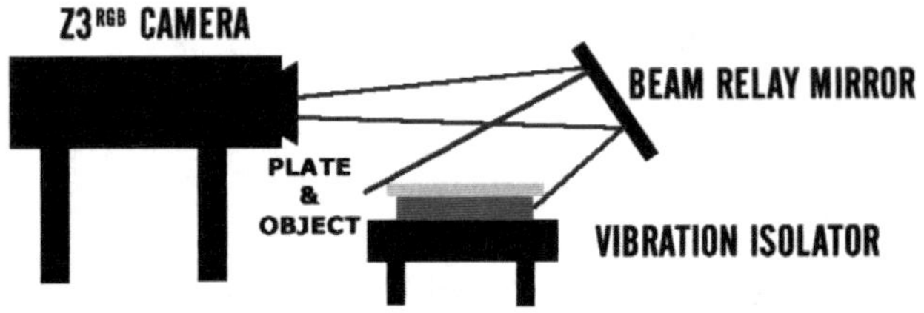

Figure 7. Setup for the 'Chios Epitaph' hologram recording.

The hologram was recorded on 30x40 cm 'Ultimate' 08 plate with an exposure time of 12 seconds. The object has been previously left to acclimatize for one day. Standard 'Ultimate' developer and bleach were used and the plate was left to air dry after soaking in distilled water in which a few drops of PhotoFlo were diluted. The hologram exhibits saturated colors and fine details under HoLoFoS illumination (figure 8).

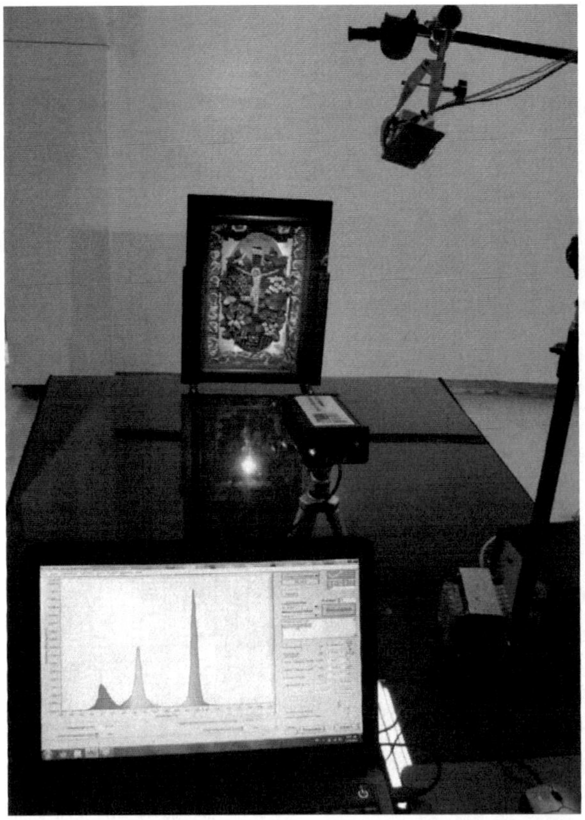

Figure 8. Spectrum from the 'Chios Epitaph' hologram, under HoLoFoS illumination, measured with our Jeti Specbos 1211 spectro-radiometer.

6. Conclusions

We have designed and tested a transportable camera for the in situ recording of color holograms of cultural heritage artifacts and an RGB LEDs illuminating device to illuminate the holograms.

The $Z3^{RGB}$ is stable under adverse environmental conditions and is capable of exposing 30x40 cm 'Ultimate' plates with short exposures of 6 to 12 sec. The HoLoFoS provides narrow bandwidth quasi-point source illumination that enhances colors and reduces blur.

Our experience with the 'Ultimate' emulsion is limited and we have to further quantitatively investigate the emulsion's characteristics to ensure repeatable results using spectro-radiometric measurements. At the moment we get remarkable results though not consistently due to the emulsion's high sensitivity to humidity.

Finally, we have to investigate the option to add more laser lines to enhance color reproduction.

7. Epilogue

Since its inception, display holography offered the highest hopes for recording objects of cultural interest and using the resulting holograms as *Virtual Artifacts*. However, mostly on account of the limitations of the technology (such as monochromaticity, unavailability of consistent commercially available emulsions, size and performance of LASERs, etc.) such hopes never really materialized in a publicly acceptable way despite the early promising applications and the occasional successful implementation of holograms as a medium for creating optical replicas of the original artifacts.

In our view, these efforts also lacked an important angle to their approach to Museum curators, conservation scientists and decision-makers: the proposed holograms were considered mainly on their properties for replicating quasi-realistically the selected objects or were assessed on basis of their subjective aesthetic and artistic value. Our proposal is that the new generation of ultra-realistic display holograms or ***Optically Cloned Artifacts*** - as termed by our Institute- should also serve as a means of **visually documenting** the cultural artifacts in addition to latest tendencies in museology such as digital modeling, multi-perspective imaging, digital virtual data-basing etc.

Within this framework, the novel approach of the Hellenic Institute of Holography to incorporate its techniques for creating display holograms with the scientifically-proven proprietary equipment for the analysis and diagnosis of works of art developed by the Institute of Electronic Structure & Lasers (Foundation for Research & Technology-Hellas) offers not only the necessary enhanced scientific credentials but also a complementary integrated approach described by the connoted term ***Hyper-Documentation***^{TM} [17]. Such services, as marketed by **ArtGnomon**, a spin-off of FORTH-Crete and TAURUS SecureSolutionS Ltd., open a completely new approach to the study of cultural artifacts in which ultra-realistic display holograms have an important function to play.

References

[1] Leith EN, Upatnieks J. 1965 *Scientific American* 212 6 24-35.
[2] Saxby G., *Practical holography* 2004 3rd edn Institute of Physics Publishing London.
[3] Denisyuk Y.N., 'On the reproduction of the optical properties of an object by the wave field of its scattered radiation', *Optical Spectroscopy* (USSR) 1963 **14** 279–284.
[4] Hans Bjelkhagen and Jill Cook, 'Colour holography of the oldest known work of art from Wales', *The British Museum Technical Research Bulletin* 2010 **4** 1-9.
[5] Anon., 'Greece plans Holograms of Antiquities', *Holography News* 2009 **23** 11 1-8.
[6] Bjelkhagen, H.I. and Mirlis, E., 'Color holography to produce highly realistic three-dimensional images', 2008 *Applied Optics* **47** 4 A123–A133.
[7] K. Bazargan, 'Techniques in display holography', 1986 *Phd Thesis* Optics Section Blackett Laboratory Imperial College of Science and Technology London.
[8] Arno Klein, 'Dispersion Compensation for Reflection Holography', 1996 *M.S Thesis* in Media Arts and Sciences M.I.T
[9] W. A. Thornton, 1971 *JOSA* **61** 1155-1163.
[10] K. Bazargan, 1989 *Proc. SPIE* **1051** 6.

[11] P.M. Hubel and A.A. Ward, 1989 *Proc. SPIE* **1051** 6.
[12] P.M. Hubel and L. Solymar,1991 *Appl. Opt.* **30** 4190-4203.
[13] T. Kubota, E. Takabayashi, T. Kashiwagi, M. Watanabe, and K. Ueda, 'Color reflection holography using four recording wavelengths' in Practical Holography XV and Holographic Materials VII, S. A. Benton, S. H. Stevenson, and T. J. Trout,eds., 2001 *Proc. SPIE* 4296 126-133.
[14] M. S. Peercy and L. Hesselink, 'Wavelength selection for true-color holography' 1994 *Appl. Opt.* **33** 6811-6817.
[15] Y. Gentet and P. Gentet,“"'Ultimate'" emulsion and its applications: a laboratory-made silver halide emulsion of optimized quality for monochromatic pulsed and full color holography" 2000 Proc. SPIE 4149 56-62.
[16] Hans Bjelkhagen, 'Silver-Halide Recording Materials',1995 2nd edn Springer.
[17] Fotakis C., D. Anglos, V.Zafiropulos, S. Georgiou, V. Tornari, Ed. R. G. W. Brown, E. R. Pike, 'Lasers in the Preservation of Cultural Heritage; Principles and applications' 2006 *Taylor and Francis, New York.*

A HOLOGRAPHIC COLOR CAMERA FOR RECORDING ARTIFACTS

Abhay Jith
School of Physics, University of Hyderabad, Gachibowli, Hyderabad, 500046, Andhra Pradesh, India.

Email: abhayjith@gmail.com

Abstract: Advent of 3D televisions has created a new wave of public interest in images with depth. Though these technologies create moving pictures with apparent depth, it lacks the visual appeal and a set of other positive aspects of color holographic images. The above new wave of interest in 3D will definitely help to fuel popularity of holograms. In view of this, a low cost and handy color holography camera is designed for recording color holograms of artifacts. It is believed that such cameras will help to record medium format color holograms outside conventional holography laboratories and to popularize color holography. The paper discusses the design and the results obtained.

1. Introduction: There are several holographic color camera designs proposed in the scientific literature available. However, a quick search of the commercial holography websites reveals that there are no off-the shelf availability of cost effective and handy holographic color cameras that help to meet the requirements of an inquisitive student or a hobbyist. This paper sketches my preliminary efforts to build a compact holographic color camera that can record color holograms of small and medium format. The effort was fueled by enthusiasm of a student rather than the hardcore knowledge level of an experienced optical engineer. Building such a camera not only popularizes holography among students but among holography hobbyists also.

2. Experiment: The basic objective here is to build a compact color holographic camera that can record holograms of about 15 x 10 cm^2size. Three diode lasers (TEM_{00}) of wavelength 446 nm, 532 nm and 660 nm were used to form a white source of out power of about 60 milli watts. The Power Supply (Figure 1) and the laser head unit (Figure 2) were made compact and were own built, under the guidance of the engineers of Light Logics Holography and Optics, as part of my student project in the Company.

Figure1: Power Supply

Figure2: Laser Head Unit

Aluminium coated Beam Splitters of 50% reflectivity on BK7 glass substrates were used as Beam Combiners for the green and the blue lasers. The red laser, properly oriented, was injected from the non-coated side of the first beam combiner, as shown in Figure 3. Power Supply electronics were adjusted so that the final output of each of the beams from the second beam combiner was about 20 mW. A 20X microscope objective and pinhole arrangement was used to do the combined spatial filtering. Good white light illumination was obtained through precise adjustments of the opto-mechanics. The entire setup was mounted on a 90cm x 60cm breadboard with small pneumatic pads, used for the vibration isolation. An electronic shutter unit was applied to control the exposure time.

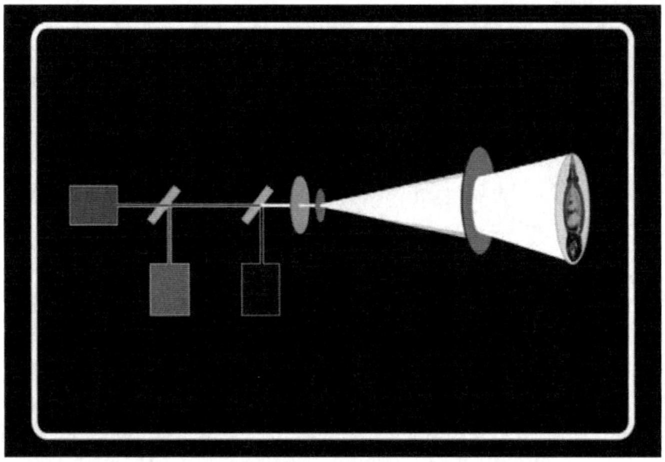

Figure3: Schematic diagram of the setup

A modified model of '*Kathakali* face, a traditional art form of Kerala was used as the subject. Due to the presence of all three primary colours and limited depth, this formed an ideal object for recording a color hologram.

Initially, holograms of the component colors were recorded to have the basic system adjustments. A fresh pack of PFG-03C color plates and standard chemistry recommended by Slavich were used to record the holograms. Photographs of the monochromatic illumination of the object are given in Figure 4 (a), 4(b) and 4(c).

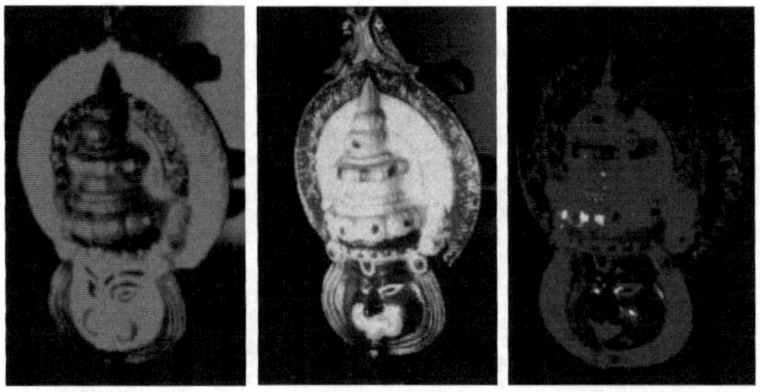

Figure 4(a) Figure 4(b) Figure4(c)

3. Result: Color holograms of relatively good image quality were recorded (Figure 5) and the reconstruction yielded an apparent depth of about 5cm. The experiment proved that conventional TEM_{00}

diodes can be effectively used to make low cost holographic color cameras for recording holograms of about 15cm x 10 cm size. Experiments on color control, diffraction efficiency optimization and further improvement by using dichroic mirrors and a 40X microscope objective are progressing. Positive results in this direction will yield a more compact and handy holographic color camera for various applications. Another advantage of this camera is that by selectively switching the lasers and adjusting the subject, medium format creative holograms of distinct color and dimensional effects can be recorded. Commercialization efforts to make the camera available off the shelf for creative artists, students and hobbyists are also progressing.

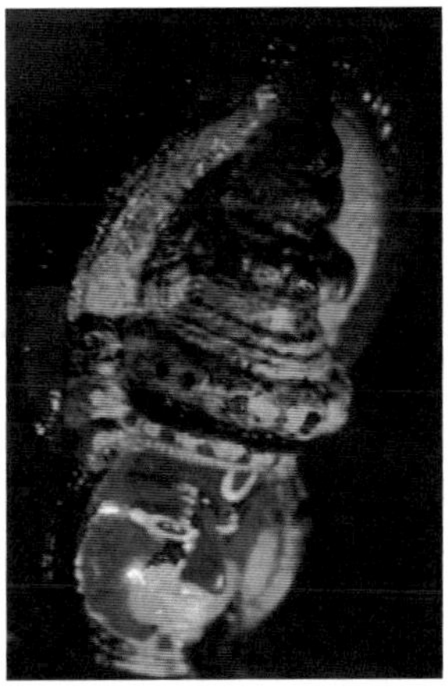

Figure 5: Image of the hologram recorded.

4. Acknowledgement: I would like to mention here my indebtedness to the Dean and the Professors of the School of Physics, University of Hyderabad, for the support and encouragement given to me. Also, I owe a lot to Light Logics Holography and Optics for the equipment, technical and the financial support.

5. References:

[1] P. Hariharan, 1987, "*Optical Holography Principles, Techniques and Applications*" Cambridge University Press, Cambridge, U.K, 131-143.

[2] J. Ludman, H.J Claulfield, J. Riccobono, 2002, "*Holography for the New Millennium*", Springer-Verlag, New York, Inc. 13-35.

[3] H.I. Bjelkhagen, 1993, "*Silver Halide Recording Materials for Holography and Their Processng"* Springer Series in Optical Sciences, **Vol.66** , New York.

[4] H. I. Bjelkhagen, T. H. Jeong, and D. Vukièeviæ, 1996, "*Color reflection holograms recorded in a Ultra high-resolution single-layer silver halide emulsion,*" *J. Imaging Sci. Technol.* **40** 134-146.

[5] P. M. Hubel and L. Solymar 1991 "Color reflection holography: Theory and experiment," *Appl. Opt.* **30** 4190-4203.

Journal of Physics: Conference Series **415** (2013) 012026 doi:10.1088/1742-6596/415/1/012026

Cave Holography – Out of the lab and under the ground

J Klayer

Hobby Holographer, Bonaire, Georgia USA

klayer@bellsouth.net

Abstract. This paper describes the combination of my hobbies, caving and holography. Most traditional holography involves bringing the objects to a lab with all the necessary holography equipment mounted on a stable table. I instead bring all the equipment assembled as a portable unit to the natural formations in a cave with the cave itself being the stable table. The first successes were Denisyuks made with a HeNe or laser diode and spatial filter mounted on a tripod. For greater depth, transmission holograms were made with a DPSS laser in several configurations sometimes using fiber optics to route the reference beam and sometimes a spatial filter and mirrors. The cave environment presents unique obstacles that have been overcome as evidenced by the beautiful holograms made.

1. Beginning

I developed an interest in caving and holography about the same time in my life so naturally I tried to combine the two hobbies. Unfortunately the low power HeNe lasers and homemade power supplies available in the 1970s were not up to the task of doing the split beam transmission holograms I was attempting in a cave. Discouraged, I put the cave holography on hold but continued my other holography interests on and off as I could afford over the years.

2. Starting Again

In the 1990s I revived my cave holography interest due to the availability of efficient DC HeNe power supplies and inexpensive long coherence length red laser pointers and diodes. These lasers allowed me to make Denisyuks of limited depth using the HeNe or pointer with a spatial filter mounted on a tripod and the hologram plate simply leaning on the cave formation (Figures 1, 2, 3).

HeNe laser powers were limited by the available DC power supplies and diodes much larger than 5mW tend to mode hop thus again frustrating my desire to do split beam work in the cave environment.

Figure 1. A typical cave Denisyuk.

Figure 2. A HeNe and spatial filter mounted on a tripod. The PVC pipe protected the laser as it was dragged into the cave.

Figure 3. A diode laser and spatial filter being adjusted for a Denisyuk. The white square on the floor is to center the beam on the formation.

3. Going Green

High power green DPSS lasers with long coherence lengths are on the surplus market now making split beam holograms possible with no need to match path lengths. Experiment found that multimode fibers can carry the reference beam thus greatly simplifying the portable apparatus being hauled into the cave but some grain is introduced into the image (Figure 4).

Split beam transmission holograms are desirable because they have extreme depth of field and parallax, viewable from all angles and get better looking as the observer gets closer.

The first DPSS setup I made used a fixed beam splitter sending the object beam through a diffuser to the object and the reference through a multimode fiber to the plate (Figure 5). No holograms resulted from the first apparatus. Though somewhat discouraged, I built a second smaller apparatus that used a variable density microscope slide beam splitter (Figures 6, 7). That effort yielded some decent holograms (Figure 8). I later substituted a ½ wave plate and polarizing cube beam splitter for the slide to improve the reference to object ratio adjustment.

My latest apparatus uses a spatial filter and mirrors in place of the fiber to eliminate some of the grain in the final image (Figures 9, 10).

Setting the reference beam at a 70 degree angle allows the apparatus to be placed much closer to the objects and upon reconstruction greatly reduces the amount of laser light directed toward the observer.

Figure 4. The experiment demonstrating that a multimode fiber can carry the reference beam.

Figure 5. The first apparatus using a fixed beam splitter.

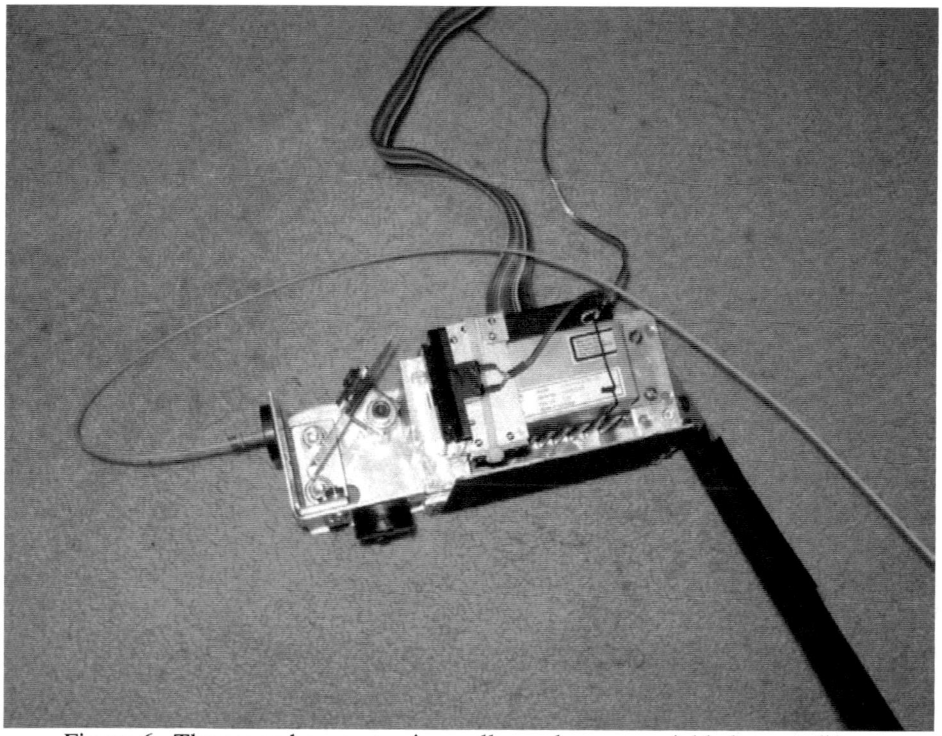

Figure 6. The second apparatus is smaller and uses a variable beam splitter.

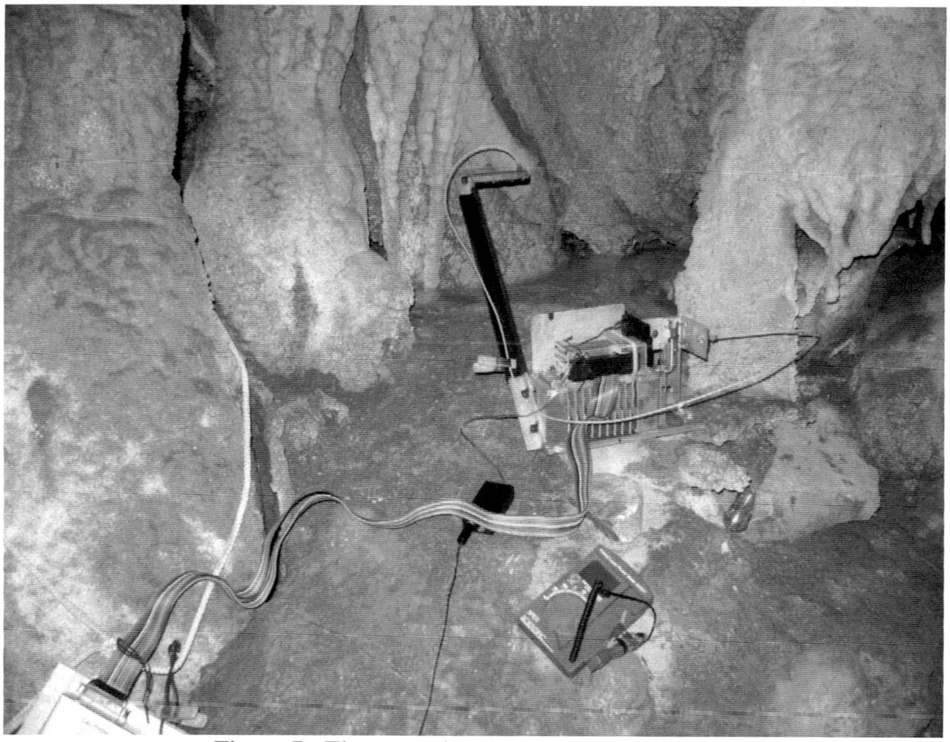

Figure 7. The second apparatus set up in a cave.

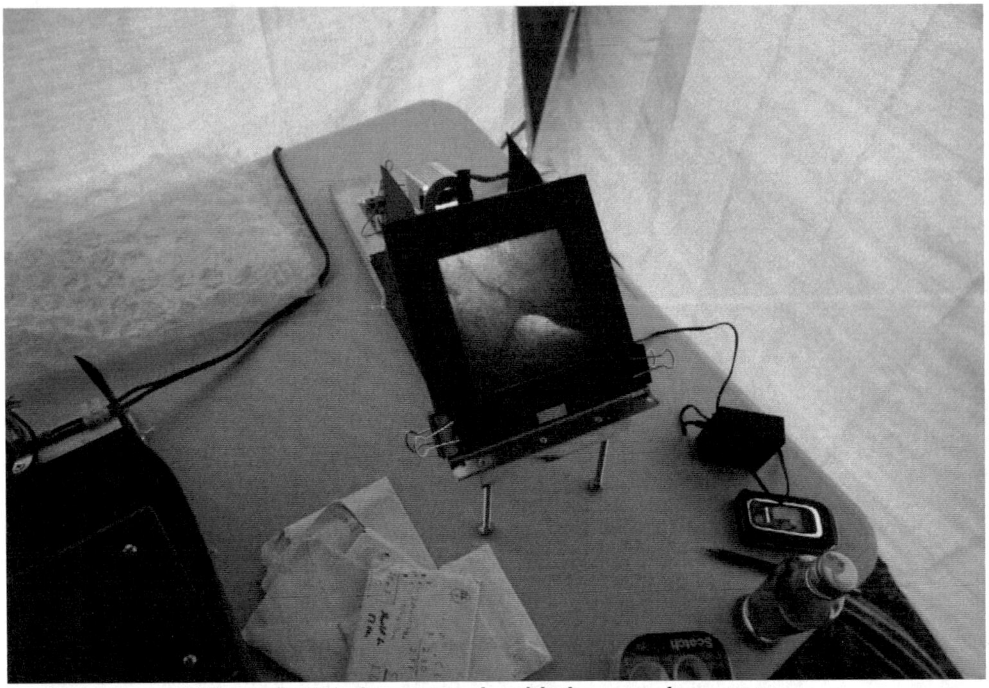

Figure 8. A hologram made with the second apparatus.

Figure 9. The newest apparatus uses a spatial filter and mirrors.

Figure 10. A cave hologram shot with the newest apparatus and the reference set at 70 degrees.

4. Obstacles Overcome

My first several split beam attempts were failures which were very frustrating because interferometer tests confirmed the cave to be rock solid. Finally the last shot of one set had an image so I figured a certain waiting time is needed for temperature stabilization. Now I routinely place the apparatus in the cave to temperature soak for a few hours before returning to do the shots.

Among other minor problems are bats flying about; the solution is to not set up during the morning and evening while they are active. The occasional salamander will crawl around and insects fly. A mouse once tried to steal a storage bag. Weather changes sometimes causing a breeze through the cave. The optics can become fogged by the steam from my body so I use a long wire for the shutter release. The high humidity causes dripping from the ceiling. Condensation forms on the apparatus when it is removed from the cold cave back to the outside air.

5. Future plans

I would like to produce larger holograms and find an inexpensive yellow laser for reconstruction or perhaps produce some H2 reflection copies. I will keep making cave holograms (Figure 11).

Journal of Physics: Conference Series **415** (2013) 012026 doi:10.1088/1742-6596/415/1/012026

Figure 11. Entering a cave with the holography equipment in protective packaging.

A survey of lasers at the birth of holography

Robert A Hess
Point Source Productions, San Jose, California, USA

Abstract. The 50th anniversary of the first hologram made with laser light is an appropriate time to shine some light on the lasers that made it possible. Dubbed "a solution looking for a problem", the laser emerged from industrial research labs as a new kind of light source for any application requiring coherence. Immediately commercialized by both start-up and well-established corporations, lasers sold in their first year of production left much to be desired but made a good first impression. Their second year of production saw technical improvements that met the requirements for practical holography, and by the end of 1963 were good enough to make holograms of three dimensional, deep and diffuse imagery. Mostly obsolete by the end of the 1960s, very few of the Kennedy era lasers have survived. This survey is intended to identify and celebrate them, so more may be preserved as they are found, to be admired in the future.

1. Introduction

When Ted Maiman's assistant Irnee D'Haenens first proclaimed the laser to be "a solution in search of a problem", he probably wasn't aware of two researchers at the University of Michigan's Willow Run Labs who were exploring Denis Gabor's technique of "wavefront reconstruction", eventually called holography. The laser would however, quickly progress from a device anticipated to be a "death ray" as visualized in pulp fiction throughout the first half of the century, to a commercial product for a wide range of practical applications. Holography is one a technique made practical by the laser. On the 50th anniversary of the first hologram made with laser light, it is appropriate to review the technology of the time that led to its success.

The first period of laser development was dominated by pulsed ruby, and Helium-Neon gas lasers operating in the infrared. Between July 1960 and the end of 1961, lasers were mostly built by those who wanted to study or use them. Starting in early 1962, commercial lasers became increasingly available and by the end of 1963 three dozen models from at least sixteen companies were in production.

It was the continuous visible beam from the He-Ne laser that really enabled holography, but little is known about the actual lasers used by the early holographers. This survey highlights commercial laser development against the background of pre-laser holography, and celebrates the wonderful little devices that made it all possible.

2. Masers to Lasers

The story of lasers begins with masers. Conceived in 1951 by Charles Townes, the first gas maser operated in April 1954 at Columbia University [1]. This was followed by the first solid-state maser at Bell Labs and the first solid-state maser amplifier at MIT's Lincoln Labs [2]. Chihiro Kikuchi's group at the Willow Run Labs started using pink ruby in their maser research in 1957 [3], and Theodore Maiman built a very compact ruby maser at Hughes Research shortly thereafter [4]. The 1959 conference at the Shawanga Lodge brought many maser researchers together, and with a paper by Arthur Schawlow on "Infrared Optical Masers", the stage was set for the laser to emerge.

Journal of Physics: Conference Series **415** (2013) 012027 doi:10.1088/1742-6596/415/1/012027

When Hughes announced operation of the first laser in July 1960, researchers at Willow Run were primed to jump right on it along with all the other groups doing maser research including Bell Labs, TRG, Raytheon, and RCA. Around the same time they all began work to repeat the results of Ted Maiman, Juris Upatnieks started working with Emmet Leith (also at Willow Run) to repeat the results of Denis Gabor [5]. Many had early success repeating Maiman's design. TRG and Bell Labs did it by August (using the larger flashlamp pictured with Maiman in the Hughes press release) [6] and RCA by November [7]. Raytheon went a step further immediately afterward however, by proposing an elliptical pump cavity with the ruby rod and flashlamp at the two foci [8].

Raytheon built a number of laboratory prototype lasers in November 1960 to demonstrate their improvement [9], and one of these lasers was pictured in a number of early books about lasers [10] [11]. The one shown here (Figure 1) differs from the one in previously published pictures by the cylindrical rather than conical rod support on the output end, and placement of the pump cavity reflector seam on the lamp side instead of rod side of the cavity. The rod ends are coated with mirrors to form the resonator as in Maiman's design, and a spring plunger at the rear holds the rod in place (Figure 2). Originally there was a trigger button at the rear end of the housing and a few more electronic components inside a cylindrical housing [12].

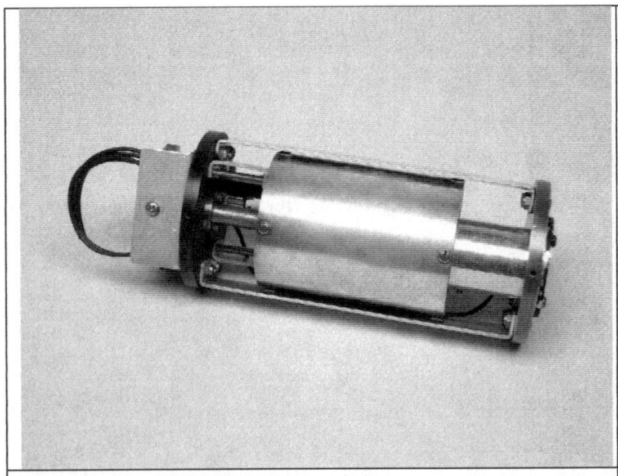

Figure 1. 1960 Raytheon laboratory prototype ruby laser head with elliptical pump cavity.

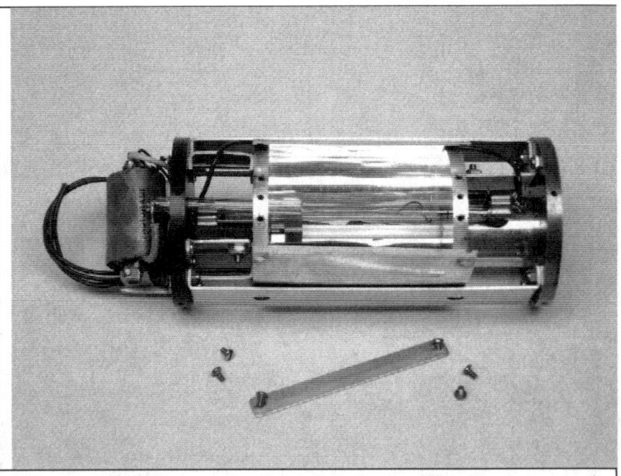

Figure 2. Chrome plated brass reflector opened to show linear flashlamp and ruby rod.

Meanwhile, Bell Labs continued on with dark ruby for pulsed lasers, and invigorated their program toward gas lasers [13]. The latter reached fruition at 4:20 on December 12, 1960 when Ali Javan suggested that Donald Herriot "just wiggle the mirrors" and the first Helium-Neon laser flashed to life [14]. Operating with RF excitation and flat resonator mirrors inside the plasma tube, this laser had an infrared output at 1150nm. But it was a continuous beam, and that's the type of light source researchers working with optical data processing needed.

3. Commercial Pulsed Lasers

By February 1961, Leith and Upatnieks had demonstrated their off-axis technique for holography by using a 100W GE high-pressure mercury arc lamp, a grating based recording system, and a wire for the image [5]. But then Juris Upatnieks went into military service, putting further work on hold for the next year and a half. Across the field at Willow Run however, work with lasers was just beginning. After a presentation on the ruby laser there by Robert Collins of Bell Labs, Lloyd G. Cross and Don Gillespie from Kikuchi's maser group (already in business together to make ruby masers) started Trion Instruments to make ruby lasers instead [15].

Trion was the first company formed specifically to make pulsed lasers commercially, but it was a few months until they had their first model LS-1 ready for sale. One of their first systems was rented by Peter Franken at the University of Michigan for his first generation of optical harmonics with laser light [16] (Figure 3). Trion ruby lasers were characterized by their large dewar for a liquid nitrogen cooling system, and were bought by many large organizations near the end of its first year in operation. In the middle of 1962 Lear Siegler acquired them, but despite being a major supplier of lasers throughout the early 1960s, the operation didn't survive the decade.

The first commercial laser sold however, was built by Raytheon and sold at the March 1961 meeting of the Institute of Radio Engineers in New York City [8]. Their model LH-1 (Figure 4) was a pulsed ruby laser similar to the laboratory prototype described above. It featured a housing designed for easy replacement of the rod and lamp, iron sights on top of the head, and a price of $5850 [17]. This laser operated at room temperature like the prototype before it, but three other models soon followed featuring the capability of cooling with liquid nitrogen, and pumping with multiple lamps bringing output energies up to 50J by the end of the year [18].

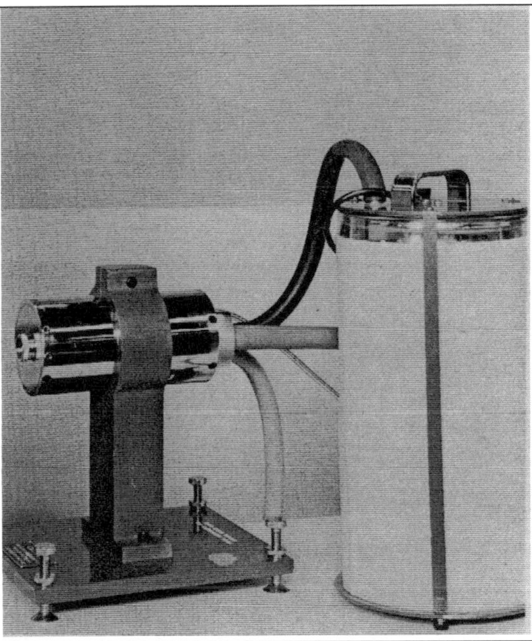

Figure 3. 1961 Trion Instruments model LS-2 pulsed ruby laser with liquid nitrogen dewar. The earlier model LS-1 was similar. Photo from Trion brochure.

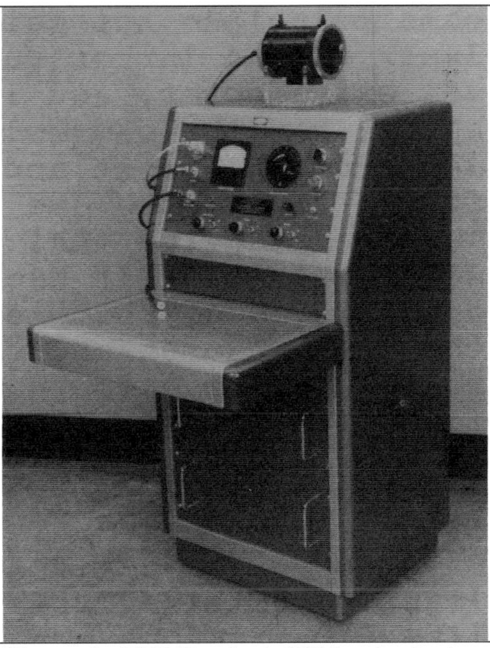

Figure 4. 1961 Raytheon LH-1 pulsed ruby laser system, the first laser model sold commercially. Photo from Raytheon brochure.

Early 1961 also saw the announcement of the q-switch, invented by Robert Hellwarth at Hughes Research [19]. This device eliminates the spiking that occurs in pulsed ruby lasers, and dramatically increases the peak powers achieved by them. Many techniques have been employed for q-switching including the Kerr cell, a rotating mirror, the Pockels cell, and saturable absorbing dyes. Q-switched lasers would eventually be necessary for many successful applications of holography. Another device of great significance to holography that was developed in the middle of 1961 was the intracavity etalon for single longitudinal mode selection, by D.A. Kleinman and P.P. Kisliuk [20].

Other pulsed ruby lasers available for sale near the end of 1961 included the Trion Instruments models LS-2 and LS-4 (which had a 30J output), the Raytheon model LH-2, the Maser Optics model 500, and the model 100 from Optics Technology Incorporated. While Raytheon and Trion lasers were characterized by their high-energy outputs and liquid nitrogen cooling, the initial offerings from Maser Optics and Optics Technology were smaller lab lasers intended for basic experimentation with laser light.

Maser Optics was a major supplier of lasers throughout the 1960s. Headed by Harry E. Franks, they offered ruby lasers, laser photocoagulators, educational kits, He-Ne lasers, and laser accessories from 1961 to 1968. In 1964 they produced a ruby laser rifle for the army, and were reported to be the world's largest consumer of synthetic rubies [21]. Throughout the second half of the '60s however, they were experiencing problems with the SEC, and in April 1968 became Laser Nucleonics (also headed by Franks) [22].

Optics Technology Incorporated was formed to be a modern optics research operation in January 1961, when a seven-man group headed by Dr. Narinder Kapany was acquired by Spectracoat Inc., a manufacturer of optical coatings since 1958. Along with coatings, the company produced ruby and He-Ne lasers, power meters and interferometers, and photocoagulators [23]. Similar to the model 100, their model 120 ruby laser system (Figure 5) featured an easily accessible pump cavity, a 3 x 1/4-inch rod, 0.1-J output energy, an iris diaphragm on the output bezel, and threads on the front end for a microscope objective. OTI was a major manufacturer of lasers throughout the 1960s, but by 1973 had fallen on difficult times and ceased all laser production [24].

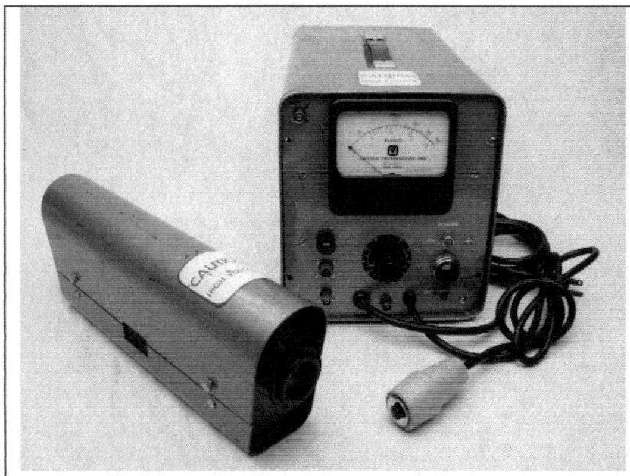

Figure 5. c1962 Optics Technology Incorporated model 120 ruby laser system with the model 100 power supply.

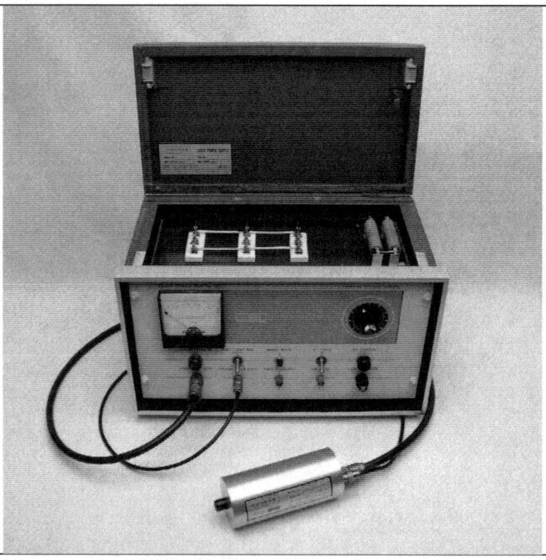

Figure 6. 1962 Hughes Aircraft Company model 200 ruby laser system, fully operational with all original components.

In 1961 however, Helium-Neon lasers were still built in-house by and/or for researchers at Bell Labs, Hughes and Raytheon (for example), but that would soon end. In September, Spectra-Physics was founded in a deal with Perkin-Elmer making it the first commercial continuous wave laser company. The deal included labeling all lasers with both names, and was to be terminated in two years or upon the sale of the 75th laser, whichever came first [25].

4. The Continuous Wave

In February 1962, Yuri Denisyuk made volume reflection holograms of a convex mirror and a micrometer scale using a high-pressure mercury arc lamp, as was being used by Leith and Upatnieks at Willow Run. Juris Upatnieks was still away in service, but laser technology was quickly expanding.

In March at the west coast meeting of the IEEE, Perkin-Elmer Spectra-Physics (P-E S-P) introduced their model 100 Helium-Neon laser with an infrared output, the first commercial laser with a continuous beam [26]. Its head consisted of an electronics filled box with the plasma tube protruding out both sides. The resonator mirrors were attached to the ends of the plasma tube [27]. Other He-Ne lasers with infrared outputs that became commercially available shortly afterward were the Sylvania model GL-6211 and one from Raytheon. All of these early commercial gas lasers were RF excited with only a few milliWatts of output power, and very delicate.

In May of that year, Hughes Aircraft Company introduced their first commercial laser despite having built the first laser two years earlier [28]. The model 200 pulsed ruby laser was a commercialized version of Maiman's first device, but with a longer rod having dielectric mirrors coated on its ends (Figure 6). There were two or three other models built after the model 200 and similar to it, then external mirrors were employed, and finally a Kerr cell q-switch to achieve higher energies in later devices [29].

More significant to holography however, was the continuous wave He-Ne gas laser. In August 1962, Alan White and J. Dane Rigdon of Bell Labs achieved laser action in He-Ne at the visible wavelength of 632.8nm. A month later, P-E S-P introduced the model 110, the first commercial laser with a continuous visible beam [30] (Figures 7). Mated with the model 71 power supply, this laser featured a metal resonator rod surrounded by a heat shield to hold the mirrors, a smaller diameter plasma tube, an output of 1.0mW from both ends multimode (with a confocal resonator), 0.3mW output from both ends single mode (with a hemispherical resonator), and cost $7900 new [31] (Figure 8). This was the breakthrough laser that many working with communications, optical processing and optics in general were waiting for. One of those researchers was Anthony VanderLugt at Willow Run, who acquired one immediately for his work on spatial filtering.

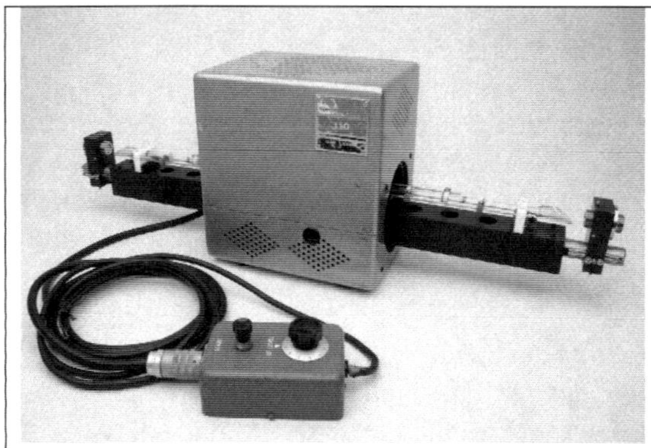

Figure 7. 1962 P-E S-P model 110 Helium-Neon laser head, without the cylindrical aluminum covers for both ends.

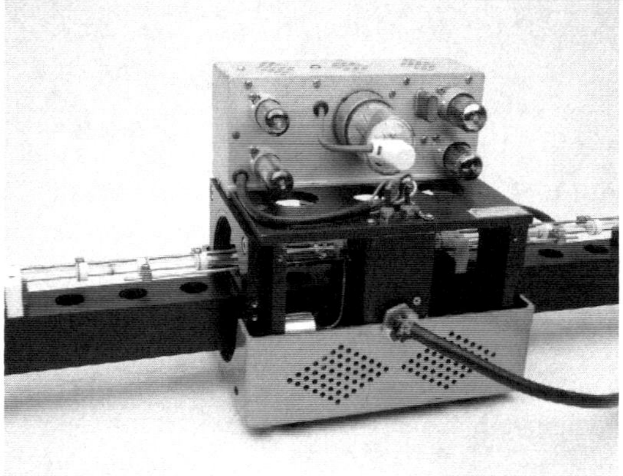

Figure 8. Uncovered P-E S-P model 110 head showing the RF exciter above the plasma tube.

When Juris Upatnieks returned from military service near the end of 1962, he and Emmet Leith immediately got back to work on wavefront reconstruction of binary images with the mercury arc lamp. Near the end of December however, they borrowed the beam from VanderLugt's P-E S-P model 110 laser and on Christmas Eve made the world's first hologram with laser light and 2D imagery [5].

5. The Birth of Laser Holography

The first commercial laser with a continuously visible beam left much to be desired for making holograms. The output power was very low, especially in single mode operation. The Brewster windows were attached with epoxy instead of being hard sealed, limiting tube lifetime. Mirror coatings were delicate and easily damaged. And finally, the mechanical design of the resonator structure and mirror mounts required frequent adjustment to keep them lasing.

In the spring of 1963, Leith and Upatnieks acquired a He-Ne laser for their own lab. This could have been a device made by Raytheon, a P-E S-P model 111 (Figure 9), a P-E S-P model 112, and/or the Semi-Elements model SEOG-MK1, all of which were also introduced near the end of the previous year. The P-E S-P model 111 was essentially a longer version of the model 110 with the same mirror mounts and resonator rod (Figure 8), and the model 200 RF exciter powered it. It had an output power of 30mW multimode (with long radius mirrors) and 10mW single mode (with a hemispherical resonator) [31]. The model 112 had the same plasma tube as the model 111, but a larger resonator rod and better mirror mounts. In March and with their larger laser, Upatnieks made an unsuccessful attempt to record a hologram of three-dimensional objects (steel rods), then spent the next five months improving their techniques and testing better equipment [5].

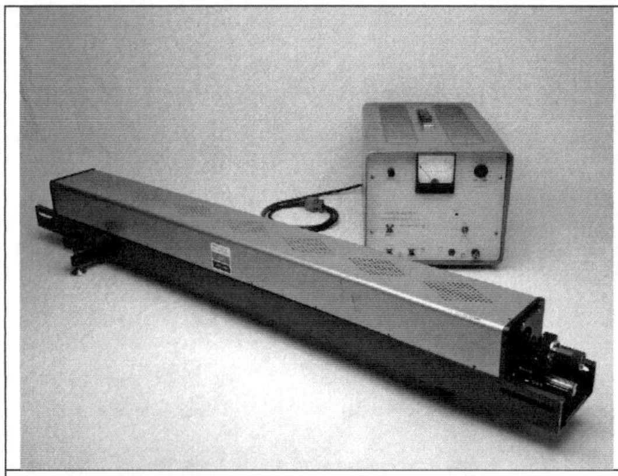

Figure 9. 1962 Perkin-Elmer Spectra-Physics model 111 Helium-Neon gas laser system with model 200 power supply.

Figure 10. 1963 Spectra-Physics model 115 Helium-Neon gas laser head. This head is about half as long as the model 111.

By June, Perkin-Elmer had sold the 75th laser under its deal with Spectra-Physics and the two companies separated. Spectra-Physics continued production with the model 112 and also introduced the models 115 and 130. The model 115 (Figure 10) was an RF excited improved version of the model 110. It featured better mirror mounts and adjusters, a cavity length adjustment knob, a much larger gas reservoir tube, 1mW output power in single mode, and a price of $4650 [32].

The model 130 (Figure 11) was the first laser "pointer", as it was the first continuous wave laser with a visible beam and a self-contained power supply. This was possible because it was also the first commercial laser with DC excitation. It had an output power of approximately 0.5mW, weighed ten pounds, was 13" long, and cost $1525 [33]. The model 130 was introduced with a hot cathode, but it was apparently replaced with a cold cathode as seen below. This laser was also designed with its resonator structure also serving as the outer case, a feature that may be described as an exoresonator. A ten-pound, 0.5mW laser pointer with a handle attached to the resonator!

Perkin-Elmer introduced their model 5200 He-Ne in 1963 as their first laser product, and it became very popular throughout the rest of the decade (Figure 12). It was the first He-Ne with the now familiar cylindrical head shape, and featured a hot cathode, a double walled plasma tube for increased lifetime inside an exoresonator, and an output power of 0.5mW [34]. This laser was designed to be rugged for reliable operation in factory and construction site settings, and had a very "James Bond" style to its design.

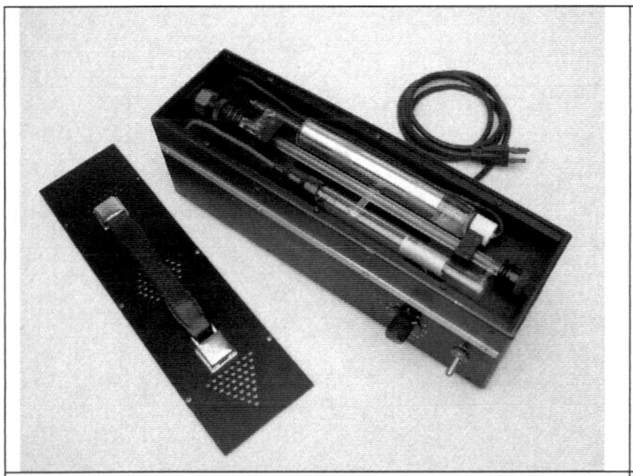

Figure 11. 1964 Spectra-Physics model 130B. The model 130 was the first commercial laser with an integrated DC power supply, and designed to be used as a "laser pointer".

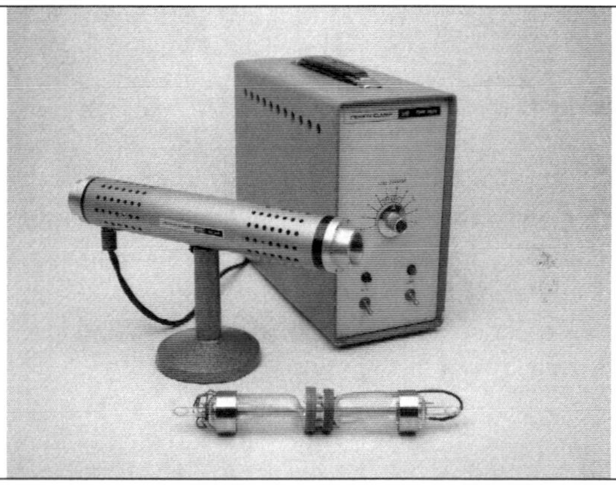

Figure 12. Perkin-Elmer model 5200 Helium-Neon gas laser system with a model 5202 power supply. Also shown is the plasma tube.

Meanwhile, other researchers applied the laser to holography. Ralph Wuerker learned about Gabor's wavefront reconstruction technique from Professor Paul Kirkpatrick while at Stanford (at the same time as Ted Maiman). While at Quantatron with Maiman, he helped design their first pulsed ruby and q-switched lasers, but left there to go to TRW in late 1962. There, he built his own q-switched ruby laser to record Gabor holograms for the examination of spray fields from a colloid propulsion system [35]. His recordings were probably the first holograms recorded with a pulsed laser.

Brian Thompson at the University of Rochester was, like Leith and Upatnieks, using coherent light with an optical processing technique, but he was measuring the size of aerosol particles. He chose the pulsed ruby laser for its ability to freeze particle motion for imaging, and because it also had a visible output. He also used a spatial filtering technique to reconstruct his recordings and didn't credit Gabor's technique when describing his work initially, but they were nonetheless Gabor holograms made with a pulsed ruby laser. As for Wuerker, the multiple images reconstructed by Gabor's in-line geometry were not a problem with their application because the depth of focus for the reconstruction system was much smaller than the distances between them. Coherence requirements were also minimal.

The laser Thompson used was q-switched, had a pulse length of about 2 microseconds and a repetition rate of 4-6 seconds [36]. It was probably a Korad model K1-Q, but may also have been made by Hughes or Trion/Lear Siegler, both of which are known to have offered q-switched lasers at the time. Other pulsed laser manufacturers in 1963 included CSF in France, EG&G, TRG, and Semi-Elements.

Korad was formed when Quantatron, the company started by Ted Maiman after leaving Hughes in early 1961, was acquired by Union Carbide in November 1962. Their first products were the models K1 and K1-Q, the first commercial q-switched laser [37]. With Maiman at the helm, Korad was one of the major suppliers of commercial solid-state lasers for the rest of the decade. It was eventually acquired by Hadron in the early 1970s.

Trion/Lear Siegler offered a q-switch for their lasers in early 1963 called the Optul. It was a spinning mirror device that could be used with their models LS-1 and LS-4, and was also sold as a separate component.

The third quarter of 1963 also saw the introduction of the Gallium Arsenide semiconductor injection laser to the marketplace. This type of laser was first made to operate a year earlier by Robert Hall's group at General Electric, followed immediately by groups at IBM and MIT's Lincoln Labs. Nick Holonyak Jr. at GE Syracuse operated the first semiconductor laser with a visible output shortly thereafter. The first commercially available GaAs laser however, was the model ILS-1, introduced at the end of that year by ITT Industrial Labs [38]. It was followed within a few months by the Korad model K-S1, the Philco model GAE-404, and the Texas Instruments model SNX-110. All of these early GaAs lasers were cryogenically cooled.

By September however, Leith and Upatnieks had optimized their film and apparatus for making holograms with laser light, and introduced diffuse illumination to their hologram recording systems [5]. Diffuse illumination of 2D objects removed the coherent noise and diffraction ring patterns caused by particles on the optics in the system. Ironically, these artifacts of highly coherent light in optical systems of the day were essentially what Brian Thompson was using to measure the size of the offending particles!

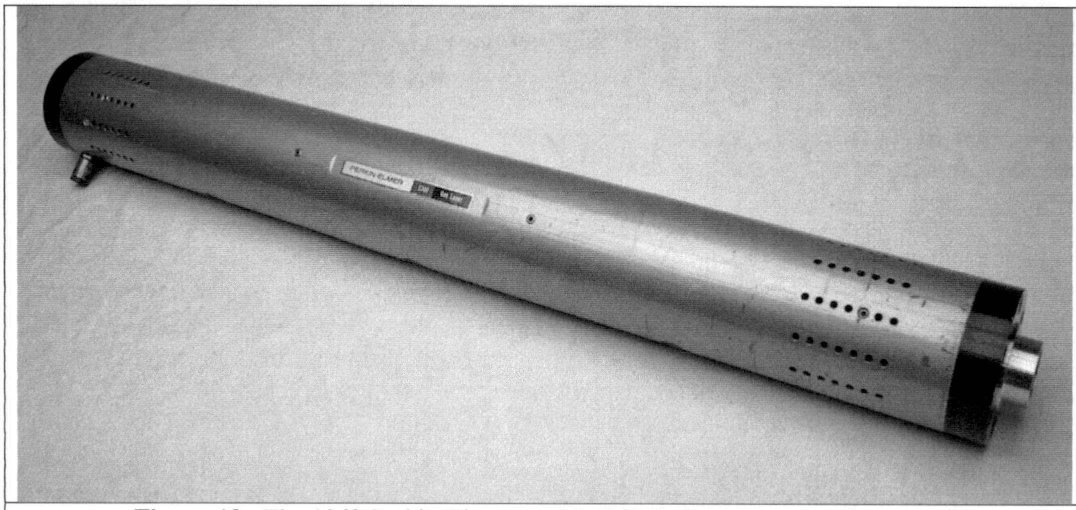

Figure 13. The 1963 Perkin-Elmer model 5300 Helium-Neon gas laser head.

Around that same time, Perkin-Elmer introduced their second and larger He-Ne laser, the model 5300 (Figure 13). This laser had an output power of 8mW single mode, 15mW multimode, and a DC excited plasma tube with a hot cathode inside an exoresonator (a 35 pound, 33" long metal tube with 1" walls) [39]. It was apparently more stable and easier to keep aligned than the Spectra-Physics lasers, and became the laser of choice for their next step. So finally, on December 22, 1963 and using the Perkin-Elmer model 5300 Helium-Neon gas laser and a model train for the object, they recorded the world's first hologram of diffuse three dimensional imagery [40].

6. Epilogue

The hologram of that little train made a big splash at the spring meeting of the Optical Society of America in early April, 1964. Its presentation was just a few months after publication of their work with off-axis holography with two dimensional greyscale images, and the phrase "lensless photography" that was coined to describe it [41].

Commercial laser development continued at a rapid pace. Many companies joined those already mentioned making ruby and He-Ne lasers, and they all contributed in greatly increasing the output power, stability, ease of use, and lifetime of them. Spectra-Physics introduced the models 131 and 116 He-Ne lasers near the end of 1963. The model 116 was the first wavelength tunable laser available commercially [42]. Mated with the model 250 exciter, it was DC excited but also used a small amount of RF to quiet and stabilize the discharge, had an output of 25mW [43], and cost a whopping $13,250.

A significant technical improvement that enabled many applications of He-Ne lasers was the invention of the oxide-coated aluminum cold cathode with a large surface area, by Hochuli and Haldeman in 1965 [44]. This advance greatly increased the lifetime of this laser type, and was probably first introduced in the "Stabilite" series of lasers from Spectra-Physics the following year. Another big improvement to He-Ne lasers was the incorporation of internal mirrors, first available on the Optics Technology model 170 also in 1966 [23].

Other laser types made commercially available in the mid 1960s included Nd:Glass and Nd:YAG solid-state lasers, the carbon dioxide molecular laser, the Helium-Cadmium metal vapor laser, and Argon and Krypton ion lasers. Of these, the He-Cd and ion lasers had the most dramatic impact on holography because of their many visible wavelengths and high output powers. First to market with the He-Cd laser late in 1969 was Spectra-Physics with their model 185. It produced 50mW of blue light from a tube specified to last 3000 hours [45]. Raytheon introduced the first commercial Argon ion laser in November 1964, the model LG-12 [8].

This first generation of lasers was obsolete before the end of the 1960s, and are now highly collectable [46]. The Spectra-Physics model 125 was introduced in 1966 with an output power of 50-100mW at 633nm. Despite its six-foot length, it became the He-Ne laser of choice for holographers until the end of the century. The vast majority of the Kennedy-era lasers however, were scrapped long ago. But holography found its way into many and varied applications as lasers and recording materials improved, and it truly was one of the problems that lasers solved.

7. Acknowledgements

The author gratefully acknowledges the very helpful correspondence and discussions with pioneering scientists who graciously shared their experiences, and helped identify the "cabinet grannies" shown here. They include Irnee D'Haenens and Doug Buddenhagen from Hughes, Colin Bowness, Michel Ciftan and Michael Holt from Raytheon, Donald Gillespie, Clarke Harris and Don Broadbent from Trion, Narinder Kapany from Optics Technology, Rod Waters from Korad, John Ekstrand from Spectra-Physics, Brian J. Thompson, Tony Siegman, Ralph Wuerker, Jeff Hecht and of course and with great respect, Juris Upatnieks. Special thanks also to William S. Alcott, the industry analyst who saved all his copies of The Laser Letter, Laser Weekly, and Laser Newsletter, and to the executors of his estate who put them all up on Ebay to be found and used extensively to reconstruct the very early days of the laser industry.

8. References

[1] Townes C 1999 *How the Laser Happened* (Oxford: Oxford University Press) p 66
[2] Meyer J 25Apr1958 The Solid State Maser A Supercooled Amplifier *Electronics* **31** 66
[3] Bromberg J 1991 *The Laser in America* (Cambridge: MIT Press) p 40
[4] Townes C 1960 *Quantum Electronics* (New York: Columbia University Press) p 330
[5] Johnston S 2006 *Holographic Visions* (Oxford: Oxford University Press) pp 96-108
[6] Hecht J 2005 *Beam* (Oxford: Oxford University Press) p 199
[7] Shore B 1962 Lasers: New Power from Light *Electronic Age* **21** 10
[8] Bushor W (ed) 15Jun1966 *Laser Focus* **2** 11
[9] Ciftan M, Luck C, Shafer C and Statz H May 1961 *Proc. I. R. E.* **49** 960
[10] Klein H 1963 *Masers and Lasers* (Philadelphia: Lippincott) p 133
[11] Carroll J 1964 *The Story of the Laser* (New York: E.P. Dutton) p 167
[12] Hoaglin C 25Jun2010 Private communication
[13] Maiman T 2000 *The Laser Odyssey* (Blaine: Laser Press) p 143
[14] Hecht J 1992 *Laser Pioneers* (San Diego: Academic Press) p 160
[15] Johnston S 2006 *Holographic Visions* (Oxford: Oxford University Press) p170
[16] Bromberg J 1991 *The Laser in America* (Cambridge: MIT Press) p 117
[17] Luck C July/August1961 Optical Masers *Electronic Progress* **6** 13
[18] Lehr C 8May1962 *Principles and Applications of Lasers* (Lexington: Raytheon Company) 10
[19] Smith G 1984 *IEEE J Quant Elect* **QE-20** 579
[20] Kaminow I, Siegman A (ed) 1973 *Laser Devices and Applications* (New York: IEEE Press) 190
[21] Bushor W (ed) June1964 *The Laser Letter* 23
[22] Pasek R (ed) July1968 *Laser Newsletter* **5** 1
[23] Bushor W (ed) 15Aug1966 *Laser Focus* **2** 7-16
[24] Bushor W (ed) March1973 *Laser Focus* **9** 30
[25] Bushor W (ed) 15Jul1966 *Laser Focus* **2** 9
[26] Bromberg J 1991 *The Laser in America* (Cambridge: MIT Press) p 121
[27] Levine A October 1962 *The Exchange* **23** 4
[28] Staff 1May1962 Hughes Aircraft Adding New Line *New York Times* 49
[29] Buddenhagen D 26Nov2006 Private communication
[30] Stocklin W (ed) December1962 *Electronics World* **68** 37
[31] Ekstrand J 2Jul2010 Private communication, owners manual
[32] Kuper H (ed) June1963 *Rev. Sci. Instr.* **34** 727
[33] Kuper H (ed) August1963 *Rev. Sci. Instr.* **34** 953
[34] Bushor W (ed) 1Aug1966 *Laser Focus* **2** 14
[35] Wuerker R 21June2012 Private communication
[36] Thompson B 1963 *J. Soc. Photo-Opt. Instr. Eng.* **2** p 43
[37] Wixom C (ed) May1966 *The Laser Letter* **3** 6
[38] Abelson P (ed) 27Sep1963 *Science* **141** 1227
[39] Bushor W July/August1964 *The Laser Letter* **1** 21
[40] Upatnieks J November2011 Private communication
[41] Berland T March1964 *Popular Mechanics* **121** 105
[42] Kuper H (ed) December1963 *Rev. Sci. Instr.* **34** 1464
[43] Ekstrand J June2012 Private communication
[44] Hochuli U and Haldeman P 1965 *Rev. Sci. Instr.* **36** 1493-4

[45] Pasek R (ed) 1Sept1969 *The Laser Weekly* 4
[46] Hess R December2010 *Photonics Spectra* **44** 51

Journal of Physics: Conference Series **415** (2013) 012028 doi:10.1088/1742-6596/415/1/012028

What are the ideal wavelengths for full color holography?

Kaveh Bazargan

holographer.org, Unit 1, 43 Hereford Road, London W2 5AH

E-mail: kaveh@holographer.org

Abstract. One of the holy grails in display holography is the production of natural color holographic images. Various sets of wavelengths for recording have been suggested, some favoring three wavelengths, some four, and even more. I will argue that the choice of recording wavelengths is completely independent of the holographic process; it was in fact was solved once and for all by scientists working in general lighting in the 1970s. I will suggest an ideal set of wavelengths which will produce color rendition equal to better than conventional photographic processes.

1. Introduction

Color science is an extremely complex subject. It is a subtle combination of many disciplines, including physics, chemistry, physiology and psychology. Under normal circumstances, our perception of colors derives from the spectral power distribution of the illuminating source (normally continuous, e.g. the sun), the spectral reflectivity of the object being viewed, and finally our eye/brain combination. The fundamental problem in color holography is that the illuminating source (2 or more laser sources) is not continuous, but discrete. Hence the only information we have on the spectral reflectivity is at those discrete wavelengths. The problem, then, is which wavelengths we choose to get the best color rendition of the object.

In this paper I shall follow the following reasoning:

- When the loci of the recording wavelengths are plotted in the CIE Color Chromaticity Chart, it is a widely held view that the geometric recording wavelengths must cover as large an area of the diagram as possible; this assumption is incorrect;
- There are good reasons for *not* covering the largest area referred to above;
- The choice of recording wavelengths has little if anything to do with holography, and has been addressed and solved by researchers working in the area of color and illumination.

(Most of the content contained herein was first published in 1989 by Bazargan [1].)

2. the CIE Color Chromaticity Chart

Many systems have been developed to represent systematically all the colors we perceive. All the systems are three-dimensional, i.e. need three variables in order to define all visible colors. Examples of these variables are RGB, HSB, LUV, etc – see figure 1.

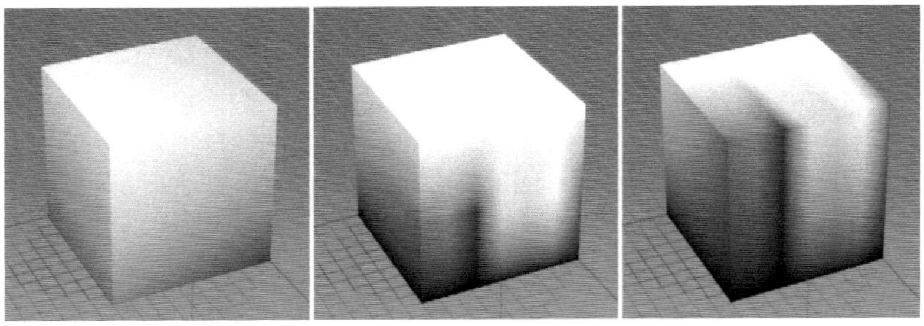

Figure 1. Examples of models for representing all visible colors in 3D. From left: RGB, HSL, HSV. (*Source: Wikimedia*)

But a three-dimensional model is hard to work with, as it cannot be easily represented on paper. In order to have a convenient 2D model, one of the axes needs to be ignored. The CIE color model does this by ignoring the Lightness scale. The 1931 CIE Color Chromaticity Chart (figure 2) is still the most popular chart for plotting colors, but the 1976 chart (figure 3), which is a geometric transformation of the original chart, and which has a more realistic distribution of colors, is recommended and should be used instead. The saturated colors are represented by points near the periphery of the shape, with the pure spectral colors being represented by the curve on the left and top of the shape. Approximate positions for three wavelengths are shown in figures 2 and 3.

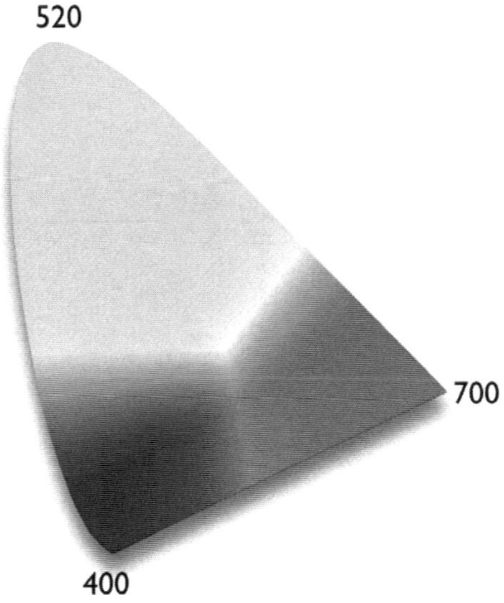

Figure 2. The 1931 CIE Color Chromaticity Chart allows a two-dimensional representation of all distinguishable colors, by ignoring the third dimension of 'lightness'.

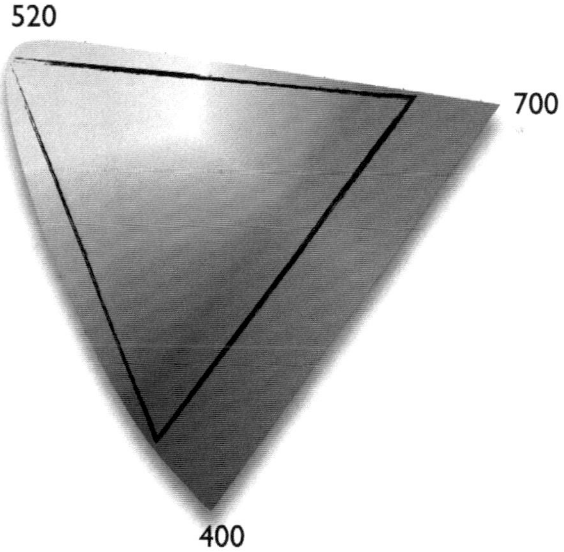

Figure 3. The 1976 CIE Color Chromaticity Chart gives a better distribution of colors and should be used in place of the original 1931 chart.

A useful property of the CIE color chart is that it allows the prediction of which colors can be matched by the additive mixture of several other colors. Let us consider the black triangle in figure 3. The vertices of the triangle represent three pure spectral colors, as for instance generated by a laser. According to the geometric principle of the chart, all colors within the triangle can be perfectly matched by mixing the three spectral colors in the correct proportions.

If we look at publications on color holography since the early days, it is almost always assumed that the choice of laser colors has to be such that the greatest area of the CIE chart is covered, in order that any conceivable color can be reproduced in the hologram. But this is a misconception, as I will show below.

3. Why we don't need the biggest triangle

The key point that holographers have been missing is that the CIE chart represents all colors that the eye can perceive, *not* all colors occurring in everyday objects. Let us plot some real object colors onto the chart to get an idea of the area that we need to cover. In figure 4 we have plotted some common everyday colors and a few saturated colors from the Macbeth Color Checker Chart (B, G, Y, R). We can see that they are all clustered in an area away from the periphery of the chart. The irregular triangular shape enclosing these colors is the gamut of *all* natural and man-made object colors that could be found by Wintringham [2] in 1951. In 1980, Pointer [3] extended the gamut slighly, mainly to include more recent artificial materials. But we are unlikely to come across any colors in everyday life that lie outside Wintrigham's original gamut, and most triangles formed by RGB laser primaries will cover almost the entire gamut of everyday colors.

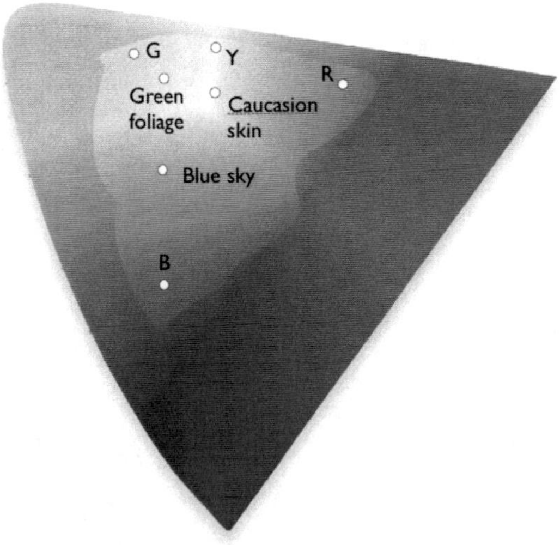

Figure 4. The 1976 CIE Color Chromaticity Chart gives a better distribution of colors and should be used in place of the original 1931 chart.

4. Reason for *not* choosing a large triangle

Notwithstanding the above argument, one might be tempted to use a larger polygon or triangle of colors, on the basis that there is no disadvantage. But in fact there are good reasons *not* to choose wavelengths at the extremes of the spectrum. Figure 5 depicts the *Luminous efficiency* curve, which shows the sensitivity of the eye to the visible spectrum. It can be seen that for the same power of a laser beam, colors near the centre of the spectrum seem brighter than those

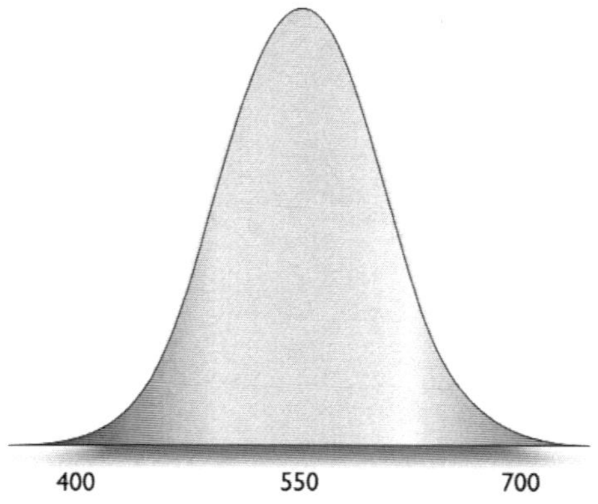

Figure 5. The luminous efficiency curve, showing that the eye is most sensitive to the centre of the visible spectrum, i.e. around 540 nm, and falls off to zero around 400 and 700 nm. So the same power emitted by the laser beam will look less bright towards the extremes of the visible spectrum.

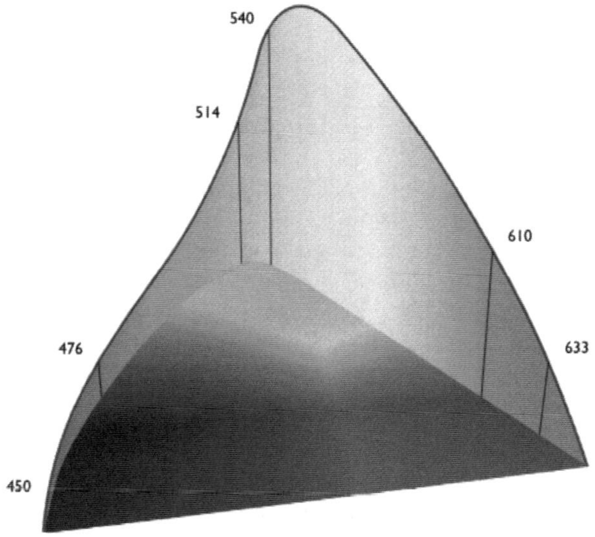

Figure 6. The luminous efficiency curve plotted on the locus of spectral colors of the CIE chart.

near the extremes. Colors below ~350 nm and above ~750 nm will be completely invisible to the eye.

Let us now plot this curve onto the periphery of the CIE curve (figure 6). We can see that as the blue and red wavelengths approach the extremes of the visible spectrum. In the limit of trying to obtain the largest polygon or triangle, the red and blue wavelengths will be invisible, or appear to be of very low brightness.

So all else being equal, this is a good reason *not* to try and cover the entire CIE chart with the recording wavelengths.

5. So which wavelengths *should* we use?

The conclusion from the above arguments must be that the CIE chart is of little use for choosing wavelengths for color holography.

Let us now look at how we perceive colors in everyday life. Figure 7 shows a simplistic diagram of an observer viewing an object. The colors that we perceive depend essentially on two parameters:

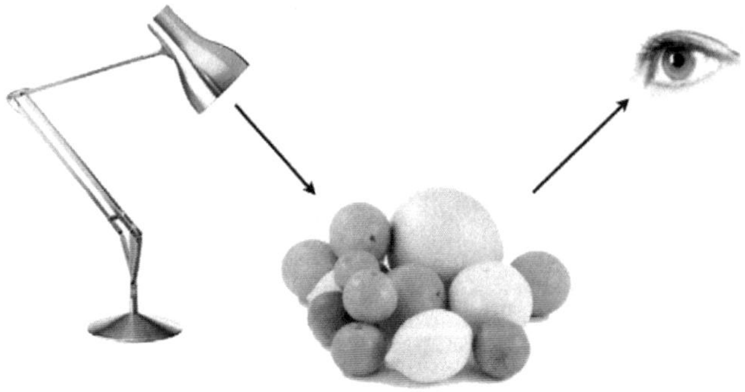

Figure 7. How we see objects depends on the properties of the object, and those of the illuminating source.

- Spectral reflectivity of the object;
- Spectral power distribution (SPD) of the illuminating source.

Figure 8 is a schematic diagram to show how we perceive colors under normal conditions. The light has a spectral power distribution (which, for example, determines whether it is a warm or cool light, etc). Each object has a spectral reflectivity curve which determines how much of each wavelength it reflects. The combination of these two spectral functions create a composite spectral distribution which determines the color the eye perceives. The essential point here is that the SPD of the light is in general continuous, so the object is 'tested' at all visible wavelengths.

When recording a color hologram, we are only using a small set of wavelengths (let us say three). For the purposes of discussion we can in fact forget the fact that we are making a hologram. The problem is one of the illumination source. So effectively the SPD of the source is three discrete lines in the spectrum, as illustrated in figure 9. So in contrast with normal conditions whereby the eye receives information about the reflectivity of the object throughout the spectrum, now only three discrete values are received.

So here is the question. Is it possible to choose an illuminating source comprising three wavelengths, such that everyday objects look to the eye the same color as they would under broadband illumination? It would seem an impossible task as the data received by the eye is so small. But there is a key publication which can shed light on the problem. In 1971 William Thornton [4] published his results on a comprehensive examination of the Color Rendering Index (CRI) of white light. He found that there are three areas in the spectrum (around 450, 540, and 610 nm) which seem to have a disproportionate 'pull' in the CRI. In fact, his conclusion was that a light comprising solely of these three narrow wavelengths had a higher CRI than *any* other combination of wavelengths, whether discrete or continuous.

Thornton named these the *Prime-color wavelengths.* How these wavelengths relate to the absorption dyes in the eye is still debated today, but as far as color holography is concerned, I believe that these 'magic' wavelengths are the ideal for holography, and will deliver color rendering as good as any photographic process.

6. Conclusions

The CIE chart is of little use in determining the wavelengths to use in full color holography. The choice of wavelengths is not related to holography at all, but to color science. Thornton's prime-color wavelengths (around 450, 540, and 610 nm) are the ideal set for holography.

Journal of Physics: Conference Series **415** (2013) 012028 doi:10.1088/1742-6596/415/1/012028

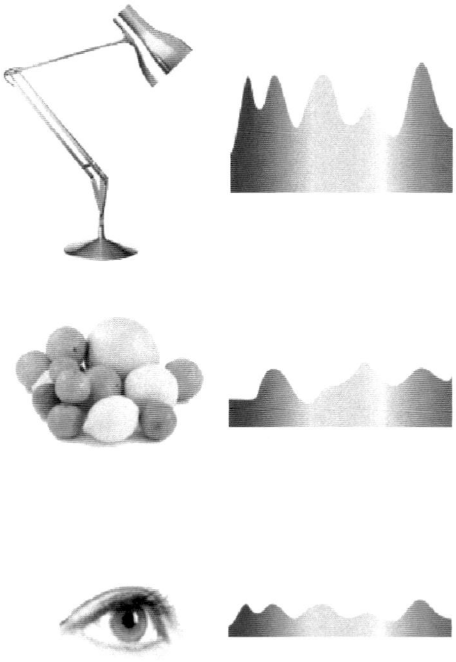

Figure 8. Viewing a scene under normal conditions. The spectral power distribution of the illuminating source is continuous, so the reflectivity of the object is tested throughout the entire spectrum.

Figure 9. Viewing a scene using discrete light sources, e.g. three laser wavelengths. They eye receives a tiny proportion of the information regarding the reflectivity of the object. Is it possible to find three wavelengths that can emulate a continuous source?

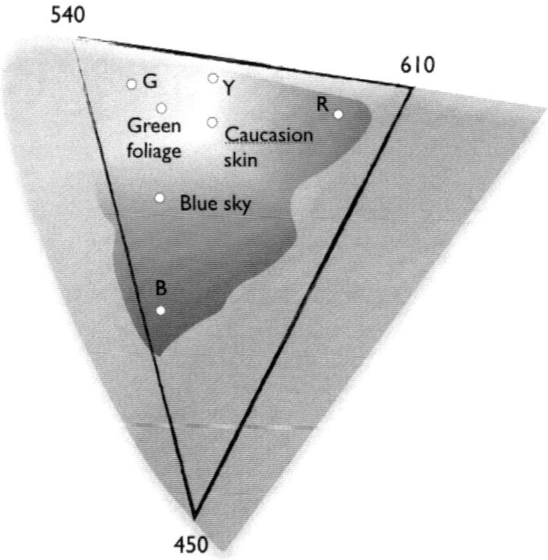

Figure 10. Thornton's 'prime-color wavelengths' plotted on the CIE diagram.

[1] Bazargan K 1989 *Second International Conference on Holographic Systems, Components and Applications* Proc. IEE (Bath, 1989) 49–50
[2] Wintringham W 1951 *Proceedings of the IRE* **39** 1135–1172
[3] Pointer M R 1980 *Color Research and Application* **5** 145–155
[4] Thornton W A 1971 *JOSA* **61** 1155–1163

New recording material with AgBr nanoparticles for optical holography

M Květoň[1], M Škereň[1], P Fiala[1] and A Havránek[2]

[1] Department of Physical Electronics, Faculty of Nuclear Sciences and Physical Engineering, Czech Technical University in Prague, 115 19 Prague 1, Czech Republic

[2] Department of Macromolecular Physics, Faculty of Mathematics and Physics, Charles University, 180 00 Prague 8, Czech Republic

E-mail: milan.kveton@fjfi.cvut.cz

Abstract. A new recording material with silver bromide nanoparticles and photopolymerization system has been prepared and tested as a recording medium for optical holography. Through the method of the real-time measurement of a diffraction grating growth, we have shown that an efficient volume phase grating is formed during a holographic exposure. The material response on different exposure parameters has been measured and obtained results have been discussed. The transfer of nanoparticles within the recording layer, which is the main cause of the grating growth, has been experimentally verified through the direct observation of the grating micro-structure with the scanning electron microscope.

1. Introduction

Photopolymerizable recording materials with nanoparticles are the subject of an intensive study in last ten years (e.g. [1–5]). In comparison with common photopolymers, they contain inert inorganic particles with a high refractive index. Similarly as in the case of common photopolymers, the materials are self-developing and no additional wet chemical developing process is needed. The refractive index modulation, which is formed already during a holographic exposure, is relatively high and stable. Hence, the materials are attractive for applications such as image holography, for fabrication of diffractive optical elements, optical security elements, integrated optics, etc. The development of our photopolymer with silver bromide (AgBr) nanoparticles was inspired by the phenomenon of light induced mass transport in volume phase recording materials which can lead to redistribution of components (polymers and nanoparticles) within the medium and hence formation of a refractive index modulation. It is known that a transport of mass is the main cause of local changes of the refractive index in all efficient holographic recording materials. However, the mechanisms of molecules or particles redistribution differ.

In photopolymers, during a holographic exposure, polymer chains are formed in bright regions of the interference field. As the mobility of growing polymers is much lower than that of monomers, the polymers remain at their places of origin and so the modulation of polymer density is formed. The value of the refractive index modulation caused by the modulation of polymer density is relatively low. It has been recently shown that addition of inert particle into the photopolymerizable medium can increase the refractive index modulation. TiO_2, SiO_2, or ZrO_2 nanoparticles with different refractive indexes from that of the host material are used most often. The particles are spatially redistributed due to the inhomogeneous polymerization initiated by a holographic exposure.

Mass transport is running also in silver-halide emulsions (SHE) when the process of the rehalogenization bleach is used for conversion of an amplitude hologram into a phase hologram [6]. SHE are composed of gelatin matrix in which grains (nanoparticles) of silver halide are dispersed in a high concentration. With the rehalogenization bleach technique, the developed metallic silver particles are oxidized into silver ions which became a part of the bleaching solution. The bleaching solution contains also halide ions. Due to the affinity of silver and halide ions to original grains in the unexposed areas, the grains grow at their expense and migration of silver ions occurs from exposed to unexposed areas. The formed refractive index modulation is high as the difference between the host gelatin matrix and the bulk silver halide material is high. However, the rehalogenization technique causes that the size of grains increases and so does the scattering of light which is an unwanted effect.

The basic idea of our concept is to add a suitable photopolymerizable system into the gelatin matrix with silver halide nanoparticles. When the material is holographically exposed, the photopolymerization process will cause the redistribution of nanoparticles. The advantage of the recording system is that silver halide nanoparticles are redistributed without additional developing processes and remain of the same size after the holographic exposure. AgBr nanoparticles have been chosen because of the fact that the refractive index of AgBr material is relatively high (about 2.3 for visible light) in comparison with gelatin or polymers and the preparation procedure of AgBr nanoparticles in a gelatin matrix is well developed for production of SHE.

2. Mechanism of polymer recording

In recent years we have formulated the immobilization-diffusion theory (ID-theory) of photopolymer holographic recording [7; 8] where we described why the growing photopolymer was located in the illuminated parts of the recording medium and why its position was practically fixed after the exposure. Details of the argumentation are thoroughly given in the cited references. The location of the photopolymer remains the same in the recording materials with nanoparticles as it has been in the previously studied acrylamide-based recording materials. However, as the nanoparticles are not compatible with the polymer, they are extruded out of the illuminated parts of the recording material and create an inverse distribution with respect to the interference field and polymer density. In a simple way, the mechanism of polymer recording is described in Fig. 1.

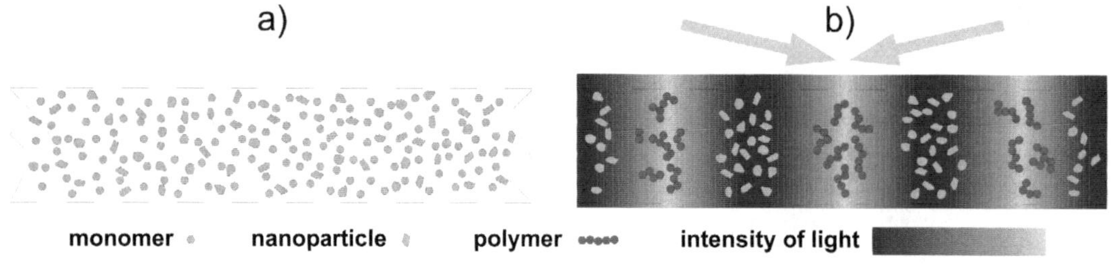

Figure 1. Distribution of monomers, nanoparticles, and polymers before (case a) and at the end (case b) of a holographic exposure within the recording layer.

The extrusion may be more seriously described by physical chemistry of polymers, namely by the Flory-Huggins theory of polymer solutions (e.g. [9–11]). Some attempts to solve this problem have been done in the work [12]. According to our opinion, neglecting the Flory-Huggins interaction parameter χ in their calculations is misleading. Namely, this parameter describes the polymer-monomer interaction in a solution and we think that the idea may be widen also to the polymer-nanoparticles mixture. More detailed discussion on the physical chemistry description of the grating growth in composite systems lies beyond the capacity of this short article. In practical evaluation of our experimental results, we continue in our previous techniques and for the specific features combined with nanoparticle systems, we were inspired by the works [1; 13].

3. Experimental

The recording material is composed of a gelatin binder with AgBr nanoparticles and acrylic monomers with a photoinitiation system. Typically, the diameter of nanoparticles is about 30 nm and they have relatively narrow size distribution. The preparation procedure of a recording layer consists of several steps which involve controlled precipitation of AgBr grains in gelatin solution, addition of a photopolymerization system, and coating thin layers on glass plates. The prepared layers are typically 20 μm thick and are deposited on a glass plates. The absorption maximum of the material is at a 550 nm wavelength (green light), its optimal exposition dose ranges from 5 to 10 mJ/cm^2, and the diffraction efficiency of a formed grating (volume phase transmission grating) reaches 80 %. Unique self-developing characteristics of photopolymers enable the real-time measurement of a forming diffraction grating. With the method, a response of the recording material on exposure of the harmonic interference field is studied for different recording parameters such as recording intensity, grating spatial period, or exposure time. The redistribution of AgBr nanonoparticles within the recording layer is also studied with a scanning electron microscope (SEM) JEOL JSM750.

3.1. Material composition and preparation

The initial ingredients for preparation of fine grains in a gelatin binder are water solutions of silver nitrate ($AgNO_3$) and potassium bromide (KBr) with a small amount of potassium iodide and a growth stabilizer. At first, a water solution is homogenized in a paddle-agitated reaction vessel which is temperature controlled. At second, solutions of $AgNO_3$ and KBr are supplied through two vertical mixing jets to the agitated gelatin solution. In the vessel, the AgBr grains of a controlled size are precipitated and fall out of the solution. The gelatin acts as a stabilizer for the individual grains, preventing them from coagulation to form larger aggregates. After the precipitation, the gelatin solution contains a number of soluble by-products which are removed by washing as they may adversely affect the shelf-life or performance of the emulsion.

The radical photopolymerizable system is added to the melted gelatin-nanoparticles mixture. The system is composed of two liquid monomers, namely acrylic acid and N,N-Dimethylacrylamide, photoinitiator H-Nu 535 (2,4,5,7-tetraiodo-6-hydroxy- 3-fluorone), and co-initiator borate V (butyryl choline butyltriphenylborate). The prepared mixture is poured over a leveled glass plate, spread with a stick or other coating tool and gravity settled. The amount of solution is a function of the desired thickness of the sensitive layer. For example, if we want to obtain a 20 μm dry layer, we are to take 0.02 cm^3 of the solution on each 1 cm^2 of a glass plate. After deposition, the layer is cooled down and dried in dark for about 4 hours. To prevent evaporation of liquid monomers, the sticky gel layer is covered with a thin PET foil. Concentrations of components of the prepared recording layer are summarized in Tab. 1.

3.2. Detection of a diffraction grating growth

With the detection method, a response of the recording material on exposure of the harmonic interference field is tested. A volume phase diffraction grating is formed already during the exposure and so its diffraction efficiency can be continuously measured. The diffraction grating is characterized with the diffraction efficiency, but the main quantity of the recording material

Table 1. Typical chemical composition of the photopolymer with nanoparticles.

Chemical	function	weight fraction
Gelatin	matrix	20 %
AgBr	nanoparticle	15 %
Acrylic acid	liquid monomer	57 %
N,N-Dimethylacrylamide	liquid monomer	7 %
H-Nu 535	photoinitiator	0.2 %
Borate V	co-initiator	0.8 %

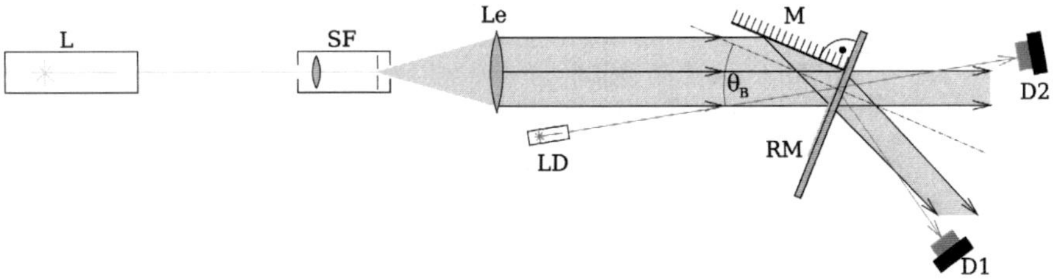

Figure 2. Schema of the experimental setup. Legend: L - recording laser (532 nm), SF - spatial filter, Le - collimating lens, M - plane mirror, RM - recording material, LD - laser diode (656 nm), D1, D2 - photo-detectors.

is the refractive index modulation. It can be evaluated from the measurement of the diffraction efficiency with Eq. (1) derived in the coupled wave theory [14]. The material parameter refractive index is connected with the chemical composition of the recording material. So measured changes of the refractive index may also be used for characterization of recording processes (polymerization, diffusion) and their further optimization.

In Fig. 2, a schema of the recording and detection setup is given. With the setup, an unslanted transmission grating is recorded with two "s" polarized waves. For the exposure, the diode-pumped, frequency-doubled Nd:YAG laser, that emits at a 532 nm wavelength, is used (Coherent model 532-400). A narrow beam of the detection laser diode, which wavelength is out of the absorption band of the material, is used for the measurement of the diffraction efficiency. The beam is adjusted at the Bragg angle of the recorded grating. The length of the exposure is controlled with an electronic shutter which also triggers the real-time measurement of the diffraction efficiency $\eta(t)$. The refractive index modulation $n_1(t)$ is evaluated as an inverse function from the expression which was derived in the coupled wave theory [14]

$$\eta(t) = \sin^2 \frac{\pi n_1(t) d}{\lambda_r \cos\theta_B}, \tag{1}$$

where d is the thickness of a recording layer, λ_r is the reconstruction wavelength and θ_B is the Bragg angle. The obtained dependence of the refractive index modulation on time we call the grow-curve and is used for characterization of self-developing recording materials. More details about the detection method and experimental setup can be found in our recent paper [8] which deals with the acrylamide-based photopolymer.

4. Results and discussion

In this section, an experimental verification of transfer of nanoparticles during holographic exposure is given through the measurement of a grating growth and direct observation of the

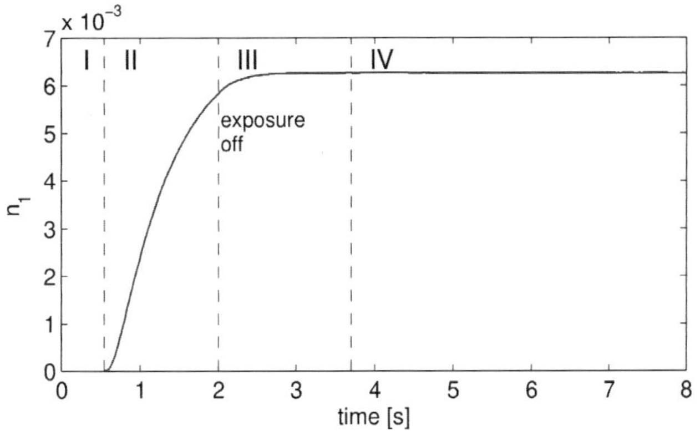

Figure 3. Grow-curve with four formation phases: I – induction period, II – exposure growth, III – post-exposure growth, IV – final period.

microscopic structure of the grating.

4.1. Formation of a refractive index grating

The first experiments verified the ability of the recording material to form a volume phase diffraction grating. The process of a grating formation was measured with the method described in the previous section. A typical grow-curve obtained from the real-time measurement is presented in Fig. 3. Within several seconds, a grating with diffraction efficiency of 60 % was formed in the layer of the photopolymer which was about 20 μm thick. The recording intensities, exposure time, and period of the recording interference field used in the experiment were 1.8 mW/cm^2, 2 s, and 700 nm respectively.

The grow-curve describes the process of the diffraction grating formation through the parameter n_1. At the beginning of the exposure, a short induction period is observed. In spite of no detectable change of the refractive index, some initiation processes may be running in the recording material. After the induction period, the growth of the grating begins. This phase is interrupted when the recording laser is switched off. The rate of a grating formation depends on exposure parameters and chemical composition of the material. Typically, the formation rate increases in the beginning of this phase. The grow-curve goes through its inflexion point and the formation rate decreases at the end of the exposure phase. An additional increase of the refractive index modulation occurs even if the exposure is switched off. It runs out successively and saturates as the formation rate decreases to zero. This phase also strongly depends on exposure parameters. In the final period, the refractive index modulation remains stable or it may slowly decrease in time. The decrease is caused by degradation processes and its rate depends on the value of n_1 and chemical composition of the material. In fact, the degradation process may already be running from the beginning of the grating growth, but it is observed only at later exposure times when the grow rate decreases.

4.2. Influence of exposure parameters

The grow-curve depends on exposure parameters which are mainly recording intensities I_1 and I_2, spatial period of the interference field Λ, and exposure time t_e. An influence of the recording intensities was measured at first and the results are given in Fig. 4. In the case of a higher recording intensity, a higher recording rate is detected than that in the case of a lower intensity. The effect of intensity can be theoretically explained by the fact that the polymerization rate

Journal of Physics: Conference Series **415** (2013) 012029 doi:10.1088/1742-6596/415/1/012029

increases with increasing intensity of light as more radicals are produced. The polymerization process is assumed as the driving mechanism and so the higher polymerization rate causes faster formation of a grating. Next to the rate of the grating growth, the final value of the refractive index modulation depends also on the recording intensity. This effect is still not very clear although it is also observed in other types of photopolymer recording materials.

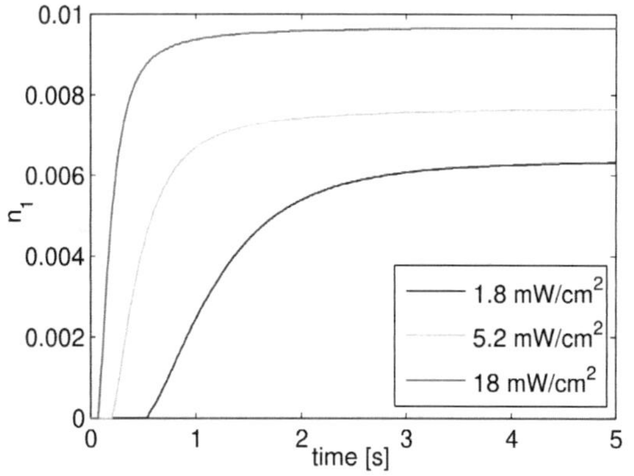

Figure 4. Influence of recording intensities on the process of a grating formation ($\Lambda = 700$ nm and $t_e = 5$ s for all curves).

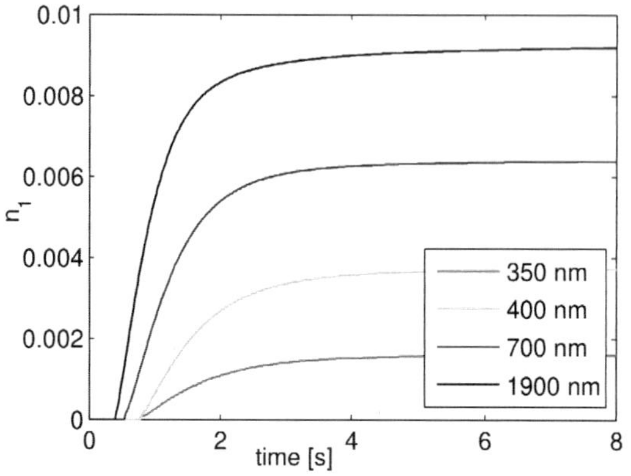

Figure 5. Influence of spatial period Λ on the process of a grating formation ($I_1 = I_2 = 1.8\,\mathrm{mW/cm^2}$ and $t_e = 8$ s for all curves).

The grating growth depends strongly on the spatial period of the recording interference field (see Fig. 5). In the case of a long period, the grating grows up rapidly and a high value of n_1 is reached. On the other hand, a short spatial period is difficult to record and a low value of n_1 is reached. We assume that this effect of the limited resolution is caused by the size and mobility of the growing polymers. Hence their positions are not located only in bright areas of the interference field, which are very narrow in this case, and the separation process of nanoparticles and polymers does not proceed optimally.

4.3. SEM observations
The direct observations with SEM microscope give evidences of redistribution of nanoparticles within the layer. With SEM, only AgBr nanoparticles can be observed; the other components such as monomer, polymers, or gelatin matrix are not visible. Before a holographic exposure, the distribution of nanoparticles in an unexposed material is homogeneous. SEM images of cross section of an exposed material (grating) in the thickness direction are shown in Fig. 6. In the figure, a grating, which follows the intensity distribution of the recording interference field, can be seen. When the higher magnification of the microscope is applied, a periodic distribution of nanoparticles can be seen. The observation has this verified that the refractive index grating is formed due to the transfer of nanoparticles caused by the inhomogeneous polymerization.

5. Conclusions
We have introduced a new photopolymer recording material with AgBr nanoparticles. We have demonstrated that a volume phase grating can be formed in the recording material when it is exposed with the harmonic interference field. The performance of the material has been tested at different exposure conditions. It has been found that high values of the refractive index modulation are reached for broader spatial periods and higher recording intensities. The evidence, that the grating is produced due to redistribution of nanoparticles during a holographic exposure, has been given from SEM observations of the grating microstructure.

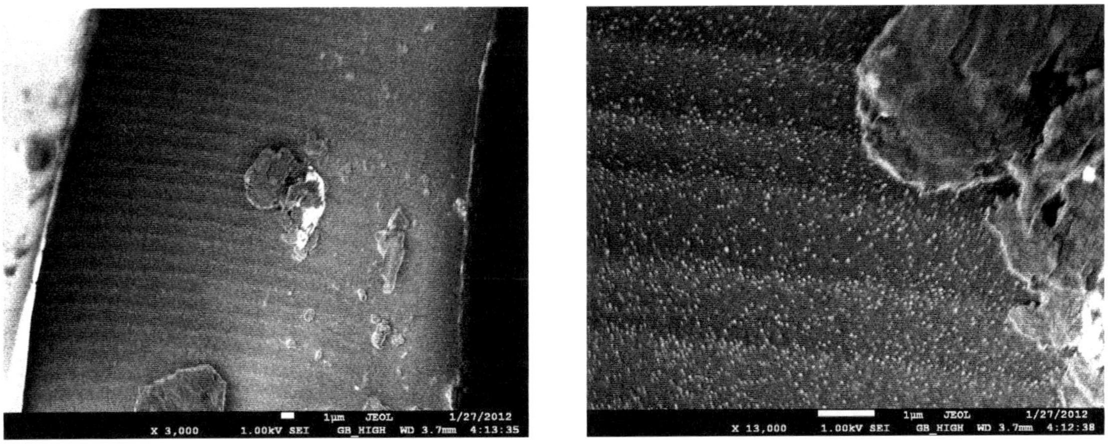

Figure 6. SEM images of the cross section of a grating recorded in photopolymer with AgBr nanoparticles. The grating microstructure is magnified 3000x (image on the left) and 13000x (image on the right).

Acknowledgments
This work was supported by the Grant Agency of Academy of Sciences of the Czech Republic under the project KAN 401220801.

References

[1] Smirnova T N, Kokhtich L M, Sakhno O V and Stumpe J 2011 *Opt. Spectrosc.* **110** 129–136

[2] Suzuki N and Tomita Y 2004 *Appl. Optics* **43** 2125–2129

[3] Tomita Y, Suzuki N and Chikama K 2005 *Opt. Lett.* **30** 839–841

[4] Juhl A T, Busbee J D, Koval J J, Natarajan L V, Tondiglia V P, Vaia R A, Bunning T J and Braun P V 2010 *ACS Nano* **4** 5953–5961

[5] Naydenova I, Leite E, Babeva T, Pandey N, Baron T, Yovcheva T, Sainov S, Martin S, Mintova S and Toal V 2011 *J. Phys.* **13**

[6] Hariharan P 1996 *Optical Holography: Principle, techniques, and applications* 2nd ed (Cambridge UK: Cambridge University Press)

[7] Havranek A, Kveton M and Havrankova J 2007 *Pol. Bull.* **58** 261–269 ISSN 0170-0839

[8] Kveton M, Fiala P and Havranek A 2011 *Holography, Research and Technologies* ed Rosen J (Intech) pp 58 – 82

[9] Flory P 1953 *Principles of polymer chemistry* 1st ed (Ithaca, N.Y.: Cornell University Press)

[10] Tanford C 1961 *Physical chemistry of macromolecules* 1st ed (New York: Wiley)

[11] Yamakawa H 1971 *Modern theory of polymer solutions* 1st ed (New York: Harper and Row)

[12] Leewis C, de Jong A, van IJzendoorn L and Broer D 2004 *J. Appl. Phys.* **95** 4125–4139

[13] Smirnova T N, Kokhtich L M, Sakhno O V and Stumpe J 2011 *Opt. Spectrosc.* **110** 137–144

[14] Kogelnik H 1969 *Bell Syst. Tech. J.* **48** 2909–2947

Holographic Metrology and Basic Physics

Nils H. Abramson, Professor *Emeritus*
Industrial Metrology, Royal Institute of Technology
Stockholm 10044, Sweden

Abstract. A short pulse of light is emitted from one point followed by a short observation from another point separated in space and time from the first. Even if space is full of scattering particles no sphere of expanding light is seen from outside by the observer, instead he finds himself inside an ellipsoid of light. We use this ellipsoid for measurement and in a graphic way to explain and evaluate optical resolution, gated viewing, radar, holography, 3-D interferometry and Special Relativity. In the later case the Lorentz Contraction together with the Time Dilation are explained as results of the eccentricity of the measuring ellipsoid, caused by its velocity. Finally, the extremely thin ellipsoid of the very first light appears as a beam aimed directly at the observer which might explain the wave or ray duality of light and entanglement.

Keywords: Gated viewing, ultrashort pulses, holography, optical resolution, interferometry, relativity, duality of light

1. Appearance of the sphere of observation

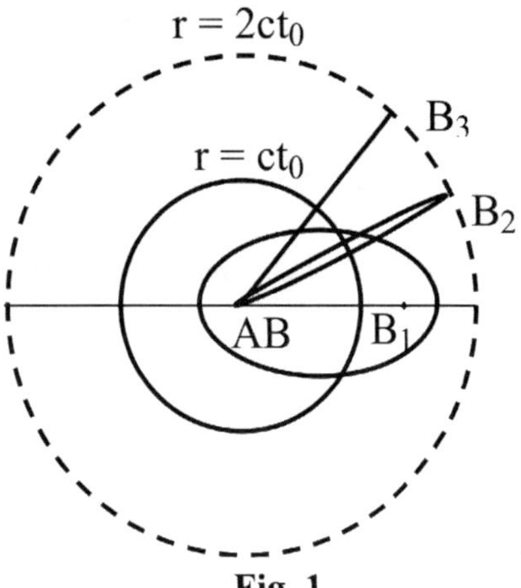

Fig. 1

(A) emits a sphere of light with radius (R) which is observed by (B).The separation A-B = 2f.
2f = 0 : Sphere with radius r.
0 < 2f < R : Ellipsoid (B_1 B_2)
2f = R : linje (B_3)
2f > R : no observation

In a space filled by scattering particles (C) a point source (A) emits a short pulse of light of the length of 3mm (c:a 10^{-11} sec). After the time t_0 pulse has expanded into a sphere of two meters radius . (1 meter is the distance light travels in 1/299, 792,458 of a second.). It is now photographed by camera (B) with such a short exposure time that it freezes the motion of the light. Recorded will be only those particles or those objects that at that moment are illuminated by the thin spherical shell of light. If the distance A-B is zero, a sphere with a radius of only one meter is recorded, because the light has to go the distance A-C-B where C is an illuminated point. This sphere we name "*the sphere of information*". If the distance A-B is nonzero, but less than two meters the recording will be that of an ellipsoid with (A) and (B) as focal points because light still has to travel the two meter distance A-C-B. If A-B is greater than two meter nothing will be recorded because *the sphere can not be seen from outside*. Finally, if A-B is exactly two meter the ellipsoid has such high eccentricity that it has become just a line from (A) to (B). (Fig.1).Could this deformation in absurdum also be used to visualize the apparent concentration of the sphere into beams of energy?

Thus, we have shown that the sphere of information changes from a sphere to an ellipsoid to a line solely because of the existence and the position of an observer (Fig.2).

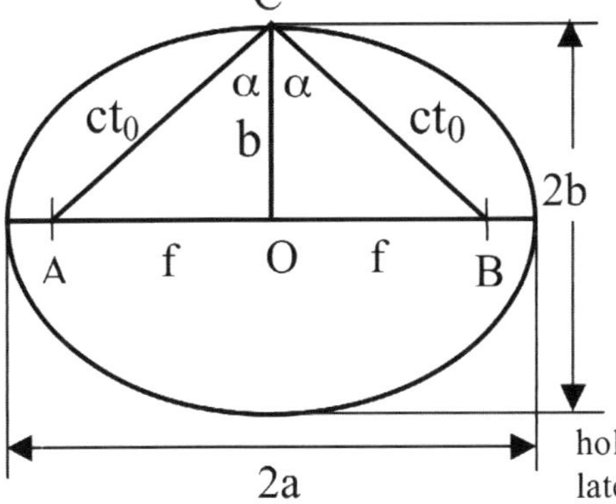

Fig. 2
The sphere of information transformed into ellipsoid:

$$\frac{b}{a} = \frac{b}{ct_0} = \cos\alpha = 1/k = \sqrt{1-\frac{f^2}{c^2 t_0^2}}$$

Usually these changes are of no importance when light is used for measurement such as classic interferometry, RADAR, LIDAR where A-B is zero. However, for holographic interferomety[1] and for Special Relativity the situation is different because A-B are separated by static distance respective by velocity. Thus, if intersections of the ellipsoids are used for measurements of object deformations the same correction factor has to be introduced in holographic interferometry and in special relativity as shown later on. This correction can be visualised by the holodiagram (Fig.6). The deformation of the sphere into an ellipsoid might not be known resulting in measuring errors. One reason to this fact is that any intersection of a plane surface by an ellipsoid to an observer at one focal point appears as an intersection by a sphere[2].

2. Moiré analogy

In Fig.3 we see the moiré analogy to the ellipses[3]. Two sets of circles are added to each other, the centre of one represents a light source (A) the other (B) an observer. The moiré pattern formed by the two sets of circles consists of one set of ellipses and one set of hyperbolas, which are normal to each other. The ellipses are stationary if one set of circles is moving outwards (source) while the other is moving inwards (observer) at the same velocity. The ellipses also represents the interferometric

Fig. 3
The moiré analogy to the holodiagram. Ellipses represent interferometric, hyperbolas transverse, resolution.

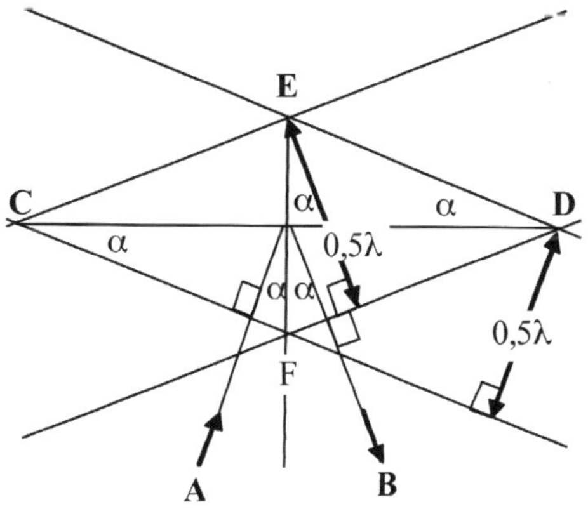

Fig. 4
EF = 0,5 λ/cosα; CD = 0,5 λ/sinα
One of the rhombs of Fig.3. Where EF represents resolution of holo- graphic interferometry and CD represents eg the resolution of a lens.

measuring sensitivity e.g in holographic interferometry where the hyperbolas represent sensitivity direction. If both circles move outwards or inwards the hyperbola represent stationary interference fringes eg visualizing the resolution of a lens having A-B as diameter.

If A-B is the diameter of a lens (D) the separation of the hyperbolas represents the resolution limit of that lens[4]. If A-B instead represents the diameter of an object the separation of the hyperbolas represent the lens diameter needed to resolve that object. From Fig.3 we see that the interferometric sensitivity is maximal ($\lambda/2$) where the ellipsoids are closest together, which is along the axis outside A-B and at infinite distances. The transversal sensitivity is maximal where the hyperboloids are closest, which is closest to the lens while it is zero at infinite distances. These are some of the reasons why interferometry is useful for long distances, while microscopes are used for as short distances as possible.
The hyperbolas also represent the information carries in holographic recordings (as described by Tung H. Jeong)[5].

3. Holographic interferometry

Let (A) and (B) be sources of coherent light that are phase locked by simply arriving from the same laser. In that case stationary interference fringes are projected corresponding to the hyperbolas of Fig.3. The separation (d_{hyp}) of these fringes is $d_{hyp} = \lambda/2\sin\alpha$. The ellipses are formed when (A) is the source of coherent light while (B) is a point of observation made phase locked to (A) eg being observed through a hologram plate that was exposed and reconstructed by light arriving from (A). The separation of these fringes used foor holographic interferometry are $d_{ellips} = \lambda/2\cos\alpha$.

The ellipsoids, which we refer to as "ellipsoid of observation" can berecorded and studied using a method named "light-in-flight recording by holography". A hologram is recorded when two beams of coherent light (the object beam and the reference beam) intersect at the holographic plate. *Interference fringes can be formed on this only if the two components illuminate the plate at the same time.*[1] In Fig. 5 one single picosecond pulse is emitted by (A) and a mirror (C) reflects the pulse to the holographic plate (B). Thus, the reference beam works as a picosecond shutter that moves over the hologram plate with a velocity higher than light, recording on each point of the plate only objects that are situated on the ellipse that intersects the reference mirror [6,7].

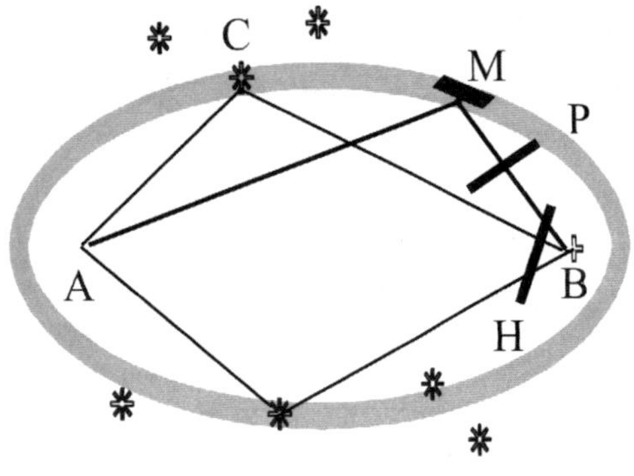

Fig. 5
Light-in-flight recording by holography. Referens pulse (P) from (A) to mirror (M) to hologram plate (H). Only objects intersected by the ellipse are recorded.

If we let the steps by which we increase the (ct_0) of Fig.2 be very small, e.g. only the wavelength of light (λ), we will find that (b) increases by $\lambda/2\cos\alpha$. The result is that the wavelength along the y-axis appears to be increased to $1/2\cos\alpha = k$. An ordinary example of such apparent increase of λ is that a rough surface appears more mirror-like the more oblique is the observation. In interferometric measurements, classic or holographic, the number of interference fringes has to be multiplied by 0,5 $k\lambda$ to find the displacement causing the fringes (Fig.3). If we do not make any corrections, measured distances appear shorter where the separation is larger (Fig.3).

Journal of Physics: Conference Series **415** (2013) 012030 doi:10.1088/1742-6596/415/1/012030

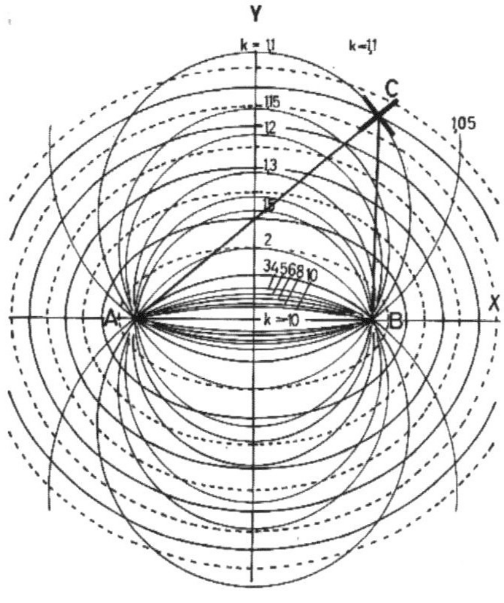

Fig.6
The original holo-diagram used to calculate deformations (d) and vibrations from number of interference fringes (n). α = ACB

The "holodiagram" (Fig.6) is based on the ellipsoids of Fig.3 and constructed to evaluate the interference fringes in holographic interferometry[1]. The focalpoint (A) is the point source of light, while focalpoint (B) is the point of observation, behind the holographic plate. In double exposed holography the deformation (d) at any point (C) is calculated: d = n k 0,5λ where (n) is the number of interference fringes, (k) is 1/cos α where α is half the angle ACB. The angle α is also constant along circular arcs through (A), (B) and (C). The distance A-C-B is ct where (c) is the velocity of light and (t) is time. The interferometric sensitivity vector is everywhere perpendicular to the ellipsoids. On the x-axis in between (A) and (B) the k-value is infinite, while outside (A) and (B) it is unity. In Fig.4 we study the k-value at the y-axis.

The calculations made in Fig.2 reveals that the k-value used in holographic interferometry (eq. 1) is strikingly similar to equations in Special Relativity which was one of the facts that inspired to use the ellipsoids also in that field[8].

$$k = \frac{1}{\cos\alpha} = \frac{a}{b} = \frac{f}{ct_0} = \frac{1}{\sqrt{1 - \frac{f^2}{c^2 t_0^{\,2}}}} \qquad \text{Eq.1}$$

4. Special relativity

What happens if the separation from (A) to (B) of Fig.2 is dynamic instead of static, if the one who performs the measurement at high velocity passes the object to be measured (C)? At (A) the pulse is emitted and at (B) it is observed. The sphere of observation is transformed into an ellipsoid just as in Fig.2 but in this case the distance (f) is represented by vt.

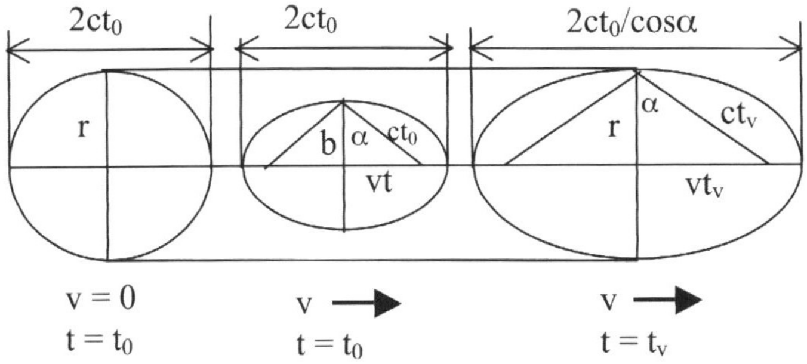

Fig. 7
Relativistic time dilation: To restore the decreased b= rcosα, the time has to be increased by the time dilation factor $t_v = t_0/\cos\alpha$

The elongation of the sphere of observation by 1/cos α results in that objects measured with this ellipsoid appears shortened in the direction of velocity by the factor cos α. (Fig.6). In this way the Lorentz contraction is explained as caused by the elongation of the measuring tool, the sphere into the ellipse (Fig.7). This fact is further confirmed by the fact that our basic measuring tool the Meter is by definition elongated by velocity. Whether the Meter is defined by lines on a ruler, by wavelengths or by (ct_0) it is elongated along the line of velocity by the factor 1/cosα caused by time dilation.. There is however one problem.

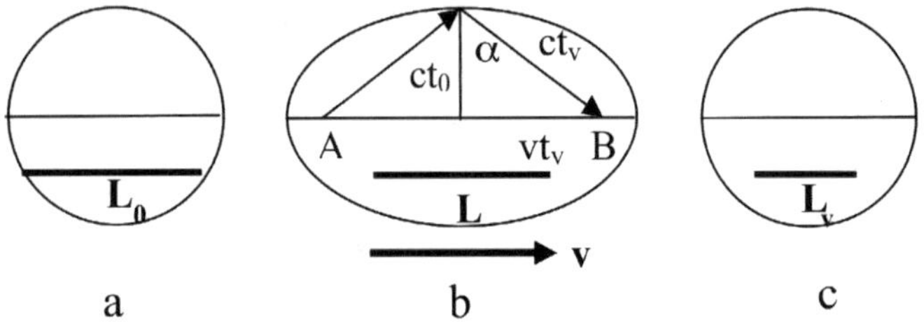

Fig. 8

The velocity elongates the sphere (a) by a factor 1/cosα to the ellipse (b). This effect is invisible to the observer B who thus measures the object L foreshortened (c) by the factor cosα.

We have up till now studied how the thickness 2b of the ellipsoid has decreased from ct to ct cosα as its length increased by the factor 1/cosα (eq.of Fig.2). When the ellipsoid is used for measurement the result would be that object dimensions transverse to the velocity would be influenced. Such a result is impossible as stated by Einstein: "A fast moving object could not pass through a hole but the fast moving hole could pass over the object". The situation is solved by the Time Dilation γ introduced by Einstein[9] where time is increased by the factor 1/cosα resulting in that ct and vt are elongated by the same factor so that (b) is no longer influenced by velocity (v). Thus we have by using the holodiagram visualized Einsteins two equations:

$$\frac{t_v}{t_0} = 1/\cos\alpha = k = \frac{1}{\sqrt{1-\frac{(vt_v)^2}{(ct_v)^2}}} = \frac{1}{\sqrt{1-\frac{v^2}{c^2}}} = \gamma \qquad \text{Eq.2}$$

$$\frac{L_v}{L_0} = \frac{b}{a} = \cos\alpha = 1/k = \sqrt{1-\frac{v^2 t_0^{\,2}}{c^2 t_0^{\,2}}} = \sqrt{1-\frac{v^2}{c^2}} = \frac{1}{\gamma} \qquad \text{Eq.3}$$

Going back to the rhomb of the holodigram (Fig.4) we find that Eq.2 is represented by the diagonal (EF) if the fixed distance A-B is exchanged by the increasing distance vt, resulting in all ellipses of the original holodiagram having the same eccentricity.

5. Conclusion and consequences

What is light? What does it look like? Is it made up of waves or of particles? Why is it so that when the sphere of light expands and its energy is diluted, it still can be absorbed by atoms as if its energy was concentrated to beams with needle-point accuracy. We explained this fact with our alternative graphic derivation of relativity using the deformation of spheres into ellipsoids. Thus, we have shown that just only the existence of an observer causes the ellipsoid and that the position and velocity of this observer influences its eccentricity. The extremely thin ellipsoid of the very first light appears as a beam aimed directly at the observer, which might explain the duality of light which sometimes appears in the form of waves sometimes as rays. When the original sphere consisted of the energy of one single photon the first light arrives in the form of a ray and if it is absorbed no collapse of the sphere is needed for explanation.

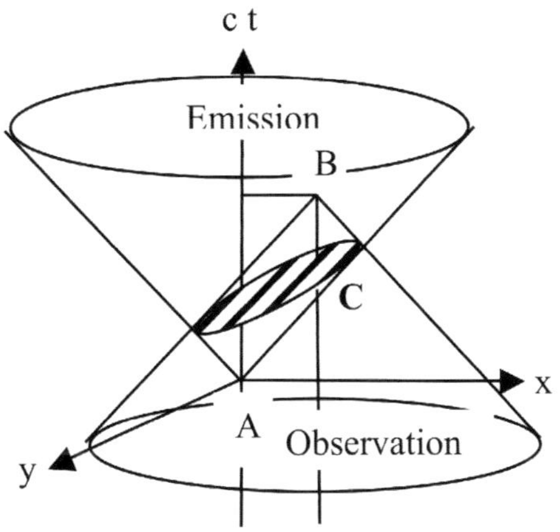

Fig. 9

The Minkowski cones, modified by moving one cone in relation to the other along x- and t-axis.

The Sphere of Observation is also a sphere of apparent Simultaneity when observed from its center. When this sphere is elongated the loci of apparent simultaneity will be on the surface of the ellipsoid, which when the velocity approaches that of light becomes a line of stationary time. The fact that (A) and (B) are the focal points of the same ellipsoid results in that they share the same apparent Sphere of Observation, and the same point of time which probably could explain some of the spooky effects of entanglement[11] where information appears to pass from (A) to (B) instantly

A velocity higher than light is however forbidden by Einstein who states that nothing can travel faster than light. Phenomena connected to entanglement appear at first to be exceptions. My explanation is that in those cases the information has no mass or energy. It is not transported by any carrier. Such "naked" information can not be reconstructed until energy is later sent in the form of correlation using ordinary information at the velocity of light. In entanglement we see that even if the naked information can not be detected directly because its luck of energy it still can influence what happens at random, because in Quantum Physics there is by definition no energy difference between two states that happen randomly.

The collapse of the sphere can be explained in an alternative way by using the four-dimensions of relativity as described by the Minkowski cones [6], which we have modified by moving one cone in relation to the other along x- and t-axis [5,10] (Fig.10). The illumination cone represents the expanding sphere with its centre at (A). The observation cone represents the sphere contracting towards the point of observation (B). Different intersections of the cones produce all the ellipses (ellipsoids) of Fig.1. When the two cones just touch each other the tangential line represents the concentrated beam of the "first light". Finally, the null result of the Michelson Morley experiment is with our method confirmed by the fact that the light path ct = A-C-B is constant independent wherever on the ellipsoidal surface (C) is positioned. Fig 2 also illustrates the impossibility of a velocity (v) greater than that of light (c).

References

[1] N. Abramson, "The Holo-diagram: a practical device for the making and evaluation of holograms," Engineering Uses of Holography,Strathclyde (1968) and *Appl. Opt.* **8,** 1235-1240 (1969).
[2] N. Abramson, J. Boman, B. Bonnevier, Plane Intersections of rotational ellipsoids, *Amer. Math. Monthly* 119 (4), 336-339 (2006)
[3] N Abramson, "Moiré patterns and hologram interferometry," *Nature* 231, 65-67 (1971)
[4] N. Abramson, Book: "Light in Flight or The Holodiagram, The Columbi Egg of Optics" Academic Press, Resolution of a lens, page 74, 1996.
[5] Tung H. Jeong, "Geometric Model for Holography." American Journal of Physics, Vol. 43, No. 8, August 1975, pp. 714-722

[6] D.I. Staselko, Y.N. Denisyuk and A.G. Smirnow,"Holographic recording of the time-coherence pattern of a wave train from a pulsed laser source," *Opt.Spectros.* **26**,413 (1969)
[7] N. Abramson, "Light-in-flight Recording: high-speed holographic motion picture of ultrafast phenomena," *Appl. Opt.* **22**, 215-232 (1983).
[8] N. Abramson, "Light-in-flight recording. 4: Visualizing optical relativistic phenomena," *Appl. Opt.* **24**, 3323-3329 (1985). Fig 7 at 3327
[9] A.Einstein et al, *The Principle of Relativity* (Dover, New York, 1952)
[10] N. Abramson, " Minkowski diagram in relativity and holography", *Appl. Opt.* **27**, 1825-1832 (1988)
[11]. Alisa Bokulich, Gregg Jaeger, *Philosophy of Quantum Information and Entanglement*, Cambridge University Press, 2010, xv.

Aberrations in Holography

D Padiyar, J Padiyar

1070 Commerce St suite A, San Marcos, CA 92078

dinesh@triple-take.com joy@triple-take.com

Abstract. The Seidel aberrations are described as they apply to holography. Methods of recognition of an aberrated holographic reconstruction are described, as well as a recognition of the type of aberration. Experimental and theoretical strategies to minimize aberrations are discussed, including geometric considerations in the recording and reconstruction of holograms. Aberrations due to the recording of a hologram with one wavelength and reconstructing of it in another is also examined.

1. Introduction

An aberration in an imaging system is defined as any distortion of an image due to imperfections in the imaging system. In particular, the monochromatic aberrations, known as the Seidel aberrations, are aberrations irrespective of dispersion in the optical imaging system. In a wave picture, the Gaussian image point is considered to be the center of a spherical wavefront, known as the reference sphere, with the distance from the Gaussian image point to the exit pupil as the radius. The phase mismatch between actual wavefront and the ideal reference sphere is then the aberration. The parameters of the aberration function are the height of a point on the wavefront from the axis, or principal ray, denoted by r, the azimuthal angle of a point on the wavefront, denoted by ϕ and the height of the Gaussian image point from the axis, denoted by σ

Eq.1 $$W = {}_0C_{40}(r^4) + {}_1C_{31}(\sigma)(r^3)\cos(\phi) + {}_2C_{22}(\sigma)^2(r^2)\cos^2(\phi) + {}_2C_{20}(\sigma)^2(r^2) + {}_3C_{11}(\sigma)^3(r)\cos(\phi)$$

The aberrations described by the five terms in the aberration function above are known as the five Seidel aberrations of: spherical aberration, coma, astigmatism, petzval curvature and distortion. The first three terms lead to a blurring of the image point and the next two lead to a distortion of the image.

2. Analysis of hologram formation and aberrations

The wave picture may be used to create an analogue of the Seidel aberrations in holography by determining the variation of a wavefront generated by the reconstruction of a hologram to an ideal wavefront that should be generated by the same hologram. It will be found that an aberration function, W, can be derived under these conditions that predict the same parameters as may be predicted by using the lens aberration function. In particular, the parameters of the position of an image, it's lateral, angular and longitudinal magnification and its five Seidel aberrations may be predicted. The basic theory assumes a point source as an object and so does not take into account extended objects such as planes.

The analysis of the construction and reconstruction of an image follows Meier [1] and

Champagne [2]. It assumes a point source as the object and a point source for the reference (Fig 1). The analysis then assumes a different point source for the origin of the reconstruction source. From the analysis, an image point (X_i, Y_i, Z_i) is derived using the coordinates of the object point (x_o, y_o, z_o), the original reference point (x_r, y_r, z_r) and the reconstruction point (x_c, y_c, z_c). Seidel aberrations are then determined by the phase mismatch between the image point as reconstructed by the original reference point source, the ideal image point, and the image point as reconstructed from a different point. The hologram itself is positioned in the x-y plane in this coordinate system with z=0. The analysis takes into account the fact that both the original reference and the reconstruction source may be collimated, i.e. the point of origin of these beams may be infinity. In this situation it is not possible to determine the exact coordinates of the original reference wave and the reconstruction wave and so only the directions of the reference and the reconstruction beams are the parameters on which the aberration is based. Furthermore, changes in wavelength between recording and reconstruction are also incorporated. Thus, if the recording wavelength is λo and the reconstruction wavelength is λc, then the ratio of the recording wavelength to the reconstruction wavelength is $\mu = (\lambda c / \lambda o)$. Both Meier and Champagne essentially use the same techniques, but Meier analyses the paraxial case and Champagne the non-paraxial case.

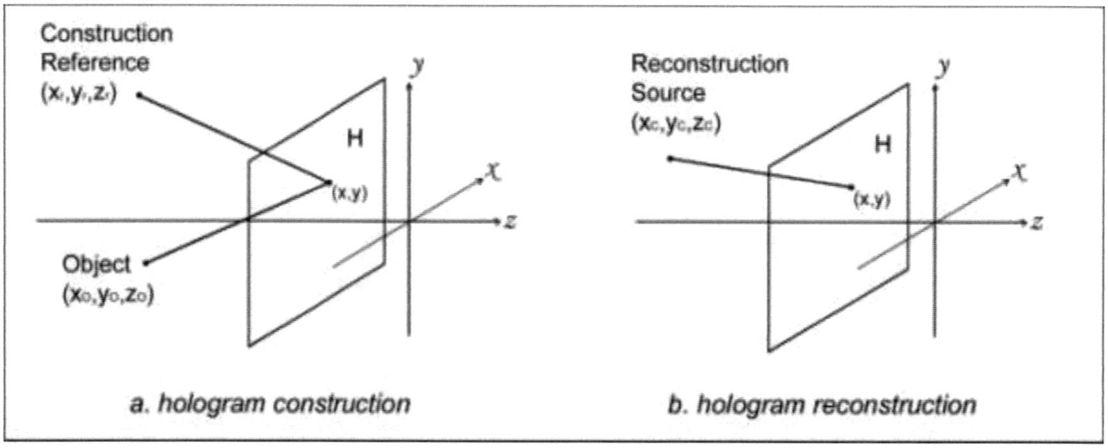

Fig 1. The geometry of recording and reconstruction of a hologram in this analysis.

The reconstructed phase wavefront emerging from the hologram at (x,y) in the hologram plane is then calculated as the sum of the recorded phases, ϕr and ϕo and the phase of the reconstructed wave, ϕc. Thus the total phase at the hologram, on reconstruction, is

Eq.2 $$\Phi = \phi c +/- \phi r +/- \phi o$$

where the signs are dependent on whether or not the reconstruction was carried out with the original reference beam or its conjugate and whether or not the desired image is the original (virtual) image or its (real) conjugate. The general expression for the phase of a spherical wave, from a general point (x_g, y_g, z_g) to a point (x, y) is given exactly by:

Eq.3 $$\Phi = 2\pi/\lambda\{[(x - x_g)^2 + (y-y_g)^2 + z_g^{\ 2}]^{1/2} - [x_g^{\ 2} + y_g^{\ 2} + z_g^{\ 2}]^{1/2}\}$$

If now Eq.3 is substituted into Eq.2 with the specific coordinates of the object, reference and reconstruction coordinates replacing the general coordinates, an exact expression is obtained for the phase Φ at the hologram plane H. However, a computationally simpler expression may be derived by expanding the exact phase expansion in a Taylor expansion about the z coordinates, assuming that $z_g^{\ 2}$

> $x_g^2 + y_g^2$. To determine the position of the image point and the magnifications of the image, a first order expansion is sufficient. Under these circumstances, the phase of the reconstructed wave, i.e. the reconstructed wavefront, to first order, is:

$$\Phi^{(1)} = (\pi/\lambda c)\{(x^2 + y^2)(1/z_c + \mu/z_o - \mu/z_r) - 2x(x_c/z_c + \mu x_o/z_o - \mu x_r/z_r) - 2y(y_c/z_c + \mu y_o/z_o) - \mu y_r/z_r\}$$

This wave, ideally reconstructed, should be a spherical wavefront converging towards an image point at (Xi, Yi, Zi). By comparison with the ideal spherical wavefront, expressions are derived for the image position (here the Yi expression is not expressly stated but is exactly the same as the Xi expression except to replace all the x values with the appropriate y values):

Eq.4 $$Zi = z_o z_c z_r/(z_o z_r + \mu z_c z_r - \mu z_c z_o)$$

Eq.5 $$Xi = (x_c z_o z_r + \mu x_o z_c z_r - \mu x_r z_c z_o)/(z_o z_r + \mu z_c z_r - \mu z_c z_o)$$

It can be seen that if (x_c, y_c, z_c) is equal to (x_r, y_r, z_r), i.e. the reconstruction point is identical to the reference point, and $\mu = 1$, i.e. the hologram is reconstructed with the same wavelength with which it was recorded, then $Zi = z_o$ and $Xi = x_o$. That is, the image coordinates are the same as the object coordinates and the image appears in the same position as the original object with no magnification or aberrations.

3. Resolution

The resolution of a hologram is often brought up and compared against the resolution of electronic displays. The validity of such a comparison is doubtful, since a hologram is a diffractive effect whereas in an electronic display, the resolution of the display is determine by the pixel density. However, a plausible figure may be given for the resolution of a hologram by determining the variation of the size of the image point with respect to the reconstruction source size. While this calculation may be carried out more rigorously using the Van Cittert–Zernike theorem and so calculating the degree of spatial coherence of the source, a simpler back-of-an-envelope approach may be used. The variation of size of an image with respect to the variation in the source size, both laterally and longitudinally, may be given by $\partial X_i/\partial x_c$ and $\partial Z_i/\partial z_c$. To this end, from Eq.4 and Eq.5 we derive

Eq.6a $$\partial X_i/\partial x_c = (z_o z_r)/(z_o z_r + \mu(z_o z_r - z_o z_c))$$

Eq.6b $$\partial Z_i/\partial z_c = (z_o z_r)/(z_o z_r + \mu(z_c z_r - z_c z_o))$$

It can be seen that in Eq.6a, the lateral resolution, if the z coordinate of the reconstruction point is identical with that of the reference point, then the resolution of the image is equal to the size of the source, regardless of any change in wavelength. However, for the longitudinal case, Eq.6b, this is not so. If the z coordinates are now identical depends on both the z coordinate of the object position and the change in wavelength.

4. The Holographic Model

In order to show a correlation between holography and conventional lenses, an equivalent "focal length", f_R for the hologram may be derived from the reconstructed phase wavefront. This is allowable insofar as the reconstructed phase wavefront is also a converging spherical wavefront and maybe thought of as the "reference sphere". Thus, for the object beam coordinate, z_o, and the equivalent image position Zi, under the assumption of a spherical wavefront, we may construct

$$(1/z_o) + (1/Zi) = 1/f_R$$
$$= 1/z_o +/- (\mu)(1/z_o + 1/z_r)$$

by comparing the standard lens equation $1/s_i + 1/s_o = 1/f$
and also Champagne's derivation of image distance $1/Rf = 1/Rc +/- (\mu)(1/Ro + 1/Rr)$

The +/- refers to the either the virtual image or the real image.

If the reference and/or the reconstruction beam is collimated, i.e. z_r and/or $z_c = \infty$, then there is no unique position of the reference and reconstruction beams. In this case, the ratio of the x coordinate to the z coordinate may be used to determine the beam direction with reference to the recording surface normal, as opposed to a unique position. These ratios are the sin of the angles of the beams, thus, the image is defined by its direction:

$$Xi/Zi = x_c/z_c +/- \mu(x_o/z_o - x_r/z_r)$$

Which leads to,

$$\text{Sin}\alpha_\iota = \text{Sin}\alpha_c + \mu(\sin\alpha_o - \sin\alpha_r)$$

Where α_ι is the direction of the image position, α_o is the direction of the object position, α_r is the original reference direction and α_c is the reconstruction beam direction. The Meier case is concerned with the paraxial case and poses no boundaries on the values of Xi and Yi by use of the paraxial approximation. While the non-paraxial case analysed by Champagne has the restriction that, for image formation

$$(Xi/Zi)^2 + (Yi/Zi)^2 < 1$$

If this condition is not satisfied, then the image field becomes evanescent and the hologram cannot be reconstructed.

5. Magnification in holography

There are three different kinds of magnification that must be taken into account: lateral magnification, M_T, longitudinal magnification, M_L, and angular magnification, M_A. Lateral magnification, M_T, is the increase in the lateral size of the image, i.e. its increase in the xy direction. This is given by the ratio of the object's original distance to the image's distance as the original distance varies. This is

$$M_T = dX_i/dx_o = 1/(1 +/- z_o/\mu z_c - z_o/z_r) \quad \text{(Meyer; paraxial)}$$
$$M_T = (\cos\alpha_o/\cos\alpha_i)\{1/(1 +/- z_o/\mu z_c - z_o/z_r)\} \quad \text{(Champagne; non-paraxial)}$$

Three interesting facts emerge:

1. If there is no change in wavelength between the recording wavelength and the reconstruction wavelength ($\mu = 1$), and the reconstruction distance is the same as the reference distance ($z_c = z_r$), then the magnification is unity, i.e. there is no magnification.
2. If the reconstruction beam is collimated ($z_c = \infty$), then the μ factor drops out. That is, the magnification of an image by change in wavelength occurs only for diverging or converging reconstruction beams. A collimated reconstruction beam will result in unit magnification.
3. Conventional magnification by a lens of focal length f is given by $M = f/(z_o - f)$. If the equivalent holographic focal length f_R, as derived above, is used as f in this expression, then the conventional magnification is different from the holographic magnification by a factor $1/\mu$. This shows the limitation of the lens metaphor for the hologram; the lens and holographic

magnifications are only equal if the reconstruction wavelength is identical to the recording wavelength. So, a hologram is the equivalent of a lens only insofar as there is equivalence in wavelengths.

Longitudinal magnification is a magnification along the axis of the hologram. That is, the variation of the image z coordinate, Z_i, to the original object's z coordinate, z_o, as z_o varies. This is:

$$M_L = dZ_i/dz_o = -(1/\mu)\{1/(1-z_o[(1/\mu z_c) + (1/_{zr})\}2 = -(1/\mu)M_T$$

Whereas in conventional imaging systems by lenses $M_L = -M_T{}^2$, the above shows that a holographic image is compressed along the axis by a factor of $(1/\mu)$ from conventional imaging systems [3]. Note also that if the reconstruction beam is collimated, then $M_T = 1$ (no lateral magnification), but M_L is still reduced by a factor of $(1/\mu)$. In other words, for a collimated reconstruction beam, there is longitudinal magnification with no transverse magnification. The angular magnification is given by the ratio of the angle subtended by the image to the angle subtended by the original object at the recording surface:

$M_A = d(X_i/Z_i)/d(x_o/z_o) = \mu$ (Meier; paraxial)
$M_A = \mu(\cos\alpha_o/\cos\alpha_i)$ (Champagne; nonparaxial)

Note that this magnification is independent of all the parameters of recording and reconstruction and depends only on the wavelength change. If there is no change in the wavelength, there is no angular magnification.

6. 3rd Order Seidel Aberrations in Holography

Seidel-equivalent aberrations for holography may be derived in the same manner as for lenses. The actual wavefront is compared against the ideal reference sphere and the mismatch between the wavefront and the reference sphere is taken as the aberration. However, in holography, the aberrated wavefront is calculated by carrying out a third order expansion of Eq.3, to determine the ideal, spherical wavefront, then substituting Eq.3 into Eq.2 and expanding to third order to obtain the third order expansion of the actual phase wavefront. The mismatch between the reference sphere and the third order expansion terms now gives the Seidel aberrations.

Transforming to a circular coordinate system, this gives for the Seidel aberration function:

W =	$2\pi/\lambda_c\,[(-1/8)\,\rho^4 S$	Spherical Aberration
	$+(1/2)\,\rho^3(C_x\cos\theta\ C_y\sin\theta)$	Coma
	$-(1/2)\,\rho^2(A_x\cos^2\theta+A_y\sin^2\theta+2A_{xy}\cos\theta\sin\theta$	Astigmatism
	$+1(1/4)\,\rho^2 F$	Petzval Curvature
	$+(1/2)\,r(D_x\cos\theta + D_y\sin\theta)]$	Distortion

7. Discussion

The spherical aberration, S, is:

$$S = \mu[(\mu^2 - 1)(1/z_o^3 - 1/z_r^3) - (3\,\mu/z_c)(1/z_o^2 + 1/z_r^2) + 3(1/z_c^2 - (\mu^2/(z_oz_r))(1/z_o - 1/z_r) + 6(\mu/(z_oz_rz_c))]$$

If both z_r and z_c are infinite, i.e. both reference and reconstructions beams are collimated, then

$$S - \mu[\mu^2 - 1]/z_o^3$$

This becomes zero for $\mu = 1$. Thus, if the hologram is reconstructed with the same wavelength as the original reference and both reference and reconstruction beams are collimated, spherical aberration disappears. However, if either the reference or the reconstruction is not collimated, or reconstruction occurs with a different wavelength then spherical aberration will exist.

Coma, Cx, is:

$$Cx = (x_c/z_c)[(1/z_c^2) - (1/z_c - \mu/z_o + \mu/z_r)^2] - (\mu x_o/z_o)[(1/z_o^2) - (1/z_c - \mu/z_o + \mu/z_r)^2] + (\mu x_r/z_r)[1/z_r^2 - (1/z_c - \mu/z_o + \mu/zr)^2]$$

If both reference and reconstruction beams are collimated ($z_c = z_r = \infty$), then the ratios x_c/z_c and x_r/z_r are used, where the ratios are the tangents of angles of the beams with respect to the recording normal. Hence,

$$Cx = -(\mu/z_o^2)[\mu\tan\theta_c - (\mu^2 - 1)\tan\theta_o + \mu^2\tan\theta_r]$$

Coma can be made to disappear if both reference and reconstruction beams are collimated, the object is directly in front of the medium (resulting in $\theta_o = 0$ and so $\tan\theta_o = 0$), and $\tan\theta_c = -\mu\tan\theta_r$. Thus, if the c and r beams are collimated, the coma can be made to disappear if the tan of the reconstruction angle is modulated by the ratio of the recording wavelength to the reconstruction wavelength. If both these wavelengths are the same and the reconstruction angle is the conjugate of the recording angle, with both being collimated, coma can be made to disappear. In addition, if both wavelengths are the same, the condition on the object positions vanishes.

Astigmatism, Ax, is:

$$A_x = (x_c^2/z_c^3) - (\mu x_o^2/z_o^3) + (\mu x_r^2/z_r^3) - (x_c/z_c - \mu x_o/z_o + \mu x_r/z_r)^2(1/z_c - \mu/z_o + \mu/z_r)$$

For a collimated reference and reconstruction ($z_c = z_r = \infty$), the ratio x/z is taken and tangents of the various angles are used, as above. The astigmatism reduces to:

$$Ax = (\mu/z_o)[(\tan^2\theta_o)(\mu^2 - 1) - 2\mu\tan\theta_o(\tan\theta_c + \mu\tan\theta_r) + (\tan\theta_c + \mu\tan\theta r)^2]$$

Here the same three conditions that made coma disappear can also be used to make astigmatism disappear. That is, astigmatism disappears if $\tan\theta_c = -\mu\tan\theta_r$.

Petzval Curvature, F, is:

$$F = (x_c^2 + y_c^2)/z_c^2 - \mu(x_o^2 + y_o^2)/z_o^3 + \mu(x_r^2 + y_r^2)/zr^3 - (Xi^2 + Y_i^2)/Zi^3$$

The same conditions as those of astigmatism and coma will also make field curvature disappear. However, if the reconstruction beam is not collimated and/or μ is not equal to 1, then the image will be seen to bulge outward giving the image a "rolling" look.

Distortion, Dx, is:

$$Dx = \mu[(x_o/z_o)^3(\mu^2 - 1) - 3\mu(x_o/z_o)^2(x_c/z_c + \mu x_r/z_r) + 3(x_o/z_o)(x_c/z_c + \mu x_r/z_r)^2 - (x_r/z_r)^3(\mu^2 - 1) - 3(x_o/z_o)(x_r/z_r)(x_c/z_c + \mu x_r/z_r) + x_o(y_o/z_o)^2(\mu^2 - 1)]$$

Once more, the same conditions will make distortion disappear.

8. Experimental Results

An planar image was recorded at 514nm with a collimated reference at 30 degrees to the plate normal at 13in from the recording plate.

The hologram was then reconstructed under a variety of beam geometries:
1. A collimated beam at various angles relative to the plate normal.
2. A range of diverging and converging beams. The direction of the reference was also changed by +/- 10 degrees from the original collimated beam direction.
3. The above two sets of experiments repeated with a wavelength of 633nm.

Table 1 gives the results for the first set of tests reconstructing at both 514nm (the original recording wavelength) and at 633nm. The angles are all relative to plate normal, with 30 degrees being the angle at which the hologram was recorded. The displacement refers to the displacement of the image in a plane parallel to the plane of the image. If a reference system is set up with the image reconstructed the x-y plane (z=0) and the z axis pointing away from the recording plate, the signs of the displacements are the signs of the x value. Thus, for a viewer standing behind the image, right is positive.
Table 2 and 3 give the displacement and twist of the image as a result of reconstructing the image with a diverging and converging beams. The divergence and convergence were accomplished by moving the spatial filter +/- 10 in from the focal position.

Angle	15	20	25	30	35	40	45
Displacement at 514	-3.0	-1.8	- 0.8	+ 0.3	+ 1.3	+ 2.3	+ 3.1
Displacement at 633		-2.8		- 1.25		+ 0.25	

Table 1. Lateral image displacement for reconstructions with collimated at various angles

θ	ϕ ~ -10 (divergent)		0 (collimated)		ϕ ~ +10 (convergent)	
	Image plane	twist	Image plane	twist	Image plane	twist
20o	14.5	+ 5.28o	13.13	-6.28o	11.56	-7.97o
30o	14.31	flat	13.06	flat	11.69	+ 1.86o
40o	14.25	-5.28o *	13.0	-1.72o	11.44	-4.43o

Table 2. Displacement and twist of the image for $\lambda = 514$. The θ figure gives the direction of the reconstruction beam.

θ	ϕ ~ -10 divergent		collimated		ϕ ~ +10 convergent	
	Image plane	twist	Image plane	twist	Image plane	twist
20o	11.31	-3.57o	10.31	-2.72o	9.5	-5.43o
30o	11.56	-1.86o	10.5	- .86o	9.5	-1.72o
40o	11.44	- .86o	10.63	- .86o	9.81	+ 1.72o

Table 3. Displacement and twist of the image for $\lambda = 633$. The θ figure gives the direction of the reconstruction beam.

9. Discussion of experimental data

From table 1, it can be seen that the lateral displacement follows the direction of the change in the (collimated) reference beam, being negative when the reconstruction beam is less than the recording angle and positive when it's greater. The displacement should be zero for the reference angle (30 degrees) when reconstructed at 514, but is measured at 0.3. This is due to the swelling caused by processing. Note that the reconstruction angle for the hologram recorded at 30 and reconstructed at 633 is 38, which is reflected in the results; the displacement for reconstruction at 633 is at a minimum at 40 degrees. Thus the displacement caused by altering the reconstruction angle by 50% at either wavelength causes an equal lateral displacement at both wavelengths.

In tables 2 and 3, the image plane refers to the distance between the image plane and the recording plate in the z direction (normal to the plate). Tables 2 and 3 both show that if the hologram is reconstructed with a divergence different to that of the recording divergence, there is a twist of the image and a normal (z direction) displacement. The amount of displacement also depends on the reconstruction wavelength. The object plane was recorded at 13.0 in from the recording plane, hence ideal reconstruction with the original parameters would have reconstructed the image at 13 in from the plate and with zero twist. This is reflected in column 3 of table 2, where the result is shown for a reconstruction with a collimated beam at 30 degrees (the original reference). The results also show that a small variation of the beam divergence at the correct angle shows very little twist; the amount of twist on the 30 degree row in table 2 is zero, or close to it. However, when the beam direction changes by +/- 10 degrees, there is significant twist and displacement, in the z direction, of the image. In the case of reconstruction at 633, the correct parameters for reconstruction are collimated at 40 degrees. It can be seen from table 3 that at this angle and at zero divergence, there is little twist but a significant displacement of the image in the z direction. Hence, if the image is reconstructed with a different angle than that at which it was exposed then there is a displacement away from the ideal image plane, despite being reconstructed at the correct angle as per the grating equation.

[1] "Magnification and Third Order Aberrations in Holography", Reinhard W. Meier, JOSA 1965, vol 55, 8, pp 987- 992

[2] "Nonparaxial Imaging, Magnification, and Aberration Properties in Holography", Edwin B. Champagne, JOSA 1967, vol 57, 1, pp 51 – 55

[3] D. Gabor, Proc. Roy. Soc. (London) A197, 454 (1949)

Quantitative measurement of holographic image quality using Adobe Photoshop

E Wesly

Harrington College of Design

Studio of E. Wesly & Sons

eweslystudio@gmail.com

Abstract. Measurement of the characteristics of image holograms in regards to diffraction efficiency and signal to noise ratio are demonstrated, using readily available digital cameras and image editing software. Illustrations and case studies, using currently available holographic recording materials, are presented.

1. Introduction

While working at an embossed hologram company over a decade ago, it was recognized that there was a need to evaluate the brightness of different generations of holographic images throughout the various stages of the process. Machine vision consultants were called in, quotes were quoted, but nothing happened. But it would have been a great addition to the ISO 9000 paperwork.

Thanks to the take over of digital photography, instead of using specialty machines or software, image analysis tools available in Adobe Photoshop and other bitmap image-editing programs can be utilized to give relative readings of diffraction efficiency and with proper calibration absolute readings.

A methodology of evaluating brightness and signal to noise ratios using readily available contemporary digital photography equipment is presented, and examples of case studies of the evaluation of reflection holograms recorded on a variety of contemporary materials is presented.

2. Prior art

Measuring the brightness of the elemental hologram, recorded by interfering two beams, is not that difficult. In this set up an undiverged laser beam impinges on the grating, which is tilted in it holder until the dimmest zero order and brightest first order appear at the Bragg Angle.

A measurement of efficiency could be calculated by dividing the intensity of the extinguished reference beam (zero order) by the incident intensity and subtracting that from 100%, assuming that all the rest of the light is diffracted into inverse and higher orders. This is more expedient than chasing down all the rest of the diffracted orders.

However this calculation is not entirely accurate, as interrogation in this manner of two samples had the same amount of zero-order extinction, but the power delivered into the first orders are way off. Sphere-S GEO-3 and Harman Technology Holo FX red-sensitive plates exposed to Helium-Neon 633 nm beams with 30 degrees were processed in a variety of manners and the pick of the litter were selected. They both extinguished the zero order down to 1.2 mW from a 6.2 mW undiverged beam the set up depicted. Measuring the + 1st order beam (which was maximized when the 0th was minimized) yielded 3.8 mW from the GEO-3, but only 2.4 from the Harman. It is not the diffracted light that depletes the straight through beam, but a significant amount of scatter in the higher speed and noisier Harman plates.

This concept could be applied to reflection holograms processed to replay at the same wavelength as the recording laser's. The zero order depletion could be measured by placing a detector in the shadows formed by the different exposure quadrants in figure 1, although it too can be fooled by light not diffracted but scattered. In the past it has been used by some manufacturers of photopolymers which gave quite optimistic readings.

Figure 1. Shadow due to reconstruction depleting zero order beam.

These type of direct power measurements are fine for gratings but not for pictorial holographic images. A method of measuring the intensity of each point of an image would require a multitude of tiny power meters. If a holographic image were photographed, the resulting density changes in photographic film could be measured to form a map of intensity of the image.

The concept of photographing holographic images was implemented as early as 1980 by Joly and Vanhorebeek in their investigation of development effects in white light reflection holography.[1] They photographed the image of a USAF resolution target they had recorded on 8E56HD plates, and measured the density of the bright and clear bars on the resultant negative on Agfa Scientia 23D56 film. Knowing the d-log e curve of the film, the relative brightness and signal to noise of the exposure and developer combinations and permutations can be deduced by comparing density measurements of the similar points, having made certain to standardize lighting of the hologram and processing of the photographic recording film.

The subject of this paper is to use a similar technique using digital photography. The sensor in the image plane of the lens is filled with millions of tiny light meters. Making sense of their signals requires some preliminary calibration of the equipment and software.

3. Equipment requirements

3.1. Camera and lens

A high quality digital Single Lens Reflex (dSLR) was used for these demonstrations, a Nikon D3X. Most any "prosumer" (a contraction of professional/consumer) camera could be implemented as long as it can be used in a manual mode. The camera was set to its lowest ISO rating, 200, to minimize any amplifier noise. White balance was set to 2856 K to match the replay light. Shooting in the raw file format is essential.

The lens was a 60 mm f/2.8 Micro- Nikkor macro or close up lens, a type of lens recommended for holographic analysis, since most images are usually quite small. The samples in this study are around 65 mm square. A standard focus distance should be adopted for consistency, as magnification affects f/number. This lens not only has focus distance in metric and American units, but also the magnification ratio on its barrel, and it is set to 1:3 when photographing these specimens.

Since the lens is set to a fixed reproduction ratio, the camera is raised and lowered by the elevator on a sturdy tripod until the image is in focus. A remote release eliminates camera shake when initiating exposure.

Exposures were made with the lens set to f/5.6 in all cases. This is in part because of the conventional photographic wisdom that a lens is at its sharpest at 2 or 3 stops from its largest opening, plus it has enough depth of field for all of the depth of the standard object to be in focus, and doesn't make large speckle when photographing under laser illumination.

The camera's internal meter operating mode is then set to aperture priority, and it decides what would be the appropriate shutter speed for optimum exposure.

3.2. Computers

Most contemporary computers have enough processing speed to perform image editing. But the files to be evaluated, especially those of the full format or FX sensor type (6048 by 4032 pixels, > 12 Megapixels), will be huge. For instance, the raw NEF (Nikon Electronic File) weighs in at 20.0 Mb, the jpg that the camera processed along with the raw is 8.65 Mb, while the TIFF used in the evaluation step that Lightroom processed from the NEF is a whopping 69.7 Mb! Lots of hard drive space, plus some external hard drives, and loaded to the gills with RAM would not be a bad operating mode. An Apple iMac did most of the drudgery for these evaluations.

3.3. Software

Most bitmap, or Paint style programs usually provide a pixel sampling tool which reads out intensities in values of red, green and blue. Since Adobe Photoshop in its variety of incarnations is ubiquitous, the technique is illustrated using this program.

In addition to Adobe Photoshop, Adobe Lightroom would be a welcome addition to the holographic assessor's desktop as this image database software's keywording feature can organize holograms based on the recording material, wavelength, etc. It can also batch process many image files simultaneously, exporting the camera's native raw format as TIFF's.

3.3.1. *File Formats*. The images should be shot in the camera raw file format, not the jpg option to eliminate any compression artifacts. The raw information should be exported same pixel size to the clunky but accurate TIFF format files, so there is nothing lost in translation from the camera specific raw file to something all machines can understand. As the files will be quite large, especially for the FX or full-format chip, it would be OK to use LZW compression as it is not lossy, it, but steer clear of the JPG option. Don't forget that the advantage of compressed files is that they occupy smaller disc area, but it may take longer to open thanks to the de-compressing operation.

Here is where Lightroom really shines, being able to open a batch of raw files and exporting them as TIFF's, ready to be read by Photoshop for evaluation. Photoshop can open raw files, but only one at a time.

Working with files resized to take up smaller disk space may give erroneous readings due to the sampling and averaging of the resizing process. A smaller DX sensor might speed up the work flow compared to the full format FX and still deliver enough information for quality readings.

3.4. Holographic test object

To illustrate this evaluation technique single beam reflection holograms were recorded of the tried and true standard object, which is an antique waffle iron mold sprayed with Krylon brand #1401 Bright Silver spray paint. This paint uses aluminum flakes for its pigment, which in addition to being highly reflective, preserves polarization vectors, very important in forming he highest contrast fringes.

The waffle iron is a not too deep object, with a homogenous background. A series of 4 exposures is typically made on one 65 mm square plate with each quadrant having an identical scene. Three ball bearings support the holographic plate so that it doesn't rock. Plus when viewed pseudoscopically it looks like the waffle itself!

4. Calibration

The eyedropper tool in Photoshop reads out the light intensity that a pixel in the camera saw on a scale from 0 to 255. To get a feel for what do those numbers represent, a Kodak Gray Card was photographed alongside a white piece of paper and a black velvet cloth to determine the dynamic range of the camera. The card was photographed starting at about 5 stops brighter than the built-n meter's recommendation to 5 stops under, shortening each exposure by 1/3 of a stop, which is the finest division that the camera and lens combination allowed. At the extremes of exposure the gray card appeared as bright as the paper on the overexposed side and as dark as the velvet on the underexposed end. The grey card's value went from the minimum of 0 to the maximum of 255 over 21 third-stops, meaning the dynamic range of the sensor is 7 full stops, or 2^7 or 128.

Table 1 shows the data compiled for the digital characteristic curve for this particular Nikon D-3X with the 60 mm f/2.8 Micro-Nikkor. It compares exposure versus the Photoshop numbers. Each consecutive exposure value differs by 1/3 stop, so 3 exposures values are one full stop, for instance going from value 4 to 7 means the light intensity at the film plane has been doubled.

Table 1. Eyedropper values for each exposure.

Exposure #	Red Value	Green Value	Blue Value
1	0	0	0
2	9	9	9
3	23	23	24
4	30	30	32
5	40	39	41
6	52	50	53
7	63	61	65
8	75	73	77
9	87	86	90
10	103	101	106
11	121	119	125
12	140	137	143
13	159	157	162
14	178	175	180
15	195	192	197
16	209	206	211
17	223	220	225
18	235	232	236

19	244	242	245
20	251	250	252
21	255	255	255

EV 11 looks like the center value, with the grey card weighing in closely at the median values of 128 for RGB.

Figure 1 is the graph of the data. The curve is not as straight of a line as would be anticipated; it has a toe and shoulder not unlike silver halide based photographic film. Its dynamic range is not as wide as that of film.

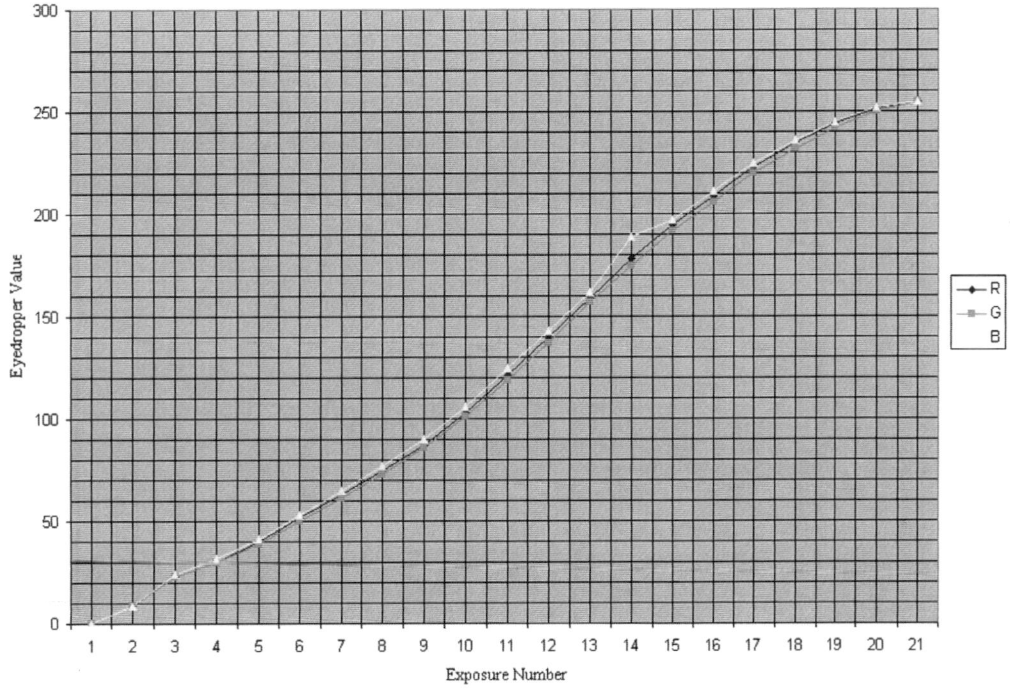

Figure 2. Characteristic curve of Nikon D3X.

It can be deduced from this data is that for every 1/3 stop of exposure, the RGB values change by 19 in the linear part of the curve. This is the Rosetta Stone for interpreting the relative brightness of image areas. 57 eyedropper units are equivalent to doubling or halving the light intensity in the linear part of the curve.

5. Set up

Repeatability of the set up is essential to be able to compare holograms made days or even years apart. Replay light distance and angle should match that recording reference beam. To do so most accurately, the shadow of a nail gnomon in a piece of cardboard placed in the plateholder can be traced, and then placed in the white light replay beam, whose tilt is finagled until the shadow at this stage matches that of the recording phase. Manipulation is made easier if the hologram lies flat on a table.

For unpainted holograms, a very dead black background to lay them on is necessary. Black velour attached to cardboard can be rendered photographically almost absolutely dead black when the calibrating gray card is in mid-scale. However the little bristles can trap not only light but dirt, which will be apparent in the photograph of some of the samples.

The camera points downward and sights directly along the normal of the hologram. Care should be taken to prevent direct light from impinging directly on the camera lighting it up, so that the camera does not see a reflection of itself.

The camera lens was set to a certain distance, make it up if you have to, For this ongoing experiment, every time a new batch of holograms were photographed, the macro lens was set to a reproduction ratio of 1:3; the image height and width are one third of the object's. The camera itself was raised up and down to focus at the standardized reproduction ratio.

For white light reconstruction, the camera's color balance needs to match that of the light source. The typical halogen lamps for replay are not necessarily 3200K like that of photographic studio floods associated with the light bulb default icon. Custom white balance might be in order, or set the camera to the 2856 typical color temp.

For consistent results, a calibration target in the form of the Kodak Neutral Gray Test Card was positioned where the holograms were to be photographed to be metered by the camera. In a suspenders and belt approach, an incident light meter, a Sekonic Studio Deluxe, was also placed in the hologram's position to correlate the exposure. Surprisingly they almost always agreed!

The gray card was always included beside the hologram in the photograph. It can then be sampled in the evaluation stage to assure consistency between different shooting sessions.

6. Methodology

Once the hologram's TIFF file is opened, the eyedropper tool needs to be engaged. Its Info Palette should be open, and its sampling size should be set appropriate to the task at hand, never single point as that would give some touchy readings, usually run with an average of many, no smaller than the 5 by 5 Sample Size. This will alleviate errors caused by dust and dirt and laser speckle.

Also visible in the Info Palette are the X and Y coordinates of the cursor, which indicates exactly which pixels are being interrogated. Very accurate relative readings can be made of exactly the same object point in a variety of holograms, especially if the Rulers' Units are set to pixels. Guides dragged out of the Rulers can provide landmarks for exact same sample locations.

6.1. Example 1: Brightness versus exposure

The Color Sampler Tool, a recent addition to PhotoShop, is a quartet of Eyedroppers. It is found in the same fly-out as the Eyedropper. Samples can be taken of 4 points, and their readings are displayed simultaneously. Figure 2 shows the readings for 4 similar object points in four differently exposed quadrants on a test strip. The hologram was shot on Sphere-S GEO-3 emulsion, with exposures of 800, 1600, and 3200 and 6400 microJoules per square centimeter. In photographic parlance, the exposures increased one stop in each fourth.

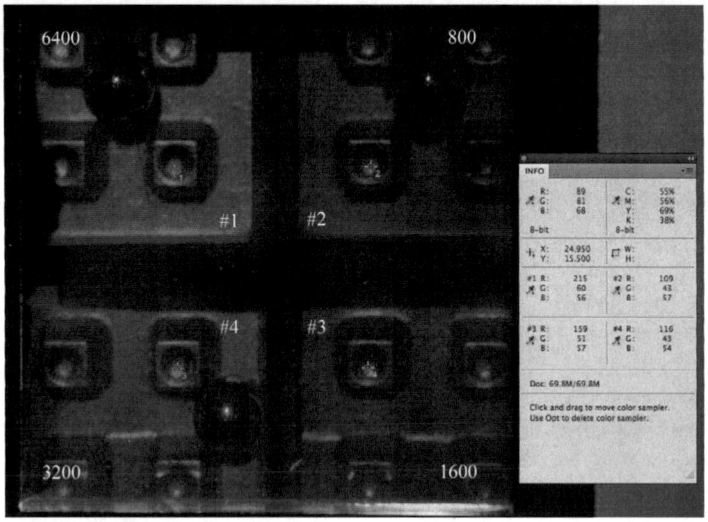

Figure 3. RGB values of varying exposures

Correspondence between Color Sampler Info Palette placement is by position; #1 corresponds to 6400, #2 to 800, #3 to 3200 and #4 to 1600. By finding the difference between any two exposures' Eyedropper numbers, and dividing by the correlation coefficient of 19 points per 1/3 stop, we can see increase of brightness with each stop of exposure. The increase in red from 800 going to 1600 is 116 - 109 = 7, about a sixth of a stop; then an increase of 43, ¾ of a stop, almost twice as bright, then the last jump of 56, a full stop, twice as bright as its predecessor, and four times as bright as the first exposure. The increasing differences in brightness shows that the material is working just over threshold, on the toe of its characteristic curve.

The blue reading is constant, which is good. Film scatter noise is dominated by the blue end of the spectrum, and these numbers show that it is not increasing with exposure. The green is increasing, along with the red, as there is a green component to the bandwidth of the reconstruction under white light, which can be seen peaking through in the highlight of the ball bearing support in the upper left quadrant.

6.2. Example 2: Signal to noise calculations

In this screen grab, the brightest and noisiest areas are sampled on a pair of Ultimate U08 plates. The one on the left received 800 uJ/cm^2 of exposure at 532 nm, while the one on the right received twice as much, seriously overexposing it. Both were developed simultaneously for the recommended time in the Ultimate Developer and bleached in the Ultimate Bleach.

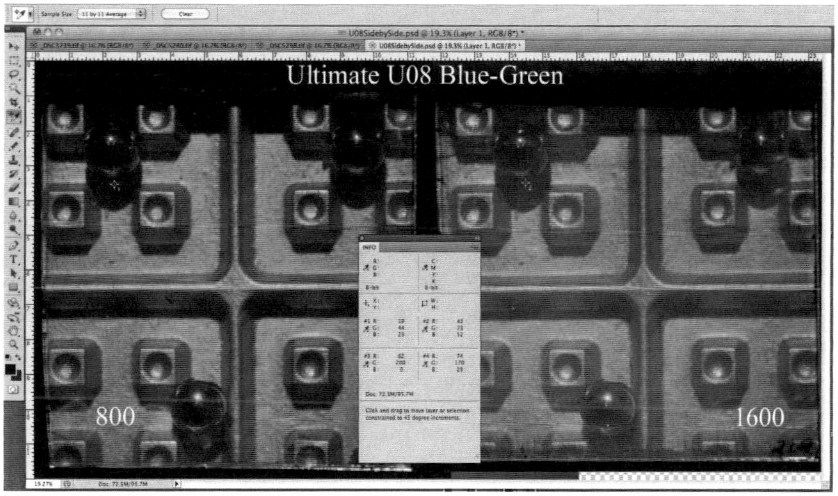

Figure 4. Signal to noise study.

The Color Sampled Readings are as follows:

- #1 (Noise in 800): R = 19, G = 44, B = 23.
- #2 (Noise in 1600): R = 43, G = 73, B = 52.
- #1 (Signal in 800): R = 62, G = 200, B = 0.
- #1 (Signal in 1600): R = 74, G = 170, B = 23.

The section of the image sampled for the noise was in the shadow of a ball bearing that supports the holographic plate during exposure. Ideally there should be no light seen from that part of the image, but there is enough light from other parts of the object slightly illuminating the umbra that there is an image background on top of the noise background caused by intermodulation and scatter due to the material and its processing.

The total luminosity of the noise is the sum of the RGB readings. For the 800 exposure, this is 19 + 44 + 23 = 86, for 1600 it's 43 + 73 + 52 = 168. The ratio of these two sums are almost doubled, but the relative brightness is found by subtracting these two numbers, 168 – 86 = 82, then dividing 82 by 57, the number found in the characteristic curve of the camera which is equal to one full stop of brightness change, to find the amount of stops difference in brightness, 82/57 = 1.44, and to find how many times brighter need to take 2 to the 1.44 power using the y^x key on the calculator, which equals 2.7 time brighter.

Finding the signal to noise ratio is done in a similar manner. The total luminosity of the bright area of the 800 is 62 + 200 = 262. This is actually a signal plus noise reading; its noise reading is 86 from the above. The difference of these two is 176 which is the signal reading; dividing 176 by 57 = 3.0, 2^3 = 8, so the signal to noise ratio of this sample is 8 to 1.

The noisier one has a signal plus noise to noise difference of 267 – 168 = 99, which is divided by 57 = 1.7, 2^1.7 = 3.2 for its signal to noise ratio.

Usually a noisier hologram appears brighter than a more contrasty, low noise one, and theswe numbers seem to bear this out. The lower exposure's total luminosity is 262 versus the higher one's 267; slightly more light arrives at the camera or the observer's eyes. But the amount of signal, as was just calculated, is 176 units on the 800 versus 99 on the 1600, a difference of 77, about 1 1/3 stops, or 2 ½ times as bright! The higher noise fills in the shadow areas, making the image appear "underwater".

6.3. Example 3: Bandwidth and shrinkage

This method can be used for checking the accuracy of the replay wavelength and its bandwidth. A baseline reading of the laser wavelength is taken, and the replay is compared to it. A gray card was photographed under spatially filtered 532 nm laser light, with exposures bracketed above and below that recommended by the camera's built-in meter. Surprisingly the eyedropper's values were not only reading green content but also some blue, even though the illumination was monochromatic! This is because the Bayer filter on top of the digital sensor, which analyzes the RGB content of the scene, acts like the human eye, whose RGB cones have overlapping sensitivities so that different hues of green for instance, can be discriminated.

Blue and green values for 3 exposures bracketed by 1/3rd stops:

- G = 165 185 202
- B = 85 95 105
- B/G = .52 .51 .52

From this calibration is can be inferred that 532 nm laser light is represented by 2 parts G and 1 part B. This is our baseline in determining color truthfulness to the laser lambda.

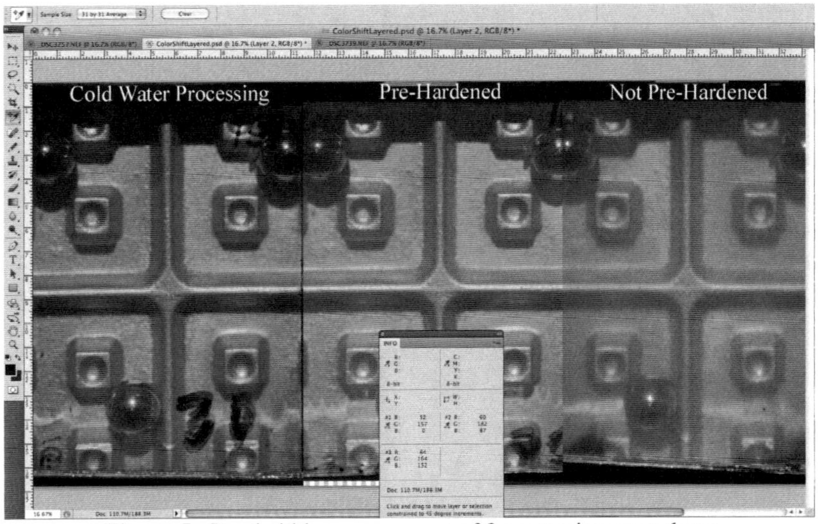

Figure 5. Bandwidth measurement of 3 processing procedures

Here is a comparison of some Sphere-S GEO-3 plates. All three were exposed for 6400 microJoules per square centimeter at 532 nm. The one on the left was processed in cold water, 15C, the middle one was dunked in formaldehyde pre-hardener and processed at room temperature, 20C, and the one on the right was processed simultaneously with the middle except it was not pre-hardened. (See Appendix for more details.) The cross piece in the middle of the object was sampled.

- Cold process: R = 52, G = 157, B = 0.
- Pre-hardened: R = 60, G = 182, B = 87.
- Not pre-hardened: R = 64, G = 164, B = 152.

All 3 of the green readings are quite similar. But the cold process plate, although not giving the same ratio of G to B under reconstruction with white light, has the narrowest bandwidth, with no B value. This is also the least noisy. The pre-hardened plate is noisier with a wider bandwidth visually, and the numbers bear that out. Obviously the hardener is a necessary evil, as without it there is considerable shrinkage to the shorter wavelength and the replay bandwidth is wider.

6.4. Example 4: Absolute efficiency

If the eyedroppered values of the holographic image were identical to that of the object, then the hologram would be 100% efficient, if we subscribe to the basic tenet that the hologram is an exact optical double of the object.

The Standard Object was photographed with 458 nm illumination. This same exposure was used to photograph two samples of Slavich PFG-03C exposed in that same location, one for 12.8 milliJoules per square centimeter, the other for 25.6. Both were processed in the same cold regimen as the GEO-3. They reconstructed quite brightly when replaced on the object, so well that real-time interference fringes could be observed. A piece of black velvet was placed under the holograms so that this effect could not be seen when photographing them.

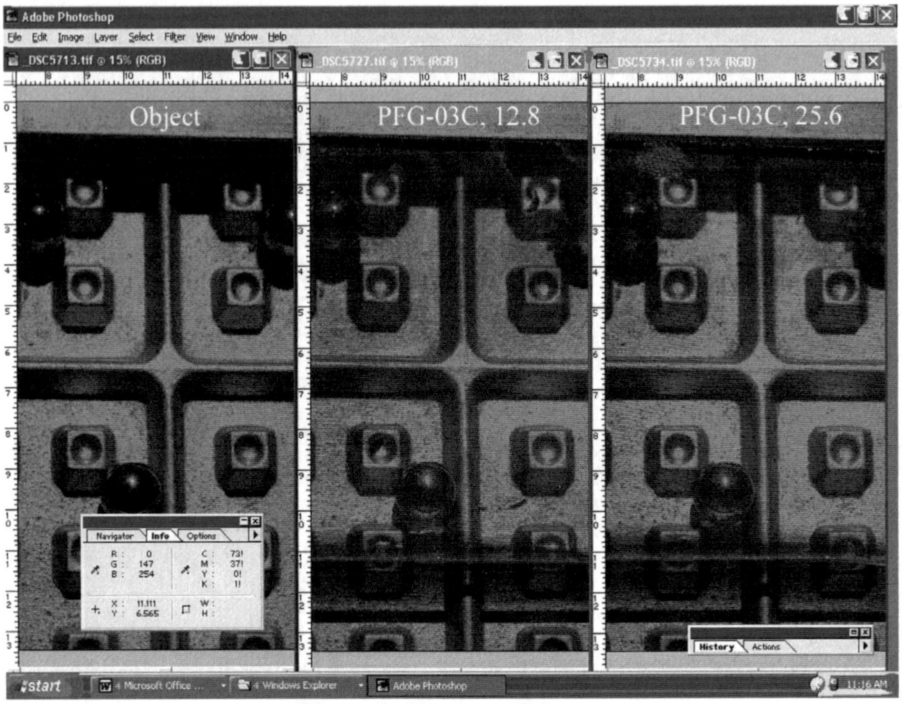

Figure 6. Absolute efficiency.

Readings at the crossbar in the center of the iron were:

- Object: R = 0, G = 150, B = 254.
- 12.8: R = 0, G = 138, B = 251.
- 25.6: R = 0, G = 150, B = 254.

It looks like 25.6 has achieved 100% efficiency! However the hologram readings contain signal plus noise, so a shadow area under a ball bearing was sampled for each scene to get the noise reading. There are some values in the object's reading, as this background is caused by diffuse reflections from other object points filling in the shadows. The hologram should record this information, but it does add its own noise to that.

- Object: R = 0, G = 19, B = 45.
- 12.8: R = 0, G = 29, B = 86.
- 25.6: R = 0, G = 29, B = 86

The noise of both holograms added 10 units of G and 41 units of B to the background, so subtracting that out from the preliminary readings gives:

- Object: R = 0, G = 150, B = 254.
- 12.8: R = 0, G = 128, B = 210.
- 25.6: R = 0, G = 140, B = 213.

Now the difference between the brighter hologram and the object is 254 -213 = 41, 41/57 = .7 of a stop, 2^.7 = 1.6, meaning that the object is 1.6 times as bright, or the hologram is 60% efficient. Which is not too shabby considering the recording wavelength!

7. Acknowledgements

I would like to send a shout out to Stas at Geola for sending me samples of the Sphere-S GEO-3 emulsion; Stephen Brierley of Harman Technology for sending me samples of their holographic emulsion; Arturo Perez Mulas from the holography forum for gifting me with an Ultimate Sampler pack; and Jesus Lopez who lent me some of the best PFG-03C that he ever used. And of course the krewe at Harrington College of Design, especially Joe Byrnes who let me take home one of the most expensive cameras in the Digital Photography Department's Cage; Photoshop guru extraordinaire Tim Arroyo, who contributed significant discussions and shortcuts in the understanding of the numbers; and, Dirk Fletcher, just because he was the first Department Chair to be wise enough to hire me full-time.

References

[1] Joly I and Vanhorebeek R 1980 Development Effects in white light reflection holography *Photographic Science and Engineering*.**24(2)** 108-13
[2] nlutie.com/ewesly for more information on cold water processing and trials of holographic materials.

9th International Symposium on Display Holography (ISDH 2012) IOP Publishing
Journal of Physics: Conference Series **415** (2013) 012033 doi:10.1088/1742-6596/415/1/012033

Drawing Light-fields: Hand-drawn Approaches to Abrasion Holography

T Duke

Museum of Jurassic Technology and Optics Divison of the Metabolic Studio

Email: tristanwduke@gmail.com

Abstract. Abrasion holography has received little attention since the technique was described by William J. Beaty in the early 1990's. In this paper the limitations of abrasion holography are explored, and new approaches are presented which expand the possibilities of the medium. New tools presented here offer new possibilities to the artist wishing to draw holograms by hand. Methods are described by which complex curves and organic forms can be constructed by hand more easily and intuitively than previously described. In an analysis of reconstruction lighting and viewing geometries, new solutions to reduce or eliminate distortions are suggested. Various tools, materials, and scratch geometries are considered for optimum 3D illusion. A new class of abrasion holograms is presented that use elliptical, hypotrochoidal, and epitrochoidal scratch geometries, exhibiting novel animation effects. In conclusion, a method for embossing abrasion holograms with the aid of an etching press is described.

1. Introduction

The depth illusion apparent in light reflected from circular scratch patterns has been noted independently by many commentators since the 1930's.[1,2,3,4,5] In the early 1990's William Beaty compared this illusion to holography and formalized a technique for creating 3D drawings by hand, which he called "scratchograms" or "abrasion holography."[6] Abramson expanded on the technique, with a method for geometric construction of a cube.[7] Several recent publications explore computer-aided methods for producing abrasion-type holograms, using CNC engravers, and milling machines.[8,9,10] Since Abramson and Beaty however very little has been published in the way of expanding the techniques available for *hand drawing* abrasion holograms.

I will assume familiarity with the basic technique of abrasion holography, and with the basic optics of their reconstruction. I will use the words *hologram* and *holographic* throughout this paper to refer to 3D images created by the scratches in an abrasion hologram. The question of whether or not these images should be considered a hologram has been treated by Beaty and Abramson and is outside the scope of this paper. For those unfamiliar with the basic techniques of abrasion holography and the optics of their reconstruction Beaty and Abramson each offer a thorough treatment of these topics.

I have spent the last three years making abrasion holograms by hand and exploring the limits of the medium. Here I present techniques, tools, and conceptual models useful to the artist who wishes to draw abrasion holograms by hand.

2. Materials

Many investigators have reported using Plexiglas or a similar plastic as the plate on which to inscribe abrasion holograms. I have found the image quality to be much better if metal is used. Metal plates are also more durable, and finished holograms on metal are less prone to being damaged by scratching. Copper plates produce a bright image with a very smooth scratch. As copper is quite expensive, aluminum plates are a good substitute. I have also found zinc plates to work quite well. Large sheets of 1/16 inch thick aluminum can be bought and cut down on a table saw.

A sturdy compass (dividers) with a steel or diamond tip scribe is used to make the scratches. When working with metal the scratch should be made with a firm, consistent pressure applied throughout the revolution of the compass. I have also employed drafting templates to good effect, especially where scratch circumference is small and use of a compass becomes difficult. Plates can be blackened by soft anodizing (or more easily with black permanent marker) before scratching; this greatly enhances the visibility of the holographic image by increasing the contrast between the plate surface and the hologram.

3. As Above So Below

In abrasion holography a single circular scratch represents two holographic points, one behind the plate (virtual image) and one in front of the plate (conjugate image). This presents a problem because any single holographic point we may wish to describe will be reflected pseudoscopically, greatly limiting the kinds of images which can be drawn.

Beaty overcame this limitation by constraining the scratch-arc to semicircular, thereby suppressing the pseudoscopic image, but as a result limiting the viewing angle.[6] In the case of a wall-mounted vertical presentation illuminated from above, Beaty's solution is passable since most of the normal range of view is covered. Beaty also described how shorter scratch-arcs could be employed to create animations or objects which disappear behind other objects. However, when presented supine (as in a table-top presentation) with illumination normal to the plate, or when the plate is presented in rotation, these techniques prove to be unsatisfactory with each single holographic point replaying pseudoscopically from the opposite angle of view.

Abramson presented another solution where full circle scratches may be used.[7] In Abramson's cube construction (fig. 5a) the holographic object is symmetrical and reflected through the plane of the plate. Abramson's cube construction can be generalized to any form with bilateral symmetry. Both real and virtual images contribute to a single holographic image with a very wide field of view. This arrangement is particularly well suited for a tabletop plate orientation with illumination at normal to the plate surface –the many advantages of which will return to shortly.

However, when presented in a wall-mounted display illuminated form above, Abramson's cube appears skewed, taking the form of a rhombic prism (a distortion which we shall return to shortly). This solution is also limited to forms with bilateral symmetry (spheres, cones, cubes, etc); the symmetrical object must be reflected across the plane of the plate (i.e. a cone can be presented where it's axis lies on the plane of the plate, however it is not possible to present a cone with its axis at any other angle to the plate, or for that matter with an axis above or below the plate).

It is clear that a method for separating "above" and "below" images is needed. This is accomplished by altering the profile of the scratch to be asymmetrical. A 90-degree conical diamond "drag engraving" tool is used in a specially designed compass (fig. 1a) which allows the tool to be tilted. Where the tool is oriented at approximately normal to the plate, the scratch profile will be symmetrical and the usual two holographic image points will appear above and below the plate (fig. 1b). If the tool is angled to tip inward toward the center of the circle, the scratch will become asymmetrical and only the holographic point "below" the plate appears (fig. 1c). Where the tool is tipped outward, only the holographic point "above" the plate appears (fig. 1d). With this arrangement the desired image point is reflected while the spurious one is completely suppressed. The tilt of the tool must be great, so that one edge of the diamond cone approaches normal to the plate. By adjusting the tilt of the tool a single scratch can be made to represent holographic points "above," "below," or both "above and below" the plate. These effects can be combined to create off axis, asymmetrical holographic objects.

With this arrangement, the requirement for symmetry and reflection across the plane of the plate has been overcome; all manner of holographic forms can be drawn with a full range of view.

Journal of Physics: Conference Series **415** (2013) 012033 doi:10.1088/1742-6596/415/1/012033

This technique also allows for Beaty's method of employing shorter arcs to be used to better effect: animations, solid objects, or objects which occlude other objects may be drawn with consistency across the visible field and without spurious artifacts.

4. Distortion and Optimum Viewing Angle

Several commentators have noted the "swinging" distortion apparent in abrasion holograms. In a wall mounted presentation with the plate illuminated from above, as the viewer moves laterally across the angle of view, the holographic object appears to "swing" dipping downward as the viewer is centered in relation to the plate and back up as the viewer moves to either side. This is remedied slightly by moving the source of illumination to an angle other than directly above the plate (with the added benefit, noted by Abramson[7], of increasing vertical parallax). A thorough examination of the optics of reconstruction reveal new solutions to this distortion.

4.1. Tabletop Display

This distortion can be avoided altogether if a tabletop plate orientation is acceptable (fig 2). Plate P is lying flat –as on a table. A circular scratch S is made on the plate with center point C and radius r. Point source illumination R is provided from directly above point C, normal to the plate and is reflected in the plane of the plate at R'. A viewer O is positioned at angle v to the reflected image of the light source. If we now draw a line connecting the reflected image R' with center point C (in this case congruent with the axis of illumination) where this line crosses with the viewers line of sight to the scratch points of light H and H' will appear. In this situation H will appear to the viewer as if above the plate, directly above C and at height Z; while H' will appear as if below C at depth Z'. We can see that where v and v' are 45 degrees then Z, Z' and r are all equal.

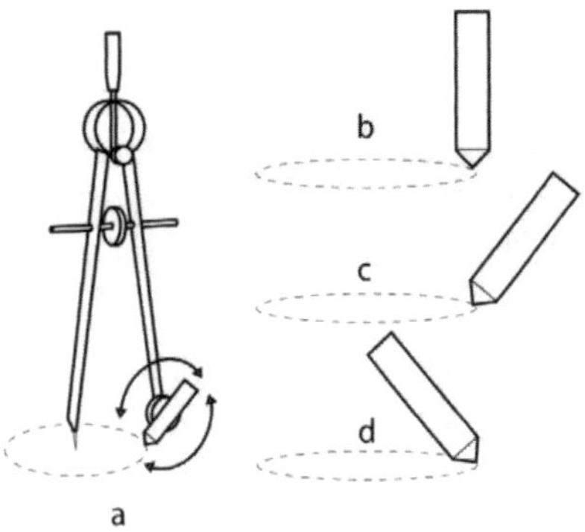

Fig. 1: Compass (a) fitted with adjustable diamond scribe which may be inclined for: "above and below" (b); "below" only (c); and "above" only (d) images.

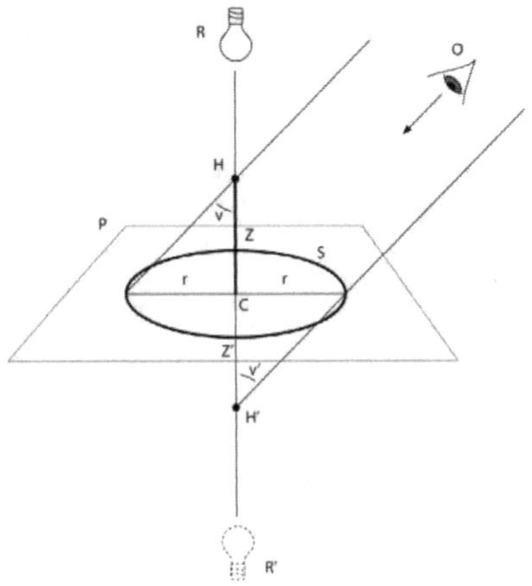

Fig. 2: Tabletop display

If we scratch a series of concentric circles centered on C, the resultant holographic image points will all fall on the line of the reflected path of illumination, in this case describing a line normal to the plate and passing through its surface at C.

In the foregoing we assume that the lines of sight are parallel; in actual practice of course they must converge at O. This will cause angle v' to be slightly diminished and v to be slightly enlarged with the result that H will appear lower and H' will appear deeper (in other words Z' will be longer than Z). This distortion becomes more pronounced as the viewer moves closer to the plate. At normal viewing distances of a few feet or more the distortion is so reduced as to be unnoticeable.

Journal of Physics: Conference Series **415** (2013) 012033 doi:10.1088/1742-6596/415/1/012033

The same can be said of the illumination source, which is treated here as a point source at and infinite distance.

With this arrangement the holographic image is dimensionally consistent as long as the angle of view relative to surface normal remains constant. However, a new "stretch" and "squish" distortion is introduced; as the viewing angle in relation to normal is varied the apparent holographic depth changes: growing more exaggerated as the viewing angle moves upward (approaching normal), causing the image to stretch; and diminishing as the viewing angle moves downward (toward the plane of the plate), causing the image to squish (fig 3). This means that as the viewer walks forward, toward the plate the apparent depth of the object increases and conversely as the viewer moves away the apparent depth is reduced. This distortion can be minimized by selecting a natural viewing angle (such as 45 degrees), and by establishing a viewing distance in the display (such as with a railing or the edge of the table).

This arrangement has the added benefit of being visible without distortion from 360 degrees around the plate (at a given viewing angle) allowing a large number of viewers, but also enabling viewers to walk all the way around the plate. This fact also means that the plate may be rotated, as on a turntable, without distortion –the holographic object appearing to rotate as if fixed to the plate. This proves to be quite desirable as the depth illusion benefits from the added motion parallax introduced by the rotation, and stationary viewers may appreciate all sides of the holographic object.

Note that in the tabletop display with illumination at normal, if the viewing angle is made to be 45 degrees, the radius of a given scratch is equal to the apparent height of the resultant point of light. This proves to be highly useful to the artist constructing an abrasion hologram by hand –as the height can be easily and intuitively determined. Here measurements can be taken directly from elevation drawings or even from physical models.

These advantages, added to those mentioned above, make the tabletop plate orientation, illuminated along surface normal and designed to be viewed at 45 degrees to normal a very useful arrangement for hand-drawn abrasion holograms.

4.2. Wall-mounted Display

Instead of adjusting our angle of view, let the view remains fixed as we hinge the plate up beginning from the tabletop orientation (as described in fig. 2) so that the surface normal approaches the line of sight of the viewer. As we do so, the reflected angle of the light source is no longer reflected back along the path of illumination but begins to shift towards the viewer. Consequently, holographic point H will travel up the line of sight of the viewer, appearing to gain altitude from the plate and move towards the viewer; while holographic point H' travels down the line of sight moving deeper into the plate and away from the viewer.

As the plate approaches 22.5 degrees from the plane of the table, holographic point H has traveled all the way up the line of sight and the reflected path of illumination is directed at the viewer.

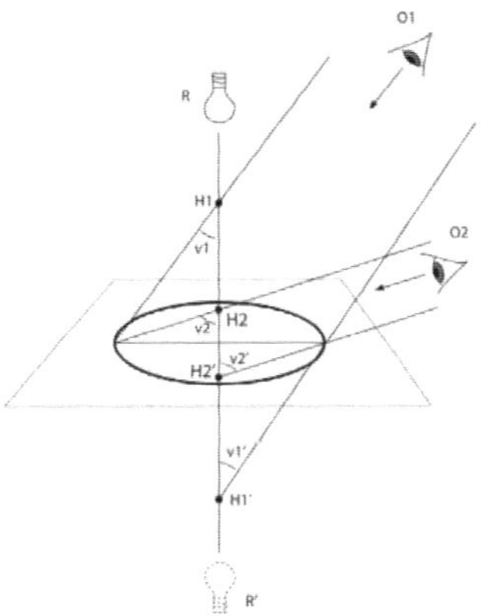

Fig. 3: Distortion as the viewing angle changes

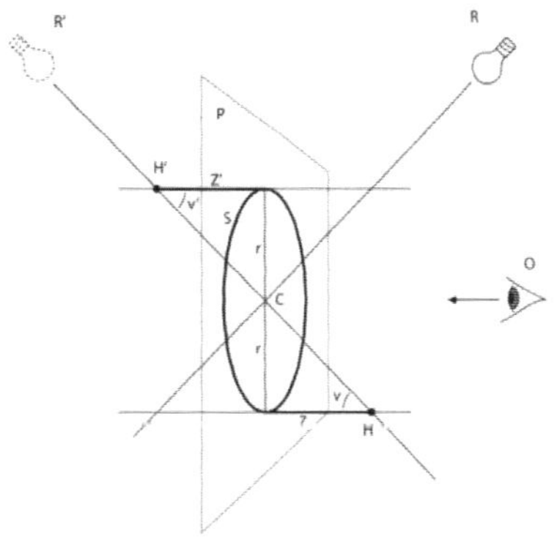

Fig. 4: Wall mounted display

The holographic image is highly distorted and obscured by the image of the light source as described by Abramson.[7]

If we continue to hinge the plate upwards however, the reflected path of illumination will continue its rotation past the line of sight of the viewer, and holographic point H will begin to travel back down the line of sight while H' will travel back up. As the plate reaches 45 degrees from the original tabletop, the line of sight becomes normal to the plate and points of light H and H' have returned to their original altitudes (as when the plate was flat on the table) only now they are no longer centered above C but appear as if directly normal to points above and below the line of the scratch (fig. 4). This is the arrangement for a wall-mounted display, with viewer normal to the plate and illumination from 45 degrees above.

Just as with the table-top display, if v is 45 degrees, then Z and r are equal (as well as Z' and r'). If we scratch a series of concentric circles centered on C, the resultant holographic image points will all fall on the line of the reflected path of illumination -in this case describing a line at 45 degrees to the plate and passing through its surface at C.

4.3. Corrected Cube

Let us consider the cube construction proposed by Abramson (fig. 5a).[7] In the table-top display the concentric circles represent lines normal to the plate, these connect two squares in planes parallel to the plate surface, one above and one below. Where the viewing angle v is 45 degrees the circles should be enlarged so that r is equal to Z (fig. 5b).

In the wall mounted presentation this construction is no longer satisfactory for an image of a cube; The two squares which are in planes parallel to the plate surface have shifted, the one above the plate having moved down, and the one behind having moved up. The concentric circles now represent lines at 45 degrees to normal. The image is no longer a cube but a rhombic prism (with the front and back faces of the cube parallel to the plane of the plate and the top and bottom faces tipped downward away from surface normal, making the remaining two sides rhombuses).

In the wall-mounted display, if we wish to describe two holographic points reflected across the plane of the plate, unit scratches will no longer be circles, but arcs tangent to each other (fig. 5c). If the "above" and "below" cutting tools are used, these may be two full circles tangent to each other (though the upper arcs will extend outside the visible area). As the viewer position approaches normal, the two points of light become coincident appearing as a single dot at the tangent point of the arcs. We

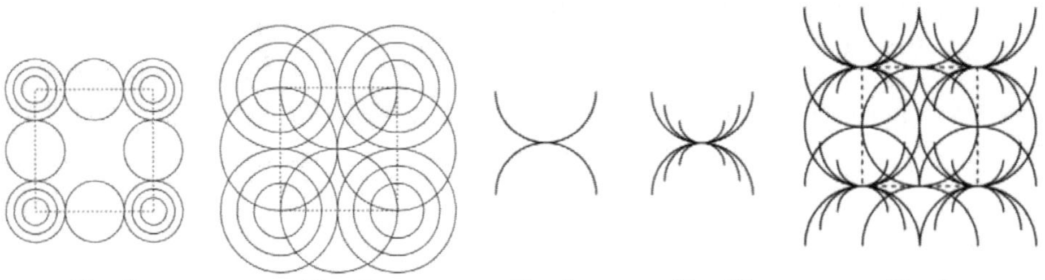

Fig. 5a Fig. 5b Fig. 5c Fig. 5d Fig. 5e

can now construct the scratch pattern for holographic points on a line extending normal to the plate (fig 5d); arcs curving upwards represent points behind the plate, arcs curving downward represent points in front of the plate. From this we can construct a corrected cube (fig. 5e), with four holographic lines normal to the plate, connecting the corners of two squares in planes parallel to the plate. The cube now appears normal in the wall-mounted display, but (as might be expected) is distorted into a rhombic prism when viewed in the tabletop display. In this case the square faces are perpendicular to the plane of the plate.

Using this corrected geometry any arbitrary holographic shape can be constructed for wall-mounted viewing. This method greatly reduces the "swinging" distortion in wall mounted display which was described above.

5. Object and Reference Projections

Let us return to the tabletop orientation described in (fig. 2). Imagine that a tiny bead or other such arbitrarily small object X is made to float at point H (at distance Z above C); the bead casts a shadow C' onto the plate at point C. Now imagine that the viewer O is replaced with a second light source.

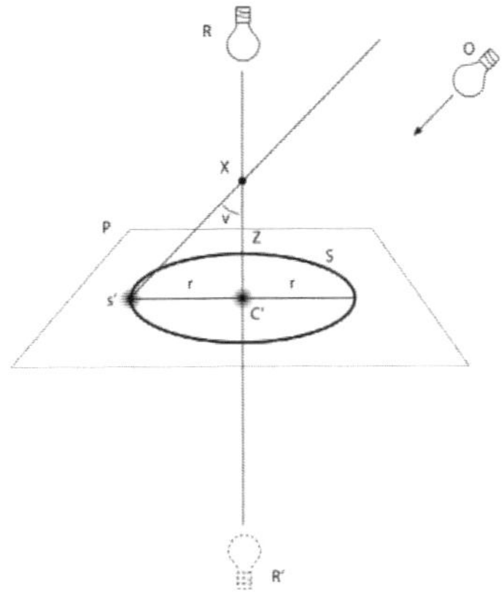

Fig. 6

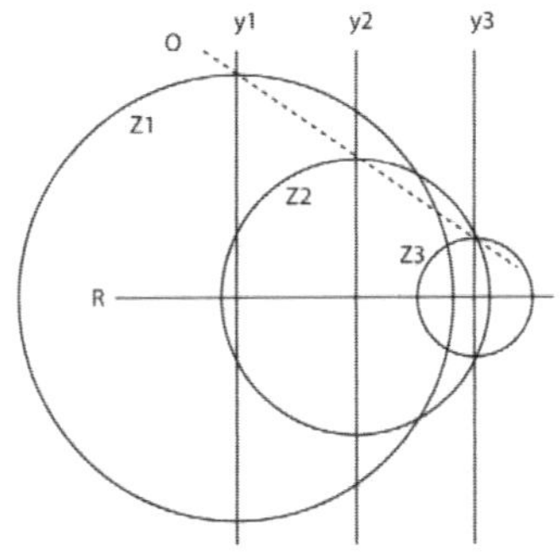

Fig. 7a

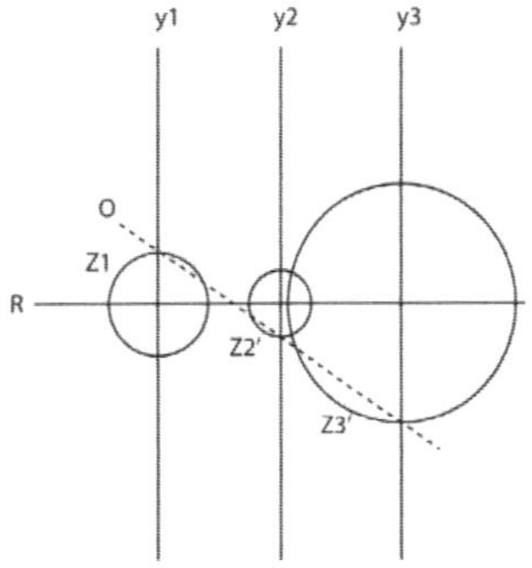

Fig. 7b

The bead now casts a second shadow S' at distance r from point C' (see fig. 6). As the plate is rotated (or as light source O is rotated around the plate), shadow S' remains a constant distance r from C' – tracing the path of circle S.

Here we can see that shadow C' corresponds to the stationary arm of the compass, while shadow S' corresponds to the moving arm. This demonstrates how the scratch pattern of an abrasion hologram can be understood in terms two projections onto the plane of the plate: one at the desired angle of illumination (R) which we shall call the *reference projection* (C'), and the other at the desired viewing angle (O) which we shall call the *object projection* (S'). Any single holographic point can be drawn by determining an object projection and a reference projection, and scratching the circle thus described.

As the bead is raised straight upwards the reference projection C' remains stationary (since the illumination is directly above) while the object projection S' moves outward away from C'. This describes a series of circles centered at C' and with increasing radii. As the bead is lowered the distance decreases, until finally the bead rests on the plane of the plate and the two shadows are congruent. This describes circles centered at C' with decreasing radii. If the bead is lowered beneath the plate the reference projection is from the direction of the reflected image of O.

By using this model we can quickly construct abrasion holograms without needing to calculate individual points.

Projective method for drawing a line (see fig. 7a):

1. Draw reference line R on the plate
2. Draw any object line O above R
3. Draw an elevation line y, at any point perpendicular to R and intersecting O
4. Place the stationary point of the compass on reference line R at the intersection of y; extend the other arm of the compass to the point where y intersects the object line; and scratch a circle z with the "above" cutting tool.
5. Repeat steps 3 and 4 for as many points on O as desired

If the object line crosses the reference line, it will pass through the plate and the "below" cutting tool should be used (fig 7b). Neither the reference line nor object line need to be straight –complicated curves can be drawn quickly with this method.

If for our object line we draw a circle above the reference line, we will have a holographic image of a circle on a plane 90 degrees from the surface of the plate. As before, if the object-circle is lowered so that part of it crosses the reference line, those scratches should be incised using the "below" cutting tool. If the object-circle is lowered to the point where its center intersects the reference line, the center point of our holographic circle will be on the surface of the plate; in this case the cutting tool can be oriented to reflect both "above" and "below" images –since the image is now symmetrical!

If we want to draw a holographic image of a circle on a plane 45 degrees to that of the plate, the reference projection will be a 45-degree ellipse while the object projection will be a straight line. If the object line is lowered so that it is coincident with the major axis of the reference ellipse, the center of our holographic circle will appear to be on the surface of the plate –however "above" and "below" cutting tools must be used, as the figure is not symmetrical across the plane of the plate. The envelope of the scratches for this form is a compressed nephroid.

If both object and reference lines are circles, then all scratches will be of the same diameter and the holographic image will be of a circle on a plane parallel to the plate. As the object circle is lowered it will eventually be congruent with the reference circle; the scratch diameter will be infinitely small and the circle will be on the surface of the plate.

In the case of the wall mounted display, object and reference projections will be reversed. The object projection will not be normal to the plate while the reference will be at 45 degrees. By this method we have a powerful tool for imagining and plotting abrasion holograms.

6. Noncircular Abrasions

If instead of circular scratches we use elliptical, hypotrochoidal, and epitrochoidal scratches we will also create holographic effects. These curves can be drawn by tracing a point on a circle as it rolls along the edge of another circle. For experiments with these curves I used a Spirograph drawing toy, a steel scribe, and aluminum plates. A set of draftsman's ellipse templates can also be used.

As may be expected, the elliptical scratch responds in a similar way to a circle, with one point appearing above and one point below the plate surface. Unlike the circle however, the curvature of the ellipse is not consistent but becomes tighter at the major axis. This causes the holographic image point to appear, when the plate is rotated, not at a fixed altitude above a fixed point on the plate, but rather as a moving point orbiting around the reflected path of the light source.

An ellipse is a special case of a hypotrochoid, where a point on a circle of a given circumference is traced as it rolls inside a circle of twice that circumference (such as Spirograph wheel #48 rolling inside Ring #96). A circle, in turn, is a special case of an ellipse; if we trace the center point of the small wheel our resulting scratch is a circle. As we have shown, the circular scratch, when illuminated from normal to the plate presents an image of a stationary point of light –even when the angle of view is rotated. If we choose a point slightly off center (such as hole #17 on Spirograph wheel #48), the resulting scratch becomes slightly elliptical, and the corresponding holographic point begins to have a slightly eccentric orbit. As we move further away from the center of wheel #48 our elliptical scratches become more and more elongated approaching the limit of a straight line. The rotation of these holographic points describe increasingly large rings of orbit around the original circular scratch. The orbit of these points is one rotation per each half-rotation of the plate –this can be understood as a function of the ratio of the circumferences of the two circles which formed the hypotrochoid (which in in the special case of the ellipse will always be 1:2).

If a series of similar ellipses are scratched into the plate, centered on the same point but each rotated by degrees to create a rosette-like pattern, then each identical ellipse will display a holographic point above and below the plate at a different stage of its orbit and each orbit will be traced as a ring of lights one above and one below the plate. When the plate is rotated the points which make up this ring rotate as described above. If a series of similar but progressively smaller ellipses are drawn on a single axis centered around the same center point, the holographic points will describe a line with an eccentric rotation as the viewing angle rotates around the plate. The rotation of this line shows the twin rings to be bases of a double cone with their shared apex on the surface of the plate at the shared center point of the ellipses. By scratching a series of concentric circles at this center point we describe holographic points on a line tracing the axis of rotation of our double cone.

If we select a slightly larger circumference for our inner wheel (such as Spirograph wheel #72), the holographic image formed is similar, though the orbits traced by the points become less elliptical and more circular as if tipped downward toward the viewer. As the inner wheel grows larger, the orbit of the holographic points slows down. In the case of wheel #72 with ring #96, a single holographic point will orbit once for every ¾ turn of the plate.

Epitrochoids also produce organized holographic images. These appear as dented or warped ellipses with some forms resembling hyperbolic paraboloids. The geometry of these images deserves further study and will be the topic of another paper.

The Spirograph drawing toy offers and interesting approach to drawing abrasion holograms. Complicated images, which would otherwise be difficult or impossible to draw by hand can be created easily. For example an approximation of a sphere can be created by using ring #96 with wheel #40, hole 13; wheel #50, hole 13; wheel #63, hole 13; and wheel #84, hole 13.

7. Printing

In collaboration with printmaker Richard Nielsen of Untitled Prints and Editions in Los Angeles, we are developing a technique for reproducing abrasion holograms using an etching press. The abrasion hologram plate is prepared as normal, by scratching in metal though with a cutting tool designed to raise a maximum burr (metal displaced by the scratch and protruding from the surface of the plate). Where a copper plate is used the burr can be reinforced by nickel plating. Using the press this burr is then embossed into a reflective surface such as a thin foil to create a "print" of the hologram -which exhibits full dimensional effects. Using this technique multiple images can be created from one master plate. The details of this technique will be the subject of another paper.

8. Conclusion

In the field of hand-drawn abrasion holography we have, so to speak, only scratched the surface of what is possible. In the short space allowed for this paper, I have only been able to introduce these methods, tools, and conceptual models. As a medium, hand-drawn abrasion holography offers many interesting and as-yet unexplored possibilities. Many of the distortions observed in abrasion holograms can be corrected by using the correct scratch geometry for the display setting. Conceptual models, such as the projective model, allow holograms to be drawn quickly and intuitively. New tools presented in this paper, such as "above" and "below" cutting tools, ellipse templates, and the Spirograph drawing toy offer exciting possibilities. The printing process only mentioned here, makes possible reproductions of holograms, making more labor-intensive images to be reproduced from a single master. It is my hope that the investigations presented here will inspire further exploration of this unique medium.

---

[1] Weil HH 1934 Improvements in Advertising and Like Signs *UK Patent* **37208/34**

[2] Lott JB 1963 Reflections on a Gramophone Record *Mathematical Gazette* **47** 360

[3] Garfield E 1981 ISI's World Brain by Gabriel Liebermann: The World's First Holographic Engraving *Essays of an Information Scientist* **5**

[4] Walker J 1989 What do Phonographic Records have in Common with Windshield Wipers? *Scientific American* **216** 1

[5] Plummer W and L Gardener 1992 A Mechanically Generated Hologram? *Applied Optics* **31** 31

[6] Beaty W 2003 Drawing Holograms by Hand *Proc. SPIE* **5005** 156-167

[7] Abramson NH, 2000 Incoherent Holography *Proc. SPIE* **4149** 1

[8] Brand M, 2011 Specular Holography *Applied Optics* **50** 25

[9] Regg C et al 2010 *ACM Transactions on Graphics* **29** 6

[10] Augier ÁG and R. B. Sánchez, 2011 Hologravure as a Computer-generated and Laser Engraved Scratch Hologram *Optics Communications,* **284** 112-117

A New Type of Coupled Wave Theory Capable of Analytically Describing Diffraction in Polychromatic Gratings and Holograms

David Brotherton-Ratcliffe

Geola Technologies Ltd, Sussex Innovation Centre, Science Park Square, Falmer, East Sussex, BN1 9SB, United Kingdom

dbr@geola.co.uk

Abstract. A new type of coupled wave theory is described which is capable, in a very natural way, of analytically describing polychromatic gratings. In contrast to the well known and extremely successful coupled wave theory of Kogelnik, the new theory is based on a differential formulation of the process of Fresnel reflection within the grating. The fundamental coupled wave equations, which are an exact solution of Maxwell's equations for the case of the un-slanted reflection grating, can be analytically solved with minimal approximation. The equations may also be solved in a rotated frame of reference to provide useful formulae for the diffractive efficiency of the general polychromatic slanted grating in three dimensions. The new theory is compared with Kogelnik's theory where extremely good agreement is found for most cases. The theory has also been compared to a rigorous computational chain matrix simulation of the un-slanted grating with excellent agreement for cases typical to display holography. In contrast, Kogelnik's theory shows small discrepancies away from Bragg resonance. The new coupled wave theory may easily be extended to an N-coupled wave theory for the case of the multiplexed polychromatic grating and indeed for the purposes of analytically describing diffraction in the colour hologram. In the simple case of a monochromatic spatially-multiplexed grating at Bragg resonance the theory is in exact agreement with the predictions of conventional N-coupled wave theory.

1. Introduction

Kogelnik's coupled wave theory [1] of the volume holographic grating has been very successful. Published in 1969 it still constitutes today a useful tool for the rapid computation of the diffractive response of general sinusoidal volume phase or mixed volume phase-amplitude gratings. However Kogelnik's theory is not rigorous. To arrive at the fundamental set of first-order coupled partial differential equations starting from Maxwell's equations, no fewer than five major assumptions must be made. Various authors have offered alternative models [e.g.2-6] or have extended Kogelnik's work [e.g.6,7].

Moharam and Gaylord [8] reported in 1981 a rigorous coupled wave theory. This theory provided an accurate computational solution to the problem of diffraction from planar gratings. It was shown that Kogelnik's analytic theory [1] provided a rather good description of the un-slanted reflection grating but that at high index modulations it overestimated somewhat the diffraction efficiency of the slanted grating. Differences in the transmission grating were shown to appear at a rather lower modulation.

Gratings produced using multilayer dielectric stacks have been traditionally analysed in a rather different manner to the analysis methods used to describe volume holographic gratings. Here Maxwell's equations are solved by analysing the individual Fresnel reflections from each layer in the stack. In 1937 Rouard [9] introduced a method by which the complex electric field amplitude reflected by a general stack could be calculated. In 1950 Abeles [10,11] introduced his highly successful chain matrix formulism which allowed a more efficient calculation of the reflective properties of a stratified medium.

Moharam and Gaylord [12] used the chain matrix model in 1982 to computationally analyse the un-slanted volume reflection holographic grating. The results were compared with both rigorous coupled wave theory [8] and with Kogelnik's theory [1]. Ning [13] also developed a polychromatic coupled wave theory of the volume reflection phase grating and compared numerical calculations of this theory with corresponding chain matrix calculations for the un-slanted case. In addition Diehl and George [14] compared the chain matrix approach to a method based on a solution of the Hill's matrix. Finally Ludman [15] has derived approximate estimations for the bandwidth and diffraction efficiency of the thick slanted transmission phase grating using the concept of Fresnel reflections.

In this paper we present a differential representation of the chain matrix model which is capable of accurately describing the general volume holographic grating (we refer for convenience to this model as the PSM model - short for "Parallel Stacked Mirrors"). The model effectively constitutes a new type of coupled-wave theory capable of describing volume phase and mixed phase-amplitude gratings of both the transmission and reflection type. Rather than seeking an approximate solution to the Helmholtz equation as Kogelnik's theory does, the PSM model is based upon a detailed analytic description of the process of Fresnel reflection within the grating.

The PSM theory produces a mixture of analytical expressions and simple differential equations which can be used to effectively characterise the general multi-colour holographic grating. In the limit of zero slant and normal incidence, the basic differential formulation is rigorous. Many of the single-colour formulae which are derived from this formulation give expressions identical or extremely close to Kogelnik's. Like Kogelnik's theory the PSM model can be generalised to an N-coupled wave theory. In the simple case of the monochromatic spatially-multiplexed volume reflection grating at Bragg resonance the generalised PSM equations are identical to the N-coupled wave equations of Solymar and Cooke [6-7]. For a full and concise mathematical exposition of the PSM theory the reader is referred to [16-18]. Here we shall restrict ourselves to a brief and simplified introduction to PSM using the least possible mathematics.

2. A Brief and Very Simplified Review of Kogelnik's Theory

The starting point for Kogelnik's theory is the Helmholtz equation. This is an equation which describes how light propagates within a medium. Mathematically, and in its most simple form, the Helmholtz equation may be written as

$$\frac{d^2u}{dy^2} - \gamma^2 u = 0 \tag{1}$$

Here the variable u measures the electric field of the light wave in a given direction, y measures the distance along the path of the light wave and γ is a variable which describes the medium and the wavelength of light travelling through that medium. In words, this equation simply states that the rate at which the rate of change of the electric field changes is proportional to the electric field itself. When the parameter γ is a constant - for example in air - then the Helmholtz equation predicts simple oscillatory or wave motion. In optics this is what we call a plane wave. The same equation arises throughout physics whenever wave motion is present.

Kogelnik's theory actually requires a slightly more complicated version of the Helmholtz equation. This is known as the two-dimensional wave equation

$$\frac{\partial^2 u}{\partial x^2}+\frac{\partial^2 u}{\partial y^2}-\gamma^2 u=0 \qquad (2)$$

The only difference here is that our wave is now not limited to travelling in a single straight line but can have a slightly more complex behaviour within the (x, y) plane.

Kogelnik's coupled wave theory starts off by assuming that the propagation parameter, γ in equation (2) is composed of two parts. The first and largest part just corresponds to a constant index of refraction. But added to this is a small oscillatory part which models the holographic grating.

Next Kogelnik makes the assumption that an impinging monochromatic plane light wave which illuminates the grating evenly will propagate in a rather special way within the grating. Specifically it is assumed that there will always be two waves within the grating - the driving wave, *R* and a diffracted wave, *S*. Both these waves are mathematically written as standard plane waves. But their amplitude is assumed to vary only with their depth of penetration within the grating (see Fig.1).

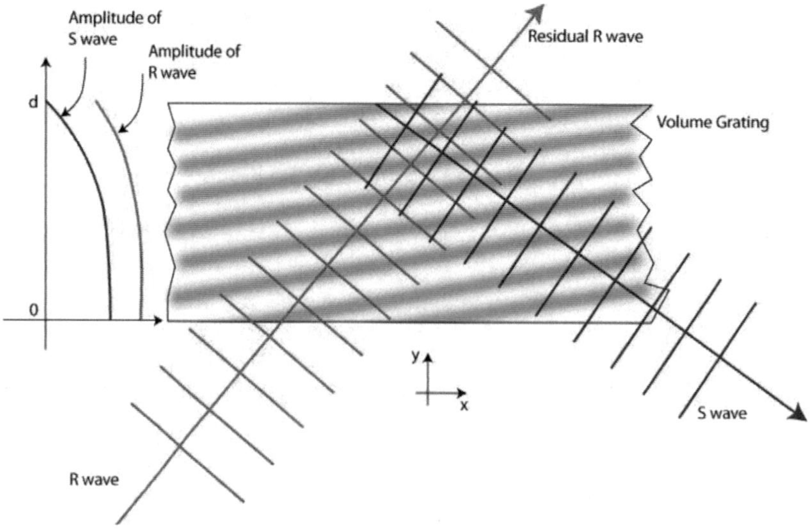

Fig.1 Kogelnik's Coupled Wave Theory assumes that two waves only propagate within the grating: The reference or R wave and the diffracted signal wave or S wave. Both waves are assumed to have the mathematical form of plane waves but with amplitudes which vary exclusively in the y dimension. The amplitude of both waves in the x direction is assumed to be constant. The example here is of a reflection grating.

By substituting the assumed form of the index distribution of the grating (or equivalently, γ) and the assumed form for the *R* and *S* electric fields into equation (2), Kogelnik derived a relatively complicated equation relating the electric fields within the grating. This equation cannot itself be solved however and needs further simplification.

The term $\gamma^2 u$ in equation (2) couples the oscillatory form of the grating (described by γ) with the oscillatory form of the two propagating waves. The first assumption is therefore to simply ignore any coupling which produces higher modes of oscillation. But this still leaves an equation containing several types of oscillating modes. Kogelnik therefore argues that each type of mode (there are two of them) must satisfy its own equation independently. This process then leads to two second order differential equations for *R* and *S*. Finally Kogelnik argues that since the amplitude of *R* and *S* must be expected to vary slowly with grating depth their second order rate of change must be small and so should be ignored. This then leads to two first order differential equations which at Bragg resonance define *R* and *S* within the grating, given the details of the illuminating wave.

Away from Bragg resonance, Kogelnik needs to make one further somewhat special assumption concerning the propagation wavevectors of the *R* and *S* waves within the grating. Specifically he assumes that the *R* wave propagates "normally" - that is that it propagates in the same direction and with the same free-space wavelength as it does outside the grating (note that the average index inside and outside the grating are the same so Snell's law doesn't bend the ray). However he is obliged to make the assumption that the *S* wave propagates in a rather special fashion in that in general its free-space wavelength is not equal to the free-space wavelength of the driving *R* wave and also its direction is different from the ray direction at Bragg resonance.

One can perhaps appreciate that Kogelnik's model, despite its great practical success in providing useful formulae for evaluating the diffractive efficiency of gratings, doesn't therefore constitute a completely transparent model of what is actually occurring inside a volume grating during the process of diffractive replay. It was with this thought in mind that I became motivated to develop an alternative coupled wave description of diffraction which is based on the concept of Fresnel reflection.

3. A Coupled-Wave Theory based on Fresnel Reflection - The "PSM" Model

When a plane light wave propagates across the boundary of two regions of constant but differing refractive index, a portion of the light wave is reflected and a portion is transmitted. In the case of a purely real index there is no absorption and the transmitted and reflected energies add up to the incident energy. The laws describing how the amplitude of the transmitted and reflected waves depends on the indices of the two regions are known as Fresnel's laws. In the case of normal incidence they are trivially simple. If the wave passes from index n_1 to index n_2 then the ratio of the reflected to incident amplitude is simply given by

$$r = \frac{n_2 - n_1}{n_2 + n_1} \tag{3}$$

Likewise the ratio of the transmitted amplitude to the incident amplitude is given by

$$t = \frac{2n_1}{n_2 + n_1} \tag{4}$$

The PSM theory models a volume grating as many thin parallel stacked layers, each with a slightly different index. Overall the grating looks just like a normal grating except that under the microscope the index does not vary smoothly from one depth to another but rather it makes small jumps. However, by making the distance between such jumps vanishingly small, we won't know the difference!

When an illuminating wave strikes a jump between one index and the next, we can use equations (3) and (4) to write down the amount of light transmitted and reflected. Of course these equations are only valid for the case of normal incidence but we can generalise them easily enough to incidence at any angle. Since the index layers are assumed flat and parallel in the simplest PSM model we can also use the well-known law that the angle of incidence is equal to the angle of reflection. If we then call our (monochromatic) illuminating wave, *R*, it is immediately obvious that at each index jump, *R* is slightly depleted and a new wave, which we can call *S*, is created by Fresnel reflection.

3.1. The Simplest Possible PSM Theory - The Normal-Incidence Volume Reflection Phase Grating

We will now present the simplest possible mathematical derivation of the simplest example of a PSM coupled-wave theory. This is the case of the sinusoidal un-slanted volume reflection phase grating under illumination at normal incidence. Fig.2 shows a diagram of the set-up. We start by modelling the grating as a stack of parallel slices, each having a slightly differing refractive index. At this stage we don't make any assumption about the form of the index distribution other than it being composed of slices! Note that each slice will have a small but finite thickness and in the other two dimensions it will have the form of an infinite rectangular plane.

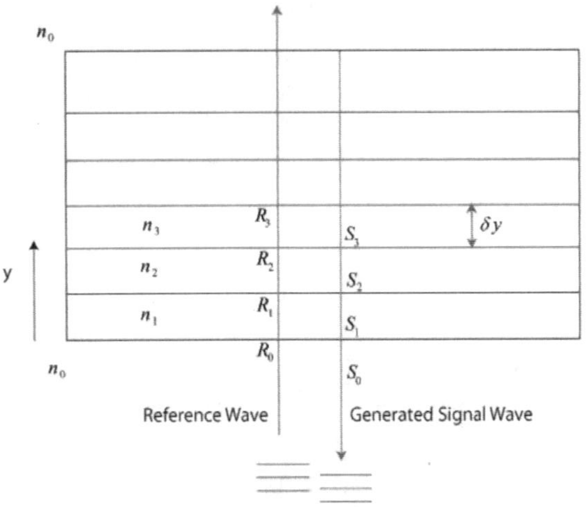

Fig.2 The simplest PSM model: The un-slanted volume reflection phase grating illuminated at normal incidence.

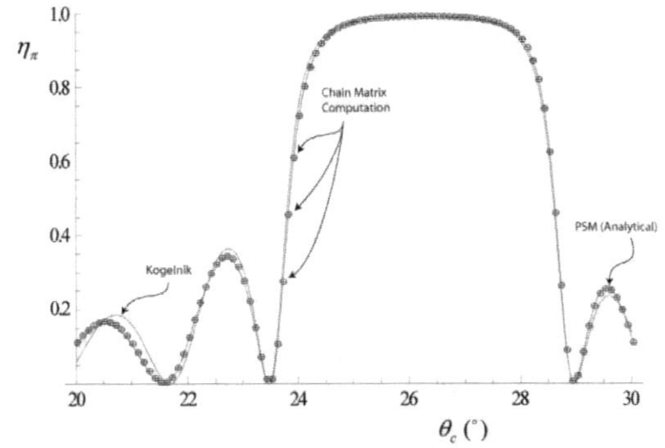

Fig.3. Comparison of the PSM and Kogelnik models with a rigorous chain matrix computation for the π-polarisation diffractive efficiency versus replay angle. The 12μm un-slanted reflection grating (n_0=1.5, n_1=0.065) has been recorded at 532nm with a recording angle of 30°.The replay wavelength is 550nm.

We now assume that a monochromatic plane wave illuminates the grating. Mathematically we state this by the equation

$$R^{ext} = e^{i\beta y} \tag{5}$$

where

$$\beta = \frac{2\pi n_0}{\lambda_c} \tag{6}$$

Here n_0 is the average value of the index inside and outside the grating and the parameter λ_c represents the free-space wavelength of the wave. If we now label each one of our "slices" by the integer, J and let the index of the Jth slice be n_J, then we can use the Fresnel relations to write down equations for how the *S* wave is generated by the *R* wave and how the *S* wave gives rise in turn to a reverse contribution to the *R* wave. Mathematically this can be written down as

$$\begin{aligned} R_J &= 2e^{i\beta n\delta y/n_0}\left\{\frac{n_{J-1}}{n_J+n_{J-1}}\right\}R_{J-1} + e^{i\beta n\delta y/n_0}\left\{\frac{n_{J-1}-n_J}{n_J+n_{J-1}}\right\}S_J \\ S_J &= 2e^{i\beta n\delta y/n_0}\left\{\frac{n_{J+1}}{n_{J+1}+n_J}\right\}S_{J+1} + e^{i\beta n\delta y/n_0}\left\{\frac{n_{J+1}-n_J}{n_{J+1}+n_J}\right\}R_J \end{aligned} \tag{7}$$

This is actually very simple! The first equation just describes how the *R* wave at the Jth position is generated by the transmission of the *R* wave at the J-1 position and by the reflection of the *S* wave at the Jth position. Likewise the second equation defines how the *S* wave at the Jth position is formed by the transmission of the *S* wave at the J+1 position and by the reflection of the *R* wave at the Jth position. You will recognise no doubt the terms in curly brackets as just being the Fresnel amplitude transmission and reflection expressions. The exponential terms take care of the phase of the waves. As each wave propagates so its phase changes and this must be accounted for properly.

The next stage is to use the laws of calculus and to let the thickness of the slices, δy tend to zero. This allows us to convert these "discrete difference" equations into the form of differential equations:

$$\begin{aligned} \frac{dR}{dy} &= \frac{R}{2}\left(2i\beta\frac{n}{n_0} - \frac{1}{n}\frac{dn}{dy}\right) - \frac{1}{2n}\frac{dn}{dy}S \\ \frac{dS}{dy} &= -\frac{S}{2}\left(\frac{1}{n}\frac{dn}{dy} + 2i\beta\frac{n}{n_0}\right) - \frac{1}{2n}\frac{dn}{dy}R \end{aligned} \tag{8}$$

Here n_J has been replaced by the continuous index variable, n. These are the PSM coupled wave equations for a normal incidence un-slanted reflection volume phase grating. Unlike Kogelnik's coupled wave equations they are actually exact. Or in other words they constitute a different way of writing down the Helmholtz equation!

In order to solve the PSM coupled-wave equations we need to define the grating in terms of an index profile. To investigate the behaviour of the simple sinusoidal grating we use the following definition

$$n = n_0 + n_1 \cos\left(\frac{4\pi n_0}{\lambda_r} y\right) = n_0 + \frac{n_1}{2}\left(e^{\frac{4i\pi n_0}{\lambda_r} y} + e^{-\frac{4i\pi n_0}{\lambda_r} y}\right) \tag{9}$$

This just says that the index profile of the grating is made up of two parts - a constant part, n_0 and a small sinusoidally varying part, n_1. We then make a mathematical transformation

$$R \to R'(y)e^{i\beta y} \quad ; \quad S \to S'(y)e^{-i\beta y} \tag{10}$$

Due to the symmetry of the problem at hand and the assumed form of the illuminating wave, (5) this is only a mathematical tool - it is not an assumption. In fact the only assumption that we need to make is that the primed terms are slowly varying compared to the exponential. And this will nearly always be true. Under this one assumption, we can write

$$\langle R' \rangle \sim R \quad ; \quad \langle S' \rangle \sim S \tag{11}$$

where the operator $\langle\ \rangle$ takes an average over several cycles of the wave. Then simple substitution of (9) and (10) in (8) yields the following much simplified coupled wave equations:

$$\frac{dR}{dy} = -i\alpha\kappa S\, e^{2i\beta y(\alpha-1)} \quad ; \quad \frac{dS}{dy} = i\alpha\kappa R\, e^{-2i\beta y(\alpha-1)} \tag{12}$$

where

$$\alpha = \frac{\lambda_c}{\lambda_r} \tag{13}$$

is the ratio of the playback to recording wavelengths. If we choose appropriate boundary conditions for the reflection grating ($R(0)=1; S(d)=0$) these equations actually have analytic solutions which are characterised by a diffractive response which is almost exactly the same as Kogelnik's theory. We can use a mathematical trick which makes this rather more obvious. This is the introduction of the pseudo-field in place of the real electric field, *S*. In fact, this is what Kogelnik does implicitly. You will remember that he assumed that the *S* field propagated in a very special manner...

The pseudo-field is defined as

$$\hat{S} = S\,\mathrm{e}^{2\beta i y(\alpha-1)} \tag{14}$$

If we substitute this expression into (12) then we arrive at Kogelnik's coupled wave equations:

$$c_R \frac{dR}{dy} = -i\kappa\hat{S} \quad ; \quad c_S \frac{d\hat{S}}{dy} = -i\vartheta\hat{S} - i\kappa R \tag{15}$$

However the constants c_R, c_S and ϑ are slightly different from Kogelnik's values. The PSM theory gives

$$c_R = \frac{1}{\alpha} \quad ; \quad c_S = -\frac{1}{\alpha} \quad ; \quad \vartheta = 2\frac{\beta}{\alpha}(1-\alpha) \tag{16}$$

whereas Kogelnik's theory gives

$$c_R = 1 \quad ; \quad c_S = (2\alpha-1) \quad ; \quad \vartheta = 2\alpha\beta(1-\alpha) \tag{17}$$

As usual Kogenik's constant is defined as

$$\kappa = \frac{\pi n_1}{\lambda_c} \tag{18}$$

One immediate advantage of the PSM theory is that it can be extended trivially to cover multi-colour gratings. We simply replace (9) with a sum of different recorded index profiles, one for each colour. Mathematically we write this as

$$\begin{aligned} n &= n_0 + n_1\cos(2\alpha_1\beta y) + n_2\cos(2\alpha_2\beta y) + \ldots \\ &= n_0 + \frac{n_1}{2}\left\{e^{2i\beta\alpha y} + e^{-2i\beta\alpha y}\right\} + \frac{n_2}{2}\left\{e^{2i\beta\alpha_2 y} + e^{-2i\beta\alpha_2 y}\right\} + \ldots \end{aligned} \tag{19}$$

Then (12) just becomes

$$\frac{dR}{dy} = -S\sum_{j=1}^{N} i\kappa_j\alpha_j\,\mathrm{e}^{2i\beta y(\alpha_j-1)} \quad ; \quad \frac{dS}{dy} = R\sum_{j=1}^{N} i\kappa_j\alpha_j\,\mathrm{e}^{-2i\beta y(\alpha_j-1)} \tag{20}$$

where the integer index, j sums the colours. As long as the colours are not too close together these equations give the following useful expression for the diffraction efficiency of the polychromatic un-slanted volume phase grating

$$\eta = \sum_{j=1}^{N} \frac{\alpha_j^{\,2}\kappa_j^{\,2}}{\beta^2(1-\alpha_j)^2 + (\alpha_j^{\,2}\kappa_j^{\,2} - \beta^2(1-\alpha_j)^2)coth^2\left\{d\sqrt{\alpha_j\kappa_j^{\,2} - \beta^2(1-\alpha_j)^2}\right\}} \tag{21}$$

3.2 Angled Incidence, Slanted Gratings and Holograms

The PSM theory can easily be generalised to the case of a tilted illumination beam [16,18]. The equations can then be solved in a tilted frame of reference to give analytic expressions for the diffraction efficiency of the polychromatic slanted reflection volume phase grating (for both σ and π-polarisations). The final results are very similar to Kogelnik's but the PSM model can usually provide a somewhat superior estimation of the off-Bragg behaviour - see for instance Fig.3. By superposing many differently tilted gratings the PSM equations can be extended to an analytic description of the spatially multiplexed grating and by employing an infinity of such gratings, to the case of the volume hologram [17,18].

The extended PSM theory directly shows that the lossless polychromatic reflection volume phase hologram is theoretically capable of perfect diffractive replay at each and every wavelength. In other words, if such a hologram is co-illuminated by P co-propagating reference plane waves, each of a different wavelength, then the hologram is capable of producing a perfect diffractive response to each of these P waves simultaneously. The all-important parameter here is the effective index modulation achievable in a given material at a given wavelength. If this is low then a thicker grating will be needed to achieve a bright hologram. The PSM model can be further extended to model lossy holograms by consideration of a complex index. In this case the strategy of simply making a reflection hologram thicker to produce a brighter diffractive response does not work as the further the reference wave penetrates into the hologram the more it is absorbed. This leads to the lossy polychromatic reflection hologram being characterised by a maximum attainable diffractive response which is dependent on the maximum achievable modulation in a given material. As such the PSM theory shows clearly that both the achievable index modulation and the intrinsic loss of a given photosensitive material must be expected to constitute the key parameters in determining the maximum practical diffractive efficiency of full-colour reflection holograms.

4. References

[1] H.Kogelnik, "Coupled Wave Theory for Thick Hologram Gratings", Bell Syst. Tech. J. **48**, 2909-47 (1969).

[2] E.J Saccocio, "Application of the dynamical theory of X-ray diffraction to holography", J.Appl. Phys. **38**, 3994-8 (1967).

[3] V.V Aristov and V.Sh. Sheekhtman, "Properties of three-dimensional holograms", Sov. Phys. Uspekhi **14**, 263-77 (1971).

[4] C.J.R Sheppard, "The application of the dynamical theory of X-ray diffraction to thick hologram gratings", Int. J.Electronics **41**, 365-73 (1976).

[5] R.S. Chu and J.A.Kong, "Modal Theory of spatially periodic media", Trans IEEE MTT-25, 18-24 (1977).

[6] L.Solymar,"Two-Dimensional N-Coupled-Wave Theory for Volume Holograms", Opt.Comm. **23**, 199-202 (1977).

[7] D.J.Cooke and L.Solymar, "Comparison of Two-Wave Geometrical Optics and N-Wave Theories for Volume Phase Holograms", J.Opt.Soc.Am.**70**, 1631 (1980).

[8] M.G.Moharam and T.K Gaylord, "Rigorous Coupled Wave Analysis of Planar Grating Diffraction", JOSA, Vol. 71, Issue 7, pp. 811-818 (1981).

[9] M. P. Rouard, "Etudes des propriétés optiques des lames métalliques très minces," Ann. Phys. (Paris) Ser. II **7**, 291–384 (1937).

[10] F. Abeles, "Recherches sur la propagation des ondes électromagnétiques sinusoïdales dans les milieux stratifiés. Application aux couches minces," Ann. Phys. (Paris) **5,** 596–640(1950).

[11] R. Jacobson, "Light reflection from films of continuously varying refractive index", *Progr in Optics*, E. Wolf, ed. North-Holland, Amsterdam, Vol. 5, Chap. 5, pp. 247–286.(1966).

[12] M. G. Moharam and T. K. Gaylord, "Chain-matrix analysis of arbitrary-thickness dielectric reflection gratings," J. Opt. Soc.Am. **72,** 187–190 (1982).

[13] X. Ning, "Analysis of multiplexed-reflection holographic gratings," J. Opt. Soc. Am. A **7,** 1436–1440 (1990).

[14] D.W. Diehl and N.George, "Analysis of Multitone Holographic Interference Filters by Use of a Sparse Hill Matrix Method," Appl. Opt. 43, 88-96 (2004).

[15] J.E.Ludman," Approximate bandwidth and diffraction efficiency in thick holograms" Am. J. Phys., 50, 244, (1982).

[16] D.Brotherton-Ratcliffe, "A Treatment of the General Volume Holographic Grating as an Array of Parallel Stacked Mirrors", Submitted 24th December 2011 and Accepted for publication 14th May 2012 to J.Modern Optics.

[17] D.Brotherton-Ratcliffe, "An Analytical Treatment of the Polychromatic Spatially Multiplexed Volume Holographic Grating", Submitted 18th May 2012 to Applied Optics.

[18] H.Bjelkhagen and D.Brotherton-Ratcliffe, "Ultra-Realistic Imaging - Advanced Techniques in Analogue and Digital Colour Holography", CRC Press, Taylor and Francis Group, ISBN: 978-1-4398-2799-4, December (2012).

Journal of Physics: Conference Series **415** (2013) 012035 doi:10.1088/1742-6596/415/1/012035

Are stereograms holograms? A human perception analysis of sampled perspective holography

Pierre St. Hilaire, Pierre-Alexandre Blanche, Cory Christenson, Ram Voorakaranam, Lloyd LaComb, Brittany Lynn, Nasser Peyghambarian

College of Optical Sciences, The University of Arizona, Tucson, Arizona 85721, USA

philaire@optics.arizona.edu

Abstract. In this paper, we analyze the optical performance of a commonly used 1-step recording geometry for stereograms and compare it to the fully fringe rendered case, taking a published optical model of the human eye into account. We compare our results to a model that conserves wavefront curvature. We then demonstrate how to optimize hogel sampling parameters as a function of image depth, size, and viewer distance. We conclude that stereograms suffer from little degradation from a viewing distance larger than 2 meters, but that nearer field images can significantly benefit by adding a second order phase correction.

1. Introduction

There historically have been two approaches for the rendition of holographic images, whether for real-life scenes or synthetic images. In the conventional method an entire scene is phase-encoded on the recording medium, either through interference from an actual object wavefront or by computation of the corresponding fringes. An alternative approach is to capture or calculate a number of discrete perspectives, which are then multiplexed after proper processing[1]. This stereographic approach makes it much easier to capture real life scenes using a conventional camera, is computationally less intensive in the case of synthetic images, and is commercially available. However, previous work on 2-step stereograms has demonstrated that discarding the second order phase information results in a loss of resolution in the case of deep images[2].

The goal of the current paper is to extend the modulation transfer function (MTF) analysis of references[2] and [3] to the case of 1-step hogel-based displays, some of which we have built at the College of Optical Sciences. These so called 1-step stereograms currently have largely supplanted the 2-step master / copy process pioneered by DeBitetto[4] and Benton[5] since they directly generate an in-plane hologram. The additional computational cost of the 1 step technique is no longer prohibitive for full parallax images and the technology has been deployed commercially.

2. 1-step image plane stereograms.

In the case of 1-step stereograms (sometimes referred as hogel-based stereograms), the direction of rays coming from a scene is discretely encoded within each individual picture element[1]. Each picture element, called a hogel, is printed separately in the geometry shown in Fig. 1. Although the stereogram is encoded interferometrically with a reference beam, the process does not conserve the 2nd order phase information (wavefront curvature) of the original scene. Instead, the wavefront curvature is approximated piecewise from the diversity of rays coming from a collection of hogels.

This loss of wavefront curvature leads to a decreased resolution in stereograms for picture elements away from the image plane, and also to aliasing artifacts if the image is improperly processed. A depth-dependent filter can be implemented to remove the artifacts. However, the loss of resolution can only be alleviated by encoding the wavefront curvature. Holographic stereograms that encode the second-order curvature information have been published by Bove's group at MIT[6]. They have been experimentally demonstrated to yield better images at a reasonable additional computational cost by using a look up table for the wavefront curvature. The principal inconvenient of the technique is that it places considerably more constraints of the spatial light modulator (SLM), which in that case has to be used as a phase modulator. With SLM's increases in performance and speed we will likely see more use of this method in coming years.

3. Wavefront errors in 1-step stereograms

In this analysis we will consider the wavefront description to be separable in X and Y axes, which is usually the case in practice due to the use of square and rectangular hogels and SLM pixels. We can then separately calculate the X and Y modulation transfer functions and express the system MTF by their product. It should be noted that our analysis does not account for astigmatism, which is a third order aberration. However, in the case of full parallax images, the astigmatic error is generally significantly smaller than wavefront curvature and can be ignored. In the case of horizontal parallax only (HPO) holograms the effect of astigmatism becomes more pronounced for deep images, but our model still provides a good approximation as we will see in the experimental section.

1-step holographic stereograms are well described in the literature so we will only succinctly describe the process in this article. The reader is referred to ref. for a more detailed explanation.

3.1. Recording

Fig. 1 gives a schematic representation of the recording of a single hogel for a transmission geometry. First, the scene of interest is calculated in a representation that corresponds to the 1-step recording geometry. This results in a 2x2 matrix representing the discrete vectors corresponding to the rays coming from the scene. Such a matrix is calculated for each hogel in the hologram. In the case of natural scenes, this process requires re-sampling the pixels corresponding to successive image perspectives in a process often referred to as *light field rendering*.

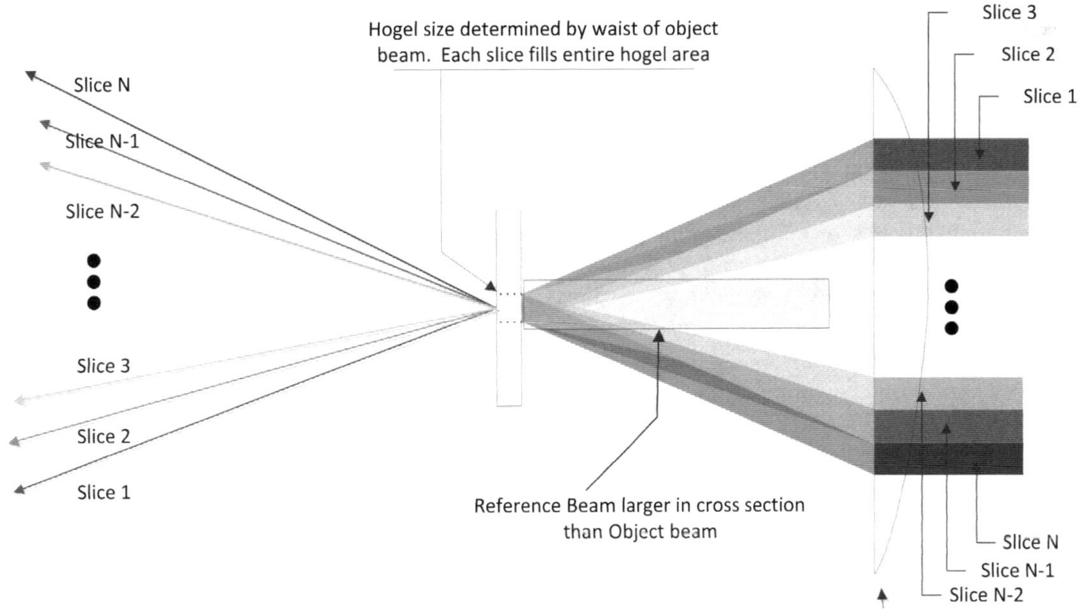

Figure 1. Hogel recording geometry.

After rendering, the matrix elements corresponding to the intensity of individual rays for a hogel are transferred onto the SLM as grayscale values, and the hogel is projected onto the recording medium where it interferes with a reference beam. Usually, the SLM is situated close to the Fourier plane of the objective lens, so that its conjugate image upon reconstruction lies in the far field. The view zone of the hologram will then be determined by the f/number of the objective lens, often requiring the use of custom made low f/number optics.

In practical systems the SLM is often relay-imaged onto the front focal plane of lens L1 where a diffuser is added to generate square uniform pixels. The effect of such a diffuser is not considered in the present paper. The effect of the diffuser is to spread the intensity of the ray bundle at the hogel plane in order to exploit the complete dynamic range of the medium, at the cost of a small decrease in resolution for deep images.

3.2. Reconstruction

The usual reconstruction consists in illuminating the stereogram with a conjugate of the reference beam, although a non-conjugate illumination can also be used (which is the case in our work at UA)[7]. Each ray direction (k-vector) then matches the recording geometry, but the radius of curvature is fixed by the focal length of the recording lens.

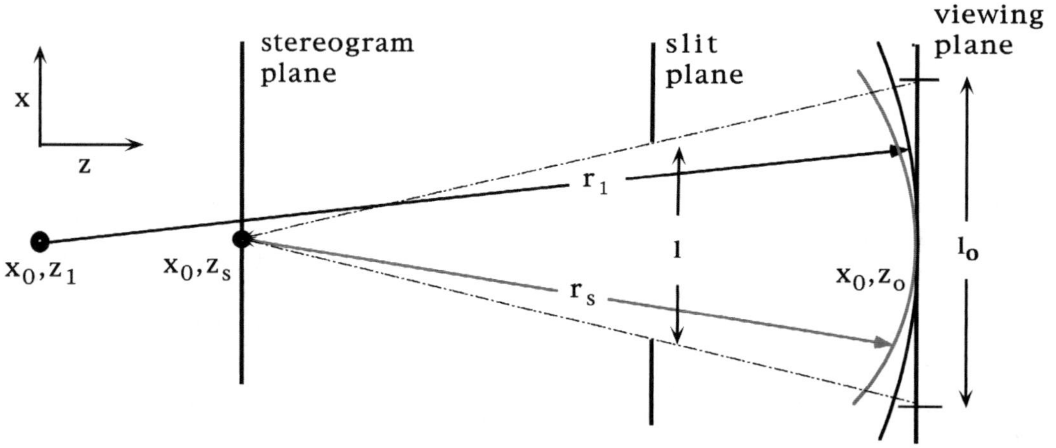

Figure 2. Wavefront curvature error. In the 1-step process the slit plane coincides with the objective lens position in Fig. 1, and l represents the extent of a SLM pixel.

The 1-step recording process thus creates errors during reconstruction because of this wavefront curvature mismatch. This discrepancy, which is not present in conventional holograms, is equivalent to a defocusing of the scene components away from the stereogram plane. The image degradation resulting from defocusing is well known and can be characterized by its effect on the modulation transfer function[8]. Such an analysis was previously applied to the analysis of conventional 2-step stereograms and Ultragrams (a generalization of the 2-step geometry for arbitrary slit position)[3]. Extending the results to the 1-step case is straightforward if we refer to Fig. 2.

The wavefront curvature error between the stereogram plane and the original object when measured at the viewer plane has the expression[3]:

$$\Delta r \approx (z_1 - z_s) + \frac{(z_s - z_1)}{2(z_1 + z_o)(z_s - z_o)}(x - x_0)^2 \tag{1}$$

And the real part of the pupil function as calculated from the viewer location is determined by the projection of each ray onto the viewer plane:

$$P(x) = rect\left[\frac{(x-x_0)z_s}{l(z_s-z_o)}\right] \tag{2}$$

where l is the spatial extent of a single perspective at the SLM plane. Note the sign convention: since we chose the origin to be at the Fourier transform lens plane, z1 and zs are generally *negative*.

Evaluating the optical transfer function (OTF) integral in an identical treatment to that of ref.[8] then leads to the result:

$$H\left(f_x^s\right) = \Lambda\left(\frac{\lambda\left|z_s f_x^s\right|}{l}\right)\operatorname{sinc}\left[\frac{(z_s-z_1)(z_s-z_o)l}{z_s(z_1-z_o)}f_x^s\left(1-\frac{\lambda\left|z_s f_x^s\right|}{l}\right)\right] \tag{3}$$

where the sinc function is defined as sinc(x) = sin(πx) / πx and the triangle function L is defined as:

$$\Lambda(x) = \begin{cases} 1-|x| & |x|\le 1 \\ 0 & \text{otherwise} \end{cases} \tag{4}$$

The modulus |*H*| of the OTF is referred as the modulation transfer function and is an important measure of image quality in optical systems.

Equation (3) can be more conveniently be expressed in terms of the image depth D, SLM size S (or its image at the objective Fourier plane if relay optics are used), number of perspectives N, and observer position Z_0 with respect to the hologram plane as:

$$H\left(f_x^s, D\right) = \Lambda\left(\frac{\lambda N\left|z_s f_x^s\right|}{S}\right)\operatorname{sinc}\left[\frac{D z_o}{z_s(D-z_o)}\frac{S}{N}f_x^s\left(1-\frac{\lambda\left|z_s f_x^s\right|N}{S}\right)\right] \tag{5}$$

where z_s now denotes the focal length of the objective lens L_1. In equation (5) we have adopted the convention that D is positive in front of the hologram for non-conjugate illumination, and negative for conjugate illumination, while the viewer position z_o is always positive. The maximum spatial frequency in the reconstruction is bounded by:

$$f_x^{s\max} = 1/2H \tag{6}$$

where H is the size of a single hogel.

As an example, let us consider an image that straddles the hologram plane and has a depth ranging from 20 cm in front of the hologram to 20 cm behind with a hogel size of 0.25 mm and view zone of 30 degrees. What would be the MTF vs. depth of the image for a viewer 0.8 m away with 100 and 250 rendered individual perspectives (or equivalently number of angular samples per hogel)?

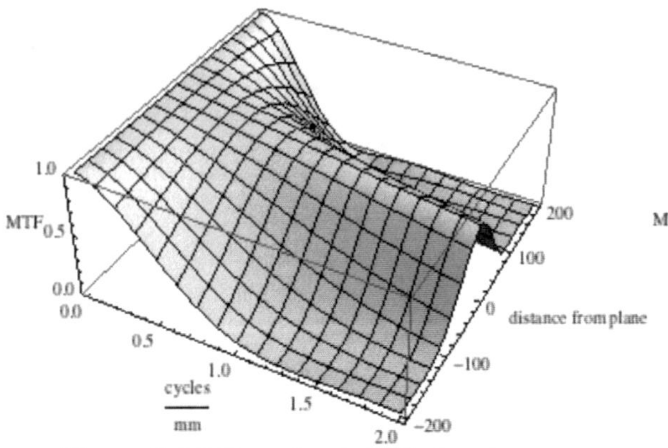

Figure 3. MTF vs. depth (100 views).

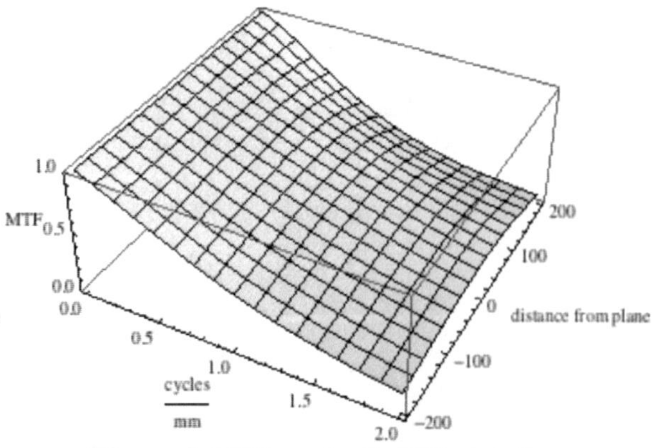

Figure 4. MTF vs. depth (250 views).

As we can see, increasing the number of perspective to 250 results in a much sharper image away from the plane of the hologram at the expense of a loss of contrast closer to the plane.

It is interesting to compute the image sharpness at maximum depth versus the number of perspectives N. We get the following curves for depths of 50, 100, and 200 mm for a spatial frequency of 2 cycles/mm.

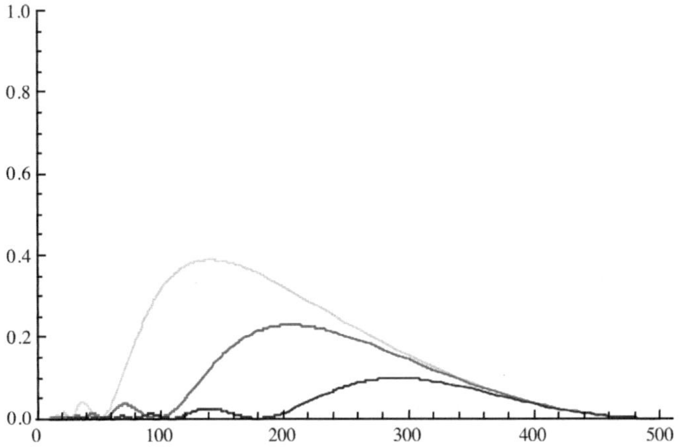

Figure 5. MTF curves for depths or 50, 100, and 200 mm *vs.* number of perspectives.

Clearly more perspectives result in a higher image resolution for deep images, and the number of perspective views should be chosen to match the depth of the reconstructed image and size of the hogels.

For an ideal optical hologram, the equivalent MTF would have an almost constant value close to 1, demonstrating that the preservation of local phase has a profound effect on hologram reconstruction. However we are now led to ask how much this matters for the perception of the image by the observer's visual system. Answering this question requires a physical optics model of human visual perception.

4. Human visual system MTF

It is essential to consider the response of the visual system when comparing different modalities of hologram recording such as phase vs. non-phase conserving displays. In this section we will use a model of the human eye MTF derived by Artal and Navarro, who used a double pass system to directly measure the contrast of images projected onto the retina[9]. The measurements were then fitted to an empirical double exponential curve. We will summarize the salient point of this description as following:

- The human eye is not diffraction limited. Only for a pupil size of 1.5 mm or less does it approximates a diffraction-limited system.
- Acuity goes down as pupil increases in size (up to 8 mm), so display brightness affects performance. Brighter displays look crisper.
- The numerical models of the eye MTF that have been published do not completely agree, but their differences are minor with respect to this work.
- Higher order aberrations extend the depth of field for larger pupil sizes, a natural example of wavefront coding.

It should be noted that here is a difference between the objective eye MTF (contrast measured on the retina by an ophtalmoscope, such as in our model), and the subjectively perceived MTF (also called "neural" MTF), which includes cognition and neural processing and is determined by psychophysical measurements. Using the retinal MTF as we do generally puts stricter requirements on the display, since the human visual system is good at "filling out" details. However it fits better this analysis, which is all based on physical optics.

As an example, figure 6 shows the object space MTF for a viewer 800 mm away for pupil sizes of 2.5 and 7 mm.

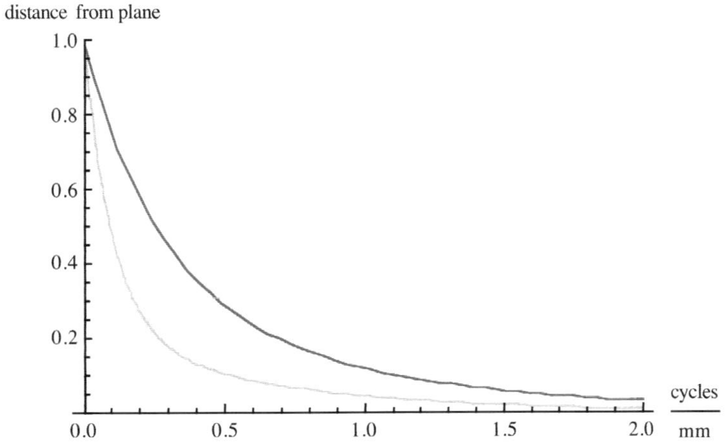

Figure 6. Modeled eye MTF for pupil diameters of 2.5 (red), and 7 mm(green).

Having obtained numerical models for both the display and visual system, we can now calculate the MTF of the optical system comprising both the display and human eye with the formula:

$$MTF_{system}\left(f_x^s, D\right) = MTF_{display}\left(f_x^s, D\right) \times MTF_{eye}\left(f_x^s, D\right) \tag{7}$$

In which MTF_{system} is given by equation 5 and MTF_{eye} corresponds to the numerical approximation of ref. [9]

For the rest of this article we assume a viewer pupil size of 2.5 mm, which is a realistic value when working with information displays. If we now plot MTF_{system} with the same parameters used in fig. 4 and 6, we get the following plots for a hogel-rendered display and a curvature preserving display:

Journal of Physics: Conference Series **415** (2013) 012035 doi:10.1088/1742-6596/415/1/012035

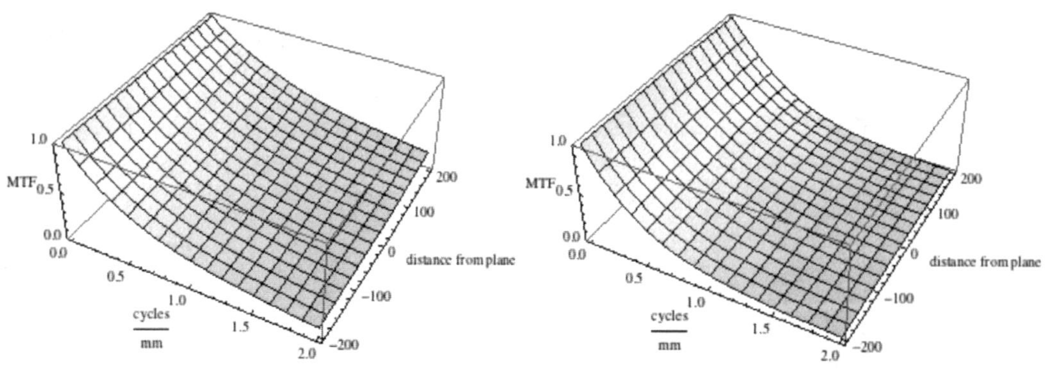

Figure 7. System MTF that includes the observer visual system using (left) full fringe rendering. (Right) Hogel based rendering. Eye pupil size is 2.5 mm.

As we can see, the limited MTF of the human eye considerably levels the playing field between phases preserving displays and hogel-based displays, although there is still is a noticeable difference.

The last parameter we have yet to explore concerns the viewer position with respect to the hologram plane. As an example, lets consider the case of observer positions 500, 800, and 2000 mm away. We get the following curves for reconstruction 200 mm in front of the hologram:

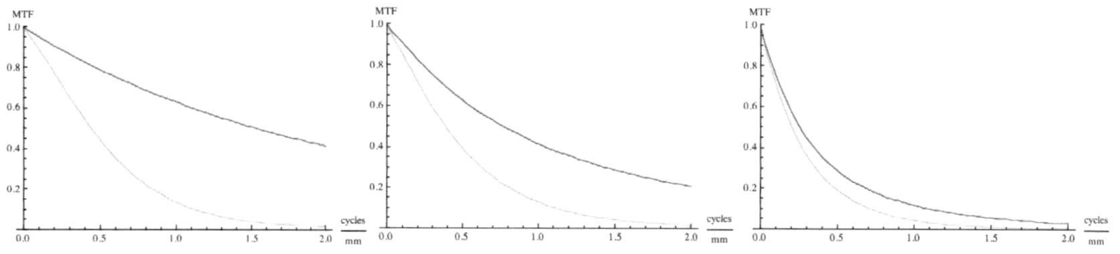

Figure 8. System MTF for a viewer distance of (right) 500, (center) 800, and (left) 2000 mm. The fringe-rendered MTF is in red and the hogel based MTF in green. Eye pupil size is 2.5 mm.

Fig. 8 illustrates how for deep images the hogel-based reconstruction dramatically degrades with respect to a coherently-rendered image as the viewer closes in to the reconstruction.

5. Discussion

The model was implemented in *mathlab* and *mathematica*, and was used to simulate the displays that our group built at the College of Optical Sciences. Having a working model allowed us to optimize both the computation and the optics of our systems, by comparing the experimental measurements to the numerical predictions. As a result, we were eventually able to write images in our photorefractive polymer that approximated the modeled MTF to within 30%, despite using off the shelf optical components (with the exception of the holographic diffusers that were fabricated in-house). As an example, figure displays the Air Force test chart written on our material at the focal plane and 90 mm in front. In this case the hogels had a 250 micrometer extent and 200 views were rendered for a total viewing angle of 26 degrees. Getting close to the predicted performance took a considerable amount of effort. Besides the design and careful alignment of optics, we had to fabricate our own LED illuminators and volume holographic band-limited diffusers. A more detailed description of our experimental work is also published these proceedings.

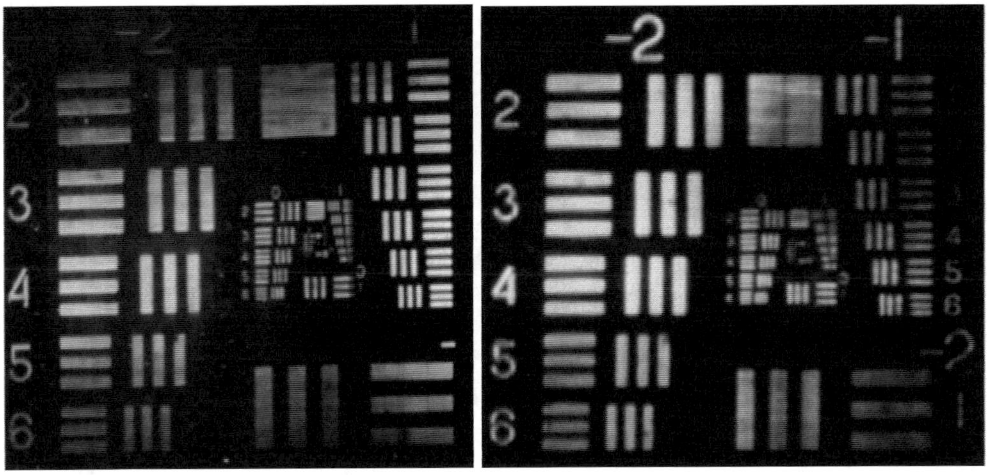

Figure 9. Air force test chart imaged (a) at the hologram plane and (b) 90 mm in front. 100 views were calculated over a total view zone of 26 degrees. Individual hogels have a 250 micron extent.

6. Conclusion

Are holographic stereograms truly holograms? We answer the question by saying that it is all in the position of the beholder. For situations equivalent to working on a desktop display or tablet computer, or for displaying medical data such as CAT scans or MRIs, the functionality of stereograms might be hampered because their resolution is reduced for deep images, despite their accurate rendition of occlusion and stereopsis depth cues. In other situations, such as televisions, advertising signage, or computer augmented virtual environments (CAVEs), the difference is hardly meaningful and the word "hologram" can be legitimately used. This is an old debate and this article does not expect to have the last word. However, from our simulations, practical holographic workstations for CAD/CAM or medical imaging will likely require wavefront curvature preservation. It also became clear within the scope of this work that light source engineering (including the use of resonant or photonic lattice LED's) is an essential element in these novel imaging systems.

We acknowledge support from AFOSR, DARPA and the NSF ERC Center on Integrated Access Networks (CIAN).

7. References

[1] S.A. Benton and M. Bove, "*holographic Imaging*", J. Wiley ed. (2011).

[2] Pierre St Hilaire, "Modulation transfer function and optimum sampling of holographic stereograms", Applied Optics, Vol. 33, No 5, Feb 1994, pp. 768-774.

[3] P. St. Hilaire, "Optimum sampling parameters for generalized holographic stereograms", SPIE Proceedings #3011 "Practical Holography XI". S.A Benton editor, pp. 96-104.

[4] D. J. De Bitetto, "Holographic Panoramic Stereograms Synthesized from White Light Recordings," Applied Optics, Vol. 8-8, pp. 1740-1741 (August 1969).

[5] S. A. Benton. "Achromatic Holographic Stereograms", J. Opt. Soc. Amer. , Vol. 71-12. p 1568A (December 1981).

[6] Q. Y. J. Smithwick, J. Barabas, D. E. Smalley, and V. M. Bove, Jr., "Interactive Holographic Stereograms with Accommodation Cues," *Proc. SPIE Practical Holography XXIV*, v. 7619, 2010.

[7] S. Tay *et al.* "An updatable holographic three-dimensional display", *Nature* 451, 694-698 (2008).

[8] J. W. Goodman, "Introduction to Fourier Optics", (Mc Graw Hill Book Co., San Francisco, 1968), pp. 101-133.

[9] P. Artal and R. Navarro, J. Opt. Soc. Am. A, Vol. 11, No. 1, 1994.

9th International Symposium on Display Holography (ISDH 2012) IOP Publishing
Journal of Physics: Conference Series **415** (2013) 012036 doi:10.1088/1742-6596/415/1/012036

Depth of Field Analysis for Multilayer Automultiscopic Displays

D. Lanman, G. Wetzstein, M. Hirsch, and R. Raskar
MIT Media Lab, 75 Amherst Street, Cambridge MA 02139, USA
E-mail: dlanman@media.mit.edu

Abstract. With the re-emergence of stereoscopic displays, through polarized glasses for theatrical presentations and shuttered liquid crystal eyewear in the home, automultiscopic displays have received increased attention. Commercial efforts have predominantly focused on parallax barrier and lenticular architectures applied to LCD panels. Such designs suffer from reduced resolution and brightness. Recently, multilayer LCDs have emerged as an alternative supporting full-resolution imagery with enhanced brightness and depth of field. We present a signal-processing framework for comparing the depth of field for conventional automultiscopic displays and emerging architectures comprising multiple light-attenuating layers. We derive upper bounds for the depths of field, indicating the potential of multilayer configurations to significantly enhance resolution and depth of field, relative to conventional designs.

1. Introduction

Thin displays that present the illusion of depth have become a driving force in the consumer electronics and entertainment industries, offering a differentiating feature in a market where the utility of increasing 2D resolution has brought diminishing returns. In such systems, binocular depth cues are achieved by presenting different images to each of the viewer's eyes. Automultiscopic displays present such view-dependent imagery without requiring special eyewear, providing both binocular and motion parallax cues. Conventional designs use either light-attenuating parallax barriers, introduced by Ives in 1903 [1], or refracting lens arrays (also known as integral imaging), introduced by Lippmann in 1908 [2]. In both cases, an underlying 2D display is covered with a second optical element. For barrier designs, a second light-attenuating layer is placed slightly in front of the first, whereas lens arrays are directly affixed to the underlying display. In most commercial systems either parallax barriers or lenticular sheets are used for horizontal-only parallax. However, pinhole arrays or integral lens sheets can achieve simultaneous vertical and horizontal parallax. While automultiscopic displays realized with lenses are brighter than barrier designs, lens-based designs reduce spatial resolution; as a result, we seek to address the limitations of automultiscopic displays, particularly diminished brightness, reduced spatial resolution, and limited depth of field using emerging multilayer architectures.

Depth of field is a key metric for judging the relative performance of various automultiscopic display technologies. As described by Zwicker *et al.* [3], the *depth of field* of an automultiscopic display is an expression describing the maximum spatial frequency that can be depicted, without aliasing, in a virtual plane oriented parallel to, and located a known distance from, the display surface. In this paper, we derive upper bounds on the spatio-angular bandwidth for any

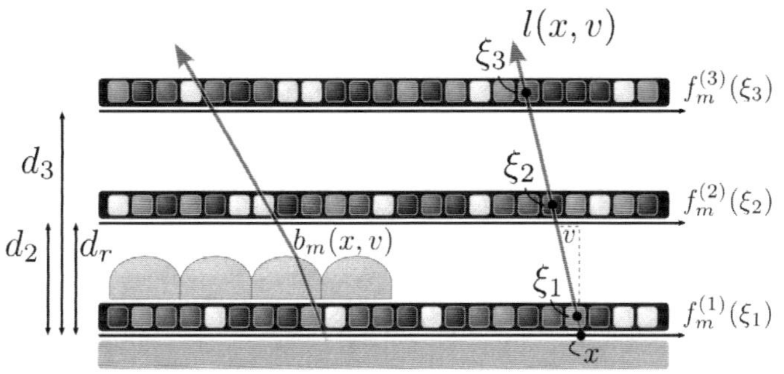

Figure 1. Tensor displays. A stack of N light-attenuating layers is illuminated by a uniform or directional backlight (e.g., depicted above as a lenslet array affixed to the rear display layer). A time-multiplexed sequence of M patterns, with transmittance $f_m^{(n)}(\xi_n)$ at point ξ_n on layer n during frame m, are rapidly displayed such than an observer perceives the time average. The emitted light field $l(x, v)$ is defined using a relative two-plane parameterization, where v denotes the point of intersection of the ray (x, v) with a plane located a distance d_r from the x-axis. The directional backlight emits a low-resolution light field $b_m(x, v)$ during frame m.

multilayer, attenuation-based display incorporating either uniform or directional backlighting [4]. This upper bound is shown, for two layers, to encompass prior depth of field expressions for parallax barriers and integral imaging. Most significantly, the upper bound indicates multilayer, attenuation-based displays can significantly increase the spatial resolution for virtual objects located close to the display surface, as compared to current architectures.

Automultiscopic display systems must optimize their performance by considering three components: multiview content, display elements, and viewers. Prior works consider the benefits of pre-filtering multiview content for a particular display device [5]. Others consider adapting display elements (e.g., the spacing within parallax barriers) depending on viewer position [6, 7]. We propose four generalizations of conventional parallax barriers: *High-Rank 3D (HR3D)* [8], *Layered 3D* [9], *Polarization Fields* [10], and *Tensor Displays* [11], for which display elements are optimized for the multiview content. As shown in Figure 1 and recently reviewed by Lanman *et al.* [12], such displays comprise stacks of time-multiplexed, light-attenuating (or polarization-rotating) layers (e.g., multilayer LCDs). The resulting generalized barriers significantly differ from existing heuristics (e.g., grids of slits or pinholes) and achieve increased optical transmission, enhanced spatial resolution, greater depth of field, and higher refresh rates, while preserving the fidelity of displayed images. Most significantly, displays employing these generalized, multilayer barrier architectures can similarly benefit from pre-filtered multiview content and can adapt to the viewer position, allowing simultaneous optimization of all three system components.

In this paper, we present a signal-processing framework for comparing the depths of field for conventional automultiscopic displays and emerging architectures comprising multiple light-attenuating layers. Section 2 assesses the depth of field for conventional designs. In Sections 3 and 4, we derive upper bounds for the depth of field for any multilayer attenuation-based display illuminated with uniform or directional backlighting, respectively. In Section 5, we compare these upper bounds to formally establish the benefits of multilayer automultiscopic displays.

2. Depth of Field for Conventional Automultiscopic Displays

As described by Zwicker *et al.* [3], the depth of field of an automultiscopic display is an expression describing the maximum spatial frequency $\omega_{\xi_{\max}}$ that can be depicted, without aliasing, in a virtual plane oriented parallel to, and located a distance d_o from, the middle of the display enclosure (see Figure 1). As established in that work, the depth of field can be assessed by analyzing the spectral properties of the emitted light field. For conventional parallax barriers and integral imaging, the discrete sampling of emitted rays (x, v) produces a light field spectrum $\hat{l}(\omega_x, \omega_v)$ that is non-zero only within a rectangle, where ω_x and ω_v are the spatial and angular frequencies, respectively. As described by Chai *et al.* [13] and Durand *et al.* [14], the spectrum of a diffuse surface, located at distance d_o, corresponds to the line $\omega_v = (d_o/d_r)\omega_x$ in the frequency domain, where d_r is the distance between the x-axis and v-axis. The spatial cutoff frequency $\omega_{\xi_{\max}}$ is given by the intersection of this line with the spectral bandwidth of the display.

Consider the emitted light field spectrum for a parallax barrier or integral imaging display. Let Δx denote the spatial sampling rate (i.e., the spacing between barrier slits, pinholes, or lenslets). Similarly, let Δv denote the angular sampling rate; for a conventional automultiscopic display with field of view α and A distinct views, the angular sampling rate $\Delta v = (2d_r/A)\tan(\alpha/2)$. Thus, the light field spectrum for a conventional automultiscopic display is non-zero only for $|\omega_x| \leq 1/(2\Delta x)$ and $|\omega_v| \leq 1/(2\Delta v)$. Intersecting the line $\omega_v = (d_o/d_r)\omega_x$ with this rectangular region yields the following expression for the depth of field.

$$\omega_{\xi_{\max}}(d_o) = \begin{cases} \frac{1}{2\Delta x} & \text{for } |d_o| \leq d_r\left(\frac{\Delta x}{\Delta v}\right), \\ \frac{d_r}{2|d_o|\Delta v} & \text{otherwise} \end{cases} \tag{1}$$

This expression supports the following intuition: near the display (i.e., for $|d_o| \leq d_r(\Delta x/\Delta v)$), the maximum spatial frequency that can be depicted in a virtual plane, separated by d_o, is limited by the spacing Δx between slits, pinholes, or lenslets. Far from the display, however, the maximum spatial frequency is limited by the angular sampling rate Δv and is independent of the spacing between the slits, pinholes, or lenslets.

3. Depth of Field for Multilayer Displays with Uniform Backlighting

The upper bound on the depth of field for a *static* multilayer display with uniform backlighting (i.e., one for which the layer patterns do not vary over time) is similarly assessed by considering the maximum spectral bandwidth, for all possible layer patterns. The light field $l(x, v)$ emitted by an N-layer display with uniform backlighting is given by

$$l(x, v) = \prod_{n=1}^{N} f^{(n)}(x + (d_n/d_r)v), \tag{2}$$

where $f^{(n)}(\xi_n) \in [0, 1]$ is the transmittance at the point ξ_n of layer n, separated a distance d_n from the x-axis (see Figure 1). The 2D Fourier transform of this expression is given by

$$\hat{l}(\omega_x, \omega_v) = \int_{-\infty}^{\infty} \int_{-\infty}^{\infty} \prod_{n=1}^{N} f^{(n)}(x + (d_n/d_r)v)\, e^{-2\pi j \omega_x x}\, e^{-2\pi j \omega_v v}\, dx\, dv. \tag{3}$$

By the convolution property of Fourier transforms [15], this expression reduces to a *repeated convolution* of the individual layer spectra $\hat{f}^{(n)}(\omega_x, \omega_v)$, such that the light field spectrum $\hat{l}(\omega_x, \omega_v)$ is given by

$$\hat{l}(\omega_x, \omega_v) = \mathop{\ast}_{n=1}^{N} \hat{f}^{(n)}(\omega_x)\, \delta(\omega_v - (d_n/d_r)\omega_x), \tag{4}$$

where $*$ denotes convolution and the repeated convolution operator is defined as

$$\mathop{\ast}_{n=1}^{N} \hat{f}^{(n)}(\omega_x, \omega_v) \equiv \hat{f}^{(1)}(\omega_x, \omega_v) * \hat{f}^{(2)}(\omega_x, \omega_v) * \cdots * \hat{f}^{(N)}(\omega_x, \omega_v). \tag{5}$$

Note that each layer produces a spectrum $\hat{f}^{(n)}(\omega_x, \omega_v) = \hat{f}^{(n)}(\omega_x)\,\delta(\omega_v - (d_n/d_r)\omega_x)$ that lies along the line $\omega_v = (d_n/d_r)\omega_x$, following Chai *at al.* [13]. Since each layer has a finite resolution, the layer spectra $\hat{f}^{(n)}(\omega_x)$ are non-zero only for $|\omega_x| \leq \omega_0 = 1/(2p)$, where p is the pixel pitch.

In the preceding analysis we have focused on static multilayer displays. A *dynamic* multilayer display exploits rapid temporal modulation, such that an observer perceives the average of an M-frame sequence. Generalizing Equation 2, the emitted light field $l(x, v)$ is given by

$$l(x, v) = \frac{1}{M} \sum_{m=1}^{M} \prod_{n=1}^{N} f_m^{(n)}(x + (d_n/d_r)v), \tag{6}$$

where $f_m^{(n)}(\xi_n)$ is the transmittance at the point ξ_n of layer n during frame m. Note that, by the linearity property of the Fourier transform, the inclusion of time multiplexing does not alter the maximum spectral support for a given multilayer display configuration. That is, both static and dynamic multilayer displays have identical spectral bandwidths. Yet, as established by Wetzstein *et al.* [11], the added degrees of freedom afforded by temporal multiplexing allow dynamic multilayer displays to more closely approach the upper bound on the depth of field.

3.1. Two-Layer Displays with Uniform Backlighting

Consider a two-layer display with uniform backlighting, with the layers separated by a distance Δd and $\omega_0 = 1/(2p)$ denoting the spatial cutoff frequency for each layer with pixel pitch p. Equation 4 yields the following analytic expression for the light field spectrum $\hat{l}(\omega_x, \omega_v)$.

$$\hat{l}(\omega_x, \omega_v) = \hat{f}^{(1)}(\omega_x)\,\delta(\omega_v + (\Delta d/(2d_r))\omega_x) * \hat{f}^{(2)}(\omega_x)\,\delta(\omega_v - (\Delta d/(2d_r))\omega_x) \tag{7}$$

As shown in Figure 2, a diamond-shaped region encloses the non-zero spectral support for any two-layer display. This region is bounded by a rectangle, such that $|\omega_x| \leq 2\omega_0$ and $|\omega_v| \leq (\Delta d/d_r)\omega_0$. The spatial cutoff frequency $\omega_{\xi_{\max}}$ is again found by intersecting the line $\omega_v = (d_o/d_r)\omega_x$ with the boundary of the maximum-achievable spectral support. This geometric construction yields the following upper bound on the depth of field for any two-layer display.

$$\omega_{\xi_{\max}}(d_o) = \left(\frac{2\Delta d}{\Delta d + 2|d_0|}\right)\omega_0 \tag{8}$$

Note that this expression is equivalent to that previously derived for two-layer displays by Wetzstein *et al.* [9].

3.2. Three-Layer Displays with Uniform Backlighting

Equation 4 yields the following analytic expression for the light field spectrum $\hat{l}(\omega_x, \omega_v)$ for any three-layer display, with layers separated by a distance Δd.

$$\hat{l}(\omega_x, \omega_v) = \hat{f}^{(1)}(\omega_x)\,\delta(\omega_v + (\Delta d/d_r)\omega_x) * \hat{f}^{(2)}(\omega_x)\,\delta(\omega_v) * \hat{f}^{(3)}(\omega_x)\,\delta(\omega_v - (\Delta d/d_r)\omega_x) \tag{9}$$

As shown in Figure 2, a hexagonal region encloses the non-zero spectral support for any three-layer display. This region is bounded by a rectangle, such that $|\omega_x| \leq 3\omega_0$ and $|\omega_v| \leq 2(\Delta d/d_r)\omega_0$. Intersecting the line $\omega_v = (d_o/d_r)\omega_x$ with the boundary of the maximum-achievable spectral support again yields the following upper bound on the depth of field.

$$\omega_{\xi_{\max}}(d_o) = \begin{cases} \left(\frac{3\Delta d}{\Delta d + |d_0|}\right)\omega_0 & \text{for } |d_o| \leq 2\Delta d, \\ \left(\frac{2\Delta d}{|d_o|}\right)\omega_0 & \text{otherwise} \end{cases} \tag{10}$$

As shown in Figure 2, the bandwidth for three-layer displays exceeds that for either conventional or two-layer architectures, motivating the development of multilayer displays for extended depth of field. Note that, using the geometric construction outlined for two-layer and three-layers displays, one may derive an analytic upper bound for an arbitrary number of layers.

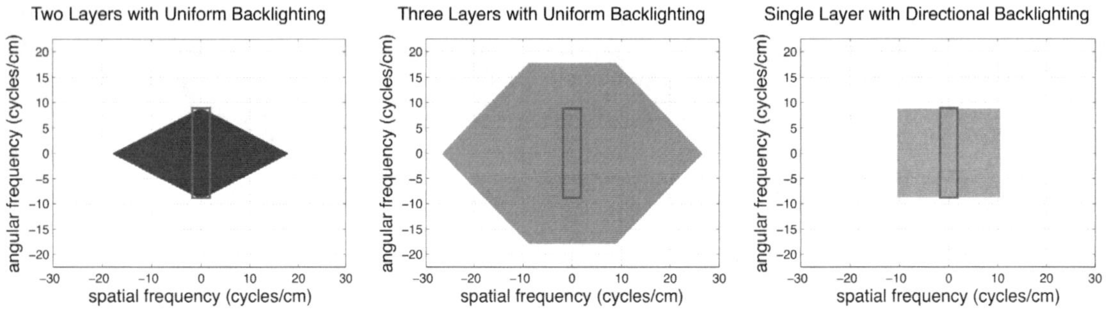

Figure 2. Spectral support for multilayer displays with uniform and directional backlighting. The spectral support is illustrated for two-layer and three-layer displays with uniform backlighting, shaded blue and green, on the left and in the middle, respectively. The spectral support for a single-layer display with directional backlighting is shaded yellow on the right. The spectral support for a conventional light field display (e.g., parallax barriers or integral imaging) is denoted by the red box. Display parameters correspond to those used for the prototype *Tensor Displays* described by Wetzstein *et al.* [11]. Specifically, multiple LCD panels, each with pixel pitch $p = 0.6$ mm, are separated by a distance of $\Delta d = 8$ mm.

3.3. General Multilayer Displays with Uniform Backlighting

For an N-layer display, a direct geometric construction for the upper bound on the depth of field becomes cumbersome. As an alternative, we apply the central limit theorem [16] to obtain an approximate expression for the spectral support due to the repeated convolutions in Equation 4; as derived by Chaudhury et al. [17], the repeated convolution of N two-dimensional spectra, each with mean $\mu_n = [0, 0]$ and covariance matrix $\mathbf{\Sigma}_n$, tends to a bivariate Gaussian distribution:

$$\hat{l}(\omega_x, \omega_v) \approx \left(\frac{1}{2\pi |\mathbf{\Sigma}|^{\frac{1}{2}}} \right) \exp\left(-\frac{1}{2} \omega^\top \mathbf{\Sigma}^{-1} \omega \right), \text{ for } \mathbf{\Sigma} = \sum_{n=1}^{N} \mathbf{\Sigma}_n, \tag{11}$$

where $\omega = [\omega_x, \omega_v]$ is a given spatio-angular frequency.

In the following analysis, we define the the maximum bandwidth for any given layer as follows.

$$\hat{f}^{(n)}(\omega_x, \omega_v) \equiv \begin{cases} \frac{1}{2\epsilon\omega_0} & \text{for } |\omega_x| \leq \omega_0 \text{ and } |\omega_v - (d_n/d_r)\omega_x| \leq \frac{\epsilon}{2} \\ 0 & \text{otherwise} \end{cases} \tag{12}$$

This definition corresponds to a thin layer with a spatially-varying transmittance characterized by a white noise process (i.e., a uniform spectral energy density) [18]. Note that the spectrum of a layer located at a distance d_n will be non-zero along a single line, with slope d_n/d_r, only in the limit as ϵ tends to zero. Under this definition for the patterns depicted with an N-layer display, the cumulative mean spatio-angular frequency $\mu = \mu_n = 0$, since each layer spectrum is symmetric about the origin. Similarly, the covariance matrix $\mathbf{\Sigma}_n$ for each layer is given by

$$\mathbf{\Sigma}_n = \begin{bmatrix} \int_{-\infty}^{\infty}\int_{-\infty}^{\infty} \omega_x^2 \hat{f}^{(n)}(\omega_x, \omega_v)\, d\omega_x\, d\omega_v & \int_{-\infty}^{\infty}\int_{-\infty}^{\infty} \omega_x\omega_v \hat{f}^{(n)}(\omega_x, \omega_v)\, d\omega_x\, d\omega_v \\ \int_{-\infty}^{\infty}\int_{-\infty}^{\infty} \omega_x\omega_v \hat{f}^{(n)}(\omega_x, \omega_v)\, d\omega_x\, d\omega_v & \int_{-\infty}^{\infty}\int_{-\infty}^{\infty} \omega_v^2 \hat{f}^{(n)}(\omega_x, \omega_v)\, d\omega_x\, d\omega_v \end{bmatrix}. \tag{13}$$

Substituting Equation 12 yields the covariance matrix:

$$\lim_{\epsilon \to 0} \mathbf{\Sigma}_n = \frac{\omega_0^2}{3} \begin{bmatrix} 1 & \frac{d_n}{d_r} \\ \frac{d_n}{d_r} & \left(\frac{d_n}{d_r}\right)^2 \end{bmatrix}, \tag{14}$$

where the limit of ϵ tending to zero corresponds to an infinitesimally-thin layer. Thus, the cumulative covariance matrix $\mathbf{\Sigma}$ of the light field spectrum is approximated by the summation:

$$\mathbf{\Sigma} = \sum_{n=1}^{N} \mathbf{\Sigma}_n = \begin{bmatrix} \sigma_{\omega_x}^2 & 0 \\ 0 & \sigma_{\omega_v}^2 \end{bmatrix} = \frac{\omega_0^2}{3} \begin{bmatrix} N & 0 \\ 0 & \left(\frac{N(N^2-1)}{12}\right)\left(\frac{\Delta d}{d_r}\right)^2 \end{bmatrix}, \tag{15}$$

for layers uniformly-spaced on the interval $d_n \in [-(N-1)\Delta d/2, (N-1)\Delta d/2]$. Substituting into Equation 11 yields the following approximation for the spectrum of an N-layer display.

$$\hat{l}(\omega_x, \omega_v) \approx \left(\frac{1}{2\pi\sigma_{\omega_x}\sigma_{\omega_v}}\right) \exp\left(-\frac{\omega_x^2}{2\sigma_{\omega_x}^2} - \frac{\omega_v^2}{2\sigma_{\omega_v}^2}\right) \tag{16}$$

Equation 16 can be used to approximate the upper bound on the depth of field, depending on the number of layers N, the separation distance Δd between layers, and the layer cutoff frequency $\omega_0 = 1/2p$. Under this model, the spectral support of a multilayer display is approximated by a bivariate Gaussian; curves defining spatio-angular frequencies with equal modulation energies correspond to ellipses, such that

$$\frac{\omega_x^2}{\sigma_{\omega_x}^2} + \frac{\omega_v^2}{\sigma_{\omega_v}^2} = \lambda^2, \tag{17}$$

where $\pm\lambda\sigma_{\omega_x}$ and $\pm\lambda\sigma_{\omega_v}$ are the points of intersection with the ω_x-axis and ω_v-axis, respectively. One can obtain an approximate upper bound on the depth of field, for an arbitrary number of layers, by finding the intersection of the line $\omega_v = (d_o/d_r)\omega_x$ with the ellipse corresponding to the highest spatial frequency achievable by an N-layer display. From Equations 4 and 12, the repeated convolution of N layers, each extending over $\pm\omega_0$ along the ω_x-axis, will produce non-zero spatial frequencies within the region $|\omega_x| \leq N\omega_0$. Thus, the ellipse with $\lambda = N\omega_0/\sigma_{\omega_x}$ provides the following approximate upper bound on the depth of field for an N-layer display.

$$|f_\xi| \leq N\omega_0\sqrt{\frac{(N^2-1)\Delta d^2}{(N^2-1)\Delta d^2 + 12{d_o}^2}} \tag{18}$$

This expression approximates the upper bound on the depth of field for two-layer and three-layer displays, given exactly by Equations 8 and 10, respectively. A comparison of the upper bound for multilayer vs. conventional automultiscopic displays is shown in Figure 3. Note that additional layers significantly increase the upper bound on the achievable spatial resolution, which is expected due to repeated convolution of layer spectra via Equation 4. In summary, these upper bounds indicate the theoretical origin for the increased depth of field observed with recent multilayer display architectures, including *HR3D* [8], *Layered 3D* [9], *Polarization Fields* [10], and *Tensor Displays* [11].

4. Depth of Field for Multilayer Displays with Directional Backlighting

The upper bound on the depth of field for a multilayer display with directional backlighting is assessed, similar to Section 3, by considering the maximum spectral bandwidth, for all possible layer and backlight illumination patterns. Generalizing Equation 2 yields a model for the emitted light field $l(x, v)$ of an N-layer display with directional backlighting, such that

$$l(x, v) = b(x, v) \prod_{n=1}^{N} f^{(n)}(x + (d_n/d_r)v), \tag{19}$$

where $b(x, v)$ denotes the light field emitted by the directional backlight. We consider directional backlighting to be equivalent to any low-resolution light field display placed behind a stack of

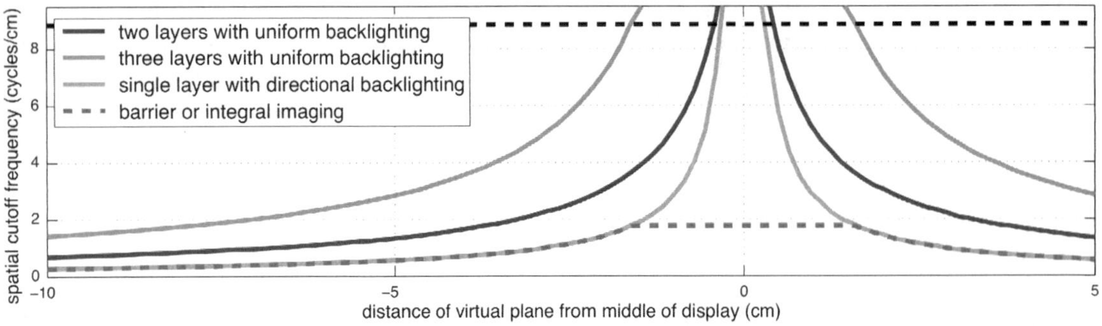

Figure 3. Comparison of the upper bound on the depth of field for conventional light field displays (red), two-layer (blue) and three-layer (green) displays with uniform backlighting, and single-layer displays with directional backlighting (yellow). The dashed black line denotes the spatial cutoff frequency ω_0 for each layer. Display parameters correspond to those in Figure 2.

light-attenuating layers. For example. as shown in Figure 1, such directional backlighting may comprise an integral imaging display, for which one desires to enhance the spatial resolution of the emitted light field. As before, the 2D Fourier transform of this expression is given by

$$\hat{l}(\omega_x, \omega_v) = \int_{-\infty}^{\infty}\int_{-\infty}^{\infty} b(x,v) \prod_{n=1}^{N} f^{(n)}(x + (d_n/d_r)v)\, e^{-2\pi j \omega_x x}\, e^{-2\pi j \omega_v v}\, dx\, dv, \tag{20}$$

yielding the following emitted light field spectrum:

$$\hat{l}(\omega_x, \omega_v) = \hat{b}(\omega_x, \omega_v) * \left[\mathop{\ast}_{n=1}^{N} \hat{f}^{(n)}(\omega_x)\, \delta(\omega_v - (d_n/d_r)\omega_x) \right]. \tag{21}$$

4.1. Single-Layer Displays with Directional Backlighting

In this section, we consider a single-layer display with directional backlighting; for example, Wetzstein *et al.* [11] describe a prototype consisting of a single LCD panel placed directly on top of an integral imaging display, comprising a lenslet array affixed to a second LCD panel. Consistent with that prototype, we assume that the directional backlight implements a low-resolution light field display, such that $\hat{b}(\omega_x, \omega_v)$ has non-zero support for $|\omega_x| \leq 1/(2\Delta x)$ and $|\omega_v| \leq 1/(2\Delta v)$ (i.e., the red boxes in Figure 2). Equation 21 yields the following analytic expression for the light field spectrum.

$$\hat{l}(\omega_x, \omega_v) = \hat{b}(\omega_x, \omega_v) * \hat{f}(\omega_x)\, \delta(\omega_v) \tag{22}$$

Note that the layer spectrum $\hat{f}(\omega_x, \omega_v) = \hat{f}(\omega_x)\,\delta(\omega_v)$ is constrained to a horizontal line of width $|\omega_x| \leq w_0$. Thus, convolution with the directional backlight spectrum results in an extended rectangular spectral support exceeding that of a conventional light field display; as shown in Figure 2, the rectangular region is given by $|\omega_x| \leq 1/(2\Delta x) + \omega_0$ and $|\omega_v| \leq 1/(2\Delta v)$. Note that placing a light-attenuating layer directly on top of a conventional light field display only increases the spatial resolution; the angular resolution of the underlying low-resolution light field display is preserved. Intersecting the line $\omega_v = (d_o/d_r)\omega_x$ with the boundary of the maximum-achievable spectral support yields the following upper bound on the depth of field.

$$\omega_{\xi_{\max}}(d_o) = \begin{cases} \frac{1}{2\Delta x} + \omega_0 & \text{for } |d_o| \leq d_r\left(\frac{\Delta x}{\Delta v + 2\Delta x \Delta v \omega_0}\right), \\ \frac{d_r}{2|d_o|\Delta v} & \text{otherwise} \end{cases} \tag{23}$$

As shown in Figure 3, the addition of a single light-attenuating layer significantly increases the spatial resolution near the display surface, as compared to a conventional parallax barrier or integral imaging display. However, far from the display, the upper bound on the depth of field is identical to these conventional automultiscopic displays.

5. Conclusion

Our analysis indicates a promising application for multilayer automultiscopic displays: increased depth of field can be achieved by covering any low-resolution light field display with time-multiplexed, light-attenuating layers. In this analysis, we assume continuously-varying layer transmittances; a promising research direction is to characterize the upper bounds with discrete pixels. However, with the presented analysis, we observe that static and dynamic multilayer displays have identical spectral supports (i.e., averaging over an M-frame sequence does not alter the maximum-achievable bandwidth). Yet, as observed by Lanman *et al.* [8] and Wetzstein *et al.* [11], time multiplexing significantly reduces artifacts occurring with static multilayer designs. We attribute this to the additional degrees of freedom allowed with time multiplexing. While the upper bounds may be identical, in practice this bound cannot be achieved with static methods, motivating multilayer automultiscopic displays for joint multilayer, multiframe decompositions capable of approaching the upper bound—significantly extending the depth of field achievable using current-generation display technologies, including LCD panels.

Acknowledgments

We recognize the support of the Camera Culture group and the MIT Media Lab sponsors. Douglas Lanman was supported by NSF Grant IIS-1116452 and by the DARPA MOSAIC program. Gordon Wetzstein was supported by the DARPA SCENICC program. Ramesh Raskar was supported by an Alfred P. Sloan Research Fellowship and a DARPA Young Faculty Award.

References

[1] Ives F E 1903 Parallax stereogram and process of making same U.S. Patent 725,567
[2] Lippmann G 1908 *Journal of Physics* **7** 821–825
[3] Zwicker M, Matusik W, Durand F and Pfister H 2006 *Eurographics Symposium on Rendering*
[4] Urey H, Chellappan K V, Erden E and Surman P 2011 *Proc. IEEE* **99** 540–555
[5] Zwicker M, Vetro A, Yea S, Matusik W, Pfister H and Durand F 2007 *IEEE Signal Processing* **24** 88–96
[6] Perlin K, Paxia S and Kollin J S 2000 *SIGGRAPH* (ACM) pp 319–326
[7] Peterka T, Kooima R L, Sandin D J, Johnson A, Leigh J and DeFanti T A 2008 *IEEE Transactions on Visualization and Computer Graphics* **14** 487–499
[8] Lanman D, Hirsch M, Kim Y and Raskar R 2010 *ACM Trans. Graph.* **29**(6) 163:1–163:10
[9] Wetzstein G, Lanman D, Heidrich W and Raskar R 2011 *ACM Trans. Graph.* **30**(4) 1–11
[10] Lanman D, Wetzstein G, Hirsch M, Heidrich W and Raskar R 2011 *ACM Trans. Graph.* **3**(6) 1–9
[11] Wetzstein G, Lanman D, Hirsch M and Raskar R 2012 *ACM Trans. Graph.* **31** 1–11
[12] Lanman D, Wetzstein G, Hirsch M, Heidrich W and Raskar R 2012 *SPIE Stereoscopic Displays and Applications XXIII* vol 8288 pp 82880A–1– 82880A–13
[13] Chai J X, Tong X, Chan S C and Shum H Y 2000 *SIGGRAPH* (ACM) pp 307–318
[14] Durand F, Holzschuch N, Soler C, Chan E and Sillion F X 2005 *ACM Trans. Graph.* **24** 1115–1126
[15] Bracewell R 1999 *The Fourier Transform and Its Applications* (McGraw-Hill)
[16] Peebles P 2000 *Probability, Random Variables, and Random Signal Principles* (McGraw-Hill)
[17] Chaudhury K N, Muñoz-Barrutia A and Unser M 2010 *IEEE Trans. Image* **19** 2290–2306
[18] Haykin S 2000 *Communication Systems* (Wiley)

An Innovative Tool for Fabricating Computer-Generated Holograms

H I Smith, M I Walsh, F Zhang, J Ferrera, G Hourihan, D Smith, R Light and M Jaspan

LumArray, Inc., Somerville, MA 02143, USA

http://www.lumarray.com

Abstract. Based on research at MIT, LumArray, Inc. has developed a maskless photolithography tool, the ZP-150, designed to cover an entire 15-cm substrate with a continuous, coherent high-resolution pattern, thereby avoiding the "stitching problem" and making it ideal for fabricating computer-generated holograms (CGH). No mask is required, as data is transferred directly from a computer to a spatial-light modulator that adjusts the intensity of 1000 beamlets, and directs them to 1000 diffractive-optical lenses. Patterns of arbitrary geometry, with placement precision ~1 nm, are written by scanning the stage in coordination with modulation of the beamlets by the spatial-light modulator. A fully automated proximity-effect-correction algorithm enables fine and large features to be written with equal ease, as well as the creation of 3-D structures. The ZP-150 uses stable, non-chemically amplified photoresists. Extension of resolution from the current 150 nm to the sub-100 nm domain is planned. In addition to providing rapid turn-around on designs, we envisage the ZP-150 being used in customized manufacturing of CGH's by virtue of its modest cost and low maintenance.

1. Introduction

In 1996 the MIT team began exploring an approach to micro- and nanolithography that deviated significantly from preexisting methods such as scanning-electron-beam lithography (SEBL) and optical-projection lithography (OPL), both of which employ a single projection lens. By employing instead a large array of lenses, in conjunction with a spatial-light modulator that directed beamlets to the individual lenses, one could create patterns of arbitrary geometry at high speed while circumventing the conventional coupling between lens numerical aperture and field of view [1-7]. The basic concept is shown in figure 1, with the lenses being diffractive-optical elements; hence the name "zone-plate-array lithography" (ZPAL).

The initial proposal called for using collimated 4.5 nm x-rays from an undulator and predicted resolution below 25 nm [1]. Subsequently, the use of ultraviolet radiation rather than x rays was explored in two Ph.D. theses at MIT [2-7], and a company, LumArray, Inc., founded [8].

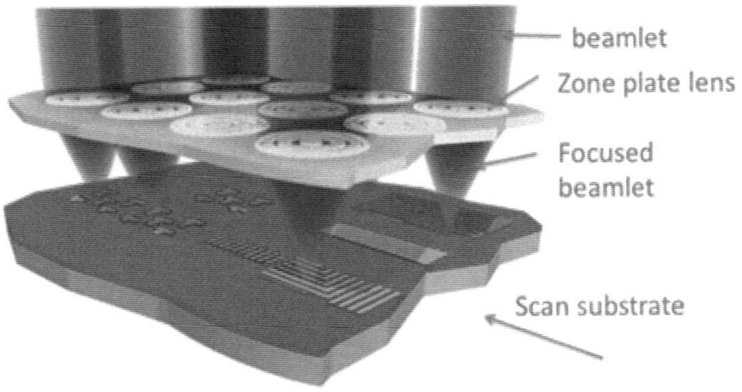

Figure 1. Schematic of the basic principle of zone-plate-array lithography (ZPAL). A spatial-light modulator directs beamlets to the individual lenses, which form focal spots on a substrate. By scanning the stage that holds the substrate, while modulating the beamlets under computer control, patterns of arbitrary geometry are written via the superposition of focal spots.

2. The ZP-150 system

Figure 2 is a schematic depicting the ZPAL configuration being commercialized by LumArray. The light source is a 405 nm GaN laser that illuminates a spatial-light modulator (SLM) from Silicon Light Machines. Each of the 1086 pixels of the SLM directs a beamlet to one of the 1086 zone plates of the array. The beamlet intensity can be controlled from zero to full on in 256 steps. This so-called gray scaling plays an important role in linewidth control, proximity-effect correction and 3D patterning.

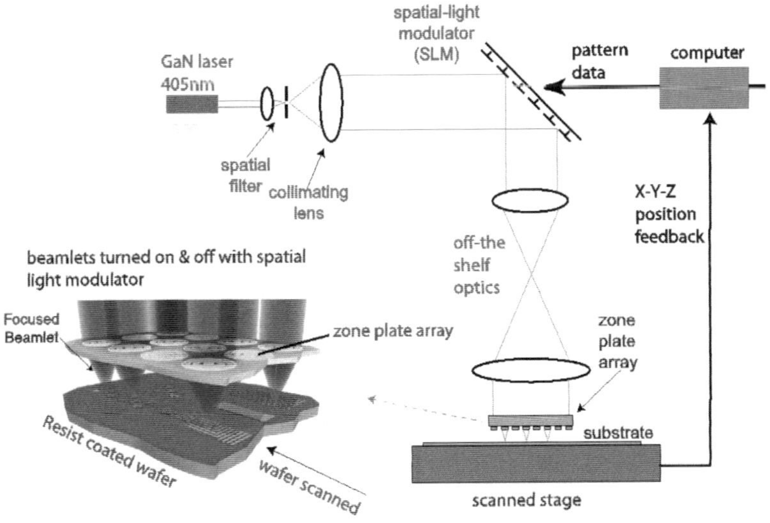

Figure 2. Schematic of zone-plate-array lithography (ZPAL) being commercialized by LumArray, Inc.

Figure 3 depicts the writing strategy. In brief, the substrate is mounted on an air-bearing stage and raster scanned across its full width. The pattern to be written is first described as a superposition of rectangles and other geometric figures. The software for the ZP-150 converts that description into commands to the spatial light modulator that are coordinated with the stage motion such that the focal spot intensity is non-zero where an element of a pattern is to be written.

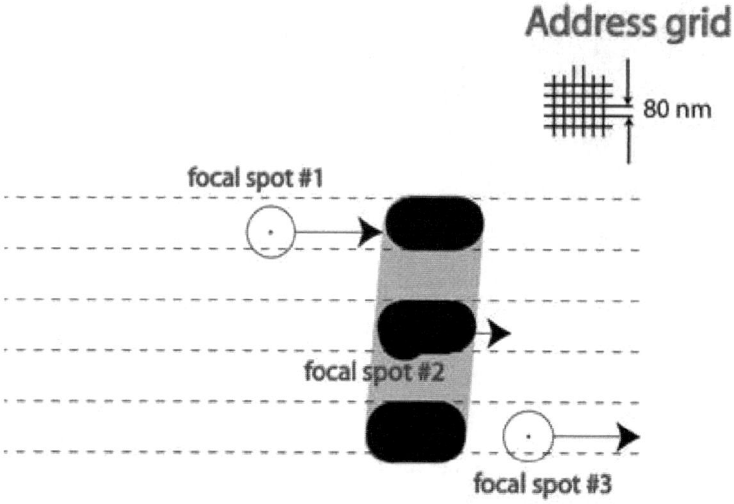

Figure 3. Depiction of the writing procedure used in LumArray's ZP-150 system at an intermediate stage in writing a particular pattern element. After a first scan in the X direction, the stage moves down in Y by one increment of the address grid (depicted as an 80 nm shift) and an X scan in the opposite direction is carried out. Three focal spots are depicted: #1 is about to make a second pass through and exposure of the pattern element, #2 is part way through its second pass, #3 has completed its second pass. The lightly shaded region shows the desired pattern element.

Figure 4 is a photograph of the ZP-150 alpha system located at the National Institute of Standards and Technology (NIST) in Gaithersburg, MD where it will be used to write large area computer-generated holograms and other nanostructured optics.

Figure 5 is a portion of a computer-generated hologram written for a potential customer. The jogs evident in the pattern are not defects but rather an intentional part of the desired hologram. In addition to writing binary type patterns, LumArray's system has also been used to write 3-dimensional patterns such as blazed diffraction gratings.

The throughput of LumArray's system (in units of area per time) is given by the following equation

$$\text{Throughput} = N\,R\,d^2, \tag{1}$$

where N is the number of lenses (equal to the number of pixels in the spatial-light modulator), R is the refresh rate of the modulator (currently 290 kHz) and d is the spatial period of the address grid. The latter is an adjustable parameter, typically set to be ½ of the desired minimum feature size. For a 150 mm-diameter substrate, covered over its entire area with minimum features of 150 nm the information content is ~4×10^{12} bytes. The writing speed in this case would be about 1.5 mm^2/sec. Patterns that are laid out on a coarser address grid (i.e., larger value of d) can generally be written more quickly.

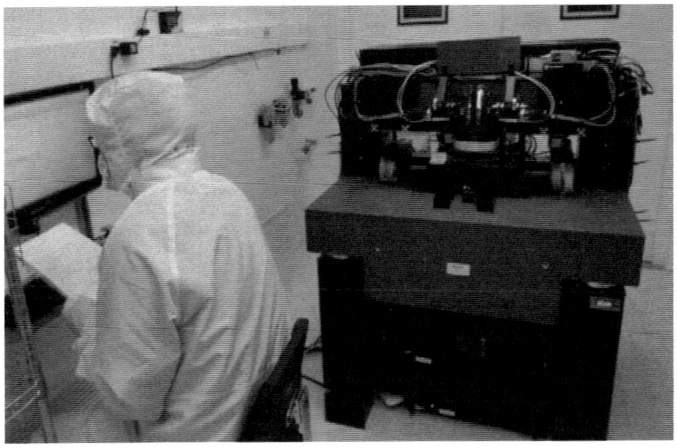

Figure 4. Photograph of LumArray's ZP-150 alpha system located within a clean room at the National Institute of Standards and Technology (NIST).

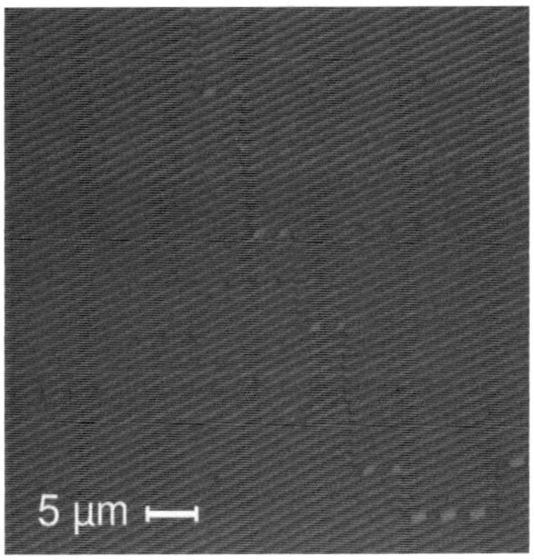

Figure 5. Optical micrograph is a portion of a computer-generated hologram written for a potential customer.

3. Innovations inherent in ZPAL

The ZPAL concept, as well as the design of LumArray's ZP-150, represent a number of innovations. We summarize them here and comment briefly on each.

a. **Maskless photolithography** – This eliminates the need for masks, enables rapid turn around and rapid convergence to optimal design
b. **Patterns are formed from overlapped focal spots** – Because the spots that form a pattern are exposed at different times, there is no coherent relationship between them and hence no troublesome interference effects. Focal spot intensities are additive and hence the mathematics of linear superposition applies.
c. **Stage is moved, not the focal spots** – The encoder that reads stage position provides nanometer-level precision. Because focal spot positions are fixed in a rigid frame, by calibrating their positions and correcting any positional errors in software, nanometer-level pattern precision should be achieved. In contrast, maskless systems that utilize beam deflection are subject to several sources of error.
d. **Multiple lenses operated in parallel** – Throughput increases directly in proportion to the number of lenses—see equation (1). Throughput in the ZP-150 is currently limited primarily by the framing rate of the spatial-light modulator.
e. **Pattern size is decoupled from its resolution** – The pattern size depends only on the range of the X-Y stage travel, while pattern resolution depends on the numerical aperture of the lenses. The two are decoupled, unlike the situation with a single projector lens. A continuous, coherent pattern can cover an entire substrate, or multiple variations on a design can be written on the same substrate.
f. **Use of diffractive-optical lenses** – Diffractive-optical lenses can be made in large arrays at very low cost using planar-fabrication techniques such as scanning-electron-beam lithography (SEBL). Moreover, with sufficiently advanced patterning techniques, focusing efficiency can approach 100%.
g. **Freedom from aberrations** – Diffractive-optical lenses focusing on axis are aberration free.
h. **Focal-spot intensity variable in 256 steps** – By controlling focal-spot intensity, so-called gray scaling, linewidths and proximity effects can be controlled, and 3-D patterns written in photoresist.

i. **Wavefront engineering** – One can engineer the progression of zones in a diffractive-optical lens, and achieve thereby focal spots other than the conventional Airy distribution. This can be advantageous in some circumstances.
j. **Circumventing the diffraction barrier** – In recent years techniques have been proposed and demonstrated that break the long-standing diffraction barrier [9-13]. In brief these techniques take advantage of wavelength-selective chemistry and optical nonlinearities. Simulations indicate resolutions below 1/20th of the optical wavelength are feasible.

4. Impact on Computer-Generated Holography

ZPAL has the potential to impact the fabrication and use of computer generated holograms for a variety of reasons:

a. **Very large area holograms** – With ZPAL, the area of a hologram is limited only by range of the stage motion, currently 150 mm x 150 mm in the ZP-150. This could be increased by employing a larger stage.
b. **Multiple designs on the same substrate** – ZPAL enables a designer to put multiple designs on the same substrate. This would facilitate a more rapid conversion to an optimal design.
c. **Customization for security** – Maskless lithography in general, and ZPAL in particular, enable one to customize holograms. If the cost and convenience are appropriate this could lead to new security markets for holograms.
d. **Higher resolution** – The ZP-150 promises higher resolution than many of the other lithography techniques currently used to make holograms. This opens the door to higher performance, higher information content and operation at shorter wavelengths.
e. **Three-dimensional holograms** – Just as a blazed grating yields higher diffraction efficiency and asymmetric beams, a capability for sculpturing the holographic medium in 3-dimensions opens the door to higher efficiency. The ZP-150 system is especially well suited to this application due to the intensity control afforded by the spatial light modulator and grayscaling.

5. Summary

When a disruptive technology comes along, its first use is generally an existing and well known application. But soon thereafter, the innovation inherent in the disruptive technology becomes more evident, giving rise to new applications and new ideas. It is our expectation that bringing zone-plate-array lithography to the attention of the holography community will open the door to new forms of holograms as well as applications not previously envisaged.

References

[1] Smith H I 1996 *J. Vac. Sci. Technol.* B **14** 4318-22
[2] Carter D J D, Gil D, Menon R, Mondol M and Smith H I 1999 *J. Vac. Sci. Technol.* B **17** 3449-52
[3] Gil D, Menon R, Carter D J D and Smith H I 2000 *J. Vac. Sci. Technol.* B **18** 2881-85
[4] Gil D, Carter D J D, Menon R, Tang X and Smith H I 2000 *J. Vac. Sci. Technol.* B **20** 2597-2601
[5] Gil D, Menon R and Smith H I 2003 *J. Vac. Sci. Technol.* B **21** 2810-2814
[6] Menon R, Patel A, Gil D and Smith H I 2005 *Materials Today* **8** 26-33
[7] Smith H I, Menon R, Patel A, Gil D, Chao D, Walsh M and Barbastathis G 2006 *Microelectron. Eng.* **83** 956-61
[8] http://www.lumarray.com
[9] Hell S W 2007 *Science* **316** 1153-58
[10] Andrew T L, Tsai H Y and Menon R 2009 *Science* **324** 917-21
[11] Brimhall N, Andrew T L, Manthena R V and Menon R 2011 *Phys. Rev. Lett.* **107** 205501
[12] Scott T F, Kowalski B A, Sullivan A C, Bowman C N and McLeod R R 2009 *Science* **324** 913-17
[13] Li L J, Gattass R R, Gershgoren E, Hwang H and Fourkas J T, 2009 *Science* **324** 910-13

Journal of Physics: Conference Series **415** (2013) 012038 doi:10.1088/1742-6596/415/1/012038

Computational hologram synthesis and representation on spatial light modulators for real-time 3D holographic imaging

Stephan Reichelt and Norbert Leister

SeeReal Technologies, Sudhausweg 5, 01099 Dresden, Germany

E-mail: sre@seereal.com

Abstract. In dynamic computer-generated holography that utilizes spatial light modulators, both hologram synthesis and hologram representation are essential in terms of fast computation and high reconstruction quality. For hologram synthesis, i.e. the computation step, Fresnel transform based or point-source based raytracing methods can be applied. In the encoding step, the complex wave-field has to be optimally represented by the SLM with its given modulation capability. For proper hologram reconstruction that implies a simultaneous and independent amplitude and phase modulation of the input wave-field by the SLM.

In this paper, we discuss full complex hologram representation methods on SLMs by considering inherent SLM parameter such as modulation type and bit depth on their reconstruction performance such as diffraction efficiency and SNR. We review the three implementation schemes of Burckhardt amplitude-only representation, phase-only macro-pixel representation, and two-phase interference representation. Besides the optical performance we address their hardware complexity and required computational load. Finally, we experimentally demonstrate holographic reconstructions of different representation schemes as obtained by functional prototypes utilizing SeeReal's viewing-window holographic display technology. The proposed hardware implementations enable a fast encoding of complex-valued hologram data and thus will pave the way for commercial real-time holographic 3D imaging in the near future.

1. Introduction

Undoubtedly, real-time display holography is one of the most promising developments for the future consumer display and TV market. Over recent years, SeeReal has developed a unique solution to real-time holography, which is ready for the display market [1, 2]. Both real-time computing and particularly a suited spatial light modulator are essential requirements for successful commercialization.

In this paper, we address different issues of hologram encoding methods for representation on customary spatial light modulators (SLM). Although a complex modulator would be most suitable to modify the complex amplitude of a wave field, realistic devices exhibit a certain modulation capability in terms of modulation type (real or amplitude-only, phase-only, coupled), modulation range (restricted phase) and bit depth (number of discrete addressable values). Hologram encoding is therefore directly related to the hardware implementation of the programmable SLM. However, each encoding scheme is subject to approximation errors either caused by the decomposition method itself or by its sensitivity to addressing deviations or quantization errors.

This paper is organized as follows. Section 2 summarizes common definitions, which are applied to evaluate the performance of holographic reconstructions from electrically driven SLMs such as liquid crystal displays. In section 3 we briefly recapitulate SeeReal's reconstruction scheme using our proprietary Viewing-Window (VW) holography and the methods for hologram calculation by means of Sub-Holograms (SH). Three different encoding schemes are discussed in section 4, including their prototype implementation on SeeReal holographic displays. We end with a comparison of the encoding schemes and concluding remarks.

2. Basic spatial light modulator model

The pixelated structure of an SLM is shown in Fig. 1. It comprises of $N_x \times N_y$ pixel with an aperture of $a_x a_y$, which are arranged at a pixel pitch of δ_x, δ_y, respectively. Within the aperture area the light can be modulated whereas the region between the pixel is inactive and usually masked for transmissive SLM. The total optical transmittance of a pixelated SLM is characterized by its fill factor $\mu = a_x a_y / \delta_x \delta_y$. Besides, the fill factor determines the distribution in the far-field diffracted light because the envelope of the reconstructed SLM diffraction orders is defined by the amplitude transmittance of the single pixel. The smaller the fill factor, the more light is distributed into higher orders. The pixelized SLM structure leads to a discrete spatial sampling of an otherwise continuous function, where aliasing has to be prevented. The amount of information that can be recorded in the hologram is directly related to the spatial resolution and the size of the SLM. This fact is represented by the dimensionless space-bandwidth product $\text{SBP} = \nu_x L_x \cdot \nu_y L_y = \frac{L_x}{2\delta_x} \cdot \frac{L_y}{2\delta_y} = \frac{1}{2}(N_x N_y)$, with ν being the maximum spatial frequency according to the sampling theorem and L the width of the modulator [3, 4].

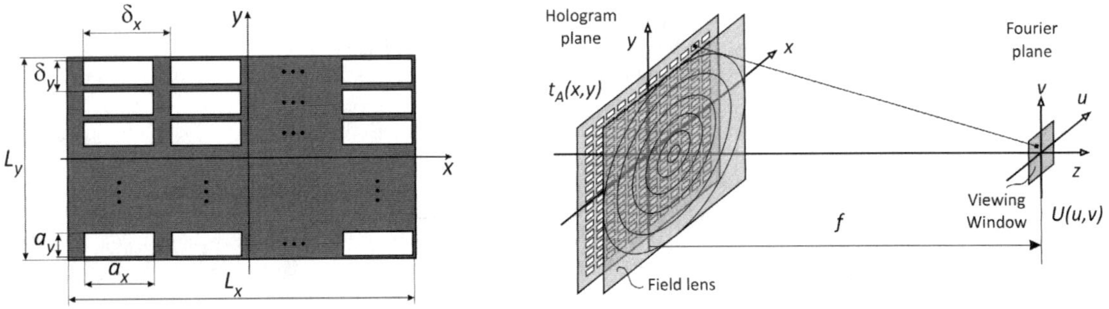

Figure 1. Geometry of the pixelated SLM and diffraction geometry.

The transmittance function $t_A(x, y)$ of an ideal pixelated SLM is given by the overall SLM aperture multiplied by the discrete complex amplitude of the pixel matrix [5]

$$t_A(x, y) = \text{rect}\left(\frac{x}{L_x}, \frac{y}{L_y}\right) \left\{ \left[\text{comb}\left(\frac{x}{\delta_x}, \frac{y}{\delta_y}\right) \cdot h(x, y) \right] \circledast \text{rect}\left(\frac{x}{a_x}, \frac{y}{y_x}\right) \right\}, \tag{1}$$

with $h(x, y)$ being the continuous complex amplitude of the hologram function. Considering SeeReal's reconstruction scheme using Viewing-Window holography, the diffracted field of the hologram in the Fourier plane is then [5]

$$U(u, v) = \frac{\exp\left[i\frac{k}{2f}\left(u^2 + v^2\right)\right]}{i\lambda f} \iint\limits_{-\infty}^{+\infty} t_O(x, y) t_A(x, y) \exp\left[-i\frac{2\pi}{\lambda f}(xu + yv)\right] \mathrm{d}x\mathrm{d}y. \tag{2}$$

where f is the focal length of the Fourier transforming lens, $k = 2\pi/\lambda$ is the wave number and $t_O(x, y)$ is the complex amplitude of the illumination wave, which is usually planar, normally incident and with uniform amplitude. Hence, the complex amplitude distribution in the focal plane is the Fourier transform of the field just behind the SLM multiplied by the quadratic phase term preceding the integral, so we can rewrite $U(u,v) = (i\lambda f)^{-1} \exp\left[i\frac{k}{2f}\left(u^2+v^2\right)\right] \mathcal{F}\{t_0(x,y)t_A(x,y)\}$.

We now consider the discrete signals existing at the SLM plane and in its Fourier plane, which are $H(x,y) = \mathrm{comb}\,(x/\delta_x, y/\delta_y) \cdot h(x,y)$ and $S(u,v) = \mathcal{F}\{H(x,y)\}$ for the hologram function and its response in the spectral domain, respectively. As a fidelity criterion of the reconstructed signal in the Fourier domain we apply a scaled signal-to-noise ratio (SNR) defined as [6]

$$\mathrm{SNR} = \frac{\iint_D |S_\mathrm{n}(u,v)|^2\,\mathrm{d}u\mathrm{d}v}{\iint_D |S_\mathrm{n}(u,v) - \beta S_\mathrm{r}(u,v)|^2\,\mathrm{d}u\mathrm{d}v}, \quad \text{with } \beta = \frac{\iint_D \mathrm{Re}\{S(u,v)\cdot S_\mathrm{r}^*(u,v)\}\mathrm{d}u\mathrm{d}v}{\iint_D |S_\mathrm{r}(u,v)|^2\,\mathrm{d}u\mathrm{d}v}, \tag{3}$$

where S_n is the desired, nominal signal and S_r is the real signal at (u,v) . The factor β represents a constant normalization factor for the power of the real signal S_r.

Another fidelity criterion to compare different encoding schemes is the encoding efficiency. It can be either defined in an absolute manner as the fraction of incident optical power on the SLM that appears in the targeting signal window (SW) in the Fourier plane or in a relative way by comparing the optical power in the signal window with those of signal and noise window (NW) together.

$$\eta_{\mathrm{E,abs}} = \frac{\iint_{\mathrm{SW}} |S(u,v)|^2\,\mathrm{d}u\mathrm{d}v}{\iint_{\mathrm{SLM}} |t_0(x,y)|^2\,\mathrm{d}x\mathrm{d}y} \quad \text{or} \quad \eta_{\mathrm{E,rel}} = \frac{\iint_{\mathrm{SW}} |S(u,v)|^2\,\mathrm{d}u\mathrm{d}v}{\iint_{\mathrm{SW}\cup\mathrm{NW}} |S(u,v)|^2\,\mathrm{d}u\mathrm{d}v} \tag{4}$$

Note that both definitions assume a 100% fill factor, consequently representing the maximum case.

3. Hologram synthesis

SeeReal's reconstruction scheme using viewing-window (VW) holography is illustrated in Fig. 1 [7]. When the SLM is illuminated by collimated coherent light, the 3D-scene is reconstructed in the frustum formed by the Fourier-transforming lens. The pixelated SLM generates a diffraction pattern in the Fourier plane. Its zero-order extension is the VW having an adapted size of $w_{x,y} = \lambda f/\delta_{x,y}$ slightly larger than the human eye pupil. Only within the VW the wave field seemingly emerging from the 3D object has to be existent when the eye position is fixed or an active observer tracking is ensured. The eye pupil itself acts as spatial filter that isolates the encoded field from nonsignal or high-order terms and thereby enabling a high SNR.

Hologram synthesis means the computational calculation of the discrete complex wave-field $H(x,y)$ at the hologram plane that reconstructs the desired 3D-scene. The 3D-scene is represented by a sufficient number n of points defined at discrete locations. Our implemented solutions for driving holographic displays with interactive or video content encoded in real-time are described in [8]. Basically, the computation methods for hologram synthesis can be grouped into direct analytic modeling by using ray-tracing and a Fourier-based modeling [9]. The first method treats each scene point $P_n(x_n, y_n, z_n)$ as a point emitter having a restricted angular spectrum according to the spatial resolution of the SLM and a defined brightness. The amplitude and phase distribution $H_n(x,y)$ of a single sub-hologram is then derived from the optical path between P_n and the hologram plane. Finally, the complex amplitudes H_n of all sub-holograms are summed up to the discrete hologram function $H(x,y) = \sum_{i=1}^{n} |H_n(x,y)| \exp\left[i\varphi_n(x,y)\right]$. In the second method the 3D-scene is sliced into layers $(L_1 \ldots L_m)$ that run parallel to the SLM plane, where the size of each layer is limited by the frustum and the sampling count is the same as for the SLM. Object points P_n are assigned to the closest layer and closest sampling point.

Each layer m is then Fresnel-propagated to the Fourier plane and its complex distributions are summed up to form the signal $S(u, v)$ in the VW-plane. To obtain the hologram function $H(x, y)$ an inverse Fourier transform is applied to the signal, $H(x, y) = \mathcal{F}^{-1}\{S(u, v)\}$.

4. Hologram representation on spatial light modulators

Numerous representation schemes have been developed since the beginning of computer-generated holography [10, 11]. The encoding step refers to the process of converting the complex amplitude of the desired reconstruction wave U_{R} into a format, which can be displayed at the SLM by addressing its pixel. In synthetic holography, a fully complex representation is most qualified. The representation techniques common to synthetic holography can be differentiate into two categories:

Direct representation: The input wave U_0 is directly modulated by the hologram in such a way that the desired reconstruction wave U_{R} emerges from the SLM. This corresponds to a multiplicative light manipulation, expressed by $U_{\mathrm{R}}(x, y) = U_0(x, y)H(x, y)$. The hologram function $H(x, y)$ is then equal to the signal function $f(x, y)$, i.e. $H(x, y) = f(x, y)$. Most prominent example of this type is the kinoform.

Encoded representation: Here, the complex amplitude of the signal function $f(x, y)$ is not equal to the hologram function $H(x, y)$ but related via an addition function to $H(x, y) = F[f(x, y)]$ with F being the encoding rule. The desired signal function must be physically decoded, which can be done by spatial filtering the signal term from noise terms in a Fourier plane of the hologram. Encoded representation methods have in common that the hologram is divided into discrete resolution cells (macro-pixel), which are further subdivided into sub-cells (sub-pixel). Usually, the signal function $f(x, y)$ is decomposed into pure phase or amplitude quantities that form the hologram function $H(x, y)$. However, this is compromised by a loss of spatial resolution. Examples of this type are detour-phase holograms [12–14] as well as double-phase holograms [15, 16].

The different encoding schemes discussed in this section are illustrated at the exemplary complex function $H(x, y)$ shown in Fig. 2, which has a sinusoidal amplitude and quadratic phase distribution.

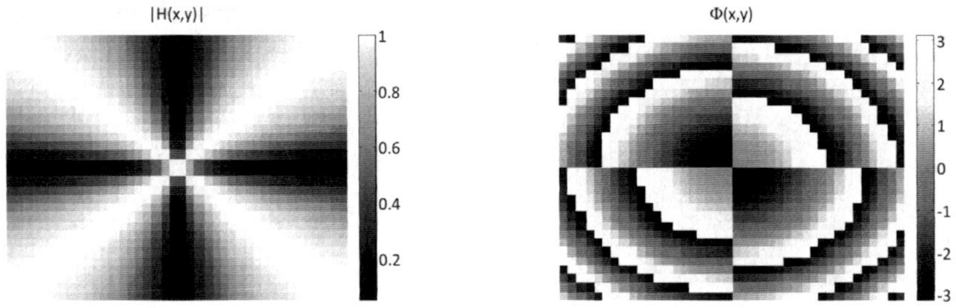

Figure 2. $H(x, y) = \sin(2\theta)\exp\left[i10\left(x^2 + y^2\right)\right]$, with $-1 \leq (x, y) \leq 1$ and $\theta = \arctan(y/x)$.

4.1. Burckhardt encoding

According to the method proposed by Burckhardt [14], a complex-valued function can be decomposed into three real and positive components. One hologram cell (macro-pixel) is laterally divided into three amplitude-modulating sub-cells, where the lateral shift between the sub-cells represents phase angles of 0°, 120°and 240°and acts as a phase offset. Thus,

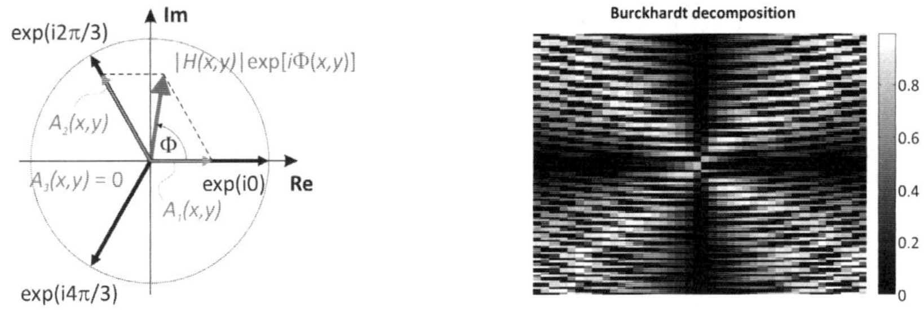

Figure 3. Burckhardt's decomposition method into three real and positive components.

Burckhardt holograms are detour-phase holograms that can be realized with an amplitude-modulating SLM. In holograms of this type, a phasor is decomposed into three vectors that run parallel to $\exp(i0) = 1$, $\exp(i2\pi/3) = -0.5+i\sqrt{3}/2$ and $\exp(i4\pi/3) = -0.5-i\sqrt{3}/2$, see Figure 3. Since the phase values are already represented by the lateral displacement of the sub-cells, any complex amplitude transmittance $H = |H|\exp(i\Phi)$ can be encoded in one macro-pixel as

$$H(x,y) = A_1(x,y)\exp\left[i0\right] + A_2(x,y)\exp\left[i\frac{2\pi}{3}\right] + A_3(x,y)\exp\left[i\frac{4\pi}{3}\right], \tag{5}$$

where the magnitude A_i of one term is always zero.

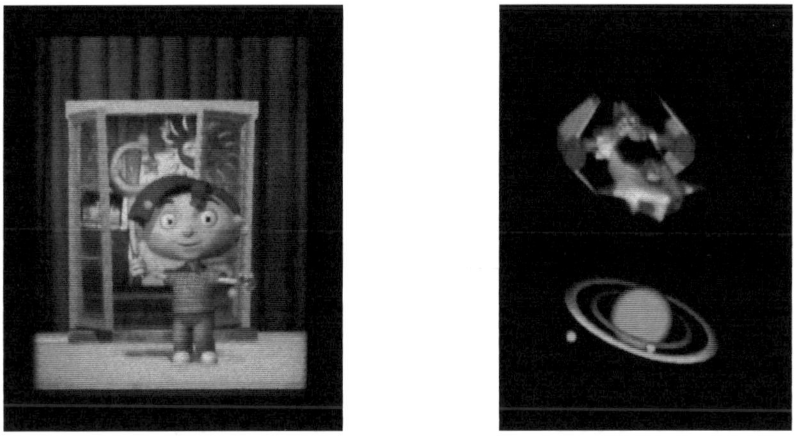

Figure 4. Photographs of holographic reconstructions from Burckhardt-encoded holograms displayed at SeeReal's 20-inch direct-view prototype "VISIO 20".

Advantage of this encoding scheme is that the decomposition can be done in an analytic way, which reduces the calculation cost. On the other hand, the diffraction efficiency of Burckhardt holograms is quite low. Notwithstanding, a fully complex modulation is achieved in a simple and easy way, i.e. by an amplitude-modulating SLM that comprises of an array of macro-pixels, albeit at the cost of reduced sampling and diffraction efficiency. Figure 4 shows reconstructions of 1D Burckhardt-encoded holograms displayed at a 20-inch direct-view color prototype from SeeReal.

4.2. Iterative phase-only encoding

Another encoded hologram representation employs the iterative Fourier transform algorithm (IFTA) [17–19], which can be adapted to not only create an intensity distribution but also

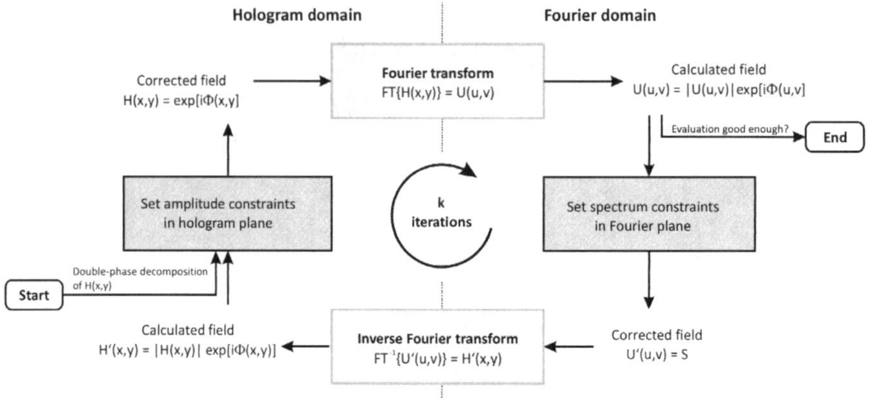

Figure 5. Flowchart of the adapted iterative Fourier transform algorithm.

a desired complex distribution in the Fourier plane of the hologram. For the hologram representation problem discussed here, the targeting complex distribution $S(u, v)$ in the Fourier domain is known, whereas a pure phase distribution in the hologram domain $H(x, y) = \exp(i\Phi)$ that can be represented by a phase-only modulating SLM is searched for. The objective of the iterative algorithm is to find a Fourier transform pair that satisfies both the constraints in the hologram as well as the Fourier plane. A flowchart of the basic design algorithm is illustrated in Figure 5.

Starting point is the nominal complex amplitude of the signal function $f(x, y)$, which is decomposed into two phase quantities by applying the double-phase hologram (DPH) approach [15, 16]. The Fourier transform of $f(x, y)$ is the nominal complex distribution $S(u, v)$ that serve as targeting function during optimization. The decomposition of $f(x, y)$ in two phase-modulating pixel doubles also the matrix size in the Fourier plane when the Fourier transform is applied. But only half of the Fourier plane carries the desired signal to be optimized, which is therefore called the signal area. The iterative algorithm works as follows: (1) Apply the Fourier transform to the initial estimate of the hologram which yields the field $U(u, v)$ in the Fourier plane; (2) Replace the resulting complex distribution within the signal area with the nominal complex distribution $S(u, v)$ (set spectrum constraints); (3) Apply the inverse Fourier transform to the modified field $U'(u, v)$ which gives a new field $H'(x, y)$ in the hologram plane; (4) Replace the

Figure 6. Holographic reconstruction from an iterative phase encoded hologram displayed at a laboratory prototype. In the left image the camera focus is set to +200 mm (in front of display), in the right image the focus is at -1000 mm (behind display).

modulus of the new calculated hologram $H'(x, y)$ with unity to provide a new hologram estimate (set hologram constraints). The iteration loop is repeated until the computed distribution in the Fourier domain satisfies the Fourier-domain constraints. Fidelity is evaluated by comparing the actual field in the signal area with the nominal signal field by use of the SNR-criterion as defined in Eq. (3).

Phase encoding has the benefits of higher efficiency and simple, single-layer SLM architecture. On the other hand, the iterative algorithm increases the computational costs and half of the native spatial resolution of the SLM is lost as one macro-pixel comprises of two sub-pixel. In Fig. 6 we demonstrate the huge depth range that can be realized by holography only. Full-parallax holograms with monochrome illumination are displayed on a laboratory prototype.

4.3. Two-phase interference encoding

By the double-phase hologram (DPH) approach, any complex number can be analytically decomposed as the sum of two pure phase values with constant magnitude as [15, 16]

$$H = |H| \exp(i\Phi) = \tfrac{1}{2}\exp[i(\Phi + \psi]) + \tfrac{1}{2}\exp[i(\Phi - \psi]) \tag{6}$$

where $\psi = \arccos |H|$ and $0 \leq \psi \leq \frac{\pi}{2}$. This holds for any arbitrary scalar complex field $H(x, y)$. Figure 7 shows the decomposition of H in the complex plane. For example, two identical phase values with $\phi_1 = \phi_2$ give a resulting vector with maximum amplitude and phase of Φ, whereas two phases with a phase difference of π result in a vector with zero amplitude with non-defined phase. Arbitrary complex values are generated by combining other phases than those of the special cases $\Delta\phi = 0$ or π.

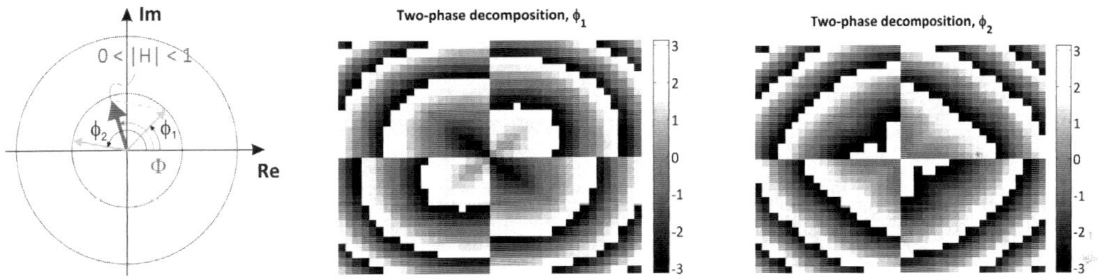

Figure 7. Geometric representation of the double-phase decomposition.

Only recently we introduced an implementation of a fully complex-modulating SLM that utilizes double-phase hologram (DPH) representation combined with two-beam interference [20]. To achieve a decoupled amplitude and phase modulation, two adjacent pixel of a phase-only modulating SLM are superimposed for interference by using polarization-sensitive components. The phase values to be addressed in neighboring pixel of the phase-SLM are $\phi_{1,2}(x, y) = \text{mod}\{[\Phi(x, y) \pm \arccos |H|(x, y)], 2\pi\}$.

Figure 8 shows the optical architecture of the complex SLM. Collimated TE-polarized light is spatially modulated in phase by an SLM and then passes through a structured half-wave plate (sHWP, isotropic/anisotropic zones), where the electric field vector of every second pixel column is rotated by 90° to TM-polarization, so that the sub-pixel to be composed to one macro-pixel have orthogonal polarization. The light fields of neighboring sub-pixel are brought to collinearity by a polarization-sensitive component (PSC), for example by a birefringent plate. The lateral displacement of the light field of every second pixel line must match the pixel pitch of the phase-modulation SLM in the selected combining direction. To bring them to interference, a subsequent linear polarizer with its transmission axis at 45° generates components of both polarizations from the two orthogonally polarized wave fields.

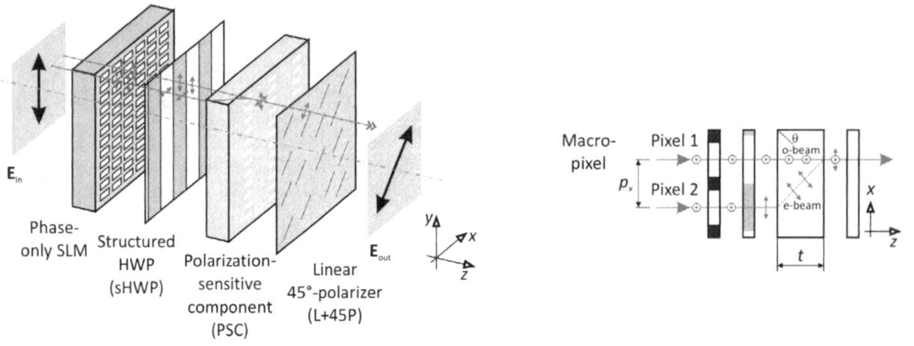

Figure 8. Optical architecture of the complex-modulating transmissive SLM sandwich.

Assuming completely polarized light and idealized components (no absorption or reflection), the Jones matrices for the two different paths through the sandwich (pixel 1, pixel 2) can be expressed as

$$\mathbf{J}_{\mathrm{P1}} = \mathbf{J}_{\mathrm{L+45P}} \cdot \mathbf{J}_{\mathrm{PSC}} \cdot \mathbf{J}_{\mathrm{sHWP_1}} \cdot \mathbf{J}_{\mathrm{LC_1}} = \tfrac{1}{2}\begin{pmatrix} 1 & 1 \\ 1 & 1 \end{pmatrix}\begin{pmatrix} e^{-i\varphi_e} & 0 \\ 0 & e^{-i\varphi_o} \end{pmatrix}\begin{pmatrix} 1 & 0 \\ 0 & 1 \end{pmatrix}\begin{pmatrix} 1 & 0 \\ 0 & e^{-i\phi_1(V)} \end{pmatrix} \tag{7}$$

$$\mathbf{J}_{\mathrm{P2}} = \mathbf{J}_{\mathrm{L+45P}} \cdot \mathbf{J}_{\mathrm{PSC}} \cdot \mathbf{J}_{\mathrm{sHWP_2}} \cdot \mathbf{J}_{\mathrm{LC_2}} = \tfrac{1}{2}\begin{pmatrix} 1 & 1 \\ 1 & 1 \end{pmatrix}\begin{pmatrix} e^{-i\varphi_e} & 0 \\ 0 & e^{-i\varphi_o} \end{pmatrix}\begin{pmatrix} 0 & -i \\ -i & 0 \end{pmatrix}\begin{pmatrix} 1 & 0 \\ 0 & e^{-i\phi_2(V)} \end{pmatrix} \tag{8}$$

where φ_o, φ_e are the phase delays for the o- and e-wave of the birefringent plate and ϕ_1, ϕ_2 denote the voltage-dependent phase shifts of the two phase-modulating LC-SLM pixel. The resulting intensity after coherent beam superposition is then $I(\phi_1, \phi_2) = \mathbf{E}_{\mathrm{in}}^{\dagger}\,(\mathbf{J}_{\mathrm{P1}} + \mathbf{J}_{\mathrm{P2}})^{\dagger} \cdot (\mathbf{J}_{\mathrm{P1}} + \mathbf{J}_{\mathrm{P2}})\,\mathbf{E}_{\mathrm{in}}$ with $\mathbf{E}_{\mathrm{in}}$ the Jones vector for linear vertically polarized light.

Besides cost reasons, advantages of two-phase interference encoding scheme are the direct optical transform of pure phase values into complex values and a high diffraction efficiency. Complex modulation with a single active panel and otherwise passive optical components is feasible, which is however, compromised by a loss of half the spatial resolution in the direction of pixel combining. The hardware implementation enables a direct encoding of complex-valued hologram data without need of time-consuming iterations.

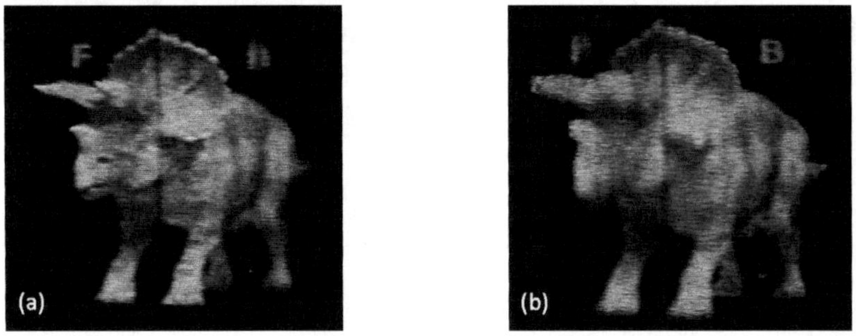

Figure 9. Holographic reconstructions taken with a digital SLR camera with focus (a) at foreground, $z = +100\,\mathrm{mm}$ (letter 'F') and (b) at background, $z = -100\,\mathrm{mm}$ (letter 'B').

Figure 9 presents holographic reconstructions from a laboratory prototype that uses a purpose-built complex-modulating SLM. The holographically generated 3D scene employs a single-parallax encoding in y-direction. Both phase and amplitude modulation is qualitatively demonstrated, which manifests in the clear depth discrimination (phase fidelity) and gray level representation (amplitude matching) of the reconstructions.

4.4. Comparison

When comparing the discussed encoding schemes, not only optical performance criteria but also SLM hardware architecture and calculation cost have to be considered. Most certainly, the intended application will determine the best choice of SLM and encoding. In some optoelectronic systems, a precomputing or delayed updating of the hologram might be acceptable. For real-time 3D imaging however, real-time capability is a must. Therefore, pixel-oriented encoding schemes that employ simple operations have a big advantage over global optimization methods. Table 1 summarizes key parameter of the encoding schemes discussed in this paper.

Table 1. Comparison of different encoding schemes.

	Modulation type	Number M of sub-pixel	SBP	maximum $\eta_{\mathrm{E,abs}}$	typical $\eta_{\mathrm{E,abs}}$	Encoding
Burckhardt	Amplitude	$M = 3$	$\frac{1}{3}(N_x N_y)$	$\leq 1.36\%$	0.67%	direct
Iterative phase	Phase	$M \geq 2$	$\frac{1}{M}(N_x N_y)$	$\leq 50\%$	50%	iterative
Two-phase interference	Complex	$M = 2$	$\frac{1}{2}(N_x N_y)$	$\leq 100\%$	8.5%	direct

We have analyzed the three encoding schemes with regard to their sensitivity to quantization and addressing errors. Simulation was done with the exemplary complex function shown in Fig. 2, which was sampled with 640 × 480 points. In Fig. 10 (left) is shown how the SNR increases with the number of bit levels that were employed for discretization of the hologram function. Two-phase interference encoding exhibits the steepest increase, whereas the pure phase or amplitude representations approach their non-discretized maximum values with increasing bit level. Note that the SNR is also dependent from the total number of sampling points. Fig.10 (right) shows the SNR degradation in dependence from a percentage error up to 5% (amplitude error for Burckhardt and phase error for the others) which was added to a 8-bit discretized hologram function. The Burckhardt encoding scheme is least sensitive to possible addressing errors. The tolerable error level depends on the on the targeting SNR.

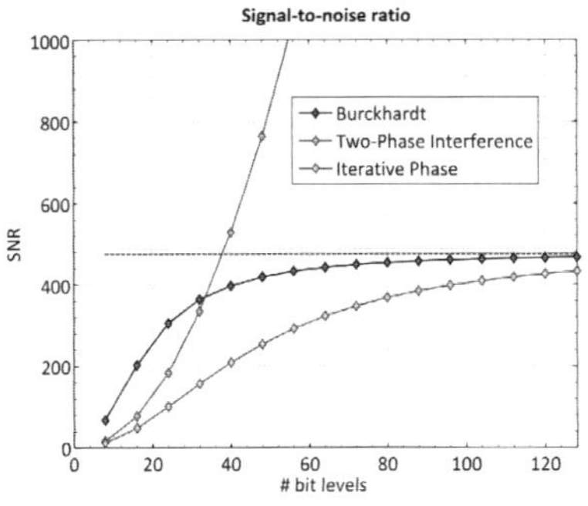

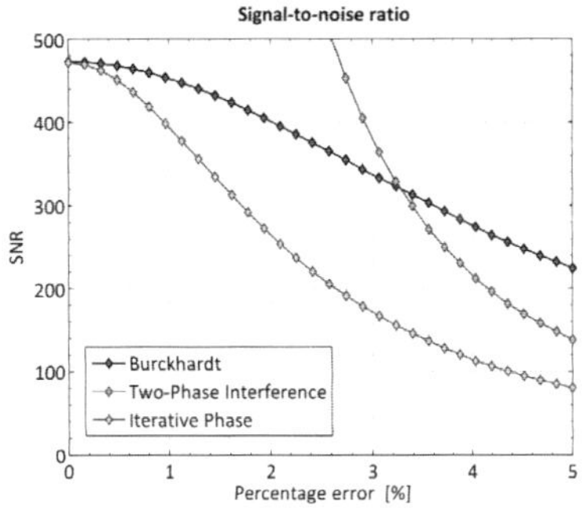

Figure 10. Sensitivity of SNR to bit depth discretization (left) and to percentage error.

5. Summary

For dynamic applications such as three-dimensional holographic display, a full-complex modulation is highly desirable, because any restricted coding domain or operating curve results in reduced efficiency, noise terms, and complementary diffraction orders. The essentials of three different encoding schemes have been discussed and their implementation on prototypes using SeeReal's proprietary Viewing-Window (VW) holography has been experimentally proven.

References

[1] A. Schwerdtner, N. Leister, and R. Häussler, "A new approach to electro-holography for TV and projection displays," in *SID-Proc.*, p. 32.3, 2007.

[2] R. Häussler, S. Reichelt, N. Leister, E. Zschau, R. Missbach, and A. Schwerdtner, "Large real-time holographic displays: from prototypes to a consumer product," *Stereoscopic Displays and Applications XX* **7237**(1), p. 72370S, SPIE, 2009.

[3] A. W. Lohmann, "The space-bandwidth product, applied to spatial filtering and holography," *Research Paper RJ-438 (IBM San Jose Research Laboratory, San Jose, Calif.)* , pp. 1–23, 1967.

[4] A. W. Lohmann, R. G. Dorsch, D. Mendlovic, Z. Zalevsky, and C. Ferreira, "Space-bandwidth product of optical signals and systems," *J. Opt. Soc. Am. A* **13**(3), pp. 470–473, 1996.

[5] J. W. Goodman, *Introduction to Fourier Optics*, McGraw-Hill, New York, 2nd ed., 1996.

[6] J. Bucklew and J. Neal C. Gallagher, "Comprehensive error models and a comparative study of some detour-phase holograms," *Appl. Opt.* **18**(16), pp. 2861–2869, 1979.

[7] S. Reichelt, R. Häussler, N. Leister, G. Fütterer, and A. Schwerdtner, "Large Holographic 3D Displays for Tomorrow's TV and Monitors – Solutions, Challenges, and Prospects," in *Proc. 21st Annual Meeting of the IEEE Lasers and Electro-Optics Society LEOS 2008*, pp. 194 – 195, 2008. (invited).

[8] E. Zschau, R. Missbach, A. Schwerdtner, and H. Stolle, "Generation, encoding, and presentation of content on holographic displays in real time," *Three-Dimensional Imaging, Visualization, and Display 2010 and Display Technologies and Applications for Defense, Security, and Avionics IV* **7690**(1), p. 76900E, SPIE, 2010.

[9] S. Reichelt, R. Häussler, N. Leister, G. Fütterer, H. Stolle, and A. Schwerdtner, *Advances in Lasers and Electro Optics*, ch. 29 "Holographic 3-D Displays – Electro-holography within the grasp of commercialization", pp. 683–710. InTech, available from www.intechopen.com, April 2010. ISBN 978-953-307-088-9.

[10] A. W. Lohmann and D. P. Paris, "Binary Fraunhofer Holograms, Generated by Computer," *Appl. Opt.* **6**, pp. 1739–1748, Oct 1967.

[11] D. Schreier, *Synthetische Holografie*, Physik-Verlag, Weinheim, 1984.

[12] B. R. Brown and A. W. Lohmann, "Complex spatial filtering with binary masks," *Appl. Opt.* **5**(6), pp. 967–969, 1966.

[13] W. H. Lee, "Sampled fourier transform hologram generated by computer," *Appl. Opt.* **9**(3), pp. 639–643, 1970.

[14] C. B. Burckhardt, "A simplification of Lee's method of generating holograms by computer," *Appl. Opt.* **9**(8), pp. 1949–1949, 1970.

[15] D. C. Chu and J. W. Goodman, "Spectrum shaping with parity sequences," *Appl. Opt.* **11**(8), pp. 1716–1724, 1972.

[16] C. K. Hsueh and A. A. Sawchuk, "Computer-generated double-phase holograms," *Appl. Opt.* **17**(24), pp. 3874–3883, 1978.

[17] P. M. Hirsch, J. A. Jordan, and L. B. Lesem, "Method of making an object dependent diffuser," 1971. US Patent 3,619,022.

[18] R. Gerchberg and W. Saxton, "A practical algorithm for the determination of phase from image and diffraction plane pictures," *Optik* **35**, pp. 237–246, 1972.

[19] N. C. Gallagher and B. Liu, "Method for computing kinoforms that reduces image reconstruction error," *Appl. Opt.* **12**, pp. 2328–2335, Oct 1973.

[20] S. Reichelt, R. Häussler, G. Fütterer, N. Leister, H. Kato, N. Usukura, and Y. Kanbayashi, "Full-range, complex spatial light modulator for real-time holography," *Optics Letters* **37**, pp. 1955–1957, Jun 2012.

Bit-mapped Holograms Using Phase Transition Mastering (PTM) and Blu-ray Disks

Donald Barnhart

Barnhart Optical Research LLC

613 W Delaware Ave, Urbana IL, 61801 USA

dhb@opticasoftware.com (www.opticasoftware.com)

Abstract. Due to recent advances made in data storage, cloud computing, and Blu-ray mastering technology, it is now straight forward to calculate, store, transfer, and print bit-mapped holograms that use terabytes of data and tera-pixels of information. This presentation reports on the potential of using the phase transition mastering (PTM) process to construct bit-mapped, computer generated holograms with spatial resolutions of 5000 line-pairs/mm (70 nm pixel width). In particular, for Blu-ray disk production, Sony has developed a complete process that could be alternately deployed in holographic applications. The PTM process uses a 405 nm laser to write phase patterns onto a layer of imperfect transition metal oxides that is deposited onto an 8 inch silicon wafer. After the master hologram has been constructed, its imprint can then be cheaply mass produced with the same process as Blu-ray disks or embossed holograms. Unlike traditional binary holograms made with expensive e-beam lithography, the PTM process has the potential for multiple phase levels using inexpensive optics similar to consumer-grade desktop Blu-ray writers. This PTM process could revolutionise holography for entertainment, industrial, and scientific applications.

1. Introduction

Computer generated holograms (CGH) are far from new. Ever since the widely referenced paper by Lohmann and Paris in 1967, the potential importance and value of CGH methods have sparked the imagination and challenged the intellect of countless scientists [1-6]. Unfortunately, until recently, limitations in methods of production have restricted their use to a narrow range of applications. In particular, not long ago, limitations in computer memory capacity and processor speeds have stifled the application of CGH to the large surface areas and high information bandwidths commonly found in traditional holograms [28, 29]. In addition, most methods of making computer generated holograms are lacking in continuous phase depth control (are binary) [7-9] or have inferior spatial resolution. These fabrication methods have been limited to very expensive processes such as e-beam lithography or lower resolution photo-reduction techniques. However, in a few narrow applications where low bandwidth can be tolerated or where cost is no object and the effects of binary levels can be tolerated, CGH methods have enjoyed a successful but narrow application. One notable application is for the interferometric testing of aspheric components [10]. Other applications include optical correlation for machine vision [11], and as phase masks in telescopes and microscopes. Along with the limitations of computer speeds and memory constraints, the wide application of CGH had been limited by the lack of any facility to inexpensively mass-produce such holograms with sufficiently high quality. While embossing methods have been around for years and have permitted the cheap mass production of holograms, the lack of phase uniformity and good surface flatness of embossing materials have limited their performance and subsequent use in many potential scientific applications such as holographic optical elements (HOE).

With the advent of compact disk and laser disk technology in the early 1980's, and DVD technology in the mid to late 1990's, computer generated holograms have been reconsidered and tested for various applications on these media [10]. However, before Blu-ray disks, the spatial resolution of normal compact disks continued to limit their use. In addition, compact disk and DVD technology continued to be exclusively binary and normal disk recorders lacked the necessary spatial encoding hardware to write holograms directly [12].

Blu-ray disk production, and phase transition mastering in particular, has at last unleashed the potential of mass-producing computer generated holograms for a wider range of applications that demand both high resolution, large surface areas and high-precision phase control [13-21]. At present, there have been few reported attempts to commercialize Blu-ray based holograms [16] and it appears that few if any of the large Blu-ray producers are actively engaged with CGH production. In spite of this lack-luster beginning, it does appear that the PTM process could be ideally suited for CGH mass-production. With the hardware technology already in place for the production of Blu-ray disks, once a CGH has been mastered with PTM, it can be immediately mass-produced in a Blu-ray factory with minimal change to the Blu-ray mass-production process. In addition, while most Blu-ray mastering technologies can only be deployed for binary phase applications, the PTM process has the potential to record multiple phase levels to approximate continuous phase tones.

2. Tera-pixel recording and Terabyte calculations

In many respects, the bit-mapped computer generated hologram is in fact just like any other digital image. The real difference between such a hologram and a standard digital image is the higher spatial resolution of a holographic image. For example, an off-axis bit-mapped hologram could contain 100 million pixels of information for every square millimeter of surface area. This corresponds to a pixel size of 0.1 microns (100 nanometers) or even less. As a result, a hologram that is 100 x 100 mm in size would require a 1 Terabyte of data if each pixel had 256 grey levels. In other words, to capture the same data using a digital camera, one would require a sensor that has one million rows by one million columns.

While 20 years ago, an optical scientist would have no chance of working with such large sums of data in any format, such data today is almost trivial to obtain and only slightly less trivial to calculate, particularly with today's graphics cards and Cloud computing options. The next step is how to print such a digital hologram? Enter the Blu-ray disk: with its high spatial resolution and high quality optical characteristics that essentially match the requirements for a good CGH. At last, with this last piece in place, perhaps the age of the CGH can finally arrive!

3. Benefits of bit-mapped computer generated holograms

For a number of years, a category of computer generated hologram has existed that does not directly use bit-mapped data, but rather depends on the local mixing of a reference beam with the object-space frequency spectrum across an array of holographic "voxels" on the hologram surface. Such a system was pioneered at MIT and has been successfully deployed by Zebra Imaging to make large format display holograms [23]. Related approaches have also been used with great success to make holographic storage by companies such as In-Phase [24]. Such systems do have some important advantages over bit-mapped holograms in terms of larger hologram formats, reduced hologram write-times, reduced image processing, and reduced memory overhead requirements. In addition, the voxel approach is able to write volume gratings that support "thick" Lippmann-type holograms in photopolymers instead of the "thin" holograms used with bit-mapped holograms. Nevertheless, the incoherent bit-mapped CGH described in this paper do offer a number of significant advantages that are not available with interference-based methods. In particular, the bit-mapped approach can offer better performance for holographic optical elements (HOE). In addition, the bit-mapped hologram can be mass produced either in the Blu-ray disk format or as an embossed hologram. Furthermore, aberrations, distortion, and diffraction efficiency can all be improved in a bit-mapped hologram.

4. Noise Reduction

There is an important advantage with incoherent writing of a hologram pixel by pixel that needs further discussion. This advantage is the reduction (and even elimination) of intermodulation noise, which is also known as cross-talk.

In classic interference-based holography, intermodulation noise occurs as a result of the interference between the different scattering points from the surface of an illuminated object. In particular, coming from an object are a very large number of wavefronts (or angular frequency components) from each scattering point in the object space. Instead of considering all possible scattered wavefronts from an arbitrary object, as shown in Figure 1, the effects of intermodulation noise can be most easily observed if we restrict our attention to three such wavefronts that we label as A, B, and C. At the hologram plane H, these wavefronts interfere with a reference wavefront, R, that creates an interference pattern which gets recorded by the holographic film and defines the holographic grating used during holographic reconstruction.

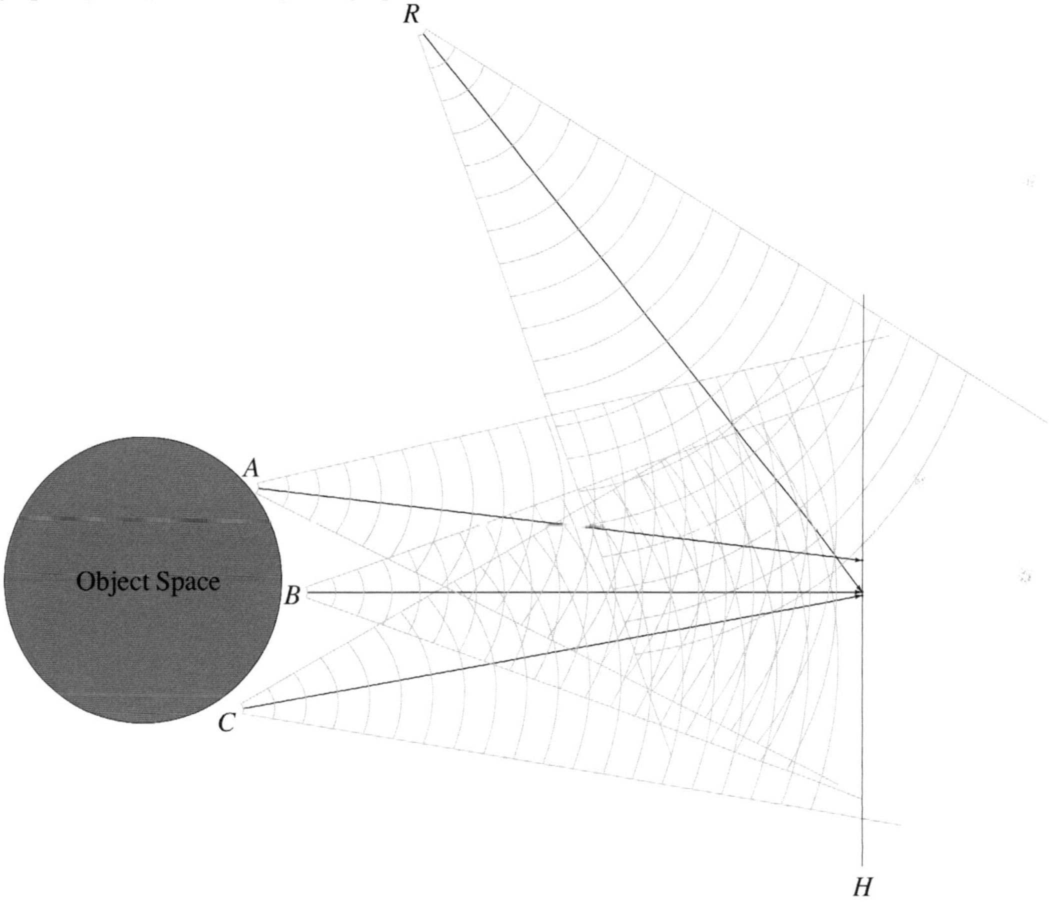

Figure 1: Interference between 3 wavefronts A, B, C and reference wavefront R at H [26].

We can next write an equation to represent this process of recording and reconstruction of a holographic image. In particular, during the recording, all of the coherent wavefronts interfere with each other at the holographic film plane, H. This interference is really the summation of the amplitudes of each wavefront multiplied by their complex conjugates that gives the intensity distribution [6]. This is shown in equation (1).

$$I = (R + A + B + C)\,(R^* + A^* + B^* + C^*) \quad (1)$$

After algebraic expansion, the result is:

$$I = RR^* + RA^* + RB^* + RC^* + AR^* + AA^* + AB^* + AC^* + BR^* + BA^* + BB^* + BC^* + CR^* + CA^* + CB^* + CC^* \quad (2)$$

Of all of these 16 terms found in equation (2), only 3 terms relate to the desired holographic image formation:

$$I[\text{image}] = A\,R^* + B\,R^* + C\,R^* \quad (3)$$

while three other terms are related to the phase conjugate image:

$$I[\text{conjugate image}] = RA^* + RB^* + RC^* \quad (4)$$

The remaining 9 terms shown in equation (5) relate to intermodulation noise and bias components:

$$I[\text{noise}] = RR^* + AA^* + CC^* + AB^* + AC^* + BB^* + BC^* + CA^* + CB^* \quad (5)$$

With classic interference-based holography, the bias and noise terms are unavoidable and inherent in the process. However, with the method of incoherent bit-mapped recording, the intermodulation noise and bias terms are not generated and it becomes possible to directly eliminate these issues. By using multiple phase levels to write a blazed hologram profile, it may also be possible to eliminate the conjugate image terms as well.

5. Polychromatic performance

With classic holographic methods based on interference, the reconstructed hologram is dependent on the wavelength of the laser used to write the hologram. The holographic recording process is formed from interference between each of the laser wavelengths present with their corresponding reference beam. After development, the fringes formed during exposure from each laser is converted into a diffraction grating at a carrier frequency that corresponds with the laser wavelength used to record it.

As such, in order for multi-wavelength reconstruction to occur for a classic hologram, it is necessary either to use multiple laser wavelengths during the recording step or a careful arrangement of the recording optics to compensate for the resulting distortion, magnification, and loss of diffraction efficiency that would otherwise occur at the different shifted wavelengths during reconstruction. In both situations fore-mentioned, the recording geometry has an increased complexity for multiple wavelength reconstruction. In contrast to this, with the CGH method of writing the hologram pixel by pixel, there is no increase in complexity or change in the recording optics to accommodate multiple wavelengths.

6. Light scattering through non-interferometric origins

With bit-mapped computer generated holograms, it is no longer necessary to create the sinusoidal grating structures found in interference-based holograms. A diffraction grating produced by classic interference methods typically has a sinusoidal shape to its surface profile. In contrast, the ruled blazed grating used with spectroscopic measurement has existed since the nineteenth century and has been historically created on a ruling machine which physically cuts the grating by a knife edge. In this case, the grating has a sawtooth profile. Such a grating has advantages for achieving a higher diffraction efficiency in a particular direction or diffraction order.

When writing bit-mapped holograms with multiple phase levels, it becomes possible to create analogous structures that are computer optimized profiles for specific applications. These shapes might have a non-smooth profile (as in the case of the blazed grating), be asymmetric or even aperiodic. The designs of such shapes need not be governed from interference considerations and may even be calculated using electromagnetic simulations of the scattering boundary conditions. Some novel surface profiles might even be fractal in nature or mimic the nano-structure on a butterfly wing to create iridescence. The key point is that light scattering profiles can have many different shapes other than the standard sinusoidal-based profile exhibited from interference recording.

7. CGH on Blu-ray disks

The Blue-ray disk technology was developed by Sony and their partners as the next step beyond DVD technology. While the DVD uses a 658 nm laser wavelength, the Blu-ray is based on the 405 nm wavelength. With a 320 nm track pitch and a 150 nm minimum pit length, a single layer Blu-ray disk is capable of storing 25 GB of data on a single layer [13]. Such numbers however do not reveal the true information density available on a Blu-ray disk. In particular, the Blu-ray disk is designed for serial digital read-out of data and is composed of a spiral of tracks in an analogous fashion to CD's, DVD's, and even the old fashioned vinyl LP records. The electronic hardware used with personal Blu-ray writers and players depend on these tracks for recording and read-out purposes.

For CGH applications, the normal tracks found on Blu-ray disks serve no purpose and only add diffraction noise to the reconstruction. In addition, these tracks are arranged with a 150 nm gap between them. These gaps are needed for serial digital read-out in order to prevent optical cross-talk between the tracks. With CGH, it becomes possible to completely fill-in the space on the disk surface and eliminate the tracks altogether. In addition, with Phase Transition Mastering (PTM), it is also possible to reduce the size of the recorded spot from 150 nm down to below 70 nm. This is due to the remarkable nonlinear behavior of the PTM process that enables spots to be recorded at super-resolution (smaller than the diffraction limit (NA = 0.85)). While the diffraction-limited spot-size of the laser focus optics is about 0.4 micron, the PTM advantage gives it less than 0.1 micron by thermal threshold. Taken together, a single Blu-ray disk layer can store a Terabyte of data or a CGH filled with a 10^12 pixels. In fact, holograms aside, Sony has already successfully produced Blu-ray ROM disks containing a Terabyte of data [27].

8. Quality control and mass production of Blu-ray disks

While Blu-ray technology enables recorded resolutions that are comparable with the best holographic films, it is the incredible perfection of the Blu-ray quality control during mass production that makes it ideally suited for CGH production. Unlike other methods used to produce CGH media, with Blu-ray mastering each pixel can be tested to accurately ensure that the intended phase level at each pixel has been obtained. In addition, once a Blu-ray master has been created, the subsequent mass production of Blu-ray disks from the master has already been carefully fine-tuned by the Blu-ray industry to deliver copies that are optically flat to a fraction of a wavelength with high fidelity reproduction of the phase information.

9. Overview of the PTM process

The PTM process was originally developed by Sony in 2003 [14]. By 2009, of the 400 million Blu-ray disks produced world-wide, 94% were created with Sony's PTR-3000 BD Mastering System.

PTM operates by depositing a layer of imperfect oxides of transition metals on an 8 inch silicon wafer substrate. More specifically, the deposited layer consists of imperfect oxides of tungsten and molybdenum. Originally, the deposited layer is amorphous. Because the material is non-transparent, it absorbs the blue laser light (405 nm). Sudden increases in temperature (where the light is absorbed) cause oxidation with oxygen from the surrounding environment. This results in a change from an amorphous to a crystalline structure. There are also chemical changes in the material. When the material is immersed in an alkaline developing solution, only the areas crystalized through exposure to the blue laser are dissolved [22].

The silicon wafer substrate also plays an important role in the mastering process by ensuring an optically flat surface as well as conducting the heat from the exposed region more quickly than the glass substrate used traditionally with DVD production. This increased heat transfer permits smaller write zones than otherwise possible by dissipating the heat away from neighboring unexposed regions.

10. Features of PTM

The Phase Transition Mastering process (PTM) is only one of several competing processes developed for Blu-ray mastering [13]. However, for the purposes of CGH mastering, it appears to be superior to other methods.

11. PTM vs e-beam lithography

With PTM, the mastering equipment costs are extremely low compared to e-beam lithography. In addition, the time required to create a BD master is greatly reduced over e-beam. In particular, e-beam mastering requires 12 processes and takes approximately 11 hours to write a 25 Gigabyte layer [21]. With PTM, there are only 5 processes and the total write time required for a single 25 GB layer can be as short as 40 minutes for a 3x write speed or as long as 80 minutes for the more standard 1.5 x write speed. When making a Terabyte CGH recording, this could translate to an overall mastering time of 24-100 hours depending on the mastering write speed [27]. Another advantage with PTM is that the areas crystallized through exposure to blue laser light are slightly raised compared with other areas. This means that error checking can be carried out at the same time as mastering, leading to 100% error-free masters [21]. Finally, e-beam lithography has traditionally been deployed as a binary process. However, e-beam writing has recently been used by researchers to create nano-structures in materials such as PMMA. It would appear that the PTM process can also be harnessed in a similar fashion.

12. Phase depth control

With PTM, the pixel depth can be controlled to deliver a maximum phase modulation of 2π radians for greatest diffraction efficiency. For reflective Blu-ray disks, this corresponds to a profile depth of 0.5 wavelength/refractive-index. This means that for the visible wavelength range, a pixel depth of less than 200 nm is needed to achieve 2π radians of phase modulation in the reflected beam. While the standard PTM mastering system is designed for binary phase operation, the Sony PTM process has the potential to support multiple phase levels. This is due to the fact that the Sony PTM process is a positive process in which the exposed surface is removed during chemical development and can increase the phase depth through increased exposure at different regions across the CGH bit-map.

While the several other PTM mastering technologies now exist that compete with Sony's PTM, these competing methods are generally lower resolution and based on negative processes in which the unexposed surface regions are removed. In contrast to positive PTM, negative PTM processes have no ability to increase the phase depth through increased exposure and the resulting spot size is larger. While positive PTM has the potential to write multiple phase depths, this property has not been exploited by the current PTM control software nor has it been reported for CGH production by other researchers. Instead, all reported CGH research using PTM has been exclusively binary phase to date [12, 14, 16]. However, PTM has been successfully applied by Sony to write different phase levels for the creation of both Blu-ray and DVD masters (which have different phase depth requirements) and there is a limited capability in some PTM control software to program the pit depth up to 100 levels. Before it can be fully exploited, further development will likely be required to add support for multiple phase levels on current PTM machines [27].

13. Electroplating from PTM

The PTM process creates a master that is electrically conductive. This enables nickel stamps to be electroplated directly on the PTM master without the additional metal sputtering step required for traditional photoresist or non-conductive, glass-substrate based disk mastering techniques, thereby simplifying the production cycle.

14. CGH with PTM

In traditional holography, the object and reference beams interfere at the holographic plate by means of standing wave patterns in intensity. This interference pattern is recorded by chemical changes in the photosensitive medium. Because the light is spread out over an area on the holographic plate, the intensity or energy density at the film surface is greatly reduced. As a consequence, the photosensitive medium must have a certain degree of sensitivity to chemically capture and amplify the impact of the interference fringe intensity. For example, traditional media will have a sensitivity that ranges between 50 micro-Joules/cm^2 for silver halide and >2 mJ/cm^2 for photopolymers.

In the case of phase transition mastering, the recording media has no chemical sensitivity toward light because the physical processes involved are phase changes due to heat. This can only happen with light that is tightly focussed to a very small spot with very high power density. The spatial scale of the heat zone can be less than 100 nanometers. As such, traditional methods of holographic recording will not work with PTM. Instead, the hologram must be synthesized pixel by pixel (or point by point) into a composite structure to diffract the light during holographic reconstruction.

15. Thin versus thick holograms

PTM holograms are generated in a single diffractive surface rather than being distributed within a diffractive volume. This is the classic difference between a "thin" hologram and a "thick" hologram. It is also the reason why holograms created with PTM can be mass-produced by the same methods as either Blu-ray disks or classic embossing. In particular, the "thin" quality of holograms made by PTM allows their features to be stamped into a plastic substrate or foil.

The thin nature of the PTM-based holograms permits diffraction to occur with equal efficiency across a broad range of wavelengths and angular frequency inputs. This property can be very useful in the application of holographic optical elements (HOE).

16. Applications

Bit-mapped holograms with the PTM process could revolutionise holography for entertainment, industrial, and scientific applications. In particular, Blu-ray mass production could enable high quality holograms to be mass-produced for the first time at an affordable cost. By using the standard Blu-ray disk mass-production process, the manufactured cost of each Blu-ray CGH can be less than one dollar.

17. HOEs

The low cost and high performance make Blu-ray HOE's viable for many new and previously unrealised applications. Today, the true potential of HOE devices has been left untapped because of insufficient performance characteristics combined with inadequate means for mass-production. With Blu-ray CGH production, this could change. In particular, with the right performance characteristics in place, HOE components could displace current optical systems that now cost many thousands (and even hundreds of thousands) of dollars. The most critical characteristics include a sufficiently high diffraction efficiency and sufficiently low aberrations and wavefront distortion. Of course the precise limits of acceptable performance varies with the application. For some applications, a diffraction efficiency approaching 100% could be necessary, while in other applications a diffraction efficiency of 50% might be perfectly acceptable. In some cases, the high diffraction efficiency may be far less critical than minimizing all distortion, while the requirements may be reversed in other cases.

Here is a partial listing of potential HOE applications:

- Laser beam-shaping
- Toys and entertainment
- Sign projections for advertising and displays
- Laser welding
- Light collection systems
- Architectural lighting systems
- Three-dimensional shape measurement and shape validation
- Solar energy concentrators
- Spectroscopy
- Bio-medical instrumentation optics
- Imaging lens optics
- Integrated circuit fabrication validation
- Aspheric and free-form optic characterization
- Head-up displays
- Three-dimensional displays
- Space-born and military optics
- Astronomy
- Microscopy.

18. Display Holograms

Display holograms have different requirements from HOE applications. In general, display holograms do not require high diffraction efficiencies and can tolerate greater aberrations and distortion than for HOE applications. Here is a partial listing of some specific display hologram applications:

- Blu-ray disk movie cover-art
- Watermarks [16]
- Anti-counterfeiting
- Embossed hologram masters
- Holographic "postcards" of venue-specific material (including landscapes and landmarks)
- Advertising and product promotion
- Artistic display holograms.

19. PTM for embossed hologram mastering

PTM mastering could revolutionize the embossed hologram industry. Many embossed hologram masters are produced one at a time by a highly skilled artist in a holographic studio. This is a delicate and expensive endeavor which limits the types of subject matter that can be produced in the hologram. While the PTM process is currently only used to create Blu-ray masters, it is naturally suited to produce embossed hologram masters since it already generates nickel replicas for stamping. With PTM-based hologram masters, the subject is always computer generated and the resulting embossed holograms can include any subject without limitation. Finally, PTM-based hologram mastering can be a turn-key process with predictable production time-lines.

20. Embossed versus photopolymer versus Blu-ray holograms

For a number of years both embossed and photopolymer materials have been used to mass-produce holograms. These methods of hologram production have different strengths from Blu-ray holograms. Both embossed and photopolymer holograms are usually created in a flexible material. With embossed holograms, a metalized-mylar film is bonded to a substrate such as paper, plastic, or card stock. Photopolymers are inherently pliable and are often produced in flexible film. Blu-ray disks, on the other hand, are rigid. This has the result of the Blu-ray disk preserving the optical phase integrity more accurately than embossed or photopolymer media. For holographic display applications, this phase flatness is not a big issue. However, for HOE applications, phase front flatness can be extremely important. For such applications, the Blu-ray disk is a far superior vehicle for CGH production.

21. Blu-ray mastering versus Blu-ray disk burning

This paper has placed a particular emphasis on the Blu-ray mastering process for making computer generated holograms on the Blu-ray disk format for mass production. In particular, Blu-ray mastering is a process that occurs in a Blu-ray factory to create a master used for the mass-production of Blu-ray disks. However, there also exists the potential of using the hardware from an inexpensive personal Blu-ray burner to make computer generated holograms. In particular, the laser optics associated with a Blu-ray writer is essentially the same as those found with Blu-ray mastering hardware. The difficulty lies in the absence of the necessary spatial encoding hardware required to accurately print the bit-map pixel locations. In contrast, the Blu-ray mastering systems normally found in a Blu-ray factory can be readily configured with optical encoders to provide positional feedback for the direct printing of bit-mapped holograms.

In addition, the recordable Blu-ray disk used in personal Blu-ray writer is based on a different material process from PTM (the details of which are beyond the scope of this paper). It appears that this recordable material could also be used to make computer generated holograms, but such disks usually get pre-imprinted with spiral recording tracks at the time of disk manufacture. For CGH, the presence of these tracks creates noise in the reconstructed image.

While it could be possible to adapt a personal Blu-ray writer for CGH production, it will require significant modifications to the writer hardware and software to make this possible. This could be worthwhile if the end-goal is to construct an inexpensive prototype holographic printer, but it will require significant costs and risks of additional research and development to adapt such hardware. By using an existing mastering facility at a Blu-ray factory, the research and development risks are greatly reduced and the resulting CGH masters made with PTM will be optimized for immediate mass production.

22. Mass-production versus custom printing

With the development of digital cameras that can capture fully three-dimensional pictures, such as the Lytro camera [25], there is likely to be a very significant market for custom printed display holograms taken by ordinary people. In theory, this could be a very good application of bit-mapped holograms printed on Blu-ray disks. However, the inherently longer process times of a bit-mapped printed hologram (24-100 hours) is much slower than alternative holographic print techniques, such as offered by Zebra Imaging. This might not be a problem for custom HOE applications since they often cannot function well in competing formats. However, for custom display hologram applications, other faster hologram print methods can be better suited.

23. Conclusion

The holography business needs a heavily invested industry to "piggy-back" onto. At the same time, with millions of dollars already invested into Blu-ray production and stiff competition from video streaming and digital downloads, the CGH market could provide a much needed source of revenue for the Blu-ray industry. The synergy between standard Blu-ray and CGH production could become an ideal partnership. In fact the technical requirements of Blu-ray already closely match CGH technical requirements and any enhancement made to the Blu-ray mastering process will immediately improve CGH production capabilities as well.

While PTM bit-mapped holograms are inherently slower to record than many other computer generated hologram methods, it appears to be the best option for the mass-production of both HOE's and display holograms on Blu-ray disks.

Acknowledgements
During the course of conducting research for this paper, the author wishes to acknowledge the helpful conversations and technical guidance of several individuals including: Masanobu Yamamoto, Seiji Kobayashi, Matt Lindsey, Tao Wu, Lon Westfall, Ed Wesley, Kevin Keilbach, Bill Wilson, and Tei Barnhart. The author wishes to thank the Sony Corporation for inventing the Blu-ray disk and PTM technology. Finally, the author wishes to thank Molly for financially supporting this effort.

References

[1] W. Lohmann and D. P. Paris, "Binary Fraunhofer holograms, generated by computer," Appl. Opt. 6, 1739-1748 (1967)
[2] L. B. Lesem, P. M. Hirsch, and J. A. Jordan, “Kinoform - a new wavefront reconstruction device,” IBM J. Res. Develop. **13(2),** 150–155 (1969).
[3] D. Leseberg and O. Bryngdahl, “Computer-generated rainbow holograms,” Appl. Opt. **23(14),** 2441–2447 (1984).
[4] G. Tricoles, “Computer generated holograms: an historical review,” Appl. Opt. **26(20),** 4351–4357 (1987).
[5] M. Clark, “Two-dimensional, three-dimensional, and gray-scale images reconstructed from computer-generated holograms designed by use of a direct-search method,” Appl. Opt. **38(25),** 5331–5337 (1999).
[6] J. Goodman, Fourier Optics, Roberts & Company, (2005)
[7] P. Chavel and J. P. Hugonin, “High-quality computer holograms - problem of phase representation,” J. Opt. Soc. Am. **66(10),** 989–996 (1976).
[8] N. C. Gallagher, Jr., J. C. Angus, F. E. Coffield, R. V. Edwards, and J. A. Mann, Jr., “Binary phase digital reflection holograms: fabrication and potential applications,” Appl. Opt. **16(2),** 413–417 (1977).
[9] J. P. Allebach, “Representation-related errors in binary digital holograms: a unified analysis,” Appl. Opt. **20(2),** 290–299 (1981).
[10] C. Pruss, S. Reichelt, H. J. Tiziani, W Osten, “Computer-generated holograms in interfereometric testing,” Optical Engineering, 43 (2), 2534-2540, (2004).
[11] D. Psaltis, M. A. Neifeld, A. Yamamura, and S. Kobayashi, “Optical memory disks in optical information processing,” Appl. Optics / Vol. 29, 2038-2057, No. 14 / 10 May 1990
[12] A. Cable, P. Mash and T. Wilkinson, "Production of computer-generated holograms on recordable compact disk media using a compact disk writer," Opt. Eng. 42 (9) pp 2514-2520 (2003)
[13] http://www.blu-raydisc.com/Assets/Downloadablefile/BD-ROMwhitepaper20070308-15270.pdf
[14] K. Osato, "Phase transition mastering for Blu-ray ROM disc," in *Optical Data Storage*, OSA Technical Digest Series (Optical Society of America, 2003), paper MD1.
[15] E. R. Meinders, R. Rastogi, M. van der Veer, P. Peeters, H. El Majdoubi, H. Bulle, A. Millet, and D. Bruls, “Phase-transition mastering of high-density optical media,” Jpn. J. Appl. Phys. **46(6B),** 3987–3992 (2007).
[16] C. Martinez, O. Lemonnier, F. L., A. Fargeix, F. Tissot and M. F. Armand, “Complementary computer generated holography for aesthetic watermarking,” Optics Express 5547, 27 February 2012, Vol. 20, No. 5
[17] C. Martinez, A. Fargeix, O. Lemonnier, B. Martin, M. Armand, and R. Templier, “Blu-Ray mastering process applied to the manufacturing of computer generated holograms,” in *Digital Holography and Three-Dimensional Imaging,* OSA Technical Digest (CD) (Optical Society of America, 2009), paper DWD5.
[18] Y. Usami, T. Watanabe, Y. Kanazawa, K. Taga, H. Kawai, and K. Ichikawa, “405 nm laser thermal lithography of 40 nm pattern using super resolution organic resist material,” Appl.

Phys. Express **2(12),** 126502 (2009).

[19] K. Kurihara, C. Rockstuhl, S. Petit, Y. Yamakawa, and J. Tominaga, "Plasmonic devices with controllable resonances--an avenue towards high-speed and mass fabrication of optical meta-materials," J. Microsc. **229(Pt** 3), 396–401 (2008).

[20] J. H. M. Neijzen, E. R. Meinders, and H. van Santen, "Liquid immersion deep-UV optical disc mastering for Blu-Ray disc read-only memory," Jpn. J. Appl. Phys. **43(7B),** 5047–5052 (2004).

[21] M. Gerspach, S. Noehte, and S. Borsmüller, "Holographic polymer data storage medium as a security label," Photonik. International. 30–32 (2009).

[22] K. Aratoani and A. Kouchiyama, "Reliable Low-cost Mass-production Technology for Blu-ray Discs," Special Interview published on Sony website, http://www.sony.net/SonyInfo/technology/interview/sp03.html#block4

[23] www.zebraimaging.com

[24] www.inphase-technologies.com

[25] www.lytro.com

[26] Illustrated by *Optica*: www.opticasoftware.com

[27] Personal communications with M. Yamamoto, M. Lindsey, and S. Kobayashi.

[28] Barnhart,D. H., Adrian, R. J. and Papen, G. C. "Phase-conjugate holographic system for high-resolution particle-image velocimetry", *Applied Optics*, October 20, vol 33, no 30, 7159-7170 (1994).

[29] Barnhart, D. H., Koek, W. D., Juchem, T. , Hampp, N. A., Coupland, J. M. , and Halliwell, N. A., "Bacteriorhodopsin (BR) as a high-resolution, high-capacity buffer for digital holographic measurements", Meas. Sci. Technol., vol 15, pp 639-646, (2004).

9th International Symposium on Display Holography (ISDH 2012) IOP Publishing
Journal of Physics: Conference Series **415** (2013) 012040 doi:10.1088/1742-6596/415/1/012040

Discrete calculation of the off-axis angular spectrum based light propagation

P Lobaz

Department of Computer Science and Engineering, University of West Bohemia, Univerzitni 8, 306 14 Plzen, Czech Republic

E-mail: `lobaz@kiv.zcu.cz`

Abstract. Light propagation in a free space is a common computational task in many computer generated holography algorithms. A solution based on the angular spectrum decomposition is used frequently. However, its correct off-axis numerical implementation is not straightforward. It is shown that for long distance propagation it is necessary to use digital low-pass filtering for transfer function calculation in order to restrict source area illumination to a finite area. It is also shown that for short distance propagation it is necessary to introduce frequency bands folding in transfer function calculation in order to simulate finite source area propagation. In both cases it is necessary to define properly interpolation filters that reconstruct continuous nature of the source area out of its sampled representation. It is also necessary to zero-pad properly source area sampling in order to avoid artifacts that stem from the periodic nature of the fast Fourier transform.

1. Introduction

To calculate coherent light propagation in a free space, scalar approximation is used frequently. A common task is such a calculation where complex amplitudes of light are given in the area *source* in a plane $z = 0$ and we look for complex amplitudes in the area *target* in a plane $z = z_0$, $z_0 > 0$. A common procedure leads to the Rayleigh-Sommerfeld integral of the first kind [1], or to its mathematically equivalent form, the angular spectrum decomposition [2]. Approximations of these formulas are used frequently, namely Fresnel and Fraunhofer approximations. However, these approximations are used in paraxial regime while in computer generated holography an off-axis solution is often needed (e. g. [3]); therefore we will not discuss them.

The problem has to be solved numerically in computer generated holography. This leads to discretization of signals. Discretization of the Rayleigh-Sommerfeld solution is not straightforward [4, 5] and discretization of the angular spectrum decomposition is tricky [6, 7]. Onural [6] describes the discretization process in general and shows that it leads to formation of signal copies in spatial domain that can be filtered out. However, he does not discuss implementation issues. Matsushima [7] deals with the implementation and solves a troublesome aliasing problem by local frequency estimation; however, he does not analyse the effect of hard frequency clipping.

This article focuses on reference calculation of light propagation between parallel planes using angular spectrum decomposition. We will point out what makes the discretization difficult and how to overcome the problems so that the method provides the same results as the Rayleigh-Sommerfeld method. We will deal with both large propagation distances discussed by Matsushima [7] and small propagation distances that were not discussed in literature yet. We will explain the meaning of "large" and "small" distance later.

Structure of the article is as follows. At first we will precisely show how to discretize the propagation calculation based on convolution (i. e. based on Rayleigh-Sommerfeld integral). We will show that it is necessary to introduce spatial limitation of a convolution kernel for successful discretization. This leads to frequency limitation of the transfer function used in the angular spectrum decomposition method. We will show that the frequency limited transfer function can be calculated using digital signal processing methods. At last we will deal with the case where this frequency limitation is actually useless due to small propagation distance. We will show that, in this case, we have to deal with exact nature of sampling and reconstruction of signals involved. We will show that the choice of reconstruction leads to introduction of artificial alias.

2. How to read the article

The mathematical explanation presented may be unpleasant to follow. Readers are therefore encouraged to go through the presentation packed as a "multimedia" attachment to this article. It shows pictures containing various problems that appear when calculating the propagation numerically. I have decided to attach these pictures as a separate media for two reasons. The first one is: the pictures show mainly problems with aliasing. It is therefore needed to control the display of these images precisely which is not possible in a PDF reader or in printed media. The second one is: separate media gives the opportunity to show much more images than any printed media allows. I should note that the presentation does not contain any information not covered by the article.

3. Convolution discretization

Let us assume that we know complex amplitudes $u(x, y, 0)$ of monochromatic coherent light of wavelength λ in the area *source* ($x_{s\min} \leq x < x_{s\max}$, $y_{s\min} \leq y < y_{s\max}$, $z = 0$) and we want to calculate complex amplitudes $u(x, y, z_0)$ in the area *target* ($x_{t\min} \leq x < x_{t\max}$, $y_{t\min} \leq y < y_{t\max}$, $z = z_0 > 0$). The solution is given by the Rayleigh-Sommerfeld integral of the first kind:

$$u(x, y, z_0) = \frac{-1}{2\pi} \iint_{-\infty}^{\infty} u(\xi, \eta, 0) \frac{\partial}{\partial z} \frac{\exp(\mathrm{j}kr)}{r} \, \mathrm{d}\xi \, \mathrm{d}\eta \tag{1}$$

where $r = \sqrt{(x-\xi)^2 + (y-\eta)^2 + z_0^2}$, $k = 2\pi/\lambda$, $\mathrm{j}^2 = -1$ and $u(\xi, \eta, 0) = 0$ for $[\xi, \eta, 0] \notin$ *source*. The second term of the multiplication inside the integral depends on $(x - \xi)$ and $(y - \eta)$ only, which means that the integral can be rewritten as the convolution with the Rayleigh-Sommerfeld kernel $h(x, y, z)$:

$$u(x, y, z_0) = u(x, y, 0) \otimes h(x, y, z_0) = \iint_{-\infty}^{\infty} u(\xi, \eta, 0) h(x - \xi, y - \eta, z_0) \, \mathrm{d}\xi \, \mathrm{d}\eta \tag{2}$$

$$h(x, y, z) = \frac{-1}{2\pi} \frac{\partial}{\partial z} \frac{\exp(\mathrm{j}kr)}{r} = \frac{-z}{2\pi} \left(\mathrm{j}k - \frac{1}{r} \right) \frac{\exp(\mathrm{j}kr)}{r^2}$$

$$r = \sqrt{x^2 + y^2 + z^2}$$

The kernel $h(x, y, z)$ can be interpreted easily: the value $h(x_d, y_d, z_d)$ describes the change of the complex amplitude of light travelling from the point $[x_s, y_s, z_s]$ to the point $[x_s + x_d, y_s + y_d, z_s + z_d]$.

To calculate $u(x, y, z_0)$ in the area *target* using (1) we have to know the kernel $h(x, y, z_0)$ for $x_{t\min} - x_{s\max} \le x < x_{t\max} - x_{s\min}$, similarly for y. The value $h(x, y, z_0)$ is not important for other x, y because in this case $u(x, y, 0) = 0$. We will use this fact in a while.

We discretize (1) easily by changing integrals to sums and differentials to differences. The sums will have finite extent of the indices thanks to (in fact) finite domain of the integration. Therefore their calculation will be easy, although the computational complexity will be high.

To reduce computational complexity, let us rewrite the equation (2):

$$u(x, y, z_0) = u(x, y, 0) \otimes h(x, y, z_0) = \mathcal{F}^{-1}\Big\{\mathcal{F}\{u(x, y, 0)\} \cdot \mathcal{F}\{h(x, y, z_0)\}\Big\} \tag{3}$$

where $\mathcal{F}$ and $\mathcal{F}^{-1}$ are 2-D Fourier and inverse Fourier transform respectively. We will try to use the discrete Fourier transform (DFT) implemented as the fast Fourier transform (FFT) after discretization.

The discrete Fourier transform can be defined if the original continuous functions are periodic before discretization (sampling); then we take into account one period of functions $u_p(x, y, z_0)$, $u_p(x, y, 0)$ and $h_p(x, y, z_0)$ derived from functions $u(x, y, z_0)$, $u(x, y, 0)$ and $h(x, y, z_0)$. We can define the DFT another way if the functions to be transformed are spatially limited; in this case we can assume just one period of the functions $u_p(x, y, z_0)$, $u_p(x, y, 0)$ a $h_p(x, y, z_0)$. Both ways lead to the same results. This means that the results of the DFT can be interpreted in both ways. We will choose the way that will be more suitable in a particular situation.

Let us briefly give a hint what is the meaning of the functions $u_p(x, y, 0)$, $u_p(x, y, z_0)$ and $h_p(x, y, z_0)$ before we define them precisely. Their period in x direction is equal to the sum of widths of the *source* and the *target* (similarly in y). One period of the function $u_p(x, y, 0)$ is composed of values of $u(x, y, 0)$ in the way that one corner of the *source* is translated to the origin. The meaning of the function $u_p(x, y, z_0)$ is similar. The $h_p(x, y, z_0)$ is restricted and shifted version of the function $h(x, y, z_0)$. The meaning of the value $h_p(0, 0, z_0)$ is the change of the complex amplitude of light travelling from the point $[x_{s\min}, y_{s\min}, 0]$ to the point $[x_{t\min}, y_{t\min}, z_0]$, see fig. 1.

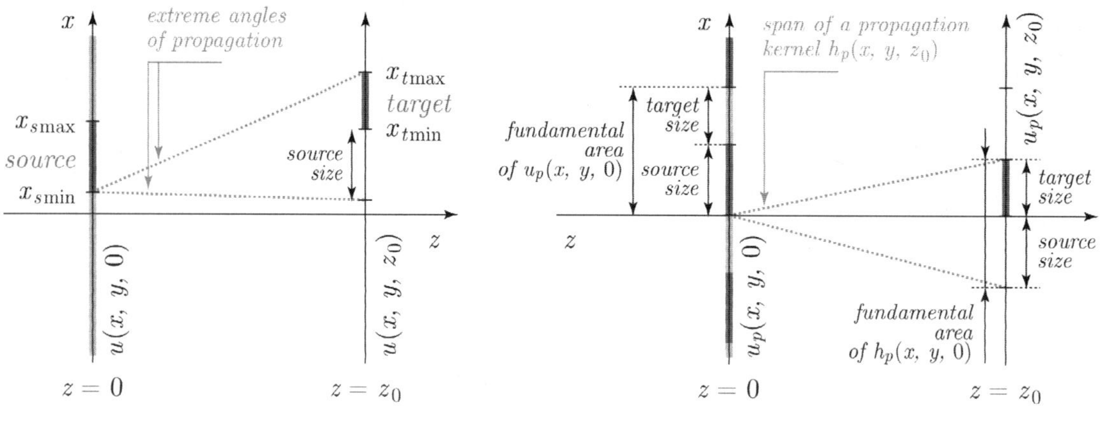

Figure 1. Geometry of the problem in xz slice. *Left image:* original setup. Red line in the source area denotes the complex amplitudes to be propagated, blue line in the same plane defines zero value. It should be clear that we need to know the values of a propagation kernel contained in a green wedge only. *Right image:* setup prepared for discretization. *Source* and *target* are shifted to the z axis, blue and red lines in the *source* plane show both zero padding and periodicity. The green wedge shows one period (fundamental area) of the function $h_p(x, y, z_0)$.

Let us define the functions precisely. The function $h_p(x, y, z_0)$ is defined in "the fundamental area" $x_{s\min} - x_{s\max} \leq x < x_{t\max} - x_{t\min}$ (similarly for y) as:

$$h_p(x, y, z_0) = h(x + x_{t\min} - x_{s\min}, y + y_{t\min} - y_{s\min}, z_0) \tag{4}$$

Definition outside the fundamental area depends on situation: it can be either zero or the function can be made periodic. However, as we will see, the values outside the fundamental area are not important, which means that the periodic nature of the function is not harmful.

Let us define the function $u_p(x, y, 0)$ in the fundamental area $0 \leq x < (x_{t\max} - x_{t\min}) + (x_{s\max} - x_{s\min})$ (similarly for y; notice that the size of the area is the same as for $h_p(x, y, z_0)$) as:

$$u_p(x, y, 0) = \begin{cases} u(x + x_{s\min}, y + y_{s\min}, 0) & \text{for } \; 0 \leq x < x_{s\max} - x_{s\min}, \\ & \qquad 0 \leq y < y_{s\max} - y_{s\min} \\ 0 & \text{elsewhere in the fundamental area} \end{cases}$$

and again let us make it periodic.

Let us calculate $u_p(x, y, z_0) = u_p(x, y, 0) \otimes h_p(x, y, z_0)$. If we consider the function $u_p(x, y, 0)$ to be periodic and the function $h_p(x, y, z_0)$ to be zero outside the fundamental area, we can easily prove that $u_p(x, y, z_0)$ will contain correct result of the propagation from the *source* to the *target* in the area $0 \leq x < x_{t\max} - x_{t\min}$ (similarly for y); the values of $h_p(x, y, z_0)$ are meaningless for other x, y as they are damaged by periodicity of the function $u_p(x, y, 0)$. The proof easily follows from the geometry of the problem.

If we consider the function $u_p(x, y, 0)$ to be spatially limited and the function $h_p(x, y, z_0)$ to be periodic, we get the same result. If we consider both functions to be periodic, the meaning of the result remains the same; however, we are in fact in the world of discrete Fourier transform.

The form of the discrete calculation follows from the aforementioned ideas. The area *source* is discretized by $M_x \times M_y$ samples, the area *target* by $N_x \times N_y$ samples. For simplicity let us assume such parameters so that the sampling period Δ is the same in both directions x and y and in both *source* and *target* areas. Its value is then e. g. $\Delta = (x_{s\max} - x_{s\min})/M_x$.

The area *source* is discretized by samples $u_0[m, n] = u_p(m\Delta, n\Delta, 0)$, the convolution kernel is discretized by samples $h[m, n] = h_p(m\Delta, n\Delta, z_0)$ for $m \in \{0, 1, \ldots, M_x + N_x - 2\}$ (similarly n, see [4]). We consider functions $u_p(x, y, 0)$ and $h_p(x, y, z_0)$ to be periodic. Then we can calculate

$$u_{z_0}[\,] = \mathbf{IDFT}\Big\{\mathbf{DFT}\{u_0[\,]\} \cdot \mathbf{DFT}\{h[\,]\}\Big\} \tag{5}$$

to get the array $u_{z_0}[\,]$ that contains complex amplitudes of the area *target* in the first $N_x \times N_y$ elements. In the equation (5), **DFT** and **IDFT** stands for forward and backward 2-D discrete Fourier transform, respectively, and $\cdot$ stands for elementwise product (Hadamard product).

We can summarize the results as follows. For the discrete calculation of light propagation, we have to use a convolution kernel spatially limited to a certain area. The *source* has to be zero-padded to span the same area. The result of their convolution contains correct values in the area of the size proportional to the size of the *target*.

4. Angular spectrum discretization for large propagation distances

The equation (1) can be written in the equivalent form called the angular spectrum decomposition [2]:

$$u(x, y, z_0) = \mathcal{F}^{-1}\{U(f_x, f_y, 0) \cdot H(f_x, f_y, z_0)\}$$

where

$$U(f_x, f_y, 0) = \mathcal{F}\{u(x, y, 0)\}$$
$$H(f_x, f_y, z_0) = \exp\left(-\mathrm{j}\, 2\pi z_0 \sqrt{\lambda^{-2} - f_x^2 - f_y^2}\right)$$

and f_x, f_y are Fourier domain variables.

It follows from (3) that

$$u(x, y, z_0) = u(x, y, 0) \otimes h(x, y, z_0) = \mathcal{F}^{-1}\left\{U(f_x, f_y, 0) \cdot \mathcal{F}\{h(x, y, z_0)\}\right\},$$

that is $H(f_x, f_y, z_0) = \mathcal{F}\{h(x, y, z_0)\}$. It seems that the propagation calculation should be advantageous using the angular spectrum decomposition compared to convolution with the Rayleigh-Sommerfeld kernel because we have one Fourier transform less – the transform of the kernel is known in the analytic form.

In the following paragraphs, we will talk about various limitations of functions in both spatial and Fourier (frequency) domain. A function $f(x, y)$ defined in spatial domain is spatially limited if it is zero outside a bounded area in the plane (x, y); it is frequency limited if $\mathcal{F}\{f(x, y)\}$ is zero outside a bounded area in the plane (f_x, f_y). Similarly, a function $F(f_x, f_y)$ defined in frequency domain is spatially limited if it is zero outside a bounded area in the plane (f_x, f_y); it is frequency limited if $\mathcal{F}^{-1}\{F(f_x, f_y)\}$ is zero outside a bounded area in the plane (x, y).

We know from the previous section that we have to introduce certain functions to discretize the calculation. We have to define the periodic function $u_p(x, y, 0)$ based on the function $u(x, y, 0)$ and to calculate its (discrete) Fourier transform; we will follow this procedure exactly. We also need to spatially restrict the function $h(x, y, z_0)$, and to make its periodic form alternatively. However, we do not know the Fourier transform of the function $h_p(x, y, z_0)$ in the analytic form.

Fortunately we can use digital signal processing tools. If the signal is spatially limited in one domain (in our case spatial domain), it is frequency limited in the other domain (in our case frequency domain). We can calculate frequency limited signal using low-pass filter $l(f_x, f_y)$. Let us assume the function $h_p(x, y, z_0)$ to be spatially limited (i. e. not periodic). For propagation calculation, we have to use the transfer function $H_p(f_x, f_y, z_0) = H(f_x, f_y, z_0) \otimes l(f_x, f_y)$.

It follows from properties of the Fourier transform that the function $H_p(f_x, f_y, z_0)$ has to be spatially unlimited. This does not matter even in numerical calculation. Since the function $u_p(x, y, 0)$ is periodic, then $U_p(f_x, f_y, 0)$ is spatially limited; and we need to calculate the product $U_p(f_x, f_y, 0)H_p(f_x, f_y, 0)$. It also follows that the value of the function $H_p(f_x, f_y, z_0)$ can be arbitrary outside the important area, and therefore we can use its periodic form to introduce discrete calculation correctly.

We cannot limit the frequency content of the function $H(f_x, f_y, z_0)$ sharply using digital low-pass filtering; the frequency limitation is approximate. In the spatial domain it means that the transition between zero and non-zero part is gradual instead of sharp. This does not matter either. All we need to do is to enlarge the fundamental area of the functions $u_p()$ and $h_p()$, i. e. zero-padding of the array $u_p[]$ will be larger than defined in section 3, so that the gradual transition will not affect the *target* area.

We will face a problem in a practical implementation. The function $H(f_x, f_y, z_0)$ is frequency unlimited – the local frequency [1, 7] in the point $[f_x, f_y]$ grows without bound as this point approaches a circle of a radius λ^{-1} centered in the origin. Moreover, if the propagation distance z_0 is large, then the local frequency will be large as well everywhere except in the origin. Therefore it is impossible to sample the function $H(f_x, f_y, z_0)$ correctly and then the low-pass filtering will not work properly.

However, we can use the same procedure as described by Matsushima [7] who realizes the aliasing problem. He evaluates local frequency when sampling the function $H(f_x, f_y, z_0)$, and if the local frequency is bigger than a half of the sampling frequency, he sets the function $H(f_x, f_y, z_0)$ to be zero. He solves the aliasing problem this way, on the other hand he introduces a spatial limitation of the function $H(f_x, f_y, z_0)$. It follows that the propagation kernel $\mathcal{F}^{-1}\{H_{\text{Matsushima}}(f_x, f_y, z_0)\}$ is then spatially unlimited which is not correct. It should be however emphasised that even though it gives remarkably good results.

The solution is easy. We can sample the function $H(f_x, f_y, z_0)$ using higher sampling frequency than desired and use Matsushima's procedure to avoid alias. Then we can filter it using $l(f_x, f_y)$ and downsample the result to get a correct sampling frequency. As the cutoff frequency of the filter $l(f_x, f_y)$ is derived from the sizes of the arrays used in the calculation, it is guaranteed that no aliasing appears when downsampling.

It follows from practical experiments that it is sufficient to sample the function $H(f_x, f_y, z_0)$ using sampling frequency twice as high as desired, and to use sinc low-pass filter with Hamming window as $l(f_x, f_y)$. The size of the filter should be chosen carefully – a long filter filters high frequencies well, but takes long to evaluate. Evaluation with a short filter is faster but slow attenuation of high frequencies has to be compensated with bigger zero-padding of the array $u_0[\,]$.

5. Discretization for short propagation distances

To calculate the propagation numerically (either using convolution approach or angular spectrum approach), we have to calculate (discrete) Fourier transform of the function $u_p(x, y, 0)$. Its result is the function $U_p(f_x, f_y, 0)$ limited to the area $A = (-1/(2\Delta), 1/(2\Delta)) \times (-1/(2\Delta), 1/(2\Delta))$, or its periodic form. We also need to calculate the function $H_p(f_x, f_y, z_0)$ in the same area, and then to calculate $\mathcal{F}^{-1}\{U_p(f_x, f_y, 0)H_p(f_x, f_y, z_0)\}$.

We can naturally ask a question: what happens if the propagation distance z_0 is so small that the low-pass filtering of the function $H(f_x, f_y, z_0)$ will not have any significant effect inside the area A? It is not easy to find the answer. Let us start with one more look into the convolution based approach described by the equation (2).

The convolution kernel $h(x, y, z_0)$ is a spatially unlimited function that is in fact frequency limited if we ignore evanescent waves. We can easily show that its local frequency grows as the point $[x, y]$ moves away from the origin. The smaler is z_0 the faster is the growth. It can therefore easily happen that the function $h_p(x, y, z_0)$ cannot be properly sampled using sampling period Δ for small propagation distances.

Physical meaning of wrong sampling is easy. Discretization is based on change of integrals to sums in the equation (1). It means that we change a continuous light field in the *source* to a number of point light sources. This replacement cannot be observed from a big distance or in on-axis case but makes a big difference close enough or in off-axis case.

We have to assume (especially in small propagation distances) that one sample of the function $u(x, y, 0)$ represents behaviour of the light field in a small neighbouring area. Let us denote the result of the sampling of the function $u(x, y, 0)$ by the function $u_s(x, y, 0) = u(x, y, 0)\,\mathrm{comb}(x/\Delta)\,\mathrm{comb}(y/\Delta)$, where $\mathrm{comb}(x)$ is the sampling function with period of samples 1 (see [1]). Then we can describe the reconstruction of the continuos form using convolution with a reconstruction kernel $r(x, y)$:

$$u(x, y, 0) \approx u_r(x, y, 0) = u_s(x, y, 0) \otimes r(x, y)$$

where $u_r(x, y, 0)$ is a continuos function reconstructed from the discrete samples. The function $u_r(x, y, 0)$ is more or less similar to the function $u(x, y, 0)$ depending on the shape of a reconstruction kernel $r(x, y)$ and size of the sampling period Δ.

We can therefore express the propagation as

$$u(x, y, z_0) \approx u_s(x, y, 0) \otimes \Big(r(x, y) \otimes h(x, y, z_0)\Big)$$

It is possible to put big parenthesses in the equation thanks to associativity of the convolution. It follows that we should not use the function $h(x, y, z_0)$ for discrete propagation calculation, but we shoud use its filtered version $r(x, y) \otimes h(x, y, z_0)$ instead.

It is most common to demand for one sample of the function $u(x, y, 0)$ to influence its close neighbourhood only. For example, the kernel $r(x, y) = \text{rect}(x/\Delta)\,\text{rect}(y/\Delta)$, where $\text{rect}(x)$ is a rectangular pulse of unity width and height centered in the origin, implies the function $u(x, y, 0)$ to be approximated with a piecewise constant function. We can construct other kernels as well that provide piecewise bilinear approximation, piecewise bicubic approximation and so on. In either case, the kernel $r(x, y)$ acts as a low-pass filter. If we use the filtered function $h(x, y, z_0) \otimes r(x, y)$ for construction of the function $h_p(x, y, z_0)$ in the equation (4), we get the result of the propagation calculation as precise as the function $u_r(x, y, 0)$ resembles the function $u(x, y, 0)$. Practical implementation of the procedure is described in [5].

We can repeat the analysis for the angular spectrum decomposition as well. We should not use the function $H(f_x, f_y, z_0)$ for construction of the function $H_p(f_x, f_y, z_0)$; we should use the Fourier transform of the filtered propagation kernel, the function $\mathcal{F}\{h(x, y, z_0\} \otimes r(x, y)\} = H(f_x, f_y, z_0)\,\mathcal{F}\{r(x, y)\}$. If we choose convenient kernel $r(x, y)$, we will know analytic form of its Fourier transform and calculation of the product will be simple. This step limits high frequencies that were caused by the discretization process.

It remains to solve the last, but important detail. In the beginning of the section we have stated that the calculations in frequency domain are done inside the area A. It is however possible that the support of the function $H_p(f_x, f_y, z_0)$ will not fit into the area A even if it was filtered with both filters $l(f_x, f_y)$ and $r(x, y)$. If we reject the values of the function $H_p(f_x, f_y, z_0)$ outside of A despite that fact, it means that the final function $u(x, y, z_0)$ was filtered by a third, still unjustified low-pass filter.

We can explain this final low-pass filter. It would be appropriate if the kernel $r(x, y)$ represents a perfect sinc low-pass filter limiting the frequency content of the function $u(x, y, z_0)$ to a range described by the area A. Then the kernel $r(x, y)$ has to be be spatially unlimited. If we do not care, the analysis is finished.

If we would rather keep the kernel $r(x, y)$ spatially limited (which is physically more natural), we have two choices to choose from. The first one is simple – we can use such a sampling period Δ/s, $s \in \mathbf{N}$ for the calculation so that the area A covers the support of $H_p(f_x, f_y, z_0)$ now. This leads to increase of time and memory demands of the calculation, of course. Moreover, if we demand sampling of the *target* to be Δ, we have to downsample the result; we have to admit that a lot of values were calculated needlessly.

The second one is a bit strange at first sight. Signal downsampling in spatial domain can be described easily in frequency domain – the frequencies f and $f + n/\Delta$ merge due to downsampling, where f is frequency f_x or f_y from the range $(-1/(2\Delta), 1/(2\Delta))$ and $n \in \mathbf{Z}$. This effect is called aliasing. We want to avoid it usually; however, if we are decided to sample the *target* with an insufficient sampling period, we have to accept aliasing. It is worth to note that aliasing needs not be harmful. If we are interested in intensities $|u(x, y, z_0)|^2$ only, aliasing in the real or imaginary part of the function $u(x, y, z_0)$ may be harmless. For example, if $u(x, y, z_0) = \cos x + \mathrm{j} \sin x$, then the intensity is always 1 regardless sampling period used, while intensity of "correctly sampled" (i. e. low-pass filtered) version can be either 1 or 0 which is not correct.

The procedure in the second case is obvious: we will perform the downsampling process in frequency domain. Let us calculate the function $H_p(f_x, f_y, z_0)$ in the area $(-s/(2\Delta), s/(2\Delta)) \times (-s/(2\Delta), s/(2\Delta))$, assume it is zero outside this area, and calculate

$$H_a(f_x, f_y, z_0) = \sum_{n_x, n_y} H_p(f_x + \frac{n_x}{\Delta}, f_y + \frac{n_y}{\Delta}, z_0)$$

for $[f_x, f_y] \in A$ and all suitable n_x, n_y. To calculate the propagation, we have to use the function $H_a(f_x, f_y, z_0)$.

6. Conclusion

The analysis shows how to numerically calculate the propagation of light using the angular spectrum decomposition method. Unlike the procedures described in the literature, it defines low-pass filters $l(f_x, f_y)$ and $r(x, y)$ that are needed to introduce for correct discretization. Then we can choose either more precise, slower calculation or faster, less precise by choosing their parameters. The analysis also shows that it is worth introducing aliasing of the transfer functions in certain situations. The results are shown in figures 2 and 3. The images may display wrong due to resampling; the reader is encouraged to look at the images in the multimedia attachment.

References

[1] Goodman J W 2004 *Introduction to Fourier Optics* 3rd ed (Roberts & Company Publishers) ISBN 0974707724
[2] Lalor E 1968 *J. Opt. Soc. Am.* **58** 1235–1237
[3] Hanák I, Janda M and Skala V 2010 *Visual Computer* **26**(2) 83–96
[4] Lobaz P 2011 *Opt. Express* **19** 32–39
[5] Lobaz P 2012 Calculation of a coherent light propagation in a free space using filtered convolution Tech. rep. University of West Bohemia URL `http://www.kiv.zcu.cz/en/research/publications/`
[6] Onural L 2007 *J. Opt. Soc. Am.* A **24** 359–367
[7] Matsushima K 2010 *Opt. Express* **18** 18453–18463

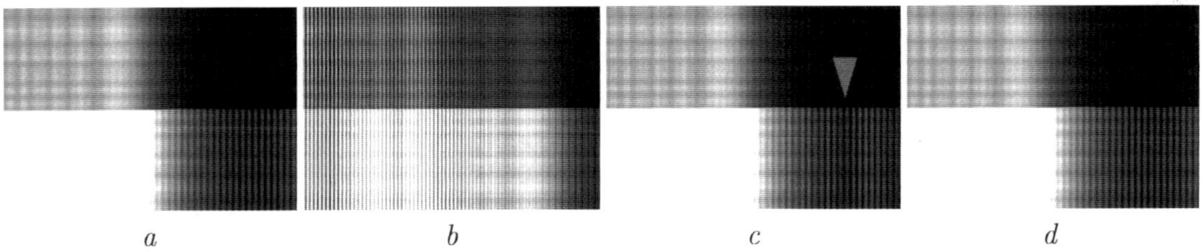

Figure 2. Light ($\lambda = 650$ nm) diffracted by a grating (3×3 mm^2) composed of vertical slits (slits distance 40 μm). Off-axis propagation to a distance of 300 mm, size of each image 3×3 mm^2. Lower half of each image is overexposed to show the details. *a)* Reference Rayleigh-Sommerfeld calculation. *b)* Angular spectrum based calculation without any modification. Notice that aliasing errors destroy the image completely. *c)* Angular spectrum based calculation with Matsushima's kernel filtering. Notice different brightness of fine stripes compared to the reference image. *d)* Angular spectrum based calculation with $l(f_x, f_y)$ sinc kernel of length 50.

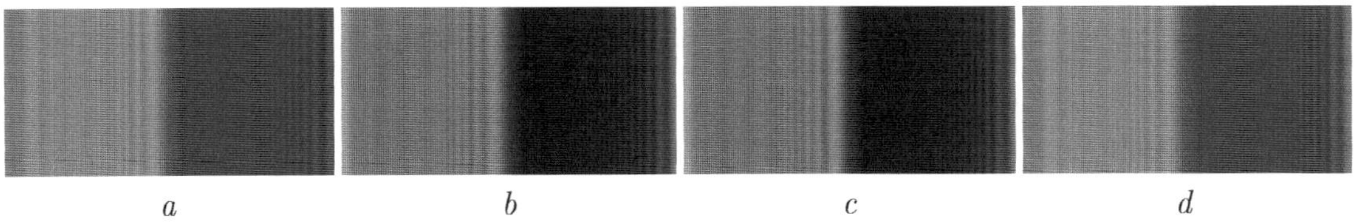

Figure 3. Light ($\lambda = 650$ nm) diffracted by a grating (3×3 mm^2) composed of vertical slits (slits distance 20 μm). Off-axis propagation to a distance of 50 mm, size of each image 3×1.5 mm^2. *a)* Reference Rayleigh-Sommerfeld calculation. *b)* Angular spectrum based calculation without any modification. Notice that the right half is much darker than in the reference image. *c)* Angular spectrum based calculation with Matsushima's kernel filtering. Notice that the image is the same as without any modification because aliasing did not occur. *d)* Angular spectrum based calculation with introduced aliasing. Notice that the image is almost the same as the reference one.

Haptic Holography/ Touching the Ethereal.

Michael Page,
P.I. PHASE Lab
OCAD University
Toronto, Ontario Canada.
Email: mpage@faculty.ocadu.ca

Abstract. Haptic Holography, was perhaps, first proposed by workers at MIT in the 90s. The Media Lab, headed up by Dr. Stephen Benton, with published papers by Wendy Plesiak and Ravi Pappuh. -1

Recent developments in both the technology of digital holography and haptics have made it practical to conduct further investigations. Haptic holography is auto-stereoscopic and provides co-axial viewing for the user.

Haptic holography may find application in medical & surgical training and as a new form of synthetic reality for artists and designers.

At OCAD's PHASE Lab (Prototypes for Holographic Art and Science Explorations) workers are exploring hybrid forms of augmented reality, that combine haptics, interactivity and auto-stereoscopic imagery.

Conventional Haptic environments, while presenting a 3D physics environment, typically provide a 2D visual work/play space. Orienteering in such an environment creates an uncertain spatial relationship for the user.

Our group creates 3d models from which we create holographic constructs. The same model is used to create the physics environment. The two models are super-imposed. The result: Holograms you can touch.

1. Introduction

As artists and designers our research is focused on human perception and how we might create multi-layered synthetic reality experiences and to successfully blur the lines between illusions to create a hyper state of synthetic reality that is also interactive.

Holography has long been the supreme 3D visioning medium. The public critique of the medium is that it is too static, not interactive. By combining 3D graphics, holography, haptics and audio, we hope to create a more immersive environment for a single user in what we have dubbed the haptic holo workstation.

Haptic derived from the Greek word (haptikos), means pertaining to the sense of touch. Our sense of touch is interpreted through a series of sensors in the brain that can be activated by movement, pressure, chemical and/or temperature changes and vibration. The parietal lob integrates sensory information from various parts of the body (fiqure 1). -2. Touch is a big part of sensory experience.

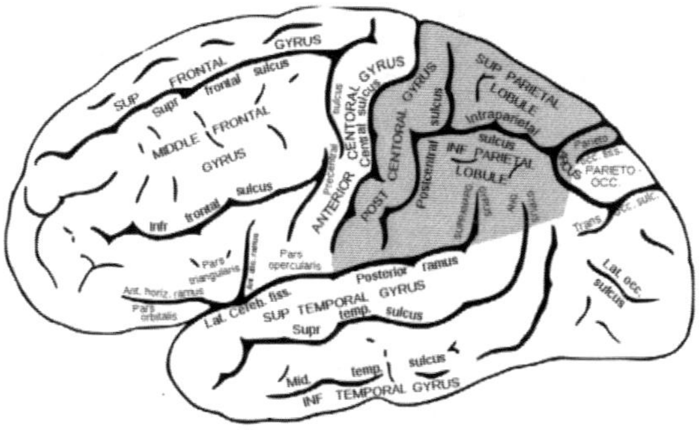

Figure 1 parietal lob integrates sensory information from various parts of the body

The application of haptics to computer technology typically invovles a joystick or stylus equipted with servo motors and encoders that "pay-back to the user and provide tactile sense that reponds to a physics-based environment wherein vitual objects are defined in three dimensional space and are given properties such as:

- deformable,
- rigid but movable,
- rigid, not movable, etc.

Haptic technology does for the sense of touch what computer graphics does for vision. -3

Our approach to creating prototype simulations is to create a 3D model which is then used to print a digital hologram. The same model is imported into the haptics software to create a 3D physical model of "touchable objects. We then super-impose the two models, add video and sound to create an interactive 3D experience. The haptic/holo workstation has the advantage of requiring a single user interface. Holograms, of course, display full parallax, or in the case of most digital holograms, horizontal parallax. In order to successfully marry 3D animation with holographic projections, head tracking can be used.

The station functions in transmission and reflection mode. In transmission mode the hologram is lit by a low powered laser diode. In reflection mode an array of bright LEDs are used to animate the hologram. This changes the holographic scene for the single viewer to, for example, zoom in on a scene or change scenes completely.

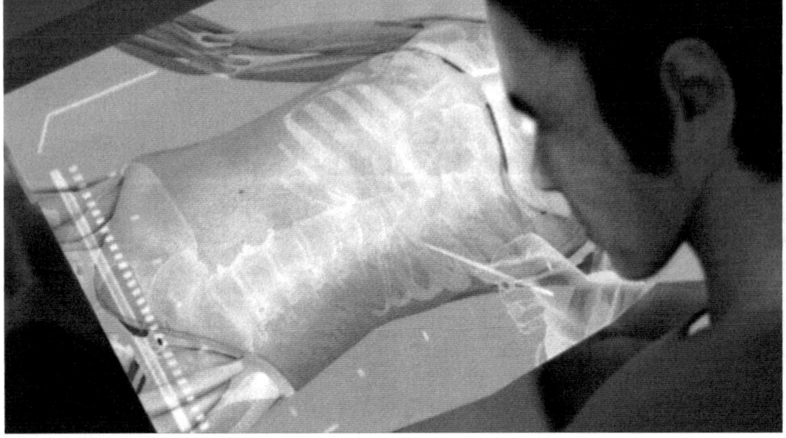

Figure 2. & 3 Visualization of haptic holo workstation illustrating the co-visual aspect of the workstation

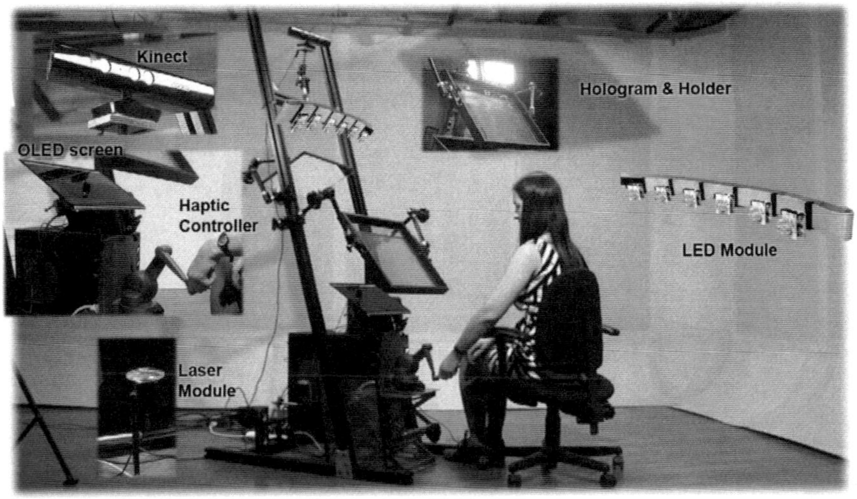

Figure 4. The haptic holo workstation and modules.

The Haptic Holo Workstation has a number of modules that are responsible for displaying imagery and creating interactivity. (Figure 4)

1. **Hologram and holder**. This device holds the hologram, transmission or reflection in a precise position, aligned with the OLED display.

2. **OLED display**. This display produces very bright and chroma saturated images* and is viewable at any angle without degradation of the image.

3. **The laser module.** This low power inexpensive laser illuminates the hologram in transmission mode.

4. **The Kinect module.** This device tracks the viewer's head position and instantly modifies the video parallax to conform to that of the hologram.

5. **The Super-bright LED module.** This device, in reflection mode, changes the reconstruction angle of the hologram, which animates the scene. The module is controlled by the computer and will change the holographic scene to accommodate "zooming in" on an image, for example.

6. **The computer** controls all of the input data, real-time processing and output data.

*This means that the video image will be as saturated as the holographic scene.

Future modifications include:

- The use of direct-write reflection holograms.
- The implementation of next generation OLED technology. These brilliant displays, only a few millimetres thick, are transparent where a normal flat screen is black. This will allow the hologram and video display to be sandwiched as one, further compressing the size of the display. (figure 5)
- LED technology continues to evolve into semi-coherent light sources that will sharpen the image in our holograms
- Haptic technology continues to advance into new areas of sensory perception.

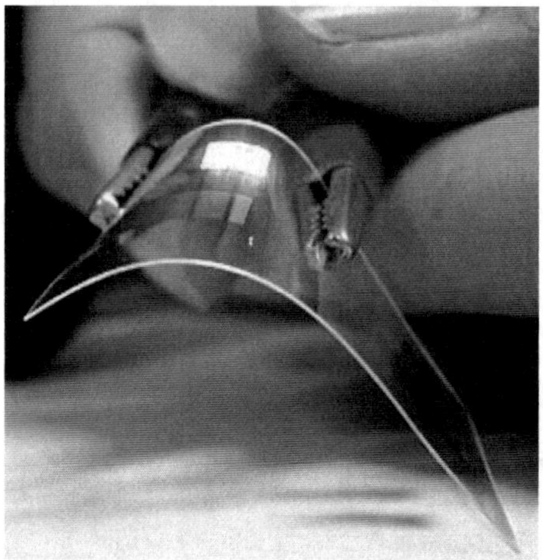

Figure 5 Flexible, thin, transparent OLED screen

Examples of other haptic research

MIT has been a world center for research in haptics, including haptics holo video. (Real-time manipulation of holographic video!) **-5** The MIT touch lab has been instrumental in bringing about research in hand therapy, intelligent prosthesis design, and the development of autonomous robots that need to perform human-like functions in unstructured environments since 1990. -4

New areas of research

Mutual touch over the net. (see figure 6)-6

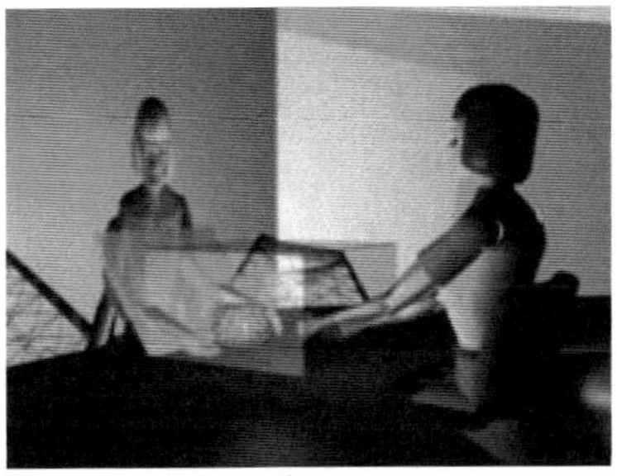

Figure 6 Mutual touch

Disney's Tactile Brush which uses localized piezoelectric devices to create patterns on the back or arms of users that "touch" the user. **-7** Disney's "Touché" haptic device can make your body into an MP3 player.**-8**

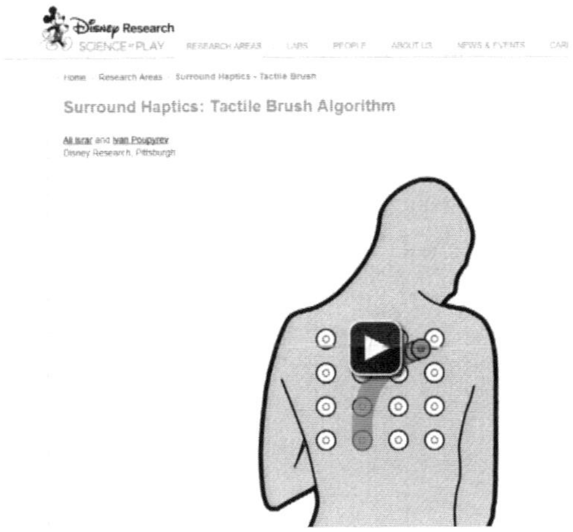

Figure 7. Disney's tactile brush. My computer is tickling me?

Journal of Physics: Conference Series **415** (2013) 012041 doi:10.1088/1742-6596/415/1/012041

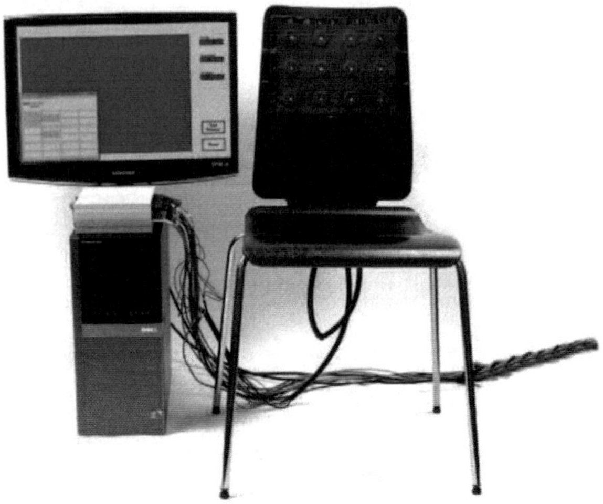

Figure 8 Disney's gaming chair

Arguments for haptic holography over conventional haptic interfaces.

The haptic interface we have constructed can be configured to be co-visual. This means that, for example, the user's hand holding the haptic controller would be in the same position as the virtual object, a scalpel or needle.

Unlike other 3D haptic displays that require head gear, goggles or glasses, holograms are auto-stereoscopic.

Haptic interfaces allow the user/medical trainee the option of "zooming in" or making the initial training less critical, then altering the resolution to a more critical state.

Text and other visual or audible communication to the user is also possible (figure 9).

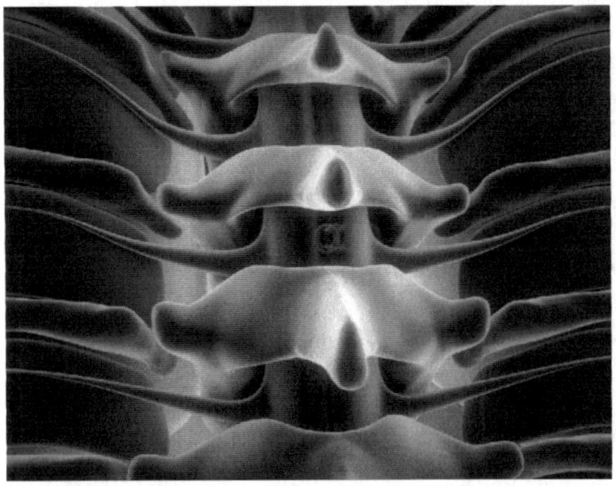

Figure 9 Image from holographic simulation showing the point of entry of the lumbar puncture as well as a graticule, guiding the angle of entry.

References

1. http://web.media.mit.edu/~wjp/pubs/thesisAbstract.pdf
2. http://www.brainline.org/multimedia/interactive_brain/the_human_brain.html?gclid=CLexj5374bACFQK5Kgod-R3W3Q
3. Gabriel Robles-De-La-Torre. "International Society for Haptics: Haptic technology, an animated explanation". Isfh.org. Retrieved 2010-02-26..
4. http://touchlab.mit.edu/
5. http://alumni.media.mit.edu/~pappu/pdfs/PappuPlesniak-SPIE-1998.pdf
6. http://imsc.usc.edu/haptics/
7. http://www.disneyresearch.com/research/projects/hci_surround_haptics_drp.htm
8. http://www.wired.com/beyond_the_beyond/2012/05/augmented-reality-disney-touche-haptics/

Challenges in using GPUs for the real-time reconstruction of digital hologram images

P R Hobson, J J Nebrensky and I D Reid

Centre for Sensors and Instrumentation, School of Engineering and Design, Brunel University, Uxbridge UB8 3PH, UK

E-mail: peter.hobson@brunel.ac.uk

Abstract. In-line holography has recently made the transition from silver-halide based recording media, with laser reconstruction, to recording with large-area pixel detectors and computer-based reconstruction. This form of holographic imaging is an established technique for the study of fine particulates, such as cloud or fuel droplets, marine plankton and alluvial sediments, and enables a true 3D object field to be recorded at high resolution over a considerable depth.

The move to digital holography promises rapid, if not instantaneous, feedback as it avoids the need for the time-consuming chemical development of plates or film film and a dedicated replay system, but with the growing use of video-rate holographic recording, and the desire to reconstruct fully every frame, the computational challenge becomes considerable. To replay a digital hologram a 2D FFT must be calculated for every depth slice desired in the replayed image volume. A typical hologram of ~100 μm particles over a depth of a few hundred millimetres will require $O(10^3)$ 2D FFT operations to be performed on a hologram of typically a few million pixels.

In this paper we discuss the technical challenges in converting our existing reconstruction code to make efficient use of NVIDIA CUDA-based GPU cards and show how near real-time video slice reconstruction can be obtained with holograms as large as 4096 by 4096 pixels. Our performance to date for a number of different NVIDIA GPU running under both Linux and Microsoft Windows is presented. The recent availability of GPU on portable computers is discussed and a new code for interactive replay of digital holograms is presented.

1. Introduction

In the last decade digital in-line holography has replaced silver-halide based techniques for the 3D recording of volumes of small particles. It has applications in a number of areas of science and engineering, for example cloud or spray droplets [1], marine plankton [2] and sediments [3]. Digital holography replaces the older technique, which primarily used high-resolution silver-halide emulsions, with direct recording onto a solid-state pixel sensor such as a CCD or CMOS array. There are a number of different mathematical techniques that can be used to calculate the replayed image volume; ours is based upon the convolution approach [4]. To replay the image volume encoded by a digital hologram requires that the two-dimensional discrete Fourier transform of the hologram is multiplied by a phase factor related to the recording beam and the distance of the replaced depth slice and then the image is created by taking the inverse discrete Fourier Transform and calculating its magnitude. A typical CCD or CMOS sensor used in digital holography has a few million pixels, although individual sensors with over eighty million pixels have been made. To fully reconstruct the volume recorded on a digital hologram requires 2D depth-slices at sub-millimeter intervals to be reconstructed over the entire recorded depth which is often a few hundred millimeters in extent. Thus for a single hologram of order one thousand individual 2D images will be replayed. The computational task, even with modern multi-core processors and efficient discrete Fourier transform algorithms such as FFTW [5] is considerable. In previous work [7], [8] we have exploited the trivially parallel nature of the hologram replay to enable us to use computational grid resources. Unfortunately the stochastic nature of submission to grid resources, via a broker, and the possibility of being queued until the validity of a job's X.509 proxy certificate expires means that the initial significant speed advantage over a single local computational node is rarely maintained and quite often not all of a volume is reconstructed. In earlier work a number of different jobs each reconstructing ten, fifty or one hundred sequential depth slices, for a total of 2200 planes for each run, were submitted to nodes on the EGEE grid [9]. An initial large advantage over a single processor was seen but the rate dramatically slowed in all cases and only a fraction of all the jobs completed, behaviour which is typical of current scientific production grids.

With the advent in recent years of powerful and increasingly inexpensive GPU chips considerable effort from ourselves [10] and many other researchers (see for example [11], [12], [13]) has been expended in porting, or developing, digital hologram replay code to run efficiently on these systems. In this paper we describe our latest work using NVIDIA Tesla and Fermi class GPU systems, the optimization techniques that we have deployed to achieve very significant replay speed, and the prospects for further improvement. Finally we speculate on whether a hybrid approach of GPU and distributed (grid or cloud) based computing may be optimum when the task of both replay and object extraction is considered.

2. Recording and replaying a digital hologram

We consider in this paper only the recording and replay of in-line (Fraunhofer) holograms recorded digitally on a pixel area sensor. Other forms of digital hologram, for example those recorded in off-axis holography, do not differ in the essential principles of replay.

2.1. Hologram recording

We recorded in-line holograms of objects in water using an 8 megapixel camera (ATMEL Camelia 8M, 2300 by 3500 10 μm-square pixels with 12-bit depth) with a collimated beam from a c.w. HeNe laser (λ=633 nm, 1 mW). As in earlier work [10] we recorded a sample volume consisting of cenospheres mostly of 100-300 μm dia. (Fillite Trelleborg Specialty Grade (High Alumina) SGHA 500) dispersed in a water tank (figure 1). A variety of other subjects, for example air bubbles in water were also used as test objects.

More recently, dynamic objects, such as digitalis seeds set in motion by breath [14], were recorded using a fast pulsed green laser (figure 2). The laser is a custom doubled Nd-YAG from Elforlight Ltd, model number LP3-10-532. This produced single longitudinal-mode pulses with a wavelength of 532 nm, duration of 2 ns and an adjustable energy of a few millijoules. The beam was then attenuated using a Newport 935-10 high-power variable attenuator and reflected, using a BK7 wedge, towards a beam expander telescope. The expanded beam then traversed a sample tank with BK7 flat windows and the resulting hologram was recorded using the same ATMEL camera.

2.2. Hologram replay

Our general approach has been summarised elsewhere [5]. The process is computationally heavy as it requires multiple 2D complex FFT to be calculated. Regarding only the reconstruction of a digital hologram, and ignoring the challenging and similarly computationally intensive task of locating "regions of interest" (ROI) within the replayed volume, it can be seen that every depth slice may be calculated in parallel. Since the initial Fourier transformed hologram is common to any depth slice calculation there are clearly potential advantages in doing this step once and making the transformed data available to all processors. With the recent availability of very powerful GPU processors and the porting of the FFTW algorithm to the NVIDIA CUDA library, most recent work in the digital holography field has made use of such GPU based systems. In the next section we describe results of our work on porting our existing replay code *HoloReco* [15] to the NVIDA CUDA environment.

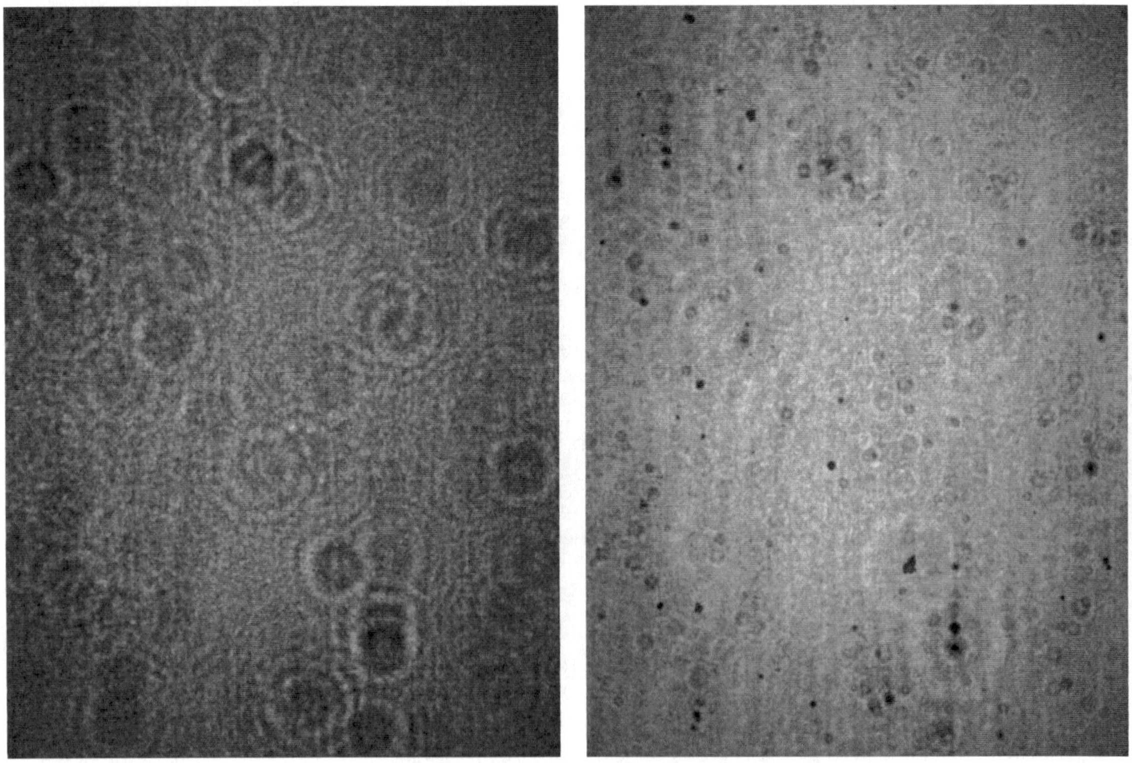

Figure 1. (Left): An extract (10 mm wide) from a digital hologram of cenospheres in water. (Right): The region of a numerically-reconstructed arbitrary plane through the sample volume. The smallest particles imaged here are approximately 30 μm in size [9].

3. Porting to the NVIDIA CUDA environment.

The original *HoloReco* code carried out reconstruction separately for every depth slice. Performing the FFT of the input hologram and maintaining the result to be used subsequently for a series of different depth slices eliminated all but one of the time intensive serial transfers from host to GPU memory.

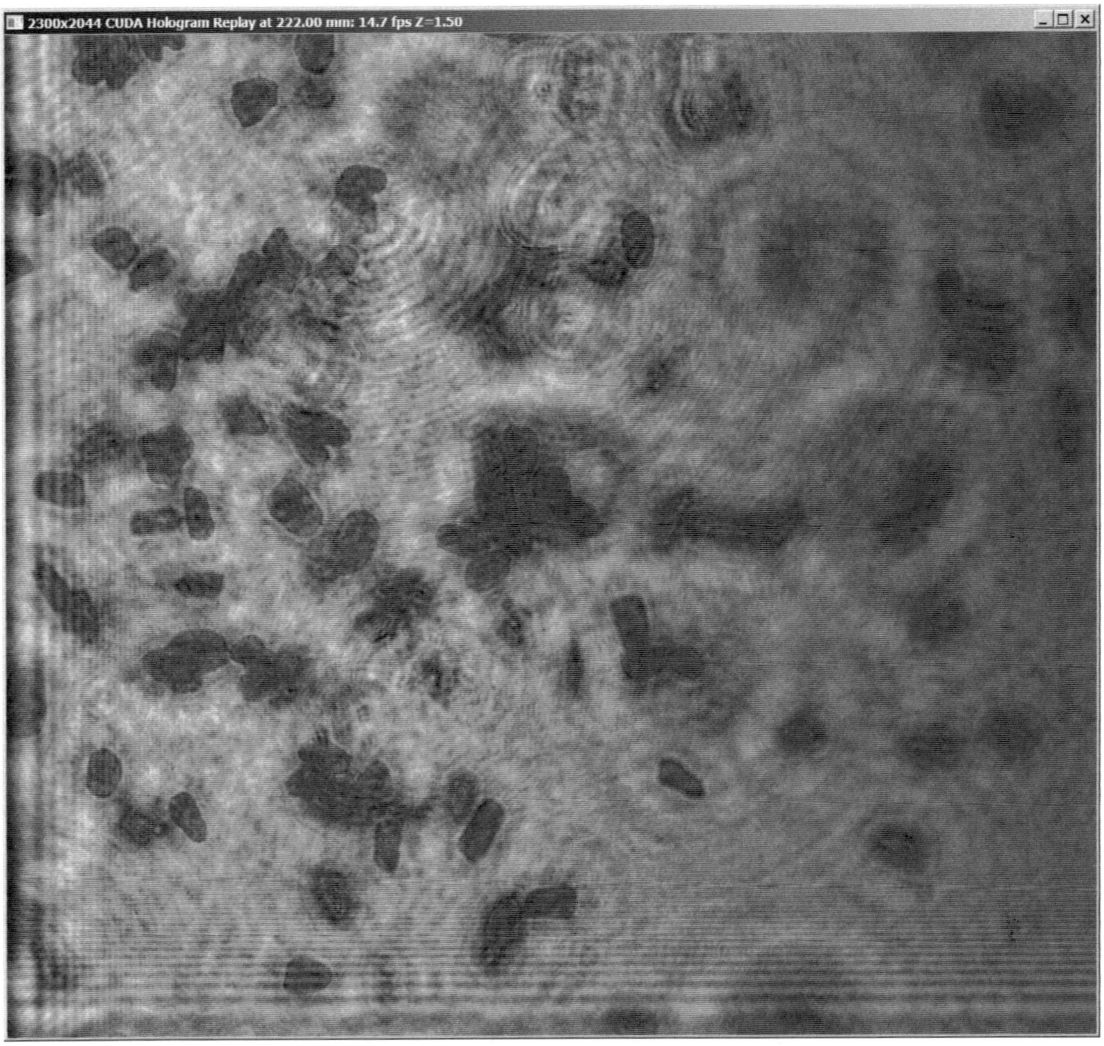

Figure 2. Window capture of the replay of a pulsed-laser hologram on a laptop computer equipped with a GT 555M GPU. The image shows digitalis seeds (~1 mm long) scattered by breath. The original hologram was cropped to 2300x2044 pixels and the image has been zoomed by a factor of 1.5 in the 1187x1080 window.

Another significant improvement to the processing speed was finding a way to parallelise the transfer function (unique to each depth plane). This allowed all computations to be carried out in the GPU and thus no bulk data movement is needed until the output stage

3.1. Tesla-based system with no video display

Using a system containing three NVIDIA Tesla S1070 (240 thread processors per GPU core, shaders clocked at 1.30 GHz) we were able to process our hologram to files at 13.6 planes/second. This should be compared to our original code running on an eight-core Xeon E5420 (2.50 GHz) based system where a speed of only 0.057 planes/second was achieved. Thus code refactoring and using three GPUs provided a factor of nearly 240 improvement in reconstruction speed. These early results used V2.2 of the CUDA libraries (including FFTW).

3.2. Fermi-based systems with video display

There are two clearly different user cases for hologram data visualisation:

- An expert user (e.g. a Marine Biologist) searches through a video stream of replayed hologram depth slices looking for "interesting" objects
- An image-processing system searches through replayed hologram image slices looking for "objects". The set of detected objects is then sent to a classifier.

A better use of the human expert is for them to locate quickly holograms that contain some objects of interest, for example a zooplankton species, and then to create a selected list of holograms to be systematically replayed in full and processed by an image-processing system and object classifier. Alternatively the expert user could look, in context, at those regions of the replayed volume in which an automated system had previously located and classified objects. Some Tesla and many Fermi-class GPUs have a video output and an OpenGL compatibility library is included in the CUDA development environment. For human visualisation this is ideal since one can efficiently copy the reconstructed hologram to the video buffer and overall performance is not compromised. As well, the OpenGL library allows simple panning and zooming of the image within the display window. The sequence of events for each replayed frame is shown in figure 3.

Table 1 lists the video rates we achieved for three different holograms (two cropped from the original) on four separate GPUs with integrated video output. Notable is the performance of the GT 555M GPU, which is designed for portable systems and in our case is part of a Dell XPS17 laptop computer which consumes just 80 W of power whilst continuously calculating and displaying the reconstructed images.

In practice, the replay software (*HoloMovie*) has a simple interface. Control and data parameters, such as laser wavelength, pixel spacing, input file-name, and reconstruction step size and depth range are read, as in *HoloReco*, from a simple text file. The programme calculates the window size which will fit the largest reconstructed image on the screen and begins the cycle of reconstruction and passing the image data to the display system. User controls are limited to manually changing the reconstruction depth or choosing to loop continuously back and forward in the range set by the parameter file, zooming the image in and out within the window, and panning the window across the image. This allows the operator to scan the reconstruction quickly and to examine closely any interesting features which are discovered.

Table 1. Reconstruction and replay rates achieved with several modern GPUs for three separate histogram sizes (zero-padded sizes in parentheses). All models are Fermi-class devices and all operating systems are 64-bit..

GPU Model	Thread Processors	Operating System	CUDA Version	Driver Version	2048x1389 (2048x2048)	2300x2044 (4096x2048)	2300x3500 (4096x4096)
GT 440	96	Windows 7	4.1	301.42	24.2 fps	13.6 fps	7.4 fps
GT 555M	144	Windows 7	4.1	301.42	25.1 fps	14.7 fps	9.3 fps
GTX 460	336	Scientific Linux 5	4.1	302.07	30.0 fps	20.0 fps	12.0 fps
C 2070	448	Ubuntu 11.10	4.2	295.53	39.8 fps	29.2 fps	15.9 fps

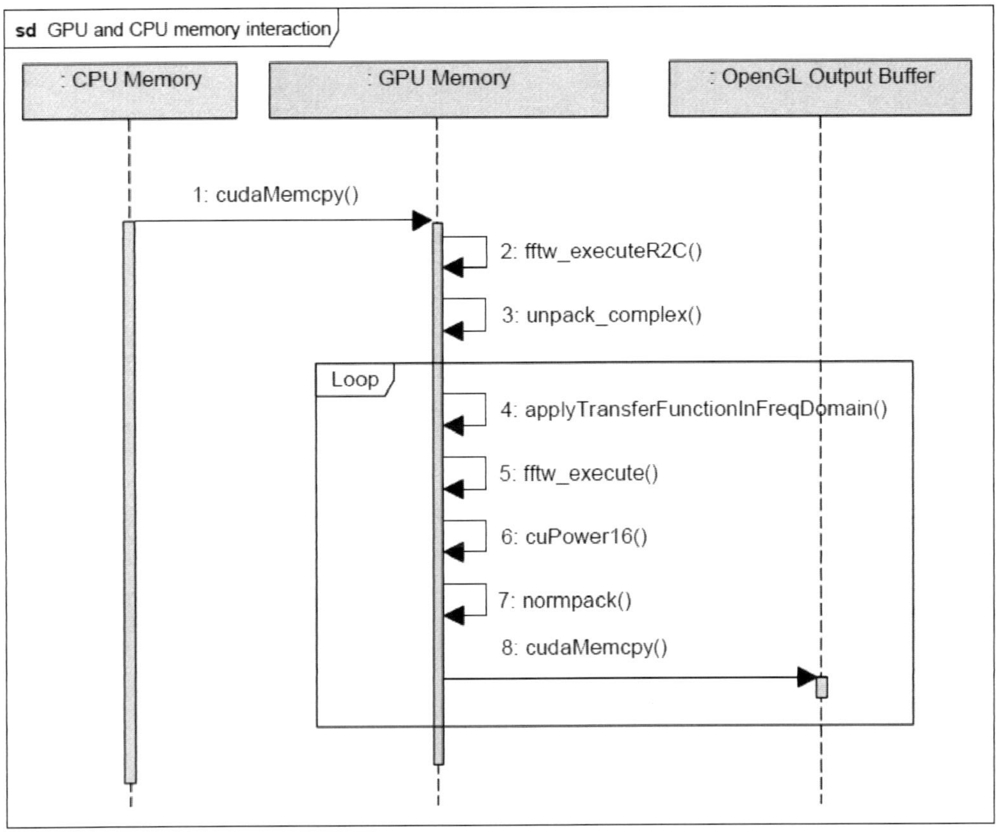

Figure 3. A UML representation of the sequence of key events in our software that occur during the processing of a digital hologram. If a number of frames from the same hologram are replayed to video then the sequence numbers 4 through 8 are repeated as often as required.

3.3. Profiling the code

The CUDA SDK also includes a profiler for mapping GPU usage which is a powerful tool for locating and understanding which operations dominate the GPU processor time. We ran the CLI version to get a timestamp snapshot of the programme running for 15 seconds on a 4-Core Xeon computer running Ubuntu 11.04 and equipped with a Fermi-based C2070 GPU. This gave the start-up and 50 iterations of the display loop (at ~4.4 frames/s; without the profiler it ran at ~16 frames/s). The hologram was 2300x3500 and zero-padded to 4096x4096 for processing. Table 2 shows information on GPU usage for this run.

The timing data show that, once the initial file is read, the majority of the GPU time is spent manipulating data – performing the transfer function to prepare for the reverse FFT and converting back from the complex transform to real magnitudes.

Table 2. (a) GPU and CPU procedure call timings and GPU occupancy for the initialisation step common to all slices. (b) GPU and CPU timings and GPU occupancy, averaged over 50 iterations, of the procedure calls that are executed for each depth slice calculated and then displayed. Note that there are two occupancy figures for the FFT because there are two separate kernels executed for each FFT; one executes once, the other twice.

(a) GPU kernel	Number of calls	GPU time (μs)	CPU time (μs)	GPU occupancy	Comment
memset32_post	3	6.7	27	0.042	
memcpyHtoD	1	12203	12313		Copy data from Host to Device
Forward FFT operations	3	5011	36	0.667/0.417	
Post-process	1	2379	5	1.000	
memcpyHtoD	1	2.0	7		Parameters for unpacking
copy_kernel	4095	11123	18524	1.000	cudaBlas unpacking FFT
insert_conj	1	15609	4	0.167	Reconstruct 2nd half of FFT
setUpX22_kernel	1	421.1	9	0.021	Calculate a lookup table for transfer

(b) GPU kernel	Number of calls/loop	Average GPU time/loop (μs)	Average CPU time/loop (μs)	Av. GPU occupancy	Comment
doInnerLoop	4096	20780	19891	0.167	Parallelized Transfer Function
Reverse FFT operations	3	9898.9	26.0	0.667/0.333	
cuPower16_kernel	1	22333	4.6	0.500	Extract amplitudes from complex & remove zero-padding
memcpyDtoH	1	2.6	40.1		Return subroutine results
memcpyDtoH	1	2.1	33		Return subroutine results
memcpyHtoD	1	1.6	5.4		Parameters for minmax to GPU
thrust::minmax_element	2	516.0	11.7	0.500	Find min & max
memcpyDtoH	1	2.0	539.0		Return subroutine results
setminmax	1	4.3	6.5	0.021	Place scaling factors in device
packfloat	1	5128.4	4.4	0.167	Create scaled, packed pixels
memcopyDtoD	1	272.4	7.8		Copy to OpenGL buffer

4. Discussion

With our code achieving a replay frame rate approaching standard video for holograms as large as 4096x4096 pixels and significantly exceeding it for 2048x2048 pixel holograms, there is no doubt that for visualisation a GPU is far superior to a conventional CPU or a computational grid. Recent advances (for example the GTX460 and GTX560 based systems) in inexpensive "desk-top Fermi" processor GPU systems, developed primarily for the high-end computer gaming community, now enable a very impressive price-to-performance ratio to be obtained for a modest monetary outlay. Even more significant are the most recent Kepler-class systems exemplified by the GTX 680 with 1536 thread processors and power consumption less than 200 W. A recently released laptop computer (Samsung Chronos 17) improves on our GT 555M system with a Kepler based GT 650M GPU, containing 384 processors for twice the performance at the same power consumption. Clearly it is now becoming feasible to take considerable processing power into the field rather than needing to bring data back to a compute farm.

Even further improvements lie on the horizon, with the upcoming CUDA 5 technology and the second-generation Kepler GPUs providing new techniques such as dynamic parallelism, bringing the need to master these concepts for maximum performance. However the picture is less clear when considering the overall task of replaying a digital hologram and then searching in 3D for in-focus images of "interesting" objects, extracting these and then classifying them.

Much existing code, including extensive image processing and classification libraries, would need to be ported to a GPU to generate a significant speed increase. Nonetheless the challenge lies not in the porting but in the code refactoring needed to make efficient use of the GPU. In our earlier work [10] the naive approach of recompiling our unmodified *HoloReco* code with the CUDA FFTW libraries resulted in only a factor of six improvement in processing speed and only extensive code refactoring coupled with detailed profiling has enabled the current factor of 240 to be achieved. In addition the task of locating objects in a replayed 3D volume is not a trivially parallel one. Several correlated depth slices need to be considered in order to locate, using some appropriate focusing metric [16], the best one so that each object may be extracted for subsequent classification. We hypothesise that a hybrid approach, with replay using a GPU and subsequent processing of the datasets using grid or commodity cloud computing may be the most appropriate combination of the two different architectures.

References

[1] Yang Yan, Kang Bo-seon 2011 Opt.Lasers Eng 49 1254–63
[2] Sun H, Benzie P W, Burns N, Hendry D C, Player M A and Watson J 2008 Phil. Trans. R. Soc. A 366 1789–806
[3] Graham G W, Smith W and Nimmo A M 2010 Limnol.Oceanogr.Meth. 82 1–15
[4] Kreis T M, Adams M and Juptner W P O 1997 Proc. of SPIE 3098 224–33
[5] Reid I D, Nebrensky J and Hobson P R (2012) Challenges in using GPUs for the reconstruction of digital hologram images Advanced Computing and Analysis Techniques in Physics Research September 5-9, 2011, Uxbridge, London to appear in *Journal of Physics, Conference Series*
[6] Frigo M and Johnson S G 2005 Proc. IEEE 93 216–31
[7] Nebrensky J J, Hobson P R and Fryer P C 2005 Proc. of SPIE 5775 285–96
[8] Nebrensky J J and Hobson P R 2006 Proc. of SPIE. 6252 62521I.1–6
[9] Nebrensky J J and Hobson P R 2009 Replay of Digitally-Recorded Holograms Using a Computational Grid [available] http://bura.brunel.ac.uk/handle/2438/3443
[10] Nebrensky J J, Hobson P R and Reid I D 2009 Digitally-Recorded Hologram Replay - a Comparison Between a Computational Grid and a GPU Based System, EOS Topical Meeting on Blue Photonics, Aberdeen, UK 18-19 August 2009
[11] Pandey N, Kelly D P, Naughton T J and Hennelly B M, 2009 Proc. of SPIE 7442 744205–1
[12] Zulfiqar A, Hyun-Eui Kim, Dongbiao Han, Jae-Hyeung Park and Nam Kim 2011 3D Res. 2 1–5
[13] Zhuqing Zhu, Min Sun, Heping Ding, Shaotong Feng and Shouping Nie 2009 Fast numerical reconstruction of digital holography based on graphic processing unit [Available] http://dx.doi.org/10.1109/CLEOPR.2009.5292249
[14] Hobson P R, Reid ID and Wilton JB 2012 Visualising breath using digital holograms, this conference
[15] [Available] http://holoreco.sourceforge.net/
[16] Meinecke T, Sabitov N and Sinzinger S 2010 Appl.Opt. 49 2446–55

Calculation method of CGH for Binocular Eyepiece-Type Electro Holography

Chanyoung Yang[1], Takuo Yoneyama[1], Yuji Sakamoto[1] and Fumio Okuyama[2]

[1] Graduate School of Information Science and Technology, Hokkaido University, Sapporo, 060-0814, Japan

[2] Faculty of Medical Engineering, Suzuka University of Medical Science, Suzuka, 510-0293, Japan

E-mail: yang.mcm@ist.hokudai.ac.jp

Abstract. We had researched about eyepiece-type electro holography to display 3-D images of larger objects at wider angle. We had enlarged visual field considering depth of object with Fourier optical system using two lenses. In this paper, we extend our system for binocular. In the binocular system, we use two different holograms for each eye. The 3-D image for left eye should be observed like the real object observed using left eye and the same for right eye. So, we propose a method of calculation of computer-generated hologram (CGH) transforming the coordinate system of the model data to make two holograms for binocular eyepiece-type electro holography. The coordinate system of original model data is called the world coordinate system. The left and the right coordinate system are transformed from the world coordinate system. We also propose the method for correcting the installation error that occurs when placing the electronic and optical devices. The installation error is calculated and the model data is corrected using the distance between measured position and setup position of the reconstructed image Optical reconstruction experiments were carried out to verify the proposed method.

1. Introduction

Holography records light waves from objects, and presents us with beautiful three-dimensional objects[1]. The light waves reflected on the surfaces of objects are called object light waves. And interference pattern of the object light waves and reference light waves, which are used to record the objects and which have its wave length equivalent to the object light waves, is a hologram.

In electro holography, electronic display device is used to display the hologram[2]. The viewing zone and the visual field in electro holography is limited into small area, because resolution of display devices is not sufficient for displaying the hologram, where the viewing zone is the area in which human can observe the image reconstructed from the hologram, and the visual field is the area in which the reconstructed image is placed. Therefore, the size of reconstructed image is limited under the size of display device. In addition, the reconstructed image needs some distance to be observed by the observer because of the characteristic of hologram. As a result, the apparent field of view, which is the angle size of the observed image from the observer, is more limited[3][4]. To make the apparent field of view wide, the Fourier transform optical system has been used[5][6]. The distance between the reconstructed image and the observer becomes shorter and the apparent field of view becomes wider in the Fourier transform system. However,

Published under licence by IOP Publishing Ltd

the size of the reconstructed image does not change and the apparent field of view is still limited in the Fourier transform system. In our previous study, we have proposed the method to enlarge the apparent field of view over 45 degree using two Fourier lens, the objective lens and the ocular lens[7]. However, the visual field has limited too narrow to observe with two eyes. In this paper, we combine two set of the previous optical system to make the reconstructed image observable with two eyes at a time. And we propose the calculation method of making the holograms with transforming the coordinate system for each eye. We also propose the method for correcting the installation error which occurs when placing the real objects, displaying devices and optical devices.

2. Theory

2.1. Monocular Eyepiece-type Optical System

The apparent field of view is extended to over 45 degrees with the monocular eyepiece-type optical system that we have proposed[7]. However, because the viewing zone is limited in very narrow region, the represented image is observed by only one eye. The monocular eyepiece-type optical system is consisted of a spatial light modulator(SLM), first lens as objective lens, a light emitting diode(LED) as reference light and second lens as ocular lens (figure 1).

At first, the primary image is reconstructed as a real image by the SLM, objective lens and reference light. And reconstruction of the real image is known as the Fourier transfer hologram. And the real image is enlarged as the virtual image by the ocular lens and the viewing zone is closed to the optical system. As a result, the apparent field of view is extended.

The monocular eyepiece-type optical system is based on the conventional theory used in the telescope or microscope. However, while the telescope and microscope consider the real image as a plane image, the real image of the monocular eyepiece-type optical system is three-dimensional image with depth. And the reconstructed image is distorted, because the magnification power is constantly varying with the depth of the reconstructed image. To remove the distortion, we proposed a calculate method reshaping the model data, which is used to simulate the hologram in CGH. When a point of model data $P(x, y, z)$ indicates the point of the virtual image, the point of the real image $P'(x', y', z')$ is expressed as

$$\begin{aligned} x' &= \frac{f_{ocular}}{f_{ocular} + z_{ocular} - z} x, \\ y' &= \frac{f_{ocular}}{f_{ocular} + z_{ocular} - z} y, \\ z' &= z_{ocular} - \frac{f_{ocular}}{f_{ocular} + z_{ocular} - z} (z_{ocular} - z), \end{aligned} \quad (1)$$

where f_{ocular} and z_{ocular} are focal length and Z-coordinate of the ocular lens. And the expanded

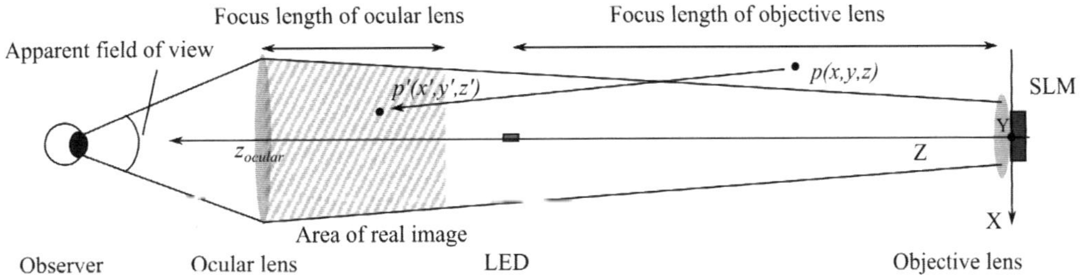

Figure 1. Monocular eyepiece-type optical system.

apparent field of view $\theta_{apparent}$ is expressed as

$$\theta_{apparent} = 2\arctan\left(-\frac{s_{ocular}}{2f_{ocular}} + \frac{s_{ocular} - s_{SLM}}{2z_{ocular}}\right), \quad (2)$$

where z_{ocular} is the Z-position of the ocular lens in the left of the right coordinate system, and s_{ocular} and s_{SLM} are the size of the ocular lens and the SLM.

2.2. Binocular Eyepiece-type Optical System

We propose a binocular optical system consisted of two monocular eyepiece-type optical system to observe the reconstructed image with two eyes. Two reconstructed image is represented for binocular observing, one image for left eye and the other image for right eye with two set of monocular eyepiece-type optical system. To be consistent the two reconstructed image, we make two coordinate systems for each eye from the coordinate system of model data. We call the original coordinate system the world coordinate system (figure 2). The right optical system slide laterally varying the distance between left and right eye from 60mm to 70mm and the world coordinate system also slide laterally half amount of the movement of the right optical system , because the distance between left and right eye is different with person.

When a point of the world coordinate is $P_W(x_W, y_W, z_W)$, the corresponding points $P_L(x_L, y_L, z_L)$ and $P_R(x_R, y_R, z_R)$ of the left coordinate system and the right coordinate system are calculated by movement and rotation transformation, which are expressed as

$$\begin{bmatrix} x_L \\ y_L \\ z_L \\ 1 \end{bmatrix} = \begin{bmatrix} \cos\theta_R & 0 & \sin\theta_R & 0 \\ 0 & 1 & 0 & 0 \\ -\sin\theta_R & 0 & \cos\theta_R & 0 \\ 0 & 0 & 0 & 1 \end{bmatrix} \begin{bmatrix} 1 & 0 & 0 & -x_{WO_L} \\ 0 & 1 & 0 & -y_{WO_L} \\ 0 & 0 & 1 & -z_{WO_L} \\ 0 & 0 & 0 & 1 \end{bmatrix} \begin{bmatrix} x_W \\ y_W \\ z_W \\ 1 \end{bmatrix},$$

$$\begin{bmatrix} x_R \\ y_R \\ z_R \\ 1 \end{bmatrix} = \begin{bmatrix} \cos(-\theta_R) & 0 & \sin(-\theta_R) & 0 \\ 0 & 1 & 0 & 0 \\ -\sin(-\theta_R) & 0 & \cos(-\theta_R) & 0 \\ 0 & 0 & 0 & 1 \end{bmatrix} \begin{bmatrix} 1 & 0 & 0 & -x_{WO_R} \\ 0 & 1 & 0 & -y_{WO_R} \\ 0 & 0 & 1 & -z_{WO_R} \\ 0 & 0 & 0 & 1 \end{bmatrix} \begin{bmatrix} x_W \\ y_W \\ z_W \\ 1 \end{bmatrix}, \quad (3)$$

where θ_R is the angle between the lines of sight and Z-axis of the world coordinate system, and

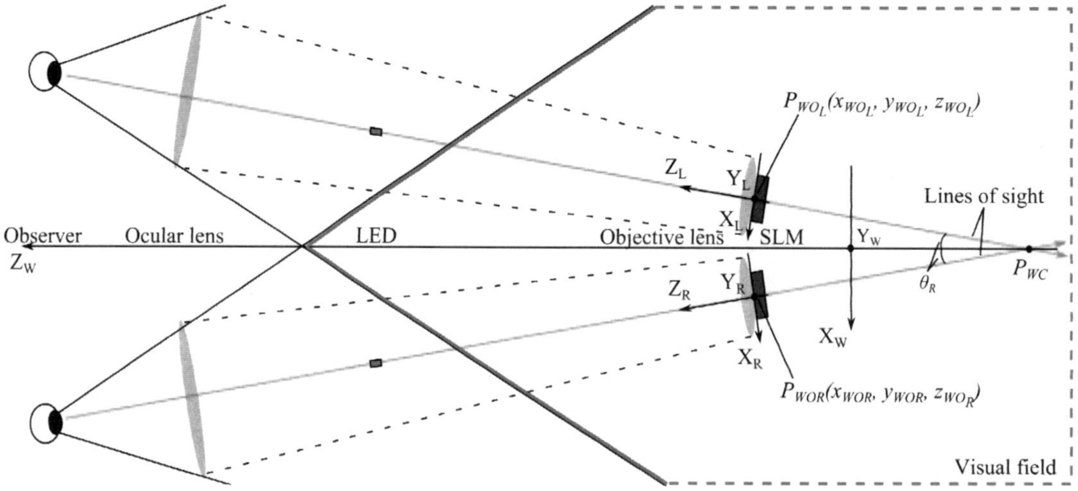

Figure 2. Binocular eyepiece-type optical system.

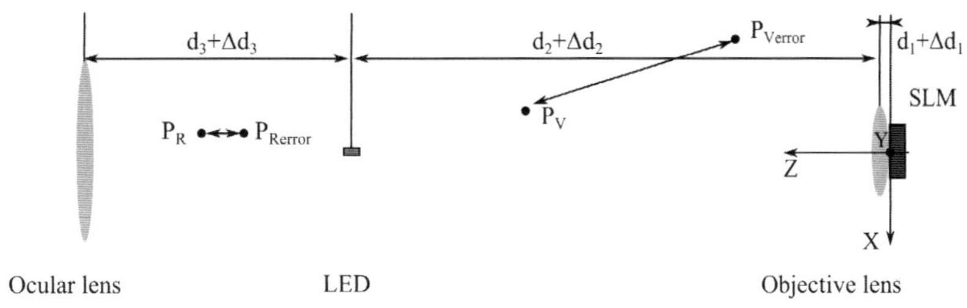

Figure 3. Installation error.

$P_{WO_L}(x_{WO_L}, y_{WO_L}, z_{WO_L})$ and $P_{WO_R}(x_{WO_R}, y_{WO_R}, z_{WO_R})$ are the coordinates of the origin points of the left and the right coordinate systems in the world coordinate system.

The slide of the right optical system makes some objects move; the right coordinate systems, the world coordinate system, P_{WO_L}, P_{WO_R} and the center point of rotation P_{WC} at which point the lines of sight cross. However, θ_R is constant.

2.3. Correcting Installation Error

The installation error of the optical system makes considerable error in reconstructed image, because the real image is enlarged to the virtual image with high-power magnification. However, making the installation error zero is very difficulty and hard work. So, we propose the method of correcting the model data to offset the error with the installation error calculated using measured error of the reconstructed image.

We make three setup parameter d_1, d_2 and d_3 to control the position of the real and the virtual images; d_1 is the distance between the SLM and the objective lens; d_2 is the distance between the objective lens and the LED and d_2 is the same with the focal length of the objective lens; d_3 is the distance between the LED and the ocular lens. The installation error Δd is consisted of Δd_1, Δd_2, Δd_3, the amount of difference in d_1, d_2, d_3 (figure 3). Δd_1 and Δd_2 make error in the position of the real image. When the point of P_{Rerror} is the point influenced by the installation error Δd_1 and Δd_2 from setup point P_R, the Z-position of P_{Rerror} is expressed as

$$z_{Rerror} = \frac{z_R f_{objective}^2 - z_R(\Delta d_1 + \Delta d_2) f_{objective} + (\Delta d_1 + \Delta d_2)^2 f_{objective}}{(\Delta d_1 + \Delta d_2) f_{objective} + f_{objective}^2 + (\Delta d_1 + \Delta d_2)^2 - z_R(\Delta d_1 + \Delta d_2)}, \quad (4)$$

where z_R is the setup Z-position of the real image point P_R and $f_{objective}$ is the focal length of the objective lens. And the Z-position of the virtual image point P_{Verror} influenced by the installation error Δd is expressed as

$$z_{error} = \frac{f_{ocular}}{f_{ocular} - (z_{Rerror} + \Delta d_3)} z_{Rerror}, \quad (5)$$

and the X-position and Y-position are calculated as same.

Figure 4 express the relationship of the installation error Δd and the Z-positions of the real image and the virtual image. The line of Z-positions moves up with increasing of the install error, which means the virtual image moves far from the observer.

When the amount of error in the reconstructed image Δd_m is measured, Δd is calculated with Δd_m by reverse of equation 5. And the model data is corrected using Δd to display the reconstructed image on the setup position. With the correction, the influence of installation error is removed and the reconstructed images becomes to have correct depth and binocular disparity.

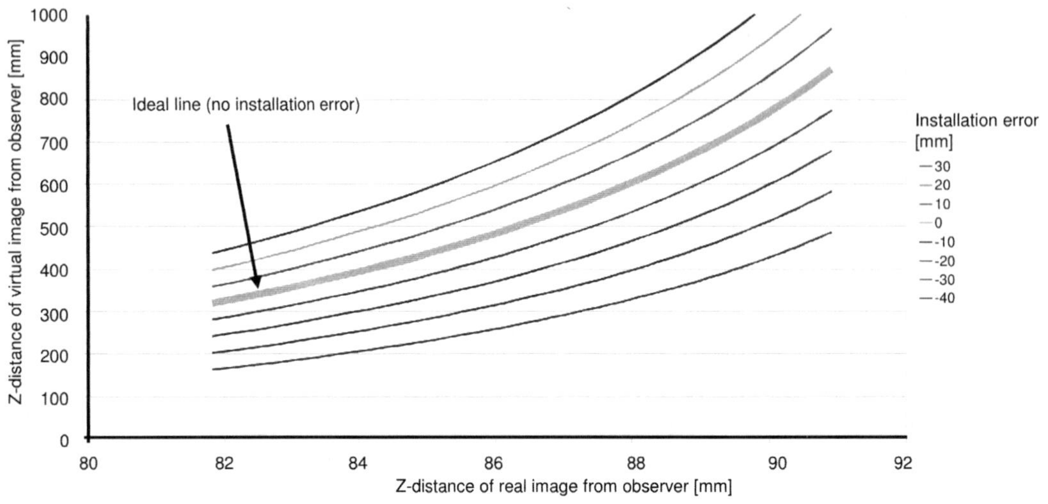

Figure 4. The Z-position of images varying by installation error.

2.4. Calculating Hologram

In the proposed method, the CGH is calculated in four steps.

(i) Read the model data and transform the model data to the left and the right data with equation 3.

(ii) Calculate the data for the real image from the data for virtual image which is the left and the right data with equation 1.

(iii) Correct the installation error from the data for the real image with the reverse equation of equation 5. The installation error z_{error} needs to be pre-calculated.

(iv) Calculate the interference pattern I of CGH from the corrected data using reference light R.

$$u_i(x_h, y_h) = \frac{a_i}{d_i} \exp\left(-j(kd_i + \phi_i)\right), \quad (6)$$

$$U(x_h, y_h) = \sum u_i(x_h, y_h), \quad (7)$$

$$I(x_h, y_h) = \mathrm{Re}[U(x_h, y_h) R(x_h, y_h)^*], \quad (8)$$

where $u_i(x_h, y_h)$ is the Fresnel-Kirchhoff diffraction on a pixel (x_h, y_h) from i-th point p_i of the corrected data which has its amplitude a_i and phase ϕ_i; d_i is distance between p_i and $(x_h, y_h, 0)$; k is the wavenumber of light; $U(x_h, y_h)$ is light wave from every point; $R(x_h, y_h)$ is light wave from the reference light;

3. Experiments

We developed an optical system for proposed binocular eyepiece-type electro holography (figure 5-6). To prevent physical interference with the SLMs and the objective lenses, we inserted a prism in the right optical system, which isolated the left and the right optical system. And whole of the right optical system was on a rail to make the right optical system slide left or right. Parameters of experiments are shown in table 1.

To make sure proposed method is correct, we measured ΔD_m with the optical system and calculated ΔD using measured ΔD_m. The shape of model is a vertical line, and the distance

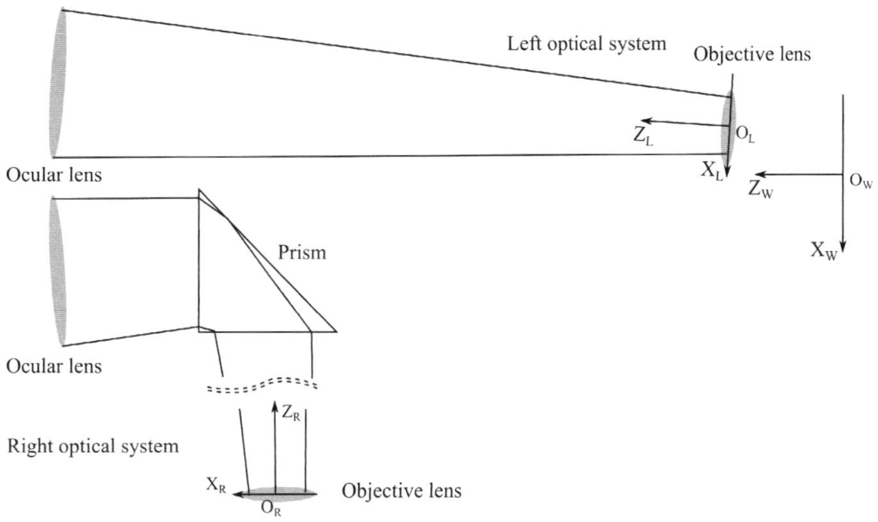

Figure 5. Implemented optical system.

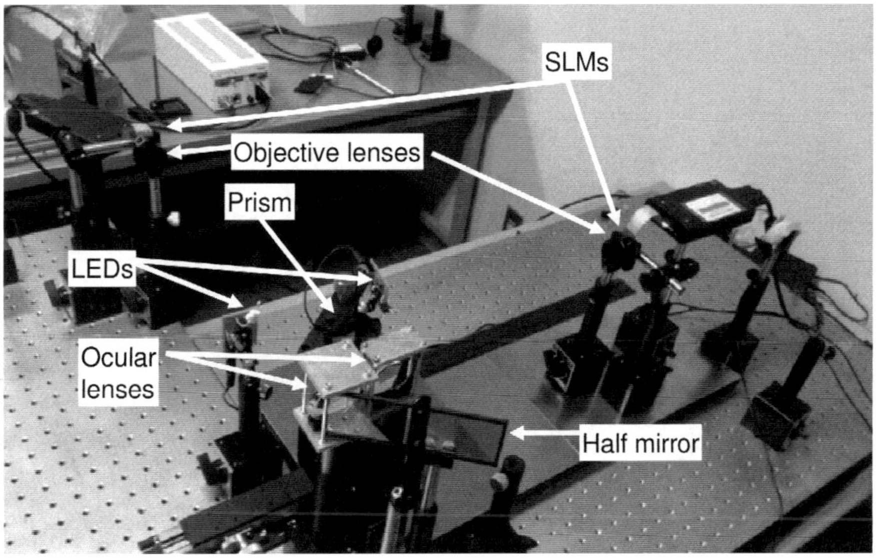

Figure 6. Implemented optical system.

of the line from observer is varying from 500mm to 900mm with 100mm of interval. And we made a hologram with proposed correction and reconstructed the image using the hologram and measured the Z-position.

3.1. Correction Installation Error

We made a hologram from model data and reconstructed the image from the hologram. And we measured 6-times, 3-times for left optical system and 3-times for right optical system, using holograms of each depth from the reconstructed image with camera focusing and comparing with real object reflected by half mirror (table 3.1). By assigning the average measured distance to equation 4, Δd was calculated to 39mm.

Table 1. Parameters of optical system.

SLM pixel pitch	8.0μm
SLM pixels	1920 x 1080
Wavelength of reference light	632nm
Focal length of objective lens	300mm
Focal length of ocular lens	100mm
Setup distance between SLM and objective lens (d_1)	10mm
Setup distance between objective lens and LED (d_2)	300mm
Setup distance between LED and ocular lens (d_3)	150mm

Table 2. Measured Z-position of the reconstructed image.

setup distance	average measured distance (left)	average measured distance (right)
500mm	351mm	330mm
600mm	416mm	388mm
700mm	470mm	460mm
800mm	522mm	496mm
900mm	557mm	592mm

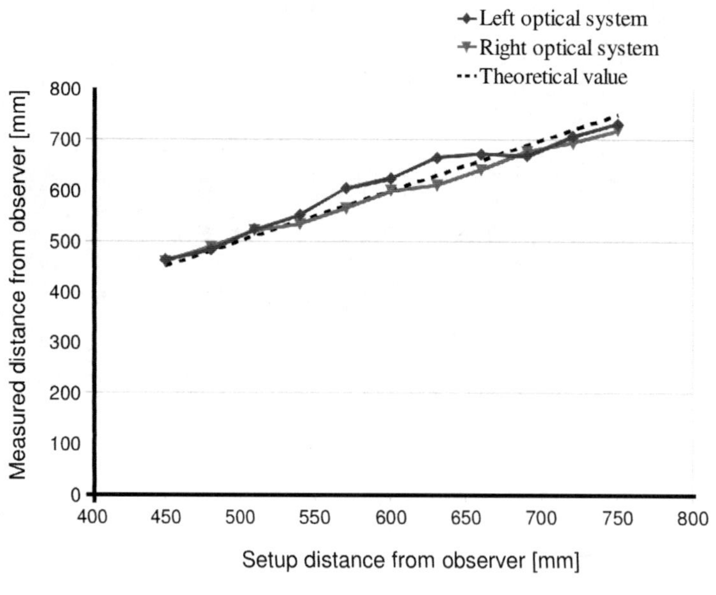

Figure 7. Measured Z-position of the corrected reconstructed image.

3.2. Optical Reconstruction

To verify the effect of correction, we simulated the holograms and measured the Z-position of the reconstructed image with line shape model corrected using Δd. To make the measure more accurately, we set the line position from 450mm to 750mm with 30mm of interval. And we

Figure 8. Reconstructed image (left).

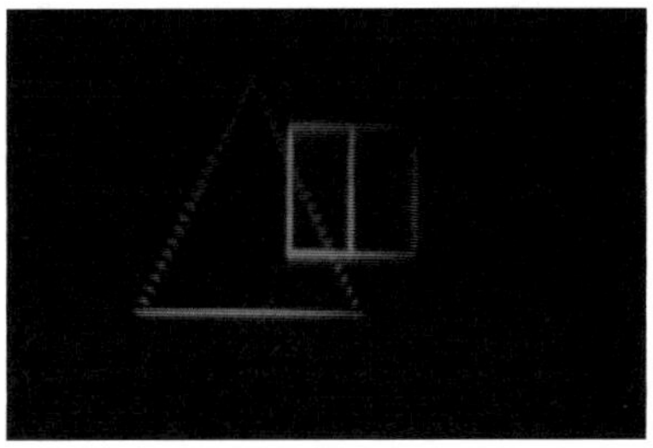

Figure 9. Reconstructed image (right).

measured the Z-position of reconstructed image 6-times as like as 3.1. In figure 7, the measured values with the left and the right optical systems seem around the line of theoretical value and the maximum error is 35mm at measured value 665mm of the left when the theoretical value is 665mm.

And we made a hologram which was simulated for reconstructing a triangle and a rectangle placed at 500mm and 1000mm from the observer (figure 8-9). The Z-position of the triangle and the rectangle was close the setup position and the binocular disparity was observed naturally.

4. Conclusion

In this paper, we proposed the calculation method of CGH for binocular eyepiece-type electro holography. Based on our previously monocular eyepiece-type optical system, we combined two optical system for each eye. And we proposed the coordinate transformation from the world coordinate system to the left and the right coordinate systems. We also proposed the correction method to remove the installation error which is too hard to make zero. The correction method is consisted of two phases, measuring the amount of error and correcting the installation error. The optical experiments showed that the correction was functioned well removing the installation error, so the setup model was shown at the setup coordinate. And the experiments also showed that the binocular images had proper binocular disparity and Z-depth, so some observers felt the images natural. However, the proposed correction method needs to be verify its validity with more experiments. And if the length of the optical system will be more short, it may be useful for a head-mount-display using electo holography.

ACKNOWLEDGMENTS

This work was supported by Grant-in-Aid for Scientific Research (B) KAKENHI(23300032).

References

[1] Benton s, Bove V and Connors-Chen E 2008 *Holographic Imaging* (New York: Wiley-Interscience)
[2] Lucente M 1994 *Diffraction-specific Fringe Computation for Electro-holography* (Boston: Institute of Technology)
[3] Sakamoto Y 2009 Proc. *Portable three-dimensional display system for cghs* (SanFransisco: SPIE Photonics WEST) vol 7233 no 3
[4] Oi R, Mishina T, Yamamoto K, and Okui M 2009 Proc. *Real-time IP-hologram conversion hardware based on floating point DSPs* (SanFransisco: SPIE Photonics WEST) vol 7233 no 5
[5] Takemori T 1997 *3-dimensional display using liquid crystal devices-fast computation of hologram* (ITE Tech. Report) vol 21 no 46 (Tokyo: Institute of Image Information and Television Engineers) p 13-19
[6] Kato A and Sakamoto Y 2010 Proc. *An electro holography using reective lcd for enlarging visual field and viewing zone with the fourier transform optical system in cgh* (SanFransisco: SPIE Photonics WEST) vol 7619 no 36
[7] Yang CY, Sakamoto Y and Okuyama F 2012 Proc. *Enlargement of visual field considering depth of object for eyepiece-type electro holography* (SanFransisco: SPIE Photonics WEST) vol 8281 no 27

A rendering method of background reflections on a specular surface for CGH

T Ichikawa and Y Sakamoto

Graduate School of Information Science and Technology, Hokkaido University, Japan

E-mail: ichikawa@ist.hokudai.ac.jp

Abstract. Computer-generated holograms (CGHs) are remarked as ideal three-dimensional displays. There are a lot of problems have to be overcome, and we focused on the rendering techniques to reconstruct realistic images. In particular, there is almost no calculation method that correctly expresses depth information. Reflected images are necessary to express complex realistic scenes. In this paper, we introduce a calculation method that generates reconstructed images with background reflections. Proposed method calculates intersections between virtual objects and rays by the ray tracing method. Then intersections are assumed as a point light source group, and light waves on a hologram plane are calculated. In the ray tracing process, when a ray hits the specular object, a ray is additionally casted to the specular direction from the intersection. If the ray hits other diffuse objects or background, the length of light path from a viewpoint to the diffuse object is calculated. By calculating light waves from a point light source on the diffuse surface distant from the light path, reflected images are expressed in the CGH. To express the quality of materials, we adopted the Cook-Torrance reflection model. In the experiment, we conducted the computer simulation to confirm that the depth of reflected images is correctly calculated. And results of optical reconstructions show that our proposed method is able to make CGHs of various qualities of material.

1. Introduction

A computer-generated hologram (CGH) is remarked as an ideal three-dimensional technology that fills human's visual features to perceive 3D images. Interference pattern on a hologram plane is calculated by simulating light propagation from virtual objects that is made in computer. In CGH, holograms are made by developing the interference pattern to the film or displaying on a spacial light modulator (SLM).

However there are some problems before CGHs will be put to practical use. First of all, the calculation of a CGH takes enormous computation time. Furthermore, rendering techniques to express realistic images like one in the field of computer graphics (CG) are scarce. To speeding up the calculation of CGH, a number of studies have been proposed already. For example, there are methods of hardware acceleration using a graphic processing unit (GPU) and special-purpose computer[1, 2, 3]. Besides, fast calculation algorithms have been proposed. In order to make the point wise method run faster, there is a method using a look-up table recording pre-calculated light wave on a hologram plane from a point light source[4]. And a technique using three-dimensional affine transforms of basic object light recording pre-calculated light waves from a basic plane has been also reported[5]. Thanks to these methods, CGH calculations can nearly work in real time.

On the other hand, studies that improve the appearance of the reconstructed images are inadequate. Rendering techniques include some kinds of problems. To improve each problem, some studies have been proposed. For example, various researchers have proposed methods that express the hidden surface removal[6, 7, 8, 9]. Moreover some methods have attempted to implement reflectance distributions that signify how an object's material affects its appearance[10, 11, 12]. However the above methods did not take into consideration the background reflection which renders reflected images mapped on metallic object's or mirror surfaces. Reflected images are necessary to express complex realistic scenes. A study [13] proposed a calculation method that expresses reflectance distributions and reflected images by modulating the phase differences on a object's surface. But this method can only apply polygon-based models and takes enormous calculation time. Some methods by using only amplitude information are existed, the depth of reflected images are not calculated correctly.

Thus we propose a calculation method of CGH which generates reconstructed images with background reflections using the ray tracing method. In the field of CG, the ray tracing method is used to express realistic images including the hidden surface removal, refraction, multi reflection and so on[14, 15] In this paper, we show the way to apply the ray tracing method to CGHs that generate reconstructed image having reflected images. In addition, accelerations of the calculation using GPU were implemented. By computer simulations and optical reconstructions, we confirmed that the reflected image with highlight was performed by proposed method.

2. Proposed method

2.1. Calculation method of CGH

We have proposed a method that calculates CGH expressing the hidden surface removal by using the ray tracing method[16]. The ray tracing method calculates intersections with objects by casting rays from a viewpoint. Our method make CGHs by treating these intersections as a point light source group. The view of this method is close to the holographic stereogram. By the ray tracing method from one viewpoint, overlapping of images with movement of a viewpoint is shown up because motion parallax is necessary in CGH. So that a hologram plane is divided into elementary holograms, and the ray tracing processes are conducted every an elementary hologram in the study [16].

First of all, a hologram plane is divided into elementary holograms (Fig. 1). Continuous parallax depends on division width of elementary holograms. The parallax between elementary holograms becomes big as the size of an elementary hologram becomes large. In this study, the size of elementary hologram is 1.216 [mm](128 [pixels]). Because it has been said that if the size is about 1[mm], there is no problem in continuous parallax by a research of CGH using multi-view images [17]. The θ in Fig. 1 signifies the angle of frequency limit. Rays are casted within the angle. Here the interval $\Delta\theta$ between rays is 1/60 degrees. Because a resolution of the human is assumed 1/60 degrees. Therefore gaps between rays is not visible by the observer.

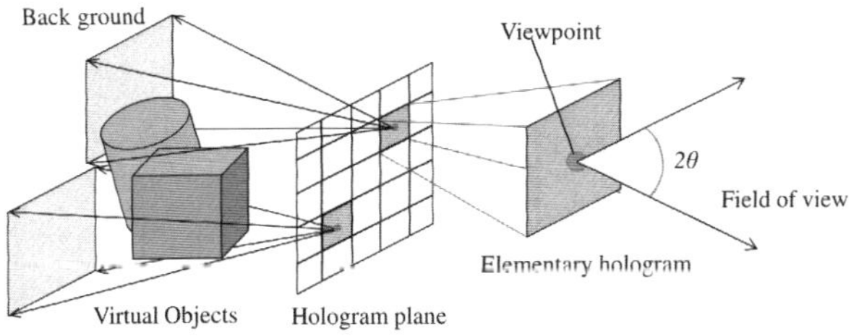

Figure 1. Division of a Hologram plane

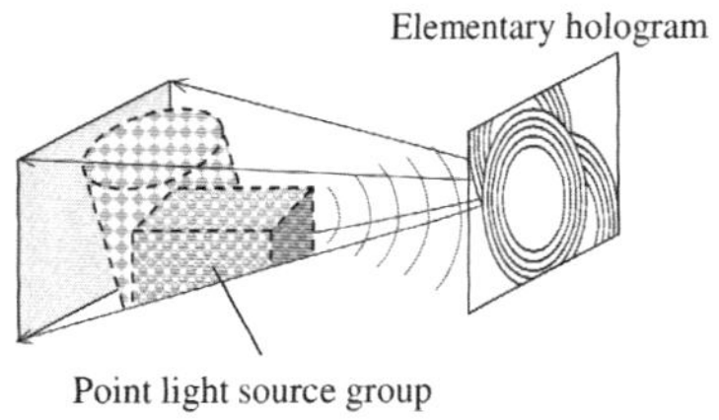

Figure 2. Light propagation

Next, every an elementary hologram, the determination of intersections is conducted with the ray tracing method. Rays are casted from the center of the elementary hologram. Each ray conducts intersection determinations with all objects (polygons) in the virtual scene. A ray obtains information of the brightness and the coordinate of the nearest intersection. Therefore, it is considered that intersections is a point light source group. The detailed computational algorithm of the ray tracing method is omitted here.

Finally, we calculate light waves on an elementary hologram. We use the point wise method that assumes virtual objects as an assemblage having three-dimensional coordinate. Here we focus on one elementary hologram (Fig. 2). This time, the calculation uses the intensities and depth information of each point light source. In the holographic stereogram, a hologram is made by combining the multi view images having only information of the intensity. On the other hand, our method is able to express depth perception in one viewpoint by using depths of the virtual objects in addition to the intensity. Complex amplitude distribution on a hologram plane $h_m(x, y)$ is calculated by

$$h_m(x,y) = \sum_{i=1}^{N_m} \frac{A_i}{r_i(x,y)} \cdot \exp\left(j\frac{2\pi}{\lambda}r_i(x,y)\right) \cdot \exp(j\phi_i) \,, \tag{1}$$

$$r_i = \sqrt{(x_i - x)^2 + (y_i - y)^2 + z_i^2} \,, \tag{2}$$

where parameters x, y and z represent the horizontal, vertical and depth components, while indices m and i respectively indicate the each elementary hologram and point light source. λ is the wavelength, and ϕ_i is the initial phase of each point light source. N_m is the number of point light sources that compose 3D objects at each elementary hologram. The intensity of the point light source A_i are added various reflectance distributions by using the normal vector of objects, position of the point light source and direction of the ray. The detailed flow of calculating the intensities is described by following paragraphs. Finally, whole light waves on a hologram plane are obtained by lining up every light waves of elementary holograms.

2.2. Representation of reflected images

Here the representation method of specular surface in CGH is described. Before I proposed a calculation method of CGH to express transparent object [18]. In a study [18], refraction on a transparent object was expressed by considering the length of light path and light attenuation in the medium. Then reflection on a surface of transparent objects is ignored. However reflected images in addition reflection on a surface are necessary to calculate the multiple reflection.

A simple graphic explanation of proposed method is shown in Fig. 3. At first, a ray (Ray_1) is casted from a center of an elementary hologram. When a ray hits the specular objects, the intensity and length $r_1 = t_1$ of a intersection P_1 is stored into a buffer of a point light

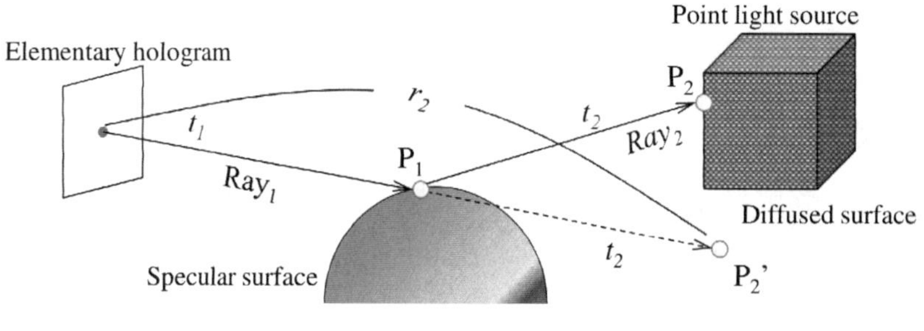

Figure 3. Expression of specular surface.

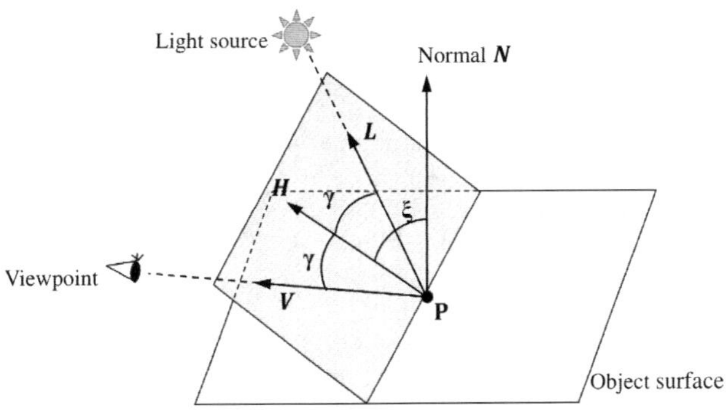

Figure 4. Cook-Torrance reflection model

source group. And more, a ray (Ray_2) is additionally casted to the specular direction from the intersection. If the ray hits other diffuse objects or background in the scene, the length of light path r_2 from a viewpoint to the diffuse object is calculated. If the ray hits the specular surface again, same process is repeated until an additional ray hits the diffuse surface or a reflection number of times reaches a setup maximum reflection number of times. So when a ray reflects I times, a length of light path is calculated by

$$r_I = \sum_{i=1}^{I} t_i \,. \tag{3}$$

In the example of Fig. 3, a intersection P_2 is the same as being arranged in the position of P'_2. By calculating light wave from a point light source on the diffuse surface distant from the light path r_2 along with a point light source on a specular surface using Eq. (1), reflected images are expressed. Furthermore the reflectance is set up on a specular surface, and the brightness of the intersection determined after a ray reflected is calculated by multiplying the reflectance. A calculation of the intersection is described by following paragraph.

2.3. Cook-Torrance reflection model

The brightness of intersections is determined with the normal vector of the surface, the viewpoint and so on. By being the brightness I_r as the intensity of the point light source in Eq. (1), various reflectance distributions are expressed. In this paper, we use a following equation to calculate the intensity of the reflected light

$$I_r = I_a\rho_a + (\mathbf{N}\cdot\mathbf{L})k_d\rho_d + (\mathbf{N}\cdot\mathbf{L})k_s\rho_s \,, \tag{4}$$

here, I_a is intensity of ambient light. k_d and k_s are respectively the ratio of Lambert light and specular light ($k_d + k_s = 1$). ρ_a, ρ_d and ρ_s are reflectance of ambient light, Lambert light and specular light. Vectors $\mathbf{N}$ and $\mathbf{L}$ are a normal unit vector and an unit vector to the light source. In proposed method, we used the Cook-Torrance model [19] to calculate the reflectance of the specular light. The Cook-Torrance reflection model is suitable for expressing metallic objects. In Cook-Torrance model, a object surface is assumed as an aggregate of minute surfaces, and it is granted that the direction of the minute surface is distributed centering on the normal direction of a surface. The Cook-Torrance model considers that reflectance distributions depend on the wavelength and incident angle of light, and the reflectance turned on materials is calculated by the Fresnel equation. These functions are composed of a half vector $\mathbf{H}$ and normal vector $\mathbf{N}$ in

Fig. 4. The reflectance by the Cook-Torrance model ρ_s is described with following equation

$$\rho_s = \frac{F}{\pi} \frac{D\,G}{(\mathbf{N} \cdot \mathbf{V})(\mathbf{N} \cdot \mathbf{L})} . \tag{5}$$

The Fresnel term F describes how light is reflected from each smooth micro facet and is given by Fresnel equations. The geometrical attenuation factor G accounts for the shadowing and masking of one facet by another. D stands for a distribution function, and is defined by Beckmann distribution with

$$D = \frac{1}{4m^2 \cos^4 \xi} \exp\left(\frac{\tan \xi}{m}\right)^2 . \tag{6}$$

Here, m is the coefficient which adjusts the roughness of a surface. As m becomes small, the surface is smooth and becomes specular. Light intensities of a intersections on a specular surface are obtained by above equations. As for diffuse surfaces, the intensity calculation is performed by the Phong model as with a study [16].

3. Experiment

3.1. Calculation system

We conducted some experiments to confirm the adequacy of the proposed method. At first, this section describes about our calculation system. The ray tracing method is avoided ever in CGH having a very high resolution because of enormous calculation cost. Then accelerations of the ray tracing algorithm and calculation of light waves by parallel computing with GPU were implemented in this study. In a calculation process of light waves, each pixel on a hologram plane can be calculated independently apparent to Eq. (1). So the calculation is suitable for a parallel calculation with GPU.

For the parallel computing of light propagation, we used the CUDA programming environment (NVIDIA). To accelerate the ray tracing algorithm, we used OptiX application acceleration engine. OptiX is real time ray tracing engine for CUDA-based video cards. A GPU that we used is an NVIDIA GeForce GTX 580, and a CPU is Intel Core i7-2600K. Tab. 1 shows detail specifications of our system.

3.2. Computer simulations

We conducted computer simulations that the reflected images are expressed in CGH. Computational reconstructed images were made by simulating the light propagation with a virtual lens from light distributions on a hologram plane. Experimental parameters in this experiment are shown in Tab. 2. At first, we verified whether the image reflected by the mirror would be reconstructed by the right position. Fig. 5 signifies the arrangement of virtual objects

Table 1. The spec sheet of our system.

Specification of Geforce GTX 580	
Number of streaming processors	512
Clock frequency	1544 [MHz]
Memory capacity	3GByte GDDR5
Specification of the host computer	
Core Clock	3.4 [GHz]
Memory capacity	8.0GByte
Operating system	Windows 7 64bit

Table 2. Setup parameters for experiment.

Number of elementary holograms	32(H) × 16(V)
Size of elementary holograms	1.216[mm] × 1.216[mm]
Number of rays of each elementary hologram	161 × 161
Number of pixels	4096 × 2048 [pixels]
Sampling pitch	9.5 × 9.5 [μm]
Wavelength	632 [nm]

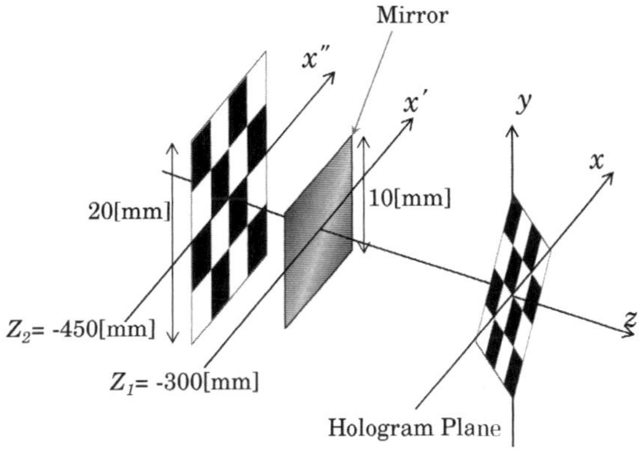

Figure 5. Geometry of the computer simulation.

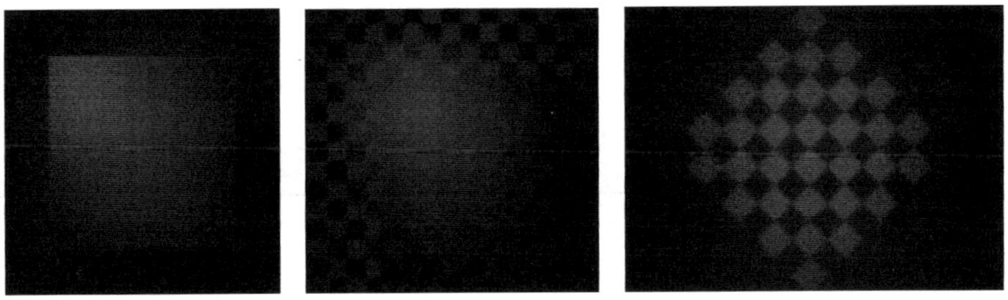

(a) $f = 30$cm. (b) $f = 45$cm. (c) $f = 60$cm.

Figure 6. Computational reconstructed images.

in the scene. The board of check pattern were made of diffuse surfaces, and a mirror had a specular surface. These objects were illuminated by a point light source. The ray tracing method was performed noting that the viewpoint was in a hologram plane ($z = 0mm$), and the hologram was calculated.

Fig. 6 show computational reconstructed images at each distance. Fig. 6 (a) was focused on the virtual mirror. Upper left part of a mirror surface seem gleaming because illuminated light hit to the surface light at an angle. Fig. 6 (b) and (c) were focused on each checker board. A reflected image by a mirror in addition to a checker board behind a mirror. It turns out that the image reflected in the mirror is also reconstructed by the right position as Fig. 5 in addition to the checkerboard in the back of a mirror. Therefore it was confirmed that the proposed method succeeded in expressing reflected images with correct depth.

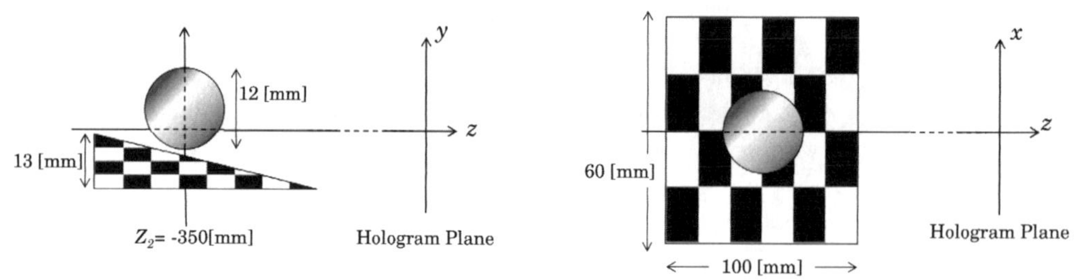

(a) y-z plane (b) x-z plane.

Figure 7. Geometry of the optical reconstruction.

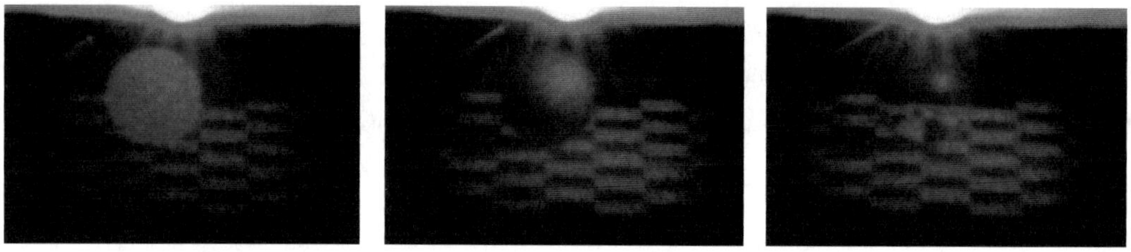

(a) Ambient light. (b) Phong model. (c) Cook-Torrance model.

Figure 8. Reconstructed images by three viewpoints.

3.3. Optical reconstructions

We conducted optical reconstruction to express the complex scene in CGH. Fig. 7 signifies the geometry of the optical reconstruction: Fig. 7 (a) is a view of y-z plane and Fig. 7 (b) is a view of x-z plane. The check pattern under the sphere was diffuse surface. The sphere was added some reflectance distributions. These objects were illuminated by a point light source in the top-right. The interference patterns were displayed on a spatial light modulator.

Results of the optical reconstructions are shown in Fig. 8. In Fig. 8 (a), the sphere was expressed with only ambient light. So the whole of a sphere is seen with the same luminance. A sphere in Fig. 8 (b) has reflectance distributions by the Phong reflection model. The sphere seems like being made out of plastic. Fig. 8 (c) was obtained by the proposed method. The sphere had specular surfaces, and the reflectance distributions by Cook-Torrance model were added. Therefore the reflected image of check pattern on a sphere is shown. Moreover the highlight by the Cook-Torrance model like a metal object appears. So this experiment shows that the proposed method can calculate CGH of various qualities of material. These results of the computational and optical reconstructed images show that the proposed method correctly expresses reflected images.

4. Discussion and Conclusion

Whole calculation of the proposed method were parallelized with GPU. Calculation times of experiments in sec. 3 are shown in Tab. 3. According to Tab. 3, it turns out that the calculation time of the wave propagation is by far longer than the ray tracing process. Our calculation algorithm of wave propagation and coding is still crude compared to the Optix ray tracing engine. So optimization and tuning the code is necessary. In the future, we aspire to becoming the real time calculation.

Table 3. Calculation times [ms].

	Setting	Ray tracing	Wave propagation	Total
sec. 3.2	2,137	2,531	16,773	21,660
sec. 3.3	2,169	2,593	17,491	22,459

In this paper, we propose a calculation method of a CGH that generates reconstructed images with background reflections using the ray tracing method. Background reflections are expressed by calculating the light path to reflected objects. And we adopted the Cook-Torrance reflection model to express metallic objects. As results of experiments, it was confirmed that the reflected image with highlight was performed by proposed method.

5. Acknowledge

This work was supported by Grant-in-Aid for Scientific Research (B) KAKENHI (23300032).

References

[1] Kang H, Yamaguchi T, Yoshikawa H, Kim S-C and Kim E-S 2008 Acceleration method of computing a compensated phase-added stereogram on a graphic processing unit *Appl. Opt.* **47** 5784-5789
[2] Nakayama H, Takada N, Ichihashi Y, Awazu S, Shimobaba T, Masuda N and Ito T 2010 Real-time color electroholography using multi graphics processing units and multi high-definition liquid-crystal display panels *Appl. Opt.* **49** 5993-5996
[3] Ichihashi Y, Nakayama H, Ito T, Masuda N, Shimobaba T, Shiraki A and Sugie T 2009 HORN-6 special-purpose clustered computing system for electroholography *Opt. Express* **17** 13895-13903
[4] Yoshikawa H, Yamaguchi T and Kitayama R 2009 Real-time generation of full color image hologram with compact distance look-up table *Digital Holography and Three-Dimensional Imaging* OSA Technical Digest(CD) (OSA **2009**) paper DWC4
[5] Sakata H and Sakamoto Y 2009 Fast computation method for a fresnel hologram using three-dimensional affine transformations in real space *Appl. Opt.* **48** H212-221
[6] Sakamoto Y 2002 An algorithm of hidden surface removal using shadow-propagation method for computer-generated hologram *IEICE Trans. Information Syst. 2002* **J85-D-II** 1832-1839 (in Japanese)
[7] Matsushima K and Nakahara S 2009 Extremely high-definition full-parallax computer-generated hologram created by the polygon-based method *Appl. Opt.* **48** H54-H63
[8] Yamaguchi T, Fujii T and Yoshikawa H 2009 Fast calculation method for computer-generated cylindrical holograms *Appl. Opt.* **47** D63-D70
[9] Chen R H-Y and Wilkinson T D 2009 Computer generated hologram with geometric occlusion using gpu-accelerated depth buffer rasterization for three-dimensional display *Appl. Opt.* **48** 4246-4255
[10] Bräuer R, Wyrowski F and Bryngdahl O 1991 Diffusers in digital holography *J. Opt. Soc. Am.* A**8** 572-578
[11] Kim H, Hahn J and Lee B 2008 Mathematical modeling of triangle-mesh-modeled three-dimensional surface objects for digital holography *Appl. Opt.* **47** D117-127
[12] Yamaguchi K, Ichikawa T and Sakamoto Y 2011 Calculation method for computer-generated holograms considering various reflectance distributions based on microfacets with various surface roughnesses *Appl. Opt.* **50** H195-H202
[13] Yamaguchi K and Sakamoto Y 2009 Computer generated hologram with characteristics of reflection: reflectance distributions and reflected images *Appl. Opt.* **48** H203-H211
[14] Appel A 1968 Some techniques for machine rendering of solids *AFIPS Joint Computer Conference* 37-45
[15] Whitted T 1980 An improved illumination model for shaded display *Comm. ACM* **23** 343-349
[16] Ichikawa T and Sakamoto Y 2012 Full parallax Computer Generated Hologram using GPU-accelerated ray tracing method *Proc. SPIE 8281* 82810E
[17] Hayashi N, Sakamoto Y and Honda Y 2011 Improvement of camera arrangement in computer-generated holograms synthesized from multi-view images *Proc. SPIE 7957* 795711
[18] Ichikawa T and Sakamoto Y 2012 Expression of Refractive Objects for Computer-Generated Hologram Using Ray Tracing Method in *Digital Holography and Three-Dimensional Imaging*, OSA Technical Digest (Optical Society of America, 2012), paper DTu4C.4.
[19] Cook R and Torrance K 1982 A Reflectance Model for Computer Graphics *ACM Transactions on Graphics* Vol. **1** No.1, pp.7-24

Real-time Image Generation for Compressive Light Field Displays

G. Wetzstein, D. Lanman, M. Hirsch, and R. Raskar
MIT Media Lab

MIT Media Lab, 75 Amherst Street, Cambridge MA 02139, USA

E-mail: gordonw@media.mit.edu

Abstract. With the invention of integral imaging and parallax barriers in the beginning of the 20th century, glasses-free 3D displays have become feasible. Only today—more than a century later—glasses-free 3D displays are finally emerging in the consumer market. The technologies being employed in current-generation devices, however, are fundamentally the same as what was invented 100 years ago. With rapid advances in optical fabrication, digital processing power, and computational perception, a new generation of display technology is emerging: compressive displays exploring the co-design of optical elements and computational processing while taking particular characteristics of the human visual system into account. In this paper, we discuss real-time implementation strategies for emerging compressive light field displays. We consider displays composed of multiple stacked layers of light-attenuating or polarization-rotating layers, such as LCDs. The involved image generation requires iterative tomographic image synthesis. We demonstrate that, for the case of light field display, computed tomographic light field synthesis maps well to operations included in the standard graphics pipeline, facilitating efficient GPU-based implementations with real-time framerates.

1. Introduction

Within recent years, glasses-free 3D displays have started to enter the consumer market; display form factors now range from hand-held devices and tablets to large-scale home theaters. However, the underlying technology has not fundamentally changed in a century. Parallax barriers [1] and integral imaging [2] are at the core of most modern displays. Parallax barriers add a light-blocking barrier mask, consisting of arrays of vertical slits or pinholes, at a slight offset in front of a conventional 2D display. An observer looking at the screen only sees a subset of the rear pixels, the rest is optically blocked. Hence, an interleaved light field is displayed on the rear device so that the viewer always perceives the correct samples of a virtual 3D scene. Integral imaging achieves a similar effect by mounting an array of lenslets in front of a conventional 2D screen. While successful in displaying glasses-free 3D imagery, both parallax barriers and integral imaging significantly reduce the image resolution; parallax barriers are also very dark. Compressive light field displays employing stacks of two [3] or more [4] high-speed LCDs, stacked transparencies [5], or stacks of polarization-rotating liquid crystal panels [6] have been proposed to overcome the limitations of conventional light field display technology by providing brighter and higher-resolution imagery within wider fields of view and larger depths of field. For the purpose of this paper, we consider a light field to be a collection of two-dimensional images, each depicting a 3D scene from slightly different perspectives [7].

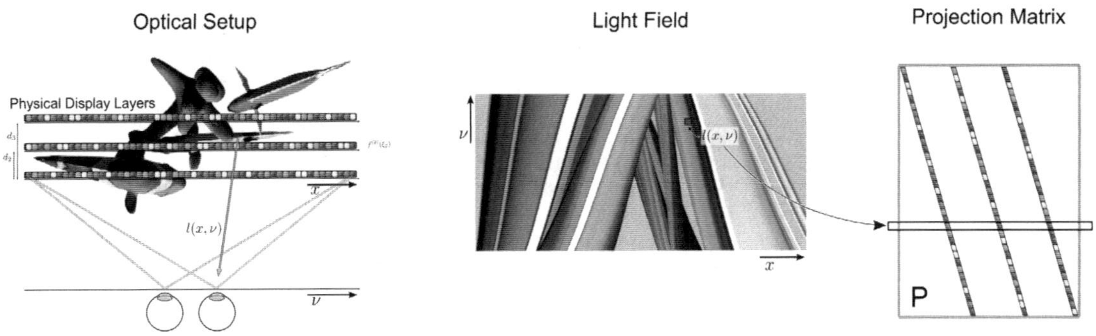

Figure 1. Compressive light field display with a three-layer display. (Left) A virtual 3D scene seen from above with physical LCD layers and viewer indicated. (Center) The corresponding light field is visualized with one spatial and one angular dimension. (Right) Tomographic light field synthesis requires the projection matrix that encodes the physical mapping from display pixels to emitted light field rays.

Compressive light field displays are enabled by the co-design of optical elements and computational processing algorithms. Rather than pursuing a "direct optical" solution (e.g., adding one more pixel to support the emission of one additional light ray), compressive displays aim to create flexible optical systems that have the capability to synthesize a compressed target light field. In effect, each pixel emits a superposition of light rays; through compression and tailored optical designs, fewer display pixels are necessary to emit a given light field, than would be demanded with a "direct optical" solution. The benefits of compressive displays, however, come at the cost of significantly increased computation. The patterns necessary to synthesize a compressed light field require inverse mathematical problems, such as computed tomography, to be solved. In this paper, we investigate efficient implementations of these algorithms and show that, for the specific case of light field synthesis, the underlying abstract mathematical routines directly map to operations supported in hardware on modern graphics processing units (GPUs). The connections between light field, virtual 3D scene, optical display, and mathematical constructs for a three-layer compressive light field display are illustrated in Figure 1.

Related work on 3D displays includes recent advances in parallax barriers, that now support viewer tracking [8] and high-speed panels allowing for time-shifted mask patterns [9]. Integral imaging today is capable of supporting 2D/3D switchable lenses [10] as well as the entire imaging pipeline including image acquisition, compression, and display [11]. Other types of glasses-free 3D displays include volumetric displays [12, 13, 14], stacks of switchable diffusers [15], directional backlighting [16], and multi-focal displays [17]. For a comprehensive review of stereoscopic and automultiscopic display technology, the interested reader is referred to [18].

2. Tomographic Light Field Synthesis

Louis Lumière, a pioneer of cinematographic image capture and display, was the first to propose multilayer display architectures in 1920 [19]. Lumière captured and exposed multiple glass film plates with a camera focused on different depths. A stack of these plates creates a three-dimensional effect, but the underlying problem actually requires an inverse problem to be solved. We show in this section how computed tomographic light field synthesis provides a powerful tool for analyzing multilayer displays and efficiently synthesizing light fields with them.

As shown in Figure 2, multilayer light field displays (e.g., stacks of transparencies or LCDs) are analyzed as general spatial light modulators that act in concert to recreate a light field by attenuating rays emitted by a uniform backlight. Since arbitrary oblique views may be

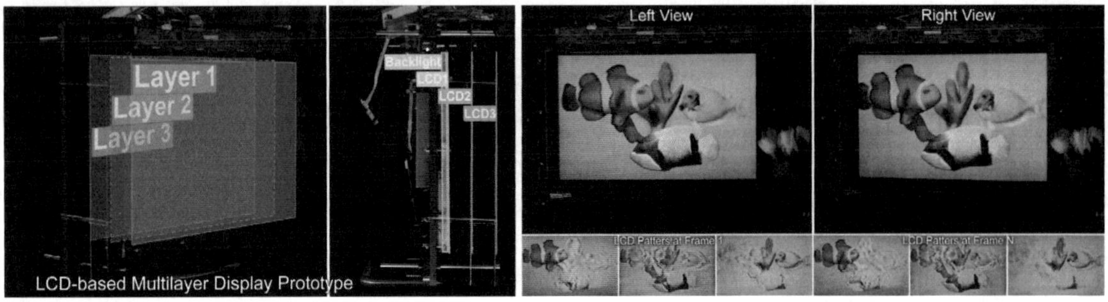

Figure 2. Wide field of view glasses-free 3D display using compressive light field displays. (Left) A prototype configured as a three-layer display; three see-through LCD panels are mounted in front of a uniform backlight. The panels are mounted on custom aluminum frames with driver electronics mounted on the frames. (Right) Photographs showing the display prototypes from two different perspectives. Multilayer light field displays support smooth motion parallax and binocular disparity at a high resolution for a large depth of field over a wide range of viewpoints.

inconsistent with any single attenuator, iterative tomographic reconstruction minimizes the difference between the emitted and target light fields, subject to physical constraints on attenuation. For 3D display, spatial resolution, depth of field, and brightness are increased, compared to conventional parallax barriers.

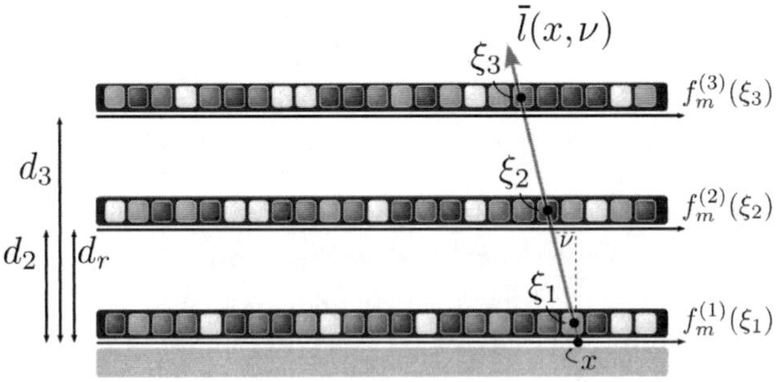

Figure 3. Ray diagram illustrating a multilayer light field display with a single, emitted light ray $\bar{l}\,(x,\nu)$ indicated.

The following analysis adopts an absolute two-plane parameterization of the 4D light field (i.e., the parameterization of rays of light propagating in free space). As shown in Figure 3, an emitted ray is parameterized by its intersection with two planes. Thus, the ray (x, y, ν, υ) intersects the "rear" plane at the point (x, y) and the front plane at the point (ν, υ), with both coordinate systems having an origin in the top-left corner and the relative coordinates $(a = \nu - x, b = \upsilon - y)$. In the following, we analyze two scenarios in "flatland" 2D with straightforward extensions to full 4D light fields and 2D layer patterns: attenuation tomography and tomographic image synthesis for layers of polarization-rotating layers.

2.1. *Tomographic Light Field Synthesis with Attenuating Layers*

An N-layer stack of optical attenuation functions (i.e., masks) is defined using the absolute light field parameterization, such that the mask on layer n is given by the transmittance function. Following Wetzstein et al. [5], the emitted 2D light field is given by the following expression:

$$l(x,\nu) = \prod_{n=1}^{N} f^{(n)}\left(x + (d_n/d_r)\,\nu\right), \tag{1}$$

where d_n is the distance of layer n from the x-axis. Taking the logarithm gives an additive model:

$$\bar{l}(x,\nu) = \sum_{n=1}^{N} \ln f^{(n)}\left(x + (d_n/d_r)\,\nu\right) = -\sum_{n=1}^{N} \alpha^{(n)}\left(x + (d_n/d_r)\,\nu\right), \tag{2}$$

where $\alpha^{(n)}(\xi) = -\ln f^{(n)}(\xi)$ is the absorbance function for layer n. We consider a discrete light field matrix $\bar{l}$ (containing the logarithm of the target intensities for each ray) and discrete attenuation vectors $\alpha^{(n)}$ for each layer n. With this notation, we formulate light field display using a multilayer attenuation-based display as the following constrained linear least-squares optimization problem:

$$\arg\min_{\alpha} \left\|\bar{l} - P\alpha\right\|^2, \quad \text{for } \alpha \geq 0. \tag{3}$$

Here, P is the projection matrix such that the total attenuation along ray is given by the summation $\sum_{j=1}^{J} P_{ij}\alpha_j$. The following section discusses a very related problem: finding an optimal set of polarization-rotation values in a stack of LCDs.

2.2. Tomographic Polarization Field Synthesis

While stacks of light-attenuating LCDs are successful in light field synthesis, they also reduce the light transmission of the display significantly. We explore optically-efficient display implementations using stacks of liquid crystal panels stripped off their polarizers [6]. The resulting *Polarization Field* display contains a stacked set of liquid crystal panels with a single pair of crossed linear polarizers. Each layer is modeled as a spatially-controllable polarization rotator, rather than a spatial light modulator.

Consider a pair of crossed linear polarizers enclosing a single liquid crystal cell; under the polarization rotator model, the transmitted intensity I is given by Malus' law:

$$I = I_0 \sin^2(\theta), \tag{4}$$

where I_0 is the intensity after passing through the first polarizer and θ is the angle of polarization after passing through the liquid crystal cell. The emitted 2D light field $l(x,\nu)$ is given by:

$$l(x,\nu) = \sin^2\left(\bar{\theta}(x,\nu)\right) = \sin^2\left(\sum_{n=1}^{N} \phi^{(n)}\left(x + (d_n/d_r)\,\nu\right)\right), \tag{5}$$

where $\phi^{(n)}(\xi)$ denotes the polarization state rotation induced at point ξ along layer n. Under this model, ray (x,ν) intersects the N layers, accumulating incremental rotations at each intersection, such that emitted polarization field approximates the target polarization field, given by

$$\theta(x,\nu) = \pm\sin^{-1}\left(\sqrt{l(u,a)}\right) \bmod \pi, \tag{6}$$

Under these assumptions, the principal value of the arcsine ranges over $[0, \pi/2]$. Note, with full generality, the target polarization field is multi-valued and periodic.

Similar to stacks of attenuating layers, light field display using multilayer polarization rotators can be formulated as a constrained linear least-square optimization problem. We consider a discrete parameterization for which the emitted polarization field is represented as a column vector $\overline{\theta}$ with I elements, each of which corresponds to the angle of polarization for a specific light field ray. Similarly, the polarization state rotations are represented as a column vector ϕ with J elements, each corresponding to a display pixel in a given layer. This yields the following linear model:

$$\overline{\theta}_i = \sum_{j=1}^{J} P_{ij}\phi_j, \tag{7}$$

where $\overline{\theta}_i$ and ϕ_j denote ray i and pixel j, respectively. Using this notation, an optimal set of polarization state rotations is found by solving the following optimization problem:

$$\arg\min_{\phi} \|\theta - P\phi\|^2, \quad \text{for } \phi_{min} \leq \phi \leq \phi_{max}. \tag{8}$$

As discussed in the following subsections, we use the simultaneous algebraic reconstruction technique (SART) to solve both Equations 3 and 8.

3. Simultaneous Algebraic Reconstruction Technique (SART)

We observe that the tomographic image synthesis problem, i.e. Equations 3 and 8, can be solved by adapting the simultaneous algebraic reconstruction technique (SART). As proposed by Andersen and Kak [20] and further described by Kak and Slaney [21], SART provides an iterative solution wherein the estimate $\phi^{(q)}$ at iteration q is given by

$$\phi^{(q)} = \phi^{(q-1)} + v \circ (P^{\top}(w \circ (\theta - P\phi^{(q-1)}))), \tag{9}$$

where $\circ$ denotes the Hadamard product for element-wise multiplication and elements of the w and v vectors are given by

$$w_m = \frac{1}{\sum_{n=1}^{N} P_{mn}} \quad \text{and} \quad v_n = \frac{1}{\sum_{m=1}^{M} P_{mn}}. \tag{10}$$

After each iteration, additional constraints on $\phi^{(q)}$ are enforced by clamping the result to the feasible rotation range. Note that ϕ can simply be replaced by α to solve the attenuation-based computed tomography problem. Building upon the Kaczmarz method for solving linear systems of equations [22], SART is shown to rapidly converge to a solution approaching the fidelity of that produced by alternative iterative methods, including trust region and conjugate gradient descent techniques [21]. The pseudo-code for the above equation is given in Table 1. In the next section, we discuss an efficient, GPU-based implementation of the SART algorithm that allows for real-time framerates in tomographic light field displays.

4. Efficient Implementation of SART

The SART algorithm introduced in the last section is well-suited for parallel processing on programmable GPUs [23]. Our code is programmed in C++, OpenGL, and Cg. Light fields are rendered in real-time using OpenGL, followed by several iterations of the GPU-based SART implementation.

Tomographic light field synthesis is not restricted to virtual, rendered scenes. Captured multiview footage, either acquired with a camera array, a depth camera such as the MS Kinect, or a single, moved camera can be displayed in the same manner. In fact, recently popularized light field cameras, such as the Lytro, capture exactly the multiview data that is required as

```
Algorithm SART - Generic

variables P, b, x, w, v, lb, ub

x = zeros();
w = 1 / (P1);
v = 1 / (P^T 1);

for all iterations k
    x = x + v ∘ (P^T (w ∘ (b − Px)));
    x = clamp(x, [lb ub]);
end
```

Table 1. SART as a generic update routine. An initial guess of the solution x, here initialized as zeros, is iteratively updated until the algorithm is converged or the maximum number of iterations is reached.

the input to our display. In the remainder of this paper, we assume that a target light field is available; in our implementation, we render it interactively using OpenGL. Each view of the light field depicts the same 3D scene from a slightly different perspective. While many possibilities of accelerating the rendering part of the implementation exist, for instance by exploiting inter-view or inter-frame correlations, this paper focuses on the tomographic part of the image generation, which equally applies to captured light fields.

As outlined in Table 1, SART requires an initial guess of the solution, which can either be all zeros or random values, and iteratively updates that. Only very few different operations are actually required: a forward matrix-vector multiplication Ax, a backprojection using the transpose matrix $A^T x$, element-wise multiplications, sums and differences as well as a clamping operation of the updated solution. Any element-wise operations such as sums, differences, Hadamard products, and the clamping operation can easily be implemented as per-pixel operations in any shader language. The critical operations for an efficient implementation are the forward and backward projections that will be discussed in the following.

4.1. The Projection Matrix

For the special case of tomographic light field synthesis with multilayer displays, the projection matrix P encodes the mapping between light field rays and layer pixels. This concept and the connections between light field rays, physical display layers, target light field, and mathematical projection matrix are illustrated in Figure 1. Each ray $l(x, \nu)$ of the target light field corresponds to one row in P, whereas the columns of P correspond to the pixels of the layers. Except for the boundary cases, each light field ray intersects each layer of the stack exactly once; hence, each matrix row has three non-zero elements. While P is extremely large, it is also very sparse. Note that SART does not require the matrix to be formed at any point as long as matrix-vector multiplications can be directly computed.

4.2. Forward Projection as Perspective Rendering

The matrix-vector product between any vector x and P basically simulates the light field that is observed when the layer pixel values encoded in x are displayed on the multilayer display. Instead of precomputing a large, sparse matrix, this operation can efficiently be implemented on the GPU by rendering each individual layer with the corresponding pattern assigned as a texture from all the perspectives of the light field. Instead of rendering each view for each layer, we initialize one offline buffer for each view. During runtime, each view buffer is activated in sequence, the corresponding projection matrix set, the accumulation buffer activated, and all layers are rendered with the depth test disabled. The accumulation buffer allows for the contribution of all layers, as seen from a single perspective, to be added up in hardware. All buffers use a precision of 16 bits to allow for non-integer values that can be larger than 255.

These operations are outlined in Table 2. Our implementation, tested on both Windows and Linux, is programmed in C++ and uses OpenGL and Cg shaders. The algorithm assumes a light field with N distinct views and K layers positioned at user-defined depths along the optical axis. Intermediate quantities, including the target light field views and temporary variables storing weights and layer patterns, are internally rendered into 16-bit offline framebuffers (i.e., Framebuffer Objects, FBOs) before the optimized patterns are displayed on the individual layers. Only three separate Cg fragment programs are required, each performing the action implied by their names.

4.3. Backprojection as Projective Texture Mapping
Another critical operation is the multiplication of the transpose projection matrix P^T with an arbitrary vector x. In this case, the vector x does not contain layer pixel values but intensity values of the light field. A backprojection smears the intensity value of each light ray back into the layer pixels and accumulates them there. In an unoptimized, sequential implementation, the intensity value of each light ray is simply added to all layer pixels that the ray intersects. An efficient implementation can perform many of these operations in parallel. Projective texture mapping, as supported in hardware by OpenGL, is one way of implementing these operations. Again, 16 bit offline buffers are initialized, this time for each layer. During runtime, each of these is activated in sequence. For each of the layer buffers, the accumulation buffer is enabled and each view of the light field encoded in x is projected using the corresponding perspective matrix set as the current texture matrix. This basically simulates a number of projectors, each placed at one viewpoint of the target light field displaying the corresponding image in x. The offline buffer for the currently active layer simply sums all the projectors' contributions via the enabled accumulation buffer. The same procedure is repeated for each layer in the display stack as outlined in Table 2.

4.4. Temporal Coherence in Dynamic Scenes
SART, as an iterative optimization algorithm, requires multiple iterations to converge to the optimal solution. For the application of tomographic light field synthesis, we experimentally determined five to ten iterations to be sufficient when the solution is initialized with zeros. For dynamic scenes, where the 3D scene or parts of it change from one frame to another, estimates for the previous frame seed the optimization for the current frame. For static scenes, this effectively implements an increasing number of SART iterations over time, while providing a suitable initialization for successive frames in a dynamic environment. In this "seeding" mode, fast moving objects observe a slight motion blur, while the image quality immediately improves when the motion stops. Figure 4 plots the performance of our GPU implementation for three different display configurations.

5. Discussion
Within the past few years, stereoscopic and automultiscopic 3D displays have started to enter the consumer market. Adoption has been slow, however, mainly due to the high bandwidth requirements, low resolution, and dim images provided by conventional multiview, glasses-free 3D displays. Compressive light field displays offer higher-resolution images to be displayed at a significantly reduced bandwidth. As demonstrated in this paper, the increase in required computational processing can be handled by efficient implementations of tomographic image synthesis on programmable graphics hardware. The underlying mathematical operations map directly to standard operations supported on hardware, such as perspective rendering, projective texture mapping, and accumulation buffering.

```
Algorithm SART - Implementing Light Field Synthesis using OpenGL

variables FBO_LF[N], FBO_LF_TMP[N], FBO_LAYERS[K], FBO_W[N], FBO_V[K], DEPTH[K]

function displayLoop()
    if not initialized
        initialize all FBO_W, FBO_V
    end
    drawLightField();
    runSART();
    drawReconstructedLayers();
end

function drawLightField()
    for all light field views i
        activate FBO_LF[i]
        set perspective i
        drawScene(); // render desired 3D scene (e.g., a teapot)
    end
end

function runSART()
    for all iterations k
        // 1. compute Ax^(k)
        for all light field views i
            activate FBO_LF_TMP[i]
            enable BLEND_MODE
            set perspective i
            for all layers l
                draw 2D plane at DEPTH[l] textured with FBO_LAYERS[l]
            end
        end
        // 2. given Ax^(k), compute W (b - Ax^(k))
        for all light field views i
            activate FBO_LF_TMP[i]
            activate CG_SHADER_MULTIPLY_SUBTRACT(FBO_W[i],FBO_LF[i],FBO_LF_TMP[i])
            draw orthographic 2D plane with normalized texcoords
        end
        // 3. given W (b - Ax^(k)), compute VA^T (W (b - Ax^(k)))
        for all layers l
            activate FBO_LAYER[l]
            activate BLEND_MODE
            for all light field views i
                set projective texture to perspective i
                activate CG_SHADER_MULTIPLY(FBO_V[l],FBO_LAYER[l])
                draw orthographic 2D plane with automatic texcoord generation
            end
        end
        // 4. enforce constraints by clamping values outside feasible range
        for all layers l
            activate FBO_LAYER[l]
            activate CG_SHADER_CLAMP(FBO_LAYER[l])
            draw orthographic 2D plane textured with FBO_LAYER[l]
        end
    end
end

function drawReconstructedLayers()
    for all layers l
        set viewport for display l
        draw orthographic 2D plane textured with FBO_LAYERS[l]
    end
end
```

Table 2. Tomographic light field synthesis using SART. In this particular application, the abstract mathematical operations required for a generic tomographic reconstruction map to standard operations such as perspective rendering and projective texture mapping.

We strongly believe that the key for next-generation, glasses-free 3D display technology is the co-design of display optics and computational processing. In all of our prototypes, compressive light field synthesis is optically facilitated by stacks of conventional displays while computationally exploited with efficient implementations of advanced algorithms. While the physical limits of display optics can only be slowly overcome, available processing power has been growing rapidly; in the near future, we expect the computational side of compressive displays to increasingly contribute to advances in display capabilities. Compressive light field displays already achieve a higher resolution, brighter images, wider viewing zones, and larger depth ranges than conventional technology. We are inspired by the promise that future generations of

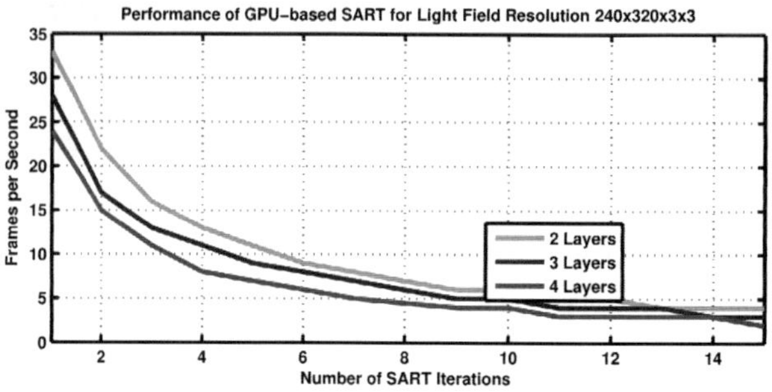

Figure 4. Performance of the GPU-based SART implementation as a function of varying numbers of SART iterations (per frame) and varying numbers of polarization-rotating or light-attenuating layers. The light field resolution is 320×240 spatial samples and 3×3 angular samples; layers have a similar spatial resolution.

compressive displays approach the realism of the physical world with technology that is available today.

Acknowledgments
We thank the Camera Culture Group for their support and also the MIT Media Lab sponsors as well as NVIDIA Research. Gordon Wetzstein was supported by the DARPA SCENICC program and by an NSERC Postdoctoral Fellowship (PDF). Douglas Lanman was supported by NSF Grant IIS-1116452 and by the DARPA MOSAIC program. Ramesh Raskar was supported by an Alfred P. Sloan Research Fellowship and a DARPA Young Faculty Award.

References

[1] Ives F E 1903 Parallax stereogram and process of making same U.S. Patent 725,567
[2] Lippmann G 1908 *Journal of Physics* **7** 821–825
[3] Lanman D, Hirsch M, Kim Y and Raskar R 2010 *ACM Trans. Graph. (SIGGRAPH Asia)* **29**(6) 163:1–163:10
[4] Wetzstein G, Lanman D, Hirsch M and Raskar R 2012 *ACM Trans. Graph. (SIGGRAPH)* **31**(4) 1–11
[5] Wetzstein G, Lanman D, Heidrich W and Raskar R 2011 *ACM Trans. Graph. (SIGGRAPH)* **30**(4) 1–11
[6] Lanman D, Wetzstein G, Hirsch M, Heidrich W and Raskar R 2011 *ACM Trans. Graph. (SIGGRAPH Asia)* **30**(6) 1–9
[7] Levoy M and Hanrahan P 1996 *ACM SGGRAPH* pp 31–42
[8] Perlin K, Paxia S and Kollin J S 2000 *ACM SIGGRAPH* pp 319–326
[9] Kim Y, Kim J, Kang J M, Jung J H, Choi H and Lee B 2007 *Optics Express* **15** 18253–18267
[10] Jacobs A, Mather J, Winlow R, Montgomery D, Jones G, Willis M, Tillin M, Hill L, Khazova M, Stevenson H and Bourhill G 2003 *Sharp Technical Journal* 15–18
[11] Matusik W and Pfister H 2004 *ACM Trans. Graph. (SIGGRAPH)* **23**(3) 814–824
[12] Jones A, McDowall I, Yamada H, Bolas M and Debevec P 2007 *ACM Trans. Graph. (SIGGRAPH)* **26**(3) 40:1–40:10
[13] Cossairt O S, Napoli J, Hill S L, Dorval R K and Favalora G E 2007 *Applied Optics* **46** 1244–1250
[14] Favalora G E 2005 *IEEE Computer* **38** 37–44
[15] Sullivan A 2003 *SID Digest* vol 32 pp 207–211
[16] Travis A, Large T, Emerton N and Bathiche S 2009 *Optics Express* **17** 19714–19719
[17] Akeley K, Watt S J, Girshick A R and Banks M S 2004 *ACM Trans. Graph. (SIGGRAPH)* **23**(3) 804–813
[18] Urey H, Chellappan K V, Erden E and Surman P 2011 *Proc. IEEE* **99** 540–555
[19] Lumière L 1920 *Comptes rendus hebdomadaires des séances de l'Académie des sciences* 891–896
[20] Andersen A and Kak A 1984 *Ultrasonic Imaging* **6** 81–94
[21] Kak A C and Slaney M 2001 *Principles of Computerized Tomographic Imaging* (Society for Industrial Mathematics)
[22] Kaczmarz S 1937 *Bull. Acad. Pol. Sci. Lett. A* **35** 335–357
[23] Keck B, Hofmann H, Scherl H, Kowarschik M and Hornegger J 2009 *SPIE* vol 7258

Limits of computational white-light holography

Sebastian Mader[1,2], Tomasz Kozacki[1] and Wayne Tompkin[2]

[1] Mechatronics Department, Warsaw University of Technology, 02-525 Warsaw, Poland
[2] Optical Technology Division, OVD Kinegram AG, 6301 Zug, Switzerland

Email: sebastian.mader@kinegram.com

Abstract. Recently, computational holograms are being used in applications, where previously conventional holograms were applied. Compared to conventional holography, computational holography is based on imaging of virtual objects instead of real objects, which renders them somewhat more flexibility. Here, computational holograms are calculated based on the superposition of point sources, which are placed at the mesh vertices of arbitrary 3D models. The computed holograms have full parallax and exhibit a problem in viewing that we have called "ghosting", which is linked to the viewing of computational holograms based on 3D models close to the image plane. Experimental white-light reconstruction of these holograms showed significant blurring, which is explained here based on simulations of the lateral as well as the axial resolution of a point image with respect to the source spectrum and image distance. In accordance with these simulations, an upper limit of the distance to the image plane is determined, which ensures high quality imaging.

1. Introduction

Conventional holography has been applied in various forms such as off-axis holography, image plane holography and rainbow holography [1]. Each of these techniques requires rather sophisticated equipment and skilled experimentalists to achieve high quality results. All these methods require further "real" objects from which the hologram will be made. These objects can be simple binary masks, gray scale masks and/or even complex 3D models.

In this work we focus on computational holograms, which can in principle replace all hologram types mentioned above. The main advantage is that one is not limited by the real objects available to the experimentalists, but only to the imagination of the hologram designer. Any 2D object, e.g. a gray scale mask or a virtual 3D object might be suitable, but we focus here on holograms based on virtual 3D mesh models, which can be reconstructed with white light. The computation is carried out on a standard desktop computer, while the writing of the hologram as a phase profile is done by a laser writer or a spatial light modulator SLM of the type Holoeye Pluto-VIS.

The main difference in terms of reconstruction quality of these writing devices is the minimum writable pixel size. The limited spatial resolution of the holograms places restrictions on the object size, the object placement, the field of view and the imaging resolution. The latter is numerically analyzed in section 4 in great detail, while in section 3 an effect is described, which we have called "ghosting". The cause of the effect is described and two suggestions to reduce it are proposed. Before commenting however on the limiting factors of computational holography, we would like to give a brief description of the point source method, which is used to compute all the holograms in this work.

2. The point source method

Several complex methods to create computer-generated holograms have been proposed in literature [2,3], but we limited ourselves to an idea in close analogy to the one presented by Yoshikawa [4]. The virtual object is assumed to be represented by a point cloud, which has been created in this work based on the vertex coordinates of a 3D mesh model, see figure 1.

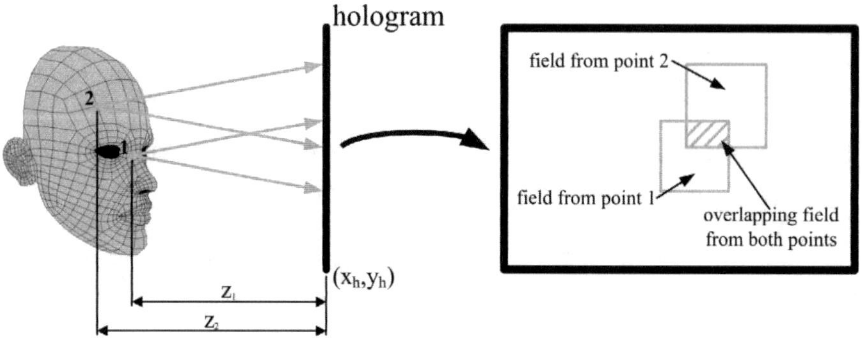

Figure 1. Visualization of the hologram generation process by the point source method

Each point of the virtual object is assumed to emit a spherical wave of unit amplitude,

$$U_i(x_{\mathrm{h}}, y_{\mathrm{h}}) = \exp(ikr)/r, \quad r = [(x_{\mathrm{h}} - x_i)^2 + (y_{\mathrm{h}} - y_i)^2 + z_i^2]^{-1/2}, \tag{1}$$

where x_{h} and y_{h} denote the spatial coordinates in the hologram plane ($z_{\mathrm{h}} = 0$) and x_i, y_i, z_i denote the coordinates of the i-th model point. The spatial extent of the field in the hologram plane is limited by the following aliasing criteria:

$$\begin{aligned} x_{\mathrm{h}} &\in \{x_{\mathrm{i}} - |z_{\mathrm{i}}| \cdot \tan(\eta_{\max}), x_{\mathrm{i}} + |z_{\mathrm{i}}| \cdot \tan(\eta_{\max})\}, \eta_{\max} = \sin^{-1}(\lambda/2\Delta_{\mathrm{x}}), \\ y_{\mathrm{h}} &\in \{y_{\mathrm{i}} - |z_{\mathrm{i}}| \cdot \tan(\theta_{\max}), y_{\mathrm{i}} + |z_{\mathrm{i}}| \cdot \tan(\theta_{\max})\}, \theta_{\max} = \sin^{-1}(\lambda/2\Delta_{\mathrm{y}}), \end{aligned} \tag{2}$$

where $\eta_{\max}$ and $\theta_{\max}$ are the maximum first-order diffraction angles and Δ_{x} and Δ_{y} are the pixel sizes of the hologram in x- and y-direction, respectively. Obeying the aliasing criteria ensures that no spatially shifted copies of the object are reconstructed since this would degrade the image visibility and the image perception. Thus one should only calculate the field values within the inner square drawn in figure 2 for each point of the model. Please note however that obeying the aliasing criteria also leads to the fact that the aperture size of each point depends on its distance to the hologram plane. Thus, points closer to the hologram plane may be reconstructed less brightly than the ones further away from it due to the difference in aperture size.

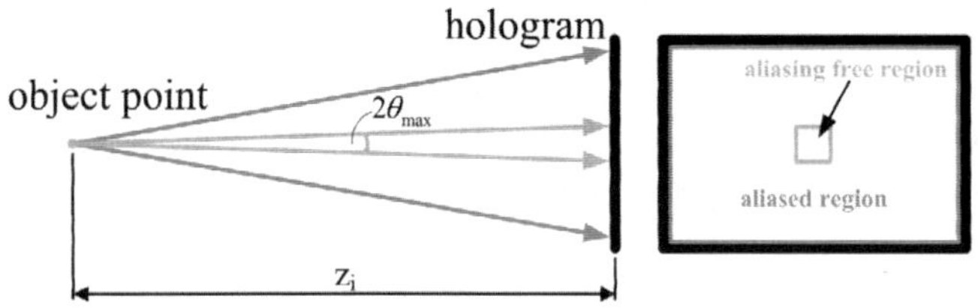

Figure 2. Scheme to show the aliasing free and the aliased region of the hologram

If one obeys the aliasing criteria, the reconstruction shows a single object point, while ignoring it causes the formation of additional and spatially shifted image points as shown in figure 3. The top row of the figure shows two holograms of an object point placed at $z_{\mathrm{i}} = 90\mathrm{mm}$. The left one was created according to the aliasing criteria (2), while the right one was not. Below each hologram, one finds the corresponding numerical reconstruction at $z_{\mathrm{i}} = 85\mathrm{mm}$. The distance 85mm was chosen, since a reconstruction at 90mm leads to point images with such a small extent that they a barely noticeable in the full field plot of the numerical reconstruction.

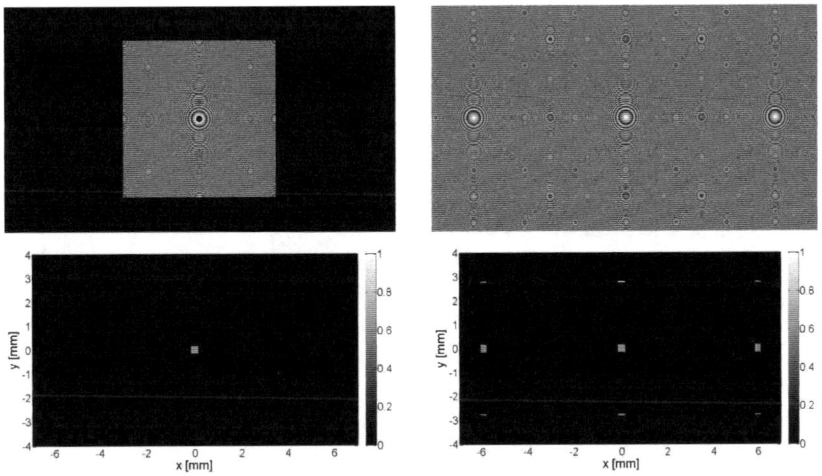

Figure 3. Holograms of an object point at $z_i = 90$mm without aliasing (top left) and with aliasing (top right) and the corresponding numerical reconstructions at 85mm (bottom row)

The total field in the hologram plane is obtained by superimposing the fields of all N_p points of the virtual object and multiplying the sum by a reference field U_r

$$U_h(x_h, y_h) = U_r^*(x_h, y_h) \sum_{i=1}^{N_p} U_i(x_h, y_h). \tag{3}$$

The field in the hologram is sampled at a resolution depending on the reconstruction device, e.g. at 8 micron if the SLM Holoeye Pluto-VIS is used for reconstruction, and the reference field may be given by a plane wave impinging under an angle θ with respect to the x_h-z-plane

$$U_r^*(x_h, y_h) = \exp[iky_h sin(\theta)]. \tag{4}$$

The angle θ should be chosen to be the opposite of the angle used during reconstruction, so that the reconstruction with a plane wave under this angle will lead to a cancelling of the reference field term and only the desired field that reconstructs the point sources will propagate. The reconstruction devices used in this work are only able to modify the phase of an impinging light wave and thus only the phase information ϕ_h can be explored

$$\phi_h(x_h, y_h) = \text{angle}[U_h(x_h, y_h)] = \tan^{-1}\{\text{Re}[U_h(x_h, y_h)]/\text{Im}[U_h(x_h, y_h)]\} \tag{5}$$

and the amplitude information at the hologram plane is ignored from this point on. Furthermore, the phase ϕ_h is approximated by a finite numbers of phase levels ph_{lev} in the range $[0, 2\pi(1 - 1/ph_{lev})]$. Thus the discrete phase values ϕ_{ch} may be computed by

$$\phi_{ch}(x_h, y_h) = \text{floor}[(ph_{lev} - 1) \cdot \phi_h(x_h, y_h)/2\pi]. \tag{6}$$

The values of ϕ_{ch} are for example all integer values in the range $[0,255]$ for $ph_{lev} = 256$, which can then be stored as a grey scale bitmap with 8-bit depth. This format is compatible to the used SLM as well as to the laser writer which was used. Equation (6) thus represents the final form of the computer-generated holograms and we can now turn our attention to a particular problem due to calculation and writing of ϕ_{ch} on a regular grid.

3. Self-ghosting due to higher order images

The cause of this problem will be explained using figure 4 (top), where the propagation of the holograms "zero-order field" created by +/-1st- order diffraction of a single image point is indicated by two solid arrows. The angle of each line with respect to the optical axis is given by θ_{max} since the aliasing condition (2) is assumed to be obeyed.

If one observes the image point at a distance of 250mm, the linear field of view is 19.3mm wide for the particular point shown in figure 6 (top) and $\Delta_x = 8\mu m$. In addition to the zero-order field, the holograms "+1st-order field" is indicated by two dashed arrows. The angles of these arrows are calculated by taking the sum of $\pm\theta_{max}$ and the +1st-order diffraction angle of the hologram, which is $3.81°$ for $\Delta_x = 8\mu m$ and $\lambda = 532nm$. The hologram thus acts as a grating since it is written on a regular pitch Δ_x. Please note that the diffraction angle of the hologram is twice as large as the maximum angle $\theta_{max} \cong 1.91°$ of the intended +/-1st-order diffraction for these parameters.

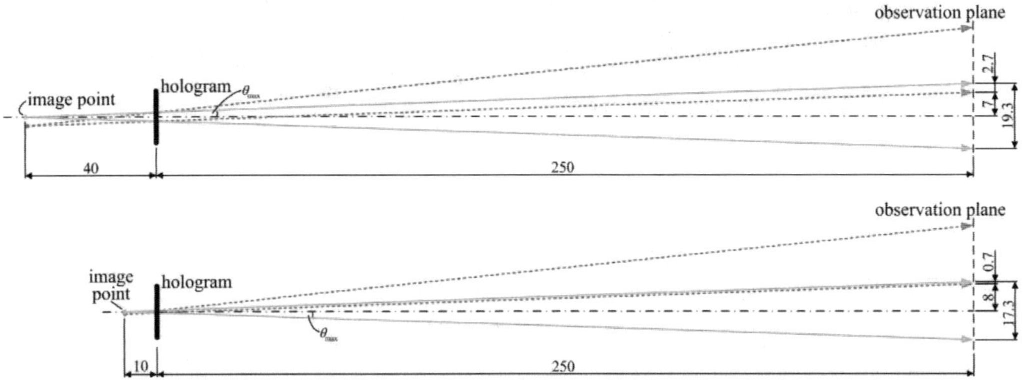

Figure 4. Scheme to visualize self-ghosting of virtual image point

One can see in figure 6 (top) that both fields intersect each other 7mm above the optical axis and the size of the overlap is approximately 2.7mm. The observer thus sees the holograms zero- and the first-order field simultaneously, an effect we have called "self-ghosting". Figure 6 (bottom) shows that the effect becomes less dominant with decreasing distance of the virtual image to the hologram plane (overlap of only 0.7mm), but due to the finite size of the observers pupil it is still visible. For real object points, the fields do not overlap, but again the effect might be observed due to observers finite pupil size. **The effect can however be reduced by limiting the field of view or avoided by writing the hologram on a non-periodic grid.** Further details and the description of two point ghosting are given in [5].

4. Imaging limitations of computational white-light holograms

The issue of imaging resolution will be discussed in the following based on a point image. In order to determine the resolution, one may first distinguish between lateral and axial resolution of an image. The latter will be related to the focal tolerance, but first the lateral resolution of a point image will be considered.

4.1. Lateral resolution

Kozacki treated in [6] the resolution question in the context of the Wigner domain and he presented a rather simple graph that showed the relation between lateral resolution, imaging distance and point position x_i in the image plane. Figure 5 was created in close analogy to figure four of his paper and it shows that the image of a point placed within the green cone will be reconstructed with full lateral resolution, while resolution and maximum intensity of a point placed outside the cone decreases linearly till the intersection of the image plane with the upper or lower arrow. The area enclosed by the arrows is the maximum range in which the object can be placed if resolution is of no importance. Our holograms explore however the full resolution at the design wavelength λ and thus the virtual objects are always placed within the green area of full resolution. This holds true for real and virtual imaging points. To determine the resolution of our holograms we used the z-position $z_{\max}$ of the last point, which can be imaged with the holograms full aperture of width $2a$, which is given by

$$z_{\max}(\Delta_x, a, \lambda) = a/\tan\eta_{\max} \cong 2a\Delta_x/\lambda, \quad (7)$$

for small ratios of $\lambda/\Lambda_{\min} = \lambda/(2\Delta_x)$. It is the only point that is imaged with full resolution and full aperture of the hologram and this special point is selected for all results described below.

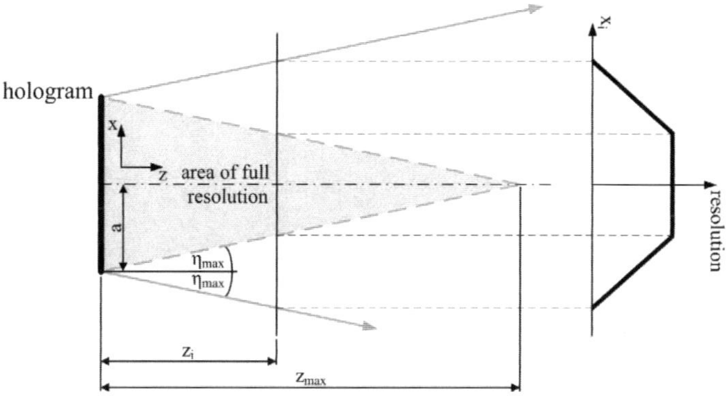

Figure 5. Lateral resolution of a point image at the image plane

In agreement with [6] the resolution of a phase-only hologram of a point image is given in Fresnel approximation at the design wavelength by

$$I_i(x_i, z_i) = (2a/\lambda z_i)^2 \mathrm{sinc}^2(2ax_i/\lambda z_i). \quad (8)$$

In our case the full aperture is contributing to the image formation process and thus the relation $z_i = z_{\max}$ holds true, where $z_{\max}$ is given by (7), which leads to

$$I_i(x_i, z_{\max}) \cong (B_f)^2 \mathrm{sinc}^2(B_f x_i), \ B_f = 1/\Delta_x, \quad (9)$$

Equation (9) shows two interesting properties of the point image. First, the lateral resolution may be defined as $2\Delta_x$, if the distance between the two first zeros of the sinc function is used to quantify resolution (other definitions see table 1) and second, the resolution does not depend on the wavelength. However, the distance $z_{\max}$ given by (7) varies with wavelength and thus also the size of the cone of full resolution. The cone is smaller for longer wavelengths and vice versa. Please note further that the hologram is designed for a specific wavelength λ and the analytic results above only hold true for this particular wavelength.

Table 1. Resolution values depending on determination criteria

Criteria	FWHM	$1/e^2$	distance between first two zeros
Resolution [Δ_x]	0.8859	1.4000	2.0000

4.2. Lateral extent of a point image by white-light reconstruction

In the following it is assumed that the spatial resolution is limited, while the phase resolution is still unlimited as in the analytical results above. The phase values are usually sampled with 8bit resolution (256 values) in the range $[0,2\pi]$ and thus the errors introduced by the finite phase resolution are assumed to be fully negligible here. The transfer function of the computational hologram T_{ch} [7] is thus given by

$$T_{ch}(x_h, z_{\max}, a) = \mathrm{rect}(x_h/2a)\exp\left[i\phi_{ch}(x_h, z_{\max})\right], \phi_{ch}(x_h, z_{\max}) = \mathrm{mod}(-kx_h^2/2z_{\max}, 2\pi) \quad (10)$$

where $z_{\max}(\Delta_x, a) = a/\tan\eta_{\max}$ as defined above. The phase values of (10) are sampled with the resolution of the hologram Δ_x and calculated at the center of each pixel and the reconstruction is performed numerically based on the angular spectrum decomposition method. One of the properties of this propagation method is that the image has the same spatial resolution as the hologram. As the simulations are performed for $\Delta_x = 1, 2,$ and 8μm the spatial resolution of the image might be insufficient to determine the resolution. Each phase value is thus resampled on N points within one pixel without interpolation to achieve a resolution better or equal to 1μm. The resampling is visualized in figure 6, where the phase value of the first pixel is resampled N times.

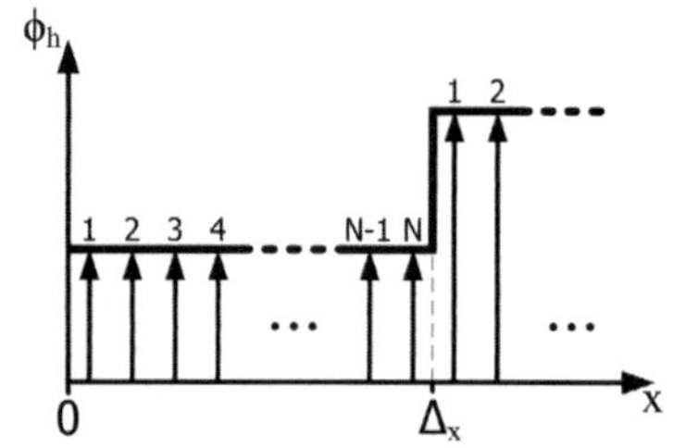

Figure 6. Scheme of the phase resampling

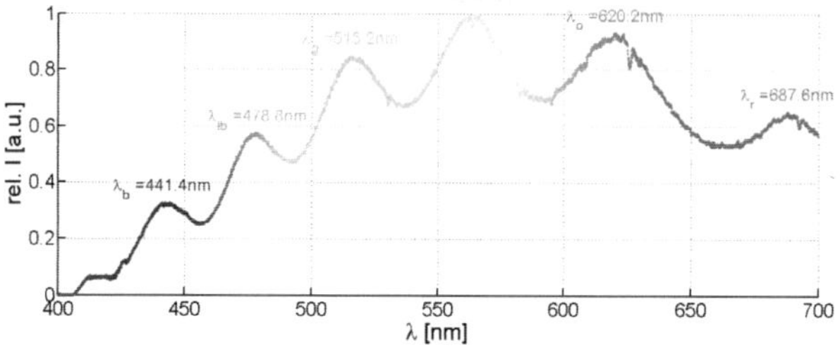

Figure 7. Measured spectrum of a halogen spot lamp

To model the decrease in resolution, it is assumed that the light source emits plane wavefronts with various wavelengths, which are totally independent of each other and normally incident on the hologram. Consequently, the intensity distribution $I_{\text{phys}}(x_i, z_i, \Delta\lambda)$ at each imaging distance z_i is the incoherent sum of the wavelength-dependent intensity distributions I_i multiplied by the wavelength-dependent emission strength $E_{\text{rel}}(\lambda)$ of the light source

$$I_{\text{phys}}(x_i, z_i, \Delta\lambda) = \sum_{\lambda} E_{\text{rel}}(\lambda) \cdot I_i(x_i, z_i, \lambda). \quad (11)$$

The modeling is based on two different spectra, a simple uniform spectrum and the spectrum of a halogen spot lamp shown in figure 7. The halogen spot lamp was used in various experiments and its spectrum was measured using the spectrometer Thorlabs CCS100 with a resolution better than 0.5nm FWHM @ 435nm. Since halogen spot lamps are quite commonly used light sources in holography and the spectrum of the sun is in "first approximation" uniform, both cases are of practical interest.

Equation (11) takes into account the wavelength-dependent emission strength of the sources, which leads to the "physical" light distribution in space I_{phys}; but to determine the observers imaging resolution, one must also take into account the wavelength-dependent sensitivity $S_{\text{rel}}(\lambda)$ of the observer itself

$$I_{\text{human}}(x_i, z_i, \Delta\lambda) = \sum_{\lambda} S_{\text{rel}}(\lambda) \cdot E_{\text{rel}}(\lambda) \cdot I_i(x_i, z_i, \lambda). \quad (12)$$

Figure 8 shows the relative sensitivity of the human eye, where the crosses mark the values given in [8] and the dots the linearly interpolated values in between them. One can infer from the plot that the human eye is most sensitive around 555nm, which is very close to the typical design wavelength $\lambda = 532\text{nm}$, which was also used here together with $\Delta_x = 1, 2$ or 8μm and $N = 1, 3$ or 9. The reconstruction wavelength was varied within the range [400nm, 700nm] in steps of 1.0nm and the reconstruction distances were varied in 10 micron steps around $z_{\max}$.

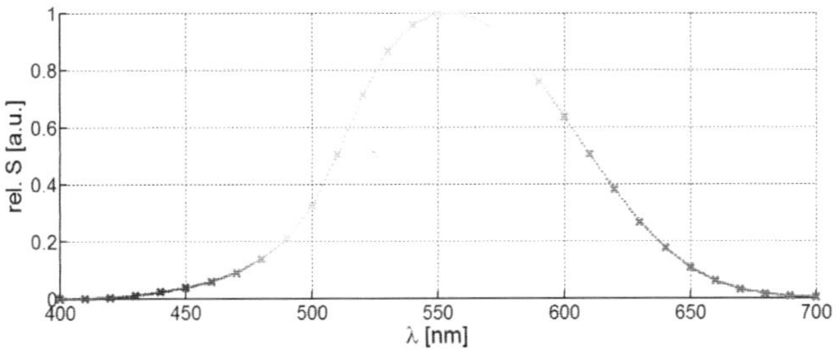

Figure 8. Relative spectral sensitivity of the human eye

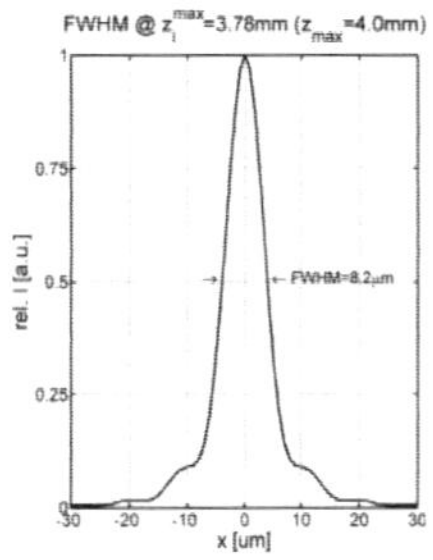

Figure 9. PSF @ $z_i^{max} = 3.78$mm

Figure 9 shows the result for $z_{max} = 4$mm based on I_{human} evaluated at $z_i^{max} = 3.78$mm. The choice of this particular imaging distance will become clear in the next section. One can infer from figure 9 that the point spread function does **not** decrease as rapidly to zero as the sinc^2-function in (9) and thus the FWHM and the 1/e^2-criteria are used to determine the lateral resolution depending on the hologram resolution and imaging distance z_{max}, see figures 10 and 11. The 1/e^2-criteria is more conservative and probably also more accurate to describe the hologram's imaging resolution since it agrees well with experimental observations.

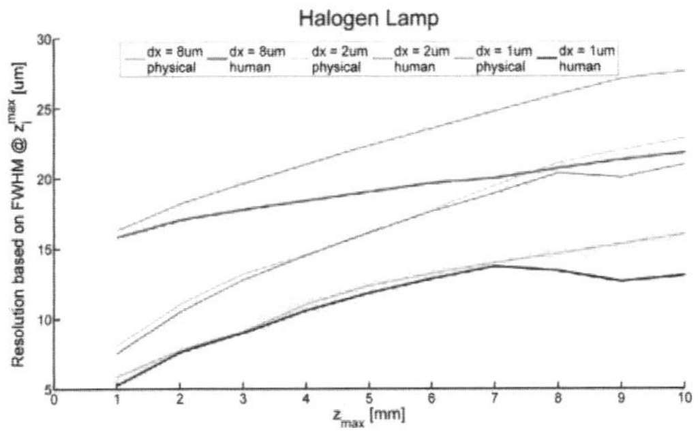

Figure 10. Resolution based on FWHM criteria

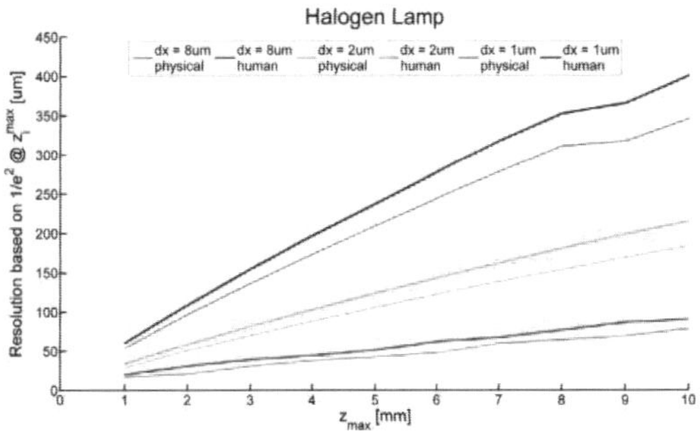

Figure 11. Resolution based on 1/e^2 criteria

Both figures show that there is a monotonic decrease in resolution with increasing imaging distance z_{max} for human as well as physical observation. Compared to the theoretical resolution value of monochromatic reconstruction $2\Delta_x$, the resolution of the reconstructions with white-light is always lower. The results show further that doubling the resolution of the hologram leads to a decrease in image resolution of about √2 for white-light illumination. High resolution imaging (<70µm or >360dpi) is thus only possible for distances <3mm (image plane holograms), if the hologram's resolution is better or equal 2µm. Please note that in case of the uniform spectrum the resolution is about 10% less in the case of the 1/e^2-criteria, but the trends are almost identical.

4.3. Axial extent of a point image by white-light illumination

According to Born and Wolf [8] the intensity distribution along the optical axis $I(z)$ is proportional to

$$I(z) = \mathrm{sinc}^2(a^2 z / 2\lambda z_{max}^2). \qquad (13)$$

It is said in [8] that a decrease of 20% from the maximum intensity is permissible to determine the focal tolerance Δz and one may thus measures the axial resolution by defining the axial extent of the focal spot $\Delta z' = 2\Delta z$. An estimation of the focal tolerance Δz is according to [8] given by

$$\Delta z = \pm\lambda/2(z_{max}/a)^2 \cong \pm 2\Delta_x^2/\lambda, \qquad (14)$$

if the relation $z_{max}(\Delta_x, a) \cong 2a\Delta_x/\lambda$ is used as given by (7), which should hold true for monochromatic illumination at the design wavelength. To determine the decrease in axial resolution due to white-light illumination, the simulations of the previous subsection were used. Figure 12 shows the evaluation of the maximum intensity I_{max} depending on the imaging distance z_i using again the parameters of figure 9. One can see that the peak intensity is not located at z_{max}, but at a different position z_i^{max} due to the white-light illumination. The amount depends on the source spectrum and can be easily corrected in the design process for a particular source since the shift is in good approximation linear with imaging distance as shown in figure 13. It also shows that the discrepancy between z_{max} and z_i^{max} as well as the focal spots axial extent increase linearly with imaging distance.

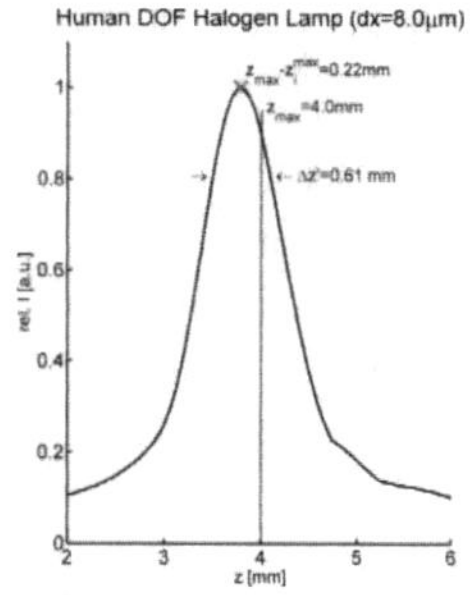

Figure 12. I_{max} vs. imaging distance z_i

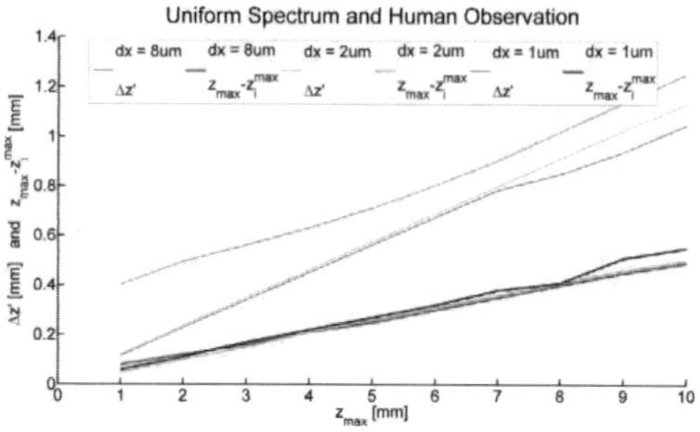

Figure 13. Evaluation of $\Delta z'$ and z_{max}-z_i^{max} vs. z_i

Please note that in the case of physical observation the discrepancy and the axial focal spot extent are approximately three times larger compared to the results shown in figure 13.

5. Conclusions

Holograms created by the point source method exhibit a problem in viewing due to writing on a regular grid, which we have termed "ghosting". The effect can be reduced by limiting the field of view or avoided by writing on a non-periodic grid. Experimental hologram reconstructions with white-light showed significant blurring, which can be explained based on image point simulations.

Numerical results based on a light source with uniform spectrum and the one of a halogen spot lamp showed that the lateral resolution is close to theoretical values if the FWHM criteria is used to determine resolution. The more conservative 1/e^2-criteria differs significantly from theory, but is in better agreement with experimental observations. The lateral extent of the image point increases in both cases linearly with distance and thus the resolution in dpi decreases proportional to 1/x. Doubling the holograms resolution leads to a factor √2 reduction of the imaging resolution for white-light illumination.

The focal spot's axial extent is related to the focal tolerance and increases linearly with distance. The position of maximum intensity differs from the design position, but the difference increases again linear with distance and thus corrected in the holograms design process for a particular light source.

References

[1] Hariharan P 2002 *Basics of Holography* (New York: Cambridge University Press) pp 1-14
[2] Matsushima K and Nakahara S 2009 *Appl. Optics* **48**, H54-63
[3] Ahrenberg L, Benzie P, Magnor M and Watson J 2008 *Appl. Optics* **47** 1567-74
[4] Yoshikawa H 2006 *Digital Holography and Three-Dimensional Display* Poon T C (US: Springer) pp 235-255
[5] Mader S 2011 *Master Thesis - Computational White-Light Holography* (Warsaw: Warsaw University of Technology) pp 58-60
[6] Kozacki T 2010 *Opt. Express* **18** 27118-29
[7] Goodman J W 1996 *Introduction to Fourier Optics* (New York: McGraw-Hill Companies Inc.)
[8] Born M and Wolf E 1986 *Principles of Optics* (New York: Pergamon Press Inc.)

Journal of Physics: Conference Series **415** (2013) 012047 doi:10.1088/1742-6596/415/1/012047

Occlusion processing for computer generated hologram by conversion between the wavefront and light-ray information

Koki Wakunami[1], Masahiro Yamaguchi[1]

Global Scientific Information and Computing Center, Tokyo Institute of Technology
4259-R2-56 Nagatsuta, Midori-ku, Yokohama, Kanagawa, 226-8503 Japan

E-mail: wakunami.k.aa@m.titech.ac.jp

Abstract. In the field of computational holography for three-dimensional (3D) display, the mutual occlusion of objects is one of the crucial issues. We propose a new mutual occlusion processing that is achieved by the conversion between the light-ray and wavefront on a virtual plane called ray-sampling (RS) plane located at near the interrupting object. The wavefront coming from background scene is converted into light-ray information at the RS plane by using Fourier transform based on the angular spectrum theory, then the converted light-rays are overwritten with those from interrupting object in the light-ray domain as an occlusion culling process. The ray information after the occlusion process is reconverted into wavefront by inverse Fourier transform at each RS point, then wave propagation from RS plane to hologram is computed by general light diffraction computation techniques. Since the light-ray information is used for the occlusion processing, our approach can realize a correct occlusion effect by a simple algorithm. In addition, high resolution 3D image can be reconstructed with wavefront-based technique. In the numerical simulation, we demonstrate that our approach for deep 3D scene with plural objects can realize a correct occlusion culling for varying observation angle and focusing distance.

1. Introduction

Holography is a superior medium for high-quality 3D displays since it reproduces the depth cues of human vision. For the electronic display of holography, the hologram pattern is calculated using the technique of computer generated hologram (CGH). Usually it is achieved by simulating physical phenomena of wavefront propagation from point/polygon sources on the object surface [1-3]. Thanks to the progress of high-speed computing, it becomes possible to define huge number of point sources in high-density on the object surface, allowing to render smooth surfaces. However, it is not easy to deal with some standard rendering technique used for photorealistic image in computer graphics, such as shading, transparency, texture mapping and light reflectance property, etc. Especially, two kinds of occlusions; self occlusion in which some surfaces are hidden by other surfaces of the same object, and mutual occlusion between different objects, are the crucial issue in CGH computation for photo-realistic reconstruction.

There have been achieved self/mutual occlusion culling by masking the wavefront at the surface of the foreground [4][5]. These approaches can be considered as an approximation of physical phenomena of light interference and diffraction on plural object's surfaces to decrease the calculation cost. However, these are still highly complicated and cannot avoid severe increase of computational cost to a large amount of light sources for higher resolution 3D scene, it is not realistic to apply for the larger hologram computation. Recently, Matsushima demonstrated the CGH computation over 4 billions of pixels for deep 3-D scene composed the plural objects [6]. In his demonstration, mutual occlusion culling is carried out by silhouette masking approach in which the wavefront from the background is interrupted by a mask of the foreground object's orthographic silhouette. Although it realizes a decreasing of a computation cost, the occlusion error should be observed from oblique direction since a silhouette mask is orthographically projected.

The other approach for occlusion culling is a depth-sorting technique for each light-ray in variable direction based on the general computer graphics (CG) techniques, such as holographic stereogram [7][8]. Computation cost of this approach is almost independent of the complexity of 3-D scene, depends on the number of sampled light-rays on the hologram plane. Because an occlusion culling process is much simpler, it is possible to decrease the computation cost as compared with the wavefront based approach. But because this approach is based on light-ray reproduction, deep 3D scene is blurred due to the light-ray sampling and light diffraction [9-11].

In previous work, we proposed a CGH computation approach using a ray-sampling (RS) plane that is located at near the object [12]. On the RS plane, light-ray information is sampled spatially and angularly at RS points as a set of projection images [see figure 1], and is transformed into a complex amplitude distribution by Fourier transform of projection images. The CGH pattern is thus obtained by calculating the wavefront propagation from RS plane to the hologram. In this approach, self occlusion effect of target object can be directly achieved by a general rendering technique on projection images.

In this paper, we show a new mutual occlusion culling method based on the conversion between light-rays and wavefront. Since light-ray domain is used for occlusion processing and the light propagation is calcutlated in wavefront domain, this approach can realize a occlusion culling with higher accuracy with simple algorithm.

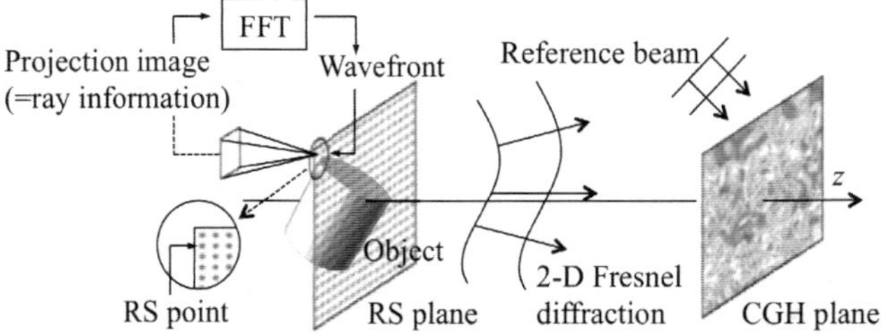

Figure 1. a scheme of CGH computation using RS plane

2. Conversion between light-rays information and wavefront on a RS plane

In this section, a principle of conversion between light-rays information and wavefront on an RS plane is briefly explained. RS plane should be set near a target object to avoid an image blur due to a ray sampling and light diffraction [12]. Light-rays passing through (i, j)th RS point on an RS plane compose a projection image $p_{i,j}$ viewing from a given RS point as shown in figure 2. Fourier transform of this projection image $p_{i,j}$ derives a complex amplitude distribution $P_{i,j}$ of a region centered at the same RS point based on the angular spectrum theory [13]. Appling this process for a set of projection images of all RS points and tiling obtained complex amplitude distributions on the RS plane, discrete wavefront on the whole RS plane is obtained. This process can be considered as a conversion from light-ray information into wavefront on an RS plane. By following an inverse process, i.e. executing an inverse Fourier transform of allotted wavefront for each RS point on a RS plane, wavefront on the RS plane can be converted into angular spectrum of each RS point corresponding to a light-ray information. This process can be considered as a conversion from wavefront into light-ray information.

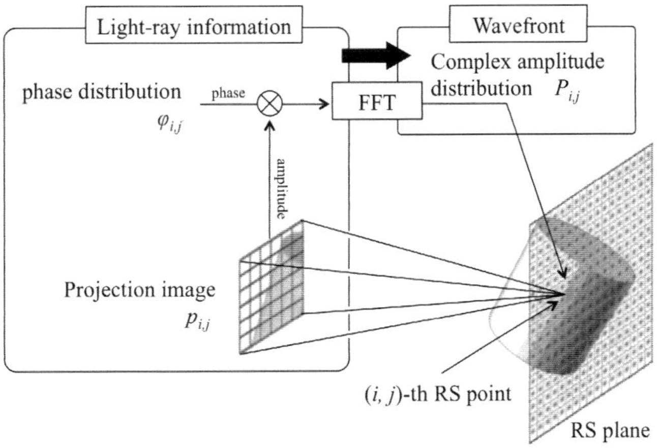

Figure 2. Conversion from light-ray information into wavefront on the RS plane

3. Mutual occlusion culling process by conversion between light-ray information and wavefront

For simplicity, we assume that a target scene consists of two objects; background and interrupting object as shown in figure 3(a). If there exist more interrupting objects in a target scene, we can apply the following way for each object. Figure 3(b) shows a flow of CGH computation with proposed occlusion process. In this approach, mutual occlusion culling process should be dealt with by defining RS planes for interrupting objects. The RS plane in figure 3(a) is set near the interrupting object, and a set of projection images p^i each RS point are rendered by ray-based rendering software in advance. Then the wavefront propagation from the background to the RS plane is calculated by Fresnel transform to derive the wavefront on the RS plane. The wavefront from background is then converted to the ray-information using inverse Fourier transform of allotted region for each RS point on the RS plane as mentioned in section 2.

Let $p^b_{i,j}$ be the light-ray information (i.e. angular spectrum) converted from wavefront of background at one RS point. Then $p^b_{i,j}$ is replaced by the light-rays from $p^i_{i,j}$ if there exists the light-rays from the interrupting object. Namely, the occlusion culling at each RS point is achieved in light-ray domain. After overwriting the rays from background with the rays from interrupting object at all RS points, combined light-ray information on the RS plane is reconverted into the wavefront by Fourier transform, and whole wavefront on the RS plane considered the occlusion culling can be obtained. Finally the Fresnel transform that simulates the light propagation leads the wavefront on the hologram plane.

Since occlusion process is achieved in light-ray domain, correct occlusion culling for variable observing angle is realized easily by substituting light-rays on the RS plane. Because each RS plane and CGH plane are set in parallel, it is also possible to implement efficient calculation for light propagation by using look up table, discrete Fresnel transform algorithm by using FFT, and/or Shifted Fresnel diffraction [14].

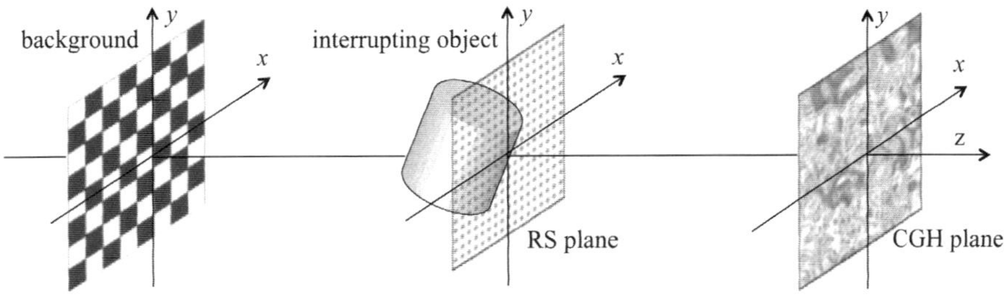

Figure 3. a scheme of occlusion culling process using RS plane

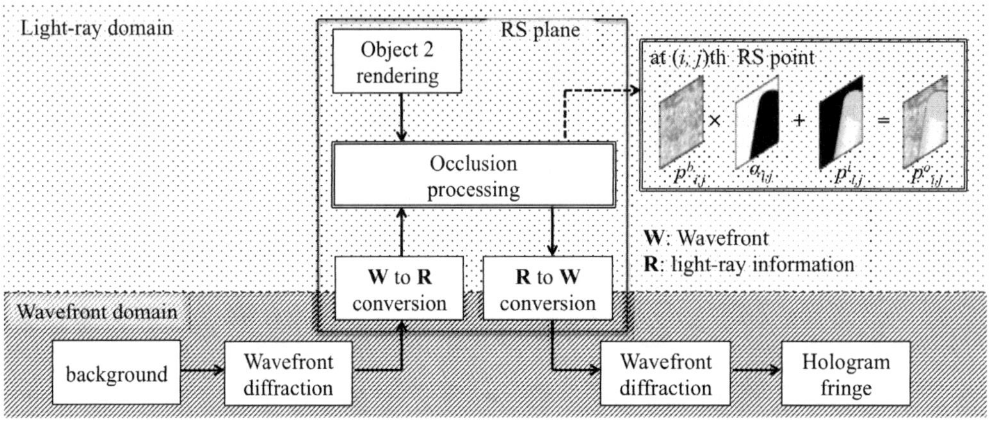

Figure 4. occlusion processing flow based on light conversion between ray information and wavefront

4. Experiment

We calculated the CGH based on the proposed method and evaluated the reconstructed image by numerical simulation. The target scene consists of two objects; a checkered panel as a background and a latticework of dice as an interrupting object located at 150mm and 100mm from CGH plane respectively. We set RS planes for both objects at 5mm in front of objects. The wavefront of background was generated by using RS plane based on the method presented in ref.12, then the occlusion process was carried out on the RS plane of interrupting object. The parameters of both RS planes and CGH plane are shown in Table 1.

Table 1. The parameters of RS planes and CGH plane

RS planes		CGH plane	
Num. of RS points	256×256	**Num. of pixels**	8192×8192
Pitch of RS points	64μm	**Pitch of pixels**	2μm
Num. of rays/RS point	32x32	**Viewing angle**	±3.6deg

CGH calculation was executed by using a PC with a CPU of Intel Westmere-EP (2.93GHz) processor and a shared memory of 24Gbytes. The computation time for occlusion process on the RS plane including a light conversion was about 14sec for 8192×8192 pixels, and two discrete Fresnel diffraction required about 192 sec(96sec×2). In a preprocessing, we needed to render a set of projection images in about 120 min. The wavelength was 532nm in CGH calculation. By applying parallel processing or GPU computing, it is possible to decrease the calculation time, especially for occlusion processing since it includes same calculation for each RS points.

In the numerical reconstruction, initially, wavefront propagation from CGH plane to the imaging lens was calculated by discrete Fresnel diffraction. The imaging lens assumed as a human eye set at 200mm from the CGH plane. The pupil size of the lens was 7mm. Then, the wavefront inside the lens pupil was multiplied by lens phase function, finally, calculate the wavefront propagation from lens pupil to image plane.

Figure 5 shows a result of numerical reconstruction. (a)-(c) are focused on different depth as in the captions of respective figures, (d)-(f) are focused on the latticework observed from left, center and right directions in 2 degrees intervals, respectively. These results show that proposed approach can reconstruct a high resolution image for deep 3D scene without occlusion error for any observing angles and focusing distances.

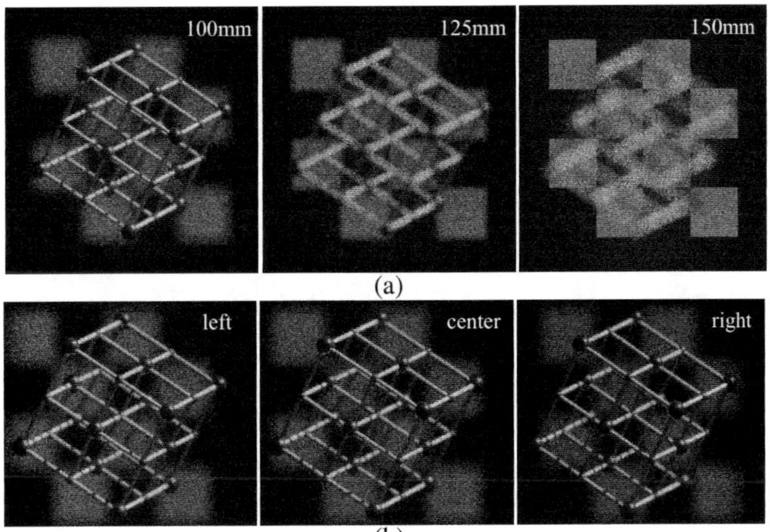

Figure 5. reconstruction images by numerical simulation (a): varying a focusing distance, (b): varying a observing angle

5. Conclusion

We proposed an occlusion processing of CGH calculation based on the conversion between the light-ray information and wavefront. Because light-ray domain is used for occlusion processing and the light propagation is calcutlated in wavefront domain, this approach can realize high accuracy occlusion culling with simple algorithm. In the simulation, we showed that our method can reproduce the image with proper occlusion for the scene there exist some objects. As a future work, it should be addressed to reduce the amount of calculation of the proposed method.

References

[1] Lucente M 1997 Optimization of Hologram Computation for Real-Time Display *Proc. SPIE* **1667** 32-43
[2] Waters J 1966 Holographic image synthesis utilizing theoretical methods *Applied Physics Letters* **9** 405-6
[3] Yamamoto K, Senoh T, Oi T and Kurita T 2010 8K4K-size computer generated hologram for 3-D visual system using rendering technology *Proc. IEEE Conf. on Universal Comm. Symp.* 193-196
[4] Matsushima K 2005 Exact hidden-surface removal in digitally synthetic full-parallax holograms *Proc. SPIE* **5742** 25-32
[5] Underkoffler J 1997 Occlusion processing and smooth surface shading for fully computed synthetic holography *Proc. SPIE* **3011** 53-60
[6] Matsushima K and Nakahara S 2009 Extremely high-definition full-parallax computer-generated hologram created by the polygon-based method *Appl. Opt.* **48** 54-63
[7] Chen R and Wilkinson T 2009 Computer generated hologram with geometric occlusion using GPU-accelerated depth buffer rasterization for three-dimensional display *Appl. Opt.* **48** 21 4246-55
[8] Smithwick Q, Barabas, J Smalley D and Bove M 2009 Real-time shader rendering of holographic stereograms *Proc. SPIE* **7233** 723302
[9] McCrickerd J 1972 Comparison of Stereograms: Pinhole, Fly's Eye, and Holographic Types *J. Opt. Soc. Am. A* **62** 1 64-70
[10] Helseh L 2006 Optical transfer function of three-dimensional display systems *J. Opt. Am. A* **23** 4 816-820
[11] Hilaire P 1994 Modulation transfer function and optimum sampling of holographic stereograms *Appl. Opt.* **33** 5 768-774

[12] Wakunami K and Yamaguchi M 2011 Calculation for computer generated hologram using ray-sampling plane *Opt. Exp.* **19** 10 9086-9101
[13] Goodman J 1996 Introduction to Fourier optics (New York: McGraw-Hill)
[14] Muffoletto R, Tyler J and Tohline J 2007 Shifted Fresnel diffraction for computational holography Optics Express **15** 9 5631-40

Full parallax three-dimensional computer generated hologram with occlusion effect using ray casting technique

Hao Zhang, Qiaofeng Tan, Guofan Jin

State Key Laboratory of Precision Measurement Technology and Instruments, Tsinghua University, Beijing, 100084, China

E-mail: zhanghaocgh@hotmail.com

Abstract. Holographic display is capable of reconstructing the whole optical wave field of a three-dimensional (3D) scene. It is the only one among all the 3D display techniques that can produce all the depth cues. With the development of computing technology and spatial light modulators, computer generated holograms (CGHs) can now be used to produce dynamic 3D images of synthetic objects. Computation holography becomes highly complicated and demanding when it is employed to produce real 3D images. Here we present a novel algorithm for generating a full parallax 3D CGH with occlusion effect, which is an important property of 3D perception, but has often been neglected in fully computed hologram synthesis. The ray casting technique, which is widely used in computer graphics, is introduced to handle the occlusion issue of CGH computation. Horizontally and vertically distributed rays are projected from each hologram sample to the 3D objects to obtain the complex amplitude distribution. The occlusion issue is handled by performing ray casting calculations to all the hologram samples. The proposed algorithm has no restriction on or approximation to the 3D objects, and hence it can produce reconstructed images with correct shading effect and no visible artifacts. Programmable graphics processing unit (GPU) is used to perform parallel calculation. This is made possible because each hologram sample belongs to an independent operation. To demonstrate the performance of our proposed algorithm, an optical experiment is performed to reconstruct the 3D scene by using a phase-only spatial light modulator. We can easily perceive the accommodation cue by focusing our eyes on different depths of the scene and the motion parallax cue with occlusion effect by moving our eyes around. The experiment result confirms that the CGHs produced by our algorithm can successfully reconstruct 3D images with all the depth cues.

1. Introduction

Holographic display is the most promising one among all the 3D display techniques, since it can reconstruct the whole optical wave field of a three-dimensional (3D) scene and produce all the depth cues.[1] The hologram plane forms a window, through which viewers can observe the 3D scene as if the object is floating behind. To generate a hologram for 3D display by calculation is not an easy task. With the development of computing technology and spatial light modulators (SLMs), computer generated holograms (CGHs) can now be used to produce dynamic 3D images of synthetic objects.[2-4]

At present, there are two main challenges in computation based holographic 3D display: the limited space bandwidth product (SBWP) of the SLM and the intensive computations required to produce the CGHs, especially those with full parallax effect.[5] The restricted SBWP of the SLM will limit the viewing angle and the image size of a 3D scene, while the computational challenge will affect both the

Published under licence by IOP Publishing Ltd

computational speed and the visual quality. Therefore, there is a need to investigate how best to calculate the CGH for 3D images in practice.

Furthermore, occlusion cue is an important property when viewing 3D scenes, but such an effect is not considered in most computational methods for CGHs. An occlusion effect is coupled with motion parallax, it should be distinguished from hidden surface removal, where the visible surfaces do not change when the viewpoint moves. In other words, for hidden surface removal, the front object does not block the back object when the viewer moves, while the occlusion effect implies that the visible surfaces change according to the movement of the viewpoint, which produces a correct depth cue for the viewers.

Most of the current algorithms for synthesizing CGHs use techniques which represent the 3D scene geometry as a collection of primitives and calculate the superimposed complex amplitude in the hologram plane. The 3D objects are commonly divided into point sources or planar segments. The point source method is more flexible in describing many kinds of 3D objects,[2, 6] while the planar segments method can perform comparatively fast calculations.[7, 8] These physically based algorithms simulate the optical transmission process from the target 3D objects to a hologram plane, which records a precise description for these 3D objects. However, it is difficult to express the occlusion effect using such techniques, since the optical transmission processes of different primitives are independent. Although the silhouette method was introduced to these algorithms for full parallax CGH with occlusion,[9] this method is specially designed with a fixed condition, namely, a 3D object with a 2D background, so it cannot produce the occlusion effect for an arbitrary 3D scene. Recently, Chen et al. proposed a method to handle the occlusion issue for a physically based algorithm, which runs visibility tests for each hologram sample using the occlude list.[10] However, this method interpolates the sampling points through a 2D interpolation kernel, which is not a rigorous description of 3D objects in general. Different from the physically based methods, the diffraction-specific approach developed by Lucente considers the reconstruction step in holography rather than the simulation of the optical transmission process from the objects. It uses functional holographic elements (hogels) to reduce the CGH computation load.[11, 12] Moreover, this algorithm is specially designed for horizontal parallax only (HPO) holograms and display systems with high horizontal resolution, such as the Holovideo display system of MIT Media Laboratory.[1] Stereogram modeling method and the methods derived from it can produce HPO CGH with occlusion effect easily because the hologram is generated through appropriately captured 2D parallax views.[13, 14] These kinds of methods are fast and they are well matched to hardware implementation. However, since the calculation of the stereogram is based on a set of 2D views, it is hard to produce all the depth cues, especially the accommodation cue.

Recently, some researchers attempted to introduce the ray casting technique from computer graphics to the calculation of CGHs. Janda et al. proposed a ray casting method to handle occlusion issue, but they restricted their 3D model to a collection of planar segments parallel to the hologram plane.[15] This approximation could cause artifacts when reconstructing 3D images experimentally. Hanak et al. made use of the graphics pipeline to generate a full parallax hologram with occlusion effect.[16] More recently, Chen et al. developed a similar algorithm to generate an HPO CGH with occlusion effect by using an OpenGL depth buffer.[17] However, because such graphics pipeline based methods produce shear transformations to the 3D objects concerned, the shading effect of the 3D scenes will be affected.

Although the occlusion effect has been addressed by some researchers, none of them has produced an optical demonstration for full parallax 3D CGH with occlusion effect. In this paper we present a ray casting based algorithm for the calculation of full parallax 3D CGH with occlusion effect and demonstrate the result in an optical experiment. In contrast to the previously mentioned methods, this algorithm has no parallel planar segments restriction or shear transformations to the 3D objects. It can, therefore, produce precise description to the 3D scene. Programmable graphics processing unit (GPU) is used to perform parallel calculation. Experiment is performed to optically reconstruct the 3D scene by using a phase-only SLM. The theoretical framework of such CGH generation is described in

Section 2. Optical reconstruction results are shown in Section 3. Finally, brief discussion and conclusion are provided in Section 4.

2. Method

2.1. Ray casting algorithm for CGH generation

Ray casting is a widely used technique in computer graphics for hidden surface removal.[18] The casting ray is projected from the eye position through each pixel of the image to find the nearest intersection object point, as shown in Figure 1(a). One can easily add light and shadow effect to a 3D scene during rendering. However, there are fundamental differences between this kind of image-order rendering and the CGH generation. The parts of a 3D scene which make a contribution to each specific hologram sample must be evaluated. This can be done by projecting a bundle of rays from each hologram sample to the 3D scene to find the nearest intersection object point for each ray, as shown in Figure 1(b). Only the nearest point provides a contribution to that hologram sample along the ray projecting direction. The occlusion issue is handled by performing ray casting calculations to all the hologram samples.

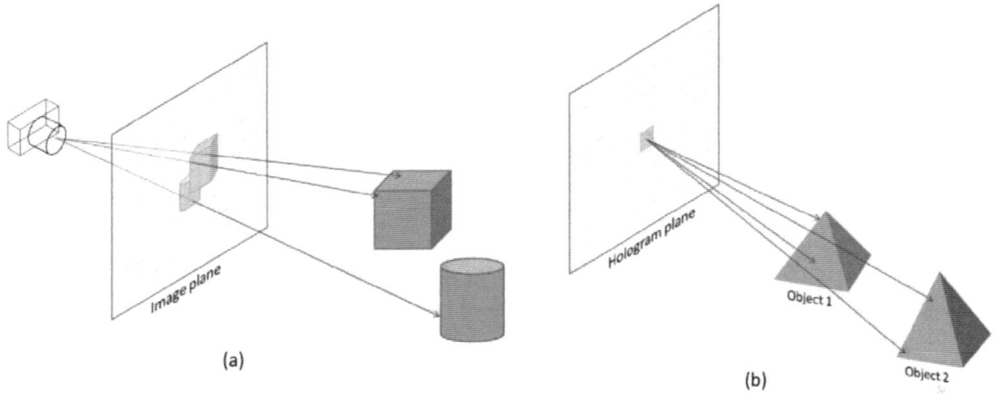

Figure 1. (a) Ray casting technique using in computer graphics, each pixel decides one casting ray. (b) Ray casting technique using in CGH generation, each pixel corresponds many casting rays.

Point source method is used in our ray casting algorithm to calculate the complex amplitude distribution of each intersection object point.[19] The optical distribution of the spherical wave which is produced from one point source can be computed as:

$$H(x,y) = \frac{A}{r}\exp[i(kr+\varphi)], \tag{1}$$

where A is the wave amplitude of the point, $k = 2\pi/\lambda$ is the wave number in free space, λ is the wavelength, φ is the initial phase (often randomized as a uniform distribution between 0 and 2π), and r is the distance between the specific point source (x_j, y_j, z_j) and the hologram sample $(x, y, 0)$:

$$r = \left[\left(x-x_j\right)^2 + \left(y-y_j\right)^2 + {z_j}^2\right]^{\frac{1}{2}}. \tag{2}$$

In order to generate full parallax CGH, the casting rays projected from each hologram sample should be horizontally and vertically distributed, as shown in Fig 2. To find the ray-scene intersection of a single casting ray, we loop over all the triangles which build up the 3D scene to find their intersections with the casting ray and keep the closest as return. The final complex amplitude of one specific hologram sample is calculated by adding contributions of all the casting rays which are projected from the sample.

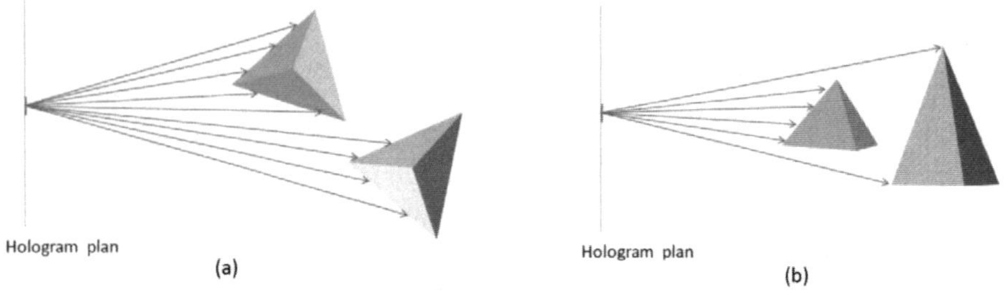

Figure 2. (a) Top view of the ray casting technique using in CGH generation, horizontal distributed casting rays are shot from each pixel. (b) Side view of the ray casting technique using in CGH generation, vertically distributed casting rays are shot from each pixel.

Lighting is an important feature when describing 3D scenes. When it comes to CGH computing, lighting affects the wave amplitude of the intersection object point A in Eq. (1). Deduced from bidirectional reflectance distribution function (BRDF),[20] which is the function that defines how light is reflected at the opaque surface of a 3D object, the intensity of each surface point p can be computed as:

$$I_p = k_a I_a + k_d I_s \max\left(\vec{L} \cdot \vec{n}, 0\right), \tag{3}$$

where k_a is the ambient reflection constant and I_a is the intensity of the ambient light, k_d is the diffuse reflection constant, I_s is the intensity of the incoming light, $-\vec{L}$ is the unit vector pointing from the light source to the surface, $\vec{n}$ is the unit normal vector at point p, and $\cdot$ is the dot product operator. So the light intensity of our model is constituted by ambient and Lambertian reflection. During computing, k_a and k_d can be customized to adjust the overall shading effect.

As the calculations performed on each of the hologram samples are independent of each other, we can explore the powerful parallel computing ability of GPU for calculation. The Computer Unified Device Architecture (CUDA) is used to perform the calculation. It is a parallel computing architecture developed by NVIDIA, which gives developers access to the virtual instruction set and memory of the parallel computational elements in CUDA GPUs. Each hologram sample is managed as a separate thread. A kernel function is executed on all the threads in order to perform the ray casting calculation, which loops over all the casting rays to compute the complex amplitudes for the hologram samples. The contribution of each casting ray can be calculated through Eq. (1). The kernel function evaluates the nearest intersection object point for each ray. Hence the distance between hologram sample and the intersection point can be calculated through Eq. (2). The wave amplitude of the intersection object point A is computed through Eq. (3). The initial phase φ can be mapped to the pre-calculated lookup table restored in the 3D texture memory.

2.2. Distribution of casting rays

The distribution of casting rays has a close relationship with the discrete SLM device to be used to display the CGH. When a collimated beam illuminates on the SLM, the diffraction beam will separate in many orders due to the pixelated structure of the modulator.[21] The maximum diffraction angle of the SLM is given by:

$$\sin\theta = f\lambda = \frac{1}{2d}\lambda, \tag{4}$$

where f is the highest spatial frequency that the SLM can display, and d is the pixel pitch of the SLM, as shown in Figure 3. The angle of the casting rays is limited by this maximum diffraction angle.

This restriction automatically solves the aliasing problem from using a physically based algorithm to generate the CGHs in the situation when 3D objects are in a short distance to the hologram plane. Only the object points which are in the diffraction region of a specific hologram sample will provide contributions to that sample. From Eq. (4), we can deduce that the diffraction angle of a given SLM is directly dependent on its spatial resolution. This angle produces the motion parallax cue during the optical reconstruction. It is obvious that an SLM of higher spatial resolution would lead to a larger viewing angle.

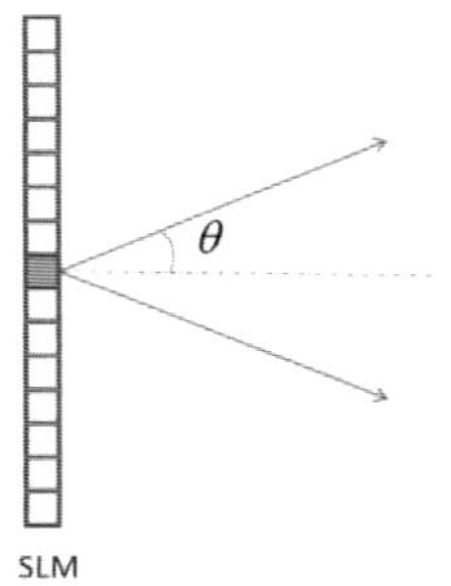

Figure 3. Angular diffraction region of the SLM.

Another parameter which needs to be considered is the angular sampling distance (β) of adjacent casting rays. This sampling distance is related to the density of the intersection object points. Dense intersection object points can give a solid surface effect to viewers. Normally we can choose $\beta = 1/60^{\circ}$, the same as the angular resolution of human eye. One may also adjust the sampling distance according to the requirements. Given the maximum diffraction angle and the angular sampling distance, the total sampling number for each hologram sample can be calculate as:

$$N = \left(\frac{2\theta}{\beta}\right)^2, \tag{5}$$

where the square is introduced because of the full parallax consideration.

3. Experimental results

To demonstrate the effect of our proposed method, an optical experiment is performed, using the HEO 1080P Phase Modulator (HOLOEYE Photonics AG). This is a liquid crystal on silicon (LCoS) analogue phase-only SLM with 1920×1080 pixels. The pixel pitch is $8\mu m$, and the phase magnitude is addressed with 8 bit gray-scale levels. The wavelength of the laser used in our experiment is 635nm.

According to the discussion in Section 2, it can be deduced that the angular diffraction region of our SLM is within $\pm 2.27^{\circ}$. To get better solid surface effect for optical reconstruction, angular sampling distance is set to 0.01° in the calculation, which means the total sampling number for each hologram plane is 454^2. The 3D scene we use to generate the CGH contains two pyramids, together with a light source at infinity. The pyramids are positioned at distances of 125mm and 175mm from the hologram plane, respectively. The calculation is performed on an NVIDIA Quadro FX 2800M graphics card, and the total computation time for this example is 80 minutes. In the reconstruction stage, we use only the phase information of the CGH displayed on the SLM, as in a kinoform. This method can maintain good quality in reconstruction for 3D diffusive objects.[7]

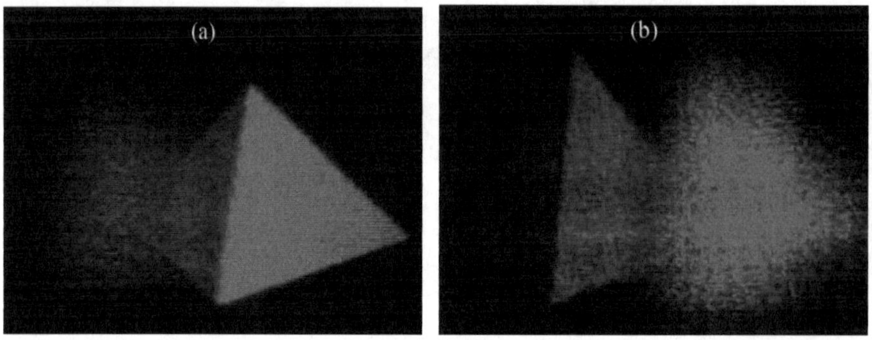

Figure 4. Optical reconstruction results when camera is focusing on (a) the front object and (b) the back object.

The reconstructed full parallax 3D scene with occlusion effect is confirmed by image recording. During which, a single lens reflex camera (Canon 500D) with a zoom lens (Tamron AF 19-200mm) is used to capture the 3D scene directly. Figure 4(a) and (b) show the reconstruction results where the camera is manually focused on the front and back objects separately (focal length: 200mm, f number: 6.3, object distance: 450mm). The front object shown in Figure 4(a) can be seen completely, while the back pyramid shown in Figure 4(b) is partially blocked by the front one. This shows that our method has successfully produced the accommodation cue with hidden surface removal effect. To demonstrate the occlusion effect of our method, the camera is placed at different viewing positions to capture the 3D objects (focal length: 200mm, f number: 20, object distance: 600mm). The results are shown in Figure 5, demonstrating that full parallax with occlusion effect can be perceived in the optical reconstruction. Long depth of field is needed when recording the images in Figure 5 because the two pyramids are displayed at different distances. This can be achieved by using a small relative aperture during image recording, but such a measure will decrease the quality of the captured images. We can also use our eyes to view the reconstructed 3D scene instead of the camera and perceive the depth cues of the scene.

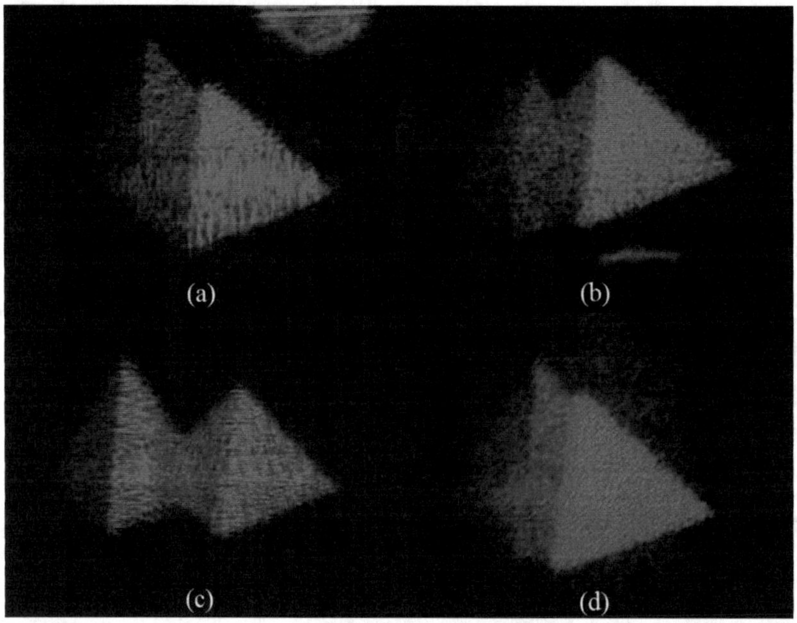

Figure 5. Optical reconstruction results when viewing from (a) top, (b) bottom, (c) left and (d) right, showing occlusion effect and full parallax.

4. Discussion and conclusion

In summary, we propose an algorithm to generate full parallax CGH with occlusion effect, and demonstrate its effectiveness by optical reconstruction. The ray casting technique used in computer graphics is introduced in this algorithm, using casting rays projected from each hologram sample to find the nearest intersection point with the objects. Our approach does not require the target objects be transformed or restricted, and it allows a rigorous description to the 3D scene with correct shading effect when it is optically reconstructed. The computational speed is highly dependent on the number of casting rays sampled. Faster calculation can be achieved through the sacrifice of visual effect by limiting the angular diffraction region or by increasing the step of sampling angle. The calculation is implemented on GPU platform for parallel computation. The optical experiment is performed using a phase-only SLM. By loading the calculated phase CGH on the SLM, we successfully reconstruct the 3D scene with full parallax and occlusion effect. Finally it should be pointed out that our method can also be used to produce large holograms with complicated 3D scenes.

References

[1] S. A. Benton and V. M. Bove, *Holographic Imaging,* Wiley, Hoboken (2007).
[2] M. Lucente, "Interactive computation of holograms using a look-up table," *J. Electron. Imaging* **2**, 28-34 (1993).
[3] F. Yara, H. Kang, and Levent Onural, "State of the Art in Holographic Display: A Survey," *J. Display Technol.* **6**, 443-454 (2010).
[4] C. W. Slinger, C. D. Cameron, S. D. Coomber, R. J. Miller, D. A. Payne, A. P. Smith, M. G. Smith, M. Stanley, and P. J. Watson, "Recent developments in computer-generated holography: toward a practical electroholography system for interactive 3D visualization," *Proc. SPIE* **5290**, 27-41 (2004).
[5] K. Matsushima, "Computer-generated holograms for three-dimensional surface objects with shade and texture," *Appl. Opt.* **44**, 4607-4614 (2005).
[6] Y. Ichihashi, H. Nakayama, T. Ito, N. Masuda, T. Shimobaba, A. Shiraki, and T. Sugie, "HORN-6 special-purpose clustered computing system for electroholography," *Opt. Express* **17**, 13895-13903 (2009)
[7] H. Zhang, J. Xie, J. Liu, and Y. Wang, "Optical reconstruction of 3D images by use of pure-phase computer-generated holograms," *Chin. Opt. Lett.* **7**, 1101-1103 (2009)
[8] L. Ahrenberg, P. Benzie, M. Magnor, and J. Watson, "Computer generated holograms from three dimensional meshes using an analytic light transport model," *Appl. Opt.* **47**, 1567-1574 (2008)
[9] K. Matsushima and S. Nakahara, "Extremely high-definition full-parallax computer-generated hologram created by the polygon-based method," *Appl. Opt.* **48**, H54-H63 (2009)
[10] R. H.-Y. Chen and T. D. Wilkinson, "Computer generated hologram from point cloud using graphics processor," *Appl. Opt.* **48**, 6841-6850 (2009).
[11] M. Lucente, "Diffraction-specific fringe computation for electro-holography," Ph. D. dissertation, Massachusetts Institute of Technology, (1994).
[12] M. Lucent, "Computational holographic bandwidth compression," *IBM Syst. J.* **35**(3&4), 349-365 (1996).
[13] M. Lucente and T. A. Galyean, "Rendering interactive holographic images," in *Proceedings of SIGGRAPH '95*, 387-394 (1995).
[14] Q. Y. J. Smithwick, J. Barabas, D. E. Smalley, and V. M. Bove, Jr., "Real-time shader rendering of holographic stereograms," *Proc. SPIE* **7233**, 723302 (2009).
[15] M. Janda, I. Hanak, and L. Onural, "Hologram synthesis for photorealistic reconstruction," *J. Opt. Soc. Am. A* **25**, 3083-3096 (2008)
[16] I. Hanak, M. Janda, V. Skala, "Full-Parallax Hologram Synthesis of Triangular Meshes using a Graphical Processing Unit," *In Proc. Of 3DTV-CON*, (2007)
[17] R. H.-Y. Chen and T. D. Wilkinson, "Computer generated hologram with geometric occlusion

using GPU-accelerated depth buffer rasterization for three-dimensional display," *Appl. Opt.* **48**, 4246-4255 (2009)

[18] K. Zhou, Q. Hou, R. Wang and B. Guo, "Real-Time KD-Tree Construction on Graphics Hardware," *ACM Trans. on Graphics (Siggraph Asia)*, **27**(5), 126 (2008).

[19] X. Xu, S. Solanki, X. Liang, S. Xu, R. B. A. Tanjung, Y. Pan, F. Farbiz, B. Xu, and T.-C. Chong, "Computer-Generated Holography for Dynamic Display of 3D Objects with Full Parallax," *The International Journal of Virtual Reality*, **8**(2), 33-38 (2009).

[20] F. E. Nicodemus, "Directional reflectance and emissivity of an opaque surface," *Appl. Opt.* **4**, 767–773 (1965).

[21] H. Zhang, J. Xie, J. Liu, and Y. Wang, "Elimination of a zero-order beam induced by a pixelated spatial light modulator for holographic projection," *Appl. Opt.* **48**, 5834-5841 (2009).

Journal of Physics: Conference Series **415** (2013) 012049 doi:10.1088/1742-6596/415/1/012049

Digital hologram generation for a real 3D object using by a depth camera

SeungHyun Lee[1], SoonChul Kwon[1], HoByung Chae[1], JiYong Park[1], HoonJong Kang[2], and James Do Kyoon Kim[3]
[1] Dept. of Holography and 3D Contents, Kwangwoon Univ.
[2] Korea Electronics Technology Institute [3] Samsung Electronics
26 Kwangwoon-Gil, Nowon-Gu, Seoul, 139-701, Korea

E-mail: shlee@kw.ac.kr

Abstract. We introduce a digital hologram generation of a real 3D object captured by a depth camera. The depth camera is employing a time multiplexed concept for color and depth image acquisition in a single chip sensor. The 3D information, point cloud, corresponding to the real scene, is extracted from taken image pairs, a gray texture and a depth map by a depth camera. The extracted point cloud is used to generate a digital hologram as input information. The experimental results are satisfactory.

1. Introduction

Holography is a unique technology that can express 3D information almost perfectly in human-friendly way[1]. Optical holography has developed into digital holography which creates interference pattern of object using digital technology and displays it through spatial light modulator.

In general, a 3D computer graphic model is being used to generate a digital hologram as the input information because the 3D information of an object can be extracted from a 3D model easily. Also there are many available methods to find the depth information including a stereo camera[2], depth camera of TOF(time of flight) method[3], 3D scanner, and Kinect of Microsoft[4].

In this paper, we introduce a digital hologram generation method of a real 3D object captured by a depth camera. The depth camera is a 3D image sensor based on the Time of Flight (TOF) principle using photon gating techniques. The basic feature incorporates real time depth imaging by capturing the shape of a light-pulse front as it is reflected from a three dimensional object. The depth camera is capable of producing RGB and D signals, where depth map D stands for the distance to each pixel. The taken RGB texture can be regarded as the intensities of the observable real object surface at the corresponding view point. However, it is still challenging to implement a single image sensor that can simultaneously acquire color and depth images. We present a 2nd-generation 2D/3D imager based on the pinned photodiode pixel structure. The time-division readout architecture for both image types (color and depth) is maintained. A complete redesign of the imager makes pixels smaller and more sensitive than before[5, 6].

The taken image set, the color texture, and the depth map are mapped to a 3D coordinate system, and are represented to a 3D point cloud in which each point is consisted with 3D coordinate values and color intensities. This extracted 3D information from a real object is used to generate a digital hologram.

2. TOF camera to obtain both color and depth images in a single sensor[5,6]

A 3D image sensor is presented employing a time multiplexed concept for color and depth image acquisition in a single chip to generate a real-time 3D image of an arbitrary scene. Fig. 1 shows the conceptual diagram of a time multiplexing architecture for obtaining color and depth images. There is no dedicated pixel for each image, but all the pixels have the same structure for both images that can be alternatively captured by switching operation mode. Because the conventional color filter is transparent to near infrared (NIR), there is no problem for using NIR emitting source in TOF depth cameras.

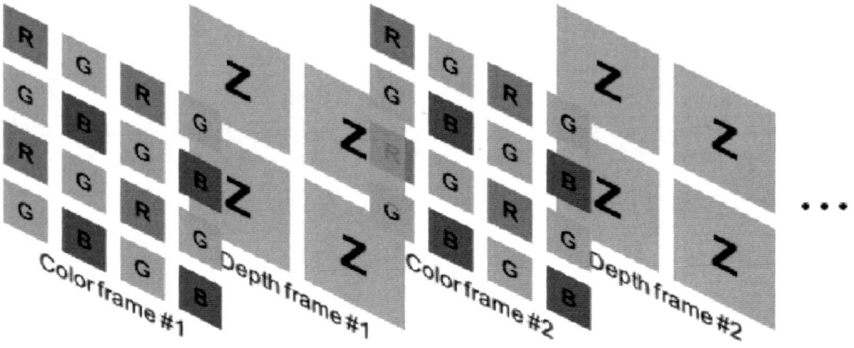

Fig.1. Conceptual diagram of a time multiplexing architecture

The sensor can operate in two alternate readout modes for color and depth images in each frame using a time-division scheme shown in Fig. 2. The IR cut or pass filter is adjusted to the next image's mode during a filter change time between both operation modes, and all the transfer gates are simultaneously activated to reset photodiodes.

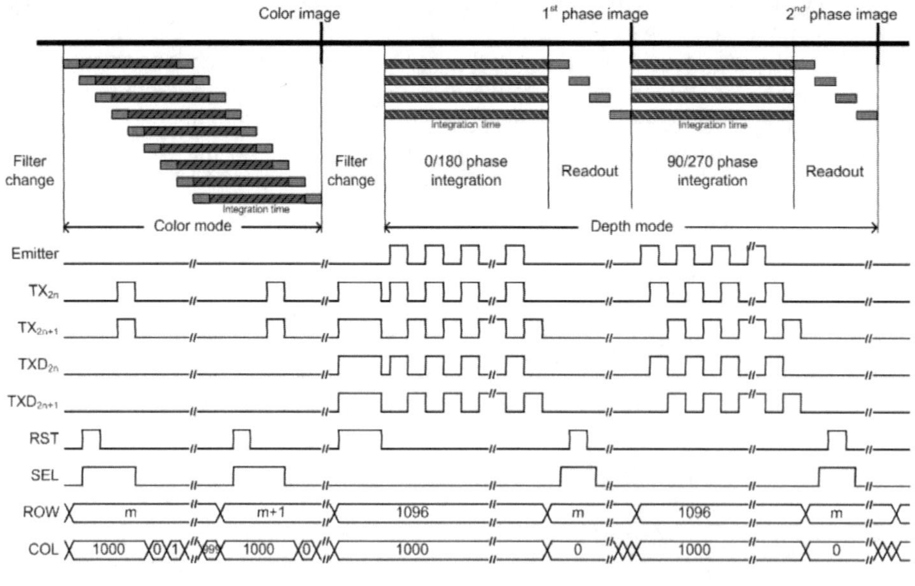

Fig. 2. Timing diagram for obtaining color and depth images

Fig. 3(a) shows a color image from the prototype sensor under typical room illumination (about 1000lux) with gain 2. The reason why the lower image is brighter than the upper is that half of pixels have small size of FD nodes for a test. Fig. 3(b) is a depth image of the same object and three flat panels located on 1.2m, 1.5m and 1.9m from the sensor are clearly distinguished by 8-bit gray level. Any image processing algorithms for the improvement of image quality are not applied to both images.

Fig. 4 shows the measured depth and accuracy of a flat white board between 1m and 3m per every 0.25m distance. The integration time is 50ms and the modulation frequency is 10MHz. The depth below 1m is distorted by pixel saturation so that the sensor cannot work properly under this condition. After calibration, the non-linearity error is about 0.8% and the depth accuracy, which means temporal non-uniformity, is lower than 62mm at 3m distance measured from small FD pixel showing better performance. Table 1 summarizes the prototype sensor characteristics and measurement environments.

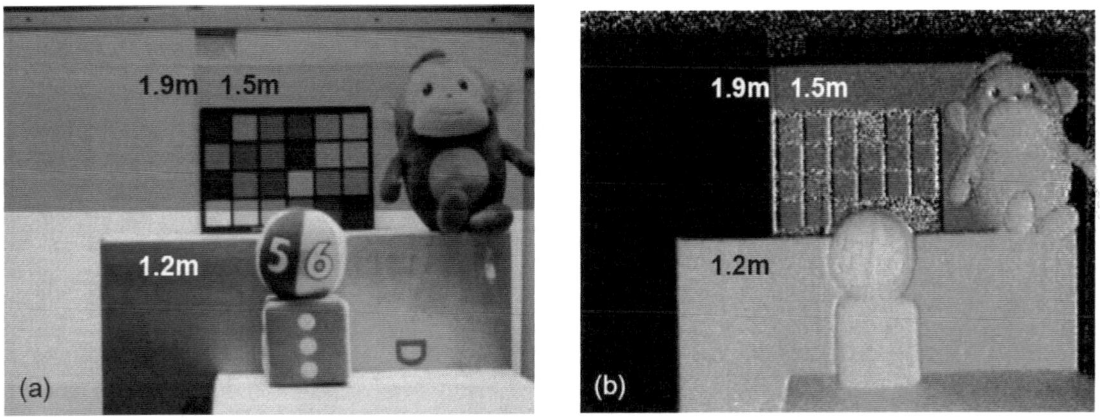

Fig. 3. (a) Color image and (b) depth image from the sensor

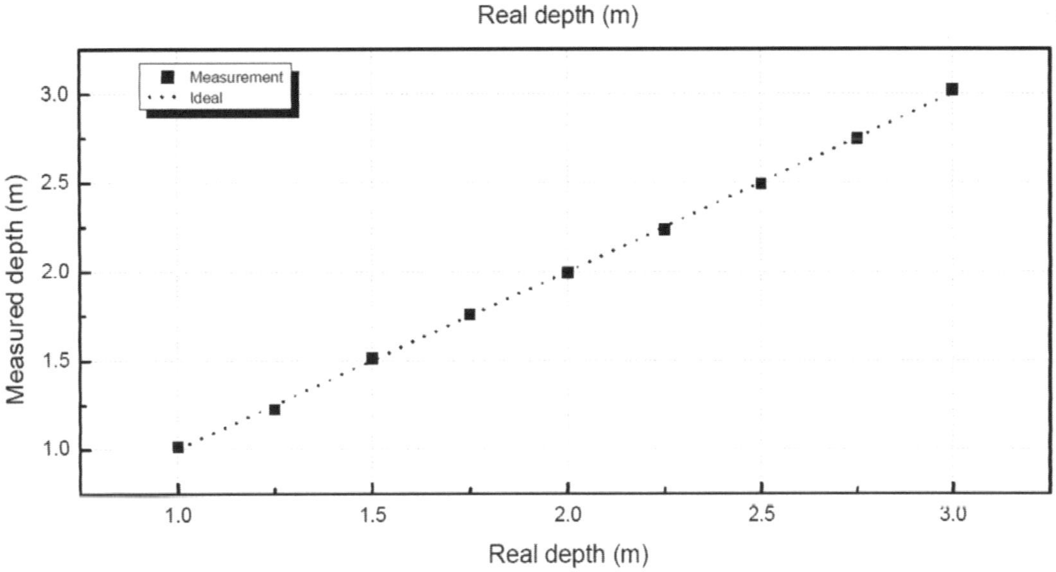

Fig. 4 Measurement data of a depth

Table 1 Prototype sensor characteristics

Sensor	Process	0.11μm 1P4M CMOS
	Fill factor	38.5%
	Die size	9.7×6.2mm^2
System	Emitter	LED with 850nm wavelength (SFH4259)
	Lens	Focal length – 16mm, F# – 1.6
	Supply voltage	3.3V
Color	Pixel size	3.65μm (2.5Tr/Pixel)
	Resolution	2000×1096 (Effective: 1920×1080)
	Frame rate	30fps (Single color mode)
Depth	Pixel size	14.6μm (4×4 binning)
	Resolution	500×274 (Effective: 480×270)
	Frame rate	11fps (Single depth mode, int. time 40ms)
	Modulation frequency	20MHz
	Measured range	0.75m ~ 4.5m (Setup limit)
	Non-linearity	0.93%
	Depth error	10mm ~ 38mm (with different int. times)

3. Experiment

By using the taken image pair which is composed with a color texture and its corresponding depth image, the 3D information can be extracted. The depth image is two-dimensional gray scale image. The intensity of the depth image is used as the depth information, and expressed by a gray scale. The depth range of the camera is dependent to the system initialization, whose parameter value is initialized by a user. Therefore, since we know the depth range and the field of view of the depth camera, the coordinate values on the three dimensional coordinate system can be extracted from the taken depth image. The taken gray image has scene information. Therefore, two-dimensional coordinate values on the coordinate system and intensity of the real object can be extracted. Accordingly, the necessary 3D information, coordinate (x, y, z) and real object intensity (A), corresponding to the real scene can be extracted as shown in Fig. 5. The extracted information is used to generate a digital hologram.

A digital hologram can be obtained by using the Rayleigh-Sommerfeld(RS) integral since such an approach gives exact complex light disturbance over a plane at a certain distance from an interesting object [7,8]. The complex function form of the RS integral in (ξ, η) plane is

$$S(\xi, \eta) = \sum_{p=1}^{N} \frac{A}{r_p} \exp(jkr_p)$$

where N is the number of object points. $A = a_p \exp(\varphi_p)$ is the complex constant, where a_p and φ_p are the amplitude and the phase of the pth object point in a point cloud, respectively. The wave number k is $2\pi/\lambda$, where λ is the free-space wavelength of the coherent light. The distance r_p

between the pth object point and the point (ξ, η) on the hologram is $\sqrt{(\xi - x_p)^2 + (\eta - y_p)^2 + z_p^2}$. The RS integral treats the propagating light from a point as a spherical wave, and the desired complex amplitude distribution on a hologram plane can be exactly determined.

The taken color image and its corresponding depth image is shown in Fig. 5. A point cloud which has 3D information is extracted from the image pair. The extracted point cloud is utilized to generate digital hologram. The number of the points in the point cloud is about 20,000 due to the used camera resolution. The generated digital hologram by the R-S integral and its numerical reconstruction are shown in Fig. 6. In addition, the optical reconstruction is shown in Fig. 7. As shown in Fig. 6 and 7, even it is only red channel, the quality of the numerical reconstruction and the optical reconstruction are acceptable. The experimental environment is shown in Table 2.

(a) (b)

Fig. 5. The utilized input images: (a) color image, (b) depth image

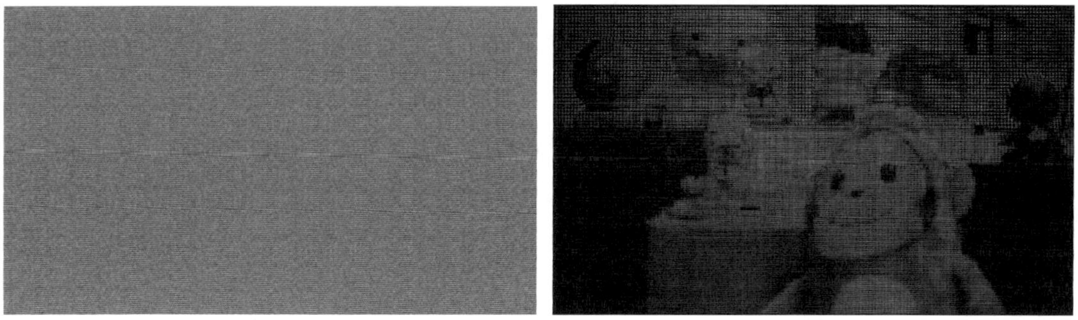

(a) (b)

Fig. 6. (a) the generated digital hologram and (b) its numerical reconstruction

Fig. 7. Optical reconstruction for the red color information

Table 2. Experimental environment

Digital hologram generation	Resolution of hologram	1920 x 1080 pixels
	Pixel interval	7um
	Object distance	500mm
Display system	Wavelength	633nm
	Resolution of SLM	1920 x 1080 pixels

4. Conclusion

Color image and depth map of an object have been necessary to create a real 3D image. We presented a time-of-flight (TOF) 3D image sensor for acquiring a depth image as well as a normal 2D image. In this work, we propose an improved 3D pixel architecture to obtain both color and depth images in a single sensor. It will be necessary to conduct further research on a depth camera and hologram generation methods for the high quality digital hologram of a real object.

Acknowledgments
This research was supported by Ministry of Knowledge Economy (MKE), Korea as a project, "The Development of Active Sensing-based 3D HD Depth Camera".

References

[1] D. Gabor, "A new microscopic principle", Nature, 161, pp.777-778, 1948
[2] Jing Zhang, Liang-Hao Wang, Dong-Xiao Li, and Ming Zhang, "High quality depth maps from stereo matching and ToF camera", Soft Computing and Pattern Recognition (SoCPaR), pp.68-72, 2011
[3] G. J. Iddan and G. Yahav, "3D imaging in the studio (and elsewhere...)," in Three-Dimensional Image Capture and Applications IV, B. D. Corner J. H. Nurre, and R. P. Pargas, eds., Proc. SPIE, vol. 4298, pp.48-55, 2001.
[4] K. Khoshelham,"Accuracy analysis of Kinect depth data", GeoInformation Science, vol.38, no. 5, 2010.
[5] Kim, S.-J., Kim, J. D. K., Han, S.-W., Kang, B., Lee, K., and Kim, C.-Y., "A 640x480 image sensor with unified pixel architecture for 2D/3D imaging in 0.11μm CMOS," in *Symposia on VLSI circuits*, pp.92-93, 2011.
[6] Ouk Choi, Hwasup Lim, Byongmin Kang, Yong Sun Kim, Keechang Lee, James D. K. Kim, and Chang-Yeong Kim, "Discrete and continuous optimizations for depth image super-resolution," Proc. SPIE, vol.8290, 2012.
[7] Kang, H., Fujii, T., Yamaguchi, T., and Yoshikawa, H., "Compensated phase-added stereogram for real-time holographic display," Optical Engineering, vol. 46, 2007.
[8] Kang, H., Yamaguchi, T., and Yoshikawa, H., "Accurate phase-added stereogram to improve the coherent stereogram," Appl. Opt., vol.47, pp.44-54, 2008.

Journal of Physics: Conference Series **415** (2013) 012050 doi:10.1088/1742-6596/415/1/012050

Recent advancements in photorefractive holographic imaging

B Lynn[1], P-A Blanche[1], A Bablumian[1], R Rankin[1], R Voorakaranam[1], P St. Hilaire[1], L LaComb, Jr.[1], M Yamamoto[2] and N Peyghambarian[1]

[1]College of Optical Science, University of Arizona, Tucson, Arizona 85721, USA
[2]Nitto Denko Technical Corporation, Oceanside, California 92054, USA

E-mail: blynn@optics.arizona.edu

Abstract. We have recently demonstrated several improvements in material properties and optical design to increase the resolution, size, brightness, and color range of updatable holograms using photorefractive materials. A compact system has been developed that is capable of producing holograms with brightness in excess of 2,500 cd/m^2 using less than 20mW of CW laser power. The size of the hologram has been increased to 300mm x 150mm with a writing time of less than 8 seconds using a 50 Hz pulse laser. Optical improvements have been implemented to reduce the hogel size to less than 200 μm. We have optimized the color gamut to extend beyond the NTSC CIE color space through a combination of spatial and polarization multiplexing. Further improvements could bring applications in telemedicine, prototyping, advertising, updatable 3D maps and entertainment.

1. Introduction

Three dimensional (3D) displays are becoming more prevalent by the day, with entertainment systems providing a semblance of three-dimensionality already flooding the consumer market. The motion picture industry has stepped in as well, integrating stereoscopic filming and post-production depth enhancement in some of the latest feature films. Standard two dimensional (2D) displays integrate many of the depth cues that a viewer uses to judge distance and relative position, such as shading, occlusion and linear perspective, yet the more subtle but equally important cues like parallax, stereopsis and convergence are omitted. By integrating this second set of depth cues into a 3D display the viewer is provided with a more physiologically natural method by which to assimilate the information, ensuring a richer viewing experience. The speed and accuracy at which a user can perform a task has been shown to increase when presented in a 3D format rather than a 2D format [1] indicating significant possible benefits for uses in fields such as medicine [2], terrain perception [3] and complex data analysis [4].

Many approaches to creating 3D displays have been seen over the recent decade, with a majority of the systems falling into one of three categories as defined by Holliman et al [5]. These are two-view systems, horizontal-parallax systems and full parallax systems in which the distinction is made based on the degree of parallax that the system is able to display at any one point in time. Head-mounted displays and those requiring eyewear to achieve the 3D effect fall into the two-view taxonomy while the horizontal and full parallax systems present multiple views of a scene, allowing the viewer to look around an object. As the number of views presented to the viewer increases, the reproduced scene begins to approach a true representation of the original scene, eliminating the vergence-accommodation conflict when the wavefront of the projected image approaches that of the object [6,7]. A hologram is the ideal truly 3D display, reproducing the amplitude and wavefront of a scene with resolution only limited by that of the recording media [8]. Computer generated holograms have also been exhibited [9], eliminating the need for direct holographic recording of an object in order to reproduce it. The main drawback to holography is that it is based on diffraction from features that are on the scale of the wavelength of the illuminating light, or around 500nm for a visible display. This pattern must be reproduced in some updatable fashion at the desired refresh rate, requiring the display media to have high-speed, sub-micron resolution with a size suitable for viewing, or on the order of 10^{12} pixels for a 20 inch display. For a 3 color, 8-bit display updating at 30Hz, this means transmission of 10^{16} bits per second. In contrast, a 1080i HDTV with the same display parameters only needs a pipeline that can transmit 10^{7} bits per second.

Due to this high bandwidth requirement, conventional computer generated holographic displays do not appear to be practical at this time, but significant effort has been expended to develop approximations retaining the realism of holographic imaging. One way to reduce the computational complexity of the system is to remove the phase information from the wavefront reproduction, leaving only the intensity and direction to be recreated (light field reproduction), as in holographic stereography [8]. This technique has been used to produce very realistic representations of scenes and is the technology upon which our updatable display is based.

2. The Holographic Printer

Holographic stereograms, also known as integral holograms, are created by holographically recording views of a scene or object taken from different angles. In our holographic printer, the necessary views of a 3D model or scene that will be reconstructed are processed into hogels, or holographic pixel elements as described in prior work from our group [10]. The hogels are uploaded to a spatial light modulator (SLM) which is used to structure the object beam of a holographic recording system. This object beam is focused onto the film plane by a lens, creating the angular subtense of the image and is there interfered with the reference beam to create an interference pattern, the basis of the hologram.

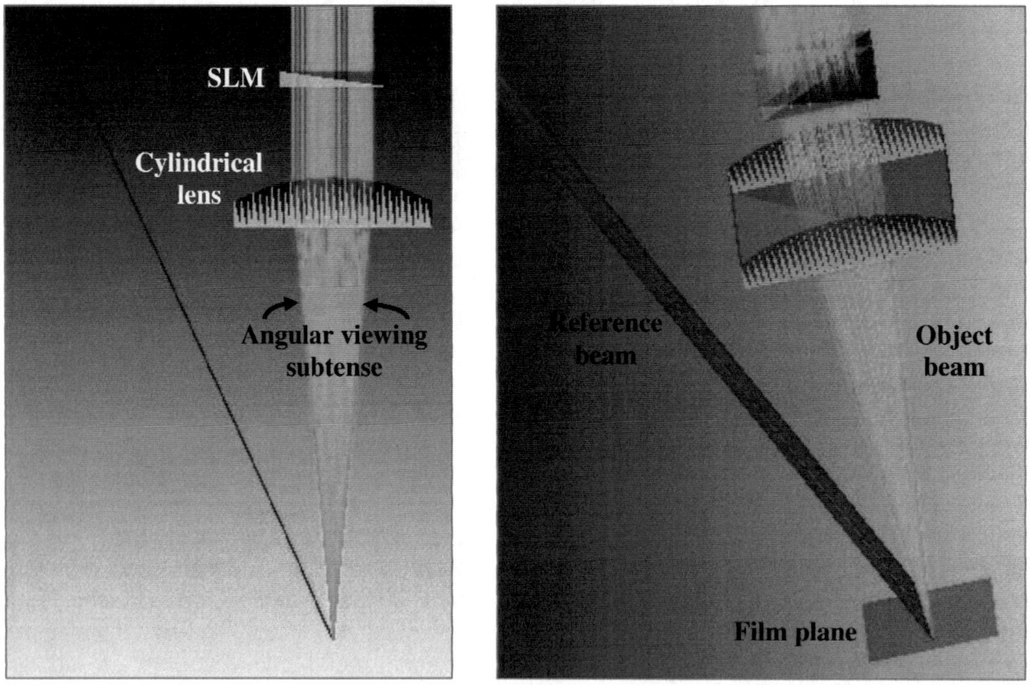

Figure 1. Representative illustrations of the main holographic printer components and organization. The reference beam is illustrated in blue to distinguish it from the object beam.

Figure 1 shows an illustration of this holographic interference concept for a horizontal parallax only system, as noted by the cylindrical lens and the line focus on the film plane. The full parallax system replaces this cylindrical lens with one that is circularly symmetric, illuminating a square hogel on the film. Illumination with an LED source from the correct angle reproduces the recorded model for viewing. In our case, the media into which we record the interference pattern differs from that used in other types of computer generated holograms.

Silver halide and photopolymers have often been used as the recording film but images in these media are permanent. In order to create a dynamic and updatable display using holographic stereography, the recording media must be able to record and display the image, but also erase and record/display a new image. Photorefractive materials fit this bill, enabling continually refreshing scenes to be holographically recorded and displayed, with photorefractive polymers being the best material for a large viewing area. The photorefractive polymer media is based on a composite of a hole transporting polymer doped with a non-linear optical chromophore, a plasticizer and an optical sensitizer. The polymer is then melt-processed between two transparent electrodes creating the photorefractive film onto which we record the holographic stereogram. The material composition and film preparation method have been described in previous literature [11]. When the interference pattern from the writing system illuminates the film, the intensity distribution is recorded nearly instantaneously into the media, resulting in a refractive index distribution and allowing us to reconstruct the 3D image by diffracting the reading illumination from the index grating.

3. State of the art performance

3.1. Brightness

Display brightness plays an important role in the comfort of the user, with low luminance levels contributing to eye strain and fatigue. In order to be a viable system, any display must be able to reproduce the illumination levels that are present in current day computer monitors and televisions. Standard computer monitors and televisions are designed to be viewed in a variety of ambient illumination levels, from a darkened room to outside viewing conditions and produce luminance levels in the range of 300-1000 cd/m^2. Our current system is illuminated with a Luminus color LED reaches a luminance of greater than 2,500 cd/m^2 in the image plane, indicating that it is more than sufficient to meet the range of desirable brightness levels any user could need.

3.2. Reduced Laser Power

The sensitivity of our current system has increased dramatically, with significantly less laser power required to record images. In previous work [10, 12] we reported a pulse laser output power of 10W in one iteration and in another iteration 100mW/cm^2 CW laser flux measured at the film surface. Now our total laser power for the CW system is less than 20mW, with the flux at the surface significantly less accounting for the various losses throughout the system. One major benefit for reducing the laser power needed to write an image is that it reduces the overall energy requirements of the display system.

3.3. Increased Display Size

Display size is another of those factors that plays an important role in usability of a system. As a display increases in size, the quantity of information that can be displayed without compromising the viewability also increases. Additionally, larger displays can more easily accommodate multiple viewers at one time. We are successfully moving toward larger displays with our system architecture and our photorefractive film material lends itself nicely to scaling to larger screen sizes. In the previous iteration of our system, we recorded and displayed an image on our 100mm x 100mm sample, taking approximately 2 seconds to record using a 6 ns pulse laser with a repetition rate of 50Hz [12]. In our current work we can write and display an image on a 300mm x 150mm sample in less than 8 seconds using the same pulse laser.

3.4. Improved Resolution

In previous work, we reported images written on the CW display with a resolution of 0.8mm hogel requiring illumination time of one second per hogel [13]. Currently, we are able to achieve hogel sizes of less than 200μm, taking only 0.4 seconds to write each hogel on the CW display. This improvement in horizontal resolution allows us to more accurately reconstruct higher frequencies in the horizontal direction. The vertical resolution is only constrained by the resolution of the SLM which for this system is on the order of 110μm. Figure 2 shows pictures of holograms taken with a digital SLR.

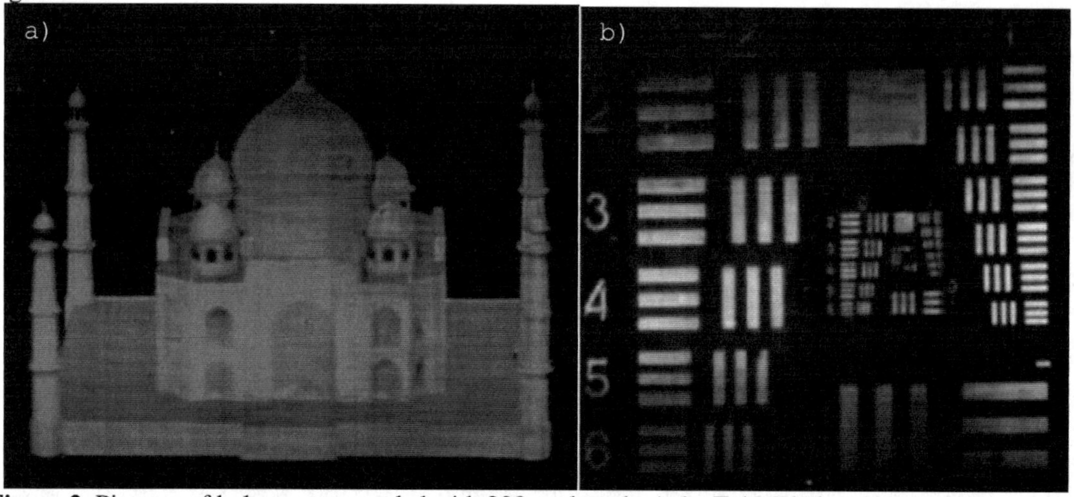

Figure 2. Pictures of holograms recorded with 200μm hogels a) the Taj Mahal and b) an Air Force test target used to measure the size of the hogels [14]

3.5. Optimized NTSC CIE Color Space

One benefit of the illumination scheme of this display is that we are able to reconstruct the image with any wavelength using LED sources. We are currently using three Luminus LEDs with nominal

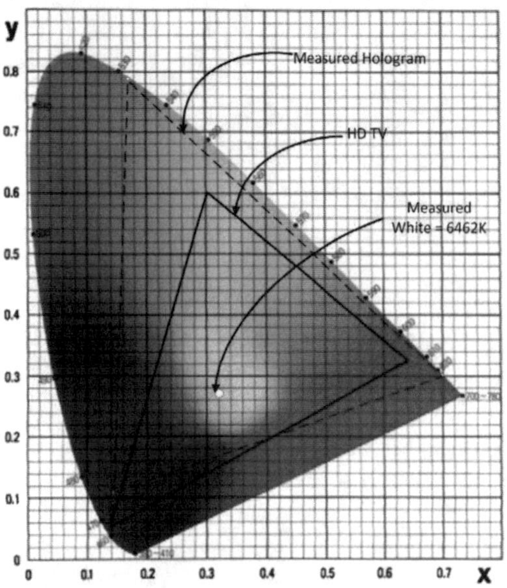

Figure 3. CIE color space comparison for a standard HDTV vs. output measured from our hologram

Figure 4. Photograph of a three color hologram written on our system

wavelengths 632nm, 525nm and 460nm, combining to provide a crisp white at 6462K. This gamut far exceeds that of the standard HDTV as shown in figure 3, allowing a larger range of colors to be reproduced by this display. Figure 4 shows a photograph of a three color hologram written by our display system and illustrates a number of vivid colors.

4. Conclusion

Our holographic stereogram display is fast approaching a near video-rate display with vastly improved brightness, display size, resolution and color gamut all while decreasing the laser power necessary for recording. Though our photorefractive polymers have shown sufficiently fast response times for video rate refreshing [15] current laser technology falls short of our need for kHz pulse repetition rates and tens of mJ per pulse. Additionally, the move toward a faster updating SLM or the like would allow us to increase the system refresh rate for the three color systems. We continue working toward increasing the sensitivity of our polymers as well as reconfiguring the optical designs, advancing our progress toward a 3D updatable video-rate display.

We acknowledge support from AFOSR, DARPA and the NSF ERC Center on Integrated Access Networks (CIAN). This research was funded by the Office of the Director of National Intelligence (ODNI), Intelligence Advanced Research Projects Activity (IARPA), through the AFRL contract FA8650-10-C-7034. All statements of fact, opinion or conclusions contained herein are those of the authors and should not be construed as representing the official views or policies of IARPA, the ODNI, or the U.S. Government.

References

[1] Martin J and Holzbach M 2008 *Air Force Research Laboratory Human Effectiveness Directorate, Warfighter Readiness Research Division* (Technical Report approved for public release via www.dtic.mil website, id: AFRL-RH-AZ-TR-2008- 0025, accession number ADA486177)
[2] Held R T and Hui T T, 2011 *Academic Radiology* **18** 1035-48
[3] Merritt J O 2005 *Proc. SPIE* vol 5800 (Orlando: SPIE) pp169-76
[4] Burwell C, Jarvis C and Tansey K 2011 *Remote Sensing Letters* **3** 481-90
[5] Holliman N S, Dodgson N A, Favalora G E and Pockett L 2011 *IEEE Transactions on Broadcasting* **57** 362-71
[6] Kim J, Hoffman D M, and Banks M S 2011 *Journal of Vision* **11** 1-29
[7] Blanche P-A 2012 *Information Display* **28**-3
[8] Benton S A and Bove V M Jr. 2008 *Holographic Imaging* (New York: Wiley-InterScience)
[9] Waters J P 1966 *Applied Physics Letters* **9** 405
[10] Tay S et al. 2008 *Nature* ***451*** 694-8
[11] Thomas J, Christenson C W, Blanche P-A, Yamamoto M, Norwood R A, and Peyghambarian N 2011 *Chemistry of Materials* **23** 416-29
[12] Blanche P-A et al, 2010 *Nature* ***468*** 80-3
[13] Blanche P-A et al. 2010 *Practical Holography XXIV: Materials and Applications* **7619** 0L-5
[14] St. Hilaire P 2012 Are stereograms holograms? A human perception analysis of sampled perspective holography *To be published in the same proceedings*
[15] Eralp M et al. 2006 *Applied Physics Letters* **89**

Holographic 3D imaging - methods and applications

J Svoboda, M Škereň, M Květoň, P Fiala

Czech Technical University in Prague, Faculty of Nuclear Sciences and Physical Engineering, Brehova 7, 11519 Prague, Czech Republic

E-mail: jakub.svoboda@fjfi.cvut.cz

Abstract. In this paper, different techniques of three-dimensional holographic imaging with respect to particular applications are discussed. The realization techniques based on image synthesis at the hologram plane and at the observer's eye pupil plane are presented and compared with classical image holography approach. Each technique is accompanied with an example of the hologram, that was created using some of the in-house developed devices.

1. Introduction

Modern 3D visualization techniques can be applied in many sectors (entertainment industry, medical applications, etc. [1, 2]). During the last decade, the non-holographic computer based techniques took the lead in the 3D displaying industry [3, 4]. Although 3D visualization using these so called stereoscopic techniques can be sometimes even more realistic than holography (true color, motion), it brings also specific problems. For example accommodation of observer's eye is often in disagreement with vergence of eye-pair, which can be significantly disturbing for some observers. Therefore, for some applications, the holographic approach still represents the only sufficient alternative.

Nowadays, the progress in holography is mostly carried by the synthetic approaches. These approaches allow us to employ and take the advantages of the stereoscopic methods in holography [5, 6]. For example it is possible to create holograms of large sceneries, holograms of living objects or even holograms of moving objects. The model for the synthetic hologram can be fully virtual, created in a 3D modeller in a computer.

First approach, that will be discussed is direct writing of the hologram structure using a lithographic device. Mostly these holograms are created using either electron beam lithography or laser beam lithography [7]. Not to say, that e-beam technology is not suitable for synthetic holography, however in the particular case of synthetic image holograms it is kind of overkill. Therefore, in this case, we will focus on laser beam lithography. The laser lithography can not achieve such a resolution as e-beam, but it has way lower operational costs and higher throughput.

Using the laser (or e-beam) lithography, one of the most important problems is joining the exposed elements into larger area with high accuracy. This puts very high demands on the positioning systems of the lithographs and restricts the area of the hologram up to a few square centimeters (considering reasonable costs and reasonable exposure times) [8]. Therefore, these techniques are not very suitable for creating larger synthetic holograms.

Synthetic holograms of larger scale can be created using the method, where the multiplexing of the stereoscopic images is not created at the plane of the hologram, but at the plane of

Published under licence by IOP Publishing Ltd

the observer's eyes. This kind of synthesis is mostly done in two-step setup. The set of 2D stereoscopic views is recorded into a master-hologram (first step) and then it is transferred into a final hologram (second step).

The image synthesis, however, can be used also in "classical" image holography. By classical, we mean recording the hologram from a real model in an optical setup on an optical table. Several different objects can be recorded into one hologram, for example to create an artistic impression. In our case, we used synthesis in classical image holography to create an object for opthalmologic system for dynamic measurement of ocular deviations [9].

All of these methods have its advantages and disadvantages and are suitable for different particular applications. A comparison will be given in the following text.

Before presenting the synthetic holography techniques, it is necessary to understand the stereographic principles.

2. Stereography

The visual sense of a 3D scene is caused by our brain interpreting the image information from both eyes. Due to the relative position of our eyes, the image from the left eye slightly differs from the image from the right eye. This is the crucial aspect causing the 3D sense and that is why it is used in hologram synthesis. There are also other cues that enter the image processing in our brain, however they are not that strong [10].

In fact it is possible to use only a stereoscopic pair of images to create the 3D sensation. The displaying methods that use only a pair of planar views are called binocular and it can be for example Anaglyphs or the modern 3D cinemas [11, 12]. The main disadvantage of these methods is the lack of movement parallax.

The movement parallax can be "added" using the autostereoscopic methods, when more than two angular views of an object are multiplexed together [13, 14]. In such a case, the displaying device creates the viewing zones and the observer, while moving, enters the different zones and sees the object from a different angle. The examples are integral photography, synthetic holography, or the classical image holography (see Figure 1).

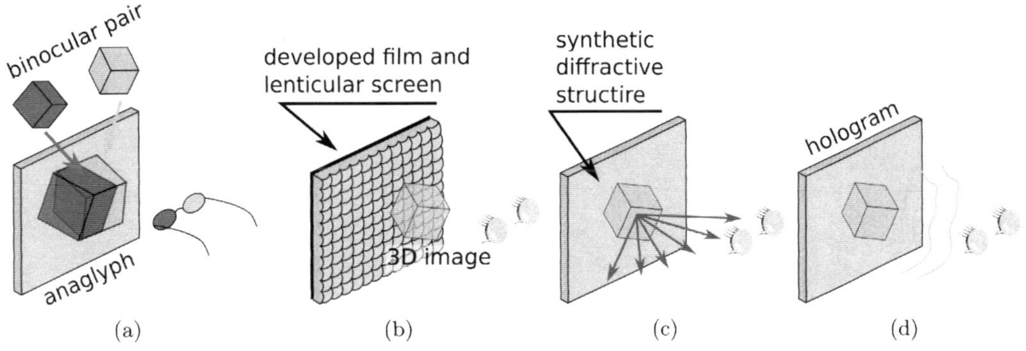

Figure 1. 3D displaying methods: **(a)** In binocular methods, only two channels are employed which overlap each other in space. In this case, an anaglyph is shown. **(b)** Integral photography is an example of non-diffractive autostereoscopic methods, in which multiple viewing zones are spatially separated. **(c)** Diffractive autostereoscopic methods create the spatially separated viewing zones as in case (b). **(d)** Hologram is capable of reconstructing the image wavefront in its full complexity. It acts as if there is infinite number of viewing zones.

To create the synthetic hologram, it is necessary to capture the set of 2D stereoscopic views. These images can be obtained using a real camera taking pictures of a real object or using a

Journal of Physics: Conference Series **415** (2013) 012051 doi:10.1088/1742-6596/415/1/012051

virtual camera capturing a virtual 3D object in a PC. The way of capturing depends on the demands on the 3D behavior of the reconstruction - if the final hologram should be composed of both parallaxes or if it should be single parallax only. Figure 2 shows one of the possible capturing setups for both the single-parallax and the full parallax case.

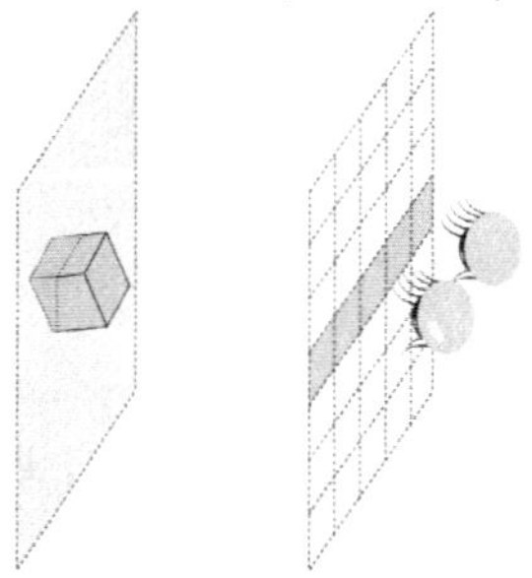

Figure 2. The positions of the viewing zones are coincident with the positions of a camera during the capturing process. Therefore the setup is the same for both cases except that the "capturing" camera lens is changed for the "viewing" eye lens. The figure shows the equidistant capturing with linear movement of the camera. The blue line represents the single parallax case - in each "square" one snapshot is taken. The whole array represents the case of capturing both parallaxes when the camera moves in two mutually perpendicular directions.

3. Synthesis at the hologram plane

Holograms of this kind are directly written using some of the lithographic techniques. The method is based on the limited resolution of the human eye in the image plane. When the hologram is observed from the conventional distance $\sim 25\,\mathrm{cm}$, the approximate resolution limit of the human eye is $\sim 0.1\,\mathrm{mm}$. From the point of view of diffractive structures such a dimension is large enough for carrying much more than just the information about a single point. As a rough estimate, if we assume an elementary diffraction grating with period $\sim 800\,\mathrm{nm}$ and area $4 \times 4\mu\,\mathrm{m}$ (which contains 5 periods of such a grating), we can place more than 600 different gratings within the area of $0.1 \times 0.1\,\mathrm{mm}^2$. The main idea of image multiplexing at the hologram plane is based exactly on this idea.

To understand the method one has to realize how the principle of stereoscopy applies for the given geometry. Let us assume the situation according to Figure 2. If the observer sees the 3D image floating in the neighborhood of the hologram plane, it means, that each point on the hologram should send different information to the different directions (meaning that the visibility of the particular point on the hologram surface in terms of color and intensity is dependent on the observation angle). The idea of realizing an angularly selective image on the base of diffraction is pretty straightforward. Therefore, each distinguishable point on the hologram surface (a macro-dot) is divided into a set of smaller units (micro-dots). Each micro-dot is filled with a regular grating with appropriate period and orientation. By changing these two parameters, it is possible to adjust the proper color and direction of diffraction. An illustration of the macro-dot/micro-dot structure can be seen in Figure 3.

Most holograms synthesized using this method are of rainbow type, thus the vertical parallax is omitted and angular views are captured only in the horizontal direction. This significantly reduces the number of necessary micro-dots. In practice, usually 9-16 angular channels are used to create a satisfactory 3D perception.

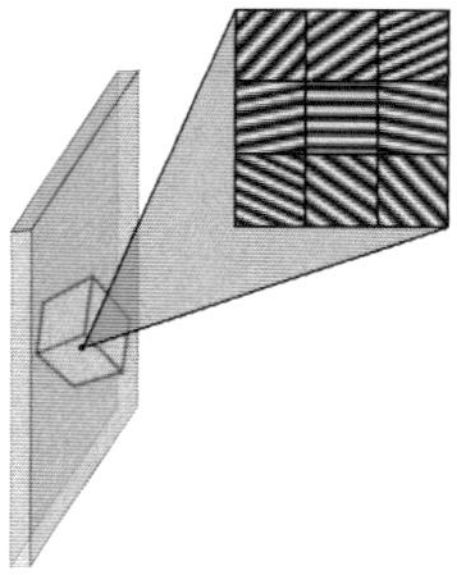

Figure 3. Basic idea of generating the spatial image channels using diffraction gratings.

The basic colors of the hologram are the spectral ones. If a non-spectral color is required (e.g. white), the additive color mixing must be introduced. The most common way of achieving a mixed color is to use two or three spectral colors, usually red, green, and blue (RGB). These colors are obtained by the means of diffraction from the gratings.

The true-color micro-dot is divided into three smaller areas, each filled with one grating. The relative areas of these three gratings control the contributions of each of the colors to the true-color mix. Furthermore, the area of the micro-dot is divided into two major parts - one filled with grating and the second empty. The ratio of these areas defines the overall luminosity of the image point. An example of this kind of hologram, created at the department is shown in Figure 4.

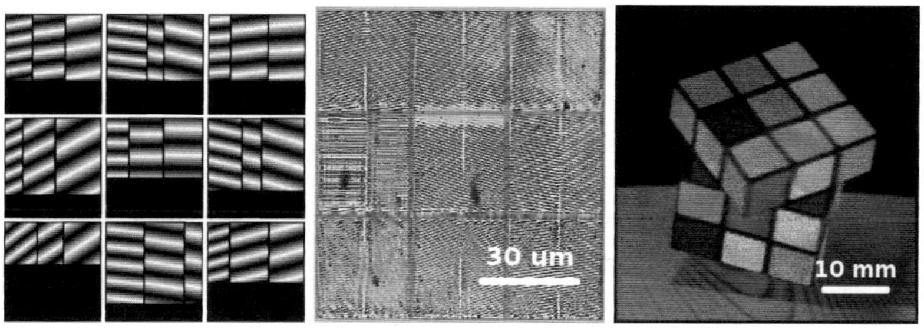

(a) (b) (c)

Figure 4. Directly written synthetic hologram: **(a)** Theoretical idea of macro-dot segmentation for RGB color mixing, **(b)** image of a real micro-structure, and **(c)** an example of the real reconstruction of the true-color RGB synthetic 3D rainbow hologram.

Advantages of the directly written holograms are mainly in the high achievable contrast, color saturation, and in the big variety of the features that can be implemented into the final hologram (especially in the field of optical document security). The biggest disadvantage is in the area limitation due to the necessity of very precise positioning of the recording material and due to low throughput for the holograms bigger than a few square centimeters.

4. Synthesis at the eye-pupil plane

In the case of synthesis at the eye-pupil plane, the final hologram is created in two steps. Recording the master-hologram and transferring it into the final one. The discretization of the spatial information is not processed at the plane of the final hologram as it was in the previous case. The micro-structure of the final hologram is continuous as in the case of classical holography. The discretization is brought in during the step of transferring the synthetic master-hologram. The master-hologram consist of many "small" separated holograms representing the particular viewing zones (as in Figure 2). These primary holograms create multiple signal beams that interfere with the reference beam during the transfer process and create the final continuous micro-structure. The name of the method is derived from the fact that the position of the former master-hologram coincides with the eye-pupil plane when viewing the final hologram. The principal method of this synthesis is shown in Figure 5.

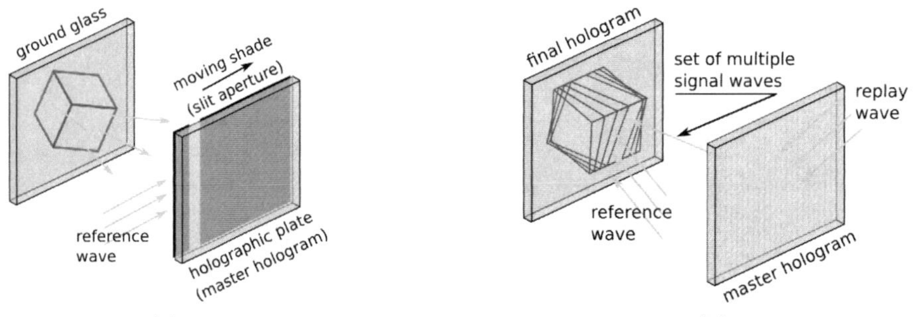

Figure 5. Hologram synthesis in the eye pupil: **(a)** The 2D views are sequentially displayed on a ground glass screen and recorded as elementary holograms in the form of narrow stripes. This means that the master-hologram is composed of a number of these elementary holograms placed next to each other. **(b)** The transfer scheme. All the stripes from the master-hologram are replayed at once using a conjugate "reference wave" from (a). All the reconstructed images overlap in the former ground glass plane and the final hologram is recorded.

In the case shown in Figure 5, there is only horizontal parallax recorded. Therefore, the final hologram is of the rainbow type. However, it is possible to modify the recording setup to create master holograms with both parallaxes. Such a master hologram can be then copied into final reflection hologram. To create the master hologram with both parallaxes, we have developed a fully automatic recording device. The scheme of this device can be seen in Figure 6.

The 2D image, representing a particular view, is uploaded to a transparent SLM. The signal wave passes through the SLM and the 2D image is focused by the writing objective onto its image plane with some magnification. Near the output pupil of the objective, the pattern of interference between the signal and reference beam is recorded onto a holographic plate. Then the plate is moved, the image on the SLM is switched for the next view and the next viewing channel is recorded, and so on. The movement of the holographic plate is compensated by moving the SLM in opposite direction (to keep all the views in register). When exposing hologram with two parallaxes, the procedure is repeated in two perpendicular axes.

Since the reference wave was collimated during the recording, the replay beam can be a wide collimated wave that reconstructs all the recorded views at once. All the reconstructed images overlap at the place where the object was situated during capturing. The final hologram plate is placed into this place and the final hologram is recorded (similar to Figure 5b).

The color mixing in this case of synthesis is processed already during the recording of the master hologram. Since the complete recording process is performed using a single laser (including the transfer process of the master-hologram), the three different color channels are achieved using three separate master-holograms. Reconstruction of these three masters during the transfer process provides three angularly separated sets of signal waves and three different "color" gratings are recorded. In practice, the master-hologram for color mixing consists of three independent vertically shifted rows of exposures. Each of the signal waves coming from the three

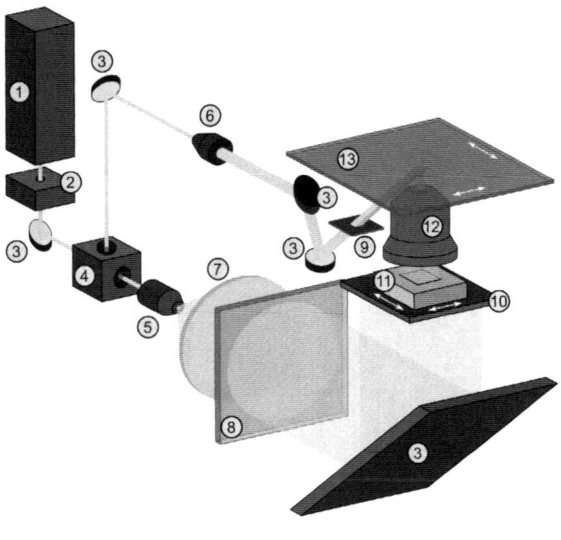

Figure 6. Scheme of the eye pupil synthesis device: **1** laser, **2** shutter, **3** mirror, **4** beam splitter, **5** microscope objective with spatial filter, **6** beam expander, **7** collimator, **8** holographic ground glass, **9** square aperture, **10** signal shade, **11** SLM, **12** special objective, **13** holographic plate.

channels carries the same spatial information about the object and the relative intensity of the waves is adjusted according to the desired color. Figure 7a shows copying of such an RGB synthetic master into the final hologram and Figure 7b shows the diffraction of white light on the final rainbow hologram.

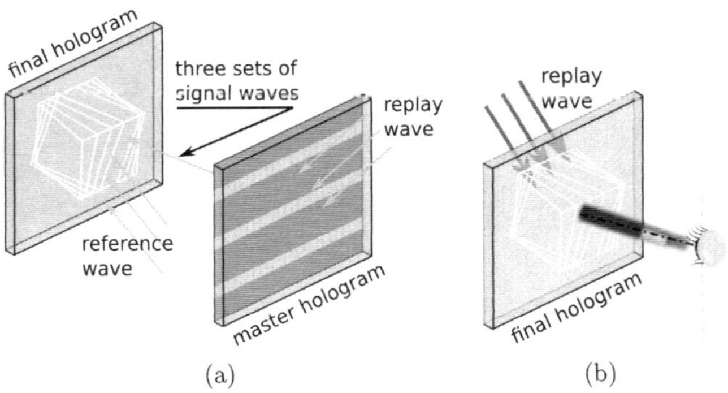

Figure 7. (a) Transferring three color masters into final RGB color rainbow hologram. **(b)** Reconstruction of the final hologram from the step (a) with white light replay beam. Only the orders that are diffracted in the direction of the observer are indicated in the figure.

In Figure 8a there is an example of the RGB color master for a rainbow hologram synthesized using our device. The final hologram (copy of the master) is displayed in Figure 8b.

The main advantage of this method over the method of synthesis at the hologram plane is in the achievable size of the final hologram - unlike the directly written synthetic holograms, there is no high demand on accuracy of the positioning of the recorded stripes. Furthermore, the size of the final image is not limited by the size of the master-hologram, but by the size of the image on the screen. The general disadvantage of this method is in lower contrast and color saturation compared to the directly written synthetic holograms.

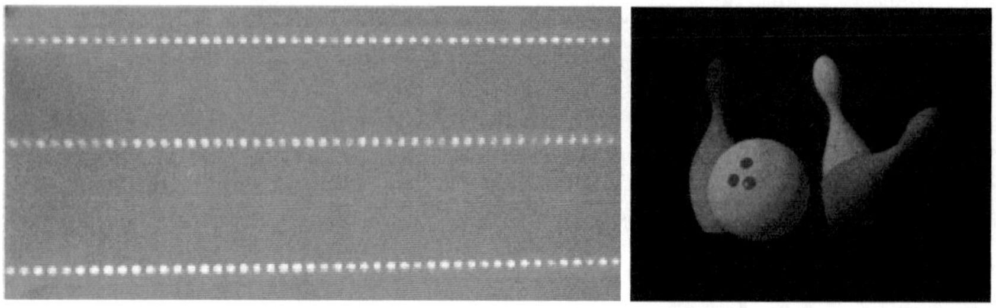

(a) (b)

Figure 8. (a) Photograph of the "RGB color" master-hologram. The three color channels can be clearly seen. **(b)** Photograph of the final color synthetic rainbow hologram, that was obtained using the master from step (a).

5. Synthesis in classical image holography

It is not that common to use hologram synthesis in classical holographic setup, however there are several areas where the synthesis of more models into one hologram may be very useful. An example of such a synthesis conducted at the department is a synthetic hologram for opthalmologic system for dynamic measurement of ocular deviations. The task was to create a hologram for the measurement according to Figure 9. The hologram includes four images of a digger in two different positions. The examination consists of rapid switching between the reconstruction sources, while the camera is observing the eyes of the patient in IR.

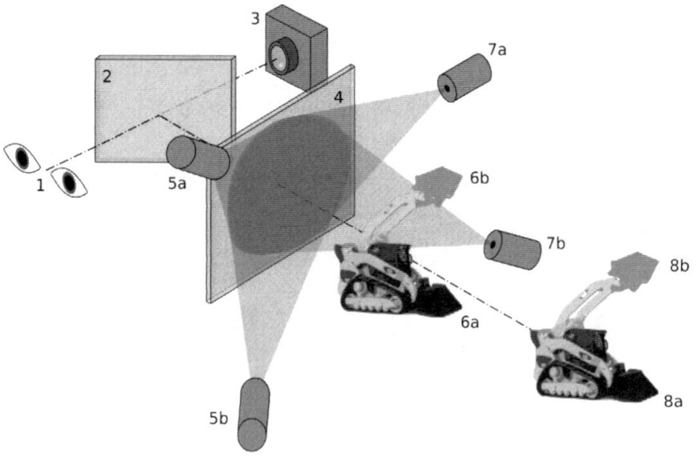

Figure 9. Dynamic measurement of ocular deviations:**1** pacient, **2** beam splitter, **3** camera, **4** synthetic hologram, **5** reconstructing lasers for closer images, **6** closer image, **7** reconstructing lasers for distant images, **8** distant images.

To make such a hologram work, all the sub-holograms should be approximately of the same efficiency. Therefore, the final hologram is composed of two plates, each for one position of the digger. Each plate than has two exposures in it. Each with different reference wave for a different position of the bucket. The two holograms are sealed together so that the emulsions are in contact.

Advantages of this method would be in the continuous change of the viewing angle, high achievable contrast and high versatility of the recording scheme. However, since everything in the process has to be "hand-made", this technique is extremely time demanding. Also, a big disadvantage is the need for a real model for the synthesis.

6. Conclusions and Acknowledgements

The three synthetic holographic methods have been presented and compared. All the methods have been accompanied by examples of holograms that have been created at the department. Each of them has its specific advantages and disadvantages which determine/limit their usage. Therefore, it is certain, that none of the methods would vanish, no matter how some of them would be prioritized.

This work has been supported by the Grant of Czech Academy of Sciences Project No. KAN401220801.

References

[1] Takanori O 1976 *Three-Dimensional Imaging Techniques* (Academic Press)
[2] O'Shea R P, Blackburn G and Ono H 1994 Contras as a depth cue. *Vision Research* **34** 1595-1604
[3] Lipton L 1982 *Foundations of the Stereoscopic Cinema - A Study in Depth* (New York: Van Nostrand Reinhold) p 56
[4] Faubert J 2002 *Three-dimensional Television, Video, and Display Technologies* (New York: Springer-Verlag) pp 359-396
[5] Benton S A 1977 *Applications of Holography and Optical Information Processing* 401-9
[6] DeBitetto D J 1969 *Applied Optics* **8** 1740-1
[7] Verheijen M J 1993 *Journal of Modern Optics* **40** 711
[8] Pease R F and Chou S Y 2008 *Proc. IEEE* **96** pp 248-70
[9] Jindra T, Dušek J, Dostálek M, Fiala P, Škereň M 2011 *Proc. SPIE* **8306** 83060M-3
[10] Javidi B and Okano F 2002 *Three-Dimensional Television, Video, and Display Technologies* (New York: Springer-Verlag)
[11] Zone R 2003 *Stereo World* **29** 11-13
[12] Turner T L and Hellbaum R F 1986 *Information Display Magazine* **2** 22-24
[13] Ferris S H 1972 *Journal of experimental psychology* **95** 258-63
[14] Halle M 1997 *Computer Graphics* **31** 58-62

Real-time Holographic Display Based on a Super Fast Response Thin Film

Hongyue Gao[1,2], Xiao Li[1], Zhenghong He[1], Yikai Su[1], and Ting-Chung Poon[2]
[1]Center for Opto-electronic Materials and Devices, National Engineering Lab of TFT-LCD Materials and Technologies, Shanghai Jiao Tong University, Shanghai, China, 200240
[2]Bradley Department of Electrical and Computer Engineering, Virginia Tech, Blacksburg, Virginia, USA, 24061

Corresponding author: Hongyue Gao
Phone Number: +1-540-629-1956, +86-13671973358
E-mail: gaohy@sjtu.edu.cn, gaohylet@gmail.com

Abstract. Real-time dynamic holographic display is obtained with super fast response in a thin film without any applied electric field. Holograms can be refreshed in the order of a millisecond and there is no cross talk between the recorded holograms because the hologram formed in the film is transient and can be completely self erased, and the hologram formation time and self-erasure time are both ~1 ms. Holographic video display is achieved, which shows the real-time holographic image display capability of the thin film, and its much higher resolution than those of commercially available spatial light modulators. Furthermore, multiplexed hologram display using two polarization directions of a recorded light and multiple color holographic display at different laser wavelengths are presented, which demonstrate the feasibility of a RGB color holographic three-dimensional display with the thin film. Because the sample is easy to be fabricated into a large size screen and needs no external applied electric field, we think that the film can be developed into a large-size, dynamic, and color holographic three-dimensional display in the future.

1. Introduction

Holography [1-5] is a true three-dimensional (3D) technique by using coherent light interference. However, it is not good enough to be applied in true 3D display by now [6]. One of the problems is that holographic response is not fast enough. Since a 3D movie "Avatar" was shown, dynamic holographic display has attracted public attention and is currently under active research. Some great results on it were obtained in recent years. In 2010, holographic 3D telepresence [7], as dynamic near-real-time holographic display, was reported in Nature by P.-A. Blanche, et al. In this paper, a holographic stereographic technique and a photorefractive polymer as holographic material were used, and the holographic display that could refresh images every two seconds was demonstrated. Recently, quasi-real-time holographic display [8] with refresh rate, 5 images per second, was presented by K. Kinashi, et al. However, these achievements are not real-time dynamic holographic display. Here, we use a super fast holographic response liquid crystal as holographic recording medium to demonstrate real-time dynamic holographic display that holograms can be refreshed in the order of a millisecond and there is no cross talk between the recorded holograms. Holographic video display and multiplexed

Published under licence by IOP Publishing Ltd

hologram display is achieved, which show the real-time holographic image display capability and the feasibility of a RGB color holographic display with the thin film.

2. Experiments and results

The sample used in our experiments is a mixing liquid crystal film with a thickness of ~50 μm, sandwiched by two ITO (Indium tin oxide) coated glass substrates. The experimental setup of holographic display is shown in Figure 1. A reference beam and a signal beam, which are derived from a Nd:YAG laser (λ=532.8 nm), are set to be p-polarization by half-wave plates. The spatial information of an image is displayed on the SLM, and then is carried on the signal beam. The SLM used in the experiments is a high resolution, pure phase modulating microdisplay with 1920x1080 pixel resolution, 8.0μm pixel pitch, and 60 Hz frame rate developed by HOLOEYE Photonics AG. A He-Ne laser beam (λ=632.8 nm) is used to probe the writing region of the sample.

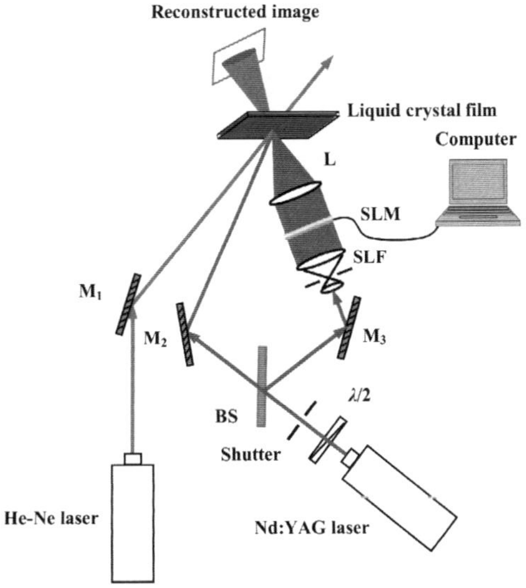

Figure 1. Experimental setup for holographic display. M's are mirrors, BS is a beam splitter, λ/2 is a half-wave plate, L is a lens with the focal length of 150 mm, SLF is a spatial light filter, and SLM is a spatial light modulator.

We study holographic display based on fast holographic response of the mixing liquid crystal film. Its holographic response is shown in Fig 2. The hologram formation time and self-erasure time are both ~1 ms in this film. Furthermore, a real-time holographic video display is achieved as illustrated in Fig. 3. We have a series of five letters inputting to the SLM sequentially and Fig. 3 shows the five reconstructions. The result turns out that this film can be used in holographic video display.

Journal of Physics: Conference Series **415** (2013) 012052 doi:10.1088/1742-6596/415/1/012052

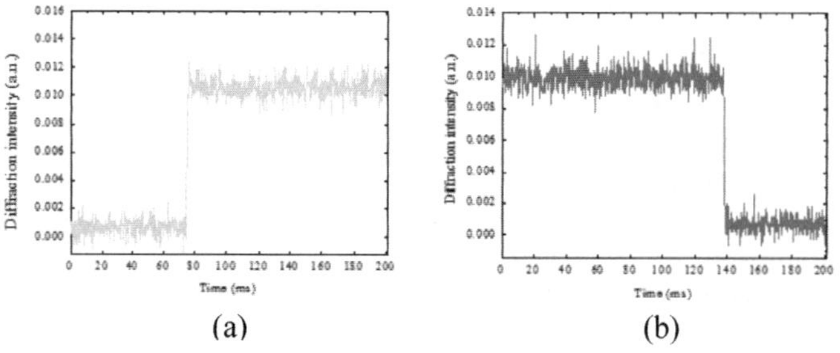

Figure 2. Hologram formation process.

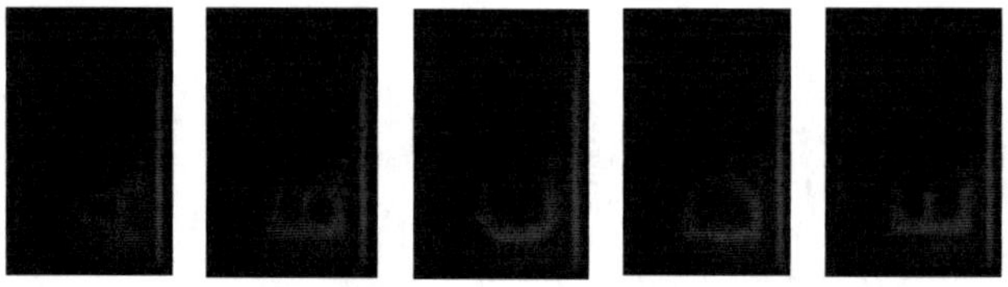

Figure 3. Several snap shots from the real-time holographic video.

Figure 4 illustrates the experimental setup for an angular multiplexed hologram display. By angular multiplexing, three holograms are written at a single location of the film at the same time. The diffractions from the angular multiplexed holograms read out by He-Ne laser are shown in Fig. 5. I_N and I'_N (N=1, 2, 3) denote the +1st and -1st diffractions from one of the three angular multiplexed holograms. This experimental result shows that it is possible to use different color lasers to record and read out the holograms.

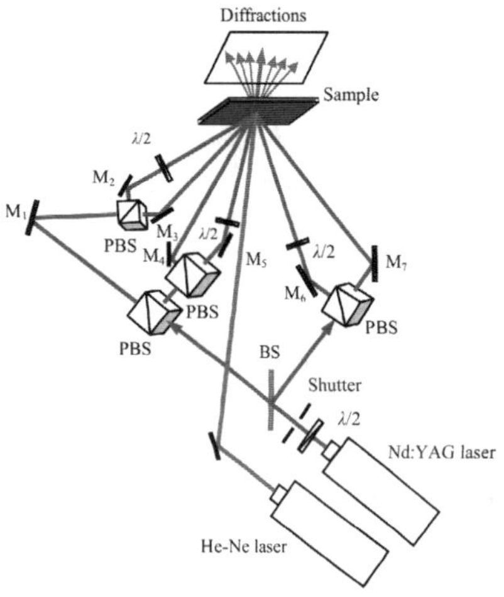

Figure 4. Experimental setup for an angular multiplexing holographic display. PBS is a polarizing beam splitter.

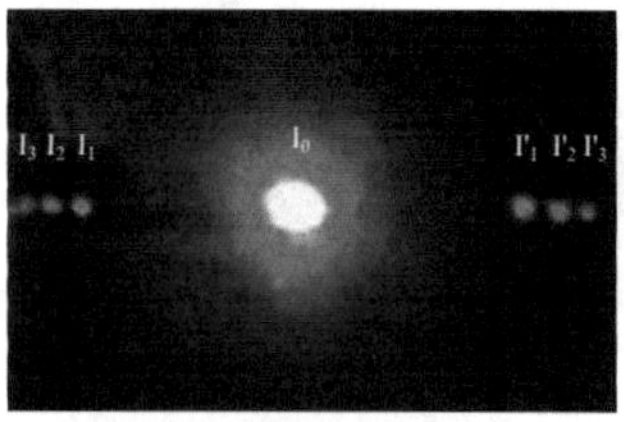

Figure 5. Diffractions from three angular multiplexed holograms.

Multiple color holographic display at two different laser wavelengths is studied. Figure 6 shows the reconstructed images from a same hologram read by He-Ne laser and Nd:YAG laser, respectively. This result means that color holographic display is possible to be realized using RGB model. Therefore, our research may be one step further towards the development of large-size, dynamic, and color holographic 3D display.

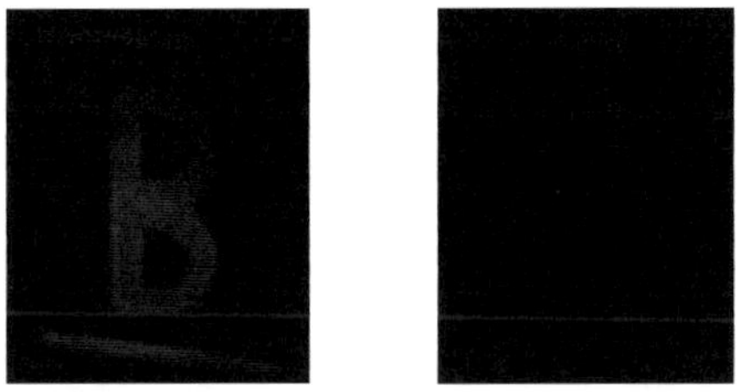

(a) (b)

Figure 6. (a) and (b) are the reconstructed images from a same hologram read by He-Ne laser and Nd:YAG laser, respectively.

3. Conclusion

We achieve real-time dynamic holographic display in the thin liquid crystal film with super fast holographic response. The holograms can be refreshed in the order of a millisecond and there is no cross talk between the recorded holograms in the film with the hologram formation time and self-erasure time, ~1 ms. We also investigate multiplexed hologram display and multiple color holographic display, which show the feasibility of a RGB color holographic three-dimensional display with the thin film. Our work turns out that the film can be developed into a large-size, dynamic, and color holographic three-dimensional display.

4. References

[1] Benton, S. A. & Bove, V. M. Jr. Holographic Imaging 2008 (Wiley-InterScience).
[2] T.-C. Poon, Ed., Digital Holography and Three-Dimensional Display 2006 (Springer).
[3] Lessard, L. A. & Bjelkhagen, H. I. (*eds*) Practical Holography XXI: Materials and Applications 2007 Proc. SPIE 6488.
[4] H. Gao, H. Pu, B. Gao, D. Yin, J. Liu, and F. Gan, 2009 Appl. Phys. Lett., **95,** 201105-201107.
[5] H. Gao, J. Liu, F. Gan, and B. Ma, 2009 Appl. Opt., **48,** 3014-3016.
[6] Savas Tay, P.-A. Blanche, R. Voorakaranam, A. V. Tunç, W. Lin, S. Rokutanda, T. Gu, D. Flores, P. Wang, G. Li, P. St Hilaire, J. Thomas, R. A. Norwood, M. Yamamoto, and N. Peyghambarian, 2008 Nature, **451,** 694-698.
[7] P.-A. Blanche, A. Bablumian, R. Voorakaranam, C. Christenson, W. Lin, T. Gu, D. Flores, P.Wang, W.-Y. Hsieh, M. Kathaperumal, B. Rachwal, O. Siddiqui, J. Thomas, R. A. Norwood, M. Yamamoto and N. Peyghambarian, 2010 Nature, **468,** 80-83.
[8] K. Kinashi, Yu Wang, Sho Tsujimura, Wataru Sakai, and Naoto Tsutsumi, 2012 Biomedical Optics and 3D Imaging OSA, JM3A.58.

Acknowledgments

This work was supported in part by NSFC (11004037), China postdoctoral Science Foundation (20110490732), Science and Technology Commission of Shanghai Municipality (09ZR1414800/11JC1405300), and Shanghai Shuguang program (09SG13).

Journal of Physics: Conference Series **415** (2013) 012053 doi:10.1088/1742-6596/415/1/012053

Computer holography: 3D digital art based on high-definition CGH

K Matsushima[1], S Nakahara[2], Y Arima[1], H Nishi[1], H Yamashita[1], Y Yoshizaki[1] and K Ogawa[1]

1 Department of Electrical and Electronic Engineering
2 Department of Mechanical Engineering
Kansai University, 3-3-35 Yamate-cho, Suita, Osaka 564-8680, Japan

E-mail: matsu@kansai-u.ac.jp

Abstract. Our recent works of high-definition computer-generated holograms (CGH) and the techniques used for the creation, such as the polygon-based method, silhouette method and digitized holography, are summarized and reviewed in this paper. The concept of computer holography is proposed in terms of integrating and crystalizing the techniques into novel digital art.

1. Introduction

In the past few years, we presented some high-definition full-parallax CGHs calculated by the polygon-based [1] and silhouette method [2,3] and fabricated by laser lithography system [4,5]. These holograms, composed of more than billions pixels, produce very fine spatial 3D images of occluded virtual scenes and objects. The optical reconstruction is comparable to that in classical holography. Strong sensation of depth caused by these high-definition CGHs has never been achieved by any conventional 3D systems and pictures other than holograms.

In addition, we have also presented a new technique called *digitized holography* [6]. In this technique, the fringe generated by optical interference between the object wave of a real existing object and a reference wave is digitally recorded by image sensors with a wide area and at a high sampling density. The recorded object wave is numerically incorporated in a virtual 3D scene constructed of CG-like 2D and 3D objects. The total field of the virtual scene, which keeps the proper occluded relation, is calculated and optically reconstructed by the technique of high-definition CGHs. This means that the whole process of classical holography, i.e. recording and reconstructing the object wave, is replaced and extended by digital counterparts. Digitized holography makes it possible to edit the 3D scene of holograms after recording object waves.

These techniques are opening the world of novel digital art, referred to as *computer holography* in this report. Various source materials can be input data in computer holography, for example, digital photos, illustrations, polygon-mesh 3D objects, multi-viewpoints images and captured fields of real existing objects. The 3D scene including these materials is designed with the technique of field-based digital-editing and optically reconstructed by a CGH as designers intended.

This paper is intended to present summary report of the concept of computer holography, technique used for creation, and some of our recent works.

Journal of Physics: Conference Series **415** (2013) 012053 doi:10.1088/1742-6596/415/1/012053

Table 1. A summary of major works in computer holography.

Title	Dimensions $W \times H$ [cm^2]	No. of pixels $W \times H$	Pixel pitches [μm] $\times$ [μm]	Creator(s)[1]	Date[2]	Ref.	Notes
The Venus	6.5 × 6.5	65,536 × 65,536	1.0 × 1.0	KM	Jan. 2009	[4, 7, 8]	The first computer hologram created by our method
Moai I	6.5 × 6.5	65,536 × 65,536	1.0 × 1.0	KM	Jun. 2009	[9]	Shifted Fresnel method [10] for numerical propagation
Moai II	6.5 × 6.5	65,536 × 65,536	1.0 × 1.0	KM	Oct. 2009	[9, 11, 12]	Shifted angular spectrum method [13] for numerical propagation
Aqua 2	10.5 × 6.5	131,072 × 65,536	0.8 × 1.0	MN	Dec. 2009	[14]	Acceleration algorithm for the silhouette method in sparse 3D scenes
The Moon	10.5 × 6.5	131,072 × 65,536	0.8 × 1.0	KM	Mar. 2010	[11, 12]	Texture-mapping
Shion	10.5 × 5.2	131,072 × 65,536	0.8 × 0.8	KM, HN	Dec. 2010	[12, 15]	Texture-mapping, Gouraud shading, Shape measuring by 3D laser scanner
Bear II	6.5 × 6.5	65,536 × 65,536	1.0 × 1.0	YA, HN	Oct. 2010	[6, 15, 16]	The first major work in digitized holography
Mountain	3.3 × 3.3	32,768 × 32,768	1.0 × 1.0	KH	Nov. 2010	[15]	Big object for mountain scenery
The Metal Venus I	6.5 × 6.5	65,536 × 65,536	1.0 × 1.0	HN	Nov. 2010	[15, 17]	Specular flat shading
The Metal Venus II	6.5 × 6.5	65,536 × 65,536	1.0 × 1.0	HN	Jan. 2011	[18, 19]	Specular smooth shading
The Hands	3.3 × 3.3	32,768 × 32,768	1.0 × 1.0	MN	Jan. 2011	[20]	Prototype using "Switch-back method"; new algorithm for accelerating silhouette method
iShion V2	5.2 × 5.2	65,536 × 65,536	0.8 × 0.8	HY	Dec. 2011	[21]	Image hologram, Reconstruction by white light
iShion V3	10.5 × 10.5	131,072 × 131,072	0.8 × 0.8	HY	May 2012		Prototype of floating hologram
Hamsters	6.5 × 6.5	65,536 × 65,536	1.0 × 1.0	DF	Apr. 2012		Resizing object in digitized holography
Parked car	6.5 × 6.5	65,536 × 65,536	1.0 × 1.0	YY	Jun. 2012		Prototype using multi-viewpoint images for background
Penguin	13.1 × 6.5	131,072 × 65,536	1.0 × 1.0	YA	Nov. 2011		Large hologram in digitized holography
Rose in Ring	10.5 × 10.5	131,072 × 131,072	0.8 × 0.8	KM, HN, KO	Nov. 2011		"Switch-back method", Specular smooth shading
Brothers	12.6 × 10.5	196,608 × 131,072	0.64 × 0.8	KM	Apr. 2012		Texture-mapping, Gouraud shading, Two live faces measured by 3D laser scanner

[1] Here, creator means programmer and/or designer. The abbreviations of creators are as follows, KM: Kyoji MATSUSHIMA, MN: Masaki NAKAMURA, HN: Hirohito NISHI, YA: Yasuaki ARIMA, KH: Kentaro HIGASHI, HY: Hiroshi YAMASHITA, YY: Yutaka YOSHIZAKI, DF: Daichi FUJITA, KO: Kohei OGAWA. All holograms are fabricated by Sumio NAKAHARA.

[2] The indication is usually the printed date of the work. There is ambiguity, because several versions are sometimes reprinted for improvement.

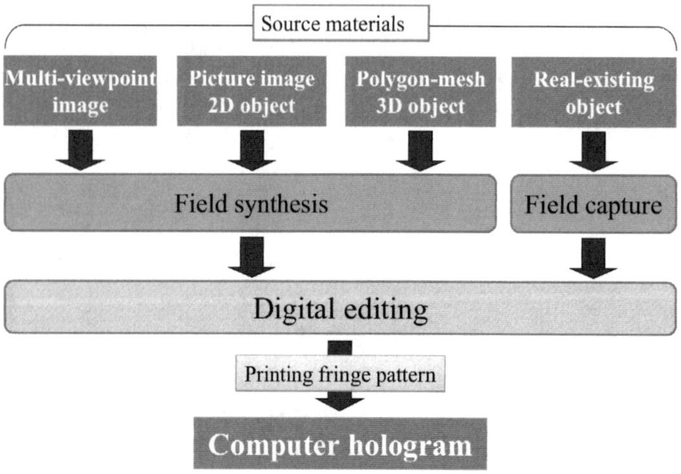

Figure 1. Conceptual illustration of the process of computer holography.

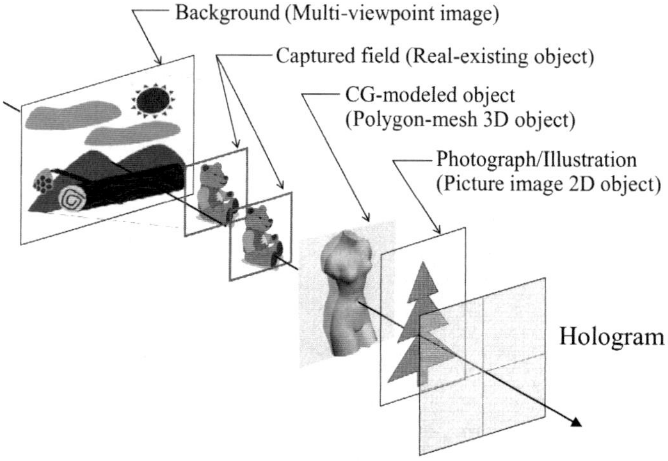

Figure 2. A typical configuration of the 3D scene in computer holography.

2. Computer holography and the materials for 3D scenes

Some major computer holograms we presented so far are summarized in Table 1. The process and typical 3D scene of computer holography are illustrated in Fig. 1 and Fig. 2, respectively. As shown in these figures, various materials can be source input in computer holography.

CG-modeled 3D objects are the most important materials for computer holography. The shapes of 3D objects are usually given by polygon-meshes. Instead of conventional point-based methods, we mainly use the polygon-based method [1, 4] for numerical synthesis, because the polygon-based method considerably reduces the computation time of full-parallax CGHs for surface-modeled objects. The point-based methods are also used complementarily in some works such as "The Moon".

Two dimensional (2D) objects such as digital illustrations or photographs are also included in the 3D scene. These digital 2D images are easily converted to distribution of complex amplitudes by oversampling and adding random or quasi-random phase patterns that give diffusiveness to the objects [4].

Although the 2D objects were used for the background of the 3D scene so far, multi-viewpoint images are more desirable for the background, because the simple 2D pictures do not produce any binocular disparity. Many special techniques have been proposed to create holograms from

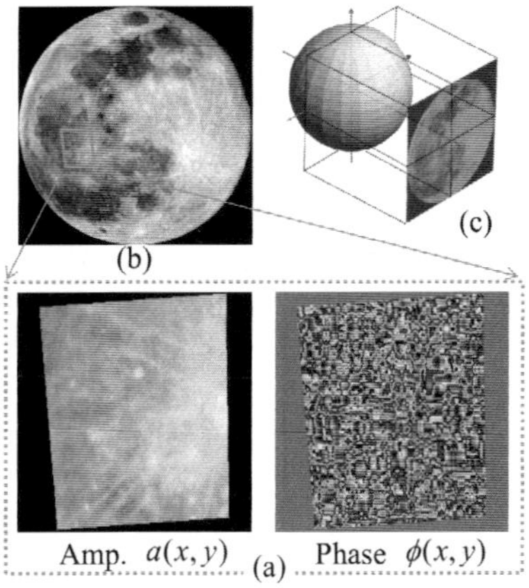

Figure 3. An example of the surface function (a) for texture-mapping of an astrophotograph (b) onto a polygon-mesh sphere (c) [12].

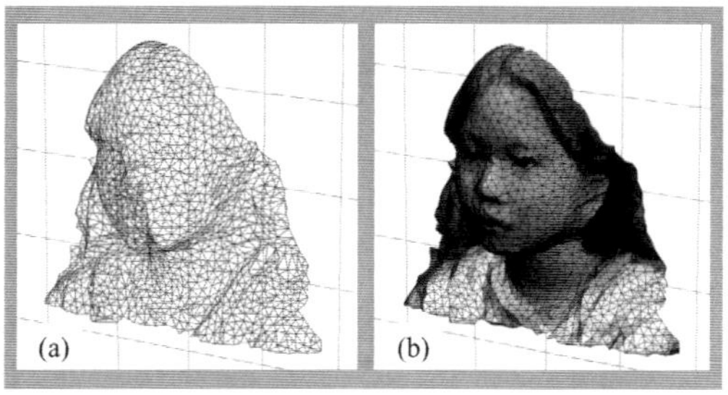

Figure 4. The polygon-mesh of "Shion" before and after texture-mapping [12].

multi-viewpoint images [22]. It is also possible to emulate conventional holographic stereograms by using technique of field propagation.

These input data of computer holography mentioned so far is not the data of the light itself. The polygon-mesh is a just shape data of 3D objects. Picture and multi-viewpoint images are also not object light but only intensity distribution of the light, even though these are of the real-existing objects. However, the object wave, i.e. the field of the real-existing objects, can be digitally captured by using image sensors. This technique is called *digital holography* in contemporary optics [23, 24] and has many applications in industry. In general, 2D images is numerically reconstructed by computers in the digital holography. In *digitized holography*, however, the captured field is incorporated in a virtual 3D scene and optically reconstructed through computer holography. The technique of digitized holography is briefly summarized in section 3.2.

All materials; 2D/3D objects, multi-viewpoint images and captured fields are arranged in a 3D scene in computer holography. The 3D scene is digitally edited in the form of wave fields. Since the occluded relations among the multiple objects should be reproduced in the optical reconstruction, we use the silhouette method and related techniques, as briefly summarized in section 3.3.

3. Basic techniques

3.1. Polygon-based method

The polygon-based method treats the object surfaces as polygonal (usually triangular or rectangular) surface sources of light [1]. A facet of the object is numerically expressed by 2D distribution of complex amplitudes referred to as the surface function. The first high-definition computer hologram "The Venus" was created in 2009 by using this technique. The parameters are shown in the top of Table 1. The 3D scene of the Venus, named after the famous statue of the Venus de Milo, included a 3D object composed of 718 polygons and a 2D background image. The computation time was approximately 45 h at that time [4]. However, the Venus is currently computed less than 3 h [17].

Another feature of the polygon-based method is similarity to conventional computer graphics (CG). This similarity makes it possible to apply CG technique to holography. Therefore,

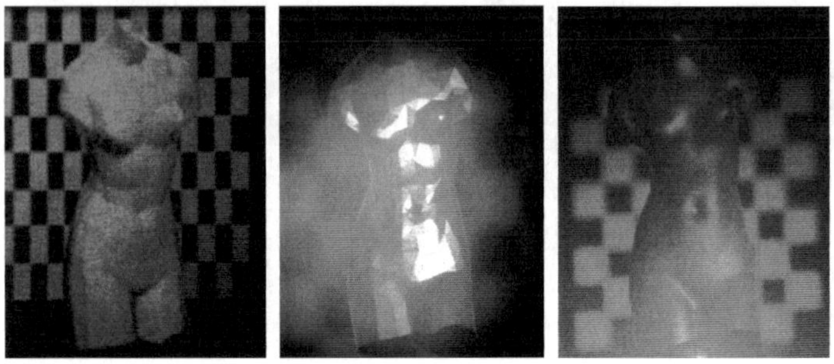

(a) The Venus (b) The Metal Venus I (c) The Metal Venus II

Figure 5. A comparison of three different techniques of wave-field rendering in the polygon-based method; (a) diffused flat, (b) specular flat and (c) specular smooth shading.

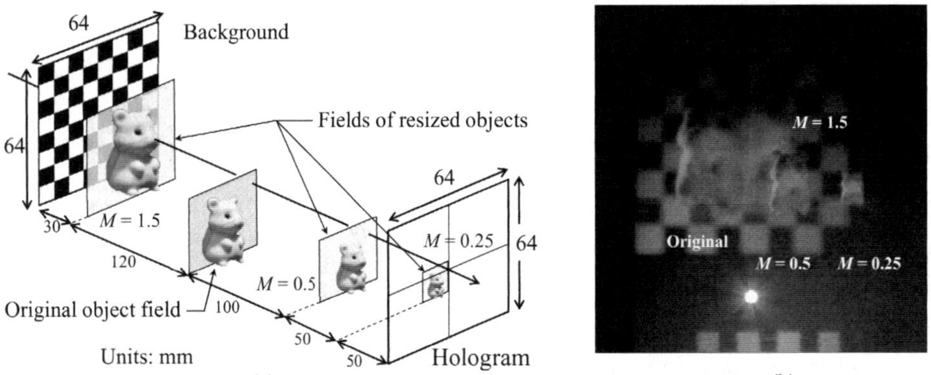

(a) (b)

Figure 6. The 3D scene (a) and optical reconstruction (b) of "Hamsters" that include multiple fields of the resized objects.

the rendering techniques in the polygon-based method are sometimes referred to as *wave-field rendering* [12]. A disadvantage of early wave-field rendering was that the reconstructed surface is not a smooth curved surface but an angular faceted surface. This issue is avoided by modulating the amplitude distribution of the surface function using techniques in CG, such as Gouraud and Phong shading. Texture-mapping in CG can also be introduced into wave-field rendering by modulating the amplitude of the surface function [12]. As shown in Fig. 3, computer holograms "The Moon" and "Shion" are created by using these shading and texture-mapping techniques.

Since all objects do not have diffused surfaces, the rendering technique of specular metallic surfaces is necessary for increasing the power of expression in computer holography. To create specular surfaces, the direction of reflection light must be limited to a narrow range. This is achieved by limiting the spatial bandwidth of the surface function. A technique based on the Phong reflection model has been proposed [17] and "The Metal Venus I", i.e. a metallic version of "The Venus", is successfully created as shown in Fig. 5 (b).

The Metal Venus I, however, reconstructs angular faceted surfaces. Tessellation of polygons is necessary to create specular smooth surface, but the tessellation most likely leads to increasing the computation time. Thus, a technique of subdividing the surface function has been proposed instead of tessellating polygons [19]. "The Metal Venus II" created by using this new technique reconstructs metal-like curved surfaces without any seams between polygons, as shown in Fig. 5 (c).

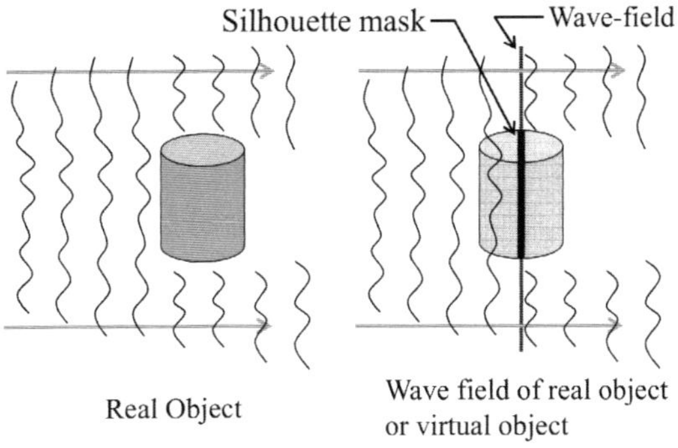

Figure 7. The principle of the silhouette method for shielding light behind objects.

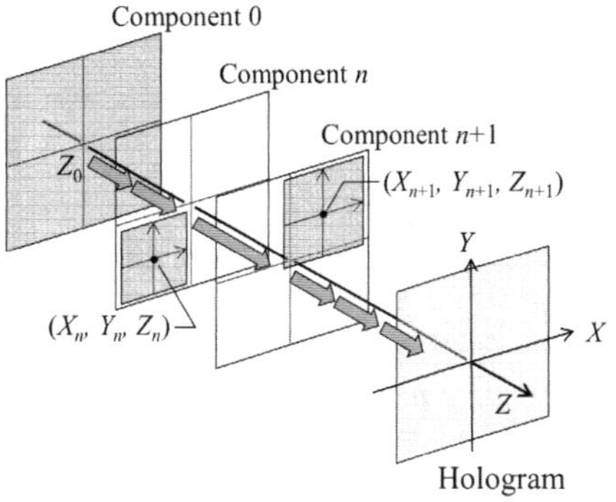

Figure 8. Field-based digital editing of 3D scenes in computer holography [6].

3.2. Digitized holography [6]

Real object fields are recorded by an image sensor using lensless-Fourier setup that uses a spherical reference field instead of a collimated field. In this technique, the object fields is given by Fourier transform of the captured wave-field. Here, phase shifting technique is used for removing the non-diffractive light and conjugate image [25]. The sampling intervals of the Fourier-transformed field is determined by the number of samplings and distance between the center of the spherical reference field and the image sensor [26]. We can adjust the sampling intervals by changing these parameters. In addition, the synthetic aperture technique [26, 27] is used for capturing large-scale fields and increasing the number of samplings.

An advantage of digitized holography is that the 3D scene can be digitally edited after recording. This is almost impossible in classical holography. In the digital editing, the captured objects are: (i) freely arranged and (ii) mixed with virtual objects in the 3D scene. In addition, (iii) a single object captured once can be duplicated in the 3D scene.

However, the captured objects cannot be simply resized unlike conventional digital images, because the recorded object field contains the phase information of the object light. Virtual optics is required to resize the objects in the 3D scene. Figure 6 shows optical reconstruction of "Hamsters" whose 3D scene includes multiple objects resized by virtual image formation using a numerical lens.

3.3. The silhouette method for light shielding in 3D scenes

In the field-based digital editing of 3D scenes, as in Fig. 2, the silhouette method plays an important roll to prevent objects from being see-through [2, 3] and awoke a strong depth perception in viewers. The principle of the silhouette method is shown in Fig. 7. Real objects shield the light behind the object. The silhouette method emulates this phenomenon by blocking the wave-field using a binary mask whose shape agrees with the silhouette of the object. This silhouette-masking is made sequentially from the deepest object to the hologram plane, as shown in Fig. 8. Note that a numerical operation for propagation is necessary for each masking.

As for CG-modeled virtual 3D objects and 2D objects, the silhouette mask is easily generated by projecting the 3D/2D shape onto the field plane. In digitized holography, the silhouette mask is produced from the numerically reconstructed 2D image of the captured object [6]

However, for both virtual and real objects, the silhouette method does not work well in some cases where the object has severe self-occlusion or the silhouette shape of the object does not fit with the cross section. In the cases, the silhouette masking must be applied polygon by polygon, i.e. the silhouette masking must be made by individual polygons. Since this is

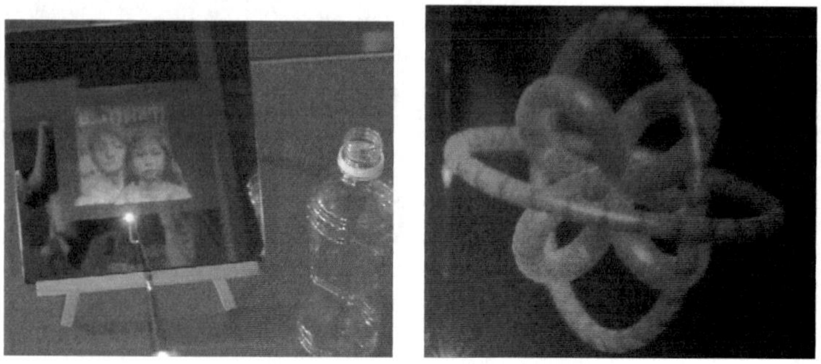

(a) Brothers (b) Rose in Ring

Figure 9. Optical reconstruction of "Brothers" (a) and "Rose in Ring" (b). The parameters are summarized in Table 1

commonly very time-consuming, a technique using Babinet's low has been proposed for speed-up the computation [14].

4. Some of recent works

The fringe patterns of the holograms were printed on photoresist coated on ordinary photomask blanks by employing DWL-66 laser lithography system made by Heidelberg Instruments GmbH, which is the equipment constructed for fabricating photomasks [5]. Since the chromium thin film coated on glass substrate forms binary amplitude fringe, the fabricated holograms are chemically very stable without suffering aging-degradation.

4.1. Brothers

This is the biggest computer hologram created with wave-field rendering up to now. The total number of pixels reaches to 25 billion pixels. The optical reconstruction is shown in Fig. 9 (a). The shapes of two live faces of a boy and girl are measured by using a 3D laser scanner. The photographs taken from the same view-point as the measurement are texture-mapped onto the polygon-meshes that are shaded with Gouraud shading.

4.2. Rose in Ring

This computer hologram is named "Rose in Ring", because the main object is generated by a mathematical 3D curve based on well-known 2D parametric curves called Rose curve. The optical reconstruction is shown in Fig. 9 (b). This model has severe self-occlusion, thus a new algorithm for accelerating the silhouette masking, named "switch-back method" is used for calculation.

4.3. Penguin

This is currently the biggest hologram created in digitized holography. The optical reconstruction is shown in Fig 10 (b). Two objects recorded using digital holography is arranged two times for each in the 3D scene, as shown in (a).

5. Conclusion

Techniques for creating high-definition CGHs and some works created so far are reviewed and summarized. The created CGHs usually make a great impact on viewers, because these synthetic holograms reconstruct continuous motion parallax in both vertical and horizontal directions and convey a strong sensation of depth like that in conventional holography. Therefore, the term "computer holography" is suitable for the techniques for creating works by high-definition CGHs. Computer holography is new digital art in 3D imaging like computer graphics in 2D imaging.

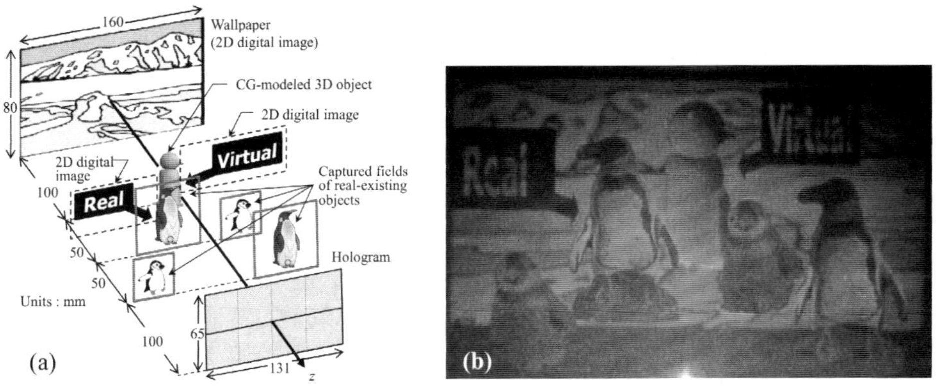

Figure 10. The 3D scene (a) and optical reconstruction (b) of "Penguin".

Acknowledgment
Authors would like to thank D. Fujita for his assistance to create Hamsters. This work was supported by JSPS. KAKENHI (21500114, 24500133) and the Kansai University Research Grants: Grant-in-Aid for Encouragement of Scientists, 2011–2012.

References
[1] Matsushima K 2005 *Appl. Opt.* **44** 4607–4614
[2] Matsushima K and Kondoh A 2004 *Proc. SPIE* vol 5290 90–97
[3] Kondoh A and Matsushima K 2007 *Systems and Computers in Japan* **38** 53–61
[4] Matsushima K and Nakahara S 2009 *Appl. Opt.* **48** H54–H63
[5] Nakahara S and Matsushima K 2012 *International Symposium on Display Holography (ISDH2012)* (IOP)
[6] Matsushima K, Arima Y and Nakahara S 2011 *Appl. Opt.* **50** H278–H284
[7] Matsushima K and Nakahara S 2009 *Proc. SPIE* vol 7233 p 72330E
[8] Matsushima K and Nakahara S 2009 *Digital Holography and Three-Dimensional Imaging* (Optical Society of America) PDWB38
[9] Matsushima K and Nakahara S 2010 *Proc. SPIE* vol 7619 p 761913
[10] Muffoletto R P, Tyler J M and Tohline J E 2007 *Opt. Express* **15** 5631–5640
[11] Matsushima K, Nakamura M and Nakahara S 2010 *Digital Holography and Three-Dimensional Imaging* (Optical Society of America) JMA10
[12] Matsushima K, Nishi H and Nakahara S 2012 *J. Electron. Imaging* **21** 023002
[13] Matsushima K 2010 *Opt. Express* **18** 18453–18463
[14] Nakamura M, Matsushima K and Nakahara S 2012 *J. Inst. Image Inform. TV. Engnr.* **66** J136–J143 (in Japanese)
[15] Nishi H, Higashi K, Arima Y, Matsushima K and Nakahara S 2011 *Proc. SPIE* vol 7957 p 79571A
[16] Arima Y, Matsushima K and Nakahara S 2011 *Digital Holography and Three-Dimensional Imaging* (Optical Society of America) DWC28
[17] Nishi H, Matsushima K and Nakahara S 2011 *Appl. Opt.* **50** H245–H252
[18] Nishi H, Matsushima K and Nakahara S 2011 *Digital Holography and Three-Dimensional Imaging* (Optical Society of America) JDWC29
[19] Nishi H, Matsushima K and Nakahara S 2011 *3DTV-Conference 2011*
[20] Matsushima K, Nishi H, Arima Y and Nakahara S 2011 *International Workshop on Holography and related technologies (IWH2011)* p 1
[21] Yamashita H, Matsushima K and Nakahara S 2012 *Proc. SPIE* vol 8281 p 82810Z
[22] Shaked N T, Katz B and Rosen J 2009 *Appl. Opt.* **48** H120–H136
[23] Goodman J W and Lawrence R W 1967 *Appl. Phys. Lett.* **11** 77–79
[24] Schnars U and Jüptner W 1994 *Appl. Opt.* **33** 179–181
[25] Yamaguchi I and Zhang T 1997 *Opt. Lett.* **22** 1268–1270
[26] Nakatsuji T and Matsushima K 2008 *Appl. Opt.* **47** D136–D143
[27] Binet R, Colineau J and Lehureau J C 2002 *Appl. Opt.* **41** 4775–4782

Direct Optical Fringe Writing of Diffraction Specific Coherent Panoramagrams in Photorefractive Polymer for Updatable Three-Dimensional Holographic Display

Sundeep Jolly and V. Michael Bove, Jr.

MIT Media Lab, Massachusetts Institute of Technology, Cambridge, Massachusetts, United States

E-mail: sjolly@media.mit.edu

Abstract. Progress in the development of an updatable holographic display system based around the direct transfer of computer-generated holographic fringe patterns from LCoS SLMs into photorefractive polymeric materials is presented. This architecture is poised as a simplifying alternative to previous demonstrations of updatable holographic displays in photorefractive polymeric materials based around conventional interference-based holographic stereogram techniques. Our system concept – comprised of fringe pattern generation on computer, fringe pattern transfer from SLM to photorefractive polymer, and spatial multiplexing for large-image generation – reintroduces accommodation cues to the resulting holographic images and represents a reduction of system footprint, complexity, and cost relative to the current interference-based systems. We present the adaptation of our Diffraction Specific Coherent Panoramagram fringe computation method – originally developed to drive AOM-based holographic displays at video rates while preserving all depth cues, including accommodation – to the current display architecture and depict methods for direct fringe transfer from SLM to photorefractive polymer. Preliminary results of horizontal parallax-only images on this display are presented and directions for performance improvements and system extensions are explored.

1. Introduction

1.1. Background

Dynamic three-dimensional holographic displays have a potential wide range of imaging applications in military, medical, and consumer entertainment contexts. Early implementations of dynamic holographic displays have employed various techniques for spatial light modulation driven with computer-generated holograms but are limited in size and resolution.

Recently, researchers at the University of Arizona's College of Optical Sciences and Nitto Denko Technical Corporation reported on the development of a photorefractive polymer for dynamic holographic recording having properties including high diffraction efficiency, long image persistence, fast writing, rapid erasure, and large area – a combination that makes the polymer especially well-suited for application in updatable 3-D holographic displays [1]. The subsequent development of a 3-D holographic display based on this material has allowed images to be written to the polymer and refreshed as desired, up to a rate of 1/2 Hz in one of the system variants [2, 3].

However, the current holographic recording process is based on the conventional holographic stereogram recording technique – optical interference of a signal beam modulated with intermediate ("hogel") representations of 3-D scenes and a reference beam – and thereby requires an optical architecture capable of producing such interference and limits flexibility in the type of hologram that can be recorded.

In contrast to optical interference-based methods for fabricating holographic gratings in photosensitive media, holographic gratings that have been generated algorithmically on computer can be directly written into a wide variety of materials. Electron-beam and optical lithographic techniques are often used for fabrication of holographic optical elements and computer-generated display holograms in massive quantity; however, their use is prohibitive in several applications due to the high cost and specialized equipment involved [4]. Direct optical writing of holographic fringe patterns into photosensitive media has been previously reported in the literature. Direct writing of computer-generated holograms onto recordable compact-disc media in a pixel-by-pixel approach has been demonstrated by Sakamoto *et al.* [5]. Furthermore, direct optical imaging of computer-generated fringe patterns displayed on spatial light modulators onto holographic film has been demonstrated in a spatial multiplexing approach by Yoshikawa *et al.* [4, 6].

Computer-based methods for holographic fringe pattern generation have matured considerably and are now widely used for creating display holograms [7]. Methods for computer-based display hologram generation include physically-based interference modeling algorithms and diffraction-specific approaches, several of which are now mature enough to support small-scale, video-rate hologram generation [8].

1.2. Motivation

Because of the combination of developments in novel photorefractive media for dynamic holography, demonstrated achievements in direct fringe printing for static holography, and fast methods of fringe calculation in computer-generated display holography, investigation into direct optical writing of holographic fringes into the novel photorefractive media is merited.

Direct writing of holographic fringes can potentially provide several benefits to a dynamic photorefractive holographic imager. In contrast to the current interference-based systems, a direct-write system can potentially simplify computational and optical architectures, reduce total system footprint, and lower cost with minimal adverse impact on overall display performance. Additionally, a direct-write system affords greater control over the type of holographic fringe pattern recorded in the material relative to the current holographic stereogram recording method thereby allowing for control over reconstructed wavefront curvature and correction for inconsistent depth cues by removing the stereogram limitation.

To this end, the current study is an ongoing assessment of the feasibility and performance of direct fringe writing in the Nitto Denko photorefractive polymer for the dynamic display application.

2. Methods

Our approach to investigating viability of direct writing is depicted in Fig. 1 and involves the computation of fringe patterns, transfer of fringe patterns from spatial light modulator (SLM) to photorefractive polymer via de-magnifying optics, and spatial multiplexing for large-image generation.

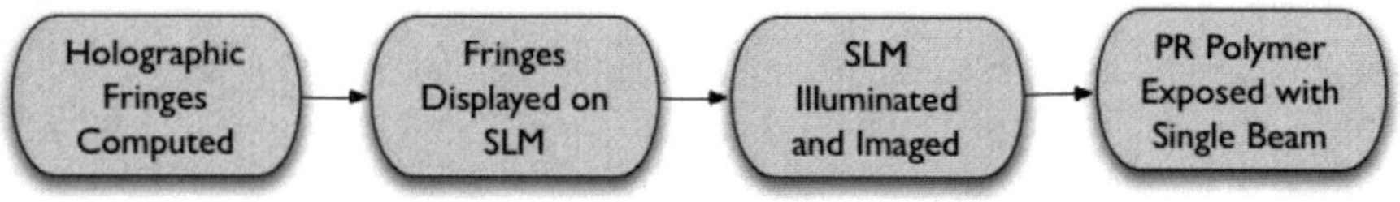

Figure 1. High-level system overview for direct optical fringe writing in photorefractive polymer.

2.1. Fringe Computation

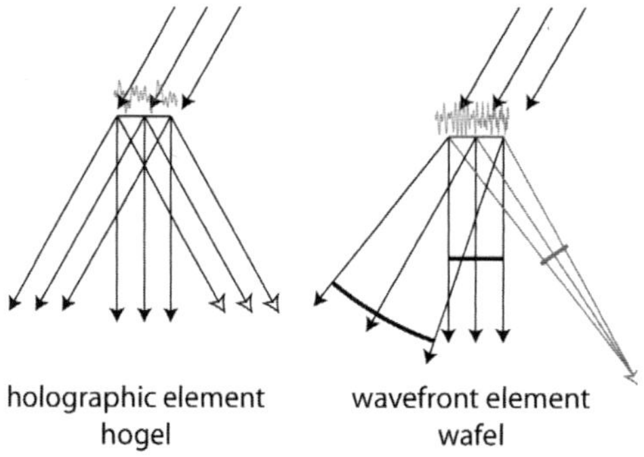

Figure 2. Comparative affordances in directional intensity and curvature for the conventional holographic element ("hogel") and the wavefront element ("wafel").

The authors' research group has previously introduced the Diffraction Specific Coherent Panoramagram (DSCP) – a diffraction-specific approach toward multi-view stereogram generation that provides smooth motion parallax and correct accommodation cues with far fewer views than required with for a normal holographic stereogram. [9, 10]. Fig. 2 depicts the affordances in wavefront curvature allowed in a conventional holographic stereogram (via hogels) and those allowed in a DSCP (via wavefront elements, "wafels"). The wafel's ability to directly generate arbitrary, positionally-variant wavefront curvature (rather than approximate such curvature through the ensemble of many constant-frequency wavelets) allows for better approximation of ideal wavefronts (and can therefore better provide accommodation cues).

In the DSCP algorithm, depth information from a 3-D scene is used to appropriately compute holographic basis functions (variable rate chirped gratings, in the current HPO implementation). The chirped gratings are then modulated with the scene's intensity information and appropriately tiled to create the complete holographic fringe pattern.

While the DSCP algorithm was originally developed for real-time generation of horizontal-parallax only (HPO) holographic fringes for output to the MIT Mark II and Mark III electro-holographic video displays at high frame rates, its current implementation for this feasibility assessment involves offline (i.e., non real-time) computation of holo-lines from synthetic 3-D models.

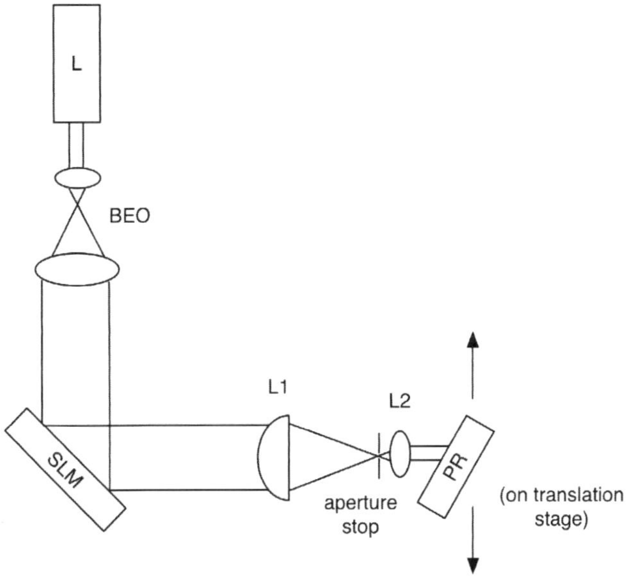

Figure 3. Schematic of optical system. L = continuous-wave laser at λ = 532 nm, BEO = beam-expanding optics, SLM = LCoS spatial light modulator, L1 = first cylindrical lens of telecentric optical system with f = 250 mm, L2 = second cylindrical lens of telecentric optical system with f = 50 mm, PR = photorefractive polymer (on translation stage for spatial multiplexing).

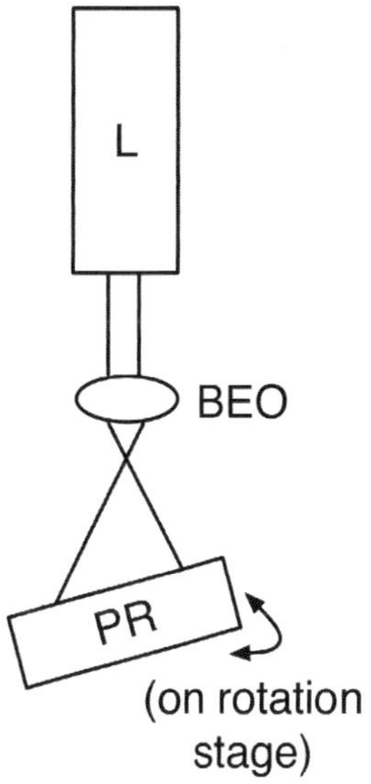

Figure 4. Schematic of illumination system. L = continuous-wave laser at λ = 632.8 nm, BEO = beam-expanding optics, PR = photorefractive polymer.

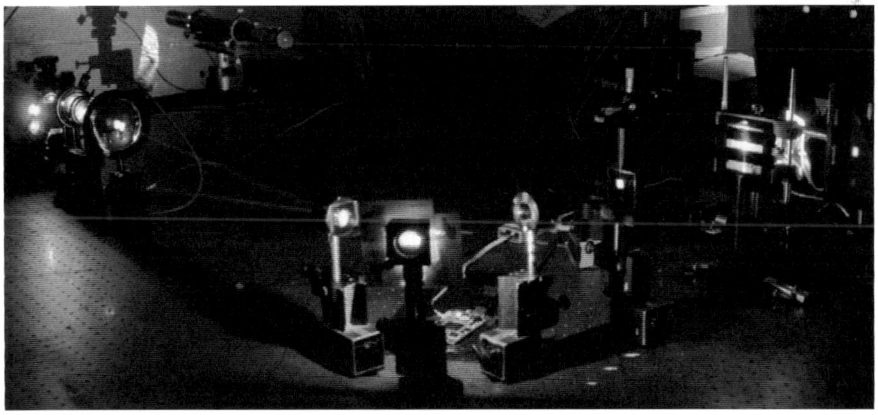

Figure 5. Photograph of the direct fringe writing setup.

2.2. High-Level Optical Setup

Direct fringe writing offers a simplified optical architecture in contrast to those necessitated by interference-based holographic recording schemes. The optical architectures used in the current writing scheme are depicted in Fig. 3 - Fig. 5. The writing system is comprised of a solid-state, diode-pumped continuous-wave laser at λ = 532 nm, beam expanding optics, liquid-crystal-on-silicon (LCoS) spatial light modulator in crossed-polarizer configuration, telecentric optical system for fringe pattern transfer and demagnification, translation stage for spatial multiplexing, and photorefractive polymer. The readout system employs an expanded beam from a continuous-wave laser at $\lambda = 632.8$ nm for illumination.

2.2.1. Spatial Light Modulation

Figure 6. Photograph of the Silicon Micro Display liquid-crystal-on-silicon module.

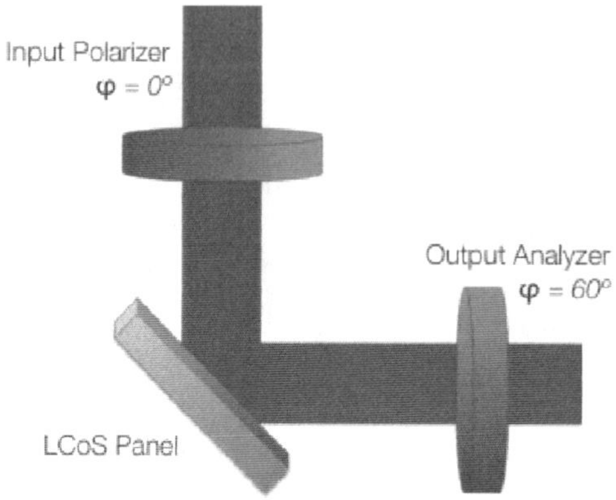

Figure 7. Cross-polarizer configuration to retrieve amplitude modulation.

For the current study, we employ liquid crystal-on-silicon (LCoS) SLM panels furnished by Silicon Micro Display. These LCoS panels have 88% fill factor and 8.5 μm pixel pitch at a resolution of 1920x1080 (see Fig. 6). For amplitude modulation, we employ the LCoS device in a cross-polarizer configuration (depicted in Fig. 7) in which a polarizer is placed before the SLM to generate the proper polarization state (with alignment parallel to the native liquid crystal axis) and an analyzer is placed after the SLM to retrieve the modulated state.

2.2.2. Optical Imaging

Because the 8.5 μm pixel pitch of the SLM is not sufficiently small for direct viewing of display holograms, an optical system for imaging fringe patterns into photorefractive material is required and must be capable of the requisite demagnification for large viewing angle. Fringe transfer from the SLM is accomplished via a telecentric optical system comprised of cylindrical lens L1 with focal length $f = 250$ mm and cylindrical lens L2 with focal length $f = 50$ mm. Because the lenses are cylindrical, demagnification is only in the horizontal axis and is given by the ratio of focal lengths (5x demagnification for the current configuration). Given the pixel size on the SLM and the nominal demagnification, the nominal pixel size resulting from the telecentric imaging is 1.7 μm.

2.2.3. Photorefractive Polymer

In photorefractive materials, incident light intensities are converted into refractive index modulation via a process of spatial charge redistribution and trapping. Because of the mechanism required for the phase hologram generation, high operating voltages of several kV are required to bias the material [1, 11, 12]. The current study employs samples with dimensions 2 in. x 2 in.

2.3. Automation and Control
NI LabVIEW 2011 is used for automation and control of SSDP laser, spatial light modulator, translation stage, and DC power supply.

3. Current Results
At the time of this writing, we have built-up a testbed for direct fringe writing experiments with the photorefractive material and have gathered preliminary results.

3.1. Validation Results

Table 1. Writing parameters for HPO teacup stereogram.

PR Polymer Bias Voltage	5.0 kV
Overall Fringe Pattern Resolution	48000 x 1080
Exposure Time per Elemental	10 s
Maximum Irradiance per Elemental	1.25 W/cm^2
Total Writing Time	300 s
Estimated Persistence per Elemental	300 s

Figure 8. Teacup model used for RIP stereogram fringe generation.

Figure 9. Resulting holographic image of teacup model.

Initial validation tests of the system were carried out using pre-computed fringes generated using an earlier diffraction specific stereogram algorithm – the Reconfigurable Image Projection (RIP) approach [13]. Fringes were computed using a model of a teacup (depicted in Fig. 8). Writing parameters and persistence observations are depicted in Table 1. The resulting holographic image in the PR material due to the direct fringe writing of these computed fringes is shown in Fig. 9.

3.2. HPO DSCP Results
Tests of the system with the DSCP algorithm were carried out using computed fringes from the Stanford Bunny model (depicted in Fig. 10). Writing parameters and persistence observations are depicted in Table 2. The resulting holographic image in the PR material due to the direct fringe writing of these computed fringes is shown in Fig. 11.

Journal of Physics: Conference Series **415** (2013) 012054 doi:10.1088/1742-6596/415/1/012054

Table 2. Writing parameters for HPO Stanford Bunny Diffraction Specific Coherent Panoramagram.

PR Polymer Bias Voltage	5.0 kV
Overall Fringe Pattern Resolution	51840 x 1080
Exposure Time per Elemental	10 s
Maximum Irradiance per Elemental	1.25 W/cm^2
Total Writing Time	350 s
Estimated Persistence per Elemental	300 s

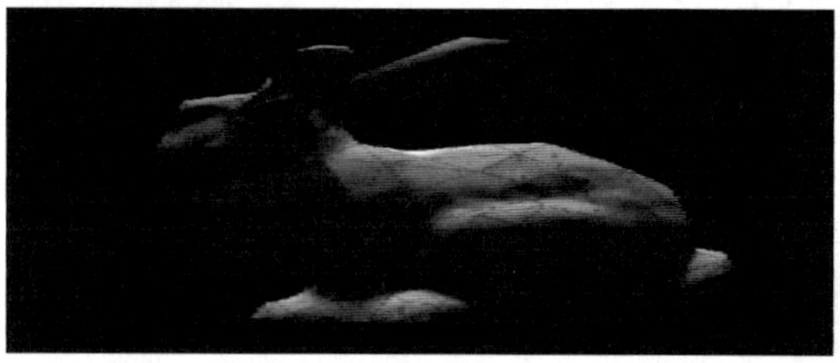

Figure 10. Stanford Bunny model used for DSCP fringe generation.

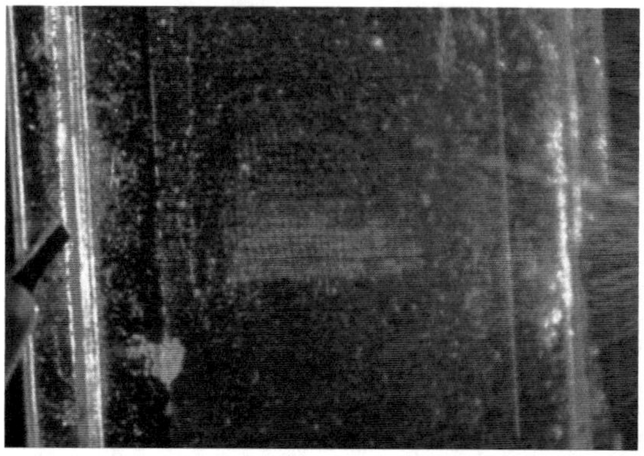

Figure 11. Resulting holographic image of Stanford Bunny model.

3.3. Observations
While the current results do indicate that holographic images can be formed via single-beam direct fringe writing as well as via optical interference, there remains much room for improvement of the display quality. At the present time, diffraction efficiency of the resulting images is poor and the images suffer from low contrast, low discriminability, and low persistence – all of which pose problems for direct viewing.

4. Future Work

4.1. Improvement of Image Quality
In order to make direct writing more practicable as a replacement for interference-based recording schemes, future work needs to address several performance issues.

4.1.1. Diffraction Efficiency
Diffraction efficiency is linked to a number of factors, including fringe visibility (i.e., modulation efficiency in the resulting refractive index distribution from recording), exposing parameters (i.e., incident irradiance and exposure time), and applied electric field strength. Improvements are likely with characterization and refinement of the imaging system MTF (up to the maximal spatial frequency indicated by its demagnification), further optimization of exposing parameters, and further characterization and optimization of the effect of the applied bias voltage.

4.1.2. Image Persistence
Image persistence is largely a function of applied electric field strength but has also been observed

to vary based on the particular sample of photorefractive material being used. Improvements are expected with the use of newer samples with longer grating holding times and optimization of the applied bias voltage.

4.1.3. Image Discriminability
Image discriminability is largely related to both diffraction efficiency and image persistence and improvements in these two areas will likely yield improvements in discriminability as well.

4.2. Extensions

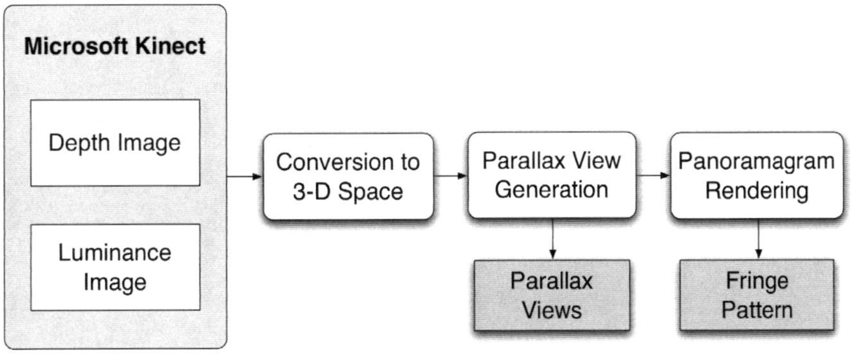

Figure 12. Pipeline for real-time generation of DSCP fringe patterns using the Microsoft Kinect peripheral.

It should be noted that while the current implementation of the DSCP fringe computation algorithm only produces "pre-rendered" fringes for recording at a later time, the algorithm can easily be integrated into the control system for real-time computation (with the caveat that such real-time computation will only be beneficial in the case of a fast writing scheme to take advantage of the fast computation). Furthermore, the authors' research group has previously demonstrated generation of DSCP fringe patterns of real scenes in real-time using consumer rangefinding cameras (Fig. 12 depicts this process). This capability can be exploited in a future holographic telepresence scheme using a direct fringe writing approach.

Direct fringe imaging for full-parallax images is a plausible extension but will require the adaptation of the DSCP algorithm, telecentric optical imaging using standard spherical optics, and spatial multiplexing in two dimensions. Alternatively, massively parallel direct writing schemes can be employed in which several areas of the overall multiplexed hologram are written simultaneously.

Acknowledgments
This work has been supported in part by consortium funding at the MIT Media Laboratory. This research was also funded in part by the Office of the Director of National Intelligence (ODNI), Intelligence Advanced Research Projects Activity (IARPA), through the AFRL contract FA8650-10-C-7034. All statements of fact, opinion or conclusions contained herein are those of the authors and should not be construed as representing the official views or policies of IARPA, the ODNI, or the U.S. Government. The authors gratefully acknowledge Silicon Micro Display for furnishing the spatial light modulators used in this work.

References

[1] S. Tay, P. Blanche, R. Voorakaranam, A. V. Tunc, W. Lin, S. Rokutanda, T. Gu, D. Flores, P. Wang, G. Li, P. Saint-Hilaire, J. Thomas, R. A. Norwood, M. Yamamoto, and N. Peyghambarian, "An updatable holographic three-dimensional display, *Nature*, vol 451, pp. 694-698, 2008.

[2] P. A. Blanche, S. Tay, R. Voorakaranam, P. Saint-Hilaire, C. Christenson, T. Gu, W. Lin, D. Flores, P. Wang, M. Yamamoto, J. Thomas, R. Norwood, and N. Peyghambarian, "An Updatable Holographic Display for 3D Visualization, *Journal of Display Technology*, vol. 4, no. 4, pp. 424430, 2008.

[3] P.-A. Blanche, A. Bablumian, R. Voorakaranam, C. Christenson, W. Lin, T. Gu, D. Flores, P. Wang, W.-Y. Hsieh, M. Kathaperumal, B. Rachwal, O. Siddiqui, J. Thomas, R. A. Norwood, M. Yamamoto, and N. Peyghambarian, "Holographic three-dimensional telepresence using large-area photorefractive polymer," *Nature*, vol. 468 (7320) pp. 80-83, 2010.

[4] H. Yoshikawa and K. Takei, "Development of a compact direct fringe printer for computer-generated holograms, *Proc. SPIE Practical Holography*, vol. 5290, no. 1, p. 114, 2004.

[5] Y. Sakamoto, M. Morishima, and A. Usui, "Computer-generated holograms on a CD-R disk, *Proceedings of SPIE Practical Holography*, vol. 5290, no. 1, pp. 4249, 2004.

[6] H. Yoshikawa and T. Yamaguchi, "Computer-generated holograms for 3D display," *Chinese Optics Letters*, vol. 7, no. 12., pp. 10791082, 2009.

[7] P. Hariharan, *Optical Holography: Principles, Techniques, and Applications*, 2nd ed. New York, NY: Cambridge University Press, 1996.

[8] S. A. Benton and V. M. Bove, Jr., *Holographic Imaging*. Hoboken, NJ: Wiley, 2008.

[9] Q. Y. J. Smithwick, J. Barabas, D. Smalley, and V. M. Bove, Jr., "Interactive Holographic Stereograms with Accommodation Cues," *Proc. SPIE Practical Holography XXIV: Materials and Applications*, 7619, 761903, 2010.

[10] J. Barabas, S. Jolly, D. E. Smalley, and V. M. Bove, Jr., "Diffraction Specific Coherent Panoramagrams of Real Scenes," *Proc. SPIE Practical Holography XXV*, v. 7957, 2011.

[11] K. Buse and E. Kratzig, "Inorganic photorefractive materials, in *Holographic Data Storage*, Eds: H. Coufal, D. Psaltis, and G. Sincerbox. Springer, New York, 2000.

[12] B. Kippelen, "Overview of photorefractive polymers for holographic data storage, in *Holographic Data Storage*, Eds: H. Coufal, D. Psaltis, and G. Sincerbox. Springer, New York, 2000.

[13] W. Plesniak, M. Halle, V. M. Bove, Jr., J. Barabas, and R. Pappu, "Reconfigurable image projection holograms," *Optical Engineering*, vol. 45, no. 11, 115801, 2006.

Holovideo for everyone: a low-cost holovideo monitor

D Smalley[1], Q Smithwick[2], J Barabas[1], V M Bove Jr.[1], S Jolly[1], C DellaSilva[1]

[1]Object-based Media Group, MIT Media Laboratory E15, 77 Mass. Ave., Cambridge, MA 02139 USA

[2]Disney Research, 1401 Flower St., Glendale, CA, 91201 USA

E-mail: telsa@media.mit.edu

Abstract. This work presents an architecture for a relatively low-cost, pc-driven holovideo monitor. The geometry uses minimal optics and is built to host a multi-channel acousto-optic modulator that can be driven by up-converted VGA signals. The display's target specifications include a standard vertical resolution (480 lines) output driven by an 18 channel acousto-optic modulator, 30Hz refresh-rate and multiple color operation. This paper reports early tests of this geometry with a single acousto-optic channel. The goal is to create a small but functional holographic display that can be readily replicated, easily driven and provide basic monitor functionality with a bill of materials in the hundreds, rather than thousands, of dollars.

1. Introduction

The holography community has a great history of resourceful individuals making sophisticated equipment with inexpensive components [1][2]. In this tradition we present this simple, relatively inexpensive holovideo display that we hope will encourage broader participation in the application and development of holographic video.

Our goal is to make the most inexpensive display possible while trying to achieve a monitor-like functionality with the following parameters

- Bill of materials of less than $500 (not including laser light source(s))
- Image extent of approximately 10cm x 5cm or larger
- Vertical resolution greater than or equal to 480 lines, HPO (Horizontal Parallax Only)
- Viewing angle greater than 15 degrees
- Frame rate greater than or equal to 30 fps
- Ability to be driven from a commodity PC with a set of three gamer graphics cards

2. Background

In order to achieve our goals we decided to build a monitor based on the MIT Mark I holovideo geometry[3]. This geometry is the most simple of the MIT architectures and allows for some flexibility in the design to allow for the substitution of low-cost parts.

2.1. Mark I Optical Path

The holographic image is created in the Mark I by descanning the aperture of the acousto-optic modulator, demagnifying the resulting long holographic line, and then vertically scanning the demagnified line to build up an image with multiple lines stacked vertically (figure 1).

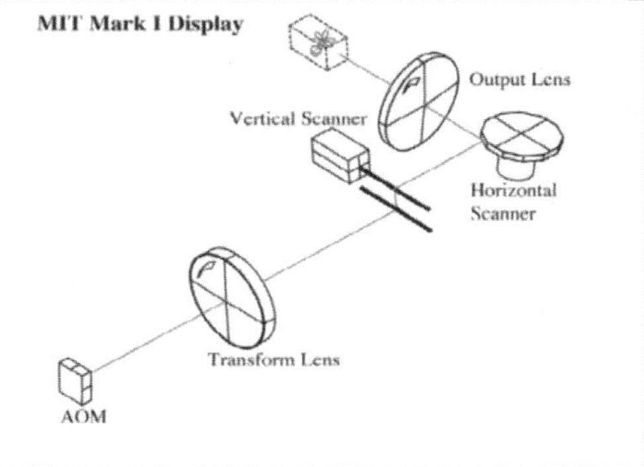

Figure 1. The Mark I geometry contains a bulk wave AOM, a two lenses in telescope configuration, a vertical scanner and a large polygon scanner.

2.2. Key Mathematical Relations

2.2.1. The Bandwidth Budget

The analog nature of the Mark I system allows fluid redistribution of display bandwidth between image extent, view angle and frame rate.
The relationship between bandwidth and system parameters is given below:

$$bandwidth \propto \frac{extent * viewangle * resolution * framerate}{\lambda} \tag{1}$$

With increased temporal bandwidth (pixels/sec) we can either increase the display extent (i.e. the display width), the view angle, the vertical resolution (lines per frame) and the frames per second or decrease the illumination wavelength, λ.

2.2.2. The Fourier Plane

The Fourier plane lies at the surface of the polygon and dictates the minimum size of the polygon facet or the maximum focal length of the transform lens[3] (equation 2).

$$\Delta x \approx \lambda f_1 \frac{\Delta \nu}{V} \tag{2}$$

In this equation, Δx, is the size of the Fourier plane, λ, is the wavelength of illumination light, f_1, is the focal length of the transform lens, Δv, is the RF bandwidth of the AOM signal and, V, is the velocity of the AOM pattern. The approximation is accurate for small angles and will suffice for the AOM which only sweeps through a range of a few degrees.

The size of the Fourier plane is also related to the size of the polygon sweep angle, Ω, and the product of the image width, d, and the image view-angle, θ[3] (equation 3).

$$\Delta x = \frac{d \sin \theta}{2\Omega} \tag{3}$$

Journal of Physics: Conference Series **415** (2013) 012055 doi:10.1088/1742-6596/415/1/012055

3. The Holovideo Monitor

Our holovideo monitor diverges from the Mark I in that it uses a smaller polygon scanner[4][5], the galvo follows the polygon instead of preceeding it, and there is a parabolic reflector instead of a lens at the output (see figures 2 and 3).

.

3.1.1. Smaller Polygon--The primary cost savings in our design is the choice of a commodity polygon scanner. The critical dimension of the facet is similar to that of the polygon scanner in the Mark I display so besides being smaller overall, the polygon also sweeps out a larger angle which, when coupled with a low f-number reflective optic, should allow for an increase in either display size and/or view angle (see equation 3).

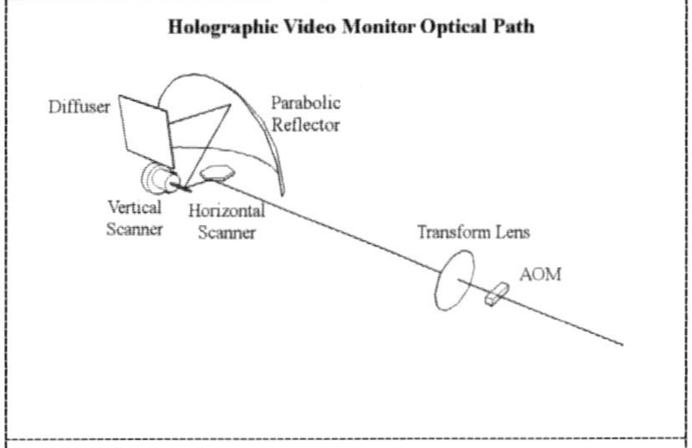

Figure 2. In contrast to the Mark I, the holographic video monitor contains a small polygon with only six facets, a vertical mirror after the polygon and a parabolic reflector.

3.1.2. Reflective Optic--The output optic is a parabolic reflector. This optic determines the size and output image of the display. A more powerful reflector will result in a smaller image with a wider view angle; a less powerful reflector will produce a larger image with the narrower view-zone.

3.1.3. Custom AOM--A bulk wave AOM is used in this paper to verify the monitor's design but bulk AOMs are expensive—several times the target cost of the display—and won't be used in the final 18 channel instantiation of the display (each channel provides 26 lines in the holographic image for a total of 480 lines using an 18 channel device). We are currently developing modulators that will provide 18 or more channels of modulated output without appreciably increasing the cost of the display. These devices will be described fully in a subsequent publication.

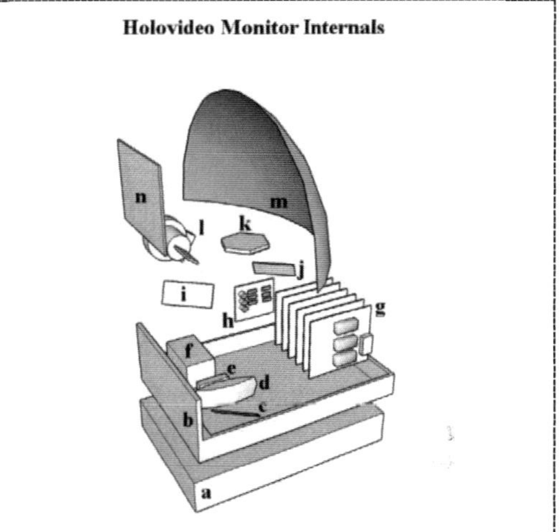

Figure 3. The monitor is composed of a power supply (a), folded aluminum chassis (b), mirrors (c, i and j), transform lens (d), AOM (e), laser source (f), bank of cards with DVI-A inputs and RF outputs (g), a phased-lock loop control to drive the polygon (k), vertical scanner (l), a parabolic reflector (m) and an anisotropic diffuser (n).

3.1.4. Chassis--Folded, waterjet-cut aluminum is used to create a structure that is both rigid and light. Extruded aluminum provides mounting locations for internal optics.

3.1.5. Electrical system—The holographic video monitor is driven by a commodity PC with three gen-locked, dual-head graphics cards. This bank of graphics cards provides 18 video signals (three color signals times six VGA outputs) which, in the final display, will each drive one of 18 AOM channels.

4. Results

4.1. Basic system verification and assembly --The optical and electrical breadboard setup confirmed that the system was descanning the video signal (figure 5). The optics and electronics were then placed into the monitor form factor (figure 6).

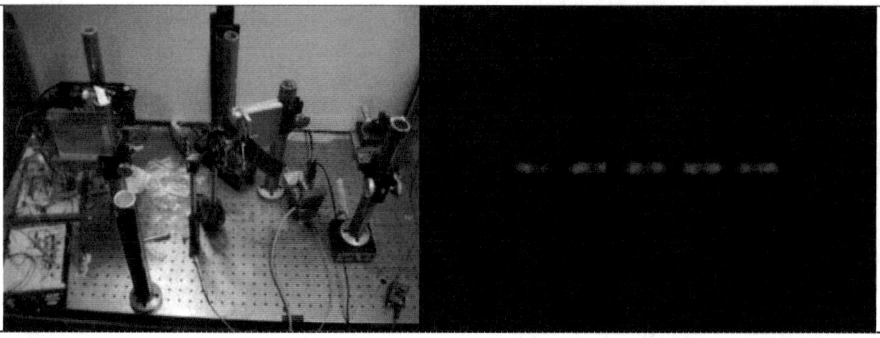

Figure 5. Breadboard prototyped display (left) and scanned output (right) to verify video descanning. The gaps in the pattern on the right are the video h-blank intervals.

Figure 6. Assembled monitor (diffuser removed).

4.2. Monitor Output

At the time of this writing we have produced a number of holographic stereograms which showcase the display's ability to present images that vary with angle. Figure 7 shows three images, each projected at a different angle.

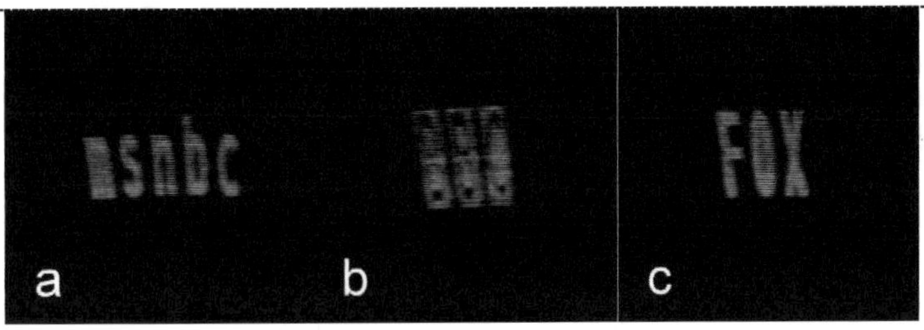

Figure 7. Holographic stereogram image from the display which shows image 'a' when viewed from the left, 'b', when viewed straight on and image, 'c', when viewed from the right.

4.3. Parameters and bill of materials

Table 1. Display Parameters

Parameter	**Design target**	**Actual to date**
Size	>10x5cm	>10x3cm
View Angle	>15 degrees	>10 degrees
Resolution	480 (18 channels)	26 (1 channel)
Frame Rate	30 frames/s	30 frames/s

Table 2. Bill of materials for holographic video monitor

Part	**Cost**[a]	**Vendor**
Polygon assembly	\$10.00	Surplus Shed
Vertical scanner	\$50.00[b]	AJP Optics
Parabolic reflector	\$15.00	Ebay
Aluminium plate	\$40.00	Admiral Metals
Extruded Aluminium	\$20.00	Ebay
Lens/mirrors	\$26.00	Surplus Shed
Boards/electronics	~\$300.00	Digikey/Advanced Circuits
TOTAL:	\$461.00	

[a] These prices are approximate.

[b] To achieve the parameters given in table 1 the vertical scanner must be refitted with a longer mirror.

5. Conclusion and future work

We have demonstrated early test images from a pc-driven holographic video monitor that is readily replicable and an order of magnitude lower in cost than earlier versions of this architecture. Next, we intend to maximize the display size, angle and resolution. This holographic monitor geometry, with relatively little modification, should be able to support a display output width approaching a maximum of 40 cm with a 10 degree viewzone (or 10 cm width with a 38 degree vie wzone). Now that the geometry has been validated, we intend to replace the low bandwidth modulator with an inexpensive, high bandwidth modulator, of our own design, capable of providing a full vertical output resolution of 480 lines. With this new modulator in place we will have developed a low cost, standard resolution, holographic desktop monitor.

6. References

[1] Saxby G., *Practical Holography* (Third Edition), Taylor and Francis, 2003
[2] Iovine J., *Homemande Holography*, Tab Books, 1990
[3] St.-Hilaire P, *Scalable Architechtures for Holographic Video*. PhD thesis, MIT, 1988
[4] Jepsen M. Holographic Video: *Design and Implementation of a Display System*, MS Thesis, MIT, 1989
[5] Kollin J., *Design and Information Considerations for Holographic Television*, MS thesis, MIT, 1988

Visual Perception and Holographic Displays

James Barabas and V. Michael Bove, Jr.

Object-Based Media Group, MIT Media Lab, Room E15-443, 20 Ames St., Cambridge, MA USA 02142-1308

E-mail: barabas@media.mit.edu, vmb@media.mit.edu

Abstract. Holographic displays have the potential to reproduce the natural parallax and focusing affordances of real scenes. Although holographic displays are still far from maturity, no other display technologies have the potential to reproduce these affordances as accurately. This paper reviews visual human-factors considerations for current and future holographic displays.

1. Introduction

Holograms can produce highly detailed images that can have an uncanny visual realism. Image appearance can change instantly and naturally with the viewers head motion, allowing viewers to experience direction-sensitive effects, like a shifting reflection on a glossy surface, or the discovery of objects hidden behind others. In addition, viewers can selectively focus their eyes on different parts of the image, allowing natural exploration of both near and far subjects, and producing natural defocus for other objects in the scene. Despite this potential, electro-holographic display systems are still in their infancy. Analog holograms that fully produce the effects above have extremely high information content, requiring digital displays to move orders of magnitude more data than traditional 2D displays in order to achieve the same visual fidelity. Current holographic displays make computation and transmission practical by reducing information content in places that have the lowest perceptual impact. Some of these techniques are adapted from traditional display holography, while others were developed solely for digital holograms. This paper reviews the capabilities of the human visual system that are of interest to designers of holographic display systems, highlighting the differences of holographic displays from other, more widely available display technologies. Although very little vision research has been conducted with holographic images, work in related areas is can also provide insight into constraints on the perception of holograms.

2. Visual Perception

2.1. Depth Perception

Human viewers estimate their distance from objects by combining information from visual and other sources. Sources of visual information that can reveal distance or relative distance are referred to as "depth cues,"

2.1.1. Pictorial information Pictorial cues are aspects of a scene that allow a viewer to discern depth using only information that appears on the retina of a single, nonmoving eye. Pictorial cues are those used when we discern thee-dimensional form depicted in two-dimensional photographs or illustrations. These cues include geometric cues, like the relative size and position of objects, blur gradients indicating deviations of objects from the plane of focus, color shifts due to atmospheric effects, and others [12]. Holographic images can provide all pictorial cues displayed by current television and film display equipment, although some prototype holographic displays currently provide limited color fidelity, contrast, or detail.

2.1.2. Motion Information Relative movement of objects within a scene can also be used to inform the viewer of scene geometry. In natural viewing, this motion can contain contributions from motion of the viewer relative to the scene, as well as motion of objects in the scene relative to each other.

Relative object motion within a reproduced scene (resulting from either moving objects, or moving camera) can convey information of relative positions and sizes of parts of a scene. These relative motions reveal spatial relationships even when viewed on a traditional two-dimensional television screen. Holographic displays are equally capable of displaying this kind of relative motion.

In natural viewing, movement of the viewers head can also provide information about scene geometry. Translational head movement allows viewers to discern distance to objects by observing the resulting induced relative motion of the objects viewed. Traditional two-dimensional and stereoscopic three-dimensional displays do not provide this cue to depth. When viewing a two-dimensional display, head motion can produce a distortion of the apparent geometry of the image of the entire scene [2]. Motion also produces the appearance of rigid uniform motion of the entire scene, which provides information about the location of the screen surface, but not about the geometry of the scene. Virtual Environment researchers have sought to overcome this limitation by electronically tracking motion of the head, and adjusting the image displayed accordingly. The addition of correct head-motion parallax to virtual environments has been found, for example, to improve time performance at a tapping task [2] by 12%.

But display changes based on head motion are generally suitable for only a single viewer, and still introduce latency between head motion and screen update [10]. This latency has been shown to degrade performance [15], and research suggests that latency of as little as 8 to 17 milliseconds can be noticed by observers [1]. Full-parallax holographic displays can however present correct pictorial information simultaneously for all eye positions within the viewing zone. Holographic displays are for this reason a natural candidate for tasks that require frequent or rapid viewer motion, and also for tasks that benefit from simultaneous inspection by multiple viewers.

Holographic displays that restrict images to HPO have appearance that can change correctly and instantly for horizontal head motion, but sacrifice correct reproduction of vertical motion parallax. In addition, HPO holograms require assumptions about the location relative to the hologram of the line connecting the viewers eyes [5]. When viewed from locations increasingly distant from this line, objects in HPO holograms experience increasing amounts of distortion. Head-tracking approaches can adjust the geometry of computer-generated HPO holograms to match viewer head position, but will still introduce delay before correct images are shown under non-horizontal head motion.

2.1.3. Ocular information Viewers also discern depth information from systems that control muscles in the eyes themselves.

Vergence

When viewing objects or displays, viewers make vergence eye movements that rotate the eyes to bring the images of points in the environment onto corresponding points on the two retinas. Distant objects are best viewed with the eyes both pointing nearly parallel to each other, while viewing close objects binocularly requires crossing the eyes inward. The vergence angle between the two eyes can be used to triangulate the absolute distance to an object. Binocular convergence is most useful as a depth cue for object distances of less than one meter [13]. For objects beyond a few meters, differences in the angle between the axes of the eyes becomes negligible.

Images on traditional two-dimensional displays produce matching images on the two retinas when the lines of sight of the eyes cross at the screen surface. Stereoscopic displays can present different images to each of the eyes, allowing the creation of image pairs that match when the eyes are verged at distances closer or farther than the screen. Vergence angle when fusing a pair of stereo images provides a cue to the absolute distance to the fused objects.

When viewing a holographic display, each of the two eyes captures a different view of the wavefronts reconstructed by the hologram. This allows each eye to receive different images as with a stereoscopic display.

Focus

Viewers actively control the refractive power of their eyes (accommodation) to keep retinal images in focus. Although viewers tend to be poor at using the accommodative state of their eyes alone to accurately gauge distances [11], accommodation can provide information when used in combination with other cues; conversely, when accommodation is not correctly provided by a display the perceptual accuracy of other depth cues can be reduced [9].

Holographic displays, unlike other stereoscopic displays, can present wavefronts that appear in focus at distances in front of, as well as behind the display surface. Designs for HPO holographic displays often use a vertical diffusing screen to allow images to be viewed from a range of vertical positions [6]. These vertical diffusers have a side effect of creating a single plane of vertical focus for light exiting the display. HPO displays can create image points with controllable horizontal focus, but vertical focus remains fixed at the diffuser. For these displays, scene points in the plane of the diffuser appear to focus and defocus naturally, but points increasingly in front of or behind the diffuser have increasingly astigmatic focus. Viewers have demonstrated the ability to mitigate some of the effects of astigmatic images, adjusting accommodation to compensate for differently oriented targets [12]. Studies evaluating the accommodative state of the eye while viewing images on an HPO holographic display show that viewers can use these images to drive an accommodative response [7]. When viewing HPO holograms from 1100 mm, viewers were able to accommodate equivalently to real-object controls when presented with targets placed up to 450 mm (1.43 Diopters) in front of the hologram plane. Despite these results, the appearance of retinal blur for out-of-focus image points on an HPO display is different than the blur seen in natural scenes, and full-parallax displays. More research is needed to explore perceptual consequences of the scene-dependent astigmatism present in HPO displays.

Holograms are most practically illuminated using one or more monochromatic light sources. This monochromatic illumination may have an impact on viewers performance at focusing an image. Some research suggests that viewers use chromatic aberration of the eye to inform changes in accommodation, and that viewers accommodate more slowly when viewing objects under monochromatic illumination [14]. Depending on the choice of wavelengths however, a holographic display illuminated by multiple different monochromatic sources would likely allow viewers to once again exploit these aberrations to some extent. Existing displays that use fluorescent, laser, or LED illumination also frequently rely on combinations of narrow-band primary colors. Viewers general ease at focusing images on these displays suggests that the narrow spectrum of holographic displays will not impose serious constraints.

Ocular Cue Interaction

Vergence and accommodation systems perform best when the cues they provide are in agreement. Stereoscopic displays that can only provide sharp focus at a single plane introduce conflicts between these cues. Since both vergence and accommodation are most active when viewing near scenes, correct presentation of these cues is especially important for images that appear close to the viewer. Although stereoscopic 3D is well suited for theatrical presentation, where images appear at long distances from the audience, home or mobile displays have more strict requirements for conflict-free viewing. Both the relatively small size of current holographic displays, and their ability to correctly reproduce both vergence and accommodation cues suggest that holographic displays will first find adoption for near viewing applications.

2.1.4. Stereopsis The difference in scene appearance on the retinas of each of the two eyes also plays a role in the perception of depth. This phenomenon is exploited by current stereoscopic 3D displays, showing images captured from two laterally displaced viewpoints, one to each of the viewers two eyes. Stereopsis takes advantage of the lateral displacements of corresponding points in retinal images of objects. Since eye movements can shift the retinas of the eyes relative to each other, retinal position (and stereopsis) can provide information about relative depth of objects, but not absolute depth [13]. The relative displacements between points in retinal images allow us to triangulate relative distances.

For stereopsis to function, the viewers eyes must first move to place the image of some object at corresponding points on the two retinas. Parts of objects at similar depths to the fixated point can also be seen clearly, but appearance degrades for objects at distance that produce large disparity in retinal image. Producers of stereoscopic 3D content use knowledge of this phenomenon to predict which parts of a scene will be clearly visible when a viewer is verged on a particular subject. Careful planning of the position of the intended subject relative to the screen can provide more clear and comfortable viewing. Developers of content for HPO holographic displays may need to make similar considerations for managing the depth of the intended subject relative to the screen surface, as HPO displays impose these stereoscopic as well as additional astigmatism-related constraints on clear viewing.

In some wide-angle binocular stereo display configurations, low horizontal resolution of displays has restricted the presentation of stereo images to only two or three distinct planes in depth [2]. Holographic displays, even HPO displays, require high spatial resolution in the horizontal direction, and therefore are unlikely to impose quantization on the stereopsis they can convey.

2.2. Image appearance

When designing holographic displays, aspects of vision relating to laser light and color vision may impose some constraints.

Although the use of coherent light allows the sharpest reproduction of holograms, the images created with coherent light are extremely sensitive to optical imperfections (in any part of the optical system, including the eye). In the presence of these imperfections, laser images appear to be modulated by a sharp, high-frequency speckled texture (laser speckle) [8]. The appearance of this speckle changes with movement of the eyes, and with constriction and dilation of the pupil. Conventional displays such as projectors that employ laser illumination can also suffer from similar speckle. Techniques for reducing speckle have been employed for these devices Error! Reference source not found., and are applicable to holographic displays as well. Depending on various characteristics of the holographic display, it may be possible to use incoherent light sources like light-emitting diodes instead of lasers.

3. Conclusions

Although holographic displays have potential for providing visual experiences that are far more natural than current 3D, much work must be done in reducing the cost of creating and transmitting content to these displays. Careful consideration of human perceptual capabilities will allow holographic displays to come to market more quickly, allowing for immersive viewing experiences that provide both comfort and realism. Despite active research on perception relating to displays in general, very little work has been done to explore the viewing of holographic images [4]. Holographic displays have unprecedented potential for reproducing many aspects of natural viewing, but also allow for creation of holograms simulating the constraints of other display systems. As such, holographic displays could be an invaluable tool for basic research in vision science. As more prototype displays become available, there are increasing opportunities for exploration of applications that take advantage of natural focus, occlusion and motion parallax. The authors see telepresence and teleoperation as application scenarios particularly well suited to these features of holograms [3], and look forward to future work in exploring this potential.

References

[1] Adelstein, B., Lee, T., and Ellis, S. Head tracking latency in virtual environments: psychophysics and a model. In *Proceedings of the Human Factors and Ergonomics Society Annual Meeting* (2003), vol. 47, SAGE Publications, pp. 2083–2087.

[2] Arsenault, R., and Ware, C. The importance of stereo and eye-coupled perspective for eye-hand coordination in fish tank vr. *Presence: Teleoper. Virtual Environ. 13* (October 2004), 549–559.

[3] Barabas, J., Jolly, S., Smalley, D., and Bove Jr, V. Diffraction specific coherent panoramagrams of real scenes. *Proc. SPIE Practical Holography XXV 7957* (2011), 795702.

[4] Barabas, J., Smithwick, Q. Y. J., and Bove, Jr, V. M. P-3: Evaluation of rendering algorithms for presenting layered information on holographic displays. *SID Symposium Digest of Technical Papers 41*, 1 (2010), 1233–1236.

[5] Benton, S. A., and Bove, Jr, V. M. *Holographic Imaging.* Wiley, Hoboken, NJ, 2008.

[6] Bove, V. M. Display holography's digital second act. *Proceedings of the IEEE 100*, 4 (april 2012), 918 -928.

[7] Charman, W. N., and Whitefoot, H. Astigmatism, accommodation, and visual instrumentation. *Appl Opt 17*, 24 (Dec 1978), 3903–10.

[8] Goodman, J. W. Some fundamental properties of speckle. *J. Opt. Soc. Am. 66*, 11 (Nov 1976), 1145–1150.

[9] Hoffman, D. M., Girshick, A. R., Akeley, K., and Banks, M. S. Vergence–accommodation conflicts hinder visual performance and cause visual fatigue. *Journal of Vision 8*, 3 (03 2008).

[10] Mania, K., Adelstein, B. D., Ellis, S. R., and Hill, M. I. Perceptual sensitivity to head tracking latency in virtual environments with varying degrees of scene complexity. In *Proceedings of the 1st Symposium on Applied perception in graphics and visualization* (New York, NY, USA, 2004), APGV '04, ACM, pp. 39–47.

[11] Mon-Williams, M., and Tresilian, J. R. Ordinal depth information from accommodation? *Ergonomics 43*, 3 (2000), 391 - 404.

[12] O'Shea, R., and Govan, D. Blur and contrast as pictorial depth cues. *Perception 26* (1997), 599–612.

[13] Palmer, S. *Vision science: Photons to phenomenology*, vol. 1. MIT press, Cambridge, MA, 1999.

[14] Takaki, Y., and Yokouchi, M. Accommodation measurements of horizontally scanning holographic display. *Opt. Express 20*, 4 (Feb 2012), 3918–3931.

[15] Watson, B., Walker, N., Woytiuk, P., and Ribarsky, W. Maintaining usability during 3d placement despite delay. In *Virtual Reality, 2003. Proceedings. IEEE* (2003), IEEE, pp. 133–140.

Modular sub-wavelength diffractive light modulator for high-definition holographic displays

Richard Stahl, Veronique Rochus, Xavier Rottenberg, Stefan Cosemans, Luc Haspeslagh, Simone Severi, Geert Van der Plas, Gauthier Lafruit, Stephane Donnay

IMEC vzw, Kapeldreef 75, 3001 Leuven, Belgium

richard.stahl@imec.be

Abstract. Holography is undoubtedly the ultimate 3D visualization technology, offering true 3D experience with all the natural depth cues, without the undesirable side-effects of current stereoscopic systems (uncomfortable glasses, strained eyes, fatiguing experience). Realization of a high-definition holographic display however requires a number of breakthroughs from existing prototypes. One of the main challenges lies in technology scaling, as holography is based on light diffraction and interference – to achieve wide viewing angles, the light-modulating pixels need to be spaced close to or below the wavelength of the used visible light. Furthermore, achieving high 3D image quality, hundreds of millions of such individually programmable pixels are needed. As a solution, we develop a modular sub-wavelength light modulator, consisting of three main sub-systems: the optical sub-system, comprising a 2D array of sub-wavelength pixels; the driver sub-system for individual pixel control, and the holographic computational engine. Based on conclusions from our state-of-the art studies, numerous experiments and holographic demonstrators, we have focused on reflective phase-modulating MEMS-based system and its scaling beyond 500nm pitch. We have devised a unique binary-programmable phase-modulating pixel architecture realizing vertical pixel displacement of up to 150nm at 500nm by 500nm pixel pitch, while sustaining low operating voltages compatible with CMOS driver circuitry. IMEC SiGe MEMS technology enables integration of the CMOS pixel-line drivers, scan-line drivers and I/O circuits underneath the 2D MEMS array, resulting in a compact and modular single-chip system design. Refresh rates of few hundred frames per second are achieved using our patented segmented driver-array architecture. Integrated circuits implementing parallel holographic computational engines can be added to the module using advanced 3D stacking technology. Herein we further report on our progress in realizing our first prototype system, featuring 2000x2000 pixels with 1.2μm pixel pitch, to demonstrate the first holographic video module with approximately 30 degree diffraction angle.

1. Introduction

Holography is the only visualization technique that can provide completely natural 3D perceptions. Digital holography combines this unique characteristic of classical holography with the flexibility of computer graphics, making it the ultimate display technology (of the future) [1].

Holographic visualization is in its essence based on two elementary properties of light propagation: light diffraction and interference of the diffracted light waves. Light diffraction is the more effective the smaller the diffractive optical elements, e.g. diffractive gratings or spatial light modulators, splitting the energy of the incident light into multiple diffraction orders (Figure 1). This means that the

Published under licence by IOP Publishing Ltd

ultimate holographic display is simply a spatial light modulator (SLM) with plenty of programmable micro-scale (or even nano-scale) pixels. Reprogramming of these pixels can so-to-speak focus the energy of diffracted light into specific points in 3D space and so create a complex 3D scene [2]. It is however mainly the physical size and pitch of the pixels that determines the overall display efficiency, i.e. the distribution of the incident light energy into ideally only one wide-angle diffraction order, the 3D projection. Such high efficiency 3D projection however requires development of SLM array with sub-wavelength pixel devices (in the range of 400 - 600nm, or smaller), as we demonstrate with our static holographic display chip (Figure 2).

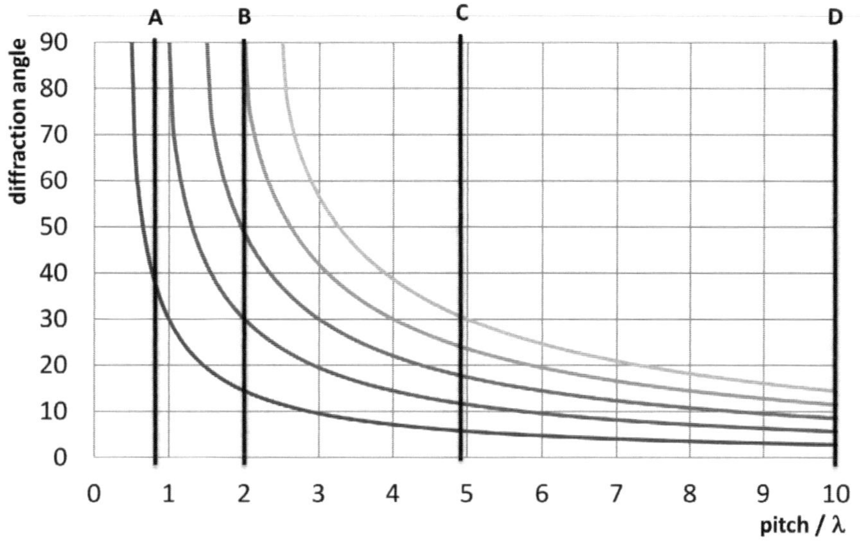

Figure 1: Diffraction angle of the first five diffraction orders as a function of relative pixel pitch (of a regular diffractive grating), wavelength λ = 635nm (red). Scaling pixels to sub-wavelength dimensions is critical for high efficiency and wide viewing angle holography. A - IMEC static demo holograms with diffraction angle of 38° (shown in Figure 2). B – IMEC prototype display with pixel pitch of 1.2μm and diffraction angle of 15° (described in Section 5). C – academic state of the art research scaling LCD-based displays to 3μm pixel pitch. D – commercially available LCD and DMD based display devices with pixel pitch down to 6 μm.

In our feasibility study we have identified the display technology scaling, and so the scaling of the light modulating pixels, as the main challenge limiting the development of high-definition digital holographic displays. As shown in Figure 1, the existing DMD [4] and LCD [5] technologies have rather limited diffractive performance, small projection angles and angular resolution. Furthermore they diffract light into multiple diffraction orders considerably reducing the overall display efficiency.

Figure 2: Photograph of IMEC static demo hologram with pixel pitch of 500nm, 38° diffraction angle, 104mm × 32mm total surface, 19° off-axis illumination, 158nm vertical pixel displacement and 3D scene of approx. 600k voxels.

In our previous work we have successfully developed and manufactured large size MEMS-based SLM for deployment in mask-less lithography [6]. Our poly-SiGe MEMS process technology provides means to achieve very high density of MEMS integration as it enables their post-processing on top of standard CMOS circuitry (Figure 3). This prior experience encourages us to further develop our MEMS platform and scale its critical dimensions down to only a few tens of nanometers, making it suitable for manufacturing of holographic display devices.

Figure 3: IMEC µ-mirror device with 8um pixel pitch, processed using low-temperature polycrystalline SiGe MEMS technology on top of standard CMOS. A 2D chip with 11 million devices has been manufactured at IMEC.

Herein we present the results of our feasibility study, consisting of the display system architecture (Section 2), sub-micron MEMS device design (Section 3), IC circuitry engineering (Section 4), as well as the latest prototype experiments (Section 5).

2. Display module overview

Our display system design is inspired by the PhD work of Mark Lucente [3], more specifically by his idea of holographic elements, the hogels: Imaging a large size holographic display composed of individual hardware modules, a sort of hardware hogels, of a size well below angular resolution of the human eye. Assuming, for example, minimal viewing distance of 2 meters, we devise a holo module of 500×500 μm^2. Such holo module is made by stacking together a display sub-module (of 1000×1000 500nm MEMS pixels with integrated CMOS drivers) and a digital-hologram generator (DHG) (Figure 4). The 500nm pixel pitch corresponds to about 30° projection angle (for green light wavelength), and the large pixel density guarantees high angular resolution. Incident light is modulated by 1-bit programmable vertical displacement of individual pixels over a distance of approximately 1/4 of the wavelength, resulting in phase modulation. The pixels are designed to have high fill-factor (of at least 80%) for good optical performance.

Operating at 50Hz in full colour mode (using red, green and blue coherent illumination) each holo module requires 150Mb/s data throughput that is easily achievable using existing CMOS technologies. Assuming each holo module encodes/projects approximately 10000 voxels (point sources of light in 3D space), it requires 1.5T holo operations per second. Each such operation is a fully pipelined custom hardware that is implementing the bipolar intensity algorithm [2]. To illustrate these computational requirements, a Radeon HD 5970 can for example reach 5TFLOPS performance, though executing relatively simple standard floating point operations. An ASIC implementation of the 1.5Tops DHG requires 750 bipolar-intensity pipelines running at 2GHz operating frequency. The DHG pipelines, implemented in 12nm CMOS technology, occupy an area of approximately $3mm^2$, which is 12-times the area of the corresponding MEMS module. It is however very likely that further algorithmic optimizations reduce the DHG computational requirements by more than the factor 12 and the DHG hardware would be implemented on silicon area the size of the corresponding display sub-module. This would enable a relatively straightforward 3D stacking of the two chips, as shown in Figure 4.

The main challenge of the system integration is the seamless tiling of the holo modules, although the ultimate challenge definitely remains the design and development of the MEMS pixel device itself.

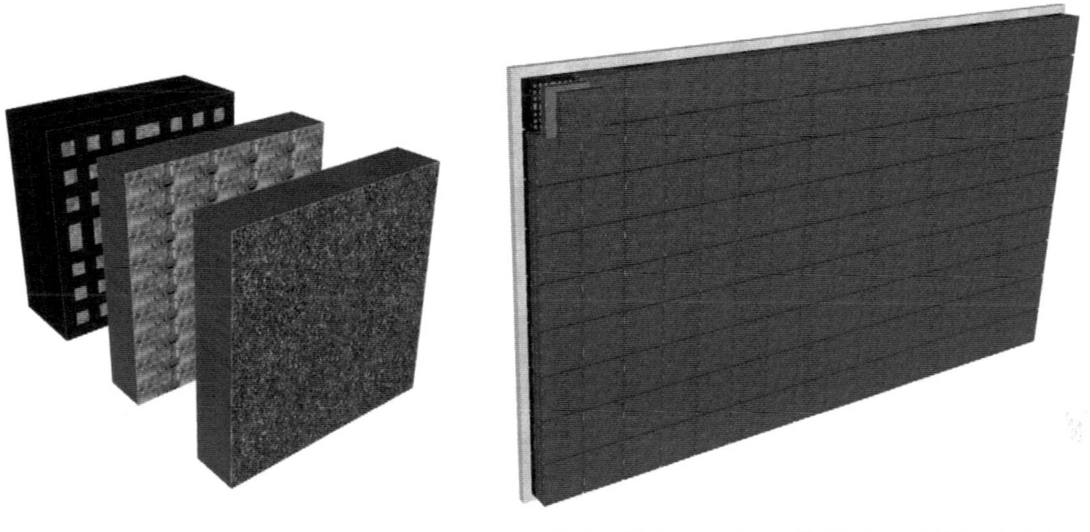

Figure 4: Modular holographic display design: Left – detail of a HW holo module with 3D stacked I/O back plate, DHG chip (containing approximately 60 full DHG pipelines) and display chip (including 1 million MEMS pixel devices of 500nm pitch, with integrated CMOS driver logic underneath). Right – example of 16×9 holo modules creating a 144M pixel holographic display with 30° projection angle and a total area of 8×4.5 mm^2.

3. Sub-wavelength MEMS design

We have explored a large number of rather unorthodox mechanical designs that could eventually deliver the 1/4 wavelength lift (of approximately 150nm in case of red illumination) within the area of just 500×500nm^2. Most of these designs are however very "MEMS processing unfriendly" as they require very small critical feature sizes. The most suitable and realistic design, which was eventually selected for later prototyping, is the staggered cantilever with three-electrode electrostatic actuation shown in Figure 5. As can be noticed, the actuator part of the pixel device is basically a considerably down-scaled version of the already proven micro-mirror design [6].

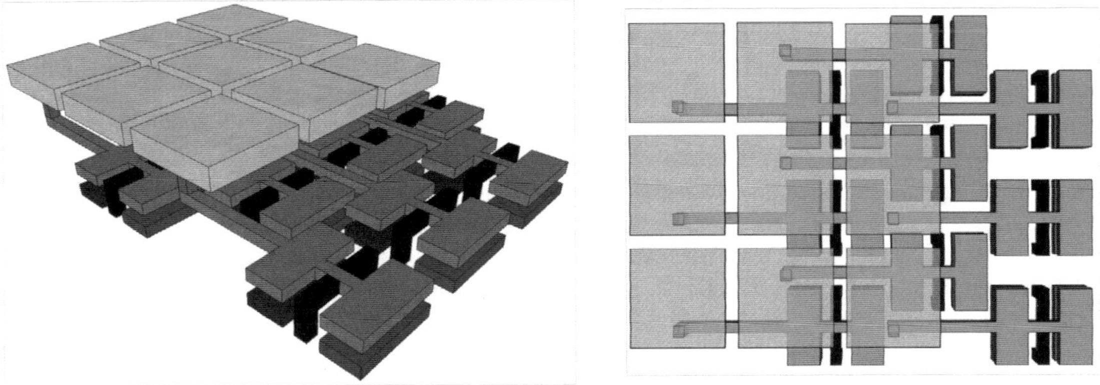

Figure 5: Conceptual design of 2D MEMS array with pixel pitch of 500nm and critical processing dimension of 50nm. The staggered MEMS actuators use 3-electrode electrostatic actuation to offset the extended cantilever. This provides an amplified vertical lift to the regularly aligned metallic mirrors atop. The metallic mirrors are designed to reach fill factor of 81% using advanced lithographic processes.

We have evaluated and refined our design using multi-physics simulation tools, taking into account mechanical and electrical properties, as well as constraints imposed by the integrated CMOS driver logic. The most limiting constraint, imposed by the standard high-voltage CMOS technology, is the 32V maximum operating voltage. This restriction considerably influences the mechanical properties of the final design (shown in scale in Figure 5): The actuator "pull-in" voltage is approximately 20V, which is well within the 32V limit, yet the critical dimension for SiGe processing is only 50nm. Resonance frequency of the whole MEMS pixel is in the range above 10MHz, which gives us more than enough time for device reprogramming and mechanical pull down, considering that each of the pixels needs to be reset only once in a few milliseconds.

Device itself exhibits an electrostatic memory effect as it operates in a so-called pull-in actuation regime. This means that we only need to apply the high voltage to pull the actuator into the programmed position, but we can hold the device in that state with much lower voltage. This allows us to simplify the programming scheme for the 2D pixel array, as described in the next section.

The main attention points of the mechanical pixel design are the stiffness of the main torsional element, together with overall robustness and stability of the extended cantilever. These issues are (to considerable extend) taken care of in the prototype design and optimization phase (Section 5).

4. Integrated driver circuitry

The main challenge of the driver logic design has been, simply said, the ever-present question of how to literally squeeze a high-voltage driver underneath every MEMS pixel to reach full array programmability. Inspired by the passive LCD display driver matrix, we have managed to combine the unique "CMOS under MEMS" feature of our MEMS technology with the electro-static memory effect of the MEMS into a so-called segmented passive driver array (Figure 6). The driver logic of the array consists of local low-voltage Pixel-Line Drivers (PLDs) and few global Scan-Line Drivers (SLDs). The 2D MEMS array is split into segments of PLDs implemented right underneath the MEMS devices they control. The PLDs in the segments run all in parallel resulting in refresh rates of few hundred Hertz for arrays as large as 100M pixels.

Reprogramming of the MEMS devices in each array segment is executed as follows (Figure 6-d): Each PLD sets the programming voltage to either "ON" (-2V) or to "OFF" (0V) state as required. A row of pixels is then selected using high-voltage SLD by pre-charging the scan line to 10V (considering for example 11V pull-in voltage). All other scan lines are kept at hold voltage (of 5V) that is well above the "pull out" voltage of the pixel devices (3V in this example). The hold voltage is applied to the scan lines for the whole duration of the image frame. This way a single row (scan line) of pixels is reprogrammed in each segment in parallel. Considering a single PLD takes approximately 9-times the area of a MEMS pixel, an example design of 100M-pixel array consist of 1111 segments driven by the 9 global SLDs. Each segment consists of 3333×3 PLDs programming a total of $3333\times3\times9$ pixels (in 9 cycles). Reprogramming of the total of 9999×9999 pixels can reach maximum refresh rate of 750Hz. Arrays of 1M pixel (such as used in the holo modules) can be in principle reprogrammed at refresh rates of few kHz.

We have further optimized the layout of the segmented array by eliminating any optically inactive chip area (SLD and I/O circuits) - we simply reduce the number of the PLD circuits and place also all the SLD and I/O circuitry neatly underneath the MEMS array. This on one hand enables seamless packaging of the module, yet on the other reduces the maximum reprogramming rate of the array.

I/O connection from the DHG to the display sub-module, implemented using standard CMOS through-silicon via technology, can reach up to 10Gb/s with only 6-pin configuration, which guarantees more than necessary data throughput between the two integrated circuits. Simulation results also show that approximately half of the total power is used by the I/O logic, while all the PLDs used only about 5%. The example 100M pixel array running at 750Hz, implemented in standard high-voltage CMOS, dissipates only approximately 0.5W of power.

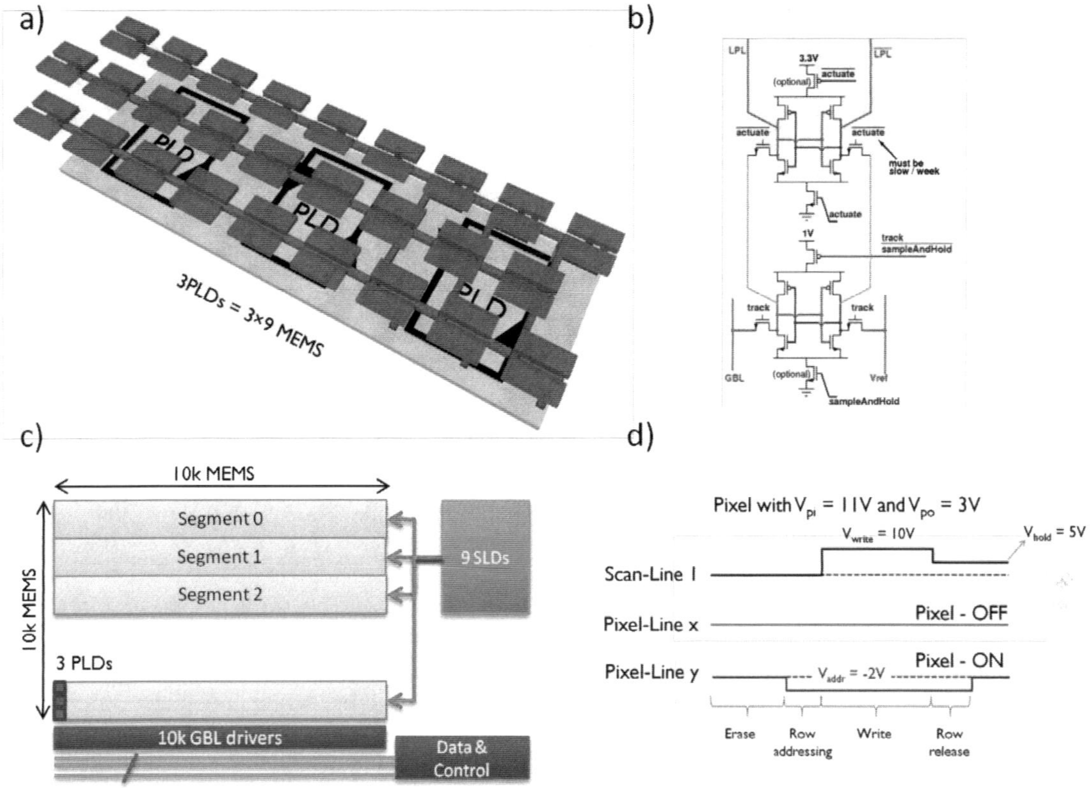

Figure 6: Example architecture of segmented passive driver array: a) 3D architecture overview with each basic segment consists of 3 Pixel-Line Driver (PLD) circuits placed underneath 3×9 MEMS devices. b) PLD circuitry details. c) 10k × 10k array layout. One scan-line (out of nine) is selected every cycle by the global Scan-Line Drivers. d) Programming scheme for pixel array with 11V pull-in voltage and 3V pull out voltage.

5. Prototype development

The original MEMS device architecture, described in Section 3, has been further optimized, taking into account practical limits of our current SiGe MEMS technology platform as well as our latest results from a project with similar processing constraints - MEMS nano-switch development for ultra low power applications [7]. The main goal of the optimizations has been improved design scalability towards the sub-wavelength dimensions.

The first prototype module is going to use a scaled down 1.5μm (instead of previously mentioned 1.2μm) pixels with greatly simplified mechanical structure (Figure 7): We have managed to double the critical processing dimension from previously required 150nm to more relaxed 300nm. Our latest experiments show that we can also reduce the vertical pixel displacement to about 50nm with only moderate influence on optical performance. We have considerably increases the mechanical stiffness of the cantilever, while keeping the pull-in voltage at approximately 26V. We further simplify the design using a two-electrode actuation instead of the previous 3-electrode proposal.

The fabricated modules will consist of 4M (2000×2000) MEMS devices with a total active area of $3\times3mm^2$. For demonstration purposes, we are also designing integrated driver logic providing 50Hz display refresh rate and 200Mb/s of input data rate. An early version of the prototype chip is foreseen with no driver circuitry underneath, focusing purely on proving the viability of the MEMS processing and experimental verification of their performance and reliability.

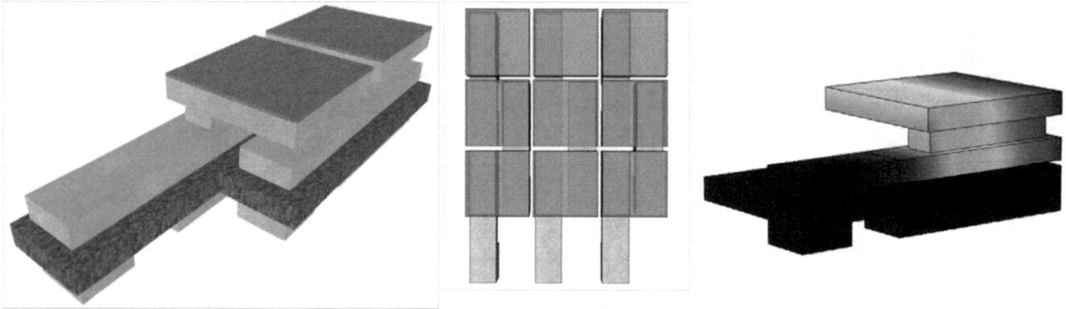

Figure 7: Optimized MEMS prototype: Left - simplified cantilever structure with 2-electrode electro-static actuation. Centre – top view illustrating the staggered cantilever alignment. Right – multi-physics finite element analysis showing amount of mechanical displacement within the MEMS structure.

6. Conclusions

We believe that the main challenge of digital holography remains the development of suitable display technologies with wide projection/diffraction angles and high angular resolution. Due to the physical principles of holography this is only really achievable with very large numbers of sub-micron diffractive light modulators. We show that MEMS technologies have reach a state where dense integration of sub-micron electro-mechanical pixels with their driver circuitry has become feasible. Moreover, we propose a holographic display module that combines the advanced MEMS spatial light modulator with digital hologram generator in a self-contained, compact and modular package. Such modules can be tiled together in large numbers to create TV sized holographic display system. The display system can reach realistic power requirements as most of the computationally intensive operations and data transfers are distributed over the many holo modules and executed locally within the DHG and MEMS sub-modules. We are working on a prototype of our holographic module to demonstrate its feasibility. Further scaling of the system will however be required to reach the true high-definition visual quality.

References

[1] S. A. Benton, V. M. Bove Jr., Holographic Imaging, Wiley-Interscience; 1 edition, 2008

[2] M. Lucente, Optimization of Hologram Computation for Real-Time Display, Proc.SPIE, No.1667, Practical Holography VI, pp. 32-43, USA, 1992

[3] M. Lucente, Diffraction-Specific Fringe Computation for Electro-Holography, Doctoral Thesis Dissertation, MIT Dept. of Electrical Engineering and Computer Science, Sept. 1994.

[4] Texas Instruments, Inc., www.ti.com

[5] HOLOEYE Photonics AG, www.holoeye.com

[6] A. Witvrouw, et. al., 11-Megapixel CMOS-Integrated SiGe Micromirror Arrays for High-End Applications, Journal of Microelectromechanical Systems, Vol. 19, No. 1, February 2010

[7] M. Ramezani et al., Effect of downscaling on the performance of a NEM switch, Proc. ICT.OPEN, 2011

Mobile glasses-free 3D using compact waveguide hologram

K Pyun [1]*, C Choi [1], A Morozov[2], A Putilin[3], I Bovsunovskiy[2], S Kim[1], J Ahn[1], H-S Lee[1], S Lee[1]
[1] SAIT, SEC, San 14, Nongseo-dong, Giheung-gu, Yongin-si, Gyeonggi-do, Korea,
[2] Russia, SMRC, SEC, Off. 1501, Bldg. 1, 12, Dvintsev St., Moscow, Russia 127018
[3] P.N.Lebedev Physical Institute of the Russian Academy of Sciences, 53, Leninskiy prospect, Moscow, Russia

E-mail: ks.pyun@samsung.com

Abstract. The exploding mobile communication devices make 3D data available anywhere anytime. However, to record and reconstruct 3D, the huge number of optical components is often required, which makes overall device size bulky and image quality degraded due to the error-prone tuning. In addition, if additional glass is required, then user experience of 3D is exhausting and unpleasant. Holography is the ultimate 3D that users experience natural 3D in every direction. For mobile glasses-free 3D experience, it is critical to make holography device that can be as compact and integrated as possible. For reliable and economical mass production, integrated optics is needed as integrated circuits in semiconductor industry. Thus, we propose mobile glasses-free 3D using compact waveguide hologram in terms of overall device sizes, quantity of elements and combined functionality of each element. The main advantages of proposed solution are as follows: First, this solution utilizes various integral optical elements, where each of them is a united not adjustable optical element, replacing separate and adjustable optical elements with various forms and configurations. Second, geometrical form of integral elements provides small sizes of whole device. Third, geometrical form of integral elements allows creating flat device. And finally, absence of adjustable elements provide rigidly of whole device. The usage of integrated optical means based on waveguide holographic elements allows creating a new type of compact and high functional devices for mobile glasses-free 3D applications such as mobile medical 3D data visualization.

1. Introduction

Recently, mobile communication markets skyrocketed due to the exploding smart-phones and tablet devices worldwide. These changes make large mass of non-technical populations experience visual data anywhere anytime. The most natural way to experience visual data is 3D since human beings perceives world as 3D. However, recording and reconstructing 3D data normally requires large quantities of optical elements for device implementation and requires cumbersome tuning for each 3D experience. Also, current CE-level 3D technology usually relies on special devices to view 3D such as external glasses, which hinders users experience natural 3D.

The holography is one of the most natural ways to view 3D without external glasses, where large diffraction efficiency can be obtained by applying Bragg diffraction of thick hologram gratings [1],

[1] To whom any correspondence should be addressed.

Journal of Physics: Conference Series **415** (2013) 012058 doi:10.1088/1742-6596/415/1/012058

Waveguide holography (WGH) is one of the unique methods of light-waves recording that uses total internal reflection on the optical media boundary for the fine separation of diffracted and non diffracted beams, which is useful for optical communications, systems, optical sensors, and display devices [2]. In semiconductor industry, integrated circuit eliminated manual assembly of discrete components, which makes its mass production reliable and economical. The same analogy can be applied in optics using integrated optics. In this aspect, we propose integrated optics using WGH for mobile and natural glasses-free 3D experience.

Important characteristic of mobile holography recording devices is to reduce the volume of device optical part without sacrificing image quality. To record full color and full parallax image, the initial laser beam is divided into two beams: signal and reference beams. The signal beam is expanded by a beam expander and after that it is modulated via a spatial light modulator in accordance with the recorded image. The signal beam passes through the focusing optical system and falls on light sensitive material. After that, the signal beam falls from the one side of light sensitive material and reference beam falls from the other side of light sensitive material.

In [3], the method and device for recording one-step, full color, full parallax, holographic stereograms is described, which consists of coherent radiated pulsed laser source, special optical system for dividing initial beam for signal and reference beams, special optical system for reference beam shaping and transformation, spatial light modulator for signal beam modulating, special optical system for hologram pixel recording into light sensitive material, and special device for measurement and controlling of spatial coherence of signal beam. All these solutions have a big quantity of different optical elements divided by air spaces. The relative positions of all these elements influence on working efficiency of whole device. In such way, these constructions have two main disadvantages: large number of separate optical elements and special units for controlling and correcting their relative positions. All those factors provide significant device complication and critical volume increase of total size dimensions.

2. RESULTS AND DISCUSSIONS

In general, each conventional device for hologram printing consists of the following main parts as shown in Figure 1: coherent laser source, beams splitter, beam expander, illumination system, spatial light modulator, and Fourier optical system, reference beam forming optical system, mechanical positioner, and light sensitive material. The precise mechanical adjustment of all optical devices makes system bulky and too heavy and requires labor-intensive tuning.

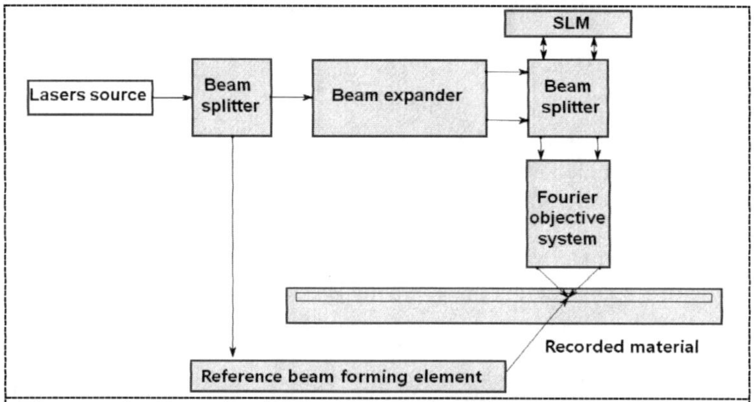

Figure 1. Conventional Hologram Recording Scheme.

Integration of optical elements and application of holographic optical elements paves the way to design compact and robust optical schemes, where the proposed devices are shown in Figure 2

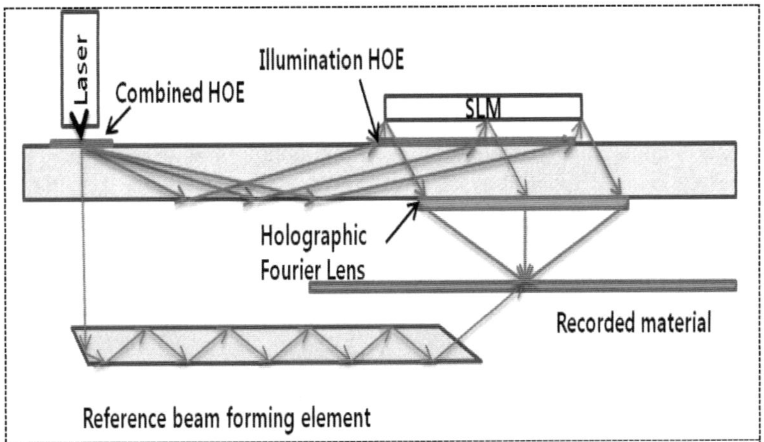

Figure 2. Proposed novel integrated optics

Figure 2 comprises the total internal reflection plate using WGH and the reference beam shaping holographic element (HOE). Signal beam element composed of combined HOE, illumination HOE, and holographic Fourier lens. The combined HOE is designed with possibility to divide beams outgoing from the coherent laser source for signal and reference beams. This combined HOE transforms the signal beam into uniform illumination on the plane of illumination HOE.

The illumination HOE can be recorded simultaneously for optimal illumination beam shaping and for required phase distribution in the plane of spatial light modulator so that it can form the virtual phase mask in the predefined plane. The holographic forming of the illumination system permits to realize the compensation of possible overall distortions or aberrations of signal beam. The holographic lens performs the Fourier transformation of the modulated signal beam in the plane of light sensitive material.

The reference beam forming scheme forms the reference beam with corresponding light distribution in the plane of optical recording media and predefined angle of incidence.

Figure 3. Sample of implemented combined and illumination HOEs

In our experiments, we implemented WGH with combined and illumination HOEs. The result of our sample WGH is shown on Figure 3. Conventional Fourier objective lens and reference beam forming elements were used as a 1st stage results. Also, we used transmission type LCD instead of LCoS as a spatial light modulator. Photopolymer was used as a recording material. And the measured characteristics of recorded holograms are summarized in Table 1. As can be seen from Table 1, we achieved state-of-the-art 3D image quality with 3D printer prototype size of 80x60x45 cm^3

	Pixel size	Angular resolution	Viewing angle	Gray level	Area Uniformity
Results	300x300 um^2	0.7°	67 °	4 bits	80%

Table 1. Experimental Results

3. Conclusion

The proposed scheme provides fundamental improvements of the basic scheme in parts of overall device sizes, quantity of elements and combined functionality of each element. The main advantages of proposed solution are as follows: First, this solution utilizes various integrated optical elements, where each of them is a united not adjustable optical element, replacing a lot of separate and adjustable optical elements with various forms and configurations. Second, geometrical form of integrated elements provides small sizes of whole device. Third, geometrical form of integrated elements allows creating flat device. And finally, absence of adjustable elements provide rigidly of the whole device. The usage of integrated optical means based on waveguide holographic elements allows creating a new type of compact and high functional devices for mobile glasses-free 3D applications such as mobile medical 3D data visualization.

4. References

[1] Kogelnik, H., "Coupled wave theory for thick hologram gratings," *Bell System Technical Journal*, **48**, pp. 2909-2949, (1969)
[2] Lin, L. H., "Edge illuminated holograms," *JOSA* **60**, 714A (1970)
[3] Klug, M, Holzbach, M., Ferdman, A., "Method and apparatus for recording one-step, full-color, full-parallax, holographic stereograms," US patent 6330088 June 17, 1998

Full Color Holographic Endoscopy

A Osanlou** *et al**.
Centre for Photonics Studies, School of Engineering, Institute of Arts, Science and Technology, Glyndŵr University, OpTIC, St Asaph, LL17 0JD, United Kingdom
ardieosanlou@aol.com *Osanlou A, Bjelkhagen H, Mirlis E, Crosby P, Shore A, Henderson P and Napier P

Abstract The ability to produce color holograms from the human tissue represents a major medical advance, specifically in the areas of diagnosis and teaching. This has been achieved at Glyndwr University. In corporation with partners at Gooch & Housego, Moor Instruments, Vivid Components and peninsula medical school, Exeter, UK, for the first time, we have produced full color holograms of human cell samples in which the cell boundary and the nuclei inside the cells could be clearly focused at different depths - something impossible with a two-dimensional photographic image. This was the main objective set by the peninsula medical school at Exeter, UK. Achieving this objective means that clinically useful images essentially indistinguishable from the object human cells could be routinely recorded. This could potentially be done at the tip of a holo-endoscopic probe inside the body. Optimised recording exposure and development processes for the holograms were defined for bulk exposures. This included the optimisation of in-house recording emulsions for coating evaluation onto polymer substrates (rather than glass plates), a key step for large volume commercial exploitation. At Glyndwr University, we also developed a new version of our in-house holographic (world-leading resolution) emulsion.

1. Introduction

In full-color interference based imaging, the produced image is essentially indistinguishable from the object. Three-dimensional holographic images of human tissue may be examined under a microscope in the same way as a tissue sample, without need for biopsy, and can resolve detail of cell nuclei. This project drew several recently developed different strands of technology namely interference based imaging techniques, recording materials and processes, fused fiber couplers, and coupled Red, Green, Blue (RGB) laser sources, for a specific application in medical imaging. Monochrome holograms could be recorded of tissue using cw laser light through a monomode fibre. Additionally, experiments using Red Green Blue (RGB) laser light through a monomode fibre were performed to illuminate recorded colour holograms of tissue samples. These preliminary experiments were aimed at investigating the influence of speckle across the illuminated field. The main focus was completely on the recording of medically useful full color holograms in a fixed rather than fibre-optic holographic setup. The fibre-optic setup will not improve the holographic image quality, but only make it possible to have a flexible recording system suitable for a medical environment.
Excellent progress was made and world first full-color holograms of human cells resolving cell nuclei were achieved. The medical assessment of the quality of the holograms was positive, as for the first time, human cell samples in which the cell boundary and the nuclei inside the cells could be focused at different depths was achieved **Figure 1.** This was the main objective and there are definite routes forward for medical applications, particularly in dermatology. However, for the mass commercialization stage, improvements are needed in both colors recording and rendering techniques as well as recording materials.

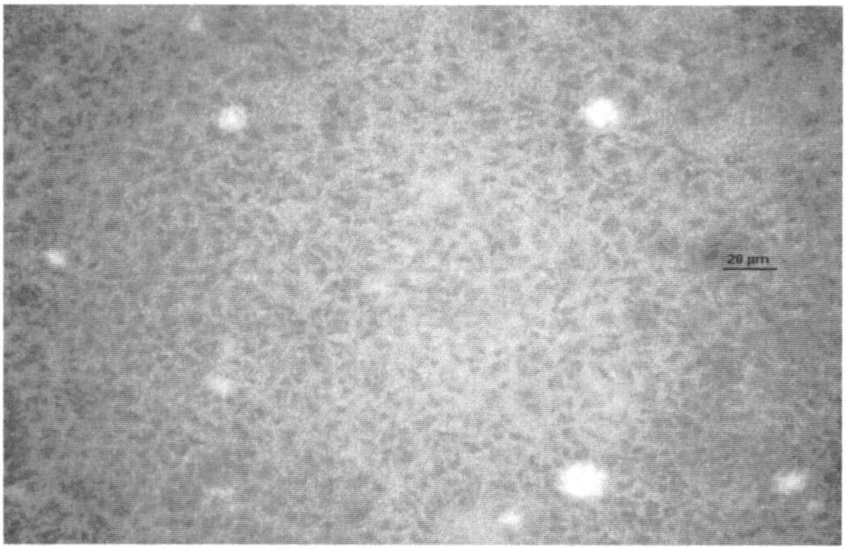

Figure 1. Hologram of human endothelial cells

2. The coherent source

2.1. Requirement for a 'white light' laser source

A fixed laboratory setup with visible CW lasers with long coherence and very stable output at 647nm, 532nm and 476nm were used to record the color holograms. However, flexible, high-quality full-color holographic endoscopy requires both visible combination and polarization-maintaining (PM) transmission integrated within a single device. Gooch and Housego fiber optics laboratories [1] with expertise in the production of laser sources, operating principally in the Red region of the visible spectrum were tasked with the development of a combination device **Figure 2(a).** with the following characteristics: (i) stable pigtailed sources, operating in the Red, Green and Blue regions, (ii) an output PER of >15dB, (iii) transmission of each visible wavelength between 70% and 90%, and (iv) Red=658nm, Green=532nm and Blue=452nm **Figure 2(b).** The choice of visible wavelengths was a compromise based on available sources most suited to holographic endoscopy.
Realization of the required tasks by Gooch & Housego is achieved by means of a relatively complex adaptation of existing technologies. Globally, this represents pioneering work and to date this has been highly successful.

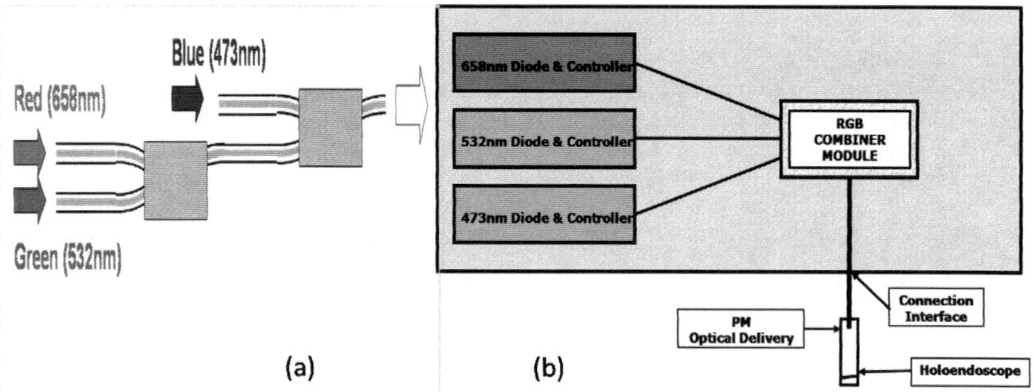

Figure 2. A flexible combination device with 'white' output(a) using RGB lasers (b)

2.2. Improved color rendering

The choice and the number of lasers used affects color rendering. Improved rendering is possible if more than the minimum three primary color laser wavelength (RGB) is used [2]. In house computer simulation **Figure 3** based on the *Macbeth Color Checker* ® assuming a perfect recording emulsion confirms this. The choice of the recording material used is also of equal importance and in this case the focus needs to be based on the use of extremely low light-scattering panchromatic materials that are fast in terms of exposure and developed. The main contenders are silver-halide emulsions such as the nano particle size emulsion developed at Glyndwr University's Centre for Photonics Studies [3] and photopolymer such as the new Bayfol®HX [4-5] from Bayer [6].

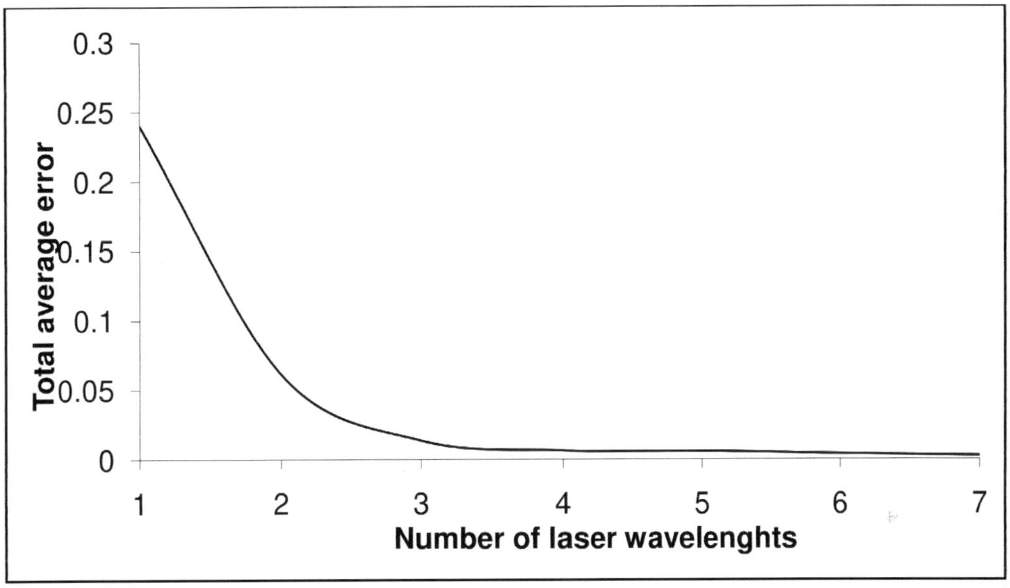

Figure 3. Decreasing total average color reproduction error with increased number of lasers

3. Recording Materials

To source the color recording material for the holo-endoscope project a number of commercial suppliers [6] [7] [8] [9] [10] [11] were approached. Despite the long list, the choice is very limited, but some panchromatic (around 450nm – 650nm) recording materials supplied on both Film and Glass substrates could be obtained from Sfera-S, Slavich and Color Holographic (Glass only). However, in order to obtain a recording materials with a diffraction efficiency of higher than 95% (Similar to that from Ultimate and Bayer) it was decided that to reproduce and optimize Glyndwr University's own ultra-fine grain (8-10 nm) recording emulsion (SilverCross) previously developed as part of an EU framework program [3]. This objective was met successfully. Additionally, for large volume commercial exploitation, optimized recording exposure and development processes for the holograms were defined. The produced emulsion has sensitivity between 4-8 mjcm^{-2} **Figure 4** and it seems to be one of highest quality color recording materials yet produced. Recording material used: - (i) Sphere-S material with an optimum recording of ~ 1.7 mJ/cm^2; (ii) In house coated material ~ 4 mJ/cm^2 and (iii) Machine coated in house recording material with ~ 6 mJ/ cm^2.

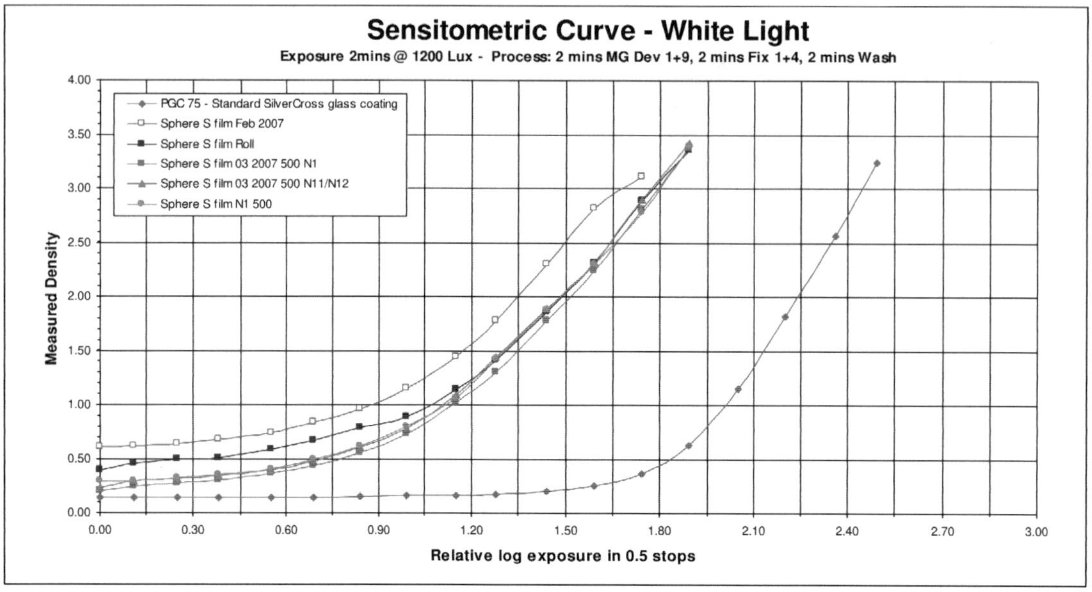

Figure 4.

4. RECORDING FIBRE-DELIVERED ENDOSCOPE HOLOGRAMS

The standard procedure was recording a reflection hologram on a silverhalide material and then chemically processing it in a darkroom.

4.1. Fibre-delivered recording

The following three Continuous Wave (CW) lasers were used to create a coherent 'white' beam to record the full color holograms: krypton ion laser (647 nm), frequency-doubled Nd:YAG (532 nm) and argon ion (476 nm). The recording setup is of reflection type using Denisyuk geometry **Figure 5.** Using multimode fiber, the setup was also used extensively for coherent holographic reconstruction. The reconstructed images were recorded using a digital microscope camera **Figure 6.**

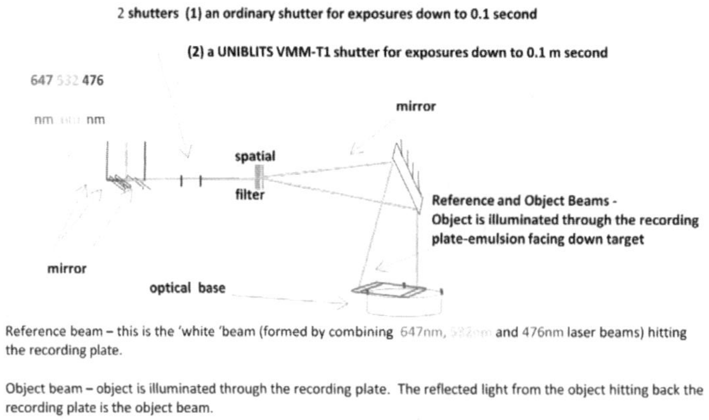

Figure 5. Holographic recording setup, also used for coherent reconstruction

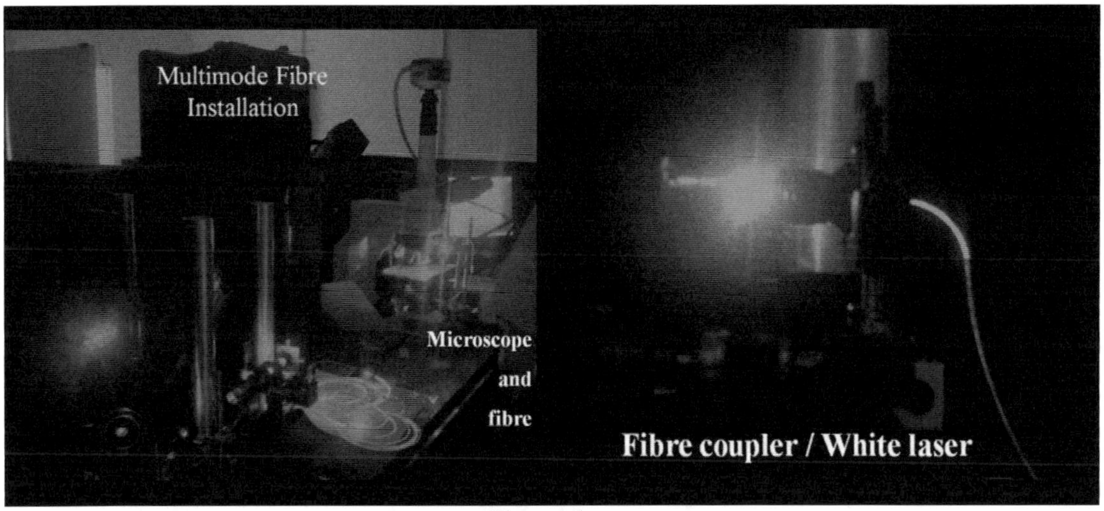

Figure 6.

The original subject matters used for recording, due to their relatively larger size were onion cells. The USAF test chart was used for calibration **Figure 7.**

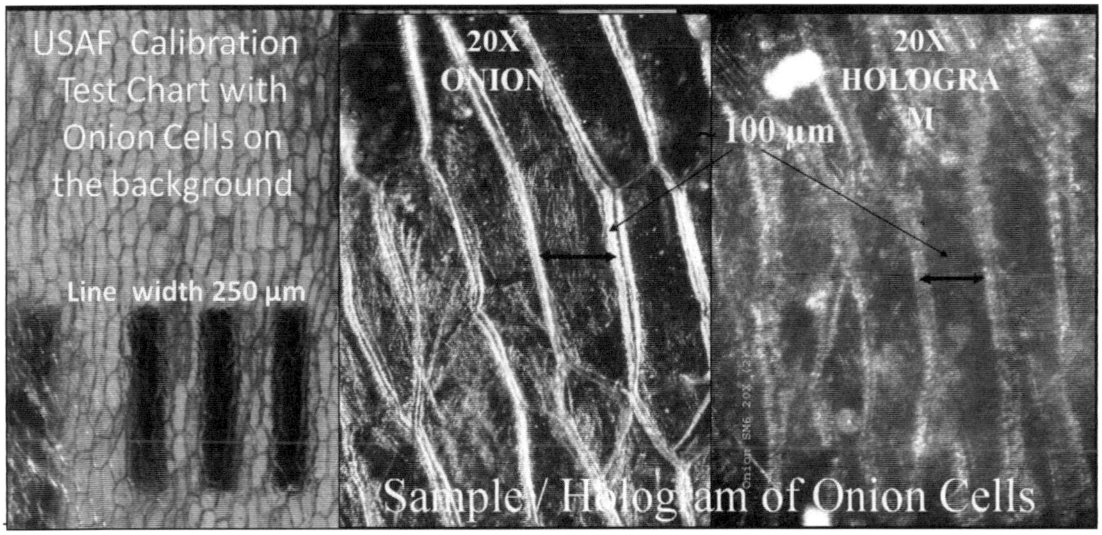

Figure 7.

4.2. Darkroom processing

The optimum darkroom processing techniques for each recording material were mostly established by trial and error in an empirical manner. Additionally, for large volume commercial exploitation, optimized recording exposure and development processes for the holograms were defined. This included the optimisation of in-house recording emulsions for coating evaluation onto polymer substrates rather than glass plates.

5. Reconstruction

The final recorded reflection holograms were reconstructed using a MEIJI Techno fibre optic (150W) light source **Figure 8.** A Nikon ECLIPSE ME600L microscope was used to observe the holograms in detail. The images were then captured using a Nikon DS-Gil 5 Mega Pixel digital camera.

Figure 8. Best image reconstruction results were only achieved through non-coherent illumination

5.1. Benchmark tests

These are standard tests looking for bright, noise free and full color holograms. **Figure 9** below, shows the spectral graph of a typical full color hologram made for this project.

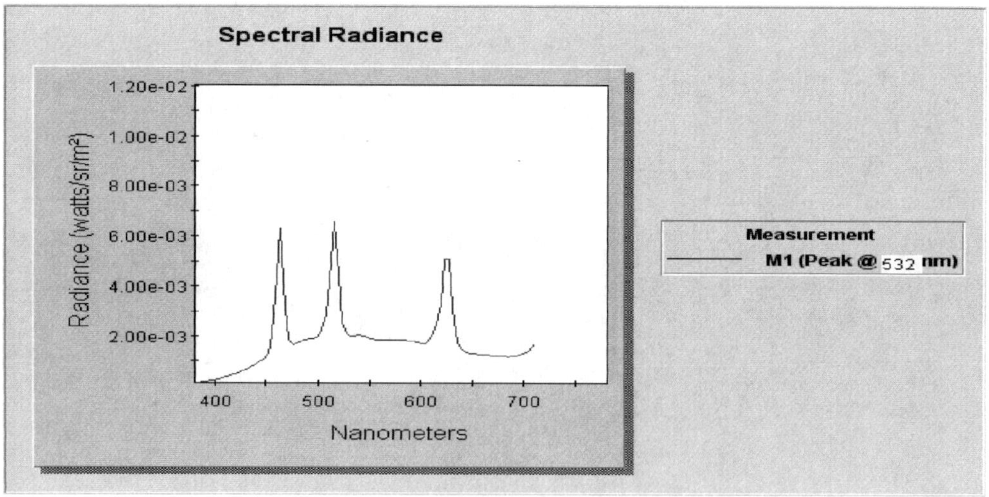

Figure 9.

The peaks on the graph should match the recording wavelengths of the three lasers used: - i.e. krypton ion (647 nm), frequency-doubled Nd: YAG (532 nm) and argon ion (476 nm). For a good color balance, the size of the peaks should be similar. All color measurements were made using the PR-650 Spectra Colourimeter from Photo Research Inc.

5.2. Image quality and speckle noise

High level of speckle noise was recorded when the single mode fibre was used for exposure. To reduce the noise level the method was abandoned in favour of the multi-mode fibre, but this method did not improve the noise level as the fibre delivery system did not match the hologram recording wavelength used. Eventually, some speckle reduction was successfully achieved by matching the recording wavelengths with the colour sensitivity of the panchromatic recording material (using chemical processing). In other words, by matching the recording color with the recording materials' color sensitivity, the speckle averaging effect is at its optimum, giving us a realistic estimate of the "best possible" colour endoscopic holograms in regard to resolution.

6. Color holograms of human endothelial

Early in the holoendoscopic investigations it was demonstrated that monochrome holograms could be recorded of tissue both directly and/or using fibre-optics. Experiments using RGB laser light through a fibre were also performed to illuminate recorded colour holograms of tissue samples. These experiments were aimed at investigating the influence of speckle across the illuminated field. Non-coherent illumination was eventually decided as the hologram reconstruction method of choice. The focus was then completely on recording and displaying of colour holograms of tissue samples without the use of fibre optics. The quality of endoholograms will be the same, independent of if they are recorded in a fixed or fibre-optic holographic setup. The fibre-optic setup will not improve the holographic image quality, but is only required for the next stage of the project to develop a flexible recording system suitable for medical endoscopic applications. The recorded colour holograms of tissue samples were investigated by Peninsula Medical School (PMC). The PMC determined that an RGB colour endohologram of cells recorded has the required quality for medical relevance **Figure 10**.

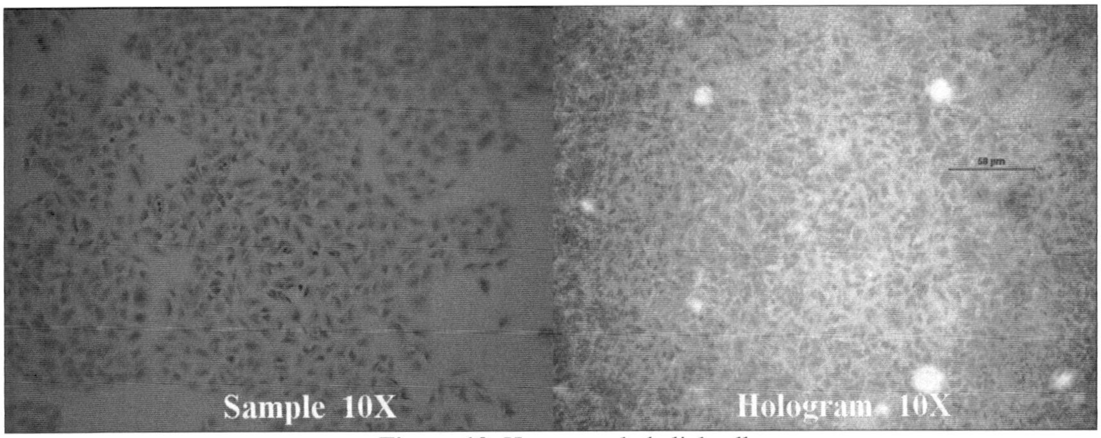

Figure 10. Human endothelial cells

7. Conclusions and the future

This project developed several different strands of technology and drew them together for a specific application in medical imaging. The main objective was to produce medically approved full color holograms of human cell samples in which the cell boundary and the nuclei inside the cells could be clearly focused at different depths. This objective has been achieved and potentially the recording could be done at the tip of a holo-endoscopic probe inside the body. Additionally, for large volume commercial exploitation, optimized recording exposure and development processes for the holograms were defined. At the Center for Photonics studies laboratories we also developed an optimised version of the Center's own nano size holographic emulsion with leading-edge characteristics for recording onto polymer rather than glass substrates, a key step for large volume commercial exploitation. However, the mobile holo-endoscope system, designed for flexible use in a medical environment, was not ready for testing in the timeframe of the project, and is not available for immediate commercialisation.

Future Work – With consideration to several advances in applicable technologies during the course of the project, here follows a list of future objectives, grouped by subject case.

A new holo-endoscope design

Analysis of promising recently developed mobile 'White' laser holographic recording equipment is of key importance [12], [13], [14].

Consideration for a new device:

- Integrated 'White' laser, with a minimum of three lasers, Red, Green, Blue, with all lasers to be fully integrated with PM fibre pigtails.
- Recording requirements - exposure dose ~1mJ/cm², nominal exposure time 1ms requires 1W.
- Full interface from Red, Green, Blue combiner to endoscope with consideration for beam profile for hologram exposure.
- Exposure time and output power balance control for all wavelengths, and shutter mechanism.

Material

- Further development on extremely low light-scattering panchromatic recording materials
- Tests on main recording materials in use today as well as newly developed photopolymer and silver-halide based emulsions.

Illumination

The illumination source to display the recorded holograms is of critical importance and newly reported illuminating system such as [15], [16], should be thoroughly investigated.

Considerations for the development of a new system:

- LEDs - choice of mixing the component LEDs in terms of bandwidth and wavelength, with controller for each LED. This is important as the color temperature of the source influences the color of the holographic image
- Source size partly determines the image resolution. Also, illumination angle and the distance from the source should be considered to avoid aberrations.

References

[1] Gooch and Housego PLC, Broomhill Way, Torquay TQ2 7QL UK

[2] Kubota T, Takabayashi E, Kashiwagi T, Watanabe M and Ueda K 2001 Color reflection holography using four recording wavelengths, in *Proc. SPIE: Practical Holography XV and Holographic Materials VII* vol 4296 ed S A Benton *et al.* pp 126-133

[3] A Glyndŵr University European CRAFT project COOP-CT-2004-5901, Centre for Photonics Studies, School of Engineering, Institute of Arts, Science and Technology, OpTIC, St Asaph Business Park, St Asaph, North Wales, LL17 0JD UK

[4] Bruder F-K, Deuber F, Fäcke T, Hagen R, Hönel D, Jurbergs D, Kogure M, Rölle T and Weiser M-S 2010 Reaction-diffusion model applied to high resolution Bayfol®HX photopolymer, in *Proc. SPIE: Practical Holography XXIIV: Materials and Applications* vol 7919 pp 0I-1-15

[5] Bruder F-K, Deuber F, Fäcke T, Hagen R, Hönel D, Jurbergs D, Kogure M, Rölle T and Weiser M-S 2009 Full-color self-processing holographic photopolymers with high sensitivity in red – the first class of instant holographic photopolymers *J. Photopolym. Sci. Technol.* **22** 257-260

[6] Bayer Material Science AG, Germany; www.bayermaterialscience.com

[7] Sazanov Y, Gradova O, Zacharovas S, Bakanas R and Gudaitis G 2006 New ultra-fine grain photofilm for pulsed colour holography, in *Advances in Display Holography, Proc. 7th Int. Symposium on Display Holography* ed (London: River Valley Press) pp 65-68

[8] Gentet Y and Gentet P 2000 "Ultimate" emulsion and its applications: a laboratory-made silver halide emulsion of optimized quality for monochromatic pulsed and full color holography, in *Proc. SPIE: HOLOGRAPHY 2000* vol 4149 ed T H Jeong and W K Sobotka pp 56-62

[9] Colourholographic Ltd UK; www.colourholographic.com
[10] Ultimate – Holography France; www.ultimate-holography.com
[11] DuPont Autentication, USA; www2.dupont.com/Authentication/en_US/
[12] Gentet Y and Shevtsov M K 2009 Mobile holographic camera for recording color holograms, *J. Opt. Technol.* **76** 399-401
[13] Sarakinos A, Lembessis and Zervos N 2012 A transportable system for the *in situ* recording of color Denisyuk holograms of Greek cultural heritage artifacts in silver halide panchromatic emulsions and an optimized illuminating device for the finished holograms *ISDH2012* IOP Publishing
[14] Jith A 2012 Holographic color camera for recording artifacts *ISDH2012* IOP Publishing
[15] AutoTech Wireless Automation, Greece; www.autotech.gr
[16] Taurus SecureSolutions Ltd. Greece; www.taurus.com.gr
**

DR ARDESHIR (ARDIE) OSANLOU ***FRPS FIET FIMA***
CENTRE FOR PHOTONICS STUDIES, GLYNDŴR UNIVERSITY

Ardie is an Accredited Senior Imaging Scientist with the Royal Photographic Society and a Chartered Scientist with the IMA of Great Britain. He is the founder and coordinator of a new MSc, MRes, MA course in Photonics Engineering and Visual Communications at the School of Engineering, Institute of Arts, Science and Technology GLYNDŴR UNIVERSITY, Wales, UK. He has two research degrees and became heavily involved in industrial contract work in applied optics since 1999.

CONTACT: Dr Ardie Osanlou, School Of Engineering, Institute , Institute Of Arts, Science And Technology, Glyndwr University, OpTIC, St Asaph Business Park, St Asaph, North Wales LL17 0JD United Kingdom
Tel: +44 (0)1745 535 131 Dir.
+44 (0)1745 535 100 Rec.
Fax:+44 (0)1745 535 101
Email a.osanlou@glyndwr.ac.uk ardieosanlou@aol.com
www.glyndwr.ac.uk/en/Ourcampusesandfacilities/GlyndwrUniversityStAsaph/

Journal of Physics: Conference Series **415** (2013) 012060 doi:10.1088/1742-6596/415/1/012060

Design and visualization of synthetic holograms for security applications

M Škereň, M Nývlt and J Svoboda

Czech Technical University in Prague, Faculty of Nuclear Sciences and Physical Engineering, Břehová 7, 115 19 Prague 1, Czech Republic

E-mail: marek.skeren@fjfi.cvut.cz

Abstract. In this paper we present a software for the design and visualization of holographic elements containing full scale of visual effects. It enables to simulate an observation of the holographic elements under general conditions including different light sources with various spectral and coherence properties and various geometries of reconstruction. Furthermore, recent technologies offer interesting possibilities for the 3D visualization such as the 3D techniques based on shutter or polarization glasses, anaglyphs, etc. The presented software is compatible with the mentioned techniques and enables an application of the 3D hardware tools for visualization. The software package can be used not only for visualization of the existing designs, but also for a fine tuning of the spatial, kinetic, and color properties of the hologram. Moreover, the holograms containing all types of the 3D effects, general color mixing, kinetic behavior, diffractive cryptograms, etc. can be translated using the software directly to a high resolution micro-structure.

1. Introduction

Today, the synthetic holograms are widely used in optical document security, where various visual effects are utilized for an evaluation of originality of the security elements. The security function is based on a very fine micro-structure with the details of order of hundreds of nanometers. It diffracts the light and creates macroscopic visual effects. The holographic security elements have been successfully used for anti-counterfeiting applications since early eighties of the twentieth century [1]. From a technical point of view, these structures are usually created as the thin diffractive relief elements in plastic foil. They are mastered in photo-resist and finally transferred to a plastic foil using the electro-forming and embossing technologies [2]. The elements are attached to the secured products either as conventional stickers or using the hot-stamping technique. Volume holograms are also used for security applications. Then they are usually recorded in a photo-polymer material.

A very important property of the holographic security element is a possibility of an easy evaluation of the originality by a common observer without any sophisticated knowledge of the optical principles and usually also without any additional inspection tools. Today, there are many different techniques capable of evaluating the originality with very high reliability (for example some spectroscopic techniques or others), but most of them require relatively sophisticated inspection equipment and trained operators. A macroscopic visual behaviour of the holograms based on a very fine micro-structure enables to avoid the complicated inspection process and to move the process of the authenticity inspection to the public. Thus the visual design plays an important role not only from an artistic point of view, but also from a point of view of the security function of the element. When the designed element contains many different visual effects, it is often difficult for the designer to simulate the complex behavior of the hologram only using conventional graphic tools and approaches. Moreover, it is almost impossible to present the hologram properties to most customers, who are not familiar with the holography at all. On the other hand, the modern computer graphics offers several interesting tools for visualization of the dynamic and three-dimensional effects created by the holographic micro-structures. A proper presentation of the visual effect can also significantly help to evaluate the security features and to reveal the potential counterfeits.

In the following sections, the most important visual effects and other security features are described, which are typically present in majority of current security holograms. Elementary physical principles of most effects are briefly discussed in order to explain the basic requirements on visualization of the designs and consecutive fabrication of the micro-structures.

2. Design and fabrication of holographic security elements

Wide variety of different holographic elements is used for security applications. Many of them are based on common holographic principles and have a similar visual behaviour [3, 4]. For most security holograms an elementary functional unit is a diffraction micro-grating [5]. Light from an illumination source (usually a white light source with some degree of spatial coherence) is diffracted from the grating and sends an information about the particular image point to a well-defined direction (see figure 1).

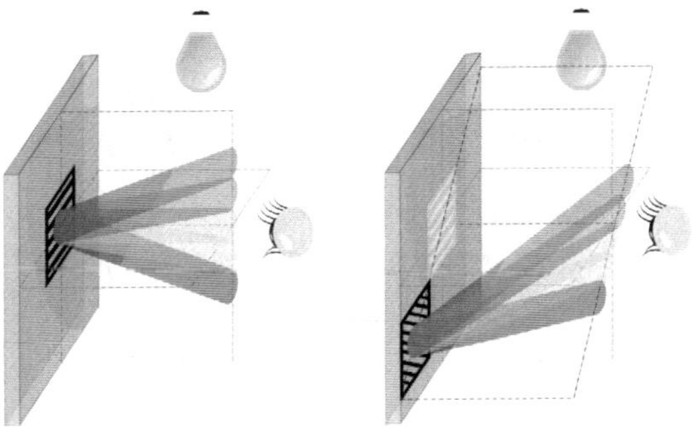

(a) (b)

Figure 1. Common geometry of reconstruction of the thin rainbow holograms for two different micro-gratings ((a) horizontal grating, (b) tilted grating). The color and angular behaviour are given by the position, periodicity, and orientation of the grating.

Usually, the thin gratings are used and thus a whole spectrum of wavelengths is diffracted, each color to a slightly different direction. By changing the grating frequency and orientation of grating lines, each direction in the half-space can be addressed with an image information. The dimensions of micro-gratings can be chosen according to the needs of a particular design, however the grating must contain at least several periods if it should diffract the light effectively. Considering the common observation geometry and the range of visible wavelengths, the grating dimension must be at least several micrometers. On the other hand, when the resolution of the observation system is not better than 0.1 mm, hundreds of different gratings can be placed next to each other within a resolvable point. It means that hundreds of different signals can be included within a single image point. This effect is the base for most of the unique visual effects of the synthetic image holograms, which use the spatial multiplexing of the images and the limited spatial resolution of the observing system. The holographic element can multiplex various images also using overlapping micro-structures, but this approach is rarely used for preparation of the synthetic embossed holograms.

2.1. Encoding of the design

The design of holographic elements differs from any other graphical artworks in the necessity to define additional parameters for description of the visual behaviour. In contrast to common graphical designs, where each point is usually defined only by color coordinates, the holographic elements can change their visual appearance depending on the geometry of reconstruction (usually the observed view depends on the direction of observation and also on the direction from which the incident light illuminates the hologram).

In fact, most holographic effects such as the dynamic behaviour, 3D properties, and variable colors are based on the ability of the holographic element to send a completely different information to the different directions from a single area of the elementary resolvable image point. Thus for each point on the surface of the holographic element, the behaviour must be defined for all possible geometries of reconstruction. A simple color image is not sufficient for such a complex description. Very often, the angle of incidence between the illuminating beam and the plane of the hologram is fixed and only the observation angle can vary. There are several approaches how to encode the image data of this kind. In our design software two image maps were chosen, one containing the information about color coordinates and the second describing the observation angle from which the particular point can be observed. When the resolution of such image maps corresponds to the elementary micro-grating dimensions, all desired effects can be represented by these two images. In figure 2, there are examples of such image maps. The color map is usually an RGB mixed image, the map describing the angular properties of the hologram is often a grayscale image (when 8-bit encoding is used, 255 different angles can be encoded).

(a) (b)

Figure 2. Simple example of the encoding of the holographic design using two raster images. The color image defines the color properties of each image point (periodicity of the corresponding micro-grating), the grayscale image carries the information about the direction of observation for each point (orientation of the grating lines).

The visualization of the design consists of simulating the different spatial views of the hologram for various light sources, observation conditions, etc., based on the prepared design image files. The design itself can be created in any graphic software which is capable of exporting the raster design images with sufficient resolution. More complicated designs can require several sets of color and grayscale images, each describing different properties of the same holographic element. All this data will be finally merged into a single pair of the design maps.

2.2. Visual effects and hidden features

Current security elements are based on two main groups of optical effects. The first group contains visual effects which can be easily observed by a naked eye under common lightning conditions with little requirements on the light source used. The security function is realized through the diffraction process which can create unique macroscopic visual behaviour based on tiny details of the micro-structure with dimensions of hundreds of nanometers or even smaller. These visual effects can serve as the most important group of security features dedicated to the first-line inspection. The typical effect from this group must create a visual perception, which is easily recognizable by a human observer and which cannot be reproduced by common printing technologies. The simplest representatives of this group are basic rainbow diffraction effects, images with a kinetic behaviour, 2D/3D effects, flip-flop effects, full 3D effects and some others. All of them are based on the ability of a holographic structure to spatially multiplex the image information [1, 5, 6].

Although the visual effects are essential for the first-line inspection of originality, it is desirable to combine them with additional security features which are hidden for a common observer. Thus the second group of the security features contains the approaches based on an encryption of some texts or images in the micro-structure. This information can be revealed either using a special geometry of observation, or by application of a specific light source (laser light, polarized light, etc.) [2, 7, 8]. The most commonly used structures of this type are the diffractive cryptograms and micro-texts. While micro-text is just an extremely miniaturized text or image, which can be observed using an optical microscope, the diffractive cryptogram is a holographic element, which forms the reconstruction of the hidden information out of the plane of the security element on the base of the diffraction of coherent light. The diffractive cryptograms are usually aperiodic, but they can be also assembled from similar micro-gratings as the visual effects.

2.3. Mastering techniques

The mastering technique strongly influences the effects, which can be used within the prepared hologram and it also influences the method how the design is prepared and encoded. Today, there are three main approaches to the mastering of the security elements. The first one is based on a classical holographic exposure of the elements, where the desired micro-structure is exposed using a natural interference process in a holographic setup. Relatively large areas are exposed at once and multiple exposures can be used to achieve various visual effects. Particular exposures can overlap or they can be separated using a system of exposure masks. The main advantage of this approach is a relative simplicity of the recording process and a possibility to expose larger holograms comparing to the direct writing approaches. On the other hand, the created color, dynamic, and three-dimensional effects are very limited. The two other approaches are both based on high-resolution direct-writing techniques which can record the details with sub-micrometer dimensions over the hologram area. Because the writing resolution is significantly higher than the resolution of the observing system (the human eye), multiplexing of large amount of image information can be achieved. If the elementary area (usually a regular micro-grating) has the dimension of only several micrometers, several hundreds of different signals can be recorded within a single point resolved by an observer. Such an approach offers huge flexibility in creating new features and enables almost any combination of effects within a single hologram. When designing complicated structures, a proper visualization of the design can be very helpful.

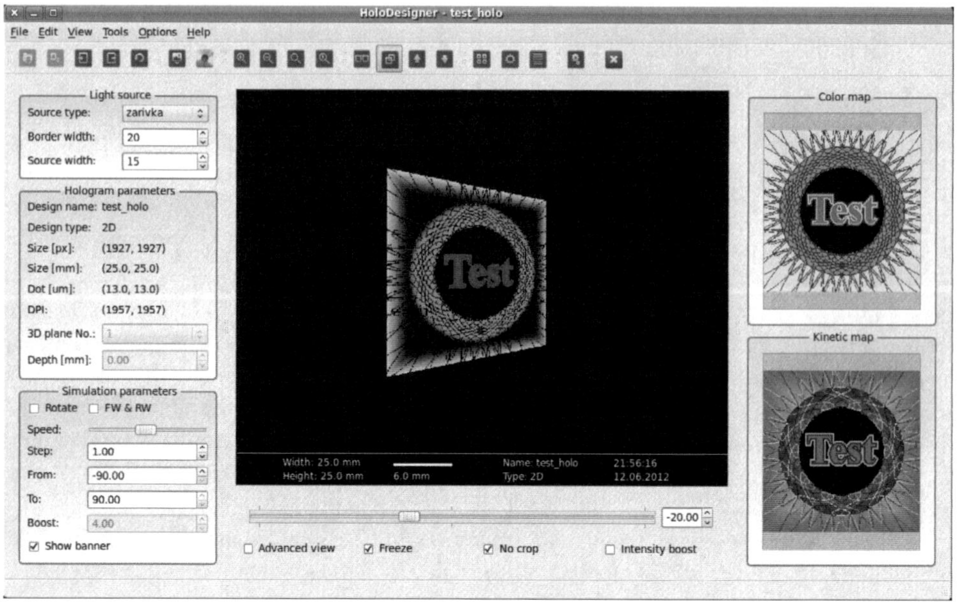

Figure 3. Main window of the HoloDesigner software. The central part is devoted to the main visualization window, on the right there are the source image maps. The left part contains basic parameters of the design, view, and light source.

3. Visualization of the security elements

The problem of visualization of the holographic elements was solved using a computer simulation of the visual behaviour of the element based on a simple design data. The most important parameters, which influence visual perception of the hologram are the geometry of reconstruction and the coherence properties of the light source. In a case of an ideal spatial coherence of the source, the rainbow effect is observed and the visualization can be relatively simple. When the source is only partially coherent (which is the case of most real sources), the visual behaviour is more complicated. Application of a partially coherent light can be described by using the convolution of the image data with a coherence function of the source.

A software tool has been built for visualization of existing designs containing wide variety of effects. In figure 3 there is a graphical interface of the HoloDesigner software. The code was written in Python language and optimized for Linux operating system. The software can generate various views of the designed hologram in real time with a possibility to interactively change full range of parameters. All kinds of designs can be loaded and also partially edited by the software.

A two dimensional preview of the holographic elements can be used for presentation of the designs in most cases, especially when the important parameters such as the viewing angle and coherence of the light source can be easily changed in the computer. However, for real three dimensional images it would by much easier to visualize the design using some of the 3D projection techniques. A combination of the interactive graphical interface with a three dimensional visualization can bring new dimension to the processes of the design and presentation of the holograms. Such a tool can be very helpful not only for previewing the future hologram prior to the expensive fabrication process, but also for evaluation of the originality of existing elements. Today, there are several techniques commonly used for the 3D visualization in the entertainment industry. For our application, two different approaches have been chosen. For a high-quality 3D visualization the two channel stereo with a temporal multiplexing of two spatial channels was used. Different hardware devices can be used for such an imaging, the output of the HoloDesigner package was optimized for application of the NVIDIA 3D Vision technology with synchronized shutter glasses. The device is based on a fast temporal switching of two spatial viewing channels with a refresh frequency 120Hz (60 Hz per channel) using a projection device such as a computer monitor or a projector with an appropriate switching frequency. The separation of spatial channels for observer is done using the synchronized shutter glasses based on the liquid crystals. The image data consists of a set of 2D views from two different observation directions. The HoloDesigner can export video preview of any holographic design in such a format. The exported data can be viewed also using another similar 3D projection systems.

The second option for the three dimensional visualization included in the software is a conventional anaglyph projection with two channels multiplexed using the pre-defined color shifts of the channels. Although this technique influences negatively the color perception of the holographic image, its main advantages are a wide availability, simplicity, and little hardware requirements (except for the passive color anaglyph glasses). In figure 4, there is an example of the anaglyph output within the main window.

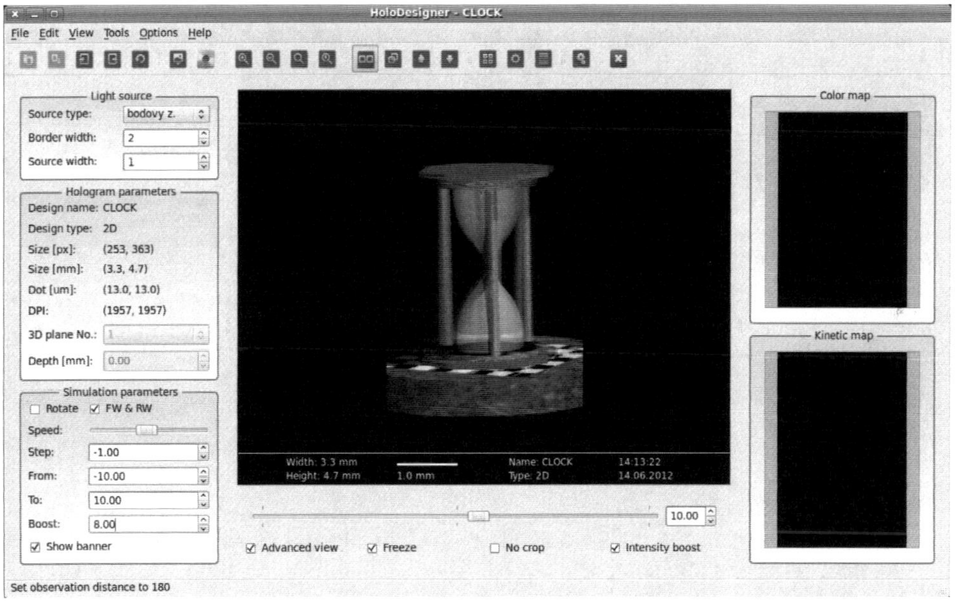

Figure 4. HoloDesigner with three-dimensional design in anaglyph mode. The view can be fully adjusted from a point of view of observation distance, position of an observer, anaglyph type, etc.

The 3D preview can be used not only for the designs containing some three dimensional objects or spatial planes in the 2D/3D design, but also for a three dimensional observation of two dimensional holograms. In figure 3 there is an example of a two dimensional design tilted in the 3D space which corresponds to the common geometry of reconstruction when the security elements are visually inspected. Such a three dimensional observation can be performed in the perspective projection mode, the anaglyph mode, and the real 3D Vision mode.

4. Processing of the design

The described software HoloDesigner can be used also as a tool for translation of a visual design to a high-resolution micro-structure. After finishing the composition, setting the 3D spatial parameters of particular objects and loading the data for the recording device, the software can merge all information to a single set of the design maps similar to those from figure 2. All viewing channels for the 2D/3D and full 3D objects are spatially multiplexed at the hologram plane, the color mixing is performed for the areas with the mixed colors, and finally the hidden features such as the micro-texts and cryptograms are included. In figure 5 there is and example of a small part of the resulting micro-structure represented as the raster image. Particular micro-gratings are clearly visible. Several micro-gratings are further segmented due to the multi-channel color mixing. The software uses 6 basic colors for mixing from which it automatically selects two or three closest colors for particular mix. The software can use gratings with various profiles such as harmonic, binary, and also more general user defined shapes. Different types of profiles can be also easily combined within a single design.

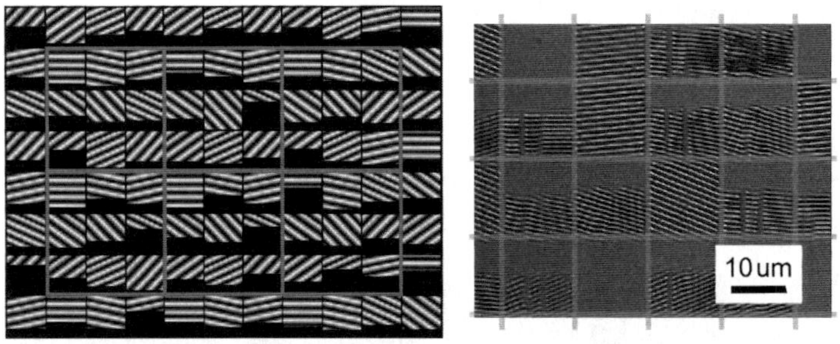

(a) (b)

Figure 5. (a) Example of the micro-structure generated by the HoloDesigner in the case of the 3D true color hologram. (b) Microscope image of the micro-structure fabricated using a matrix laser lithography device.

5. Conclusions and acknowledgements

The design and visualization software for synthetic security holograms was presented. The visualization tool has proven its importance in the design and presentation processes. Application of the recent 3D projection systems rapidly improved effectiveness of the visual design and contributed to a significant time and money savings in the holographic mastering process.

This work has been supported by the Grant Agency of Czech Academy of Sciences Project No. KAN401220801.

References

[1] Van Renesse R L 1994 *Optical document security* 2nd ed (London: Artech House)
[2] Skeren M 2006 *CTU Reports* **10** 1-181
[3] Saxby G 1994 *Practical holography* 2nd ed (Prentice Hall International)
[4] Hariharan P 1996 *Optical holography* 2nd ed (Cambridge, Cambridge University Press)
[5] Svoboda J, Skeren M and Fiala P 2011 *Advanced holography - metrology and imaging* ed Naydenova I (Intech) 209-232
[6] Girnyk V I, Grygoruk V I, Borisov I S and Kostyukevych S A 2003 *Proc. SPIE* **5290** 179-189
[7] Skeren M, Fiala P and Richter I 2006 *Appl. Optics* **45** 27-32
[8] Aggarwal A K, Kaura S K, CHihachhia D P and Sharma A K 2004 *J.Opt. A-Pure App. Opt.* **6** 278-281

Flying with Antares DLR-H2 - From Stereo images to multi view. Image making from the research of the German Aerospace Center (DLR)

Dietmar Öhlmann, MA(RCA), Holographer, 3D artist, Syn4D and Dr Josef Kallo, Group Leader Electrochemical Systems, DLR Stuttgart Germany.

do@syn4D.com www.syn4D.com and josef.kallo@dlr.de www.dlr.de

Abstract. 3-Dimensional scientific illustrations have been used by the German Aerospace Center for more than 20 years. Dietmar Öhlmann has transformed over the years scientific abstract information into visual presentations in hologram, S3D, and M§D media. The latest project Antares DLR-H2, the first emission free flying airplane in the world has been documented in 3D through stereoscopic video. Its research and progress have been documented in two and three dimensional media, a project still in progress.

1. Introduction

The digital hologram of the 4D arrival management system of flight guidance, the holographic animation of the Reichstag in Berlin, a flight with the atoms in a fuel cell for a glasses free multi barrier monitor and the first electric noiseless hydrogen driven plane of the world, the Antares H2 documented with a stereo rig: The German Aerospace Center (Deutsches Zentrum für Luft- und Raumfahrt; in this text used DLR) researches give incredible data base giving access to new vision of our realities, and new visual sensations. The artist and author Dietmar Ohlmann went through an astonishing journey to create the visualization of invisible world of information. From the scientific world into an artistic presentation, we transformed multidimensional data into animated holograms and stereo movie.

2. A timeline of 3D illustration

2.1 Reproduced Time Coordinates

Pic 1 Antares DLR-H2

Three dimensional reproductions give us the opportunity to zoom into complex process of natural science, and understand nanoscopic details. We can now follow an atom of Hydrogen on its way into a fuel cell, see the redox reaction between Anode and Cathode with Oxygen to water, watch how electrons are used to drive an engine, and watch the engine driving a car or even a plane. Modern illustration tool of 3 D gives us the freedom to move from flat plane graphics into space and observe the whole process in time. As one image can explains more than a description with thousand words, a 3 dimensional image of a hologram can show more than one thousand of singles two dimensional graphics.

2.2 Source of images making

The potential of illustrating spatial time, recording/illustration of scientific research at the DLR has been already used to illustrate a 4D arrival management system of flight guidance in 1997. Mr. Peter Gent (DLR Braunschweig)**(1)** used the hologram to illustrate the temporal relationship between space and time in the planning of runway approaches. To transform a multidimensional matrix of numbers with X-Y-Z-and time coordinates into a visual illustration we used the graphic power of an Amiga 500 and the Di-Ho System of Spatial imagine. The illustration engine of the Amiga reproduces the geography of Frankfort airport, the different landing lanes of the planes and very important the exact timing of the landing planes between 9:05 to 9:09 AM. Those images have been sent on a laser lit LCD screen and recorded for an embossed hologram. This hologram then has been printed on foil and placed on the cover of a book to illustrate the functionality of the management flight system. At this time, it was a technical novelty. Later the dot-matrix printer "lightgate" has been further developed by the team of Rob Munday and other companies. Today, digital holographic printers, like the one of Zebra imaging, or XYZ imaging and Geola give possibility to produce holographic images in the size of several meters. Power walls and cave system give feeling of being inside Michael Crichton's roman "Disclosure," and James Cameron Avatar brings back the stereogram into cinema. Media techniques reach a level of satisfaction with the public acceptance, and creative making of professionals. Now the question is: how do to get the content satisfying?

2.3 Stereo 3D for scientific illustration

Looking at scientific illustration, we find out that there are different groups of production, there are illustrations:

- Scientific correct, without esthetic and hard to understand.
- Good looking graphics, but lots of scientific errors.
- Good communication, good graphics and scientific correct.
- This later one requiring a lot of exchanges between the scientist and the artist.

Looking for new content in 3 dimensional illustrations, Dietmar Ohlmann was at first overwhelmed by the mass of information and impressions transmitted by Research Centers (i.e.DLR). The improvement in capturing images with scanning via satellite, plane, car or any other mobile recording system, provides a mass of data. These capturing devices record information from other planets, the earth, cities, and historical important places, like the building of "Schloss Neuschwanstein." When searching our ways in Google Earth, road maps, walking maps we do not imagine the quantity of data it represents. With the data from DLR and RSS,(2) Berlin- Adlershof and Pfaffenhofen (Munich) we can have a virtual walk on the planet mars, fly over the Alps, or Berlin and choose to visit any historical places in Germany. When the artist got the data for the "Thronsaal from Schloss Neuschwanstein" (historical castle), he realized that the depth was so enormous that the whole holographic image was out of focus. So he had first to reshoot them by cutting the building in slices and place the camera out of the building to shoot inside (impossible without Virtual Reality). Then he was rebuilding the room step by step to produce the visualization. Finally the format of the visual has been transformed to fit the printing parameters of the Synfogram(3), a digital holographic recording method. For one Synfogram we use between 1200 and 2000 different stereoscopic images, divided them into pixel and merged those pixels into one holopixel of 0,8mm high resolution and 1,6mm for low resolution and large scale (Geola).

Journal of Physics: Conference Series **415** (2013) 012061 doi:10.1088/1742-6596/415/1/012061

Hologram represents a good media for reproducing illustration of objects, like the "Reichstag in Berlin"(printed at Rabbit Hole), a throne Room of a castle. It is a kind of "hard copy" of the illustration, easy to carry and to show. If we compare the hologram to a stereo movie, the content has to be reduced to 2000 frames i.e. around 7 seconds' animation. Careful! There is no real time, because the animation is dependent of the movement of the spectator and/or the light. The grammar and the vocabulary of stereo recording and holographic imaging are somehow different, like between 2 dimensional graphic and stereo 3D. The choice of a media depends of the target of your communication and the quantity of information you need to pass on.
The German Aerospace Center, Stuttgart, DLR, is working on innovative energy supply systems for aviation asked for my service. The Institute of Technical Thermodynamics at the German Aerospace Center (Deutsches Zentrum für Luft- und Raumfahrt- DLR) (4) conducts research in the field of efficient energy storage systems that conserve natural resources and is expected to supply the next generation energy conversion technologies. The Electrochemical Energy Technology Department works on development of efficient electrochemical energy converters, mainly batteries, fuel cells and electrolysers; their importance for future power systems, both in stationary power supply and in electro mobility, increases continuously. The department's activities range from cell design, manufacturing processes, and diagnosis to system optimization and demonstration. The scientific and engineering challenges of electrochemical storage technology and energy conversion consist of handling the conflicting goals of efficiency, operating life, convenience, safety and costs.

There in Stuttgart an idea was born: produce the first noiseless, zero-emission aircraft and so the Antares DLR-H2 flies successfully on hydrogen fuel cell power alone. The production of a documentary 3D movie of this flight with the first noiseless hydrogen driven plane is one of the latest 3D productions. Dr Josef Kallo, Group Leader of Electrochemical Systems **(5)** at DLR Stuttgart had ordered the documentation of the flights with a JVC stereo camera. He indented to use it for visual documentation and education.

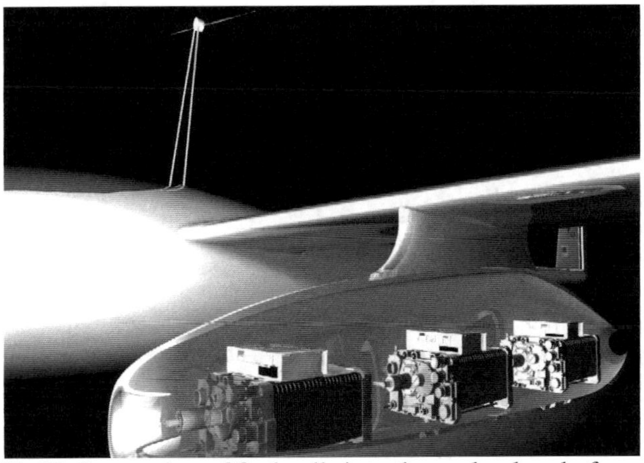
Pic02: Integration of fuel cells into the payload pod of Antares DLR H2

Before being able to see the Antares DLR-H2 flying, we need to understand what is new. A Fuel Cell produces electric energy through the reaction of hydrogen. This is fully noiseless, no sound comes out. We just hear the wind moving around the plane. This is nothing to compare with other reaction, like the explosion of a motor with petrol. And because hydrogen fuel cells only react with oxygen in ambient air, the lone byproduct is water, which has no ill effect on the environment. When flying in a straight line, the Antares DLR-H2 is capable of running at about 52% efficiency, drawing 10kW of power in the process (it's capable of delivering up to 25kW). It can go for roughly 450 miles, flying at speeds of 105 mph.
The final film shows different steps: the stereo recording of the plane Antares DLR – H2 as well as the scientific illustration of the function of the fuel cell. The 3 dimensionality allows us to show the integration of the playload pod of the Antares DLR-H2, what the interaction of producing energy is and how this is used in fly. This documentary presents a high level research between institute and Industry to produce the next generation of aircraft. More than effect and surprises, expected in 3D

Journal of Physics: Conference Series **415** (2013) 012061 doi:10.1088/1742-6596/415/1/012061

cinema, here, the three dimensionality of the film served to enhance its educative components with a virtual flight through the energy process of a flying machine.
The idea, to produce an animation giving a sensation of holographic perception, a feeling of" floating in space", while maintaining a big scale of stereoscopic depth is a big challenge for this ongoing project. Already, to shoot real life recording of a plane with a stereo rigg and to transform them later for a multi panel view presents some difficulties. The stereo-base distance of finished stereo camera, like the JVC GS TD1 or Sony HDR-TD10E, works super for most standard recording situation as used for home video. A professional rigg certainly has more possibilities, but for price –quality relationship those cameras are worth their price. If we want to interpolate the stereo images to fit on a 5 view or 9 view stereo panels, we find that there is still some spatial correction to do. Professional software like the z.l.i.c.e.3D from 3D International can help to do so. Cheap solution does not exist for this task.
The presentation has been planned to be presented at a special environmental "parcour" to a passing by public at the Stuttgart Airport. A stereo system with glasses would have been impossible because the installation should stay for a long period and in a free public access. A large size 3D panel, say a 55/60 inches wide panel, looks great with the flying Antares DLR-H2, but this size of monitor needs a long viewing distance of 5 to 8 meters to see a sharp image, not being into it. So it was not possible to place it in a gangway. To satisfy the spectator in close and far distance, a special 26 inches monitor for medical application has been used.

3. Discussion

An illustration for 3 Minutes video would need the capacity of 5 multiplexed holograms. The multiplexed recording system, first formulated by the Nobel Prize winning physicist Gabriel Lippmann in 1908 is quite difficult to use in real time recording, while stereo riggs and electronic interpolation provide good results.**(6)** The documentation and the illustration of the Antares DLR-H2 show a new need and potential for scientific documentation, education and communication given by 3D media. The floating images of a hologram is finally the most despaired wanted constructed perception, still we have to reach it on the base of physical availability and affordable bases. Legal fights between different patent, like different AVHC Format and incompatible viewing systems still reduces artistic and commercial possibilities. Without those problems, the way for S3D, M3D and digital holography is open and gives us access to new media to illustrate a growing complex world of scientific and artistic image making and discovery. Nevertheless, as fantastic as the image of S3D cinematographic production looks like, do not you think it needs further education of a proper grammar in content production? Should not we prepare all together, artists, scientists, educators, the time where 3D is not just "nice to have", but "useful to read"?

4. Conclusion

The world of 3D is able to give us access to a new perception of the micro- and the macro-cosmos. The most fascinating images are created by natural phenomena, modern instrument are able to capture. There are a few historians who mention, if the history of image making would not have had the strong focus on fine art, the images of the Hubble telescope would be the great art of our time(7). 3D has its fantastic resources, but still the difficulties of its technological imperfections: we need to wear glasses, fight with lenticular distortions or be confronted to external sources of constructing light sources and clipping windows. Still, researchers and visual creators are pushing those media to the outer limit, showing us a world of surprise and fascination. In art, holography is not accepted because it is too technical, in advertisement, it is too imperfect, for technical illustration is a true need but a too slow interface.

References

(1) Prof. Dr. Heinz Winter, Peter Gent, Dietmar Öhlmann 1997 *an intelligent solution: the four dimensional arrival management system of flight guidance*. (The first synthetic 4D hologram for the DLR flight Braunschweig)

(2) Dipl. – Geol. Frank Lehmann, Leiter der Struktureinheit Sensorkonzepte und Anwendungen 2007. Using 3D capturing and mapping to create a true colored hologram of Berlin Reichstag.. DLR Berlin. Synthetisches Hologramm

(3) Odile Meulien Dietmar Öhlmann 2007 *Synfograms: a new generation of Holographic Applications* (Proc. on CD Holopack-Holoprint, Hong Kong)

(4) Dr. N Wagner P Szabo M Lang D Öhlmann *Stereo Illustration of energy storage and extraction*. Video Production 2010 (DLR Stuttgart, Institut für Technische Thermodynamik, Elektrochemische Energietechnik. 5view)

(5) Dr. Josef Kallo, Projektleiter D Öhlmann 2011 *Stereo recording with 3D camera: Antares H2, the electric plane. „Antares"(* DLR Stuttgart, Institut für Technische Thermodynamik, Elektrochemische Energietechnik. Stereo recording and multi channel conversion for 5-view presentation). The film is in permanent presentation at Stuttgart Airport, terminal 3, on a glasses free stereo panel. DLR Press.

(6) Odile Meulien Dietmar Öhlmann 2008 *"Syn4D displays for a new generations of Holographic Applications"* (Proc. SPIE Europe, Strasbourg. Vol 7001)

(7) James Helkins, art historian, School of Art, Chicago, 2009, *"Masslose Bilder" I.Reichel, Siegel,* Wilhelm Fink Verlag.

Images

Pic 1: Antares DLR-H2, ©DLR 2010
Pic 2: Integration of PEM-FC, ©Syn4D2012, research by DLR, Lange and HYGS
Pic 3: Design of a 4D Hologram for embossing, DLR 1997, ©Dietmar Öhlmann

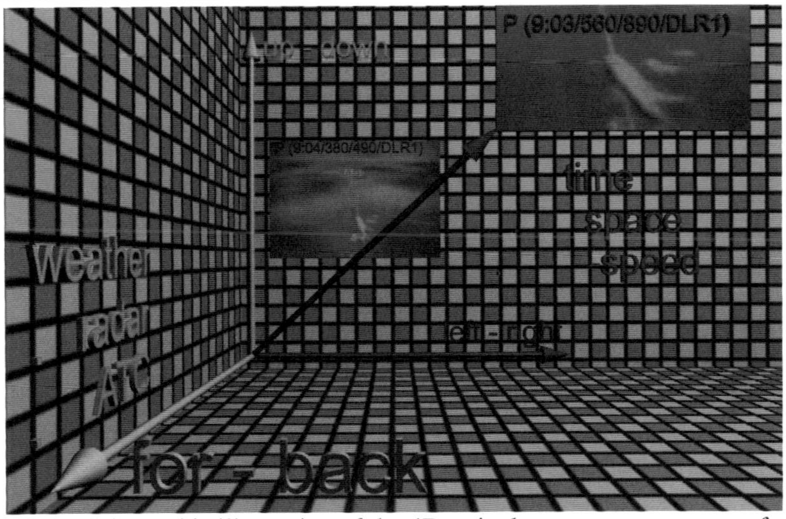

Pic03: Holographic illustration of the 4D arrival management system of flight guidance, printed as Hologram by the DiHo System, 1997.

Art-science, beauty-reason and holography

T H Jeong
Professor *Emeritus*, Lake Forest College, Department of Physics,
Lake Forest, IL 60045 USA

E-mail: tjeong@aol.com

Abstract. Display holography holds a distinction that makes it appealing to a wide audience. It can be appreciated at a deep level by people of all ages and in all fields of endeavor. It provides a unique opportunity for us to gather in an intimate location to learn, enjoy, and enlighten one another. This paper offers demonstrations to explore the relationships between art and science, esthetics and mathematics, and the dualities that exist in nature. On the practical level, a visual model for deep understanding of holography and a proposal for "making holograms that sell" will be presented. In writing this article, the author acknowledges the fact that for this symposium, a Proceeding will be published as well as a set of audio-visual recordings. With that in mind, this article represents largely the printable contents, leaving the audio-visual part as "performance" to be electronically recorded.

1. Introduction

It is my honor and privilege to give the "welcome" paper to the participants of the Ninth International Symposium on Display Holography (ISDH). Three years ago, this symposium took place in Shenzhen, China, where I delivered the opening paper entitled "Teaching Advanced Physical Concepts Using Display Holograms [1]." Today, I wish to continue on that theme but broaden it to include some relevant personal histories along with a "geometric model [9]" that is used to explain holography to a general audience. The "live" oral presentation will include material not possible to replicate here in print.

The most noticeable difference between this gathering with other symposia and conferences is the broad inclusion of artists, scientist, entrepreneurs, and other people with diverse educational backgrounds, personalities, and philosophies. I wish to stress the similarities amid this diversity, between art and science, and between beauty and reason.

For those interested in the business of display holography, it is disheartening to realize that in spite of the years of attempts and expectations, the market has not really caught on. I wish to offer my opinion on this subject and a possible solution.

Note that I have avoided the use of the word "truth" in the title. The meaning of the word has lately been contaminated by the media and we have developed doubts about its "truthiness." On the other hand, 2+2=4 is still accepted universally. It is "reasonable."

2. Why I am a teacher and who taught me how to teach

Three weeks before this symposium, my Yale University class of 1957 held its 55th reunion with a special theme. We were requested to voluntarily submit a "Life Report" that defines us as to who we are. A panel of distinguished writers held a forum to publically discuss this collection and possibly assemble it into a book form.

My own "Life Report" relates not only to "Why I am a Teacher", but in particular, a teacher of holography. I therefore include an expanded version of this report.

2.1. Why I am a teacher – a Life Report

Back in 1947, I was one of about 20 children from ages 6 to 12 and older, fortunate enough to attend school in a village near Guangzhou (Canton), China. Because of droughts, floods, plagues, war, and revolution, we never

Published under licence by IOP Publishing Ltd

had a single year of uninterrupted attendance. In one room, we had one teacher who taught the equivalence of first to eighth grades all at once. I never learned his name because everyone just called him "Teacher."

He was a poet and the most respected person in the village. We relied on him to help resolve all conflicts, public and personal. He even dictated that the pronunciation of certain words be changed so that they rhyme in Tang poetry. Most of the lessons consisted of reciting poems in unison in a sing-song manner. For art, we had calligraphy, and for math, we learned the use of the abacus.

One day, he fell ill and asked me to take over the class. I remember it well. We were studying an epic poem about an emperor and his favorite concubine and had a great deal of subtlety. I had to go over it word by word and found that there were many parts I could not understand. We agreed to ask Teacher the next day.

We did so. Teacher told us that it was fine not to understand. But if we memorized the poem and recite it to ourselves when we become older, we would come to understand it eventually. That weekend, I encountered a group of my classmates in the market place. They stopped, bowed to me in unison, and addressed me as "Little Teacher."

I was deeply moved. It was at that moment I made a life-long decision to become a teacher.

An earlier event was responsible for my interest in science in general and nuclear physics in particular. It was the ending of the Japanese occupation of our village in 1945 by two "atomic" bombs.

The choice to accept the first and only job in 1963 at Lake Forest College was based on its village-like location and the broad educational experience it offered to a relatively small number of students.

It was immediately apparent to me that nuclear physics was not a suitable subject for teaching undergraduates. Research had to be performed at an up-to-date accelerator, which was the isochronous cyclotron at the Oak Ridge National Laboratory 500 miles away. I tried to convince my wife that we could treat the summer trips as vacations near the Smoky Mountains. The commuting during the academic year was extremely difficult, however, especially after our first child was born in 1964.

Oak Ridge was a center for refining uranium for the "bomb". Because I was not a US citizen at that time (my legal status was Displaced Person), I was prevented from access to most areas outside of academic research at the cyclotron. It helped that George W. Clement, then Governor, signed a statement declaring me an Honorary Citizen of Tennessee. I was permitted to visit laser researchers at Oak Ridge and subsequently learned to build helium-neon lasers at Lake Forest College. That turned out to be a critically important experience.

Lake Forest College had a national reputation in developing new physics experiments for undergraduates. It received a government grant to host a summer conference on this subject in 1965. In searching for new topics, I discovered the publications on holography by Emmett Leith and Juris Upatnieks [2,3]. I wrote to Leith and asked for a loan of an example to be shown at our conference.

To my surprise, he not only sent me a 4x5 inch hologram of chess pieces, he asked me to keep it. His generosity literally changed my life.

Immediately after I learned how to view the hologram and studied the basic theory involved, I realized that this subject could be applied to a wide spectrum of topics from technology to art. Being visually appealing, it would be possible to teach the subject at any level to a broad audience, including artists.

I also questioned the format of the 4x5 inch hologram: If this is a 3-D medium, why are we stuck with a 2-D flat format like paintings and photographs? I immediately worked with two first-year students and published a paper entitled "360° Holography [4]" in 1966 followed by another paper called "Cylindrical Holography [5]" in 1967. This opened to me the flood gate of interests from the commercial world. I started to get grants [6], made two educational films [7,8], and later received a generous donation of an entire professional laboratory on holography and fiber optics from the Newport Research Company then led by Milton Chang. My dream of switching from nuclear physics to holography was finally realized!

During my first sabbatical leave in 1969-70, I was appointed by the NSF supported program, the Chautauqua Lectures, to tour the US on twelve university campuses, coast to coast, from Stanford to Maryland. The objective was to teach the new subject of holography to other physics professors and have them spread the new frontier. Each university had a center to recruit twenty-five other professors from nearby colleges and universities to gather for a two-day session in the fall, and another two-day session in the spring for feedback. To my knowledge, it is a broad program that is continued today, and includes many areas of new developments in science and technology.

On my second sabbatical leave in 1976-7, I was appointed Adjunct Lecturer by the School of the Art Institute of Chicago. I created a geometric model [9] for teaching holography to artists through visualization and without mathematics. During this same sabbatical, the Chinese government invited me, my entire family of five, and my wife's parents, to a month-long lecture tour in Shanghai, Beijing, and Guangzhou.

In recent years, I have become increasingly interested in reaching out to a younger audience. My latest "gig" was accepting an invitation from the Lake Forest Country Day School to be their inaugural Scientist in Residence program on April 27, 2011. After giving lectures with demonstrations to all the first to eighth graders in the morning, I taught a kindergarten class of about 40 children in the afternoon.

This *déjà vu* moment reminded me of the time sixty-four years ago when I took over a class under one roof to teach children of all ages in a village in China, and how I was inspired to become a teacher for life. On that revelation, I decided to retire completely from paid teaching appointments.

I cannot end this life-report without mentioning an incident involving an interaction with my grandchildren. Three summers ago when little Anna was three and Ian was six, I thought the time was right to take them to my lab and show them holograms.

At the entrance to the science building at Lake Forest College, they suddenly became really excited and yelled, "Rocks!" What they saw were garden variety pebbles of difference sizes and colors around the bushes and trees. They were jumping around, picking, inspecting, and could not stop playing with them. I had to pull them into my lab and finally showed them the holograms.

But it was clear that the highlight of their visit were the rocks.

Now, every time I go to my lab, I have to pass by these damned rocks. They remind me of the lesson I learned from the children, which has helped me to keep my feet firmly to the ground.

If only I were a poet, I would have written an "Ode to the Rocks."

2.2. My secret physics teacher – Richard P. Feynman

By sheer luck, my employment at Lake Forest College in 1963 coincided with the publication of the three-volume Lectures on Physics by Richard P. Feynman [10]. In fact, I was able to obtain mimeographed copies as early as 1961 from friends in Caltech. I was completely overwhelmed by the clarity and elegance of his expositions on every topic that I was to teach. But they did not include any homework problems, thus could not be used as textbooks. Therefore, I chose an existing well known standard textbook for students, and used the Feynman treatments for preparing my lectures and demonstrations. I kept this fact a secret from students otherwise they would anticipate my presentations and encouraged to skip classes.

I was so obsessed with Feynman's charm and wisdom on every topic that I not only read all books I could find written by and on him, but also watched the 16mm films of his lectures produced at Cornell University before he joined California Institute of Technology. I imitated his lecture style - roaming the floor with chalk in hand, and even the way he told jokes.

The ultimate example of his pedagogic genius is the 152-page book on quantum electrodynamics, entitled QED [11]. Without a single equation, he clearly explained this advanced subject in graduate level physics using the now universally accepted "Feynman diagrams." It was this *tour de force* that continued to inspire me.

At Lake Forest College, as far back as I can recall, the Physics Department has made it a tradition to give each physics major at commencement a complete set of the Feynman Lectures - a hefty gift both in bulk and in cost. This year, we had ten recipients. Thank goodness, the set is now available in DVD format.

I surmise that countless physics professors share my secret. We all appear smarter in doing so, thanks to Richard P. Feynman.

2.3. My not-so-secret teacher on holography – Joseph W. Goodman

As I have written in Section 2 above, I was appointed to teach a Chautauqua Program on holography during my 1969-1970 sabbatical leave. It was the most daunting experience of my teaching career, knowing that I was to teach colleagues most of whom are older and probably smarter than me. I was frantic in making preparations, and soon realized that serious readings of the Gabor and Leith-Upatnieks publications were not enough.

Then I found Prof. Joseph W. Goodman's book Introduction to Fourier Optics, now in its 3rd edition [12]. That gave me all the "ammunition" I needed, never mind that Stanford was one of the twelve universities in my program.

During the following sabbatical leave in 1976-7, the Chinese government invited me to a lecture tour that included Beijing, Shanghai, and Guangzhou (Canton). I knew only Cantonese and had to lecture in English, leading to great problems during my first program in Beijing. The interpreter appointed to translate for me had great difficulty with the technical vocabulary and phrases. In the audience was Prof. Hsu Da-Xiong from the Beijing University of Post and Telecommunication, who understood everything I was saying and volunteered to take over as translator.

It was surprising to me that Prof. Hsu's conversational English was quite poor at that time, but he was fluent in technical English, especially on the subject of Fourier optics. He later told me that he had been teaching from Goodman's textbook for years, solved every problem in it, and practically memorize much of the book. That's how he had learned English! Goodman was to him was like Feynman was to me!

Professor Hsu, who soon taught holography to students who later became the founders of many of the holographic label companies in China, is often honored as the "father of holography" in China.

2.4. The Geometric Model for Holography.

As I had mentioned in Section 2.1, during the 1976-7 sabbatical, I was appointed as Adjunct Lecturer at the School of the Art Institute of Chicago.

Here I had the opposite problem encounter with the Chautauqua program – I was to teach artists, most of whom did not have technical training in holography, and some of them downright abhorred technology and mathematics.

At that time, long before the publication of the book QED, the Feynman diagram was already famous. I thought: if Feynman can come up with simple diagrams to explain this difficult subject in physics, I should be able to do something similar on the particular subject of holography. The result was my publication of the article "Geometric Model for Holography. [9]" The reader is respectfully asked to consult the publication directly.

3. Why Big Display Holograms Don't Sell

3.1. The business of holography

My first and last interest in holography has always been its beautiful concepts. In addition, because of its visual appeal, it lends itself to become an instrument for scientific education to a broad spectrum of students. That is not to say that I have no interest in making a business of selling holograms for profit. Even in doing so, I sold holograms and the material for making them strictly for educational purposes [14].

From the moment Leith and Upatnieks published their work along with Denisyuk's discovery of Lippmann holograms [17], countless entrepreneurs have set up businesses to market them as novelties, advertising, and of course, fine art. Very few were able to make it a successful enterprise, with only one exception. Stephen A. Benton's invention [15] of "rainbow" holography drastically reduced the band-width and allowed holograms to be mass produced by embossing. This opened up a multi-billion dollar business around the world. Let us thank him and all those who have kept holography alive.

3.2. Why deep-scene holograms don't sell

In my opinion, there is a very simple reason why people don't buy beautiful, big, deep scene transmission or reflection holograms, even with full color: **It is too cumbersome and expensive to illuminate them.**

Large holograms need large space to display them. For a transmission hologram, much space is needed behind the piece. If it requires laser illumination, safety rules forbid outputs over 5 milliwatts. For a reflection holograms, a viewer trying to get a close look will be burned between the lamp and the hologram. Placing a hot lamp next to the hologram is not a solution.

3.3. Possible solutions

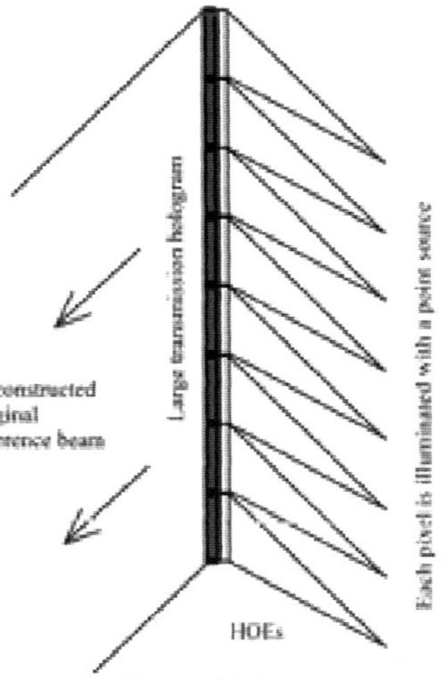

Figure 1(a)

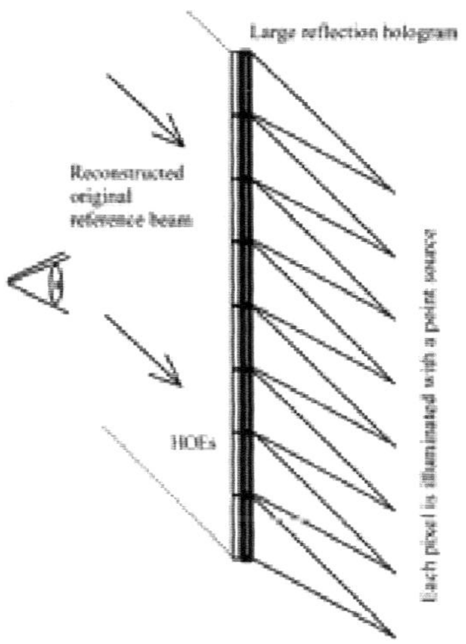

Figure 1(b)

In 2001, I presented a paper [16] proposing a scheme that can universalize display holograms into one single modular design that can meet the following goals:

- Safety from heat and radiation
- Bright enough for outdoors
- Thin, independent of hologram size
- Fixed alignment
- Free form – display in any shape
- No size limit

The basic concept of the Pixelated Holographic Display (PHD) system is to illuminate the hologram from behind with a panel of small lasers or light from the tips of fiber optics. Figure 1(a) shows a basic tile of a transmission hologram setup. It is a 30x30 cm section being illuminated 2 cm behind by an array of small diode lasers each without a collimating lens. Each individual spreading beam is directed upward at 45° at a 1 cm^2 area of a corresponding HOE in contact with the hologram. The HOE converts the spreading beam to a collimated beam going through the hologram downwards at 45°. As a whole, the entire hologram is illuminated by the equivalent of one single collimated reference beam.

Figure 1(b) depicts a similar concept for a tile-sized reflection hologram with a sheet of HOE on the left (viewer) side. The light from the single-mode fiber optics tips, fed into the fiber bundle by a single white-light or laser source far away, passes through the transparent hologram at the "wrong" angle, thereby suffering no diffraction. It then encounters the HOE that is designed to convert the spreading incident light into a reflected collimated beam going downward at 45° back through the hologram. Together, all the light from the fiber tips go through the hologram and is transformed into a single collimated light directed at the hologram from the viewer's side. In this fashion, a viewer can touch the hologram without blocking the reconstruction light.

The collective result satisfies all the bulleted points listed above. In addition, we have avoided the classical Rayleigh resolution criterion because it does not apply to Gaussian beams or any beam that fades out at the edge of an aperture.

Thus far, we have assumed fixed holograms made on "classical" materials like silver-halide or photopolymers by currently existing methods. However, PHD will apply to future "live" holograms that are composed *in situ*, and in real time. When that day comes, a PHD-tiled wall can serve as a living room "window" conveying 3-D motion pictures captured by an array of micro-cameras broadcasting from a submarine moving through an ocean or set up on top of a beautiful mountain.

Note that nothing above violates any fundamental physical laws. Therefore, this proposed scheme is "possible". Let us hope that with the rapid development of nano-technology, the dream of achieving this feat might be realized in one or two decades.

4. Abstract Symmetry and Equations as Poetry

4.1. An example of abstract symmetry

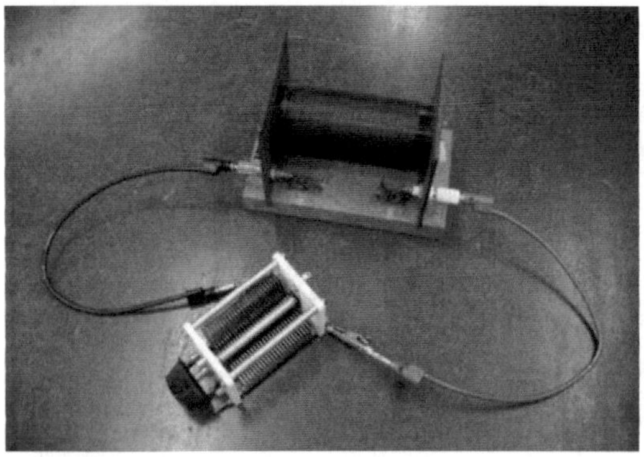

Figure 2a

Figure 2b

Observe the two figures above:

What is so "symmetrical" about them? Holographers will recognize that Figure 2b shows components of a spatial filter. There is a microscope objective, a pinhole and the adjustment screws to direct a laser beam and focus it through the pinhole.

What is in Figure 2a? It is an "inductor" made of a coil of wire, which is electrically connected to an adjustable "capacitor" made of a stack of parallel aluminum plates. The knob on the latter allows one to "dial" one row of plates into the other, varying the parallel area between the two stacks, and controlling the capacitance. This "resonant circuit" allowed us to "tune" to a radio station of our choice in the vacuum tube days.

Now imagine that the spatial filter had not been invented. How would you have come up with the idea?

Here's one way we could think of it, the explanation of which will illuminate the beauty and value of "symmetry" in science. If you expose a ring of wire to the broadcast radio waves, it will respond to all electromagnetic disturbances including thunders, spark plugs firing in car engines, and other "noise" and "cross talk" from all radio stations. To obtain good reception, one needs to invent high quality components to provide a "narrow bandpass" filter so that you can fine-tune to a desired channel and block out all of the "noise".

When a laser beam is directed by mirrors and then spread out onto your holographic plate by a lens, all of which have dirt on them, the resultant light has many distracting patterns on it. We holographers call that "noise". Using the same concept of providing a "narrow bandpass" filter in radio, we focus the beam through the pinhole so that only the original parallel beam from the laser can pass. Light that scattered off dirt does not focus through the pinhole and is blocked. That is how we "clean-up" the beam. Thus, it is a spatial filter.

Therefore, the invention of a spatial filter is accomplished by recognizing the "symmetry" between an event in the time domain and translating it into an equivalent event in the space domain. It is a beautiful symmetry between space and time.

All physical laws are fundamental; and they manifest their consequences in subtle ways when applied to difference situations. Knowing this abstract "connection" is the reward of studying science.

The 1953 Nobel Prize was awarded to Fritz Zernicke for inventing the phase contrast microscope, thereby revolutionizing biology. He used the opposite concept to the spatial filter when he placed a black dot where our pinhole is, thereby blocking the original light that illuminates transparent specimens, allowing only light that is affected by the specimen to come through. In the history of science and technology, many such beautiful ideas were discovered by recognizing the reason in symmetry.

4.2. First: A confession

It is a common notion that artists in general, are "afraid" of mathematics. In fact, many artists admit this characterization themselves. I wish to recount a humbling personal experience which offers a counterpoint to this conception.

At the beginning of my tenure as an Adjunct Lecturer for the School of the Art Institute of Chicago during my sabbatical year in 1976, I was offered a guided tour of the masterpieces in the world-famous collection. I mentioned to the appointed guide that my family and I had visited the museum several times and had "seen" the most famous works, and that I really want to learn more about "modern" art.

She led me into a large room with building material and equipment spread around the floor, apparently in the process of remodeling. We stood there for what seemed to be a very long time. I finally reminded her that I was ready for the tour.

She politely pointed out to me: "Sir, we are standing in a new "installation" by (name I have forgotten)". She then showed me work by others (names I have also forgotten), and finally the large works by Jackson Pollock. Ah-hah!! I showed-off by mentioning that I have seen a movie about him.

To artists: we are even! I am just as befuddled about modern art as you might be about the theory of holography.

4.3. Equations as poetry written in an international language

We all know that many great poems don't need to rhyme. And of course, most poems are written in languages we don't know. Therefore, we should acknowledge that mathematical equations that induce awe, wonder, and inspiration can be written in the truly international language of mathematics. Let us take some examples.

First:

$$E = mc^2$$

To me, this equation brings joy, excitement, pain, relief, and grief all at once. It recalls the moment when my adopted mother and I were hiding in caves during the Japanese occupation of our village. The sudden ending of the war gave us a new life. Then I learned it was the two "atomic" bombs dropped by the United States that ended our miseries, along with the massive death and destruction in Japan.

I was greatly mystified by the awesome power of these bombs. I tried to read about it from every source I could find, and learned that our sun uses this method to create the energy that makes life on earth possible. I was so impressed with this process that it later lead me to the study nuclear physics for my Ph.D.

Second:

$$\mathbf{F} = m\mathbf{a}$$

This is Newton's Second Law of Motion. It reads: "the total force **F** acting on a body of mass m causes it to accelerate at the rate of **a** along the direction of the force. This equation "explains" everything from electrons going around the nuclei to planets going around the sun and the motion of every mass that is subjected to a force.

It has a particular meaning to me: it gives me euphoria!

When I think of this equation, I think of windsurfing! Did you know that with a steady wind speed of 12 mph, I can be sailing at 15 mph? Seems impossible? Not so, says Newton.

If I sail in a direction perpendicular to the wind, the component of the wind force along the direction of motion will be constant regardless of my speed. Thus, if this were the only force, I would accelerate toward the speed of light. However, **F** means the "total force", and it includes the "frictional force" that is always exactly in the opposite direction of the motion. The frictional force increases with the speed of the mass. Therefore, the total forces will decrease with the net speed of the mass, until it reaches a "terminal speed" when the frictional force is exactly equal and opposite to the component of the wind force in the direction of motion. The concept is exactly analogous to any mass in a free fall in the atmosphere– it accelerates until the frictional force equals (and is opposite) to the gravitational force.

Thus, my terminal speed in windsurfing is completely dependent on my skill and the design of the equipment, as long as there is a steady wind at any speed. To experience the wind at my back while appreciating the laws of nature at work adds to the awe and thrill of the whole experience. It is poetry in motion!

I can go on and apply this Newton's second law of motion to skiing! Skiing is the art of falling down a beautiful mountain ----------. Well, I can go on and on, but have to stop myself here due to space limitation!

There are several ways we can write down mathematically the relationship between the creation of a hologram and the reconstruction of its image. To me, the most "beautiful" way is to express them is as a "Fourier transform pair". I am writing them in one dimension, x, for simplicity.

$$\mathbf{g}(x) = \int_{-\infty}^{\infty} \mathbf{G}(u)\, e^{i2\pi xu} du$$

$$\mathbf{G}(u) = \int_{-\infty}^{\infty} \mathbf{g}(x)\, e^{-i2\pi xu} dx$$

Here, $\mathbf{g}(x)$ stands for the hologram; x locates a point on the hologram, $\mathbf{G}(u)$ stands for the image of the object; and u stands for the spatial frequency of the interference pattern formed between the reference beam and light from a point on the object. Thus a hologram is created by adding up all the interference patterns caused by the interference of a reference beam with every object beam; and the image of a hologram is formed in your eye by adding up all the beams diffracted by each interference pattern recorded. In short, one is the summary of the other. Each is complimentary to the other. If one is real, so is the other.

Without any regard to the mechanism of mathematics, **just look at the equations**! There is perfect symmetry. The hologram and its image are made of each other, and in the same way.

If the earth were visited by beings from a faraway galaxy that were far more advanced than we, who no longer require eyes or ears for making observations, then they would "see" our holograms and the objects recorded in them as one and the same thing.

Finally, in university bookstores as well as in technical expositions, you may often find printed on T-shirts and shopping bags the following "message":

In the beginning, God said:

$$\nabla \cdot \mathbf{E} = \frac{\rho}{\varepsilon_0} \qquad \nabla \times \mathbf{E} = -\frac{\partial \mathbf{B}}{\partial t}$$

$$\nabla \cdot \mathbf{B} = 0 \qquad \nabla \times \mathbf{B} = \mu_0 \mathbf{J} + \mu_0 \varepsilon_0 \frac{\partial \mathbf{E}}{\partial t}$$

And there was light.

There is no better way to describe "everything" about light than these four Maxwell's Equations on electromagnetism.

(At this point, let me remind my readers who are artists that I am not taking revenge for having suffered the humility of being in the middle of a "performance" art piece and not knowing it.)

In the way in which a graphic artist can perform outstanding feats on a computer without knowing the details of how each electronic component works, we can "read" these equations and get to understand their meanings without doing calculations. Michael Faraday and Thomas Edison had something in common - neither of them had training in mathematics. They were able to "visualize" concepts that led to great inventions.

Let us just "look" at these equations. There is something ugly about them! Compared to the beautiful symmetry of the Fourier transform pair for holography, these are full of asymmetries.

I have already discussed the asymmetry between the two equations on the left side in a previous paper(1). The equation on top left reads: "the divergence of the electric field at a point is proportional to the total "electric charge density at that point." The equation below it reads: "the divergence of the magnetic field at a point is equal to zero." We can understand this based on our current knowledge: If there is a net electric charge at a point, we can visualize electric field lines going into or out of the point. We accept for a fact that for each magnet, large or small, there is always a north-pole and a south-pole. Thus, every magnetic field-line we can imagine must be a continuous loop, never beginning or ending at a point.

But nature must not favor **E** over **B**! We demand that the divergence of **B** MUST NOT BE ZERO. Therefore, we conclude that there must exist magnetic monopoles, i.e., north and south poles of magnets must exist alone just like positive and negative electric charges! We just have not found them yet.

It is through this kind of symmetry arguments many now known particles were discovered. The anti-particle is a prime example.

Now onto the pair of equations at the right: they are even uglier! The top-right equation reads: The curl of the electric field at a point is equal to the negative of the rate in which the magnetic field is changing at a point. The bottom equation reads "The curl of the magnetic field is proportional to the electric current density **J** at the point plus the rate at which the electric field is changing at the point! Why the extra term with **J** in it? This discrepancy can be remedied using what we did with the divergence equations: if we allow the existence of magnetic monopoles, then there would be a magnetic current density; and both equations would each have two terms on the right side. Now we have perfect symmetry for both the divergence and curl equations.

In free space, there are no net electric or magnetic charges. Thus all four equations will have only one term on each side of the equal, in perfect symmetry.

Michael Faraday, through this kind of "thought experiments" and without mathematical calculations, invented the **motor** and the **electric generator!**

It gets better: we now can state the following absolutely astonishing facts: In free space where there are no electric or magnetic charges, any change in the electric field around an area of space generates a magnetic field through it; and any change in the magnetic field through an area of space generates an electric field through it. In this way, the electric and magnetic fields resurrect each other through free space until they encounter a flower. There, they transfer their energy to make it bloom.

Next time you see a sunset, know that somewhere on the sun an electron changed its orbit causing a shift in its electric field, which generated a magnetic field, which created what we call "light," and began a long journey through space until it reaches your consciousness.

If only I were a choreographer, I would compose a ballet with a magnificent *pas de deux* between **E** and **B** across the celestial stage of life.

Acknowledgments

I have already acknowledged my gratitude to "Teacher," Richard P. Feynman and Joseph W. Goodman in Section 2. I belatedly wish to thank Milton Chang for his generous donation of Newport Research equipment which allowed advanced topics to be studied. I want to thank Bailey Donnally, my classmate in graduate school, for having introduced me to Lake Forest College and helping to hire me in 1963. He had begun working at LFC in 1961, and developed new experiments for the sophomore Modern Physics laboratory in relativity and mu-meson half-life measurements when I arrived. I wish to thank Michael Kash, who was my student in Modern Physics and Physical Optics courses in the 1970's. He studied and completed his Ph.D. at MIT, and "came home" to LFC where he is the current Chair of the Physics Department. He and the current staff provide many able "grand-students" for me and Bailey to do volunteer teaching. I am especially grateful to Mike for giving me valuable comments and advice in writing this paper. Finally, I wish to belatedly offer my most solemn gratitude to the greatest of all benefactors in my professional life, Emmett Leith. Without his generosity in giving to me a historic hologram and his sustained support for all my professional activities that defined my career, my life would have taken a completely different path.

References

[1] Jeong T H 2009 Teaching advanced physical concepts using display holograms; Holography-culture, art and information technology *Proceedings of the eighth international symposium on display holography* eds T H Jeong, F Fan and D Wang (Beijing: Tsinghua University Press) pp 39-50
[2] Leith E N and Upatnieks J 1962 Reconstructed wavefronts and communication theory *J. Opt. Soc. Am.* **52** 1123
[3] Leith E N and Upatnieks J 1964 Wavefront reconstruction with diffused illumination and three-dimensional objects *J. Opt. Soc. Am.* **54** 1295
[4] Jeong T H, Rudolf P and Luckett A 1966 360° holography *JOSA* **9** 1263
[5] Jeong T H 1967 Cylindrical holography and some proposed applications *J. Opt. Soc. Am.* **11** 1396
[6] My first grant was from the Gaertner Scientific Corporation in 1968 to produce a "portable" educational system for holography (the Gaertner-Jeong Holography System). Its founder in 1896 was Wilhelm Gaertner, a technician who build the interferometers for Albert Michelson (Nobel Prize in physics, 1907), and the company was owned by the University of Chicago.
[7] Jeong T H 1970 Holography – A Window to the Future (film) Veterans Administration, Medical Division, Washington DC
[8] Jeong T H 1972 Introduction to Holography (film) Encyclopedia Britannica Educational Corporation in collaboration with Albert Baez
[9] Jeong T H 1975 Geometric model for holography *Am. J. Phys.* **43** 714
[10] Feynman R P, Leighton R B and Sands M 1963-5 *The Feynman Lectures on Physics* Vols. I-III (Reading MA: Addision-Welsey)
[11] Feynman RP 1985 *QED – The Strange Theory of Light and Matter* (Princeton NJ: Princeton University Press)
[12] Goodman J W 2005 *Introduction to Fourier Optics* Third Edition (Englewood CO: Roberts and Company)
[13] Jeong T H 2009 The omnigram for teaching holography eds T H Jeong, H Bjelkhagen, F Fan and D Wang, *Proceeding of ISDH-China* (Beijing: Tsinghua University Press) pp 104-107
[14] www.integraf.com
[15] Benton S A 1969 On a Method for Reducing the Information Content of Holograms *J. Opt. Soc. Am.* **59** 1545
[16] Jeong T H 2001 Large-scale self-lit holographic display systems *SPIE Proceedings of Practical Holography XV and Holographic Materials* eds S A Benton, S H Stevenson and T J Trout **4296** pp 38-46
[17] Denisyuk Y N 1963 On the reproduction of the optical properties of an object by the wave field of its scattered radiation *Opt. Spectrose* (USSR) **14** 227

An easy physics outreach and teaching tool for holography

T. Voslion, A. Escarguel
alexandre.escarguel@univ-amu.fr
Maison des Sciences de l'Université d'Aix-Marseille,

Abstract: in the framework of scientific outreach at the "Maison des Sciences" of the Aix-Marseilles University, we created a teaching kit for holography contained in a small case. It includes all the required equipment to produce holograms almost anywhere with a simple optical assembly with a very good vibration tolerance. The fundamental principles of holography and several applications are illustrated through simple experiments: reflection Denisyuk holograms, angular multiplexing, notch filters, holographic interferometry and diffraction holographic gratings. It is possible to use this tool for several purposes: science outreach, teaching for undergraduate and graduate students and continuing education. In this article, we explain the basis of holography, how the kit works, and give some applications and results that can be done with it.

1. Introduction

D. Gabor, a Hungarian-born British physicist, invented the principle of holography in 1948 while he was looking for an electronic microscopy resolution enhancement [1]. He received the physics Nobel Prize for is work on holography in 1971. The three-dimensional pictures are produced with a coherent light source (laser) and are particularly realistic. By using three lasers, one for each of the wavelengths of the primary colors, it is impossible to differentiate the real object and its hologram, provided that the hologram is properly illuminated. Holograms intrigue and allow addressing many physical principles of varying degrees depending on the audience: laser properties, colors, coherence, interference and interferometry, HVD (Holographic Versatile Disc), etc. However, the technics to produce holograms are usually very sensitive to parasite vibrations. According to many books, it is almost impossible to produce a hologram in an environment where all sources of vibration are not perfectly controlled. Depending on the type of holographic device, a movement of a few tenths of micrometers may be enough to fail the experiment.

The kit for holography presented in this paper allows making very good quality holograms (Fig. 1) almost anywhere, if the room is dark enough and has an electricity access. The assembly is simple and optimized to be extremely robust with respect to vibrations. No need to use any device to palliate parasite vibration sources. The technics is based on reflection holograms invented by Y. Denisyuk [2], which was inspired by G. Lippmann work on interferential color photography [3]. The holograms are observable in white light and illustrate several physics principles like holographic interferometry, angular multiplexing used in HVD (Holographic Versatile Disc), "notch" filters (Raman spectroscopy) and diffraction holographic gratings (passive spectroscopy). After a short introduction on the principles of holography, we present the device included in the kit and the experiments possible to do with it, depending on the audience.

Figure 1 : A reflection hologram produced thanks to the holography case with a holographic plate Ultimate [4]. Typical exposure time: 3s.

2. Hologram production principle

Holography is based on a simple principle: the interference between two coherent light beams. On Fig. 2, the beam from the laser L is divided in two parts by an optical system S and spreads out over the object O as well as on the photosensitive plate P. The "Object Beam" OB is the beam obtained by the diffusion/reflection of the light on the object, while the "reference beam" RB comes directly from the laser. The mean angle between the object and reference beams is $\boldsymbol{\theta}$. This angle is important for angular multiplexing principles (part 4). Interferences between the two beams OB and RB occur only if their path difference is less than the coherence length. This particular length correspond to $L_c=c/\Delta\nu$, where c is the speed of light, and $\Delta\nu$ the spectral width of the laser. The experimental setup used in the holographic case only needs a coherence length larger than 1 cm. By construction, the low power HeNe lasers we use have a coherence length much larger than this.

When the object is a simple front surface mirror, a periodic modulation of the light intensity is recorded on the photosensitive plate due to the phase shift between RB and OB. If the object is not plan, beam wave front of OB is deformed, and the resulting interference pattern becomes more complex. The fundamental idea of holography is to record this phase variation relatively to the reference beam RB. Once developed, the hologram illuminated with the reference beam diffracts light just like the object. Then, unlike a photograph, a hologram contains information on the object relief. For diffusive object, it is important to note that there is information redundancy, because every point of the object sends back laser light on a large part of the photosensitive plate. People interested in a complete description of holographic technics should refer, for example, to G. Saxby's book [5].

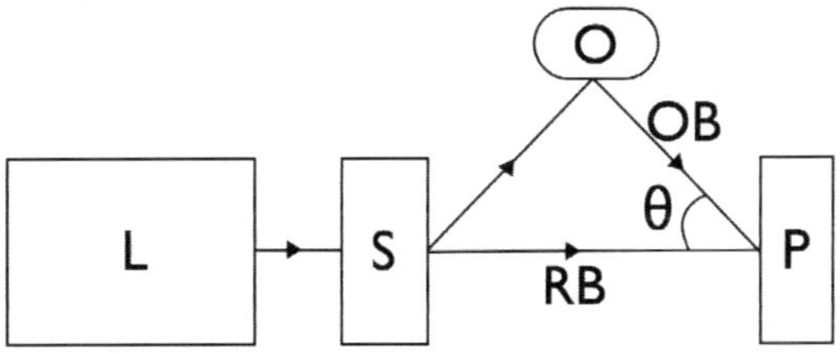

Figure 1: Usual scheme of hologram producing principle. L: Coherent light source; S: Optical system; O: object; P: photosensitive plate; RB: Reference Beam; OB: object beam; **θ**: mean angle between RB and OB. The two beams are enlarged by means of lenses not indicated in the figure, so that they cover the whole object and plate.

3. Teaching kit for holography description

The teaching kit for holography has been developed in 2010 after the CETSIS congress [6] dedicated to didactic experiments. The dimensions of the holographic case are 34cm×25cm×16cm, and includes absolutely all the necessary items to produce a hologram almost anywhere. The case includes a 0.8 mW gas laser HeNe emitting at 632.8 nm (Class IIIa). With a power under 1mW, it is possible to decrease the risk linked to highly collimated beam. This type of gas laser has a good coherence length, especially as its power is low. The laser requires no special precautions with regards to its stability. It is possible to use a laser diode, which is much cheaper. However, diodes characteristics are extremely sensitive with respect to temperature changes, which may strongly decrease the hologram success rate.

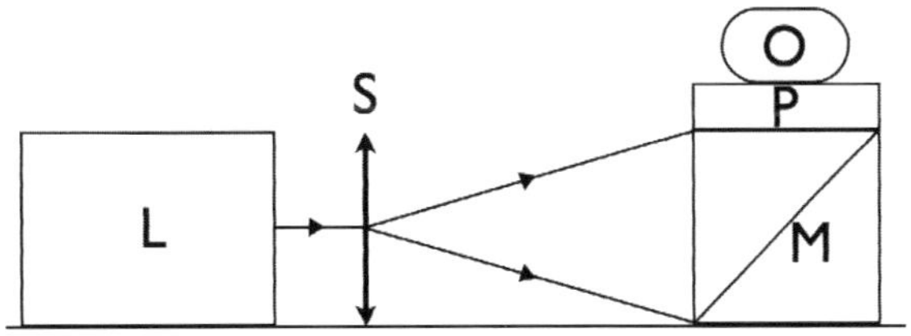

Figure 2: Schematic side view of optical assembly used in the teaching kit for holography. L: Coherent light source; M: 45° mirror; S: Optical system (diverging lens); P: Photosensitive plate, carried by the mirror mount; O: Object, carried by the plate.

The Fig. 3 shows the experimental setup, and corresponds to a single beam Denisyuk hologram setup. The laser light is spread with a lens and is directed upwards with the aid of a 45° mirror, inspired from a periscope reflective mount system. The photosensitive plate and the object are placed directly on the flat mirror mount, as indicated on Fig. 3. This mirror mount is the central part of the experimental setup, which is made by the mechanic team of the Aix-Marseilles University. Unlike the setup shown on Fig. 2, there is no need to physically divide the laser beam in two parts. The object beam is the one which is reflected/diffused by the object and which is coming back towards the plate, while the reference beam is the part of the one directly coming from the laser. In holography, stability is crucial, but only after the laser beam division [5]. In a single beam Denisyuk hologram setup, this only happens when the beam is crossing the plate. Moreover, the plate and the object are stabilized with respect to each other thanks to gravity. Thus, hologram can be produced anywhere (Classroom, hall, Kitchen, etc.), with almost any vibration condition. The produced holograms can be observed with white light spot, which is an important asset for direct observation after its completion.

4. Teaching kit for holography applications

The first aim of the teaching kit for holography is to produce simply good quality holograms, in any room with correct obscurity conditions, a power source for the laser, and water access for hologram development. Producing a hologram is an extremely rewarding experiment. Therefore, it is interesting to use it as teaching demonstration, or as practical experiment for undergraduate student.

For outreach purpose, a simple hologram can be produced during a workshop of 45 minutes. This includes basic explanations on holography physics as well as holograms production. A group of 20 primary school, junior high school and high school students can attend the workshop at the same time. To arouse their curiosity and give them information accessible to their levels, the workshop content is tailored to the level of students. With a success rate of almost 100%, students who realize a hologram are systematically proud of their production. Thus, they are in good conditions to go further in the physical notions on which holography is based.

Beyond the simple realization of a hologram, it is possible to make several experiments illustrating the applications of holography:

- Holographic interferometry, used in mechanical stress diagnostics,
- Angular multiplexing, used in data storage (Holographic Versatile Disc),
- « Notch » filter, used in Raman spectroscopy,
- Diffraction holographic grating, used in passive spectroscopy.

Holographic interferometry is the technics that introduced holography in the field of industry in 1965. The researchers found that if the object moves a few micrometers when the hologram is realized, interference fringes associated with this movement appear. It is possible to measure micrometrical deformations of any object by this technique [7]. To realize such experiment with the kit, it is necessary to make a double exposure hologram of a metallic foil F placed just above the photosensible plate and fixed at one extremity (Fig.4). Between the two exposures, the other extremity of the foil is translated of a known distance d_{mic} with a calibrated micrometer head C fixed on the mirror mount M. After the development, the two holograms shifted of a small distance will create secondary interference consisting of dark/light fringes superimposed on the foil hologram. The foil shift d_{holo} can be deduced by counting the number of dark fringes between its fixed part and the translated part. For the Nth dark fringe, we have:

$$d_{holo} = (N\text{-}1)\; l/2 + l/4 \qquad (1)$$

with λ=laser wavelength. This result can be compared to the translation of the calibrated translation device d_{mic}. The Fig. 5 shows typical results obtained by first year students at the Aix-Marseilles University. They found a good linearity between the two-measurement methods, validating the theoretical calculation.

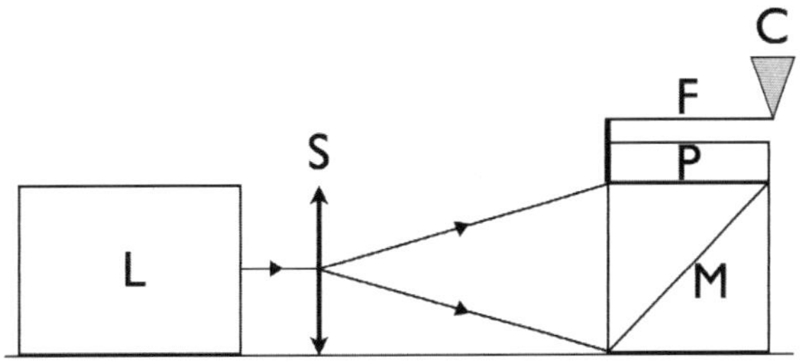

Figure 4: Table of interferometry measurement. P: Photosensitive plate. F: metallic foil, fixed on the left side. C: calibrated micrometer head, moving the foil right side.

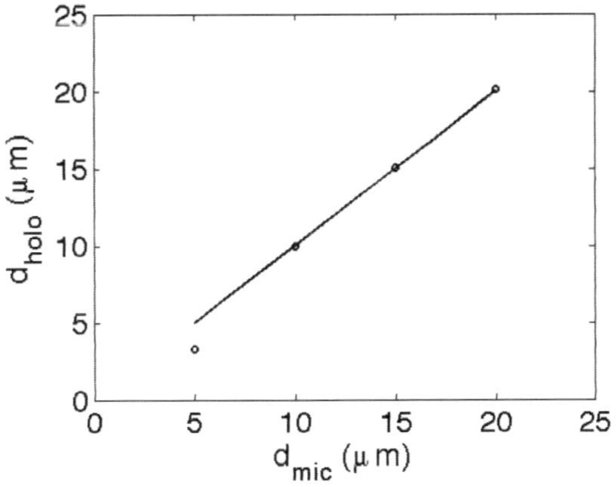

Figure 5: measured displacement d_{holo} of a metallic foil deduced from its double exposure hologram versus real displacement d_{mic} measured with a calibrated micrometer head (o). The line represents the ideal case $d_{holo} = d_{mic}$.

Information storage with semiconductor memories or disc (DVD, blu-ray, etc.) use "surface" technology. Nowadays, these technologies are well known and their capacities are close to the theoretical limit. The holography allows storing data in "volume", which increases greatly the density of information. Holographic Versatile Disc (HVD) uses this principle to store data. Moreover it is possible to increase the data density thanks to angular multiplexing. This technics rely on a singular property of holography: by changing the reference beam angle **θ** between each exposure, it is possible to record several data on the same photosensitive plate. After the development, hologram returns particular data depending on the reference beam angle used to read the disc. The kit for holography allows illustrating this angular multiplexing principle. With one photosensitive plate, two holograms are produced with two different reference beam angles. For this, a wedge is placed below one edge of the photosensible plate and a new object is put on it between two exposures to the laser.

A third application of the kit is the creation of a "Notch" filter, which is blocking a particular frequency range with much steeper fronts than conventional interference filters. These filters are used in Raman spectroscopy, where it is necessary to distinguish scattering spectra very close to the excitation source. It is possible to produce such filter with the kit, simply by producing a hologram of a front mirror put on the photosensible plate. After the development, the hologram reflects light like the object, but only on a spectral band Dl around the laser wavelength. As shown on Figure 6, it is possible to measure the transmittance of the obtained Notch filter with a white light source and a small spectrometer (not included in the kit).

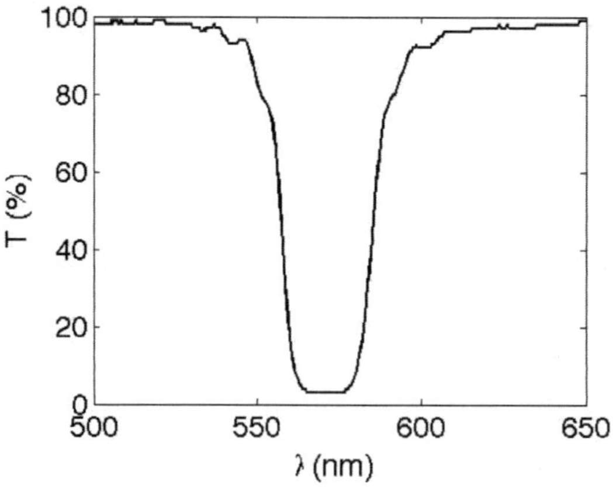

Figure 5: transmittance of a Notch filter made with the teaching kit for holography. The holographic plates are pre-swelled so that the centered wavelength of the filter is shifted of 70 nm from the initial HeNe laser wavelength (632.8 nm).

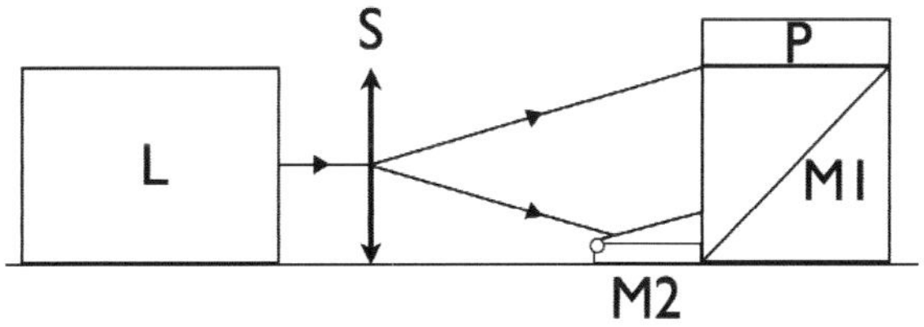

Figure 6: Table of diffraction grating. M1: main mirror. M2: additional mirror on a rotating system

A fourth application of the teaching kit for holography is the realization of holographic diffraction gratings widely used in passive spectroscopy. For this, a front surface mirror is fixed in front of the entrance hole of the mirror mount to deviate a small part of the incoming laser beam (Fig.6). A mechanical device allows its precise positioning. This part of the beam encounters the undeflected laser beam on the plate with a phase shift. This leads to the creation of interference fringes which can diffract incident ligth. The distance d_g between two successive fringes, and as a consequence, the groove spacing can be deduced from the following relation :

$$d_g = \frac{\lambda}{2\sin\left(\frac{\alpha}{2}\right)} \tag{2}$$

With a = mean angle between the two phase shifted laser beams. This experiment illustrates the principle of holographic diffraction gratings realization, and it is possible to clearly observe the first order diffraction spectrum of an incident white light. More precisely, we can calculate the mean groove spacing $d_g^{\exp}$ by measuring the first order diffraction angle b_2 of the collimated HeNe laser beam of the kit with an incident angle b_1 with respect to the normal [8]:

$$\sin(\beta_1) + \sin(\beta_2) = \frac{\lambda}{d_g^{\exp}} \tag{3}$$

Experimentally, we obtain around 1200 grooves per mm. If we use the relation (2), this corresponds to a mean angle a of 46°, which is coherent with the global orientation of the mirror. A more precise measurement of $d_g^{\exp}$ can be done by the use of other lasers at different wavelengths, for example a Blu-ray diode (λ=405 nm) and a Nd:YAG green laser module (λ=532 nm). Moreover, it is possible to study the evolution of $d_g^{\exp}$ with the mean angle a by making several diffraction gratings with different angles between the two parts of the laser beam.

These experiments have been proposed with success to physics students at the Aix-Marseille University. The experimental device was installed on common tables, without particular precautions against vibrations. They allow precise measurements and show several important current applications of holography.

5. Conclusion

The described teaching kit for holography is used since several years at the Aix-Marseille University for student's physics projects. It provides pedagogic subjects for undergraduate students with no particular need for active of passive anti-vibration mount. It is possible to develop, with educational goals, more complex experiments based on this kit: color holography (wavelength multiplexing), good quality interferential filters with central length filter control, etc. These kind of experiments may be used as a “semester project” for undergraduate as well as graduate student. In this case, students may develop the full experiment by themselves, typically one half day per week. Moreover, the optical assembly may be used for outreach purpose, like public exhibition, workshop, science festival, conferences, etc. Finally, the holographic kit may be used for junior high and high school teacher training. Thus, in an outreach point of view, teachers can teach their students how produce hologram themselves. Currently at Aix-Marseille University, two holographic kits are available for such outreach programs. Thanks to this kit, a high school student club gets qualified to the national French competition “Physics Olympiads”. They produced several live hologram in presence of a jury. An inexpensive version for massive outreach is under consideration.

Acknowledgment : The authors would like to thanks their first year students for providing the good data used in this review.

Bibliography

[1] Gabor D, 1948 A new microscopic principle *Nature* **161** pp 777-778
[2] Denisyuk Y N 1962 Photographic Reconstruction of the Optical Properties of an Object in its Own Scattered Radiation, *Dokl. Akad. Nauk. SSR* **144** pp 1275-1278
[3] Lippmann G 1891 La photographie des couleurs *Comptes Rendues des Scéances de l'Académie des Sciences* (Paris) **112** pp 274-275
[4] Gentet Y http://www.ultimate-holography.com/
[5] Saxby G 2003 *Practical holography* 3rd revised ed. (Institute of Physics ed.)
[6] Escarguel A 2010 Holographie : un outil simple pour l’enseignement du secondaire au supérieur, colloque CETSIS 2010 (Grenoble, France) / Escarguel A 2010 Journal sur l'enseignement des sciences et technologies de l'information et des systèmes **9** 0015
[7] Smigielski P 1994 *Holographie industrielle* (Teknea ed.)
[8] James J. 2007 *Spectrograph design fundamentals* (Cambridge University Press, Cambridge)

New education system for construction of optical holography setup - Tangible learning with Augmented Reality -

Takeshi Yamaguchi and Hiroshi Yoshikawa

Department of Electronics and Computer Science, College of Science and Technology, Nihon University
7-24-1 Narashinodai, Funabashi-shi, Chiba, Japan

E-mail: `yamaguchi@ecs.cst.nihon-u.ac.jp`

Abstract. In case of teaching optical system construction, it is difficult to prepare the optical components for the attendance student. However the tangible learning is very important to master the optical system construction. It helps learners understand easily to use an inexpensive learning system that provides optical experiments experiences. Therefore, we propose the new education system for construction of optical setup with the augmented reality. To use the augmented reality, the proposed system can simulate the optical system construction by the direct hand control. Also, this system only requires an inexpensive web camera, printed makers and a personal computer. Since this system does not require the darkroom and the expensive optical equipments, the learners can study anytime, anywhere when they want to do.

In this paper, we developed the system that can teach the optical system construction of the Denisyuk hologram and 2-step transmission type hologram. For the tangible learning and the easy understanding, the proposed system displays the CG objects of the optical components on the markers which are controlled by the learner's hands. The proposed system does not only display the CG object, but also display the light beam which is controlled by the optical components. To display the light beam that is hard to be seen directly, the learners can confirm about what is happening by the own manipulation. For the construction of optical holography setup, we arrange a laser, mirrors, a PBS (polarizing beam splitter), lenses, a polarizer, half-wave plates, spatial filters, an optical power meter and a recording plate. After the construction, proposed system can check optical setup correctly. In comparison with the learners who only read a book, the learners who use the system can construct the optical holography setup more quickly and correctly.

1. Introduction

Holography is well known as the technique which can record the 3D image information. By using the optical setup, the 3D image information is recorded to the recording material as the fringe pattern. For the construction of the optical setup, a user is required the precise knowledge of optical elements and the technique. However, since the beginner usually does not have the basic knowledge and the experience, it is difficult to teach the optical system construction of the hologram only with a textbook. As the education system, there are e-learning systems. Since most e-learning systems only show texts, figures and movies, it is not suitable for learning of the technique such as optical system construction.

In this paper, we proposed the new education system for the optical system construction of the hologram recording. To employ the augmented reality (AR), proposed system can provide the tangible learning[1, 2, 3]. When the user controls markers, the proposed system can make the user feel controlling virtual optical components with the direct hand. As the optical components, the proposed system has a laser, mirrors, a PBS (polarizing beam splitter), lenses, a polarizer, half-wave plates, spatial filters, an optical power meter and a recording plate. To control the virtual optical components, the user can learn not only the role of the optical components but also the setup of the Denisyuk hologram and 2-step reflection type hologram.

2. Tangible education system

Figure 1 shows the proposed education system with the augmented reality (AR). AR is a live, direct or indirect, real-world environment whose elements are augmented by the computer-generated sensory input such as the sound, the video or the graphics data. By using this technique, in this paper, we added CG object to the captured image by the camera. For the development of this system, ARToolKit[4, 5], which is a computer tracking library for creation of AR applications, is employed. ARToolKit helps to get the coordinate of the marker captured by the camera. On the basis of this coordinate information, the proposed system augments the 3D CG object and displays. As a result, the learner controls the markers, and can practice the tangible learning.

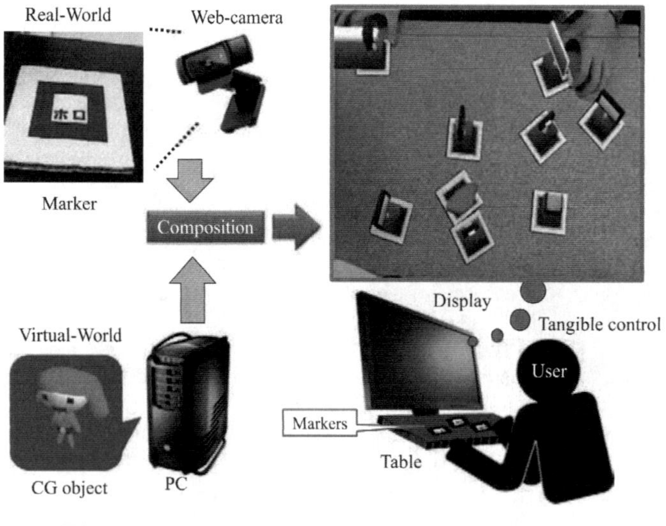

Figure 1. Proposed tangible education system.

2.1. ARToolKit

ARToolKit is a library to make vision-based AR applications[4, 5]. By using ARToolKit, the creator can make AR applications easily without the expensive censor or the high spec computer. Also, since the application with ARToolKit can employ the direct hand action, the user can easily understand how to arrange optical components.

The application with ARToolKit consists of the camera, markers, the personal computer and the monitor. Since the system is necessary to make the superimposed CG objects seem to be in real-world, the added CG objects are required the geometrical consistency with the real-world. Figure 2 shows the calculation model of the coordinate. Each marker has a black square with a unique pattern in it. To calculate the distortion from the captured marker image, the system can obtain the pose and position of the marker. The transform matrix T_{cm} from the marker coordinate system P_m to the camera coordinate system P_c is determined by

$$P_c = \begin{bmatrix} X_c \\ Y_c \\ Z_c \\ 1 \end{bmatrix} = \begin{bmatrix} r_1 & r_2 & r_3 & t_x \\ r_4 & r_5 & r_6 & t_y \\ r_7 & r_8 & r_9 & t_z \\ 0 & 0 & 0 & 1 \end{bmatrix} \begin{bmatrix} X_m \\ Y_m \\ Z_m \\ 1 \end{bmatrix} = T_{cm} \begin{bmatrix} X_m \\ Y_m \\ Z_m \\ 1 \end{bmatrix} = T_{cm} P_m, \tag{1}$$

where t is the translation component, r is the rotation component. To compare the size of the captured marker image to the registered line length of the marker, the system can display the CG object naturally.

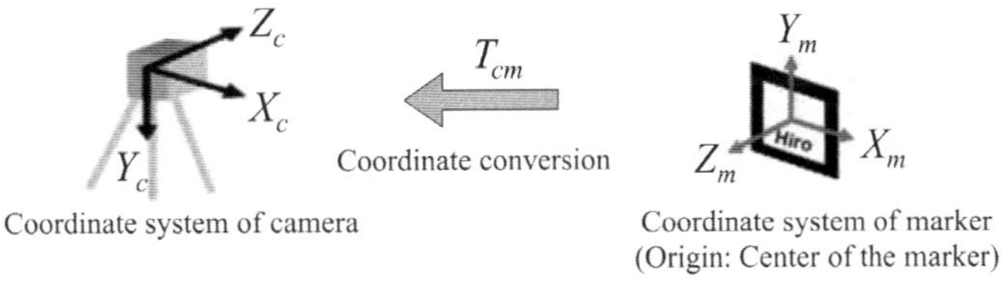

Figure 2. Coordinate conversion between the camera and the marker.

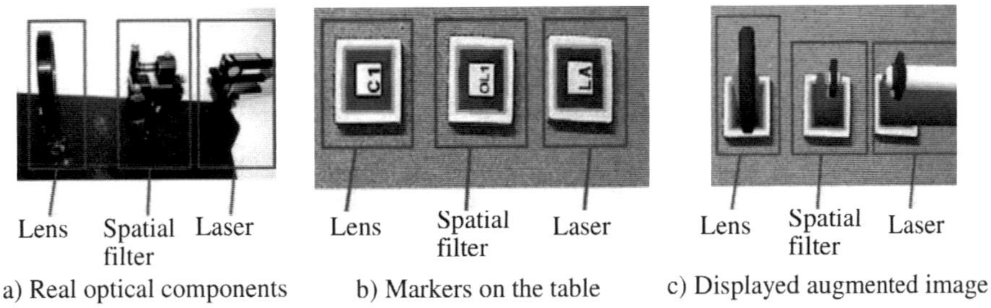

Figure 3. Real optical components and the displayed augmented images.

2.2. Education system for construction of optical holography setup
To master the optical system construction of the hologram, it is important to understand the role of the optical components and how to use it. Therefore, the proposed system makes the tangible learning possible by the control of the markers with hands. Also, to help the intuitively understanding for the role of the optical components, the proposed system displays the light path and shape formed by the optical components.

2.3. Display of the virtual optical components on the markers
Figure 3 shows the photographed real optical components, the markers on the table and the displayed augmented images. In the case of constructing the optical system shown in Fig. 3 (a), the learner arranges markers as shown in Fig. 3 (b) on the table. On the display, the captured image is augmented by the proposed system as shown in Fig. 3 (c). The augmented CG optical components can be controlled freely by the pose of the markers.

2.4. Unified coordinate system
ARToolKit only obtains the relation between the camera coordinate system and each marker coordinate system. To achieve the simulation of the optical system construction on the proposed system, each coordinate system needs to be the unified coordinate system based on the one marker. In this paper, we configure a laser marker to the base marker. The location of other markers are calculated as bellow. Figure 4 shows the conceptual diagram of the coordinate

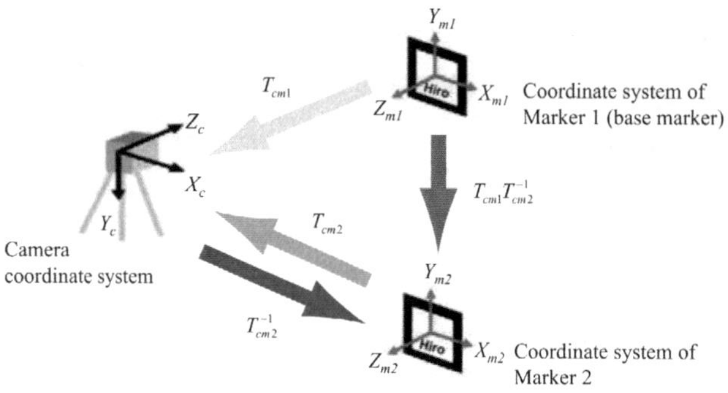

Figure 4. Real optical components and the displayed augmented images.

conversion. The relation between the camera coordinate system P_c and the marker 1 (base marker) coordinate system P_{m1} is defined as

$$P_c = T_{cm1}P_{m1}. \tag{2}$$

Also, the relation between the camera coordinate system P_c and the marker 2 coordinate system P_{m2} is defined as

$$P_{m2} = T_{cm2}^{-1}Pc. \tag{3}$$

P_{m2} can be transformed by using P_{m1} as bellow

$$P_{m2} = T_{cm2}^{-1}T_{cm1}P_{m1}. \tag{4}$$

By using Eq. (4), the proposed system calculates each marker coordinate based on the marker 1. The proposed system create the virtual optical components on the display from the calculated marker coordinates.

3. Light control with the virtual optical components

For understanding of the optical system construction to the user, it is effective to reflect changes of the light beam shape and the light path by the optical components on the display. In the case of using of the real optical components, it is difficult to see changes of the light beam by the optical components. On the other hands, since the proposed system can display the condition of the light beam, user can understand the role of the optical components in an intuitive way. Also, since the adjustment to the light intensity has the important role in the recording of the hologram, the proposed system implements the adjustment to the light intensity.

3.1. Calculation of the light path

Fist of all, to employ the calculation method described in 2.4, the proposed system obtains the coordinate of the all markers based on the laser marker as shown in Fig. 5. Then, the proposed system displays changes of the light beam by the virtual optical components which is placed on the light path of the laser beam as shown in Fig. 6. Also, the proposed system obtains the parameter of the light beam such as the light path length, the angle of deflection, the light diameter and the light power. Through this information, the user can control the light path by just using the markers.

3.2. Light path operation with laser, mirror and PBS

Figure 7 shows the laser marker, the augmented laser marker and the real laser. The CG laser and the direction of the light beam can be operated by the laser marker. Also, the light beam has the light power and the angle of deflection as the parameter.

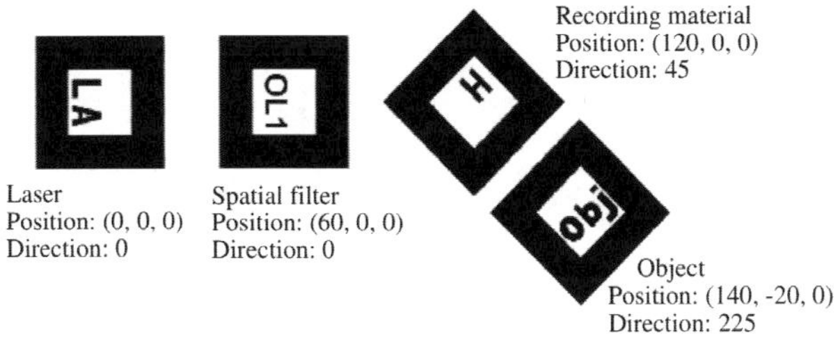

Figure 5. Acquisition of coordinate of all markers.

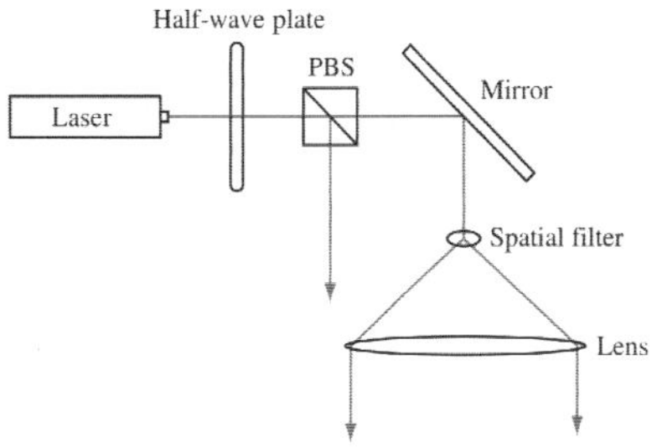

Figure 6. Drawing of the light path.

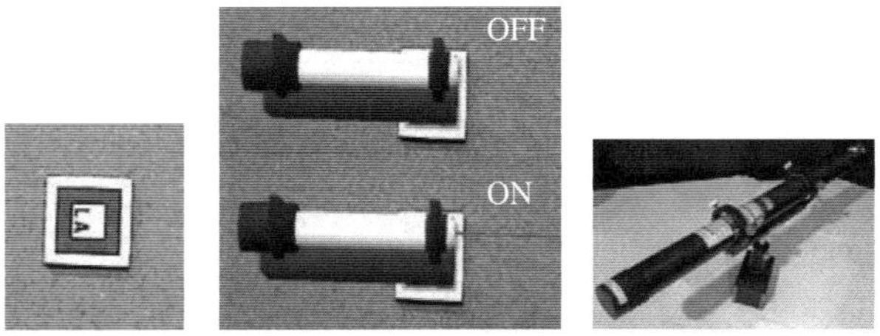

a) Laser marker b) Augmented laser marker c) Real laser

Figure 7. Laser marker and augmented laser marker.

Figure 8 shows the mirror marker, the augmented mirror markers and the real mirror. Figure 9 shows the PBS marker, the augmented PBS marker and real PBS. Mirrors can control the direction of the light beam as shown in Fig. 8 (b). The user can operate the light beam from the CG laser object by adjusting the direction of the mirror marker. The PBS can divide the laser beam depending on the deflection angle of the light beam. Therefore, user can learn to divide the reference beam and object beam.

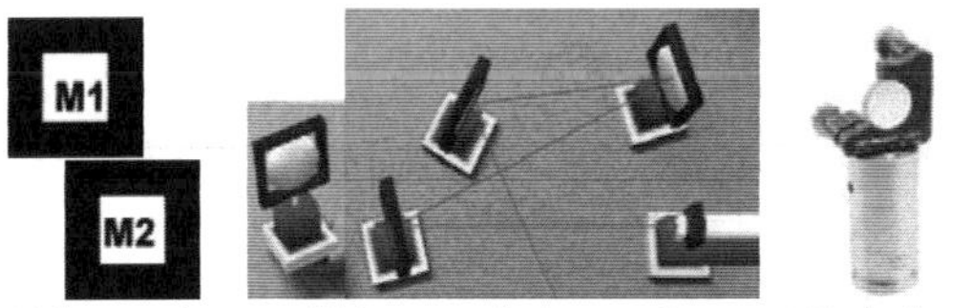

a) Mirror markers b) Augmented mirror markers c) Real mirror

Figure 8. Mirror markers and augmented mirror markers.

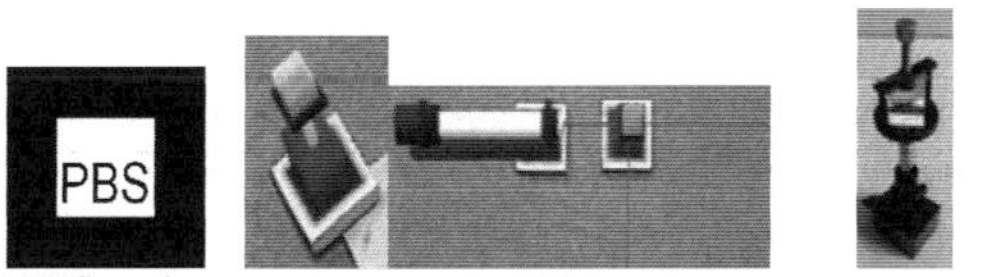

a) PBS marker b) Augmented PBS marker c) Real PBS

Figure 9. PBS marker and augmented PBS marker.

3.3. Light beam diameter operation with spatial filter and lens

Figure 10 shows the spatial filter marker, the augmented spatial filter markers and the real spatial filter. The displayed color of the expanded light beam by the spatial filter fades as the light beam expands. Also, the virtual spatial filter has the parameter of NA, and user can configure the number of spatial filter's NA according to the intended optical setup.

Figure 11 shows the lens marker, the augmented lens markers and the real lens. The CG lens objects have the parameter of the focal length. Therefore, user can control the light diameter by the position of the lens marker as shown in Fig. 11 (b).

3.4. Light power operation with power meter, half-wave plate and polarizer

Figure 12 shows the power meter marker, the augmented power meter marker and the real power meter. To set the power meter marker on the light path, the proposed system display the light power of the position where the power meter marker is put as shown in Fig. 12 (b).

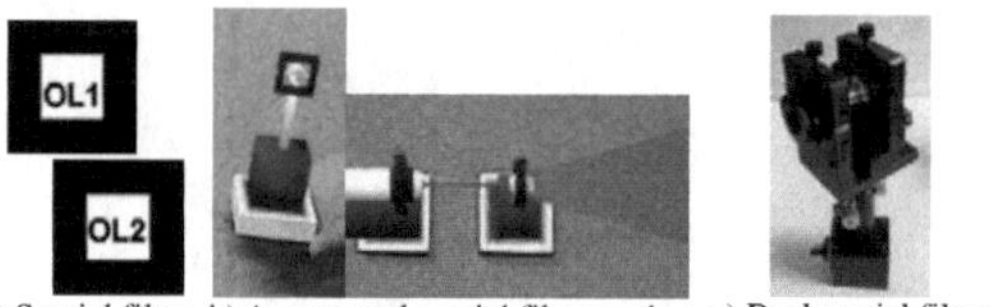

a) Spatial filter markers b) Augmented spatial filter marker c) Real spatial filter

Figure 10. Spatial filter markers and augmented spatial filter markers.

a) Lens marker b) Augmented lens marker c) Real lens

Figure 11. Lens markers and augmented lens markers.

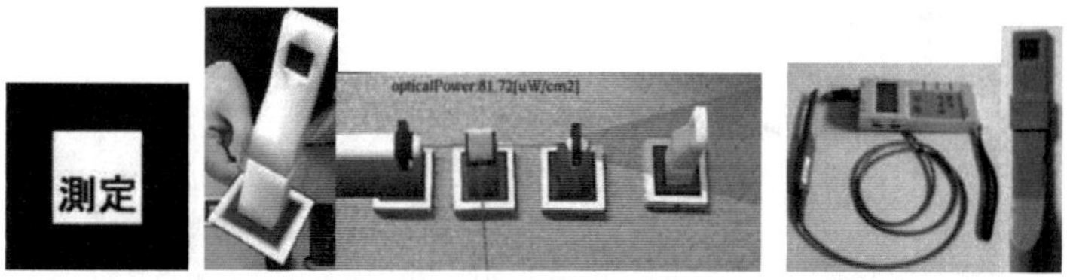

a) Power meter marker b) Augmented power meter marker c) Real power meter

Figure 12. Power meter marker and augmented power meter markers.

a) Polarizer marker b) Augmented polarizer marker c) Real polarizer

Figure 13. Polarizer marker and augmented polarizer marker.

a) Half-wave plate marker b) Augmented half-wave plate marker c) Real half-wave plate

Figure 14. Half-wave plate marker and augmented half-wave plate marker.

Figure 13 shows the polarizer marker, the augmented polarizer marker and the real polarizer. Figure 14 shows the half-wave plate marker, the augmented half-wave plate marker and the real half-wave plate. Since the displayed object is the CG object and can not be touched, we arrange the another marker to control the polarization angle. Figure 15 shows the schematic image of the polarization angle control and its results. The proposed system calculate the angular difference between the laser marker as the base marker and the control marker.

a) Control marker b) Polarizer control c) Half-wave plate control

Figure 15. Operation of the half-wave plate and the polarizer.

4. Optical system construction for the hologram recording

The proposed system support the training about the optical system construction of the Denisyuk hologram and 2-step transmission type hologram as shown in Fig. 16 and 17. The user can understand the construction of the optical setup through arranging the markers. Figures 18 and 19 show the simulation outcome with the proposed system. If the user made a mistake, the proposed system shows the reason of the mistake in the message box. In the proposed system, there are some judgment conditions such as; the difference of light paths, the incident angle of the object beam and reference beam, the light power and the polarization angle.

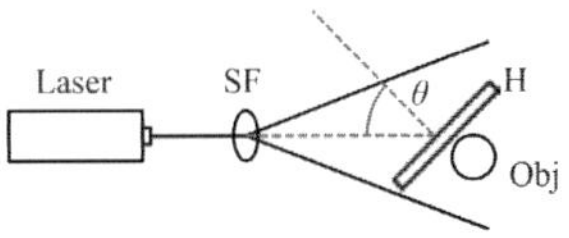

Figure 16. Schematic of the optical setup for Denisyuk hologram: SF = spatial filter, H = recording material, obj = recording object.

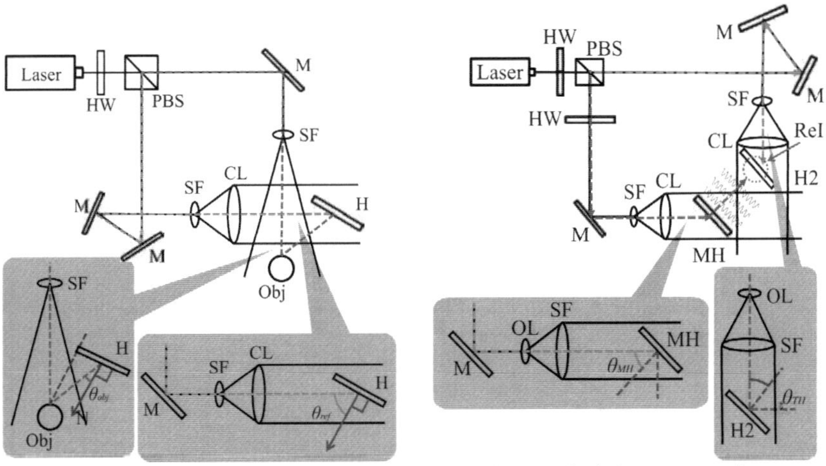

Figure 17. Schematic of the optical setup for 2-step hologram: SF = spatial filter, H, H2 = recording material, obj = recording object, PBS = polarizing beam splitter, HW = half-wave plate, MH = master hologram, CL = collimate lens, M = mirror, ReI = reconstructed image from the master hologram.

Figure 18. Simulated result about the optical setup of Denisyuk hologram.

5. Measurement of educational effect

To measure the educational effect of the proposed system, we have the proposed system used by beginners who are junior year of college and they are not familiar with the holography. The subjects were divided into 2 group which used the proposed system or only the textbook for learning. After learning, we measured the construction time of Denisyuk hologram setup for each subjects and the correctness checking. Table 1 shows the result of measurement of the educational effect. In this measurement, although there was little number of the data, but the effectiveness is confirmed.

(a) Master hologram recording (b) Transfer hologram recording

Figure 19. Simulated result about the optical setup of 2-step hologram.

Table 1. Measurement of the educational effect

With proposed system			Without proposed system		
ID	Time	Correctness checking	ID	Time	Correctness checking
a	2'03"	OK	k	3'18"	OK
b	2'18"	OK	l	8'13"	NG
c	1'54"	OK	m	3'14"	NG
d	2'11"	OK	n	3'23"	OK
e	2'18"	OK	o	4'33"	OK
f	2'27"	OK	p	8'48"	OK
g	5'35"	OK	q	6'55"	OK
h	2'41"	OK			
i	2'37"	OK			
j	3'53"	OK			
	Average time	Success rate		Average time	Success rate
	2'50"	100%		5'14"	71%

6. Conclusion

In this paper, we have proposed new education system for the construction of the optical setup. By using the augmented reality, the proposed system provide the tangible learning to the user. As a result, the learner can understand the role of the optical components and the maneuver of the optical setup construction intuitively.

[1] Nagano S, Yamaguchi T and Yoshikawa H 2009 *HODIC* **29** 27–30
[2] Shibata M, Yamaguchi T and Yoshikawa H 2010 *HODIC* **30** 17–20
[3] Shibata M, Yamaguchi T and Yoshikawa H 2011 *HODIC* **31** 18–21
[4] Kato H 2002 *IEIC Technical Report* **101** 222–232
[5] Kato H and Billinghurst M 1999 *2nd IEEE and ACM International Workshop on Augmented Reality* 85–94

9th International Symposium on Display Holography (ISDH 2012) IOP Publishing
Journal of Physics: Conference Series **415** (2013) 012065 doi:10.1088/1742-6596/415/1/012065

Going deeper: teaching more than the mechanics

R A Bruck
Adjunct Instructor
Department of Science and Mathematics
Columbia College Chicago
600 South Michigan Avenue
Chicago, IL 60605
USA
www.richardbruck.com

E-mail: rabhol@earthlink.net

Abstract. What follows is a description of an introductory holography course titled "Lasers and Holography," taught by the author at Columbia College Chicago since 1997. Because this is a science class at an arts college with an open admissions policy, these students have many different levels of education, dissimilar backgrounds, and varied fields of interest. There are few science majors. Therefore, specific learning objectives are developed. The author contends that for many of these students it is not enough to teach the physics of making holograms. To inspire and instill a lifelong appreciation for science and physics, one must go still deeper. Students need to be touched on more than just an intellectual level. Consequently, a broader approach is used. Ultimately, it may stir students to want to learn more, and to be confident they can. The paper addresses: 1) Becoming aware of one's individual state of seeing 2) Perceptual illusions: their impact on the advancement of science 3) Promoting artistic applications and exposing students to fine art holography 4) Teaching holography as an information processing, as well as an image-making technology 5) Introducing and exploring philosophical implications of holographic principles.

1.Introduction

The first version of the class this author taught was for a short summer school session for high school students. Many of the students were interested in pursuing photography at Columbia, and the similarities and dissimilarities between photography and holography, along with this instructor's background, allowed for presenting much of the course material within a context of photographic imaging.

As the course progressed to the college level the template expanded. It was clear there was fear of what physics classes were thought to be about. They were perceived to be opaque. But, it has also become clear that what inspires most of these art-oriented students, and what motivates them beyond just wanting to learn how to make holograms, is their own curiosity about the big things; their big questions, and their own grand ideas. Fortunately, the universe of physics and holography is filled with such things; things transparent and wondrously fathomless.

The philosophy for teaching is so shaped. We start with those big things. We introduce physics as thoughts: thoughts about what the universe is, about how we perceive it, about how it works, one's self and one's place in it. These things relate and resonate. This is a different physics. This can include philosophy, and it can include art. It is personal and it is passionate. It is meaningful and relevant.

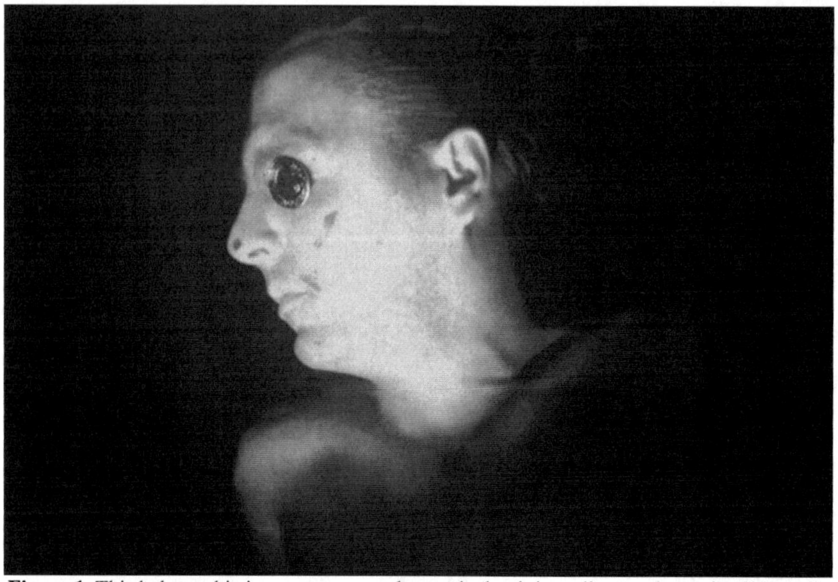

Figure 1. This holographic image appears orthoscopic, but it is really pseudoscopic. This well known cognitive illusion is thought to arise from strongly reinforced facial feature recognition which overpowers weaker, contradictory signals. *Lost in the Green River*; 50x60cm pulsed reflection hologram. By the author, 1992.

One's attention is caught and held, and sincere learning can begin. Individual desires and directions will guide the course. The smaller things, the details, can come later when needed; however named.

2. Becoming aware of one's individual state of seeing

The ultimate learning objective is to inspire a shift in a student's self-awareness: to be a catalyst that propels one to see more, to see differently; to expand one's universal view, and one's sense of place. The visual impact of holograms makes them excellent for stimulating such a shift.

To begin this process, the first class session combines lecture and demonstration to challenge students' preconceptions. Part of the lecture proceeds:

I. When looking at holograms we may be reminded that we take our visual world, what we think of as reality, for granted. We look at the universe with preconceived ideas. We assume reality.

- A. We see as if from behind a window: a little person inside this head, passively watching a world "out there."
  1. This outside world is separate from us "inside here."
  2. This outside world consists of things, objects, nouns.
  3. These objects are separate, with empty space in-between.
  4. These objects are ultimately, indivisible.
  5. Reality continues, independent of us.

II. Still, some theories in physics and some philosophies hold an opposing view:

- A. There is no person "inside here."
  1. There is no separation between the observer and the observed.
  2. The universe consists of processes and flux: not static objects.
  3. Everything is connected.
  4. We have not yet found an indivisible building block of everything.
  5. Without an observer to observe it, there is no independent reality.

The reader here may see similarities to what could be called a classical physics paradigm presented in I, and similarities to a quantum physics paradigm presented in II. But, the point is not to compare the two methods and disparage one. There is no conclusion drawn that one of these above sets of claims is superior to the other. The point though, is that there are different ways of seeing, and to bring the unaware student to such a realization. The context in which we see affects what it is we do see, and the meaning within. Here are two paradigms with which we may view the universe, reality, existence. For many, this is the first time they have encountered the idea that they see from within a context of presumptions. Many of the students in this class are in the visual arts, and not all of this is new to them. But, still they can become aware that without consciously thinking about it up to this point in their lives, they have assumed that their type of world-viewing is the only type.

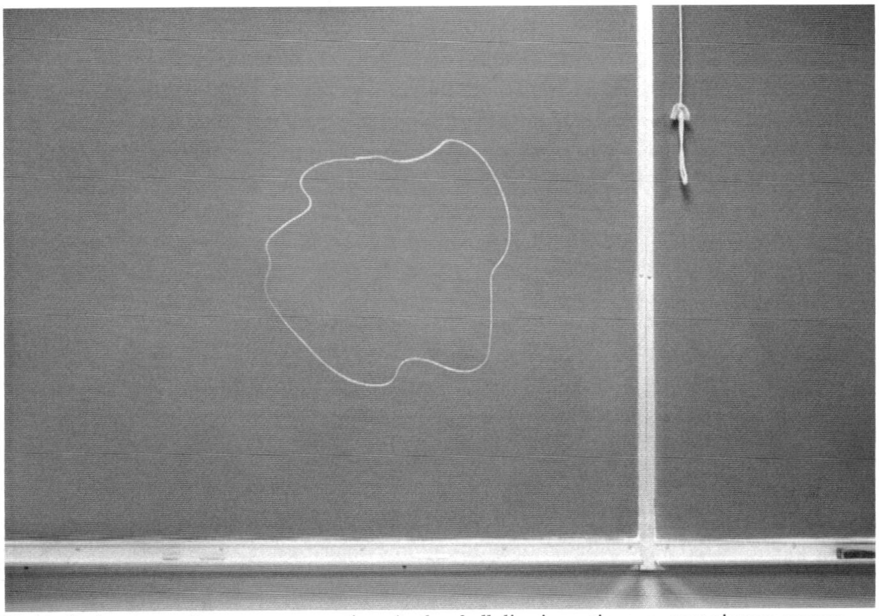

Figure 2. An assumption, or preconception. As the chalk line is continuous, most viewers assume it forms an outside border and constitutes a specific object; perhaps in this instance this is a droplet of water. However, the line could just as well be an object itself, say a rubber band. Or, it separates an interior "emptiness" from an object on the outside. Perhaps now liquid surrounds a dry spot.

Later in this first class, there is a display of large-format reflection holograms (some of which are multi-channel), a transmission master is shown full aperture with a diverged laser beam, and then its real image is projected onto a classroom wall with an undiverged beam.

For their first exercise then, students write observations of the holograms they have just viewed. They are discouraged from including subjective statements (opinions, aesthetic judgments), the broad description that holograms appear to have depth, or points that they have heard from the instructor (those that are learned and not their original thoughts). This exercise is designed for objective observation: non-judgmental viewing. This requires them to concentrate on and analyze their own authentic, visual experiences. It pushes them to consider what it means when something is said to have depth, to describe the sensation of depth without using the word itself, and what it is that makes holograms unique. It challenges them to be conscious of seeing, and to verbalize what it is they see.

As the semester progresses, there are more pragmatic exercises. One class is largely devoted to 3D imaging technologies that generally do not involve holography: Peppers' Ghost, stereo photography, anaglyphs, lenticulars, polarization, the Imax system, and some hybrid systems that employ 2D image recording (photography, cinematography, videography) with holographic synthesis, for e.g. multiplexing. Each form is briefly explained and examples discussed with the emphasis on the visual effect, and with an introduction and development of language and terms to describe the imagery. Depth cues (occlusion, parallax, etc.) are recognized along with advantages and disadvantages of these technologies. Now the fundamental point: that a hologram is not a trick. It is physics: a reconstruction of the multi-faceted and complex patterns of wavefronts emanating from something we call an object. Holograms encode patterns, and we as viewers decode those patterns.

The deep-scene transmission lab is particularly powerful for both novices and the visually astute. This is the first time a real object is used in the lab (their first hologram is a diffraction grating made only with light itself). For the deep scene though, they "light and compose" their objects. Once they have roughed-in the equipment and components, they position themselves behind a clear glass plate and look back into the scene. They are told: "Make what you see through the glass plate as you look through it as you would a window, what you hope your final holographic image will look like." The constrictions of this geometry force them to position the object precisely and to consider lighting: what it is to have an object lighted from behind or strongly from its side; how does hard laser light affect the appearance of an object, its contours, its textures, its volume; how black will shadows be; does one type of lighting mean something different from another; and more plainly, is the object well seen? As they see the laser-illuminated object for the first time, they begin to pre-visualize. In a very real sense, they are motivated to use their imaginations. After exposure and processing they compare the final holographic image with the real object. For some, this is a moving experience: its memory resounds.

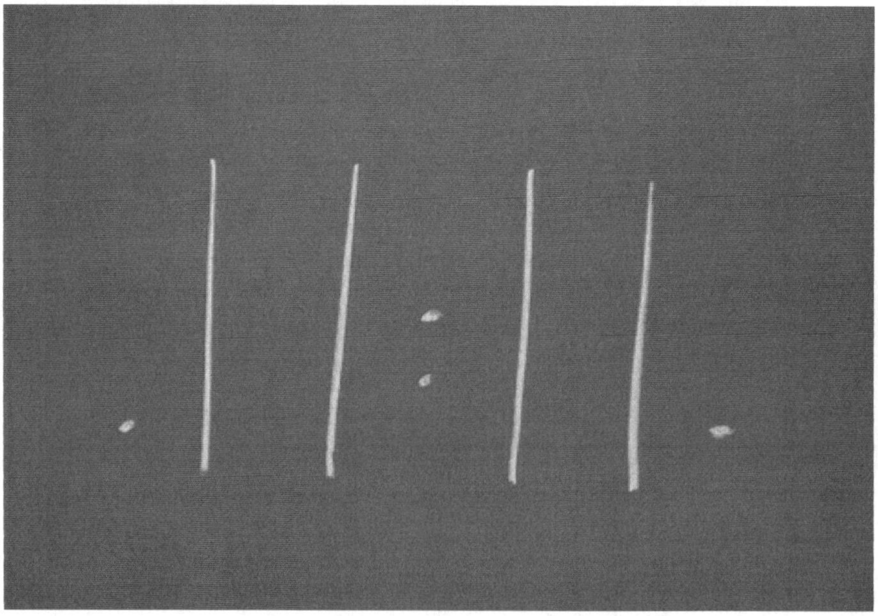

Figure 3. A visual conundrum. This pattern of dots and lines only has meaning when viewed within a context.

3. Perceptual illusions: their impact on the advancement of science

The process of perception is at least in part, a translation. But, translation however truthful, is still a form of qualification. To quote William Blake: "The eye altering, alters all." Lens aberrations deform the photon's path. Body chemistry transduces it, and the mind interprets its resulting signal. In this way, we see the world as right side up, even with our photon-inverting lenses; and we see straight lines, even as cascades of photons are caught upon our curved retinas, But, of course we still have blind spots (literally and metaphorically), and are often fooled. Our universe abounds in illusions. But, not all are absolutely, completely false; and in one sense an illusion might be considered a lesser truth.

The intent of science (and some say, art) is to see through to the truth. As we expand our vision and expose illusions, we can leapfrog from one level of awareness, one level of truth, to the next.

That we progress by dispelling one illusion with a higher level of truth conveniently dovetails with a short history of how our concept of light has evolved. It was Thomas Young's double-slit experiment that evidenced interference and the wave motion of light and saw through Newton's corpuscular theory, held for nearly a century before by most to be the ultimate truth. Works of Fresnel and Maxwell further substantiated the wave theory of light. So it was widely assumed, again for nearly a century, that in fact light is a wave. But, then Hertz perceived the photo-electric effect, and it presented inconsistencies with the light as a wave paradigm. It then took Einstein to pierce thru that veil of the time, and to reconcile the conflicting appearances with yet another paradigm: light's wave behavior is in concert with its particle (quantum) nature: wave and particle are mutual qualities. For slightly more than a century we have balanced this duality, this perspective. Perhaps it is not because this is the final truth, but because we've not seen thru to a new, more complete or more subtle model.

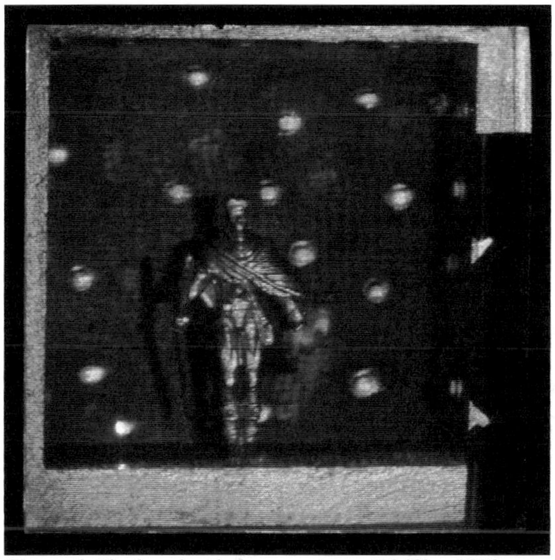

Figure 4. Final project hologram, Aaron Pfeiffer, 2012

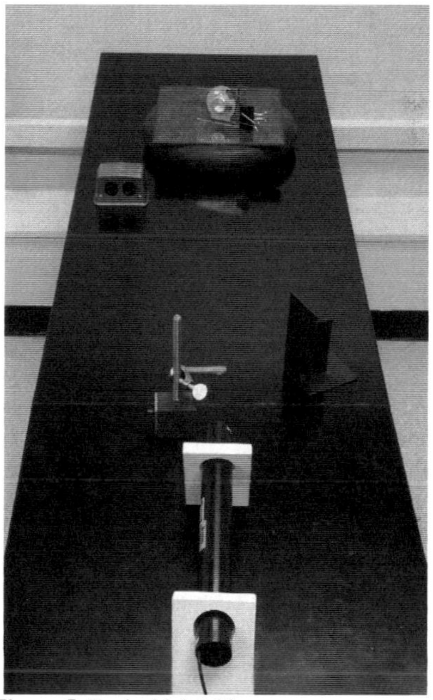

Figure 5. Typical lab table

4. Promoting artistic applications and exposing students to fine art holography

The course ardently advocates the experimental use of holography as an art form. A final project hologram which each student must conceive, design, construct, and produce, is assigned a large percentage of the total grade. Videos and slides of professional work are shown throughout the semester. Holograms are brought in for display, too. Some of these illustrate specific ideas pertaining to the final project, but still more are shown to light the imagination. The work spans the history and styles of holographic art and includes multiple artists from different regions and times. Rarely is the technical quality of the work alone discussed. Instead, the techniques are assessed as to how they contribute to the overall aesthetic of the artist and the work.

For their final project holograms, students are challenged to do more than Denisyuk arrangements with objects they have found. If they absolutely desire a found object, they need to somehow expand the project. (It is interesting that many students, especially those with backgrounds and interests from far afield, choose objects of sentimental value.) They can choose to do a transmission or reflection piece; and manipulate the color (within limits, of course) if it is in reflection mode. They can construct, combine, sculpt. They can practice with multiple exposures and reference angles, masking, pseudoscopic imagery using molds, incorporate other media, make mosaics, and use light itself as the subject. They can do anything they can imagine within the lab's capabilities.

The projects are evaluated with several criteria. The whole process is graded, from the conception, to the production in the lab, to final viewing. They are asked the rationale for their project: why this subject, why this geometry? The emphasis is on the quality of their comprehension displayed throughout, and on the intellectual and creative effort. This prevents the technical features from outweighing all others. Brightness is not the almighty arbiter; so the lucky and lazy who just happen upon bright images don't get to outshine those who have understood and worked hard but, because of the fickle finger of holography fate, or because their aesthetic doesn't demand it, don't have as bright a result.

Figure 6. Real image information is downloaded with an undiverged beam

5. Teaching holography as an information processing, as well as an image-making technology

Those of us in the field are accustomed to think of holography as a medium for recording and playing back information. But, this is a different way for students to think about "3D" pictures. Though the immediate, visceral impact of holograms may be a prime motivator for learning in this class, it is at least as important, as intimated earlier, that holography be understood and appreciated for the even larger contributions and ideas it offers. Information processing is specifically important.

So early on, it is made clear that holograms use waves as handles for carrying information. That in fact, visible light waves are not necessary to produce holograms. A definition has been carefully devised to encompass this quality: A hologram is the recorded interference pattern of a pure, coherent wave and a deformed portion of itself.

Additional points:

1. Holography is essentially an information processing technology: it records and retrieves information.
2. "Pure" holography (as distinguished from computer generated or digital hybrids) does not produce or manufacture new information; and in this sense it does not 'create' it.
3. The information is not stored locally: there is no point-to–point relationship.
4. Information can be stored throughout a volume, not just on a surface.
5. Information may even be stored inside of information.
6. This information can be accessed at the speed of light.
7. Different information is accessible from different perspectives.
8. Holographic storage volumes exceed other technologies.

"Seeing" is visual information processing: the eye gathers and transmits light information to regions of the brain which then decode and interpret it. Because of this process, there is the thought that technically we do not see objects. Instead, we recognize patterns and give them object names. In this way then, we are not just passive observers of a world 'out there', but are very active in the making of our reality.

6. Introducing and exploring philosophical implications of holography principles

Students have now moved from the introductory theories to learn the underpinnings of the physics in holography: interference, diffraction, stimulated emission. They have also made several different holograms. With added insight they are now prepared to look again at the big things. They are assigned reading on the holographic paradigm, a discussion of Alain Aspect's work on Belle's theorem, David Bohm's implicate and explicate order, and Karl Pribram's holonomic model. Some of the ideas presented:

I. A new paradox in physics has theorists suggesting a holographic model of the universe. Some say the universe actually is a hologram.

II. Researchers claim our brains may process information similarly to holographic information processing. Memory specifically may rely on interwoven neural networks like holographic interference patterns for non-local storage and rapid recall. Some say our brains actually are holograms.

III. The theories intertwine and conclude: consciousness itself might produce the appearance of reality.

These claims entice and incite strong reactions. The idea that objective reality does not exist is powerfully polarizing, with some students deeply inspired and others deeply offended (though most are inspired). Some mistake it to imply 'reality' does not exist at all, and this makes them more fervent still. Questions stream: What would the reference beam be? Where would it come from? Where is the real reality? Would there be a God? There are affirmations: It makes so much sense. Finally, there is a scientific explanation for the interconnectedness of everything. There are accusations: That is absurd: I know I exist. That does not let me control my own destiny. This is impossible to prove.

7. Conclusion

With the culminating reading, students have become fully involved; thinking and talking about the big things. Perhaps, they have seen through their earlier assumptions, and without even knowing what to call it, they now think and talk about classical physics, quantum physics, and the new physics. Then, for the last lecture, this offering:

> To understand, means literally to stand under, as we might stand under the stars. Just as we need light to enter our eyes in order to perceive it, we need to stand in reconstructing wavefronts to perceive the hologram's message. Now we have insight. As we move we gain intelligence. There are more ways to see: more to see. If we can learn to shift our perspective, our angle of view, we may learn that which we have yet to understand. Information is ubiquitous. Intelligence surrounds. Intelligence embodies.

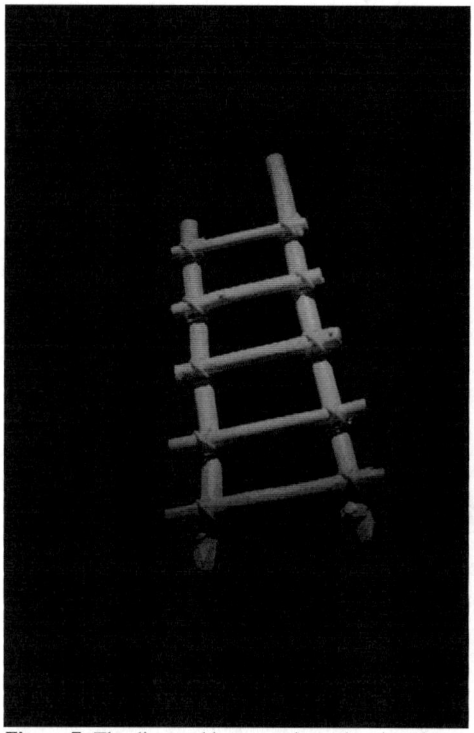

Figure 7. The diverged beam projects the virtual image of a transmission master. *Mo's Ladder*, by the author.

3D and Education

Odile Meulien Ohlmann

Odile Meulien Ohlmann is a French American, collector and sociologist. She did research on the new form of artistic expression with the art Historian and President of the French Museums René Huyghe at IPPAC in Paris, she created and managed the Museum of Holography of the Art, Science and Technology Institute, The Holography Collection ASTI in Washington DC (**6**) . She published numerous papers on the Holographic art and perception. She collaborated at the Launch of Digital Holography as a CEO of Syn4D in Germany. Now, in Strasbourg at the French/German border, she conducts research in Holography while providing exhibitions and trainings.

om@artbridge.info , www.artbridge.info

Abstract. Today the industry offers a chain of 3D products. Learning to "read" and to "create in 3D" becomes an issue of education of primary importance. 25 years professional experience in France, the United States and Germany, Odile Meulien set up a personal method of initiation to 3D creation that entails the spatial/temporal experience of the holographic visual. She will present some different tools and techniques used for this learning, their advantages and disadvantages, programs and issues of educational policies, constraints and expectations related to the development of new techniques for 3D imaging. Although the creation of display holograms is very much reduced compared to the creation of the 90ies, the holographic concept is spreading in all scientific, social, and artistic activities of our present time. She will also raise many questions: What means 3D? Is it communication? Is it perception? How the seeing and none seeing is interferes? What else has to be taken in consideration to communicate in 3D? How to handle the non visible relations of moving objects with subjects? Does this transform our model of exchange with others? What kind of interaction this has with our everyday life? Then come more practical questions: How to learn creating 3D visualization, to learn 3D grammar, 3D language, 3D thinking? What for? At what level? In which matter? for whom?

1. The present situation

The variety of media techniques available in 3D is confusing. These last 10 years the development of techniques is very significant. One development comes out, is replaced by another one, and finally just professionals keep updated. A whole list of names and products are available without reaching educators and educational institutions, as I experience it lately in France. Moreover, as I saw with children and teenagers, they have ideas of 3D, but they cannot identify techniques and the meaning of 3D except that it has relief. There is a lack of exhibition with educational goal. Moreover some companies are using word to describe their product which has nothing to do with their technical specification. The word Holography, for example, is very often used to describe the idea or even a product which do not record any object in time and space. This just sounds more attractive! This subject matter has been brought up in LinkedIn and the discussion shows that confusion in words and definition finally does not help customers' education on a long term run; even so on short term, it can bring attention to the product.

Journal of Physics: Conference Series **415** (2013) 012066 doi:10.1088/1742-6596/415/1/012066

2. New Media, new Behavior

There are a lot of engineering programs proposed at university levels in our western countries to teach the technical aspects of 3D. We can also find many trainings available in the use of 3D software packages such as cinema 4D, 3D studio max, Maya, and so on. The social aspect, the many changes implied by the multidimensionality of the medium on human behavior, and reading pattern are not included in the courses. Indeed we do not read a 2D images as a 3D one. And if we do not read it the same way, we should not write it the same way.

While working at Syn4D GmbH, we were producing synthetic digital holograms (Synfogram). Engineers could understand right away the advantage of presenting their project in a Synfogram. When coming to production we made a workshop with engineers how to create their files in time and space continuum, so that the final information could be read by engineers as well as non engineers. Engineers do visualize their prototype in 3D, with mathematical formulas, and some drawings. We were teaching them to think and create in 3D, in time and space, to understand the relation of the image to the viewer, its movement, its perception, and its knowledge. For example, we did create a synfogram of the Safety couplings for rail vehicles of express train produced by Voith Turbo. **(1)** The goal was to explain the functional safety of this device to engineers as well as deciders and politicians. The object was recorded on 12 seconds which we can see slowly or fast, as we want, with our own body movement without any other instrument. We can just bodily freeze the image, go down, back and forth, in time and space. The Synfogram indeed is a wonderful educational tool to explain machinery, medicine, biology. This means also that the reading of the image is not linear like in a 3D film it is fully multidimensional. In a 3D theater the spectator do not move in front of a 3D film, but with hologram viewer have to move. Children are the best viewer. We could even train them to repeat a movement like shows the movie done in Niederbrone les Bains in Alsace **(7)** . They repeat the movement of a football guy by moving again and again repeating the movement at their rhythm. Still the presentation and visualization of the content is extremely important not just to be exact, but also to express the hidden situation, and make it available to the eyes, what a linear film would not be able to coordinate at the same time. Nevertheless any 3D image implies movement from the viewer, to search further to look from the side, to discover the many perspectives of the image. With children I made many questionnaires and what come out is that they like 3D images for the movement in interaction with themselves. The movement for them expresses "live". Present teaching considers movement as problem and disturbing concentration. Movement is in French classroom totally forbidden.

Pic01: The use of different stereoscopic recording technique, here using a rail system to create a synfogram.

3. How and what to teach

3.1 Compositions in Time/space

3D refers to a time space/relationship which put everything in relations. Creating an image in 3D is a matter of capturing multidimensional compositions where all parameters move together; the different objects of the scenery as well as the viewer. In case of a hologram the viewer is fully interactive and capture physically, with the full body, the many information contained in the various sorts of holograms (full aperture, integral, and digital hologram/Synfogram). This multiplicity of choices in reading 3D image is resulting of the time space relationship which shows that we can never see the whole at once; always we search new perspectives to « see" more. The reading patterns of each individual differ from each others. The order is not pre-established, the individual build it up according to one's perception and interest. SO to teach we need to exchange with the different participant to let them experience the different perspectives of each one. This experience of seeing 3D imaging reinforces this ability of choices, and opens the mind towards other possibilities that one would not have thought about alone. When the participants realize the quantity of choices they access, they feel stimulated and can compose a canvas of ramifications which are far beyond their first individual approach. The necessity of searching implies a decomposition of the composite subject. Decomposing the object in relation with the whole surrounding in movement is extremely useful to create in 3D but also to evaluate different aspects of a subject, not just alone but in exchange with others. This contributes to the process of learning and should be part of any teaching concept. In 3D Stereo Cinema, this multiples information is difficult to capture as well as it is disturbing to look at. This is why cameramen use for live event, some fixed camera to identify a space and just focus on the movement of a character. The top view is very often chosen in order to give access to this frame view. A new grammar is imposed to 3D creation of movie makers and pose them problem. Holographers already know about positioning camera, setting up a space, relating image, viewer and environment. So their knowledge could be more spread and integrated in new cinema teaching programs.

3.2 The Holographic experience in the perceptive process

The physicist Regis Dutheil **(2)** explained: "it is immediately after the production of an electrical activity that the person perceives the visual sensations, having the feeling that he saw something. Between the short instant which separate the production of an electrical activity and the perception, something happens. This little gap of time between the visual sensations and the perceptive representations can be the result of the correlations existing between the information stored as references and the new information which finally give the signification." This supposes that to recollect 3D information we need to destabilize our perceptive 2D system and its codified parameters. For more than 25 years, I have been able to observe hundreds of thousands of viewers. The confrontation with the holographic image, this visual of light in volume, floating in space and changing of perspectives at the rhythms of our movement, has the advantage to destabilize our believe in a unique and materialistic reality, this invariance which we build up and use as references, separated from us. In my paper the" holographic thinking"**(3)** , I explained that Experiencing a hologram is more than seeing. Seeing is not anymore a matter of our eyes but of our full body in movement interacting with the observed subject. By creating a gap in our perception of a unique reality, the experience of the holographic image opens the door to new territories of investigations, knowledge and creativity. The neurophysiologist Karl Pibram, **(4)** who demonstrated that the brain works as a holographic model, suggests tot each the holographic paradox to children at school. Meanwhile it would contribute to give a base of observation to evaluate new structure of knowledge and be able to produce new pedagogic concept based on experience and movement!

Pic02: Laser and Master Hologram demonstration in a school class.

3.3 Virtual and New realities

If we take the example of visual perception, we now know that it is nothing else as a sensorial message resulting from the first stimulation of some cells in our retina through the transportation and processing of the signal into the brain. However, what we do not see how it interacts with other perceptive organs, for example the eyes sense with the ears. There is some experimentation where information of sound, touch and visual are experienced in different time frame or reversed or displaced producing some discomfort or dizziness. This would suppose the existence of a certain order or organization to harmonize the whole. The contents, a composition of many parts, which are proposed in many 3D films, animations and visualizations, seem to be simplified, and make the viewer smiling. When we think that our mind already simplify the compositions from the initial information received, to fit the references or invariant stored, the remaining information is very poor and not so good finally for education. Nevertheless, the access to realities created with addition and multiplication of components even so not perfect, are not produced with reduction and subtraction of information. This is already a teaching. It just needs to become clearer, and not just unfolded, hidden. 3D in comics and games avoid confrontation with our 2D representation of reality by entering an imaginary, another reality, so called virtual reality. The success of Avatar is significant. 1st the quality of images - 2nd the content , placed in an imaginary universe, opened to creativity, free from pre-conceived ideas, and capturing the essence of 3D, by entering multidimensional levels of realities. The science fiction of Avatar, such as the neurons connection at the root of the tree, kind of dreamlike life, is so strong, that the defeat of our usual material reality of confrontation is acceptable. Holographic ideas transgress too our believe in one material reality, it confronts the viewer to many questions about our representation of the world, which many often viewers are still not ready to consider and prefer to give back holography to a scientific model away from them. With holographic representation we enter into a new dimensionality of our universe which will take time before being accepted culturally and socially. This is one element of explanation in the slow immersion of the holographic concept compare to the reassuring and blooming 3D stereo-system.

4. Spreading of 3D thinking through different media

4.1 Techniques and pedagogic goals

The choice of the techniques to present educational matter need to be related to a certain pedagogic goal. 3DStereo directs our eyes in a way that we are looking at a scene behind through 2 holes of a garden fence. Instead of having a quantity of images available to our perceptive information, like in a hologram, we just have a series of 2, 6 or 9 images tricking our senses organs. If the viewer is not fully conscious about it, its perceptive awareness makes the image suspicious, creating a kind of reserve toward 3D images. With digital hologram, the content might be week; still at the production, the technical properties of the light recording, can compensate week content, as the image is always moving in relation with the viewer. Visual immersion with large scenery interacts with the body through glasses and with computer control. With this system, exploration, discovery of a historical situation or geography could be very attractive for schools, making the teaching a living experience, but schools systems want still pupils, and minimum of individual experience. This implies new educational policies, with consistent Investment not just in the material, but in training and in new pedagogic material.

4.2 Training for Teachers

New, in France is the way dissertations in literature have to be written to pass the "Bacalaureat" (final diploma of high school). The pupil should not present their idea in the traditional structure of thesis, antithesis, and synthesis but in finding a problematic by deciphering the major points, and their inter-relations. This new training which for my holographic mind, open much more possibilities to elaborate and understand the complexity of a content, is mistreated by most of teachers in French literature as they just received a usual linear training and no information and support in presenting a method they never had time to experience or learn by themselves. Finally they cannot transmit it properly to their students. This short example reflects the most difficult part in social evolution: the training, and preparation of a new generation to the significant mutation of our time.

4.3 The implicate training of the holographic thinking
Some media stimulates the approach of a kind of Dynamic linearity or pattern reading instead of a linear dualistic approach even so there are not all in 3D. These media do not suppose to train our mind in 3D so this is what I call an implicate training, inspired by the theory of David Bohm and its Implicate Order, (5) which lays in front of us hidden to our believe and perception. A good example is Internet or now most of new touch screen, phone and tablets. Make a bit of searching in Internet. What do you see?. What do you have a holographic structure where each point stands by itself, contained the whole and composed it. To read the information on the Internet screen you might have the first habit to « scan » the screen with your eyes from top left to bottom right, as we use to in our western habit of writing. Instead, the eyes make incredible movements up, down, side to side, center back left, up right and so on, the hand can move the screen in many directions, creating a pattern which has nothing to do with our usual diagonal line top left, down right , etc. The reading process of a traditional printed page is linear; this other reading forms are expressing a dynamic linearity, or a pattern reading. (3) This way of reading exercise our perception and contribute to train our mind in moving within a multidimensional net of information, in fact training us to think holographically. More than exercising perception it emphasizes individuality and develops fantasy, liberating the frustration of a compressed perception used to fit one standard. The only problem is that teaching is a matter of repetition and awareness. Repetition is there, may we need to work on the awareness.

5. My Teaching experience

5.1 Context
Since 25 years I have been teaching the public in America, in Germany and now back in France, about 3D and time space relationship. I noticed that the cultural habits changes very much the reading of visual representation, and also the teaching we need to consider. Aware of these differences, I worked on the similarity to organized holographic exhibitions and workshops. This gave me a wonderful platform of research. Lately, in Germany and in France I have introduced different other 3D techniques now available on the market, may it be 3D presentation with monitor, old and new stereoscopic presentation, anaglyph, and other printing material like lenticular.

In France, the Minister of education introduced to the program of colleges and lycées, children between 12 and 16, new programs, such as art/time and Space, and production of a project in 3D. Also the idea is very good, teachers are confronted to a field that they do not know, they did not receive any training about it and have no pedagogic material available on the subject. I have been asked by teachers to contribute with my experience to program. For example, I conducted a program for the high school Jean MONNET between January and May 2011 sponsored by the Academie de Strasbourg, and the Region Alsace. The program has been financed by ACSIMA, Groupement d'intérêt public Action Culturelle en Milieu Scolaire d'Alsace, and the DRAC.
The project, a walk around images in 3D and 4D, including holography was born of a meeting with Annick Sary, an art teacher, and on a common desire to discover the meeting of art and science.

Journal of Physics: Conference Series **415** (2013) 012066 doi:10.1088/1742-6596/415/1/012066

5.2 My Program

My program consists of 2 parts:
. **The first part** is the shock of the images floating in space that allows young people to experience physically the time space relationship. Students are encouraged to discover the exhibition organized in the college and where I provide a guided tour with a variety of small experiments and animations Each group of maximum 12 persons have the duration of a course session to discover, analyze, discuss the questions of reality, image, perspectives ... from 2D, 3D, in relation to time, the movement ..., the place of the spectator, its role and its perception ...
A quiz is provided at the end of the session
. **The second part** is a workshop where the students repeat and experiment a basic technique that can be reproduced at home. I chose the anaglyph imagery because it is easy to understand. The children can make the 2 images and combines them with a computer program. The pupils learn to:
Discover and understand the technical properties and appropriate concept
Design a project, prepare and mount it to photograph it for a 3D image (calculations of distances and adaptation of different shots required)
Shooting sessions
Data processing, printing test, visualization using special glasses.... discussion and analyze of the image. And recovery (adaptation distances photos) if the result is not conclusive!
In this way pupils can learn a technique which they can recognize outside, on newspaper, and in the film industry. I did show them also how from this basic old technique, new technologies have been developed to come to use pair of 2 images with 5, 6, 9 repetition on a screen for 3D monitor, for example. This workshop allows pupil to understand other techniques of our present time such as S3D and M3D. The teaching about geometry and light were also repeated at the same time in the courses of math, and physics. In this way pupil can evaluate that observation belongs to all matters, as well as fields apparently separated like art (often just associated to drawing or painting) science and technology.
. **Result** The teachers, the college and the pupils who participated were very happy. The program went further in a week end of science where their created anaglyph portraits.
Annick Sary, wrote : For the first part of the project, we can agree on the fact that many students enjoyed the visual and sensory experience. We would have liked to accommodate more students, open exposure to a wider audience but the time allowed and the security of the works we do not allow it.
Some pupils (Classe 3D)
«"I find it interesting that the system is to come to see a different picture, when a group is all we see an image or different colors. I think it's great. » GV
Class 3B "" I found the newspaper in 3D interesting (Das Bild 08.30.2010), I had the feeling of being in the news. I liked the hologram of the little boy with his glass and bottle and the fact of moving to see all the images. "SN

6. Conclusion

If we start thinking 3D we can imagine a different way of treating the visualization. The problem to think 3D lay in our education and habit of thinking. Educating in 3d and with 3D is a big responsibility. We are coming from a 2 dimensional world of surface, related to a universe framed by the scientific determinism, and Cartesian dualistic mind to encounter 3 -dimensional representation of new realities. It supposed a radical change in the way we conceive education and need to have trained teachers. This can be done only with new educational policies and new research on the implication of a world passing from a representation in 2D to a representation of reality in 3D. Such research needs interdisciplinary participations including social researchers. We are at a turning point, where new techniques and science open new territories of multiple possible where all is in relations with each other: The 3D techniques, contribute to show what our human perception forgot to see. And if holographic visual disappeared from exhibition, holographic philosophy is spreading through other fields.

References

[1] Odile Meulien O and Dietmar Öhlmann April 2008 *Syn4D displays for a new generations of Holographic Applications* (Proc. SPIE Europe (Strasbourg)
[2] Pr. Régis Dutheil et Brigitte Dutheil 2006 *L'Homme Superlumineux* (Sand)
[3] Odile Meulien-Ohlmann 2000 *Holographic Thinking* (Proc. SPIE 4149 p 246 St Poelten)
[4] Ken Wilber *Le paradigme Holographique*, *Interview with Karl Pibram* 1984 (le jour Editeur)
[5] David Bohm 2002 *Wholeness and the implicate Order* (Routledge, Taylor and Francis Group)

More references
[6] The Art, Science and Technology Institute, Holography Collection has been created in 1982 in DC, as tax exempt organization under section 501(c)3 and closed in 1994. It has been the first museum to present an exhibition of holograms in a completely clearly lit area and received the acclaimed of the public and the press. In 1989 it has been compared to the exhibit of Holograms presented at the National Geographic (DC) organized by the Museum of Holography of New York, "March 24 1989, Washington Post" "Holograms: Duds and Gems. It received the award of the " Best exhibit" of Washington DC in 1986 by the Washingtonian Magazine.
(7) Odile Meulien and Dietmar Ohlmann *"How to do a synfogram"*, video youtubehttp://www.youtube.com/watch?v=2Rqd00aLYu8&feature=related. Short cut of different works done with and by children, in Braunschweig (D) and Niederbronne les Bains (F) in 2007 and 2008.

Looking Back and Forward – Holography at the Academy of Media Arts, Cologne (KHM)

Martina Mrongovius and Guillermo Heinze

-1/MinusEins / Experimentallabor, Academy of Media Arts, Cologne (KHM)
Peter-Welter-Platz 2, Cologne, 50676, Germany

mlm@khm.de

Abstract. The holography laboratory at the Academy of Media Arts, Cologne (KHM) has been active since 1990. During this time numerous students and staff have created holographic artwork and experimented with various printing techniques. Several projects are presented to show how holography can be used to explore a range of artistic concepts, leading to a discussion of the holographic image in relation to creative practice and art education.

1. Introduction

Academy of Media Arts, Cologne [Kunsthochschule für Medien] known as the KHM was founded in 1990 with the aim of establishing an environment for the advancement and teaching of new media arts. The founders understood that exploring artistic questions with new media required not only technical proficiency but the cross-pollination of conceptual thinking in a number of disciplines.

One of the founding professors, Dieter Jung, established the holography laboratory – the 'hololab' – with the assistances of KHM technician Urs Fries and assistant professor Andrew Pepper. Over 22 years the hololab has seen a wide range of projects created by KHM students, many guests and a number of visiting fellows. The assistant professorship in holography is a limited term position that had been held by Michael Bleyenberg (1997–2002), Peter Schuster (2002–2008) and currently Martina Mrongovius (since 2009). In 2008 Mischa Kuball became the professor of holography, creating the -1/MinusEins / Experimentallabor and encouraged students to develop artworks with optical media and present these on the 'urban stage'[1].

The holography laboratory in -1/MinusEins currently comprises:

- A multiplex printer '*the Imager*' first installed 1991 and updated in 2003
- A dot matrix hologram printer with a resolution of 280 x 380 px installed in 2002
- A small table with HeNe laser

The multiplex and dot-matrix systems were designed by Walter Spierings of the Dutch Holographic Laboratories.

The hololab and '*the Imager*' 1994 (Left), 2000 (Centre), 2011 (Right)

This paper describes a number of conceptual explorations made with holography at the KHM to reflect on the broad scope of the medium and raise the question of where holographic imaging technology could lead next.

2. The hologram as an architectural element

Michael Bleyenberg was both a student and assistant professor at the KHM and developed holographic works that he came to consider as 'Light Architecture'[2]. Bleyenberg currently collaborates with the Cologne based company HoloPro to produce architectural installations with dot-matrix holograms.

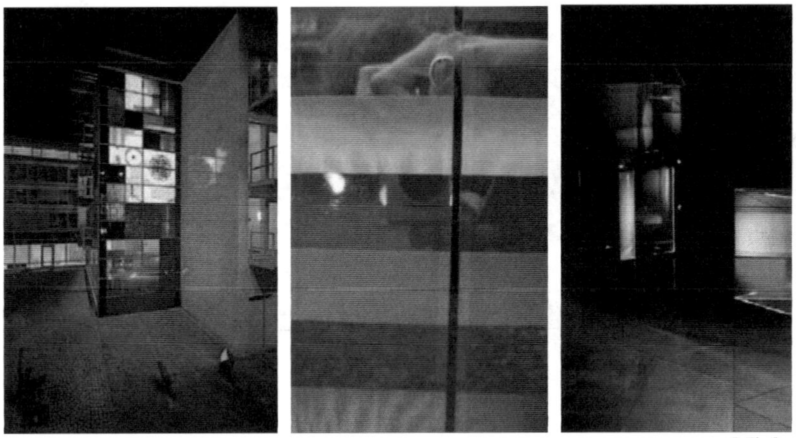

Michael Bleyenberg
Left: *EyeFire/AugenFeuer*, 2000. German Research Association (DFG) building in Bonn, 13 x 5 m
Centre: Close up of *EyeFire/AugenFeuer*, photographed 2008
Right: *Modul, 1994*

3. Animation in holographic images

The multiplex holographic printing system '*the Imager*' has been used to print both stereographic images that synthesize 3-dimensional space as well as a whole range of animated sequences.

Jean-François Moreau was a fellow from 1994–1996 and made multiplex 'video-holograms'. He also used the multiplex printing system to create 3-dimensional portraits using a camera track and through data collected by scanning the subject. Moreau was interested in how time could be caputured and viewed through holography, "the hologram introduces a new type of time... The movement in space of the viewer corresponds to a displacement in time for the image." [3]

Jean-François Moreau
Right: *Night and day*, 1994, vidéo-hologramme, 20 x 25 cm
Centre: Stereographic vidéo-hologramme portrait from camera track, 20 x 25 cm
Left: *Folie*, 1994, vidéo-hologramme, with Kirk Wollford, 20 x 25 cm

Martina Mrongovius who first came to the KHM as a guest in 2005, became interested in the way multiplex holograms provoked the viewer to move and has creating a series of works that played on the relationship between the movement of the viewer and the animation of the scene.

Martina Mrongovius, *Up the Stairs*, 2010
Installation as part of the exhibition *'-1/MinusEins im Exil'*, Sommerloch, Wuppertal

4. The hologram as a memory

The relationship of the hologram to memory was nostalgically captured by postgraduate student Stefan Sillies in his installation *Schau heimwärts!* [Look homeward] (2005). The installation consisted of six tables each with a number of small boxes, some of these boxes could be opened, some could not. Some of these boxes illuminated a hologram when opened. Other boxes when opened were empty or filled with small things. Each small box reflected a stage of life or a special event.

Stefan Sillies
Schau heimwärts! [Look homeward], 2005

A guest student from Chile, Claudia Sandoval considered the social implications of holographic images and was drawn to the disappearance of the image when not illuminated or viewed from the correct angle. Working with this quality Sandoval created a series of dot-matrix hologram portraits *'Holograms of Deportation'*:

"In *'Holograms of Deportation'*, mirrors stand blindly reflecting throughout the day on walls of Berlin. Only during a couple of hours of the day and from a certain perspective these mirrors reproduce the images of three people who have been deported from Germany, as if these dot holograms were bringing back a sort of screaming ghosts. The street art intervention is transmitted through the internet during the day. Opposite to all expectations, nothing happens to the holograms. In a city like Berlin three standing mirrors are quite invisible. As if the city itself knew they contain only ghosts, as if it was better to keep the deported as they are: Unseen."[4]

Claudia Sandoval, *'Holograms of Deportation'* 2010

5. The limited viewing of a hologram

The limited yet relational geometry of viewing white-light transmission holograms was the basis of Stuart Rosenberg's experimentation leading to *'Viewing Apparatus'*(1991–1993), in which relationships of viewer, light-source and hologram are materialised into a device. The first iteration of the project *View1* looked something like a medieval torture device: the viewer inserted their head between two planks of scaffolding wood. The physically uncomfortable restriction was reflected in the hologram designed for the installation.

Stuart Rosenberg, *'Viewing Apparatus'*
View1 and hologram for installation, 1991

Stuart Rosenberg, *'Viewing Apparatus'*
Left: *View2*, 1992. Right: *View4*, 1993

This positioning of the viewer was also the theme of Mioon's installation *Holopublikum* (2005)

> "In front of the audience there is a stadium. 200 holographs hang down within this stadium to give a clap to the audience. Moving components (consisting of ten motors) are installed behind the 200 holographs to move them. With this motors the people shown in the holographs are generated to shout and cheer. If the audience stands in front of the stadium, it can hear sounds of shouting and cheering peoples, too. Actually in our huge modern society people are standing under the influence of mass media, politicians, capital and other external influences. Historically we have seen the example of control or domination thru this so many times. So with this works I want to show this situation and the lack of critical faculty of the general public, and the lack of personality of them. Besides that, I also want to show to the people that they are controlled right now by somewhat." [5]

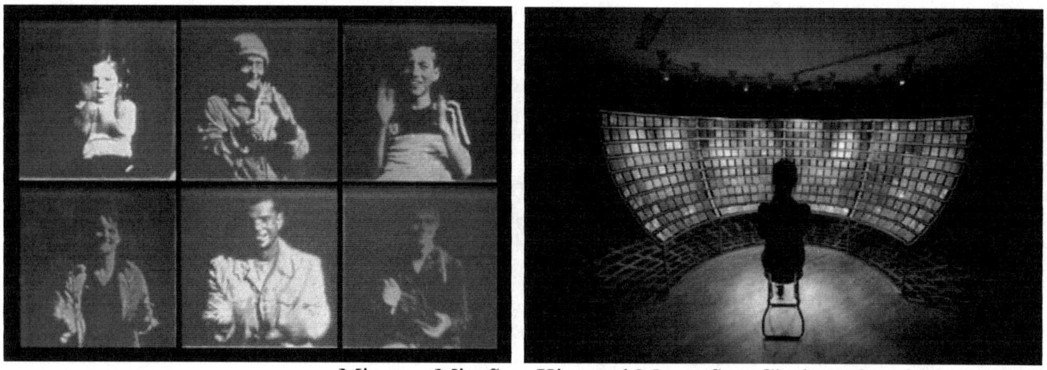

Mioon – Min-Sun Kim and Moon-Sun Choi, *Holopublikum,* 2005

6. Playing of the holographic image

Seth Riskin who was a fellow at the KHM in 1999 considered the playing of light through the hologram by using his body as a light source for performance that was then viewed through a holographic diffraction grating.

Seth Riskin, *Light Dance,* 1999

The playing of the holographic image was also explored by postgraduate student Auriel in his garden installations of dot-matrix holograms. The holograms, which he changed like during the performance, were illuminated by dappled sunlight and accompanied by a reel-to-reel recording of Auriel's music.

Auriel, hologram performance in the garden, 2010

7. Fragmentation and structure

Miri Shin, who is currently completing her post-graduate diploma, uses holography among other mediums to explore a chair as a hyper-cubist form. Shin's work fragments and combines the representation of this familiar object into a structural form, which she describes as a means "to find out another latent reality of the object by tracing aspects of changing time and space in act of looking at it".

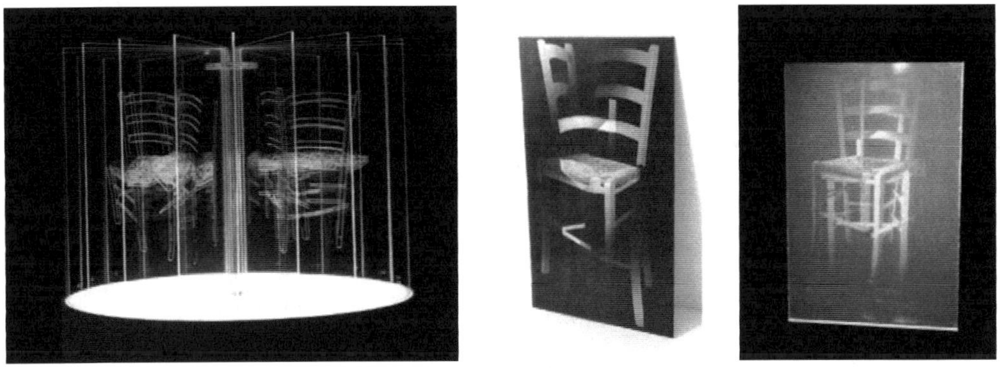

Miri Shin *'Rotating Chair'*, 2011/12

Left: 20-panel etched-image plexiglass sculpture. Centre: Lenticular. Right: 200-frame multiplex transmission hologram.

8. Visual texture

Guillermo Federico Heinze an undergraduate student, explores and uses holograms to create objects and installations with a visual texture that moves and changes colours according to the observer's interactions. In *'the hologram and the girl'*, observers could interact with a dodecahedron composed of twelve digital full color holographic stereograms. Wearing golden gloves and moving the shape around under three lights the observer can interact with the work in a ludic way. Discovering each hologram reveals that the girl inside the hologram is doing the same, wearing the same golden gloves and interacting with platonic bodies like pyramids, cubes and spheres.

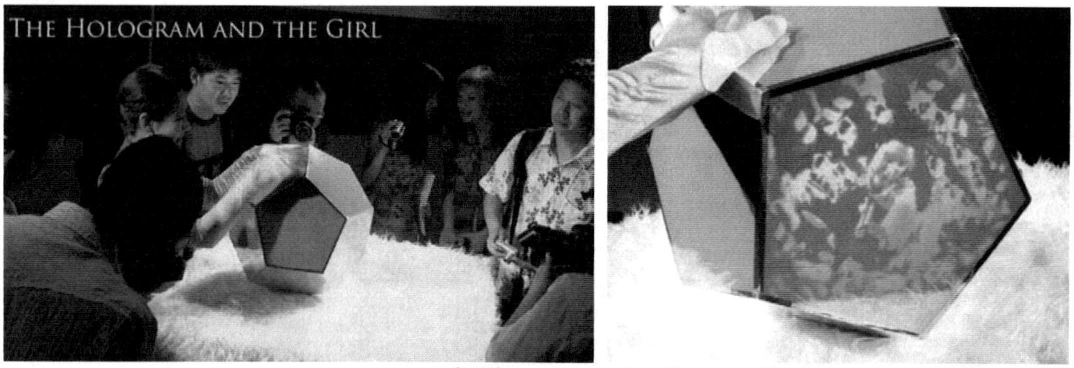

Guillermo Federico Heinze, *The hologram and the girl,* 2006

With *'In and out-sides'* Heinze's invites the observer to explore a landscape of irregular surfaces, toys and deep visual cavities of holograms on the surface of an asteroid-like sculpture. The holograms show different abstract sculptures, connected to the whole through the materials they involve. The pink surface of the sculpture is composed of Styrofoam balls, sugar and glue, designed to make the observer hungry for sweets and is reminiscent of a huge cotton candy. The kinder-surprise toys, joysticks and computer parts suggest a playful way to perceive and transform reality, something that is very usual to children.

Guillermo Federico Heinze, *Ins and outsides*, 2007

9. The collective holographic image

The seminar 'Collecting the Image' in 2009 encouraged students to compile holograms from multiple related photographs, much in the way that images are 'pooled' online.

Left and Centre: Auriel, *Holotour*, 2009. Multiplex hologram from touristic photographs taken across Europe
Right: Guillermo Federico Heinze, *mydailyfaces*, 2012

A recent work by Guillermo Federico Heinze that follows this theme is *'mydailyfaces'* – an achromatic multiplex hologram that is composed of two hundred self-portraits, taken over a two hundred day period. When viewing this hologram, it is impossible to see only one frame, instead, the viewer sees a collaged face, composed by different days, different emotional states, combined to offer a wider spectrum of the portrayed person.

10. Conclusion

These are just a few of the projects created at the Academy of Media Arts, Cologne (KHM). With each project being developed through creative dialogue and experimentation. The process of presenting concepts through installations with holographic images involves a negotiation between the possibilities of holographic conceptualisation and the recording and printing techniques available. Compared to other media in the last 22 years at the KHM, the scope for printing holograms has not changed significantly. Yet culture and technology seem poised, ready to explore new holographic installation possibilities. So the question is how to develop holographic imaging practice as a tool for future artistic expression?

References

[1] http://minus1.khm.de/
[2] Michael Blyenberg – http://www.bleyenberg-light-art.com/
[3] Jean-François Moreau. 1997. *La représentation de l'espace-temps et l'hologramme*. http://rac.chez.com/souvexp.html Translated from French: l'hologramme introduit un temps d'un type nouveau... Au déplacement dans l'espace du spectateur correspond un déplacement dans le temps pour l'image
[4] Claudia Sandoval http://www.claudiasandovalromero.com/
[5] Min-Sun Kim http://www.khm.de/pm/projekte/projekt/paview/1968/paitems/60/pasearch/mioon/

Information about holography at KHM: http://kunst.khm.de/plattformen/holographie-und-lichtkunst/

Journal of Physics: Conference Series **415** (2013) 012068 doi:10.1088/1742-6596/415/1/012068

The Center for the Holographic Arts Begins a New Artist Workshop and Residency Program in Conjunction with Ohio State University

Martina Mrongovius, Harris Kagan and Sam Moree

Center for the Holographic Arts
39-31 29th Street Long Island City, NY 11101 USA

info@holocenter.org

Abstract. This year the Center for the Holographic Arts (Holocenter) kicked off a new Artist Workshop and Residency Program in conjunction with Ohio State University. The newly renovated holography facility houses the Holocenter's pulse laser camera and two recording tables with continuous wave lasers. This facility is being utilized for undergraduate and postgraduate courses in Art and Technology as well as the Artist Workshop and Residency Program.

1. History of the program

The Center for the Holographic Arts (Holocenter) is a non-profit arts organization dedicated to developing and exhibiting holographic artwork. The Holocenter was founded in 1998 by Ana Maria Nicholson and Dan Schweitzer with four main goals – education, research, production and dissemination.

For 10 years The Center for the Holographic Arts offered an Artist In Residence program at it's Long Island City studio, with artists from all around the world coming and creating holographic artworks. These holograms were shown in museums and galleries all over the world.

Works created by 2004 Artists In Residences at the Holocenter
Left: Pepe Buitrago, *VARIABLES OCULTAS* [*HIDDEN VARIABLES*] at Catedral Gallery, Spain, 2006
Right: Eva Davidova, *'Feeding Alice'* exhibited at Center for the Holographic Arts, New York, 2007

In 2009 the Center for the Holographic Arts lost its main source of funding and could no longer afford to maintain the studio in New York City. The office and library were moved nearby to a co-working space at Flux Factory, which also has a gallery for exhibitions.

Ana Maria Nicholson *'Into the Night'*, 2009
Exhibition and artist talk at Flux Factory

The Holocenter's pulse-laser camera was loaned to Ohio State University, where Professor Harris Kagan arranged for the laboratories to be renovated in order to house the pulse-laser. In 2012 after the completion of renovations the Artist Residency and Workshop Program was launched.

2. Holography at Ohio State University

The holography program at Ohio State University is a joint initiative between the Department of Art and the Department of Physics and is taught by Professor Harris Kagan. Holography classes are a core component of The Art and Technology Program with Bachelor and Masters courses designed to enable students to develop artistic practices using science and technology. Around 140 students each year make over 700 holograms in the laboratories.

End of quarter holography lecture and exhibition at OSU, March 2012
Right: Sam Moree giving a lecture on holography to OSU Students
Center: Students looking at the exhibition of holograms
Left: A student explains holography to his family

Part of the collaboration between OSU and the Holocenter is to foster exchange projects, where experienced holographic artists can use the facilities in return for teaching. In April 2012 Betsy Connors worked with the Advanced Holography Class and also created some of her own artwork.

3. Learning in Lab

Learning holography requires both a theoretical understanding and practical knowledge. From artist proposals and questions to the Holocenter it became apparent that there was general confusion about holography. This is not surprising considering the way that 'the holographic' is represented in film and by marketing companies. The problem is that many artists want to work with holographic technologies without understanding the practicalities of the medium. We decided that the best way to tackle this was to give more artists a chance to have the hands-on experience of making a hologram.

The introductory residency is a group workshop involving five to six participants. Sam Moree and Harris Kagan teach the workshop together giving two different perspectives on holography. The workshop revolves around the creation of a pulse-laser master while a number of transfers and one-step recording techniques are also demonstrated.

Introductory workshop residency March, 2012

Through the OSU–Holocenter collaboration we can continue to offer a residency program for artists who have a base knowledge of holography and a clear concept for a holographic artwork. This year Philippe Boissonnet plans to produce a series of pulse-laser master holograms and Tristan Duke and Eva Hausam are working on a project to record the movement of beehives.

4. Future Plans

After this year's workshops and residencies the program will be revised based on feedback from the visiting artists as well as OSU students and staff. We are seeking funding to assist artists with the cost of travel to Columbus, accommodation during their stay and materials. Additional funding would also allow us to create exhibitions and symposia.

The open call for the 2013 Artist Workshop and Residency Program will be posted on the Holocenter website in October. We encourage artists to apply who either want to learn the basics of holography or have a concept for a holographic artwork.

The goal of the Artist Workshop and Residency Program is to establish a collaborative environment where artists and students can work together to develop their understanding of holography and produce holographic artwork. Conceptual expression with the medium continues to develop by enabling artists to access holography facilities. We hope that the holographic artwork produced through this program will be exhibited and allow the public to engage with the creative discourse around holography.

Further Information

[1] Center for the Holographic Arts www.holocenter.org
[2] Holography at OSU http://www.physics.ohio-state.edu/~kagan/holography/
[3] Art and Technology Program at OSU http://www.artandtech.osu.edu/

Holography and Art – Using rainbow hologram in my works

Duck Hee Cho

3434 Spectrum, Irvine, CA, 92618, US

sailhologramart@gmail.com

Abstract. Holography can be understood as one of the more sophisticated subjects within the realm of modern art. Even after several years of study and application, understanding Holography simply does not get any easier. For that reason, instead of trying to discuss both the technical and the artistic aspects of holography, the emphasis will strictly be on the latter. Just like any other paradigm or medium capable of fascinating the human hearts, holography has changed the way I encounter the world. Changed perception of the world brought me feelings of child-like delight. Among the various subdivisions of holography, I hold Rainbow Hologram exceptionally close to my heart. Not only is it related to my previous works but its bright, crisp and various color movements are also depicted wonderfully in 3-dimension. I would like to share my fascination with rainbow hologram with you.

Motivation

A new chapter of my life opened up upon moving to New Zealand in 1995, I have been a Chemistry teacher in South Korea, but I always had a passion for studying art. Moving to a new country and being exposed to a drastically different environment encouraged me to pursue what I secretly cherished.

After entering Wellington Institute of Technology, I was asked to research what fascinated me the most. I didn't know how to go about this assignment at first, but luckily I stumbled on several hologram images and my eyes were set. The hologram images were completely mind blowing and refreshing, unlike the ones I saw previously which only made me feel eerie.

However, holograms by Rudie Berkhout struck me totally in different ways. The blend of rich and refined colors he used in his works was comparable to the beauty of fine arts, especially paintings. Such strong impressions led me to write an essay on Holography and I have been captivated by holography ever since.

Fig.1 (a) Hologram I saw before.

(b) Event horizon by Rudie Berkhout.

After my first essay on Holography, I was determined to study this fascinating genre if I get a proper opportunity.

However, it took me some time before I actually started to study Holography. Although I produced various holograms such as transmission holograms, reflection holograms, stereo holograms, and color holograms, my passion was and still is for rainbow holograms.

Using rainbow in my works
It can be said that rainbow represent the very essence of my art works if not the most valuable subject. In Wellington, where I used to live, rainbows were easily spotted. Looking at the rainbows back then was like being promised of a better life in New Zealand.

The fact that rainbows are special symbols as well as religious events became my motivation and passion for dedicating myself in producing rainbow holograms.

Rainbow holograms can express various subjects very clearly.

Gorgeous colors of the world with their rich depths can be reorganized and interpreted to form diverse range of patterns and color combinations.

My rainbow holograms works

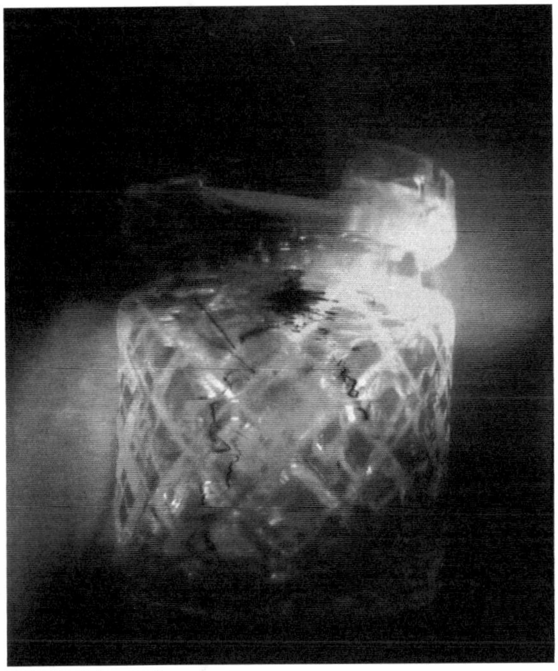

Fig2. Crystal cup

Personally, I like some crystal cups. Being transparent, these crystal cups allows us to look at both inside and outside at once.

It is interesting to both look inside from outside and vice versa. This cup is a representation of me holding my soul inside. I sometimes try to hide what I am feeling but to no avail.

The reason to such futile disguise is because of the transparency at my core. It is because I dream of cleansing and purifying my soul without hypocrisy and affectation.

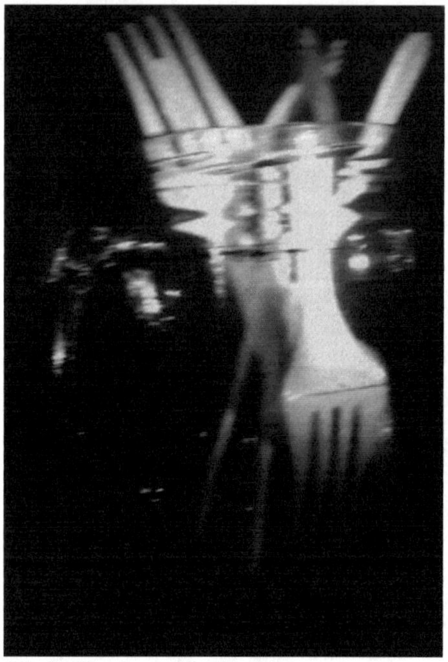

Fig3. Forks in the bottle

The beauty of being transparent is connected to how I want to live my life. Being transparent is to show everything without any filter distorting the essence. For any artists, his masterpiece is bound to be the life he yields. The white forks in the bottle can be seen as my will, the road I chose, goals I want to achieve and also the hope I hold dear to my heart. Parts of the forks showing outside the bottle represent what have been done, while the ones inside the bottle are what have yet to come.

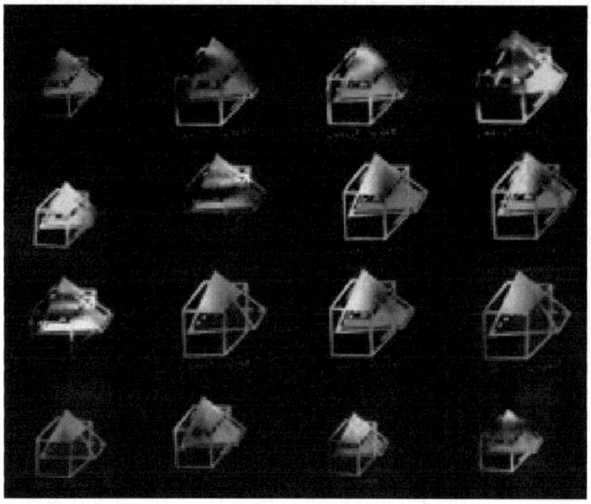

Fig4. Artist desire

I believe artists should be like a chameleon, continuously searching for what cannot be seen from the box by transforming the self into a triangle and to get out from the square, all the while painting the self with various patterns and colors. What I want is to have a desire to acquire something new endlessly.

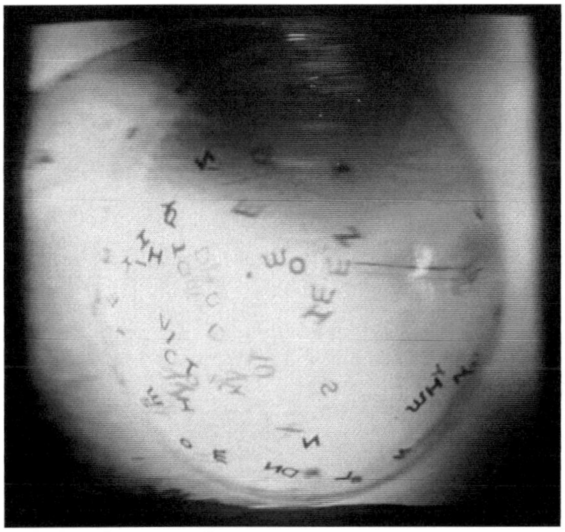

Fig 5. Floating abc

Various letters are floating around only to suggest the hardship of choosing the right word for the proper moment.

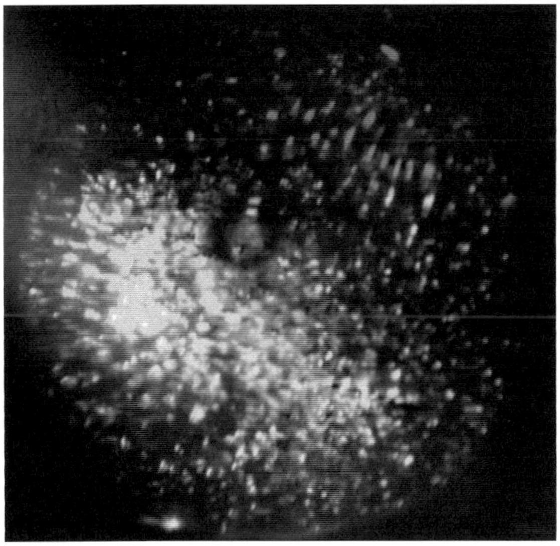

Fig 6. Big Bang

The stage where the unknown is coagulated and the desire to pinch/activate it coexists.

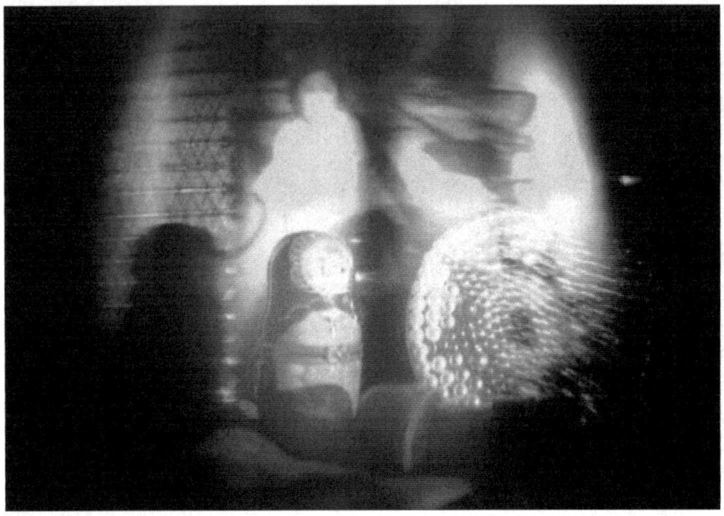

Fig 7. Reminiscence

Reminiscing on my childhood, where the most mundane become fascinating through the lens of kaleidoscope.

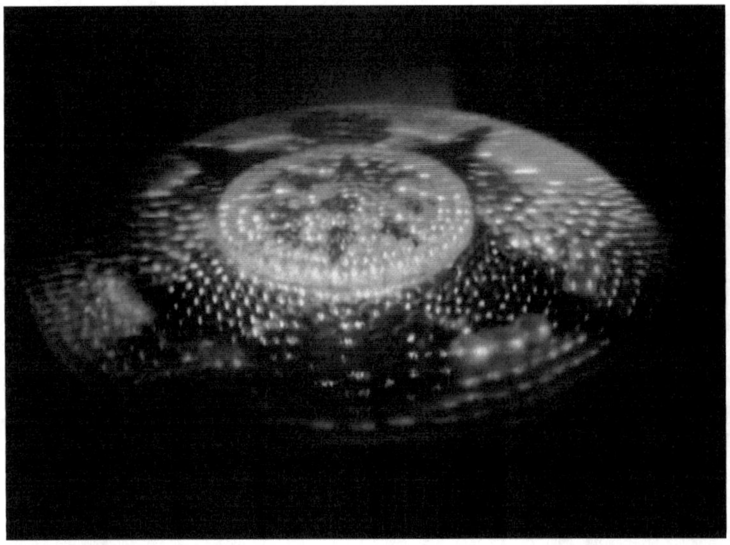

Fig 8. Wonderland

The world of few beads seen through the kaleidoscope.

Fig 9. Windy Wellington

Wellington, where I used to live is very winy city. I may shiver at a chill gust of wind against my face, but the wind does bring along a sense of freshness and bliss.

For fellow and future enthusiasts

One of the biggest obstacles people will face when wanting to become a hologram artist is to find someone from whom one can learn about holography. It is extremely hard to produce the images you want, but the chance to produce does not come by easily. In short, it is harder than you imagine to find a studio or institution where you can either learn or produce holography.

Even if you equip yourself with the proper tools, lack of knowledge for utilizinge the tools in producing holograms will prevent you from creating the image you want.

Holographic art peaked during the 1990s but has been on a downward trend ever since.

What helped me promote my work was my blog where I posted various works of holographic artists Unfortunately, due to the lack of popularity or awareness of holograms, the postings were limited to introducing various international hologram artists to Korea.

Recently, collaboration between artists utilizing different mediums to share their knowledge and insights is becoming more and more common.

I believe that it is important to accommodate artists and technicians from various backgrounds. That way, the Holography as a discipline may flourish once more.

I only started studying holography very recently. However, if I fail to find someone to whom I can offer a hand or encouragement, my interest in holograph might fade. At the same time, it is up to me to pave my own way while keeping myself motivated and passionate.
I wish there were more educational opportunities for the people wanting to learn Holography
Most hologram artists have at least two separate major areas, because holography is rather complex.

I sincerely hope that those with more knowledge and experience would be kind enough to share what they know with the novices trying their best to make a first step into the field.

"Jerry Pethick: The Missing History of the Sand Isolation Table"

Melissa Crenshaw
Research
Philips - Ledalite
Langley, BC Canada

E-mail: melissa.crenshaw@philips.com

Abstract. Jerry Pethick (1935-2003) was one of the first artists to experiment with holography in the mid 1960's. He was a close friend and collaborator with Lloyd Cross of Multiplex fame. Pethick and Cross started the School of Holography in San Francisco California on Shotwell Street. Pethick is known for building the first sand optics isolation table and writing the first holography handbook describing how to build such a system and create holograms with it. This paper offers some of the missing history of his accomplishments.

1. Introduction

Colourful would be one way to describe the people who were there at the creative forefront of holography in the 60's & 70's. One such person was Canadian sculptor, Jerry Pethick, an Associate of the Royal College of Art in London & one of the first 2 or 3 artists to explore holography. Sean F. Johnston's, *Holographic Visions A History of New Science* has provided the field of holography interesting and detailed accounts of the early days of holography as science, as an art form and an industry. Included are stories about Jerry Pethicks contribution. [1]

Many of the details in this paper & accompanying poster attempt to supply some missing history as shared with me by Jerry's widow and wife of many decades, Margaret Pethick. I have her permission to describe what she shared. This photo (Fig 1) shows Margaret in her home on Hornby Island.

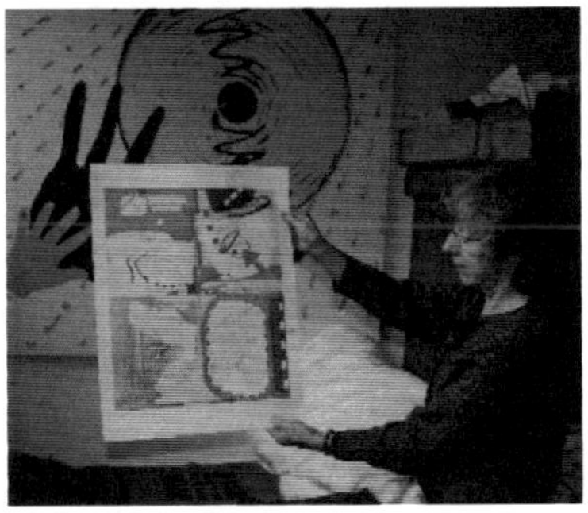

Figure 1
Margaret Pethick at her home on Hornby Island, BC Canada

Jerry's interest in holography evolved from and coincided with his interest in creating art utilizing plastics and integral photography. Jerry went on to become an internationally known artist working with lens arrays and other optical materials after transitioning on from holography. His time in the field was brief but his contribution significant as described below.

2. Jerry Pethick Time Line (quoted where noted by Margaret Pethick)

Jerry Pethick met Dennis Gabor in London around 1967, at Imperial College after seeing Gabor on BBC demonstrating Holography. Jerry decided he "had to find out how to make holograms" . It was Gabor who advised Jerry to go to Ann Arbor where more research and work in Holography was happening. Jerry also met holographic artist Margaret Benyon in London around this time and they talked often in these early days.

It was in Ann Arbor Michigan where he met & began his close collaboration with one of the great folk hero's of holography, Lloyd Cross of Multiplex fame. Lloyd Cross was working in Ann Arbor, doing advanced work in optics and holography. Jerry and Lloyd were friends and collaborators for many years. Jerry and Lloyd "hit it off immediately" according to Margaret.

Figure 2 Jerry Pethick and Lloyd Cross, date and photographer unknown.
Collection of Margaret Pethick

Jerry returned to London and Lloyd then brought Jerry his first laser, most likely a Krypton or Ruby Pulse to the studio where Jerry was setting up his sand-based, car tire tube isolation system. The studio was located over the Tube line in London and there were "horrific vibrations". Consequently, Jerry had to work at night after the tubes stopped running and before they started again. Such recording schedule considerations were emblematic of the spirit that would be the signature behaviour of future holographers.

Jerry relocated to Ann Arbor in1968 to work with Lloyd and it was there that a group of artists and scientists set up Editions Inc. to make holograms using the sand-based system. In 1969 they exhibited holograms at Cranbrook Academy in Detroit, and in 1970, *N Dimensional Space* at Finch College Museum in New York.

3. The Sand Table

There are at least two accounts of the invention of the sand isolation optics table as described in Johnston's book on pages 264 & 265. [1] There is some confusion about the exact story but there is no debate about who was there and did the inventing. The two, young holographers, Jerry Pethick and Lloyd Cross are credited in various accounts with figuring out and then building the first such systems at Jerry's London studio and at Editions, Inc in Ann Arbor Michigan around 1969.

The whole philosophy behind the sand isolation table was an accessible approach that liberated holography from the confines of well funded, off limits labs in research facilities and universities. Jerry would later write in a down to earth way about building the sand system.

"A garage or basement concrete floor is best, but anyplace away from people and traffic such as in the country is an ideal location for holographic work. Once the sand box and inner tubes have been set up, the optical components are mounted on tubes and stuck in the sand. This allows for flexibility and quick adjustments, with immediate stability as soon as the tubes are released, since the bottom portion of the tubes become filled with sand, thus lowering their centre of gravity." [2]

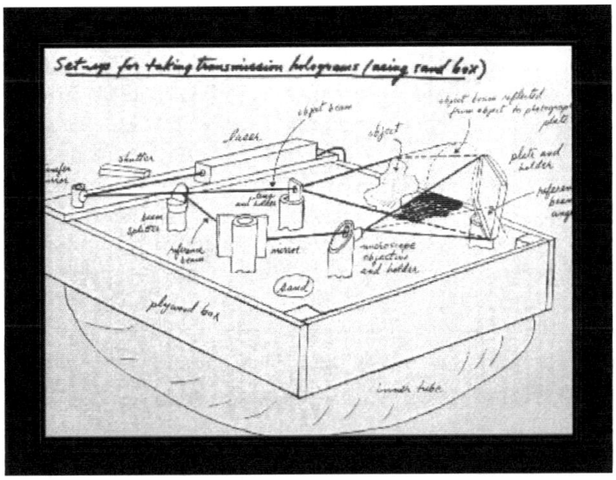

Figure 3 Jerry Pethick illustration of his Sand table isolation system

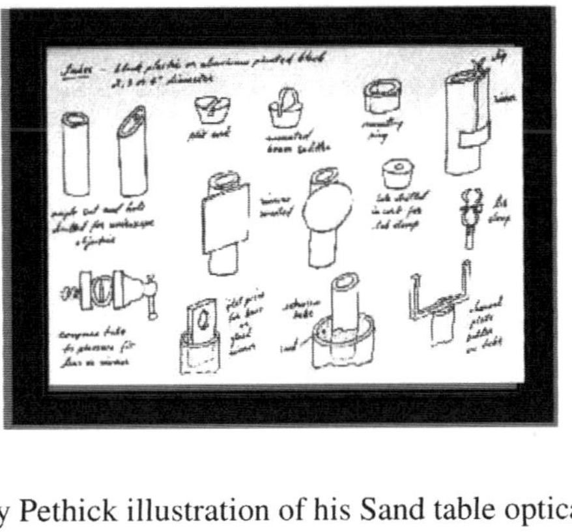

Figure 4 Jerry Pethick illustration of his Sand table optical components

The approach taken & the simple materials combined in clever ways was the basis for the garage and basement holography phenomena that survives today. Colleges and future schools of holography adapted this approach for its simplicity, portability (well, sort of) and affordability as shown in Figure 5 below.

Journal of Physics: Conference Series **415** (2013) 012070 doi:10.1088/1742-6596/415/1/012070

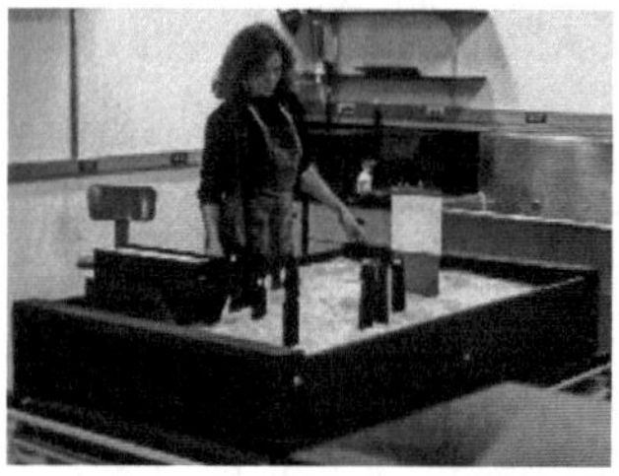

Figure 5
Evergreen State College
Olympia Washington, USA
1983 Sand table holography teaching lab
(photo M. Crenshaw)

Following the interest generated at the Cranbrook Academy and Finch College exhibitions, Jerry and Lloyd decided to start a school of holography in New York. Margaret Southcomb, Pethick's future wife moved from the UK to NY to join him. Margaret describes NY as too difficult and expensive to rent holographic space so the group looked to move on. During this time, Jerry had been doing some block teaching (sculpture) at the San Francisco Art Institute and was totally enamoured with S.F. So, in October 1970 "we all" moved from New York to San Francisco.

4. San Francisco, Project One & The School of Holography
Margaret and Jerry had been told about the former American Can factory, now called "Project One". It was just off Market Street at 9th and Howard. This square block former factory was now taken over by a group of people, trying to create an "alternative Urban society (this was late 60s!)" according to Margaret. In 1970 they moved into the basement which they rented by the square foot. They were supplied with studs and dry wall to create the space. Lloyd arrived several months later & this was the space they ultimately transformed into the first School of Holography in San Francisco.

Obtaining some type of credential acknowledgement from SF State and possibly Antioch they set up the first teaching session at the school. Now, thanks to their efforts, "there were eight more holographers in the world". Of note, is that holographic educator, author and artist Fred Unterseher was a student in this first class.

5. Holograms: How Do You Make These Things?
Jerry Pethick was the author of the first published book describing how to build a sand isolation optics table and how to make holograms with such a system. The little book with hand drawn illustrations titled "On Holography and A Way to make Holograms" was published in 1971 by Belltower Enterprises and was used for teaching at the School of Holography. [2]

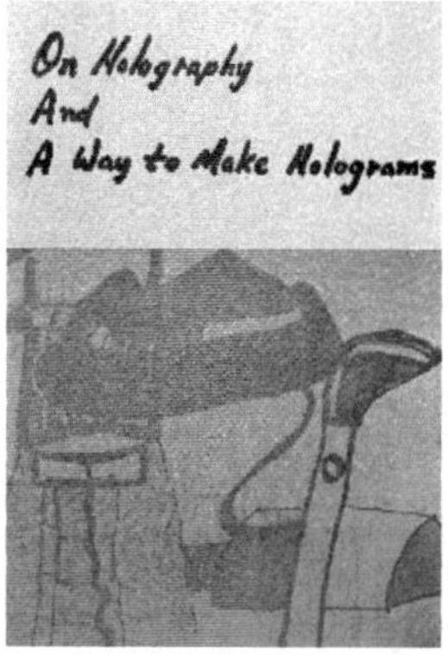

Figure 6
Cover of Pethick's
On Holography and
A Way to make Holograms

Excerpts from this book defined a language that encouraged access by anyone to this rather obscure and esoteric science/art. The language he used, although technically correct, set the tone for the intimidation-busting that was necessary for holography to open up to a wide circle of practitioners, as evidenced by the following paragraph.

"Holography is simple. Anyone with interest, basic information and minimum equipment can make a hologram. Maeterlinck said of man's use of tools (the telescope for example) that the discovery of a new star or phenomenon is not named after the tool, but after the man who discovered it. Technology has a host of new tools for those who are interested, but it will take time and talented people to realize their intrinsic value. The potential of the future is now, given imagination and the will to encounter the unknown." [2]

In these early writings by Pethick, the experimentalist view emerged that there were no wrong ways to make a hologram as long as something played back & it was even OK to break the rules. For instance you could create colour compositions resulting from the "spectral smear" from using a non coherent reconstruction source to illuminate a laser transmission hologram. And this was written in 1971!

The group continued at Project One for a couple of years and then move the school to Shotwell Street where Lloyd set up Multiplex. Shotwell became synonymous with the San Francisco School in late 1970's. More holographers than can be credited here studied at this school and many became excellent artists and leaders in the industry that was emerging at that time.

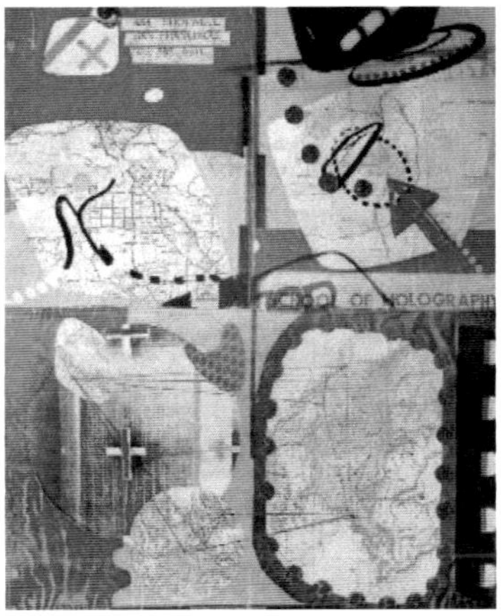

Figure 7
Shotwell Street, San Francisco
School of Holography Poster
collection of Margaret Pethick

6. Jerry Moves On

Jerry gradually moved away from holography after 1973 as he became more involved with Fly's Eye lens and integral photography, experimenting with sculptural spatial dimension and other optical materials he could fit in with the sculptural element of his work.

Margaret describes the new direction of Jerry's art as "... array pieces, multiple images taken with an instamatic camera, each from a slightly different point of view, which when viewed through an equal number of fresnel lenses at their focal length, reconstruct into one large image with dimension." At this time Jerry also sourced plastic fresnel lenses and off the shelf diffraction gratings such as Spectrafoil and used the spectral colour for pieces such as his enameled stove top works and other such found objects easily obtainable at local dumps. It's not possible in this brief history to do justice to Jerry's full life as an artist. For further study of his non holographic art works please visit the catrionajeffries.com Gallery web site.

Journal of Physics: Conference Series **415** (2013) 012070 doi:10.1088/1742-6596/415/1/012070

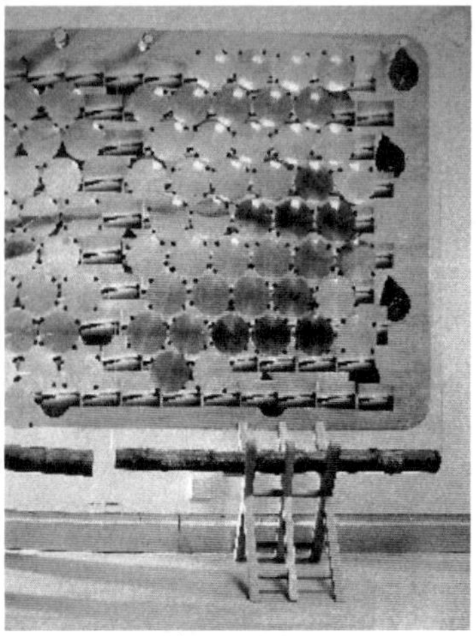

Figure 7
Detail
Notion of Nothing 1990/91
Jerry Pethick [3]

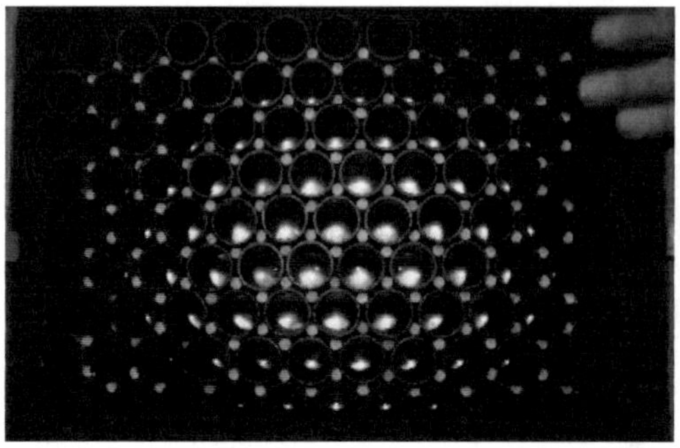

Figure 8
Lens Array Camera
1993, Jerry Pethick
Collection of Jim & Ro Allworth
(photo M. Crenshaw)

7. Annette Hurtig of the Kamloops Art Gallery wrote of Jerry Pethick:
"Jerry Pethick's innovations in aesthetics and visual imaging systems are important because they affect paradigm shifts that disrupt the visual status quo. Avoiding recourse to spectacle by using materials that are at hand, recycled, or readily and inexpensively available, Jerry Pethick intentionally maintains a modest economy of production. His work acknowledges and responds to, but also critiques, the spectacular and spectral world of commodity culture and consumerism. Favouring an enlivened, unmediated visual and cognitive experience that privileges and at the same time challenges the embodied viewer, Jerry Pethick makes art that provides (an unusual) visceral pleasure derived form visual effects that invite meandering, non-linear perception." [4]

Jerry Pethick passed away in 2003.

Figure 8 Photo of Jerry Pethick on Hornby Island
photographer unknown, collection of Margaret Pethick

8. Author's Note:

I was fortunate to meet Jerry Pethick. He was living on Hornby Island, I was living on Bowen Island and he was exhibiting his stove top series at a gallery on Granville Island. Jerry Pethick passed away in 2003. For my show of holograms and metal diffractive pieces, "Holographic Installations" in 2003, I created a metal and diffraction installation piece title "Flight of Light" (detail Fig 9) in homage to this interesting man who contributed so much. The 2007 version. "Flight of Life" (Fig 10) is now in the permanent collection of the Hornby Island Art Council.

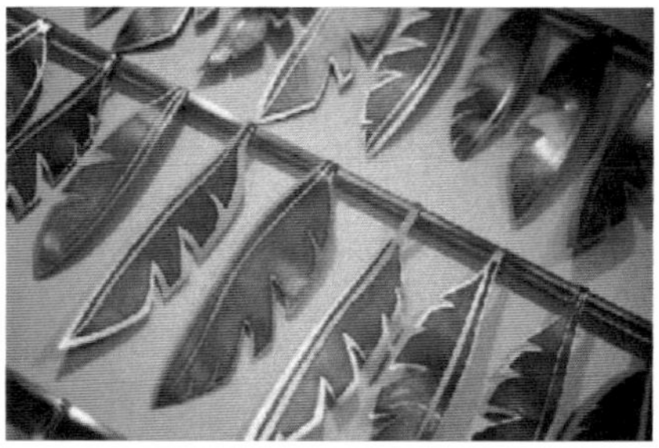

Figure 9 Detail "Flight of Light"
M. Crenshaw 2003

Figure 10 "Flight of Life" ~ 5' x5'
M. Crenshaw 2007

Acknowledgement

It was in 2007 that I met Margaret on Hornby Island. She graciously shared stories of Jerry's life in Holography and showed me Jerry's studio, posters, art works and even some very, very early holograms.

References

[1] Johnston, Sean F. 2006 *Holographic Visions A History of New Science* Oxford University Press

[2] Pethick, Jerry 1971 *On Holography and A Way to Make Holograms* Belltower Enterprises Burlington, Ontario Canada

[3] Schulz, Bernd 1994 Katalog *Jerry Pethick Notion of Nothing* Stadtgalerie Saarbrucken Kryger Druck+Verlag GmbH Dillingen/Saar

[4] Hurtig, Annette 2003 Kamloops Art Gallery, Kamloops BC Canada

Construction and Calibration of Optically Efficient LCD-based Multi-Layer Light Field Displays

Matthew Hirsch, Douglas Lanman,
Gordon Wetzstein, Ramesh Raskar
MIT Media Lab, Building E14, 75 Amherst St. Cambridge, MA, 02139, USA
E-mail: mhirsch@media.mit.edu

Abstract. Near-term commercial multi-view displays currently employ ray-based 3D or 4D light field techniques. Conventional approaches to ray-based display typically include lens arrays or heuristic barrier patterns combined with integral interlaced views on a display screen such as an LCD panel. Recent work has placed an emphasis on the co-design of optics and image formation algorithms to achieve increased frame rates, brighter images, and wider fields-of-view using optimization-in-the-loop and novel arrangements of commodity LCD panels. In this paper we examine the construction and calibration methods of computational, multi-layer LCD light field displays. We present several experimental configurations that are simple to build and can be tuned to sufficient precision to achieve a research quality light field display. We also present an analysis of moiré interference in these displays, and guidelines for diffuser placement and display alignment to reduce the effects of moiré. We describe a technique using the moiré magnifier to fine-tune the alignment of the LCD layers.

1. Introduction

Though holography may one day produce life-like 3D video displays, current state-of-the-art holographic displays face difficult trade-offs between spatial resolution, refresh rate, field-of-view, and cost. These shortcomings in the capabilities of holographic video technology, however temporary, have allowed a wide variety of ray-based 3D and 4D light field displays to find academic and limited commercial success. Conventional approaches to ray-based display typically include lens arrays [1] or heuristic barrier patterns [2] combined with integral interlaced views on a display screen such as an LCD panel.

Recent work [3] has placed an emphasis on the co-design of optics and image formation algorithms to achieve increased frame rates, brighter images, and wider fields of view using optimization-in-the-loop and novel arrangements of commodity LCD panels. We have dubbed this emerging field computational display, as it has, in part, grown out of the computational camera community. We have previously demonstrated computational displays that are compressive in nature, in that fewer parameters are configured on the display hardware than unique rays generated.

In this paper we examine the construction and calibration methods of computational, multi-layer LCD-based light field displays. We present several experimental configurations that are simple to build and can be tuned to sufficient precision to achieve a research quality light field display. The optical configurations were employed in our *Polarization Fields* [4] and *Tensor Display* [3] papers. Here we focus on their construction and calibration. We present an analysis of moiré – spatial frequency aliasing caused by the interference of pixel grids at different distances – and an analysis of diffuser placement and display alignment to reduce the effects of moiré. Finally, we describe a technique using the moiré magnifier [5] to fine-tune the alignment of the LCD layers.

2. Related Work

Conventional light field display has its roots in integral imaging [1] and parallax barriers [2], developed over a century ago. Since that time, commercial products such as the Nintendo 3DS[1] or Alioscopy glasses-free television[2] have continued to use barriers and cylindrical lens sheets, respectively, to produce stereo pairs for glasses-free viewing. These decisions are motivated by cost and the unavoidable spatial resolution loss imposed by these simple techniques.

Though switchable liquid crystal lenses[6] have begun to achieve commercial success in combining high resolution 2D display and lower resolution 3D display in a single unit, they do not address the resolution loss in integral imaging systems. Viewer-adaptive barrier patterns [7], and time sequential barrier patterns [8] can increase resolution or field-of-view for general light field display, and simple formulations for stacked LCD panels can represent limited depth at full resolution [9], but it is only recently that fast switching LCD panels and high performance graphics hardware has made it possible to address general purpose, content adaptive barrier patterns [10]. Generalizing the problem of light field display has proven a productive research direction, with recent work demonstrating both light-efficient, multi-layer designs [4, 11, 12], and formulations that incorporate multi-layer panels, directional backlighting, and temporal multiplexing [3].

3. Display Architectures

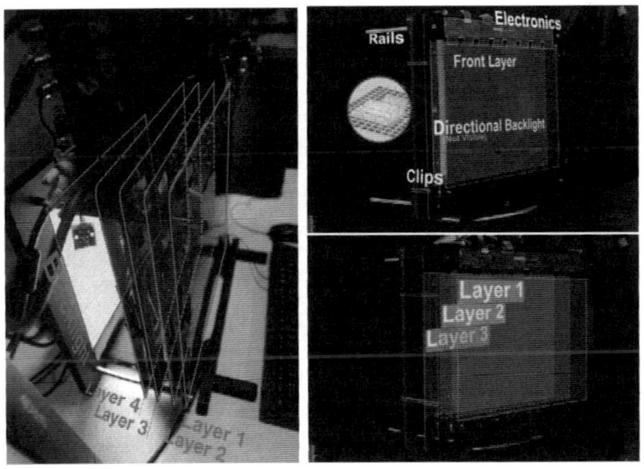

Figure 1. (Left) Polarization Field prototype. This display is composed of four linear polarization state rotators, surrounded by a pair of crossed linear polarizers. (Right) Tensor Displays. (Top) Two-layer directional backlight Tensor Display, an optically efficient design. (Bottom) Three-layer Tensor Display, composed of three high-speed attenuation-mode LCD panels.

In this section, we describe the construction of two optically efficient light field display prototypes. Both of the presented prototypes use optically efficient designs made possible by our computational image formation models, presented in our Polarization Fields [4] and Tensor Displays [3] works.

3.1. Polarization Fields

The light-efficient optical configuration of the Polarization Fields display is allowed only by the reformulation of the light field synthesis equation (Equation 2) in terms of a stack of linear

[1] http://www.nintendo.com/3ds
[2] http://www.alioscopy.com

Figure 2. We disassemble a Barco E-2320 PA panel to convert it from an attenuation based display to a linear polarization state rotating element. a) Assembled. b) Panel removed from case. c) Removing the diffusing, linear polarizing films. d) Cleaning adhesive residue with acetone and microfiber cloth. d) Stripped and cleaned panel mounted to CNC cut aluminum frame.

polarization state rotators. In turn, a stack of bare, liquid crystal panels could be constructed and modeled in many ways, but using our simplified linear polarization state rotator model, solving the image synthesis problem for this stack is amenable to fast linear methods that run in real time on modern GPUs [13].

3.1.1. Synthesis In our Layered 3D paper [14], we introduced a light field image formation model for devices composed of attenuating layers. We cast the problem of light field display from a device composed of attenuating layers as an iterative back projection problem, in which the views of the desired light field are modeled as projections through the volume of layers. In order to linearize the problem, we solve for the light field in log space, as

$$\arg\min_{\alpha} \|\bar{\mathbf{l}} + \mathbf{P}\alpha\|^2, \text{ for } \alpha \geq 0, \tag{1}$$

where $\bar{\mathbf{l}}$ denotes the log light field, $\mathbf{P}$ the projection matrix through the display stack, and α a vector of mask values. However, it is impractical to apply these methods to LCDs directly, as LCD transmissivity is low when polarizing sheets surround the LC panel.

In Polarization Fields, we reformulate our image formation model to function in polarization rotation space, modeling a bare LC panel as a linear polarization state rotator [15]. This allows us to construct a stack of optically efficient liquid crystal layers (Figure 1, (Left)). The entire stack is surrounded by one pair of crossed linear polarizers, and illuminated by a uniform backlight. As before, our formulation is posed as an iterative backprojection problem,

$$\arg\min_{\phi} \|\theta - \mathbf{P}\phi\|^2, \text{ for } \phi_{min} \leq \phi \leq \phi_{max}, \tag{2}$$

where θ denotes the linear polarization state rotation angles, per ray, necessary to create a desired light field after interaction with a linear polarizing analyzer, via Malus' law.

As in Equation 1, $\mathbf{P}$ denotes the projection matrix through the polarization fields display stack. ϕ is a vector of polarization rotation values, per layer.

3.1.2. Construction Our prototype was built using off-the-shelf components. We disassemble four Barco E-2320 PA LCD panels, removing the backlights and polarizing films (Figure 2). As these panels are medical grayscale panels, they contain no color filters. Color is theoretically restored to the display using a time sequential backlight.

The disassembled display panels are mounted CNC cut aluminum frames, which slide on precision steel rails. The rails are aligned inside a wooden, CNC machined frame. Final alignment is done using the technique described in Section 4.1.2. Light fields are not displayed at the full spatial resolution of the panel due to computational hardware constraints. This affords additional misalignment tolerance to the system.

3.2. Tensor Displays

In Tensor Displays, we further generalize the problem of light field display through attenuating layers to account for temporal modulation using high speed displays, as well as variation across layers. In addition, our Tensor Display formulation allows new refractive optical elements, such as a directional backlight.

3.2.1. Synthesis We refer readers to our Tensor Display paper [3] for complete documentation. Here, we outline the synthesis as it enables the construction of an optically-efficient light field display. The central observation of the framework is that the choice of an *absolute coordinate system* allows a light field displayed from a multi-layer device to be restricted to a Nth-order, rank-M tensor, where the order of the tensor, N, is determined by the number of layers in the multi-layer display, and the rank of the tensor, M, is determined by the number of time steps available. The layer values at each time step are then determined as the result of a Non-negative Tensor Factorization [16] (NTF). For the general case

$$\underset{\{\mathbf{F}^{(n)}\}}{\arg\min} \left\| \mathcal{L} - \mathcal{W} \circledast \tilde{\mathfrak{T}} \right\|^2, \text{ for } 0 \leq \mathbf{F}^{(n)} \leq 1, \tag{3}$$

where $\tilde{\mathfrak{T}} = [\![\mathbf{B}, \mathbf{F}^{(1)}, \mathbf{F}^{(2)}, \ldots, \mathbf{F}^{(N)}]\!]$. $\mathbf{B}$ is the light field emitted by the directional backlight, and $\mathbf{F}^{(N)}$ is a vector of attenuation values placed on the Nth layer of the display.

3.2.2. Construction The flexibility of the Tensor Display image synthesis framework allows for a variety of hardware configurations. In Figure 1 (Right), (Top) and (Bottom), our prototype is show configured in a two-layer, directional backlight mode, and a three-layer, uniform backlight mode, respectively. The Tensor Display framework does not yet handle polarization rotation, as in Polarization Fields, but allows an alternative light-efficient configuration in the form of the two-layer, directional backlight display.

Construction of the Tensor Display prototype closely follows that of the Polarization Fields prototype described in Section 3.1.2, but differs in a few key design choices. Rather than grayscale medical panels, we use high speed (120HZ) Viewsonic VX2268wm LCDs. These panels are intended for use with shutter glasses, and so have fast response times with little ghosting. Because we use these panels in attenuation mode, we place a linear polarizer at the rear of the stack, and successively crossed linear polarizers at the face of each panel.

In the light-efficient two-layer directional backlight configuration we employ a lens array to create a generic low resolution light field display for use as the directional backlight component. We implement our lens array by placing two crossed 1D lenticular sheets atop one another, such that the first sheet is rotated 90° and facing inward (Figure 1, Inset). We use a 10 lens-per-inch sheet from Microlens Technologies. The use of cheap lens sheets in this configuration is not ideal, as it results in scattering at the lens edges, creating a dark grid in the resulting images. The placement of the top lens sheet above the bottom lens sheet imposes a small astigmatism, as the top sheet is not one focal length distance from the rear LCD screen. The lenses we use are not polarization preserving, and require an additional linear polarizer between the front of the lens sheet and the rear of the top LCD screen. Each of the aforementioned issues can be resolved with more sophisticated engineering.

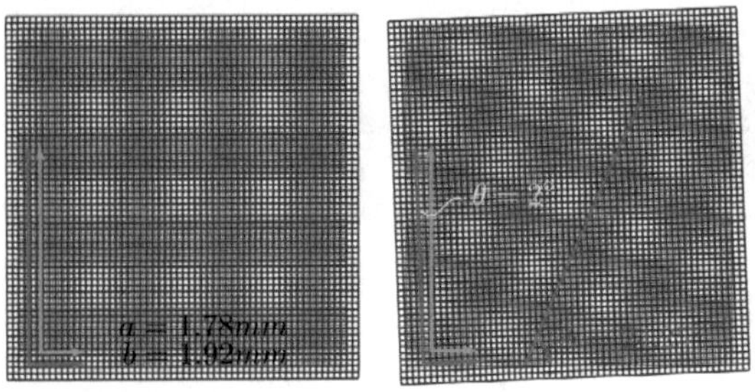

Figure 3. Moiré interference patterns caused by scaled and rotated grids. Best viewed at full resolution. The grids differ in pitch by 7%. On the right, the larger grid is rotated by $\theta = 2°$, causing an apparent rotation of the moiré fringes by $\phi = 24.25°$.

4. Considering Moiré

4.1. Calibration

Moiré fringes are observed when two patterns of different spatial frequency are multiplied. The effect is often observed in digital photography and display when patterns in a scene approach the spatial frequency of the underlying pixel grid of the image capture or display device. From a signal processing perspective, Moiré can be understood as a beat frequency between signals of similar spatial frequencies, or equivalently, spatial frequency aliasing. In this section, we demonstrate how to use moiré fringes to accurately align layered lens array and LCD systems.

4.1.1. Lenticular Alignment In many light field displays, including our two-layer Tensor Display (Section 3.2), it is necessary to align a lenticular sheet or lens array to an underlying pixel grid. Here we described a simple technique to perform rotational alignment using the moiré effect. In the case of lens sheets, this effect has been succinctly described as the *moiré magnifier* [5]. Following Hutley et. al., an expression can be obtained for the relative rotation and magnification of the image of the pixel grid of the LCD panel as viewed through the lens array. For convenience, we reproduce Hutley et. al.'s magnification, m and rotation, ϕ, here, with one minor modification: we simplify the denominator using the Pythagorean trigonometric identity.

$$m = \frac{a}{\sqrt{a^2 + b^2 - 2ab\,\cos(\theta)}} \tag{4}$$

$$\sin(\phi) = \frac{-b\,\sin(\theta)}{\sqrt{a^2 + b^2 - 2ab\,\cos(\theta)}} \tag{5}$$

As we show numerically below, for practical values of LCD and lens pitch, small rotations of the lens array will be magnified in the moiré pattern. The calibration task reduces to leveling the perceived moiré fringes by eye. If the lens pitch is nearly an integer multiple of the LCD pitch, as is the desired case, then m will be nearly infinite when $\theta = 0$. Moiré bands will not be visible under these conditions. However this calibration technique applies equally to patterns displayed on the LCD as the pixel structure of the screen itself. It is often desirable to display a pattern on the LCD to improve the contrast of the observed moiré pattern. Once the pattern has been leveled by rotating the lens sheet, the pitch of the lens sheet can be calibrated by adjusting the pitch of repeating pattern displayed on the LCD until m is infinite, or equivalently, no fringe patterns are visible. In the case of a research prototype using an imperfectly matched lens array and LCD panel, the displayed pattern may be interpolated to achieve sub-pixel alignment, with a small angular cross-talk penalty in the resulting light field display.

To determine the expected accuracy of the above method, we consider the physical values from our Tensor Display prototype. The lens pitch is $a = 2.54mm$, and LCD pixel pitch is $b = 282\mu m$. We found that the rotation of the moiré fringes could be aligned to within $\phi = 0.5°$. Substituting into Equations 5 and 4, and solving for θ and m, respectively, we get $\theta = 4°$ and $m =$ undefined. This is a result of choosing a and b as nearly integer multiples ($a/b = 9.007$), and indicates that alignment by eye will not be very accurate. To improve accuracy, we can display a linearly interpolated pattern on the LCD with a pitch of $b = 2.1mm$. Now, for $\phi = 0.5°$, $m = 5.77$ and $\theta = 0.105°$, allowing nearly $5\times$ improvement in accuracy over alignment by eye.

4.1.2. LCD Alignment Though it is possible to use the scale of moiré fringes to perform alignment in depth, we find it is much simpler in practice to use CNC machines to cut spacer clips, which can fasten to multiple layers of optical elements and space them accurately to the tolerance of the CNC machine – $0.25mm$ or less. In this section, as in Section 4.1.1, we will concentrate primarily on rotational alignment of LCD panels, to which moiré fringes are more sensitive.

Though the analysis of Section 4.1.1 was derived from the *moiré magnifier* effect of lens arrays, we observe that Equations 4 and 5 apply equally to lens arrays and grid patterns. Oster et. al. [17] use an analysis based on indical representations of curves to derive Equations 6 and 7 in their paper for moiré fringe pitch and rotation, which match our Equations 4 and 5, save for a sign difference. We show in Figure 3 that the analysis holds for a printed grid pattern.

When aligning LCD screens spaced by a distance d_s, the difference in pixel size Δp, observed by a viewer at distance d_o is due to perspective projection. By similar triangles,

$$\Delta p = \frac{pf}{d_o} - \frac{pf}{d_o + d_s}. \tag{6}$$

where f is the focal length of the human eye, accepted to be approximately $22mm$. For the physical dimensions of our three-layer Tensor Display prototype (considering the front two layers) $p = 282\mu m$, $d_o = 1m$, $d_s = 4cm$, we calculate that $\Delta p = 0.241\mu m$. Substituting the two apparent LCD pitches into Equation 5, we find that a pattern rotation of $\phi = 0.5°$ yields a screen rotation of just $\theta = 7.5 \times 10^{-6}$ degrees, indicating that aligning the LCD layers by straightening the visible moiré fringes will achieve very accurate alignment. It is useful to note that, with such a small difference in pitch, the magnification, m, will be large, making it more difficult to achieve accurate visual rotation alignment. Thus, $\phi = 0.5°$ may be an overly generous estimate.

4.2. Mitigation

While moiré is beneficial for accurate calibration, it is an unpleasant visual nuisance when observing a light field. In order to eliminate moiré, one need only prevent the multiplication of similar spatial frequency signals. We find that there are two approaches that can mitigate moiré:

- Achieve a small magnification factor, m, such that aliased copies of the signal are small relative to image features
- Implement a spatial low-pass or notch filter to remove the offending frequencies

In the case of LCD panels, the first of the above strategies implies separating the panels by a large distance. Larger separation increases Δp from Equation 6. However, a large separation distance is not always practical, and does not apply to lenticular sheets and lens arrays.

The second approach can be achieved in two-layer and lenticular devices by placing an appropriately chosen diffuser on the rear LCD layer. An appropriate diffuser choice will impose a spatial frequency cut-off such that any moiré observed will have a small magnitude or small magnification. We find that a light weight diffuser such as Grafix Matte Acetate 0.005 works well for LCD panels with a pixel pitch in the $\frac{1}{4}mm$ range.

5. Conclusion

In this work, we have described practical methods for the graphics or display researcher to construct ray-based light field displays capable of supporting, and benefiting from, the most recent work in computational and compressive display. It is our hope to inspire future research into new possibilities for glasses-free 3D displays based on simple ray optics, thanks to the increasingly low cost of high performance computing.

References

[1] Lippmann G 1908 *Journal of Physics* **7** 821–825
[2] Ives F E 1903 Parallax stereogram and process of making same U.S. Patent 725,567
[3] Wetzstein G, Lanman D, Hirsch M and Raskar R 2012 *ACM Trans. Graph.* **31** 1–11
[4] Lanman D, Wetzstein G, Hirsch M, Heidrich W and Raskar R 2011 *ACM Trans. Graph.* **3**(6) 1–9
[5] Hutley M C, Hunt R, Stevens R F and Savander P 1994 *Pure and Applied Optics: Journal of the European Optical Society Part A* **3** 133 URL `http://stacks.iop.org/0963-9659/3/i=2/a=006`
[6] Harrold J, Wilkes D and Woodgate G 2004 *Proc. IDW* vol 11 pp 1495–1496
[7] Perlin K, Paxia S and Kollin J S 2000 *ACM SIGGRAPH* pp 319–326
[8] Kim Y, Kim J, Kang J M, Jung J H, Choi H and Lee B 2007 *Optics Express* **15** 18253–18267
[9] Bell G P, Craig R, Paxton R, Wong G and Galbraith D 2008 *SID Digest* **39** 352–355
[10] Lanman D, Hirsch M, Kim Y and Raskar R 2010 *ACM Trans. Graph.* **29**(6) 163:1–163:10
[11] Gotoda H 2010 *SPIE Stereoscopic Displays and Applications XXI* vol 7524 pp 1–8
[12] Gotoda H 2011 *SPIE Stereoscopic Displays and Applications XXII* vol 7863 pp 1–7
[13] Andersen A and Kak A 1984 *Ultrasonic Imaging* **6** 81–94
[14] Wetzstein G, Lanman D, Heidrich W and Raskar R 2011 *ACM Trans. Graph.* **30**(4) 1–11
[15] Davis J A, McNamara D E, Cottrell D M and Sonehara T 2000 *Applied Optics* **39** 1549–1554
[16] Friedlander M P and Hatz K 2008 *Computational Optimization and Applications* **23** 631–647
[17] Oster G, Wasserman M and Zwerling C 1964 *JOSA* **54** 169–175

Haptic Holography: Acoustic Space and the Evolution of the Whole Message

N Logan

58 Helena Ave. Toronto, Ontario, Canada. M6G 2H2

natalogagraphy@gmail.com

Abstract. The paper argues that the Haptic Holography Work Station is an example of a medium that fit's with McLuhan's notion of Acoustic Space, that is it is a medium which stimulates more than one sense of perception at a time. As a result, the Haptic Holography Work Station transmits information about the subject much more rapidly than other media that precedes it, be it text, photography or television.

The term *hologram*, coined by its inventor, Dennis Gabor (1900-1979), derives from the Greek words "*holo-*" meaning "whole" and "*-gram*" meaning "written", "message" and "picture". Around the same time Gabor was pioneering a system of *whole writing* or a *whole message*, Canadian media theorist Marshall McLuhan (1911-1980) coined the phrase "*the medium is the message*". McLuhan's theories still resonate to this day. Though he was never able to use the Haptic Holography Work Station designed at OCAD-University with Professor Michael Page, McLuhan's theory of *acoustic space* applies to the invention. The Haptic Holography Work Station is not a new invention, but rather one that integrates current technologies of computers, robotics and holograms. I argue that the merging of these technologies creates the most complete example of a medium that creates acoustic space, that is a medium that is able to transmit an instantaneous message of information through visuals, sound and touch all at once in three dimensional space.

Media that are audio-tactile are those that engage more than one sense at a time, such as television, and they create what McLuhan calls '*acoustic space,*' where the senses are in interaction with one another. In contrast, McLuhan believed that extracting information from 2D visual sources, for example text written using a phonetic alphabet, was a linear uni-directional way of receiving information that stresses linearity, which he called '*visual space.*' McLuhan suggested that with electric media, linear uni-directional print was no longer the dominant medium and that we are currently living in a new electric age that retrieves the patterns of oral-tribal cultures. "Acoustic space is organic and integral, perceived through the simultaneous interplay of all the senses; whereas "rational" or pictorial space is uniform, sequential and continuous and creates a closed world with none of the rich resonance of the tribal echoland" (McLuhan: 1969).

With electronic media, such as television, radio, and arguably holograms, audiences receive electrically-configured information in an '*all-at-once*' or *holistic* experience similar to how auditory information is received from all directions at once. "Auditory [or acoustic] space is that sphere of simultaneous relations created by the act of hearing. We hear from all directions at the same instant (McLuhan: 1960)." In this regard, a hologram is similar to sound because it occupies acoustic space; it involves multiple perspectives of the whole where there is no centre and no linear way of looking at it. What makes a hologram "whole writing", as opposed to a photograph (which literally means *to write with light*), is that there is parallax, i.e. the hologram contains multiple perspectives of the whole. It is not, like print, uni-directional, occupying one sense at a time where you can only get the whole picture if you go in a linear sequential order starting at a beginning be it a word, a sentence or a monograph.

With our project, the Haptic Holography Work Station, one can both see the hologram and feel the outer surface of the holographic forms. Thus, the holographic display you can touch with a robotic arm provides an example of how a visual medium creates acoustic space, for there is no centre and no linear way of feeling an object and more than one sense is at play. The Haptic Holography Work Station works by merging multiple media, including programming, a computer, a haptic controller and a holographic display. A robotic arm provides dynamic resistance, allowing one to feel contours of the whole object in the holographic scene. McLuhan's description of acoustic space, namely that, "you trade an eye for an ear" when watching television applies to holography, and haptic holography as well, because your eye functions like an ear when you look at a hologram, absorbing light from different perspectives simultaneously versus directionally. When using a haptic controller, you receive information from many points in space, and there is no beginning and no end, and no correct, linear, rational way of feeling or seeing the subject. The experience is totally immersive. Another way of putting it is that a text has a beginning, a middle and an end, which is not the case with either a hologram or a haptic-hologram; they have no beginning, middle or end.

The Drum Simulation in the Haptic Holography Work Station at OCAD-U is the best example of acoustic space because of the way it affects the human *sensorium* where the three major senses: sight, sound and touch, interact with one another. The user sits before a laser-lit transmission hologram of a miniature drum set on a 30cm x 40cm plate and holds the haptic device which appears on an OLED screen behind the hologram as a 2D drum stick. The hologram is projected so that it occupies the same space as the drum stick on the screen, allowing the user to feel the drums, see the stick hitting the hologram of the drums and hear dynamic audio feedback of a drum kit being played.

CONCLUSION:

The Haptic Holography Work Station, I argue, is the evolution of the whole message, where you are not only seeing something in three dimensions but also feeling something in three dimensions; something that doesn't actually exist there. The Haptic Holography Work Station engages multiple senses at the same time, fulfilling McLuhan's notion of acoustic space which is audible, tactile and in this case, visual as well. In conclusion, the message of a Haptic Holography Work Station is the whole of the object it transmits. Its impact on the world today is that we can receive information about an object or subject faster than the amount of time it would take to read a book about on the topic and more intimately than the idea we could get about the object from still pictures or television.

References

[1] McLuhan, Marshall. "The Medium is the Message" in *Forum.* Spring 1960.
[2] McLuhan, Marshall. "The Playboy interview: Marshall McLuhan" in *Playboy Magazine.* March 1969.

Simple, complete, and novel quantitative model of holography for students of science and science education

Dale W Olson

Physics Department, University of Northern Iowa, Cedar Falls, Iowa 50614-0150

dale.olson@uni.edu

Abstract. A Moiré graphical method for predicting the directions of all permitted image waves for thin and volume holograms is described. This method has been developed in connection with a holography centred optics course for students in science, technology and science education that has evolved over several decades. A somewhat unique view of the holographic process underlies this novel method of predicting the behaviour of diffraction gratings.

1. Moiré models for wave interference reviewed: Hologram Formation

Moiré plane wave models, like those shown in figure 1, are widely used in introductory books on holography to represent wave interference and hologram formation. A comparison of figures 1a and 1b demonstrates that as the angle of the object beam relative to the reference beam *increases* in magnitude, the bright fringe spacing recorded in a thin or volume transmission hologram *decreases* in magnitude. This semi-quantitative demonstration illustrates the usefulness of Moiré diagrams. In this paper the application of Moiré diagrams will be extended to include prediction of permitted image wave directions for thin and volume holographic gratings.

Moiré plane wave diagrams not only provide a clear visual representation of the hologram recording process, but also yield geometric diagrams that lead to main holographic equations. For example, the well-known triangle diagram shown in figure 2 leads to the *light wave fringe spacing equation*, equation 1. All equation 1 symbols are explained in the caption for figure 2.

$$\Lambda_{lw} = \lambda/[2\sin\theta^*] \qquad (1)$$

Equation 1 is obtained by utilizing right triangle HGB in figure 2.

Using the symmetry between the two interfering planes waves shown in figure 2, we can know that the object and reference beam direction vectors make equal and opposite angles, θ^*, and $-\theta^*$, with the bright fringe direction. Alternately stated, the bright light fringes bisect the angle $2\theta^*$ between the two wave directions. We might call this the *fringe direction rule*.

To relate the light wave interference diagram of figure 2 to experimental arrangements involving holographic film, we can introduce angles α_O and α_R which, respectively, tell the directions of the object and reference beams relative to the film normal. The two angles α_O and α_R are shown in figure 3.

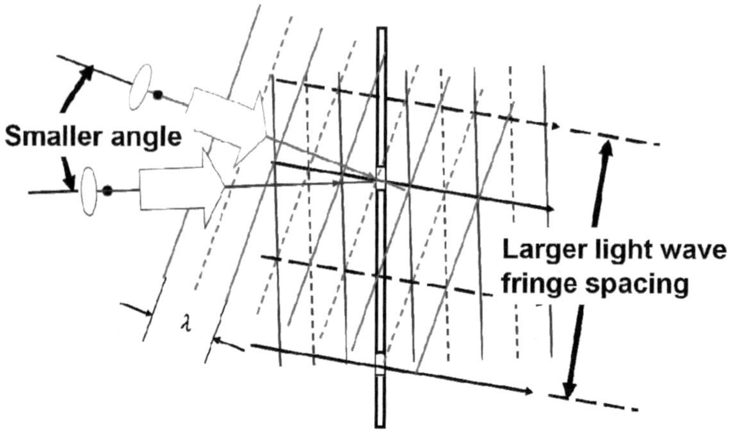

Figure 1a. Moiré pattern represents hologram formation. The reference beam angle of incidence, not marked in the film, is -20° (CW). The object beam angle of incidence is +2°, both relative to film normal.

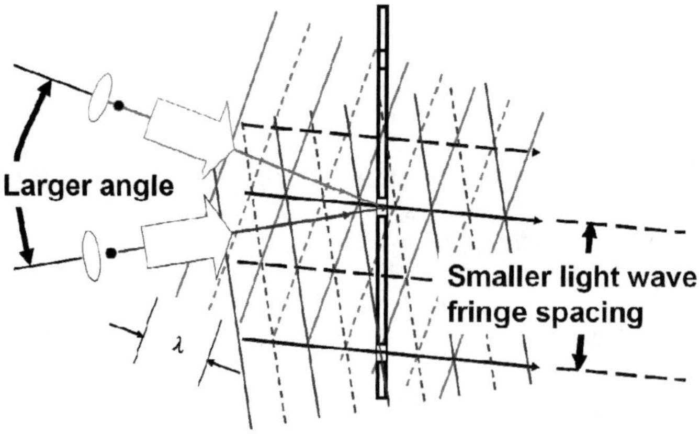

Figure 1b. Compared with Fig. 1a, the object beam angle of incidence is increased to +10°, while reference beam remains unchanged in direction at -20°.

Using the object beam and direction beam direction angles α_O and α_R shown in figure 3, we can then express the light wave fringe direction (φ) relative to the film normal by the *light wave fringe direction equation*, equation 2.

$$\varphi = [\alpha_O + \alpha_R]/2 \tag{2}$$

Also it is helpful to write equation 3 which defines θ^* in terms of α_O and φ.

$$\theta^* = \alpha O - \varphi \tag{3}$$

Then, combining equations 2 and 3, we obtain a second equation for θ^*.

$$\theta^* = [\alpha_O - \alpha_R]/2 \tag{4}$$

The inverse equations are shown as equation 5 and equation 6.

$$\alpha_O = \varphi + \theta^* \tag{5}$$

$$\alpha_R = \varphi - \theta^* \quad (6)$$

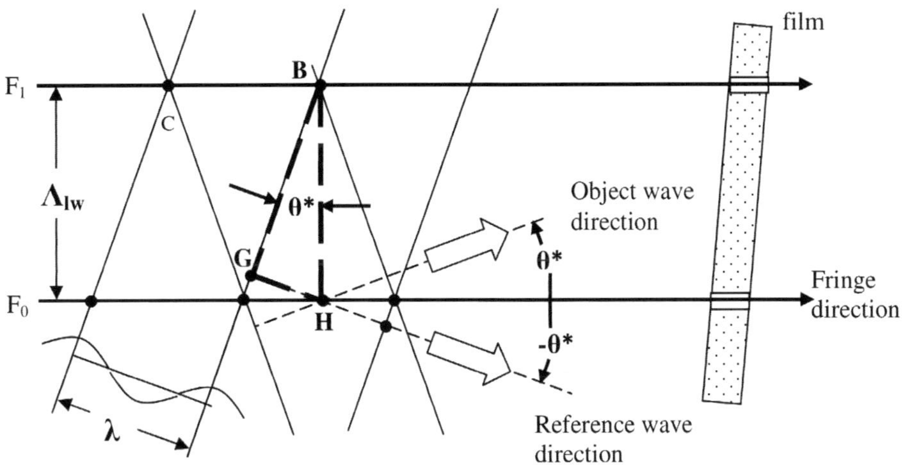

Figure 2. Figure 2 utilizes right triangle HGB to establish the relationship between light wave fringe spacing (BH = Λ_{LW}), light wavelength (GH = $\lambda/2$), and the angle ($\theta^* = \angle GBH$). θ^* is the angle by which the object wave direction is rotated out of the light wave fringe direction. We take CCW rotations as positive. Directed lines F_0 and F_1 represent bright fringes carrying wave energy. Angle θ^* relates the object and reference beam directions to the light wave fringe directions, but contain no information about wave or fringe directions relative to the film.

Equations 1 through 6 lead rather directly to two useful equations for the hologram fringe spacing (Λ_S) along the surface of a hologram. The *film surface fringe spacing equation*, version one is shown in equation 7.

$$\Lambda_S = \Lambda_{LW}/\cos\varphi \quad (7A)$$

$$\Lambda_S = \lambda/(2\sin\theta^*\cos\varphi) \quad (7B)$$

An equivalent equation, but with a form that is more closely related to the thin grating equation used to describe hologram viewing is shown in equation 8. Again, equations 7 and 8 are mathematically equivalent.

$$\Lambda_S = \lambda/(\sin\alpha_O - \sin\alpha_R) \quad (7B)$$

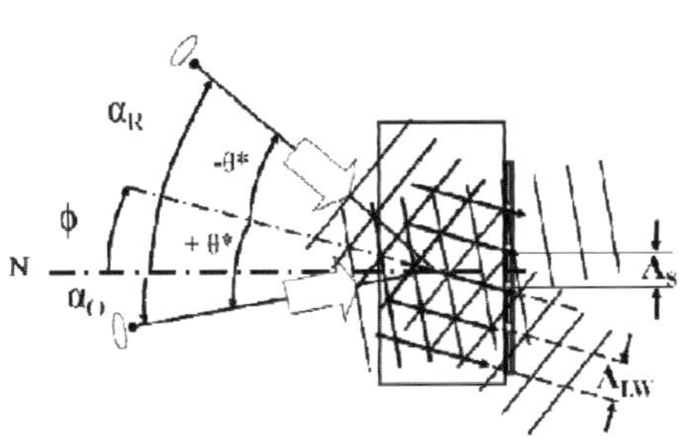

Figure 3. All important angles and fringe spacings associated with hologram formation are shown. The angles are discussed, and equations related the various angles are provided in the text.
Λ_S is the film surface fringe spacing equation, often denoted elsewhere as d, for the case of a thin grating.
Λ_{LW} is the light wave fringe spacing.

2. A novel Moiré graphical method for predicting holographic grating image wave direction

Figure 4 introduces a Moiré graphical method for locating all permitted image waves for a thin holographic grating with wavelength to grating space ratio equal to 0.5. The viewing wave direction is β_V = -20° throughout. The first and second order image directions are found using the *method of hypothetical fringes*, the fringes that would be formed by the interference of the viewing wave and the proposed image wave, should they be permitted to pass through the same region of space.

A complete set of equations for implementing the hypothetical fringe model more efficiently has been written out by the author, and will be presented elsewhere. A main point is that, because the hypothetical fringe method casts predictions about hologram image waves, i.e. hologram viewing into the 'same mold' as that used for hologram formation, all hologram viewing equations turn out to be essentially identical to those that describe hologram formation. The hologram viewing equations then become relatively easy to write down 'from scratch' and without reference to traditional diffraction grating theory. The logic underlying hologram imaging and hologram formation become essentially the same when the hypothetical fringe method is utilized. All derivations are based on the same type of Moiré diagrams, so the similarity between the hologram formation and the hologram viewing equations becomes less surprising. In particular it should be noted that the diagrams that quantitatively model hologram formation, shown in figures 1, 2 and 3 are essentially identical to those that model thin hologram imaging, shown in figure 4. Then, though volume hologram imaging has not been treated in similar detail in this essay, the explanation in figure 5 of the well known 'red shift' seen as a white light viewing beam is rotated nearer to the normal of a white-light viewable reflection indicates, correctly, that the hypothetical fringe model works equally well for thin and volume holographic gratings.

3. Discussion

In the opinion of this author a problem with traditional approaches to modeling holography is that hologram formation and hologram viewing appear to be 'disparate processes.' The fact that the hologram will form an image wave that necessarily is identical, in many ways, to the object wave used to form the hologram does not follow directly from the model or concepts introduced. Using thin amplitude transmission holograms as an example, in the traditional models, the stable wave interference pattern formed by the object and reference beam is shown to cause a pattern of microscopic clear and dark regions, called fringes, to be recorded in a light sensitive medium, such as silver halide film. Then, the chemically process film, the hologram, is interpreted as being a grating much like a traditional diffraction grating, and, correctly, traditional diffracting grating principles are applied. Again, the fact that the object wave will be spontaneously reconstructed if the hologram is correctly illuminated doesn't same natural and obvious.

This essay has demonstrated a method that the author has not seen implemented elsewhere, and which has the potential to better integrate the two parts of holography, hologram formation and hologram viewing. The conceptual reasoning used by the author to 'discover' this method will be outlined briefly here. The main point is to establish a viewpoint such that the 'magic' of the holographic process will appear as a natural outcome of wave interference principles. The goal is to retain the magic, while removing the mystery

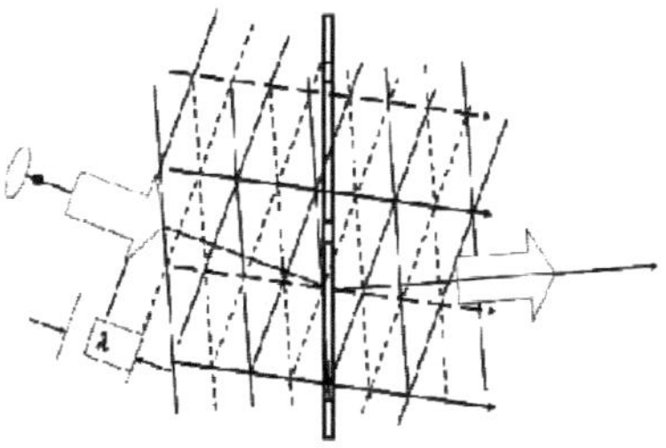

Figure 4a. Trial image wave direction $\beta_{+1} = 4°$.

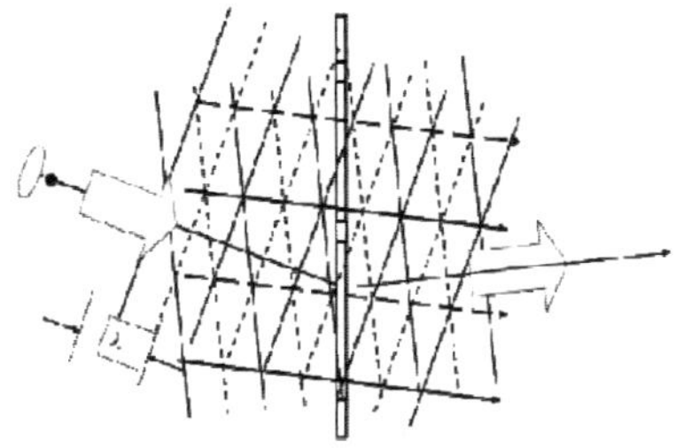

Figure 4b. Trial image wave direction $\beta_{+1} = 6°$.

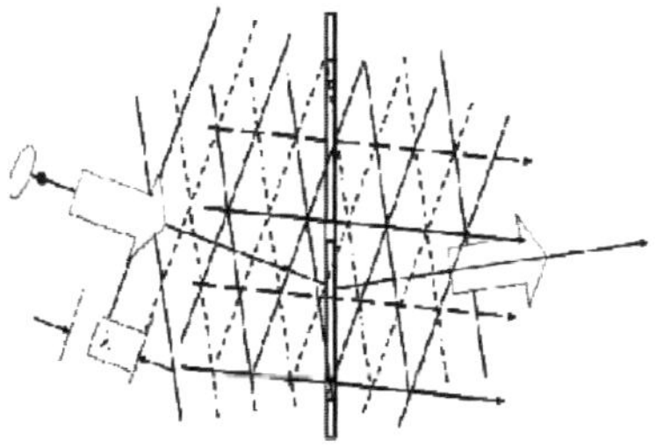

Figure 4c. Trial image wave direction $\beta_{+1} = 8°$.

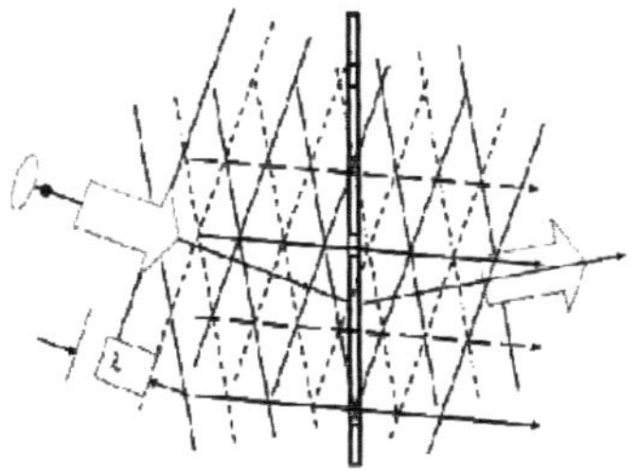

Figure 4d. Trial image wave direction $\beta_{+1} = 10°$.

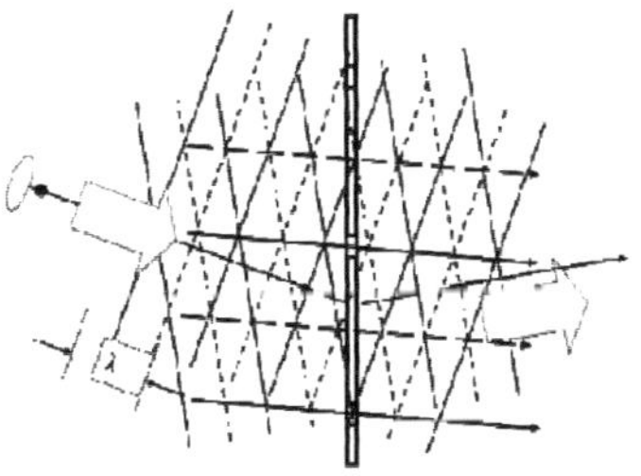

Figure 4f. $\beta_{+1} = 9°$. This is the correct direction!

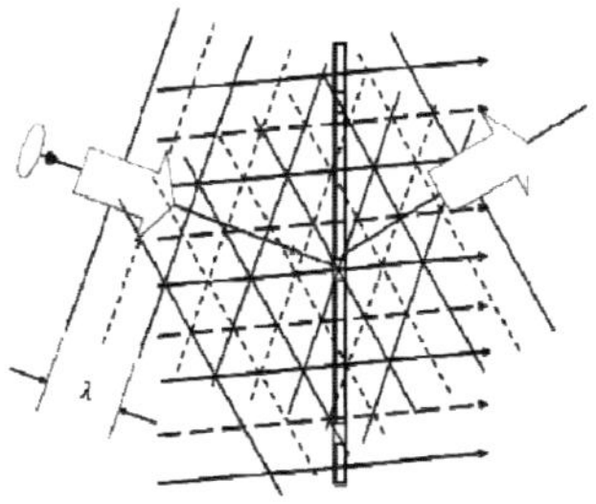

Figure 4g. Trial image wave, $\beta_{+2} = 30°$.

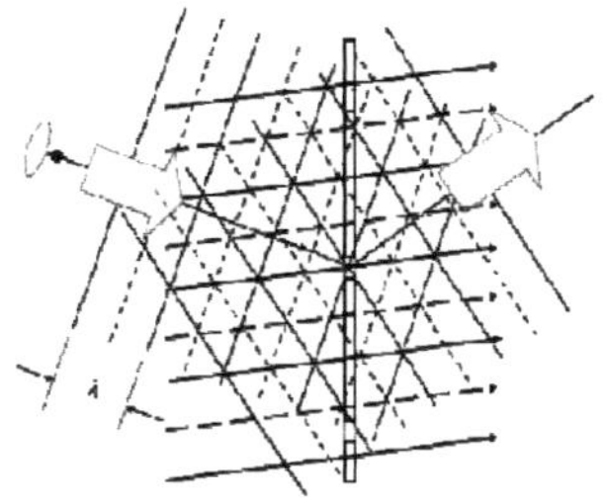

Figure 4h. Trial image wave, $\beta_{+2} = 34°$.

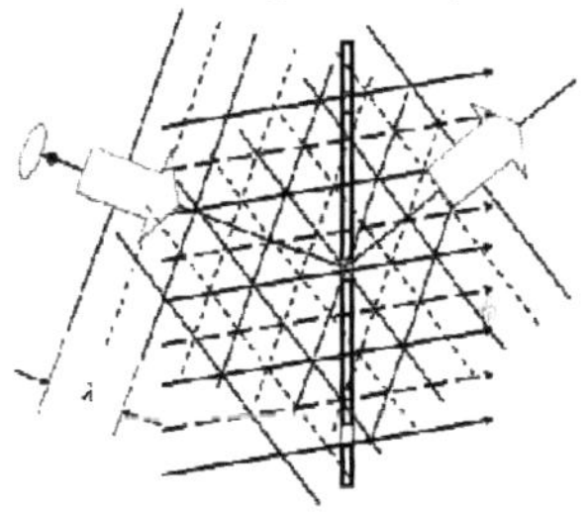

Figure 4i. Trial image wave, $\beta_{+2} = 38°$.

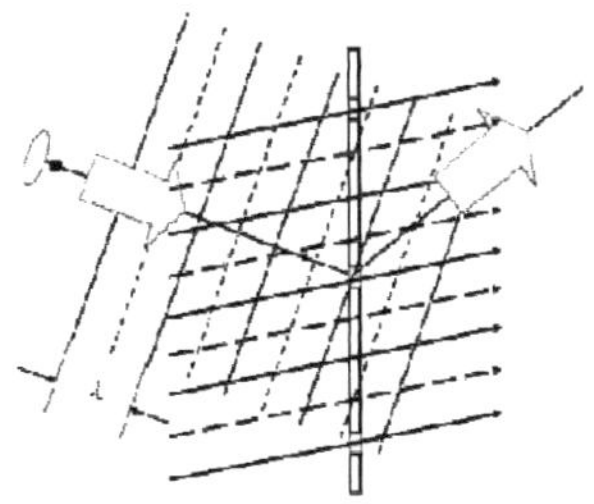

Figure 4i. Trial image wave, $\beta_{+2} = 41°$, shown immediately above. This is the correct direction for the second order (=2) image wave. A bright 'hypothetical fringe' passes through every grating opening, so this is an acceptable image wave. There are two bright fringes for every grating opening, making this the second order (m = +2) image wave.

Figure 4f shows the correct first-order (m = +1) image wave at $\beta_{+1} = 9°$.

Journal of Physics: Conference Series **415** (2013) 012073 doi:10.1088/1742-6596/415/1/012073

Figure 5 applies the method of hypothetical fringes to describe the red shift that occurs for a volume hologram, as the viewing beam is brought closer to normal incidence.

<table>
<tr><td>
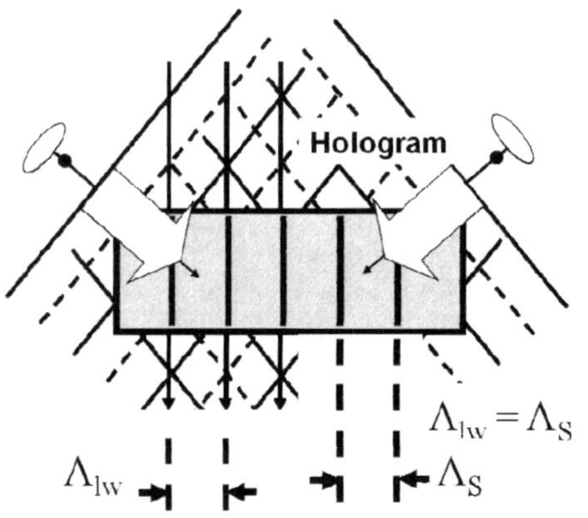

</td><td>

Figure 5a. The formation of a reflection hologram with fringes parallel to the film surface is shown at the left.

For simplicity we assume the index of refraction of the film is n = 1, so the light wave fringe structure, and the fringe structure recorded in the hologram

The fringe spacing along the film surface and in the film will be equal to the light wave fringe spacing. $\Lambda_S = \Lambda_{LW}$.
</td></tr>
<tr><td>
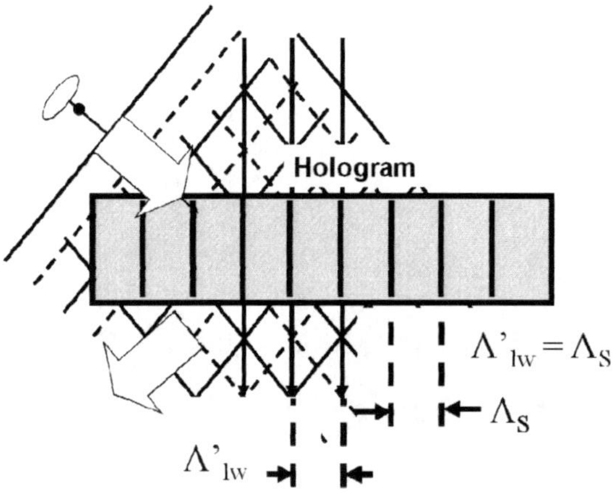

</td><td>

Figure 5b. The viewing of the volume hologram formed as in figure 5a is shown.

In this case the viewing wave is selected to have the same direction as the original reference beam used to form the hologram. $\beta_V = \alpha_R$.

The trial image wave is assumed to leave in the direction from which the original object wave arrived. $\beta_{+1} = \alpha_O$.
</td></tr>
<tr><td>
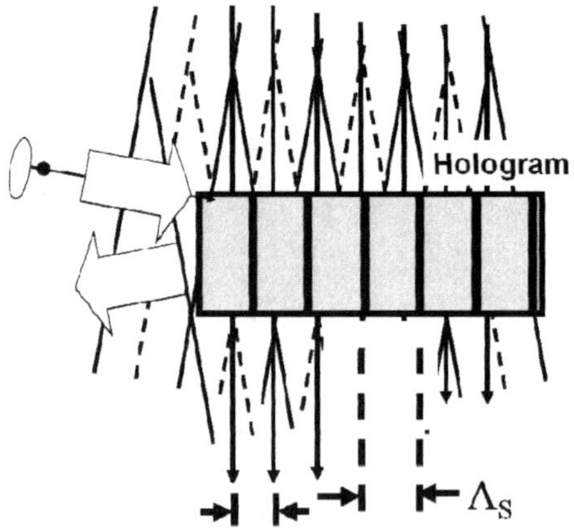

</td><td>

Figure 5b. The red-shift that occurs when a white light viewing beam is rotated closer to normal incidence is modelled.
</td></tr>
</table>

Underlying the discovery of the hypothetical fringe method is a slightly different interpretation of wave constructive and destructive wave interference that I have developed by emphasizing in my thinking the fact that holography works! To be more specific, constructive wave interference has the little noted characteristic of identifying regions of space where two equal amplitude and equal wavelength interfering plane waves are identical. The bright interference fringes formed by said plane waves mark planar regions throughout which the two wave disturbances are everywhere and always identical. Wave energy flows through these regions, the holographic film absorbs some of this energy, and, hence, holographic film automatically records locations in space where the two waves, object wave and reference wave are 'the same' It can be noted that the 'dark fringes' the locations of destructive interference mark the places where the two waves are 'most different.' In the dark regions where one wave has maximum positive disturbance, the other has maximum negative disturbance, and the two waves cancel. And, of course, in these regions, the holographic film remains clear.

So, again, the bright fringes, whose location is recorded in the hologram, mark the places where the reference beam and the object beam are the same, and Moiré diagrams very well represent this process, and permit related equations to be written. Then, as demonstrated above, the 'hypothetical fringes' that the viewing wave and the image wave <u>would</u> form should they be permitted to exist in the same region of space do correctly predict all permitted image wave directions for a thin transmission hologram. The values for β_{+1} and β_{+2} stated in the figure 4 captions agree can be shown to agree with predictions of the thin grating equation. In fact the thin grating equation can be derived using hypothetical fringe concept demonstrated in figure 4.

The author's 'short explanation' of why the method of hypothetical fringes works is that, the hypothetical fringes identify the locations where the viewing beam and the image wave are identical in phase, just as happens with the reference and object wave during hologram formation. And, in the simplest model, this is exactly the requirement that must be met at each and every holographic grating opening. This is so, because the viewing wave is the source of the image wave. Thinking of Huygens principle, each point on the viewing wave front that exists in a particular grating opening is the source of the image wave that travels outward from the illuminated hologram. In the grating openings, the image wave is the viewing wave, and vice versa. There is one and only one wave at any point in each grating opening. Having a hypothetical fringe pass through each grating opening guarantees that the 'sameness' requirement between the viewing wave and image wave is satisfied.

The author wishes to thank the University of Northern Iowa Physics Department, and the College of Humanities, Arts and Sciences for support in preparing this report.

Light liquid: a holographic 'lake' installed on the roof of an architect's model townscape.

A Pepper

Senior Lecturer in Fine Art, Nottingham Trent University, Nottingham, UK

E-mail: andrew.pepper@ntu.ac.uk

Abstract. There has been considerable speculation about the use of holography in architecture and interior design over the past 20 years, with some spectacular examples having been realised. A number of installed works are referenced which use interior and exterior structures and spaces. Scale is considered as well as the possibility of architectural works existing within an artificial (model) environment. The visual, conceptual and critical values such an installation provokes are interrogated, with particular reference to 'Light Liquid, a holographic 'lake' installed within the 2011 Miniment[s] exhibition at Nottingham Trent University, UK. Aspects of miniature public art interventions, and whether they can have a critical validity within a contrived and artificial environment, are examined.

1. Expectation of greatness

Since the first public awareness of holography in the 1960's [1] there has been an anticipation of greatness. Popular media coverage of holography has raised expectations and blockbuster survey exhibitions of holographic images have not only proved to be extremely engaging, but appear to have raised those expectations even more. The field has seen a steady raise in interest during the 1970's and 1980's, with a dip during the 1990's [2], yet in the past few years, if reaction in the UK is indicative, visitors to these exhibitions are still dreaming of possibilities [3].

Ten years ago I raised the question of whether "...the medium will ever live up to the expectations we place on it?" [4] and, although there have been some remarkable, and truly innovative, architectural and large-scale uses of holography by pioneering artists, designers and engineers [5], it is difficult to ascertain whether the use of holography in this field is anything more than post-design decoration.

Although my own work in holography has included installations which attempt to go beyond the small scale available on a gallery wall, architectural holography has never been an area I have felt capable of engaging with. This is partly because of the scale of equipment required – I prefer to work with pieces I can 'pick up', but also because the critical and conceptual compromises which need to be made when work is increased in scale have been problematic for me. There are very few opportunities

Published under licence by IOP Publishing Ltd

to truly experiment using large-scale elements. The financial and structural realities act as an inhibitor.

2. Tiny architectural monuments

One solution is to engage with issues of scale, expectation and contextualization in miniature and that opportunity presented itself in the Miniment[s] exhibition, which took place in the UK during April 2011 (figure 1).

Artists were invited to submit proposals to this innovative exhibition, which, rather than being installed in a traditional gallery setting, was mounted in an architect's model townscape, located outside a gallery in the School of Art and Design at Nottingham Trent University. This transitory space within the school is used by students, staff and visitors from several departments, who walk through the same part of the building every day. One of the aspects of the exhibition was to include new additions of work to the townscape each week over the duration of the exhibition, so providing visitors with a durational series of interventions.

Figure 1. Installation view of Miniment[s] exhibition showing the entire model townscape installed at Nottingham Trent University, UK.

Not being in a gallery was a significant premise for this exploration of how monumental work might be conceived, developed and presented. From the open call for submissions, eight artists were selected to contribute a series of (mini) major artworks and interventions to the exhibition. [6]

Selection was made by a curatorial group of practising artists [7] who identified a variety of proposals including sculpture, found objects, sonic interventions and frozen paint. Each week a new tiny monument, or intervention, was installed to change the landscape of the 'town' by locating what are, in effect, monumental gestures within the miniature townscape. Over the course of the exhibition, the previous installation remained and each week a new piece was added to the expanding series of 'public artworks'.

Exhibited works included: tiny posters promoting the Tag Group and each new artwork; a full-size cardboard radio whose aerial extended skywards and passed through the protective transparent roof of the display vitrine; a sonic installation permeating through the pavements of the town, an exploded church roof revealing interior blue sky; a designer lemon squeezer perfectly located above the town's curved seating area; a holographic roof covering a building with the shadow of undulating liquid and, to close the exhibition, two buildings replaced with their exact volume of frozen white paint which, over the course of several hours, melted and enveloped the streets and their tiny inhabitants.

3. Light Liquid

'Light Liquid' (figure 2) is a holographic surface installed onto one of the flat-roofed buildings within the miniature townscape. A glass reflection hologram displaying the three-dimensional shadow of liquid was cut to the exact size of the building's roof. Using a Denisyuk recording method and displayed pseudoscopically, these shadows, which are reminiscent of drawn marks, appeared just above the glass surface and therefore 'hovered' slightly over the top of the building.

Figure 2. 'Light Liquid' 2011. Holographic intervention installed on model building's roof.

3.1. The kinetic effect

By recording information holographically, using a shallow amount of depth and then rotating the holographic plate around its horizontal axis, to display the image pseudoscopically, the information on display appears to exist just in front of the glass plate. This 'space' is the 'property' of the viewer – it is their space, the area they move around the work within. By presenting abstract marks in this location, a visual and perceptual 'friction' is generated. We 'know' that images cannot simply 'float' in space due to our engagement with physical objects in the real world. We also understand that shadows are 'cast' because of the path of light falling on an opaque object and that, subsequently, the shadow is always 'connected' to the object which cast it. By using this simple holographic recording technique, it is possible to visually disengage the shadow from its object, leaving it to exist in a different visual and conceptual volume (figure 3).

Figure 3. Installation details showing 'Light Liquid' (foreground) in relationship to other townscape buildings and artist interventions.

In this example, as visitors move around the townscape and view the holographic roof from a variety of angles, when the holographic information is visible (and at some points it simply appears as a gloss black covering), the shadow marks appear to be visually located ON the surface of the roof, not above it, and an enhanced kinetic movement takes place, causing the marks to drift and shift. This kinetic animation has been used by the author in a number of works, using reflection holograms, dating back to 1989. [8]

3.2. The model as legitimate object

It is certainly not unusual to test out proposals using models as a way of speculating how an intervention, in a full-scale environment, might function, and also as a visual indicator for prospective investors or planners. However, in this case, 'Light Liquid' was not a scaled proposal. It existed as a sculptural work, in its own right, within the miniature narrative generated by the model townscape and the curatorial premise of the exhibition.

If logic is applied to this holographic roof, it becomes ridiculous within a framework of real-world architectural structures. It is this breakdown in the reality of the situation which allows this installation to be observed and investigated using a fictitious narrative generated by the model environment. It is conceivably possible to actually construct this roof using sections of holographic material joined together to form a large surface area. There is a precedent in large-scale architectural structures by a number of artists who have successfully used this technique. [9] However, the result would inevitably be a mosaic of surfaces. 'Light Liquid' functions within the model narrative because it is a single sheet and because of the apparent expansiveness of its roof covering.

3.3. The monumental miniature narrative

If, as visitors to this exhibition, we place ourselves in the position of the townscape residents (and some of them were present during the entire month of the exhibition), from street level we would be completely unaware of the apparent 'lake' of undulating, three-dimensional liquid on top of the building. All that would be visible would be a consistent black line (the edge of the holographic plate) around the top of the roof. To view the holographically generated marks, residents would need to be at an observation point high above the town; a position occupied by the exhibition curator, visitor or fictional town deity. The roof intervention therefore functions on two distinct levels of engagement, one providing more visual access than the other. There is a parallel here to more traditional exhibitions where there is often a separation between how work is interpreted by the 'expert', the critic and the curious visitor. The situation also puts pressure on a visitor's conceptual engagement with the work.

3.4. Concept beyond practice

The installation of this elevated holographic 'lake' has parallels to early work produced by artist Rick Silberman who, in 1982, engineered a massive installation using reflection holograms spanning the United States of America. In 'Artifacts from Descartes' Dreams', Silberman traveled around North America, installing 8 x 10 inch reflection holograms within the landscape at pre-determined off-road locations which could be safely reached by motorcycle. Each contained a three-dimensional shadow of the image of a 'wire' stretched taught by a large tent hook. The shadow of a tether extended from the hook, out of the holographic plate, and into the space occupied by the viewer. Theoretically, this 'line' continues to extend outwards to infinity.

Each of the holograms which made up the installation was orientated so that at a certain time of day it would be possible for each of them to be illuminated by sunlight and activate their extended line to coincide at a high point over America. "The conceptual construct of the tethers rising and vanishing into space seemed as natural as a reflection off the surface of rippling water, or the after haze of a moist afternoon." [10] Assuming there was someone, at a position high above America, to actually receive the light/image emitted from the holographic plates, would they actually 'see' this extended structure and would it be in three dimensions? This is possibly a redundant question, as the conceptual structure of the work takes precedent over the physical and optical reality. A narrative is created, which allows the work to exist in a framework beyond a designated gallery exhibition and offers a different platform for interrogation of the artist's practice beyond its holographic 'cleverness'.

Acknowledgements
I would like to thank the Miniment[s] exhibition selection committee for speculating on the inclusion of a fictitious architectural 'lake' in their miniature townscape; Nottingham Trent University School of Art and Design, and its research team, for supporting the writing of this paper and my colleagues in fine art, for providing an environment where practice-lead research is considered a viable pedagogic imperative.

References and Footnotes

[1] Leith E and Upatnieks J June 1965 Photography by laser *Scientific American* vol 212 no 6
[2] Pepper A 2009 The definitive holography exhibitions list - filling the data gap - assessing a trend Proc *8th International Symposium on Display Holography, Shenzhen, China,* p 65-70
[3] British collector and gallarist, Jonathan Ross, has toured survey exhibitions: Holograms, The first 60 years and Holograms 2, to major museums and exhibition venues throught the UK which have each had high attendance from visitors who are 'seeing' holography for the first time.
[4] Pepper A 2002 Architectural holography: building with light, decorating with space, *Royal Photographic Society Holography Group conference on Holography, Art and Design.* Royal College of Art, London, UK. 23rd March
[5] Pepper A 2001 Building with light: holography, glass and architecture. *This Side Up* no 15 pp 2-4 ISSN 1389 1707, September
[6] Tag (Joseph Cerski and Bobby Sayer), Susie Cochrane, Rob Squirrell, Debra Swann, Gerard Williams, Andrew Pepper and Mark Lowe.
[7] Danica Maier, Geoff Litherland, Ben Judd, Craig Fisher and Terry Shave.
[8] Early work by the author to incorporate enhanced optical kinetic effects was produced as part of a PhD research initiative, Department of Fine Art, University of Reading, UK , during 1989
[9] Although not a definitive list, large-scale works produced by artists include: Rick Silberman, Artifacts from Descartes' Dreams, 1982; Paula Dawson, There's No Place Like Home, 1985; Sally Weber, Flight of Fish, 1990; Melissa Crenshaw, interior Canadian Pavilion Expo 1992; Shu-Min Lin, Glass Ceiling, 1996; Michael Bleyenberg, EyeFire/AugenFeuer, 2000; Dieter Jung, Perpetuum Mobile, 2001; Hiro Yamagata, Quantum Field-X3, 2004
[10] Silberman R Artifacts from Descartes' Dreams, a continent-wide holographic journey and the largest holographic installation to date. Unpublished document supplied by the artist, 11 April 2012

Journal of Physics: Conference Series **415** (2013) 012075 doi:10.1088/1742-6596/415/1/012075

Fabrication of synthetic diffractive elements using advanced matrix laser lithography

M Škereň, J Svoboda, M Květoň and P Fiala

Czech Technical University in Prague, Faculty of Nuclear Sciences and Physical Engineering, Břehová 7, 115 19 Prague 1, Czech Republic

E-mail: marek.skeren@fjfi.cvut.cz

Abstract. In this paper we present a matrix laser writing device based on a demagnified projection of a micro-structure from a computer driven spatial light modulator. The device is capable of writing completely aperiodic micro-structures with resolution higher than 200 000 DPI. An optical system is combined with ultra high precision piezoelectric stages with an elementary step $\sim$ 4 nm. The device operates in a normal environment, which significantly decreases the costs compared to competitive technologies. Simultaneously, large areas can be exposed up to 100 cm2. The capabilities of the constructed device will be demonstrated on particular elements fabricated for real applications. The optical document security is the first interesting field, where the synthetic image holograms are often combined with sophisticated aperiodic micro-structures. The proposed technology can easily write simple micro-gratings creating the color and kinetic visual effects, but also the diffractive cryptograms, waveguide couplers, and other structures recently used in the field of optical security. A general beam shaping elements and special photonic micro-structures are another important applications which will be discussed in this paper.

1. Introduction

Synthetic diffractive structures offer interesting properties in many application fields. In comparison with conventional refractive optical elements, the diffractive solution is very often much more flexible, lightweight, and in a case of fabrication of large series of elements also significantly cheaper. Considering the recent progress in the design and fabrication technologies, the number of applications of the synthetic diffractive elements is rapidly growing. However, because an extremely high spatial resolution is usually needed, the fabrication of the primary structure can be still relatively expensive. Thus there is a need for a fabrication technology, which can fulfill the demands for high resolution for reasonable costs and speed of the writing process.

The direct laser beam writing techniques represent an interesting alternative to the focused electron and ion beam writing approaches for fabrication of photonic micro-structures. Although the resolution of the laser beam writers is strongly limited by the Rayleigh diffraction limit, it is not always necessary to create the features significantly smaller than the recording wavelength. In such cases, a laser exposure can be used which can often achieve considerably higher exposure speed and better versatility of the recording process.

The resolution limit of the laser lithography can be extended not only by using a shorter wavelength (the deep UV lithography, etc.) [1]. Recently, the techniques based on a two-photon absorption or a stimulated-emission-depletion [2, 3, 4, 5] have been demonstrated, which can lead to the impressive feature size reduction [6]. Most current laser writing approaches use a single-point writing with a highly focused laser beam [7, 8]. Such techniques are especially demanding from a point of view of exposure times. An exposure of larger areas is well mastered in the dot-matrix and general matrix laser lithographic systems [9, 10], which are recently used in the field of synthetic image holograms for applications in optical document security. Some modifications of these approaches have been presented, which use multiple foci from a micro-lens array [11] or from a synthetic hologram displayed on a spatial light modulator [12, 13] to speed up the single-focus process. There are also few commercially available devices such as the single point laser writer for nano-structures produced by Nanoscribe GmbH [14, 7] or the matrix laser writer Kinemax^{TM} from Polish Holographic Systems [15]. The common drawbacks of the commercial devices are their lower versatility and the lack of some specific features.

In this paper a matrix laser writing technique is presented which significantly improves the writing speed and also the robustness of a positioning system and enables us to expose fine micro-structures over large areas with high speed and moderate costs. Instead of a single spot, a larger area is exposed within a single exposure (typically tens of thousands square micrometres). The experimental setup is discussed in detail. Comparing to the current devices, the stage positioning system and the optical focusing system were improved together with the driving software. Several samples are also presented, which were prepared using the developed device.

2. Matrix laser lithography

In the field of synthetic image holography so-called dot-matrix devices are commonly used to create a system of regular micro-gratings. If the elementary exposed area contains only a regular grating, a natural process of interference of two focused laser beams can be used. When an arbitrary number of interfering beams is assumed, the recorded structure can be general. More exactly, the decomposition of the exposure field to a system of plane waves can be done by the Fourier transform of the desired shape of the structure.

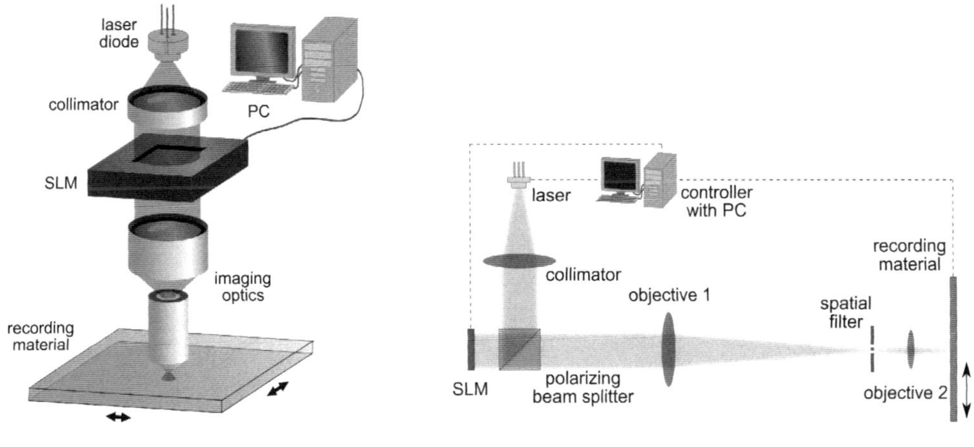

(a) (b)

Figure 1. (a) Basic idea of the advanced matrix laser writing system. The recorded structure is projected from a computer driven spatial light modulator using a de-magnifying optical system directly on the surface of a recording material. (b) Optical setup of the matrix writing device. A collimated, linearly polarized laser beam illuminates the image on the modulator which is further de-magnified using a system of objectives.

However, it can be technically complicated to set up such a system of writing beams. When the advanced matrix approach is used, the recording beams are created using a diffraction from a spatial light modulator. In fact, whole process can be interpreted as a projection of the recorded element from a micro-display to a recording material accompanied with a strong de-magnification. The basic idea of such a projection is depicted in figure 1a.

The optical resolution of the writing system is limited by several parameters. Again, the Rayleigh diffraction limit restricts the imaging process. However, the recorded micro-structure is also influenced by the elementary pixel size of the spatial light modulator and by the magnification of the imaging system. The final dimensions of a single projected addressable pixel from the modulator can be well beyond the diffraction limit (if the pixel size on the modulator is sufficiently small and the de-magnification of the optical system is high). Although a single pixel cannot be actually imaged, it can be still meaningful to use such a configuration, as the projected pixel size determines simultaneously also a positioning precision of the elementary features over the recorded element's area. As the object on the modulator is mechanically stable, a relative positioning within the exposure field is very precise (the only important distortion can be caused by imperfections of the optical imaging system). It can be also useful to expose the same area more than once with different patterns. When elements with fine details are exposed, all subsequent exposures must be perfectly aligned. This can be easily done with the presented setup.

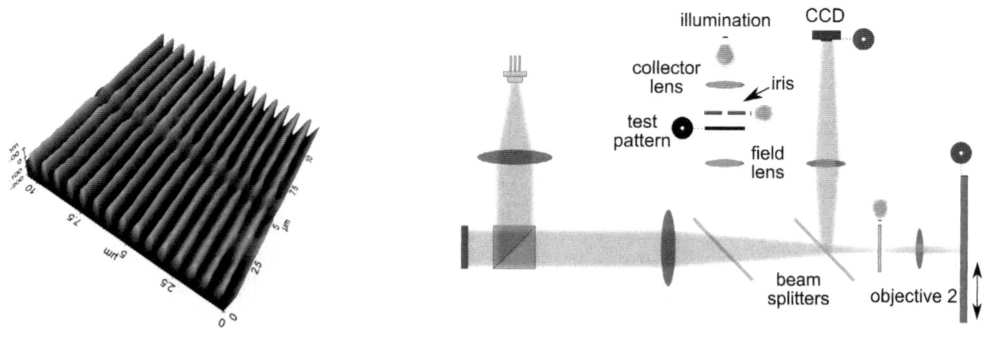

(a) (b)

Figure 2. (a) Demonstration of the stitching accuracy of the positioning system. An AFM scan of two adjacent gratings with the period 600 nm. The second grating was exposed after a long range travel of both stages from the position of the first grating and back. The stitching error is small comparing to the periodicity of the grating and also much smaller than the image positioning accuracy given by the projected resolution of the modulator. (b) The dynamic focusing is based on an auxiliary projection of a test pattern from an illumination system to the surface of a recording material. The test pattern (for example a pinhole aperture) is projected to the recording plane and the projection is observed by a CCD camera. During the exposure, the image of the pinhole is used as a reference for the focusing system and any distortion in the projection is compensated by moving the objective 2.

3. Experimental setup

The matrix writing device, which is based on the presented principles, has been built at the department. Because the main purpose of the system is the preparation of various micro-structures with high flexibility of their shapes and sizes, the main focus was held on parameters influencing the precision of the system, particularly the optical setup, the positioning system, and the focusing system.

The optical setup of the device is displayed in figure 1b. As a light source, the laser diode Nichia NDV4313 was used operating on the wavelength 405 nm with a typical optical output power 120 mW. The operating temperature of the diode is stabilized with the Thorlabs TCLDM9 cooled diode mount and the diode is powered with the Thorlabs ITC110 controller. The light from the laser is collimated and it illuminates the spatial light modulator. The Holoeye LC-R 1080 spatial light modulator was used for projecting the input data. The modulator is a reflective liquid crystal on silicon (LCoS) based device with a high resolution (1920 × 1080 pixels) and a high contrast operating in an amplitude modulation regime. The elementary pixel size is 8.1 μm and the fill factor of the display is 90%. The device operates with a refresh rate 60Hz and is addressed through a digital visual interface (DVI). Because of the operation in a reflective regime, there is a polarizing beam splitter cube in front of the modulator, which separates the incident and the diffracted beams. Finally, the image from the modulator is de-magnified in an optical system which consists of two objectives. The first objective is the photographic lens Carl Zeiss Sonnar with a focal length 300 mm and a maximum aperture 4. The second objective is the microscope objective Mitutoyo M Plan APO HR 50× with the numerical aperture 0.75 and the working distance 5.2 mm. The total de-magnification of the system is given by the ratio of the focal lengths of both objectives (75×). The Rayleigh limit for the microscope objective and the used wavelength is ~300 nm, the theoretical projected elementary pixel size at the recording material's surface is ~100 nm.

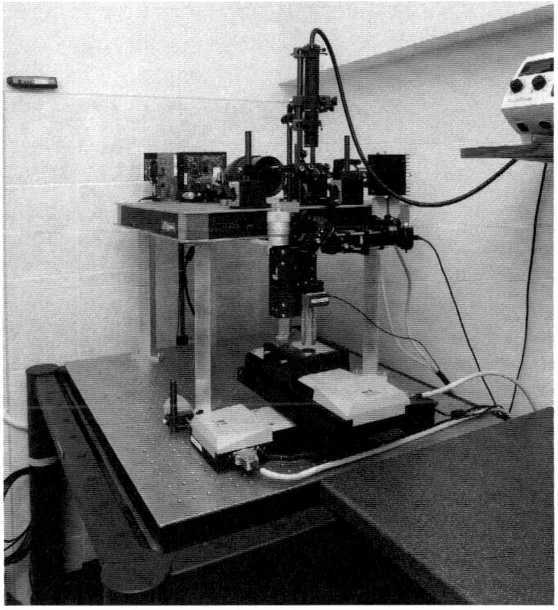

Figure 3. Photography of the device. Whole assembly is placed on a dynamic vibration isolation table. The device works in normal room environment.

Because of the positioning precision ~100 nm within a single exposure area, two M-511.HD Ultra High Resolution stages from Physik Instrumente were used to reach a comparable value also for the mechanical stitching. The stages are based on a combination of an electric servomotor with a piezoelectric actuator. The incremental step of the piezoelectric actuator is 4 nm. The linear encoder with the resolution 2 nm gives the accuracy below 50 nm with the repeatability

10 nm. These parameters can be used within a travel range of 100 mm with a maximum velocity 125 mm/s. Two stages with these specifications were crossed to form a two dimensional xy positioning system with a total covered area 100 cm^2. The positioning system was tested in an experiment which is described in figure 2a.

The theoretical limits for the elementary feature size were derived under the assumption of perfect focusing of the imaging system. However, the proper focusing is a tricky task, especially when large area structures are exposed. The dynamic focusing system, which is used in the constructed device, is described in figure 2b. The system uses an independent light source for a projection of a test pattern, which can be observed continuously during the exposure. The microscope objective is attached to a piezo-electric actuator which is connected to a feedback loop together with a CCD camera and a driving software. One of the important advantages of the laser writing techniques is a possibility to expose the three dimensional structures. The precise focusing system can be simultaneously used for the positioning of the focused exposure field in the three dimensional space.

In figure 3 there is an image of the system. The device consists of two platforms. The upper platform contains most of the optical elements, the lower platform includes only the stage and the positioning system. Between the platforms there are the microscope objective and the focusing system. Whole setup is placed on an active vibration isolation system ScienceDesk from Thorlabs. The exposure is driven directly from a PC via the custom designed software interface. The exposure speed is mainly limited by the sensitivity of the recording material and the necessary relaxation times. During the experiments, the exposure speed about 4 cm^2/hour was reached. This value is more or less independent of the micro-structure and can be further improved by increasing the laser power and by optimizing the driving software.

4. Applications and examples

The developed device can write wide range of micro-structures for various photonic applications. Because of a very precise positioning of the details within the modulator area, a nanometer precision in mechanical movement of the stages, and a general shape of the exposing field, this technique can produce very interesting patterns. Several samples of micro-structures were prepared for particular applications. They are described in figures 4-8.

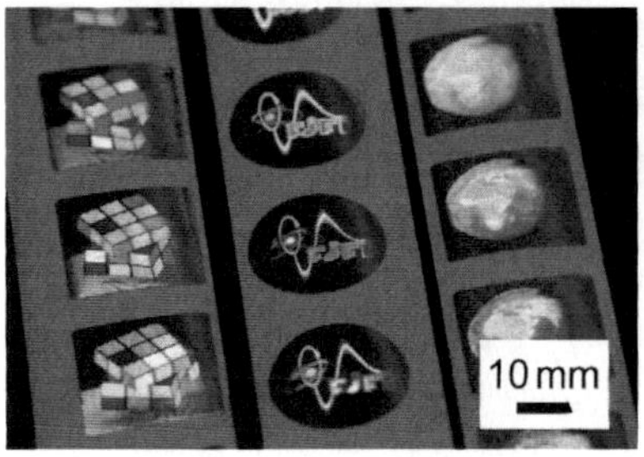

(a)

(b)

Figure 4. An example of the synthetic image holograms for application in optical document security. (a) The holographic stickers with relief holograms embossed in a plastic foil. (b) The synthetic true color 3D hologram exposed in a photo-resist material.

5. Conclusions and acknowledgements

The matrix laser lithography approach was presented which uses the projection of the desired element or its parts from the computer driven spatial light modulator on the recording material.

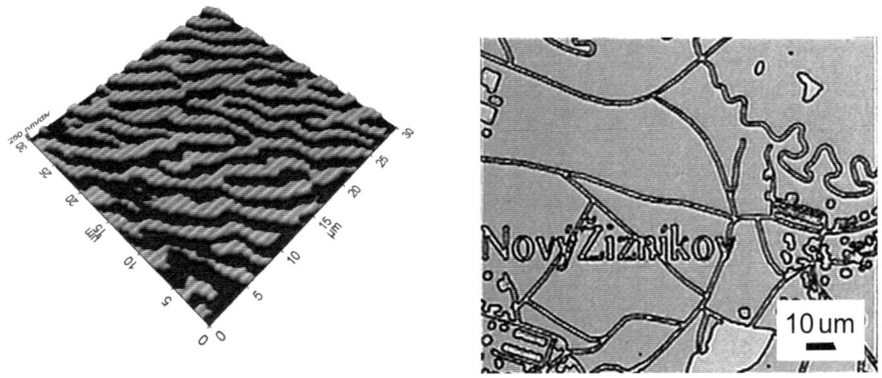

(a) (b)

Figure 5. An example of the aperiodic structures created using the matrix writing system. (a) The AFM scan of the special synthetic hologram combined with a regular grating. (b) The image from an optical microscope of the micro-map.

(a) (b)

Figure 6. An example of application of the synthetic diffractive structure for a white light beam shaping. The element was used for correction of the chromatic aberration of the car headlight system with a LED source. (a) The setup of the imaging system in the headlight and comparison of the illuminating beams without correction and with the diffractive corrector. (b) Micro-structure of an element exposed using the matrix writing system.

The main principle of the technique, which was first used in synthetic image holography, was further developed in order to improve the overall precision of the device and to enhance the flexibility of the writing process. Several examples were also presented which demonstrate the ability of the system to expose interesting elements for applications in optics, photonics, and another related fields.

Authors would like to acknowledge Jan Beránek for his help with the development and programming of electronic controllers. This work has been supported by the Grant Agency of Czech Academy of Sciences Project No. KAN401220801.

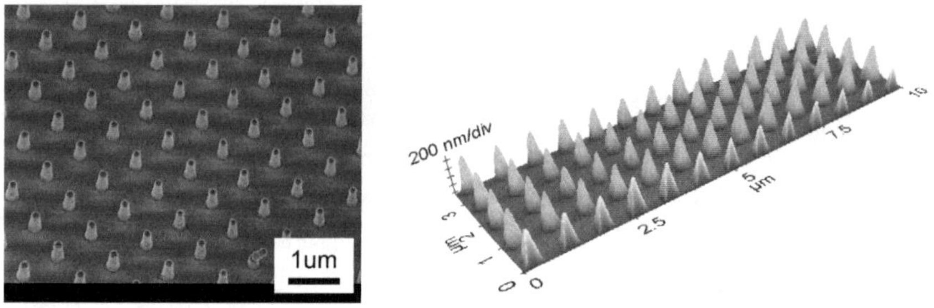

(a) (b)

Figure 7. An example of the periodic structures created using the matrix writing system. (a) The SEM image of micro-cylinders in resist material exposed using the device. (b) The AFM scan of the micro-tips array exposed using the laser matrix lithography.

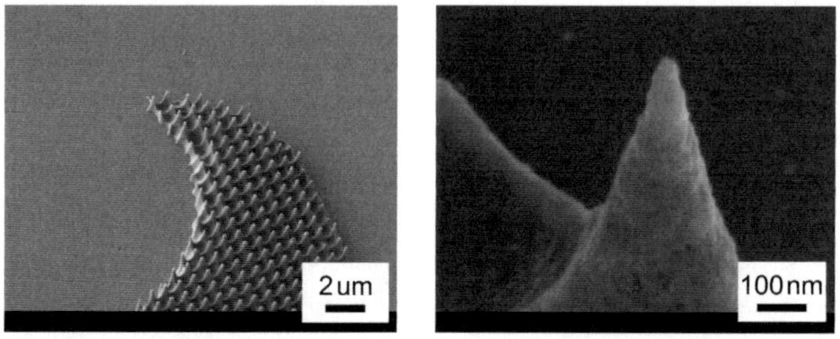

(a) (b)

Figure 8. An example of the micro-channel substrates with micro-tips for application in optical micro-manipulations. The structure was exposed using the laser matrix device.

References

[1] Selvaraja S K, Jaenen P, Bogaerts W, Van Thourhout D, Dumon P and Baets R 2009 *J. Lightwave Technol.* **0** 1-8
[2] Sun H B, Matsuo S and Misawa H 1999 *Appl. Phys. Lett.* **74** 786-788
[3] Fischer J and Wegener M 2011 *Opt. Mat. Express* **1** 614-624
[4] Haske W, Chen V W, Hales J M, Dong W, Barlow S, Marder S R and Perry J W 2007 *Opt. Express* **15** 3426-3436
[5] Fischer J, von Freymann G and Wegener M 2010 *Adv. Mater.* **22** 3578-3582
[6] Li L, Gattass R R, Gershgoren E, Hwang H and Fourkas J T 2009 *Science* **324** 910-913
[7] Ledermann A, Wegener M and von Freymann G 2010 *Adv. Mater.* **22** 2363-2366
[8] von Freymann G, Ledermann A, Thiel M, Staude I, Essig S, Busch K and Wegener M 2010 *Adv. Funct. Mater.* **20** 1038-1052
[9] Poleshchuk A G, Kutanov A A, Bessmeltsev V P, Korolkov V P, Shimanskii R V, Malyshev A I, Matochkin A E, Goloshevskii N V, Makarov K V and Makarov V P 2010 *Optoel. Instr. and Data Processing* **46** 171-180
[10] Bulanovs A, Tamanis E and Mihailova I 2011 *Latvian Journal of Phys. and Tech. Sciences* **48** 60-68
[11] Tang M, Chen Z C, Huang Z Q, Choo Y S and Hong M H 2011 *Appl. Optics* **50** 6536-6542
[12] Obata K, Koch J, Hinze U and Chichkov B N 2010 *Opt. Express* **18** 17193-17200
[13] Gittard S D, Nguyen A, Obata K, Koroleva A, Narayan R J and Chichkov B N 2011 *Biomed. Opt. Express* **2** 3167-3178
[14] Gansel J K, Thiel M, Rill M S, Decker M, Bade K, Saile V, von Freymann G, Linden S and Wegener M 2009 *Science* **325** 1513-1515
[15] Pizzanelli D 2004 *Technical review at holographer.org*

Holograms as Teaching Agents

Robin A. Walker

Program Coordinator/Interdisciplinary Doctoral Student, Teachers College,

Columbia University, New York, USA

E-mail: raw2142@tc.columbia.edu

Abstract. Hungarian physicist Dennis Gabor won the Pulitzer Prize for his 1947 introduction of basic holographic principles, but it was not until the invention of the laser in 1960 that research scientists, physicians, technologists and the general public began to seriously consider the interdisciplinary potentiality of holography. Questions around whether and when Three-Dimensional (3-D) images and systems would impact American entertainment and the arts would be answered before educators, instructional designers and students would discover how much Three-Dimensional Hologram Technology (3DHT) would affect teaching practices and learning environments. In the following International Symposium on Display Holograms (ISDH) poster presentation, the author features a traditional board game as well as a reflection hologram to illustrate conventional and evolving Three-Dimensional representations and technology for education. Using elements from the American children's toy Operation® (Hasbro, 2005) as well as a reflection hologram of a human brain (Ko, 1998), this poster design highlights the pedagogical effects of 3-D images, games and systems on learning science. As teaching agents, holograms can be considered substitutes for real objects, (human beings, organs, and animated characters) as well as agents (pedagogical, avatars, reflective) in various learning environments using many systems (direct, emergent, augmented reality) and electronic tools (cellphones, computers, tablets, television). In order to understand the particular importance of utilizing holography in school, clinical and public settings, the author identifies advantages and benefits of using 3-D images and technology as instructional tools.

1. Background: Holography and Education

What interdisciplinary research documents the effects of holography as a learning tool? Does holographic movement and stimuli impact children differently than adults? Which teaching assessments and Three-Dimensional (3D0 learning environments might support particular learning gains? The ISDH poster presentation, Holograms as Teaching Agents, is designed to prompt viewers to consider these and other questions about the influence of Three-Dimensional images, games and systems on learning science. Three-Dimensional Hologram Technology (3DHT), avatars and images have dramatically impacted the field of education, affecting the design of instructional activities and environments across all disciplines and content areas, particularly in the domain of science. Emergent hologram technology can represent diverse interactive educational platforms far different from traditional learning settings in conceptual context, time and space.

Holography is the only technique that can record the full tri-dimensional quality of an object, and allow the observer to easily see this as an image that is truly 3-D (Ko, 1998). Holograms of human organs and other figures can be used to teach adults and children simple dissections and health protocols as well as instruct medical students and physicians with the latest surgical techniques, innovative tools and environments. For learners of all ages and backgrounds, tri-dimensional games and systems offer stimulating educational training that can make complex scientific information and health content not only meaningful and instructive but more engaging.

Three-Dimensional medical avatars have been instrumental in driving development, innovation and knowledge around how patients, doctors and students learn about the human body, access medical data and record health conditions. With real-time, full color, 3-D animations and illustrations of the human anatomy, IBM™ , Apple™ and Blausen Medical, ™ have taken the lead in designing medical avatars as agents in knowledge-based instruction. Performing as science coaches in immersed learning environments, avatar apps use medical data and instructional embodiment to guide scientific learning. Students' experience and develop different sensory modalities, memories and perceptions as their brains interact with and in three-dimensional learning settings (Lu, Black, and Huang, 2005).

Holorad™ produces standard and customized sets of transparent Voxgrams®, 3-D images that literally hang in space, to teach as well as test student knowledge and understanding of anatomy. At leading U.S. medical schools, the company's new instructional tools efficiently provide "x-ray vision" for surgeons, patients and educators, giving a clear visual understanding of the critical relationships between and within anatomical and pathological structures (Holorad, 2012). Educators report that teaching is greatly enhanced with the intuitive "3-D X-Ray" viewpoint of Voxgrams. In addition to patient imaging data, Holorad has also developed popular 3-D entertainment and educational experiences at Disney's Epcot Center as well as Outer Space Holograms® that are a part of science education programs at national museums, airports and other public spaces (Voxel, 2012).

As educators, students, physicians and patients interact with various 3-D designs, programming cognitive tasks and mental representations in real-time applications, they develop science skills and abilities, building their intellectual capacity for learning science in synchronized stages. Providing immediate 3-D information, holograms uniquely facilitate the spontaneous understanding of human neuroanatomical relationships which cannot be efficiently learned with photographs or diagrams: This points the way toward future educational and biomedical applications of this emerging technology (Ko & Webster, 1995).

2. The Pulsed Brain Hologram: Superimposed Holographic Image-Guided Neurosurgery

The 8 x10-inch reflection hologram at the center of this ISDH poster design presents a Three-Dimensional image of the brain using a pulsed wave (PW) laser technique produced by neurosurgeon Dr. Kathyrn Ko (1995, 1998) in collaboration with Holographics, Inc. This instructional design employs 3-D models and imagery as instructional tools to demonstrate simple and complex scientific information in an interactive setting. As viewers observe the hologram teaching agent and engage in multimodal and multidimensional activities with the poster design, their understanding of human anatomy is enhanced. Subjects demonstrate basic cognitive skills while playing with the game box, and form mental representations when analyzing the holographic brain image.

In this reflection hologram, three-dimensional biomedical data is apparent. Applying this premise to neurosurgery, a sterilized radiological hologram of the brain is superimposed over the skill as a guiding template between the surgeon and subject.

Ko explains the surgical advantages for doctors and patients of representing biomedical data as a tri-dimensional visual guide: Holography is a departure from stereo-viewing in that the image is auto-stereoscopic and 3-D without the need of stereo pairs of stereo pairs or visual appliances. Holograms present multiple digital sets in a unified volumetric and interactive format. The accuracy of the 3-D image position and viewing angle with respect to the patient's pathological characteristics was checked in the office, leading to a higher confidence level in the use of this technology in surgery. All patients and their families preferred having their own hologram used as the visual educational aid during explanations of the procedure (Ko, 1998).

Since conceiving these preoperative simulations over 20 years ago, Dr. Ko has employed superimposed holographic images of human organs as an interactive visual map in the operative field. The neurosurgeon posits the potentiality of in vivo use of holography in diverse medical practices is enormous: Holography is not so much a jump into the third dimension, but a logical step along the progressive continuum of representing the human body; from drawing to photography, to x-ray studies, to CT and MR imaging in two dimensions and finally in three dimensions (Ko, 1998).

3. Poster Element Summary: Operation® the Game

Employing holographic images of the human brain as well as elements from the American children's game Operation®, this ISDH poster highlights the pedagogical effects of 3-D images and games on science education. In the 1960s, American scientists and young science-fiction fans alike would fashion new ways to use 3-D imagery to first entertain and later educate the masses. For the past 50 years, Operation® has inspired generations of children and adults around the world to have "funatomy," fun learning about human anatomy while playing a simple 3-D science game (Hasbro, 2005).

It was in a University of Illinois classroom that a sophomore instructional design student named John Spinello created what would become one of the world's most iconic educational 3-D games (Walsh, 2000). Credited with invented the hugely popular Operation, Spinello's premise was designed after electrified loop games from Americana country fairs and flamer stores (Mannion, 2011). Earning the highest class grade, his winning game box measured 10 x 10-inches with a 12-volt lantern battery that triggered a 6-volt bell. Designed as a 3-D operating table to test players' fine motor skills and hand-eye coordination, Operation players inserts tweezers into metal-edged holes in an electrified box, carefully removing organs, bones and other body parts without touching the sides of a multimodal, tri-dimensional patient names Cavity Sam. In late 1962, the college student pitched his educational design to Chicago toymaker Marvin Glass: I did my prototype of my magic box and everybody liked it. At first Glass was not impressed, he inserted the probe into the game board, got dinged, jumped, guffawed, and changed his mind and loved it (Walsh, 2000). The game was sold to Glass but not by Spinello who was hired 10 years later after Glass was no longer president and after Operation had become a huge success.

Milton Bradley™ purchased it in 1965 followed by Hasbro™ which continues to produce this world-renowned game in various languages and versions featuring its original 3-D patient as well as other popular children's characters from Shrek® to SpongeBob® (Hasbro, 2012). Spinello's design also inspired the 1964 children's game show, Shenanigans, which featured a life-sized, 3-D Operation game as one of its challenges. Aside from the traditional board game, a hand-held version with a screen in Cavity Sam's stomach, and a personal computer game was produced in 1998, followed by a brain surgery version (Hasbro, 2012), which required players to pull pieces out of Sam's head. Today, Hasbro jointly promotes Operation Relay® , currently shown on the television game show Family Game Night (the Hub, 2012), where two families compete through an obstacle course taking turns pulling pieces of out of an oversized Operation board.

In both the traditional game model and the television competition, the varied interactive formats of Operation encourage players to engage some Three-Dimensional version of Cavity Sam as well as science content in contexts that resemble a game, sport or amusement rather than simply an instructional lesson. John Spinello's college project would become an international educational tool that would motivate countless players of all ages and nationalities to become surgeons: I recall wearing and Operation t-shirt one day when a woman let me know of the game's impact on her son. She said, 'Because of your game, my son became a doctor' (Mannion, 2011).

4. Previous Literature and Cognitive Research

Recent cognitive studies provide a new perspective on what it means for learners to use holograms as teaching agents; specifically how Three-Dimensional images and systems will become more a part of the way that novices and experts understand and interact with the world of science. Research indicates perceptual effects on student comprehension and other strong educational benefits when instruction is mediated by 3-D imagery in immersed learning settings. Findings show objects that move or undergo perspective transformations are potentially more informative (i.e., convey additional information, such as information for an invariant property) than objects that are stationary (Gibson, 1969; Ruff, 1980).

Neuroimaging studies also indicate perceptual effects on comprehension and behavior when learning is mediated by holography. Testing this possibility in perceptual research among adults, subjects reported the three-dimensional effect to be the most salient aspect of holograms (Nelson & Horowitz, 1983). Three-Dimensional avatars used as pedagogical agents improve learners' conceptual understanding. Data also shows that 3-D medical avatars used as pedagogical agents improve learners' conceptual understanding. As they construct a concept map, students' specify propositional, functional and procedural knowledge to the avatar and are able to learn different scientific concepts better and more quickly (Bai & Black, 2005). Acting out stories using 3-D avatars also increased children's comprehension and memory.

Through innovative technology, customized content and interactive 3-D activities, holograms as teaching agents provide effective instruction to kids and adults by applying contextual and non-verbal cues. Tri-dimensional images, games, videos, storybooks and posters can make difficult health content more meaningful and engaging. By modeling instructional content and behavior in relatable contexts, holographic characters, tools and visuals can uniquely motivate students and promote interactive engagement in learning activities. Whether as a reflection hologram teaching agent, a board game or medical avatar in a computer program, each tri-dimensional tool activates science learning while motivating learners to continue to engage this educational technology.

How learners of all ages achieve greater proficiency and skills in 3-D learning environments in the future will be based on the varying medium (hologram, agent or human instruction), type of 3-D learning system used (game, computer, avatar) as well as a perceptual analysis of holographic stimuli (static or kinetic motion), on behavior and learning. Additional research is needed to examine the degree to which motion itself is a relevant or discriminant component of learning with holograms as pedagogical agents particular to adult learning, as well as 3-D's perceptual effects on children's comprehension.

5. Poster Discussion on Holograms as Teaching Agents

This International Symposium on Display Holograms poster design is built on the supposition that learners' exposure to multidimensional and multimedia instructional tools in an immersed learning environment could be associated with specific behavioral changes. Moreover, subjects' responses to Operation® as a conventional board game as well as to the superimposed holographic brain might affect particular gains in learners' skills, attitudes and knowledge in the science domain. When players create mental simulations using the 3-D game or the reflection hologram, not only are their perceptions but their knowledge, comprehension and reasoning skills are influenced.

These perceptual and technological motivators significantly impact how children and adults respond to the mental stimuli of holography in this unique learning environment. As viewers examine and engage with this poster design, questions around whether, how and to what extent Three-Dimensional Hologram Technology, images and games mediate their mastery of science content as well as their ability to regulate information in this learning environment are weighed and considered. If as perceptual researchers suspect, holography can be used as a procedure to train young children and adults to elicit scene changes from a holographic stereogram, then educators, researchers and instructional designers must continue to assess how holography performs as a learning tool and begin refinements based on those considerations. In the future, various learning environments (school, clinical and public), systems (direct, emergent and augmented reality) and telecommunications devices (computers, tablets, cellphones, TV) will be used to measure the short and long-term effects of using Holograms as Teaching Agents for learners of all ages.

References

[1] Bai X and Black J 2005 REAL: A Generic Intelligent Tutoring System Framework In C. Crawford et al Proceedings of Society for Information Technology and Teacher Education International Conference 2005 Chesapeake VA:AACE
[2] Gibson E J 1969 Principles of Perceptual Learning & Development New York Appleton-Century-Croft
[3] Hasbro 2012 Corporate Website Operation www.hasbro.com
[4] Holorad 2012 Voxgram Images for Education www.holorad.com
[5] Ko K April 1998 Journal of Neurosurgery Superimposed Holographic Image-Guided Neurosurgery Vol 88 777-781
[6] Ko K and Webster J 1995 Surgical Neurology. Holographic Imaging of Human Neurology. Brain Preparations Vol 44:428-32
[7] Lu C Black J Kang S and Huang D 2005 the Effects of LEGO Robotics and Embodiment in Elementary Science Learning
[8] Mannion A June 24, 2011 Elmhurst Tribune Local. Operation Inventor Proud of Iconic Game: http://triblocal.com/elmhurst/2011/06/24/operation-inventor-proud-of-iconic-game/
[9] Nelson C A and Horowitz E D 1983 The Perception of Facial Expressions and Stimulus Motion by Two-and Five-Month-Old Infants Using Holographic Stimuli Child Development 1983 54 868-877 by the Society for Research in Child Development
[10] The Hub television 2012 Family Game Night: Operation Relay www.thehub.net
[11] Ruff H A 1980 The development of perception and recognition of objects Child Development 1980 981-992
[12] Voxel 2012 3-D Displays and Entertainment www.voxel.com
[13] Walsh T 2000 The Playmakers: Amazing Origins of Terrific Toys. Andres McMeel Publishing

Method for evaluation of optical stability of the holographic recording environment

P A Wilksch and D T Ziedars

School of Applied Sciences, RMIT University, Melbourne, Australia

Email: philip.wilksch@rmit.edu.au

Abstract. A simple method is described for displaying and recording drifts in the path length difference between two laser beams from a single source. A beamsplitter is used to recombine the expanded beams, producing two similar interference patterns, one on each side of the beamsplitter. A photodiode is positioned in each pattern such that the two detectors sample the pattern at points where the phase difference between beams is itself different by one quarter of a cycle. The outputs of the two photodiodes are amplified and shown on an oscilloscope in X-Y display mode. After suitable adjustments of amplifier gain and photodiode position, any optical drifts cause the oscilloscope spot to move in a circle in which the angular position of the spot directly represents the phase difference between the light beams. This movement can be tracked visually, or recorded and analyzed to produce a time record of phase drifts. The system is self-calibrated, and sensitivities of 1/100 cycle are easily obtained, on time scales ranging from the minimum response time of the photodetectors to as long as desired. Examples are shown illustrating the effects of air currents and floor vibrations on the optical stability of a holographic table. Results are compared with mechanical vibration measurements obtained with an accelerometer.

1. Introduction

The necessity for optical stability when recording holograms with CW lasers is well-known, and holography texts and handbooks carry many hints and tips on reduction of disturbances due to mechanical vibrations and drifts, thermal movements, changes in humidity, air currents, and the like. The proof of a stable setup is the quality and efficiency of the hologram produced, but the time delay between recording and viewing a developed hologram makes a trial-and-error approach to the reduction of disturbances a tedious process. Immediate, real-time means of evaluation of stability are far more useful.

It is sensible to employ a method which uses the same technique as holography itself, namely interference, to evaluate stability. An interference pattern is readily set up between two or more beams arranged in a similar configuration to the one that is to be employed in recording the hologram. There is then a need for a way of recording the stability of the fringe pattern, and quantifying it so that efforts to improve stability can be evaluated objectively.

This paper points out the advantages of an arrangement that uses two detectors to sample irradiance variations at closely-separated points in a fringe pattern, and a processing algorithm that produces a continuous record of phase difference variations between the interfering beams. The techniques themselves are not new, but our experience has shown that they can be implemented easily with inexpensive equipment and can give valuable information on both short-term and long-term stability.

Published under licence by IOP Publishing Ltd

2. Theory and Method

Figure 1 shows a typical arrangement on an optical bench where a laser beam is split into two beams which, after travelling different paths, are redirected to a common point. In a holographic recording setup, the common point would be the recording medium, and the two beams would be a reference beam and an object beam, after scattering from the object. The stability of the phase difference between the beams at the point of recombination can be investigated by studying their interference fringe pattern.

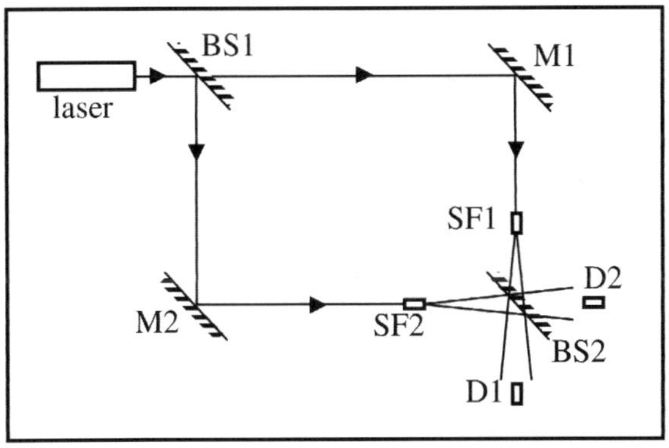

Figure 1. Recombining two laser beams to produce two interference patterns, one on each side of the final beamsplitter.
BS1, BS2: beamsplitters;
M1, M2: mirrors;
SF1, SF2: beam expander/spatial filters;
D1, D2: detectors.

To produce interference fringes with broad fringe spacing, the two beams must be brought into close parallelism. This is accomplished with another beamsplitter positioned so that the two beams make equal angles with the beamsplitter normal, but on opposite sides of the beamsplitter. Further, to spread the beams over a larger area, a beam expander/spatial filter can be used in each beam. It is an advantage if these are positioned equal distances from the point of recombination at the beamsplitter, so that the curvatures of the wavefronts are closely matched. Then by fine adjustments of the spatial filter positions and the beamsplitter tilt, the separation between the interference fringes on the far sides of the beamsplitter can be broadened, until further widening is limited by the influence of aberrations and nonuniformity in the beamsplitter.

According to the general law of interference [1], the total irradiance S_T at any particular point in the fringe pattern is given by:

$$S_T = S_1 + S_2 + 2r(S_1 + S_2)^{1/2} \cos \delta \tag{1}$$

where S_1 and S_2 are the irradiances at that point due to the two beams separately, r is the degree of coherence, and δ is the phase difference between the beams. Equation (1) can be simplified to:

$$S_T = S_{av} + S_p \cos \delta \tag{2}$$

where S_{av} and S_p are constants applying at that point. The quantity that is of interest here is the phase difference δ between the two beams. It can be expressed as:

$$\delta = 2\pi\Lambda_{12}/\lambda_0 \tag{3}$$

in which λ_0 is the free-space wavelength of the light, and Λ_{12} is the optical path difference between the two beams, measured from the point of division at the first beamsplitter to the point of recombination at the second. The optical path, in turn, is given in general by:

$$\Lambda = \int n(l)dl \tag{4}$$

where n is the refractive index of the intervening medium, and the integral is over the path between beginning and end points mentioned above. Hence Λ_{12} is sensitive to changes in the geometric path length as well as fluctuations in refractive index.

From a recording of the irradiance S_T at a single point, only the cosine of δ can be determined, so there is ambiguity in translating changes in S_T into terms of changes in Λ_{12}, particularly when such changes extend over several wavelengths. It is not possible to tell in which direction (increasing or decreasing) the change is taking place, and there is uncertainty whether or not a reversal of direction might have occurred at any point where $\cos \delta = \pm 1$.

Nevertheless, when one looks at an extended fringe pattern, it is obvious in what direction Λ_{12} is changing, because the fringes will drift in one direction or the other, and the relation between the direction of fringe drift and the increase or decrease of Λ_{12} can be determined simply by imposing a small change of known direction on the beam path, such as by light pressure on one of the mirrors. One could record movements of the fringe pattern with a digital camera, but one would still be faced with the task of analyzing the recorded images to extract quantitative measurements of optical path variations from them.

The classic technique for recovering phase from sinusoidal fringe patterns [2] is quadrature detection, where two detectors are positioned so that, if the phase difference at one of them is δ, then the phase difference at the second is $\delta - \pi/2$ rad. In (2), the irradiance at the second detector will then have a contribution proportional to $\cos(\delta - \pi/2) = \sin \delta$. With knowledge of both $\sin \delta$ and $\cos \delta$, the phase difference δ can be determined unambiguously, at least over a range of 2π rad.

This technique could be implemented by situating two detectors side by side in the fringe pattern so that their spatial separation corresponds to one quarter of the fringe separation (or any odd multiple of one quarter). Provided that the fringe separation does not change as the path lengths change, the detectors will remain in quadrature. In most cases this condition will be satisfied, because the spatial details of the fringe pattern depend only on details of the optics between the spatial filter pinholes and the detector plane, and this part of the optical system can be made compact and stable.

However, depending on the size of the detectors, it may be difficult or impossible to situate two of them close enough to sample points within one fringe spacing. One solution could be to use optical fibre pickups to collect the light. But there is another option already available in the optical configuration of figure 1, and that is the "twin" fringe pattern formed on the other side of the beamsplitter. This pattern is a mirror image of the other, and is also contrast-reversed, having bright fringes where the other has dark fringes (if the beamsplitter is non-absorbing). It is not difficult to situate one detector in each fringe pattern, and then fine-tune their positioning so that they are in spatial quadrature. A method for doing this is described in the next section.

3. Implementation

3.1. Apparatus

Measurements have been made on two pneumatically-supported optical tables in the holography laboratory at RMIT. The light source for the interference measurements was a 10 mW helium-neon laser (Melles Griot model 05-LHP-991). An arrangement similar to that in figure 1 was set up, with the beams covering a large area of the table. The two path lengths were matched to within 20 millimeters to ensure good coherence. Spatial filter assemblies were used to spread the beams from points about 100 mm before the second beamsplitter.

Figure 2 is a photograph of the typical appearance of the interference fringes. The system has been aligned to spread the fringes as far apart as possible. The residual fringe pattern shows saddle-shaped contours due to the astigmatism introduced when a spherical wavefront passes through a parallel-sided piece of glass (the beamsplitter) at an angle. A detector is then placed at the centre of each pattern where the fringes are widely separated.

The detectors were two silicon photodiodes (UDT 10DP). These have a circular sensitive area of 1.0 cm^2, but each was mounted in an enclosure with an adjustable iris diaphragm in front of the detector, so that a small area of the interference pattern could be sampled. Typically the aperture diameter was between 2 and 5 mm. The photodiode currents were amplified and converted to voltages with Terahertz Technologies PDA-750 photodiode amplifiers.

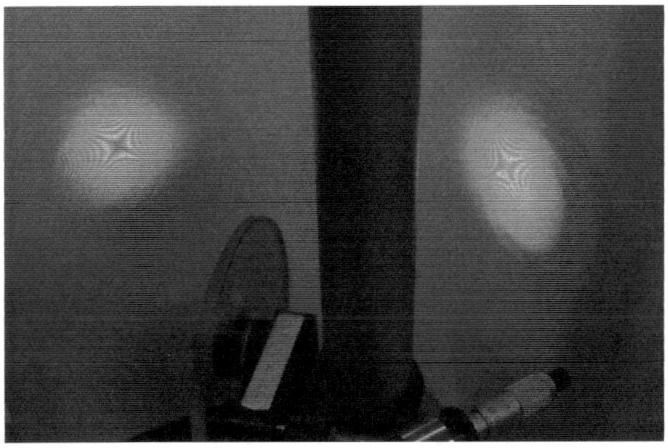

Figure 2. Twin interference patterns produced on either side of the beamsplitter/combiner. The saddle shape of the fringes is due to astigmatism introduced in the spherical waves by passage through the beamsplitter.

The two amplified signals were fed to the two inputs of an oscilloscope operating in X-Y mode. Because both signals have a component which varies sinusoidally with changes in Λ_{12}, the oscilloscope spot is observed to trace out a figure that is in general an ellipse, as the path difference changes. If the two detectors are sampling the same point in the fringe pattern, the figure will be a straight line, as the two sinusoidal components will be in phase with each other. If the detectors sample points that are one quarter of a fringe apart, the figure will be an ellipse symmetrical about the vertical and horizontal axis, or a circle if the amplitudes of both sinusoidal components are the same. The pattern varies between these extremes as the detector is displaced, as shown in figure 3.

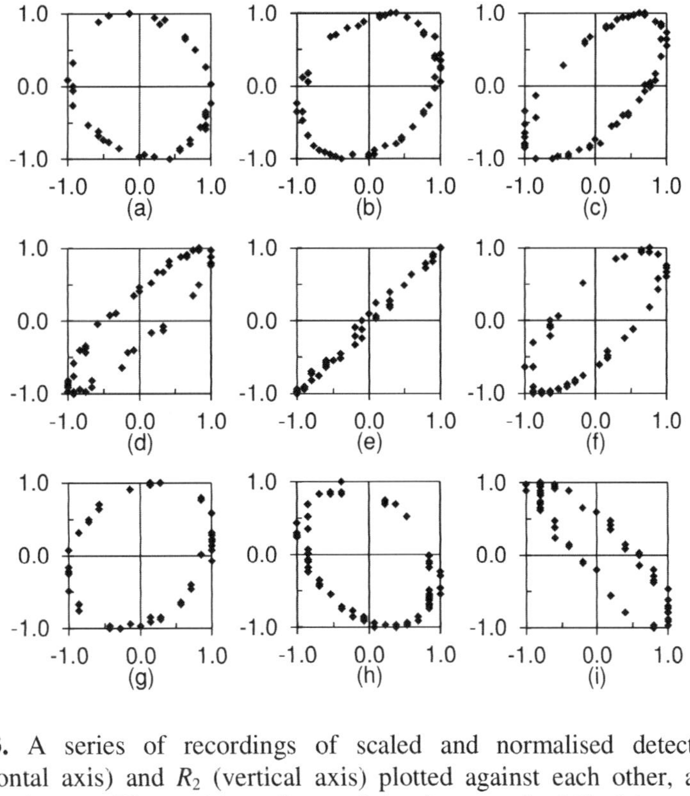

Figure 3. A series of recordings of scaled and normalised detector signals R_1 (horizontal axis) and R_2 (vertical axis) plotted against each other, as the path difference changes. The position of one of the detectors is shifted incrementally in the sequence (a) to (i). The pattern changes in shape from circular through elliptical to linear and back again as the detector is shifted across the fringe pattern.

To map out the shape of the ellipse, it is necessary to introduce a perturbation to the optical path difference between beams. This is easily done by drumming with one's fingers on the table top. The position of one of the detectors can then be adjusted until the ellipse is symmetrical about X and Y axes. The gain and offset of each channel are adjusted to produce a circular trace, centred on the oscilloscope screen. The sense of rotation of the oscilloscope spot (clockwise or anticlockwise) tells in which direction the path difference is changing. One whole rotation of the spot corresponds to a path difference change of precisely one wavelength. The system is thus self-calibrating.

To produce a permanent record of drifts, the signal outputs from the detectors must be recorded. Inexpensive digital oscilloscopes are now available which can store a time sequence of voltages and read it out to a computer. Alternatively, USB data logger units can record voltage signals over longer periods, so that long-term drifts can be studied. The series of numbers thus generated is easily converted into a record of optical path variation, as follows.

3.2. Analysis

First, each data series S is inspected to find its maximum and minimum signal values, S_{max} and S_{min}. From these, we determine $S_{av} = (S_{max} + S_{min})/2$, and $S_p = (S_{max} - S_{min})/2$. Then for each reading within the series we calculate the quantity

$$R = (S - S_{av})/S_p \tag{5}$$

which from (2) is expected to equal $\cos\delta$. It varies over the range $-1 \le R \le +1$.

Suppose that the phase difference at detector 1 is δ, and at detector 2 is $\delta - \pi/2$, and let the corresponding series of values of R be R_1 and R_2. Then we expect that:

$$\delta = \arctan(R_2/R_1) \tag{6}$$

provided that the underlying signals of S_1 and S_2 are obtained simultaneously from the two detectors.

Equation (6) gives a result over the range $-\pi \le \delta \le +\pi$. For convenience of interpretation, it is useful to convert this to a measurement in units of cycles rather than radians. Define a quantity:

$$D = (1/2\pi)\arctan(R_2/R_1) \tag{7}$$

which varies over the range $-\tfrac{1}{2} \le D \le +\tfrac{1}{2}$. Now it only remains to deal with the discontinuity that results when D crosses the boundaries of this range. Provided that successive samples differ in actual phase by less than half a cycle, this can be done by adding or subtracting one whole cycle whenever D changes by more than ½. We keep track of the number of whole cycles of phase change with an integer M, initially set to zero. At each sampling step i, we calculate the change in D, namely $\Delta D = D_i - D_{i-1}$. Then we set

$$\Delta M = \begin{cases} +1 \text{ if } \Delta D < -\tfrac{1}{2} \\ -1 \text{ if } \Delta D > +\tfrac{1}{2} \\ 0 \quad \text{otherwise} \end{cases} \tag{8}$$

and put $M_i = M_{i-1} + \Delta M$. Finally, the "unwrapped" phase P, expressed in cycles, is given by

$$P_i = M_i + D_i. \tag{9}$$

If the two detectors are not aligned exactly in quadrature, and the shape of the oscilloscope trace is slightly elliptical rather than circular, a small nonlinearity will be introduced into the relation between the calculated P_i and the actual phase difference δ, over each cycle of phase change. The discrepancy can be corrected by using the measured shape of the ellipse [3], but the correction will not be significant except in very precise applications. In any case, the fact that an integer change in P corresponds to a whole cycle of phase change continues to be exactly true.

4. Results

4.1. Long-term drift

Figure 4 shows a series of readings of phase difference taken at one second intervals over a time period of 100 seconds. At the commencement of the series (between 10 and 30 seconds on the time axis), a gradually-increasing pressure was applied to the table top, and then removed, causing the phase to change over a few cycles. From this sequence, the parameters S_{max} and S_{min} were determined for each of the two detector signals, and subsequently used in the calculation of the phase difference.

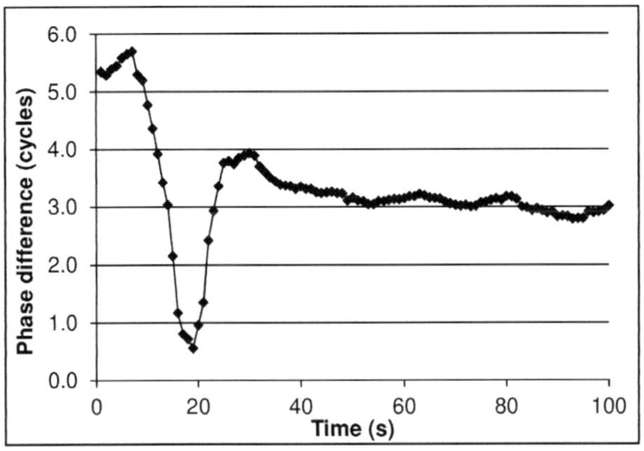

Figure 4. Variation of phase difference as a function of time, during which a phase change was induced by pressure on the table top. The phase is sampled at one second intervals.

Longer sequences of measurements are shown in figure 5, for four different conditions. The sampling period was again one second. Figure 5(a) was recorded with the room air-conditioning operating, and the table top open and unprotected. In an attempt to minimize local draughts, temperature-controlled air is introduced into the room through long, porous "socks" running overhead the length of the room. Nevertheless, random phase fluctuations extending over about half a cycle are apparent. That these are due to air currents is confirmed by figure 5(b), where the air-conditioning was switched off. The random fluctuations are seen to be reduced in amplitude, although they are still significant.

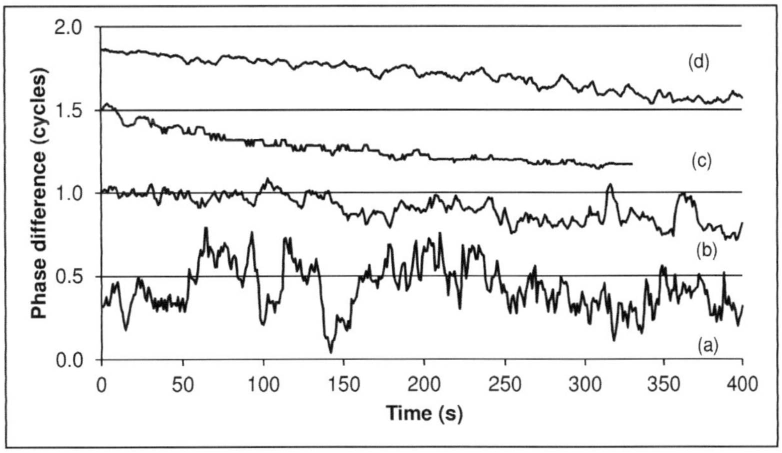

Figure 5. Recorded variations in path difference as a function of time, expressed in terms of number of cycles of phase change. The four traces are for (a) unenclosed table, air conditioning (AC) on; (b) unenclosed table, AC off; (c) enclosed table, AC on; (d) enclosed table, AC off. The phase at the starting time is set arbitrarily to offset the traces.

For the traces in figures 5(c) and (d), the table top was enclosed by walls and a roof made of acoustic insulation batts, preventing air currents from crossing the table. There is little difference between (c), with air-conditioning running, and (d), with the air conditioning off, and the random fluctuations are further reduced to less than one tenth of a cycle. However, all four traces show a steady drift in path difference over the 400-second time interval of the recordings, with some indication that the drift might be more serious when the table is covered. More measurements are needed to identify the cause of this drift.

4.2. Short-term drift

To investigate phase jitter over time periods shorter than one second, a digital oscilloscope was used to record the output of the two detectors. The output data consisted of 4096 samples taken over a period of one second, with a sampling rate close to 0.25 ms. A short segment of the processed recording is shown in figure 6.

There is a prominent periodic component in phase perturbation with a frequency close to 100 Hz, and an amplitude of about 0.01 cycle. This was present in most of the records. The ability of the technique to resolve phase changes of this magnitude is well illustrated by this example. To verify that these oscillations were not due to output power fluctuations of the laser, recordings were made of the two beams individually. No 100 Hz component was evident.

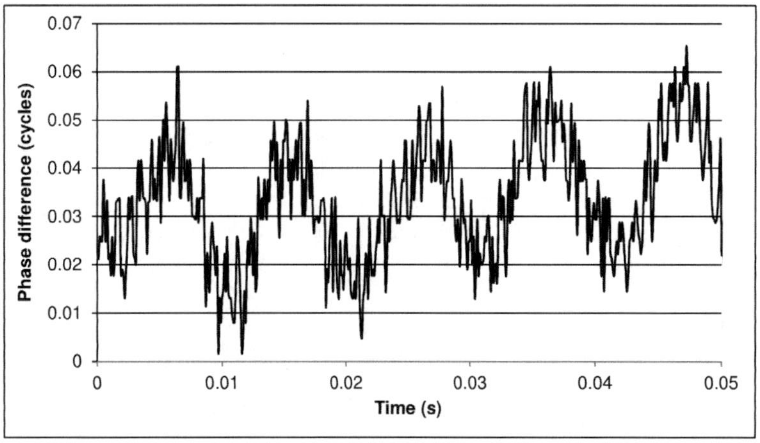

Figure 6. Phase difference variations recorded with a digital oscilloscope, with a sampling period of 0.25 ms. A strong periodic perturbation with a frequency of 100 Hz and an amplitude of about 0.01 cycle is apparent.

4.3. Accelerometer measurements

Short-term mechanical vibrations were also investigated with the use of a single-axis seismic accelerometer (Wilcoxon Research model 731A). Records of acceleration over time periods of a few seconds were Fourier-analyzed to produce measurements of the frequency components of the amplitude of displacement at one-Hertz intervals.

Measurements were made on a table next to the one on which the interferometer was mounted. Figure 7 shows the recorded vertical displacement on a logarithmic scale. The upper trace is for the accelerometer mounted on the floor under the table, and the lower trace is for the accelerometer mounted on the table top. Conversion from acceleration to displacement involves multiplying the data by the inverse square of the frequency, and results become unreliable at frequencies below about 10 Hz.

On the floor there are prominent components at 100 Hz and 200 Hz. These are attenuated by over an order of magnitude by the table suspension and damping system, but the 100 Hz vibration is still detectable on the table top. Horizontal displacements were also measured, and were found to be similar to the vertical ones but generally smaller in amplitude.

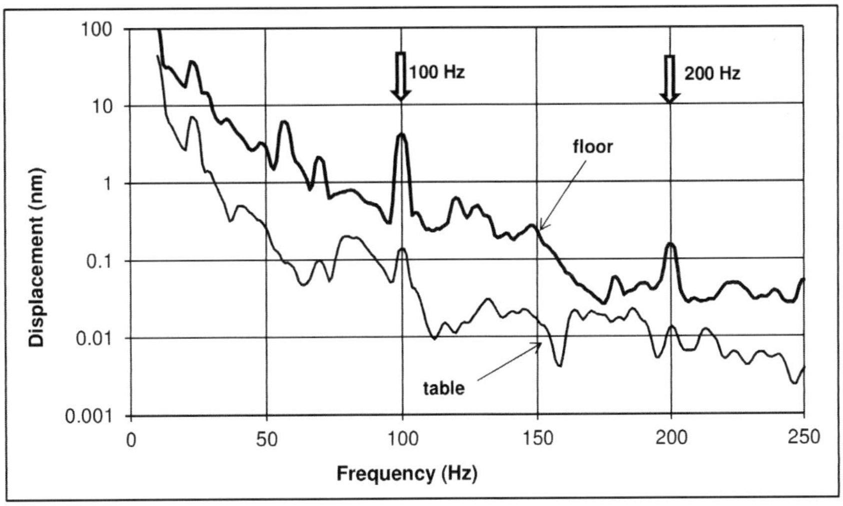

Figure 7. Frequency spectrum of mechanical vibrations, recorded with an accelerometer on the optical table top and on the floor beneath the table. The amplitude of the vertical component of displacement is shown.

The accelerometer confirms the presence of the 100 Hz vibrations found with the interferometer. It is thought that they might be caused by motors or power transformers elsewhere in the building, generating acoustic waves that are transmitted through the building structure.

5. Conclusions

A simple method has been used to measure and record phase difference variations between two interfering beams in a holographic setup. With the use of two detectors sampling an interference pattern in spatial quadrature, phase variations can be tracked over both short and long time periods without ambiguity in direction of the phase drift or discontinuities in the phase record. Phase variations can also be displayed in real time for rapid evaluation of the stability of holographic environments. Resolution is of the order of 1/100 cycle or better.

Use of the technique in the holography laboratory has illustrated the importance of good shielding of the optical table from air currents, and mechanical vibrations at 100 Hz have been identified as a source of phase perturbation.

References

[1] Hariharan P 1996 *Optical Holography: Principles, Techniques, and Applications* (Cambridge, New York: Cambridge University Press)

[2] Creath K 1988 Phase-measurement interferometry techniques *Progress in Optics* vol 26, ed E Wolf (Amsterdam: Elsevier) pp 349–93

[3] Wu C-M, Su C-S and Peng G-S 1996 Correction of nonlinearity in one-frequency optical interferometry *Meas. Sci. Technol.* **7** 520–4

Journal of Physics: Conference Series **415** (2013) 012078 doi:10.1088/1742-6596/415/1/012078

Visualizing Breath using Digital Holography

P R Hobson[1], I D Reid[1] and J B Wilton[2]

[1]Centre for Sensors & Instrumentation, School of Engineering & Design, Brunel University, Uxbridge UB8 3PH, UK
[2]Artist in Residence, Uxbridge UB8 3PH, UK
Email: peter.hobson@brunel.ac.uk, ivan.reid@brunel.ac.uk, jayne@jaynewilton.com

Abstract. Artist Jayne Wilton and physicists Peter Hobson and Ivan Reid of Brunel University are collaborating at Brunel University on a project which aims to use a range of techniques to make visible the normally invisible dynamics of the breath and the verbal and non-verbal communication it facilitates. The breath is a source of a wide range of chemical, auditory and physical exchanges with the direct environment. Digital Holography is being investigated to enable a visually stimulating articulation of the physical trajectory of the breath as it leaves the mouth. Initial findings of this research are presented. Real time digital hologram replay allows the audience to move through holographs of breath-born particles.

1. Introduction

The ability to observe organic patterns such as that created by the spent breath opens up opportunities through which to increase our understanding of communicational dynamics and the impact of the human breath. Making such dynamics visible expands our understanding of our interaction with our direct environment. Jayne Wilton works to make visible these dynamics using processes as diverse as camera-less photography and lithography. Collaboration with physicists Peter Hobson and Ivan Reid, who both have specialist experience of in-line digital holography, has facilitated an important engagement with previously inaccessible processes which has resulted in a greater technological empowerment of possible visual outcomes. In-line digital holography has not been used extensively in the making of art, so the collaborators are interested to investigate the potential for less traditional outcomes of this technique.

The creation of digital holograms of breathing patterns will allow the three-dimensional illustration and capture of the form and dynamics inherent in a range of direct and indirect communications: from the enigmatic sigh to the verbal articulation of core linguistic units or semantic primes. These holograms are played as a video, but additionally can be used interactively and navigated by the viewer so as to be able to engage with specific areas of the information captured. The three-dimensional coordinate data available from the replayed hologram images also facilitates the further potential development of the work into kinetic sculpture. The coordinates of each particle as suspended by the breath in the chamber allow the constellation to be recreated in different formats and scales and using different media.

2. Use of Digital In-Line Holography

The visualizing of breath using digital in-line holography is challenging due to the issues of the relative scale of the human breath in relation to the recording area of the hologram which is, for collimated beams, limited by the sensor area. In this collaborative project a number of approaches, including the use of spatial filtering [1], are being used to record the human breath.

2.1. Hologram Recording System

2.1.1. Continuous-wave laser. The use of holography to capture vapour bubbles is well documented and a digital hologram recorded and replayed in our laboratory is illustrated in figure 1. Bubbles clearly can be containers of breath, yet the investigators are currently also exploring the impact of breathing into an air chamber rather than through a liquid. A variety of methods have been examined ranging from breathing into a cloud generated by liquid nitrogen and breathing into a chamber of fine particles to allow the holographic visualisation of the trajectory of the spent breath.

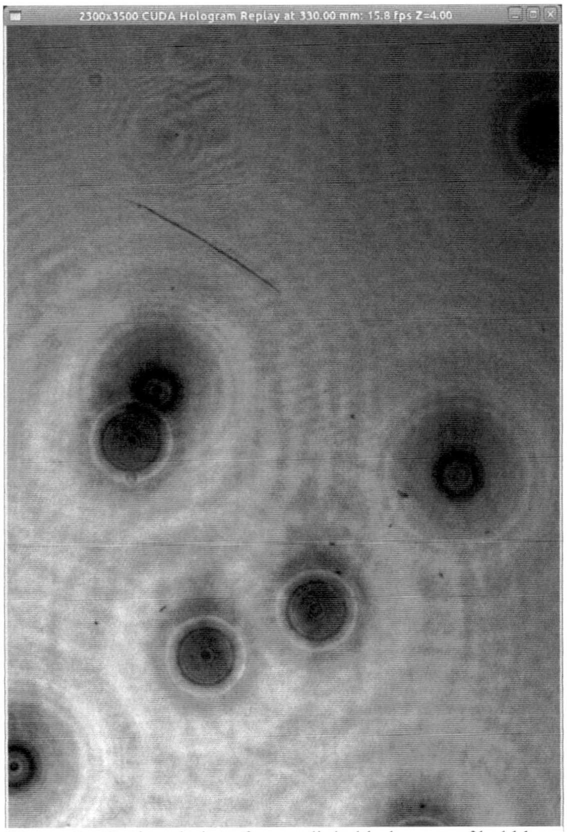

Figure 1. Replayed plane from a digital hologram of bubbles recorded using a continuous-wave red HeNe laser. The diameter of the bubbles is approximately 0.7mm.

2.1.2. Pulsed laser. In-line digital holograms were recorded using a fast pulsed green laser and a 36×24 mm^2 CCD camera (figure 2 and figure 3). The laser is a custom doubled Nd-YAG from Elforlight Ltd, model number LP3-10-532. This produced single longitudinal-mode pulses with a wavelength of 532 nm, duration of 2 ns and an adjustable energy of a few millijoules. The beam was then attenuated using a Newport 935-10 high-power variable attenuator and reflected, using a BK7 wedge, towards a beam expander telescope. The expanded beam then traverses a tank with BK7 flat windows and the resulting hologram is recorded using an ATMEL Camelia 8M CCD camera (Figure

2). The resulting holograms from the camera have 3500×2300 10 micrometre square pixels and are recorded with 12 bits of grey-scale information.

The holograms are replayed using the recently developed HoloMovie software that runs on NVIDIA GPU [2]. Our general approach to reconstruction is based on the convolution method [3] and can be summarised as follows:

- Numerically multiply the (zero-padded) hologram by a complex amplitude reference wave (in the case of collimated beam in-line holography the reference wave is 1+0i).
- Perform a 2D discrete Fast Fourier transform (FFT); real-valued data is transformed to complex-valued data.
- Multiply the result by a transfer function in the frequency domain based on the Rayleigh-Sommerfeld equation. The transfer function depends on the wavelength and the distance of the desired depth plane from the recording plane.
- Perform an inverse 2D FFT to generate the reconstructed image for a single depth plane. Extract the magnitudes, remove the zero-padding and normalise intensity.

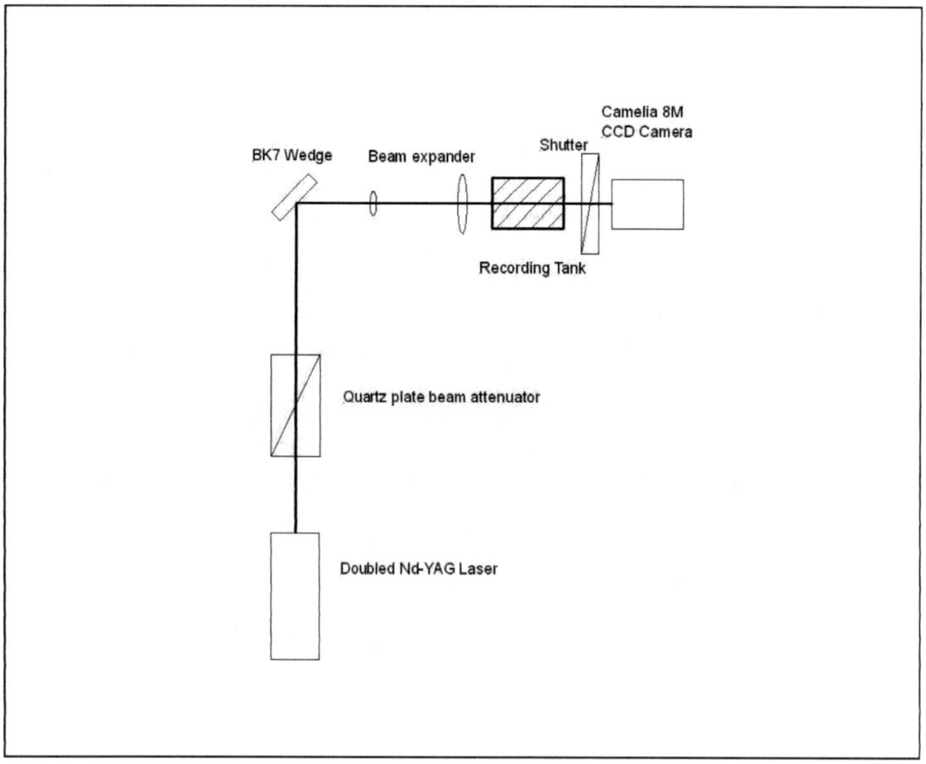

Figure 2. Diagram illustrating the optical set up for pulsed laser in-line digital holography set up.

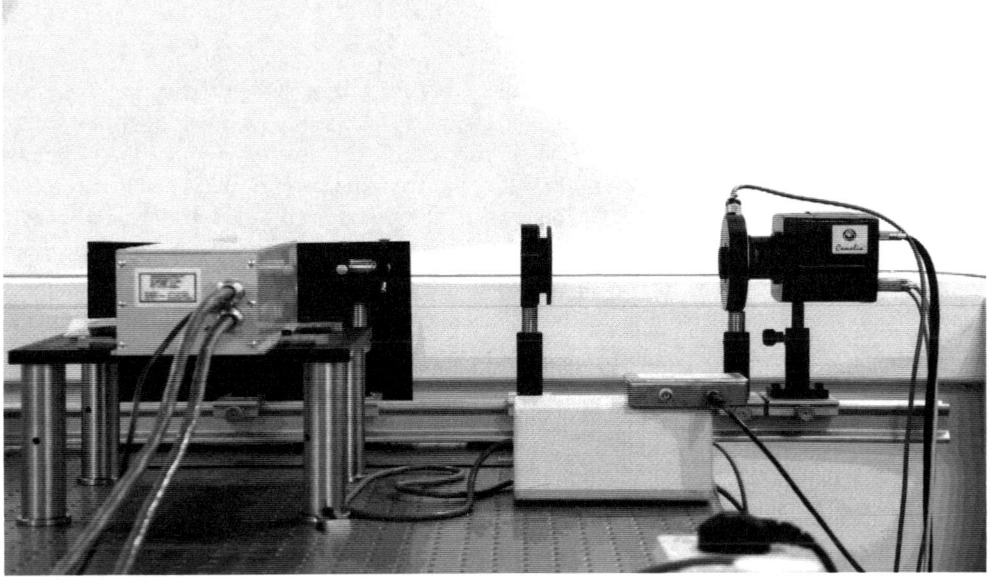

Figure 3. The configuration of the optical equipment. From left to right: the pulsed laser, the beam expander, the shutter and the camera. The variable attenuator has been removed in this picture for clarity.

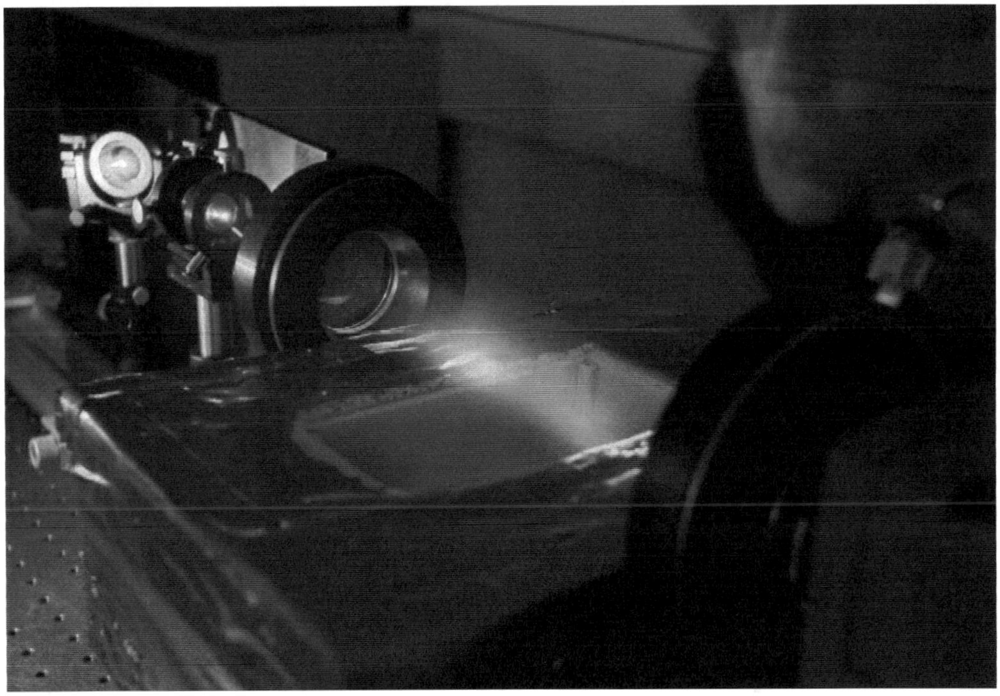

Figure 4. Recording a digital hologram of breath over liquid nitrogen.

2.1.3. Using a pulsed laser to record the dynamics of breath. Current investigations centre on the breathing of small particles such as seeds into a chamber. The subsequent holograms capture their trajectory within the vessel. A range of air-borne seeds were tested due to their aerodynamic properties. Initially poppy seeds were used. However the 1mm diameter of these seeds was too large to allow a significant cluster to be recorded. The investigators then progressed to digitalis seeds which, with a diameter of 0.4mm, allowed for a greater cloud of seeds to be recorded in the hologram. Breathing gestures such as the sigh were performed into the chamber and recorded as holograms (figures 5-7).

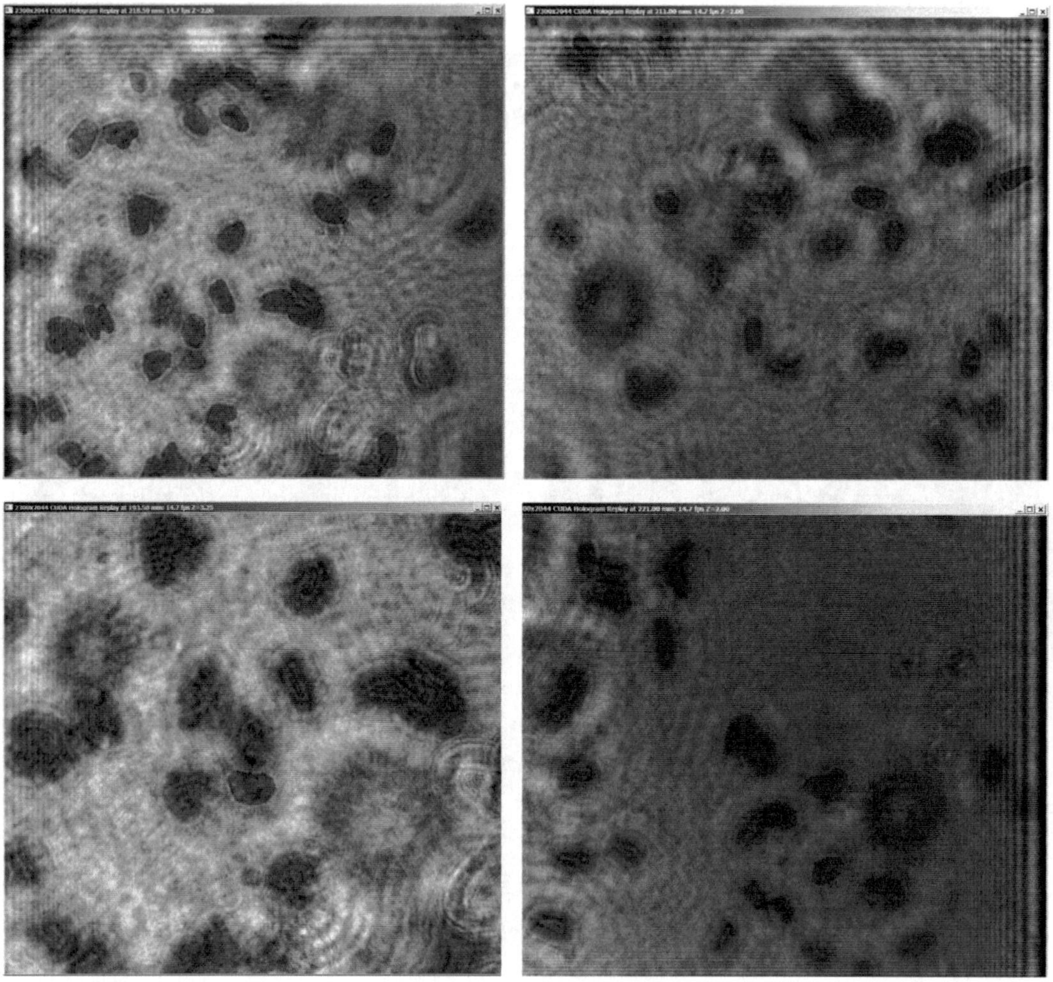

Figure 3. Series of images extracted from hologram replay of breathing gestures.

3. Hologram Replay.

The captured data is then processed using specially developed code, HoloMovie, running on NVIDIA GPU (Reid, Nebrensky & Hobson 2012[2]). Our software allows for an interactive involvement with the audience who can navigate through the depth of the replayed hologram and therefore explore the dynamic forms created by the breath. Figure 6 shows some of the dynamics available to be explored by the viewer from navigating the hologram.

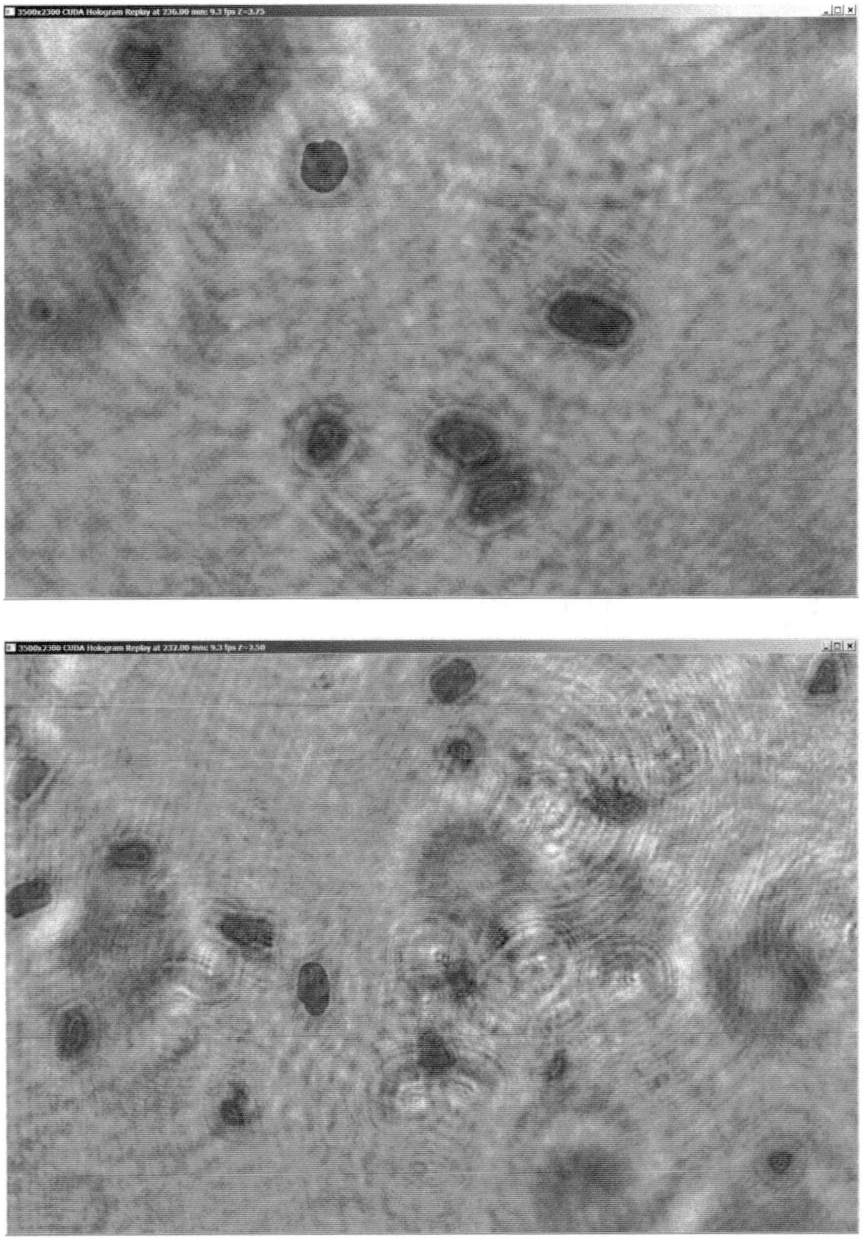

Figure 4. Images obtained from navigating through the holographic field.

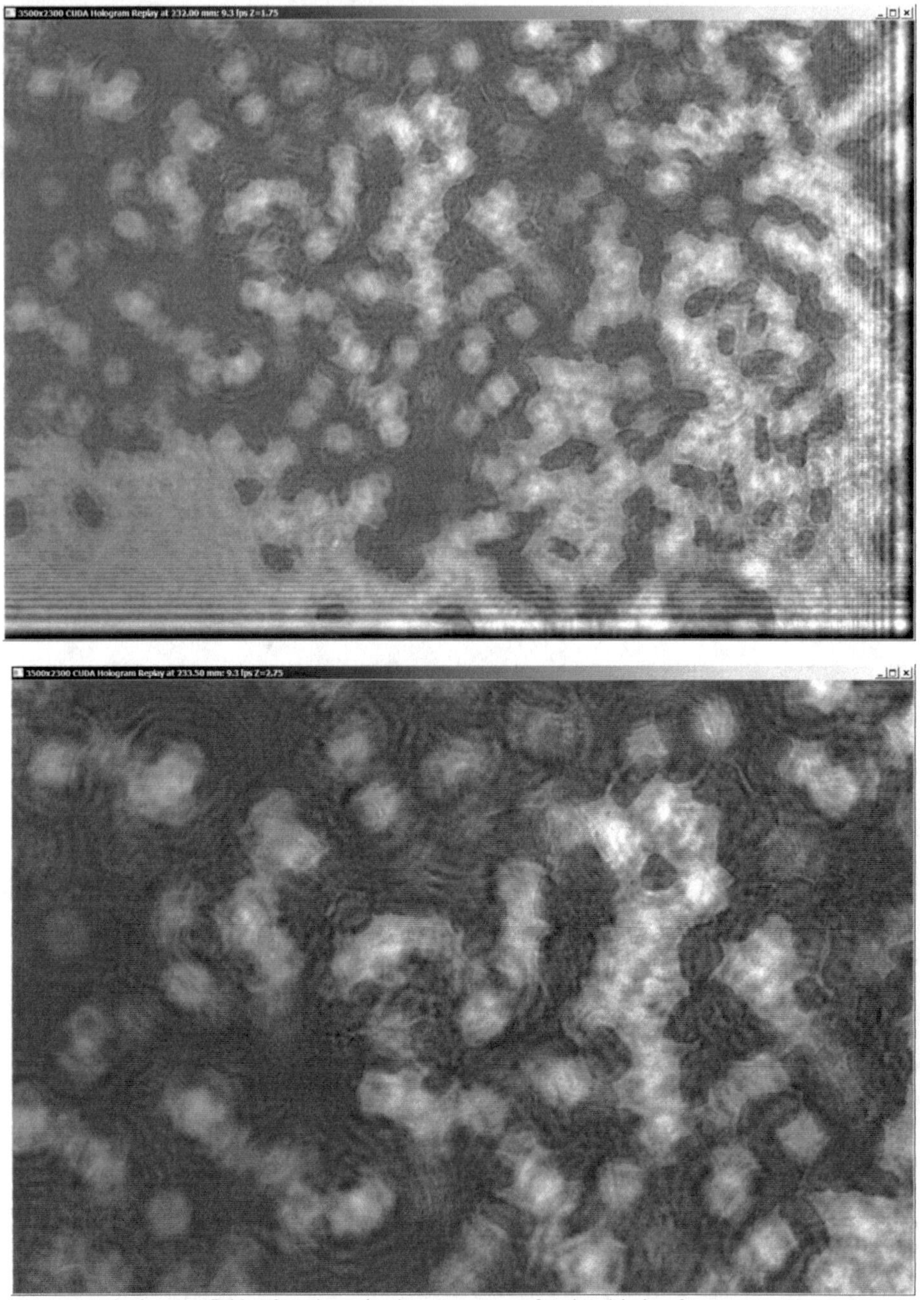

Figure 7. The intense fields of seeds evoke the movement of their initiating force.

4. Conclusion

This paper presents the first stages of investigating the use of in-line digital holography for the creation of imagery to facilitate the visualization of breath. The work is the starting point for further projects using data derived from the data capture, but the images themselves are also considered of interest. The use of seeds to render the air movement within the chamber gave an interesting view of the dynamics at play. Digitalis seeds allowed a denser cluster to be recorded on the hologram than poppy seeds and therefore allowed a more meaningful analysis of the breathing dynamic recorded. Scientifically the resulting holograms of denser clusters of seed would be seen as sub optimal. However artistically the results are of interest both aesthetically and because of information provided on the force of the expiration. The results also provide the starting point for a range of further investigations.

The collaborators will continue to explore ways of extending this initial investigation including the expiration of differing density fluids into fluid filled tanks and the creation of a schlieren field captured holograpically. At the current stage of development this work has demonstrated the ability of in-line holography to record valuable information about the movement of the breath. It is interesting to note that for the purposes of application to art we are recording interesting images with a much higher total object area to beam area ratio than would typically be regarded as usable for most scientific applications [4]. However further investigations are needed to fine tune techniques to enable comparison of dynamic breathing forms

Acknowledgements
Jayne Wilton's work on this project is made possible by a grant from The Leverhulme Trust.

References

[1] Hobson P R and Hong, J, (1988) Spatial filtering in multiplex holography. *J. Phys. D: Appl. Phys*. 21, S106
[2] Reid I D, Nebrensky J and Hobson P R (2012) Challenges in using GPUs for the reconstruction of digital hologram images *Advanced Computing and Analysis Techniques in Physics Research* September 5-9, 2011, Uxbridge, London to appear in *Journal of Physics, Conference Series*.
[3] Kreis T M, Adams M and Juptner W P O 1997 *Proc. of SPIE* **3098** 224–33
[4] Edwards P J, Hobson P R and Rodgers G J, 2003 *Journal of Modern Optics* **50** 1849-1856

A New Approach for Speckle Reduction in Holographic 3D printer

Takeru Utsugi[1] and Masahiro Yamaguchi[2]

[1]Department of Information processing, Tokyo Institute of Technology, 4259 S1-17 Nagatsuta, Midori-ku, Yokohama, Kanagawa, 226-8503 Japan
[2]Global Scientific Information and Computing Center, Tokyo Institute of Technology, 2-12-1-I7-6 O-okayama, Meguro-ku, Tokyo 152-8550 Japan

E-mail: utsugi.t.aa@m.titech.ac.jp

Abstract. A Holographic 3D printer produces a high quality 3D image reproduced by a full-color, full-parallax holographic stereogram with high-density light-ray recording. But speckle-pattern noise localized behind the reconstructed image is causing a loss of the display quality. This noise is originated from the speckle generated by a diffuser for equalizing the intensity distribution of the object light on the recording medium. We analyze some conventional ways for speckle reduction using a band-limited diffuser, and it is found that these ways cannot reduce the noise sufficiently. Then we propose two methods, one introduces a moving diffuser and the other introduces multiple exposures and a digital diffuser called as 4L-PRPS.

1. Introduction

A holographic 3D printer produces a holographic stereogram (HS) using the 3D data of a subject from a computer and outputs a hardcopy like a conventional image printer [1-3]. The image obtained by this system has full-color, full-parallax and high-density light-ray information.

In this system, by use of the optical system shown in figure 1, an array of small elementary reflection-type holograms is recorded in a thick recording material. The recording medium is translated horizontally and vertically after each exposure so that the whole surface of the hologram is exposed by an array of small elementary holograms. An aperture is placed to avoid leaking object light to the next elementary hologram. The image displayed on a spatial light modulator (SLM), in this case a liquid-crystal display (LCD) panel, is a geometrical projection of the 3D object calculated by a computer-graphic technique. In the reconstruction stage, since each elementary hologram is a volume reflection grating, full-parallax 3D image can be displayed even under ordinary white light illumination.

Since recording medium is placed on the Fourier plane, the intensity variation on the hologram plane depends on the spatial frequency component of the image displayed on the SLM. This image usually has a small high-frequency component. The phase distribution is almost dependent on the amplitude pattern since the phase is also modulated along with the amplitude by the LCD panel. Therefore the low-frequency component is dominant even in the phase pattern, and the object beam has a large intensity variation on the hologram plane and strong DC peak. As a result, two problems occur, (1) the recording of a hologram with high efficiency is difficult because dynamic range of recording medium is limited, and (2) reconstructed image gets dark because the elementary hologram

Published under licence by IOP Publishing Ltd

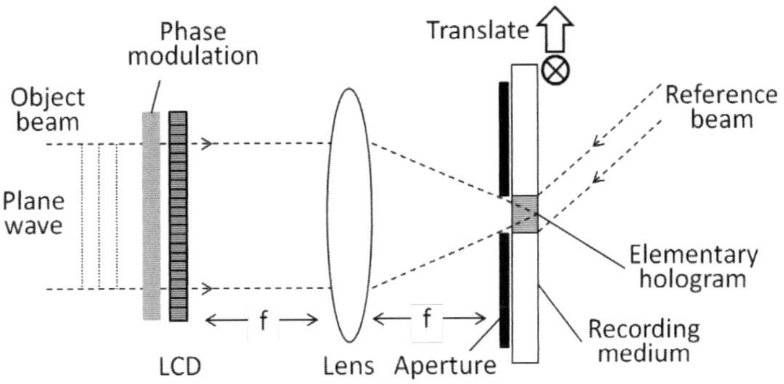

Figure 1. Optical setup for an elementary hologram recording.

has low aperture ratio.

To solve these problems, a lot of ways have been proposed. To smooth the Fourier spectrum distribution, one may simply move the recording medium slightly away from the focal plane so that the average intensity become almost uniform by defocus [4]. Then, however, the problem of low aperture ratio remains.

Another method is to use a phase modulation. Using a typical diffuser to equalize the Fourier spectrum within the aperture courses a loss of the object light and the speckle noise that appear behind the reconstructed image. Therefore various phase masks called band-limited diffuser were proposed [5-14]. However, these methods with single exposure cannot reduce the speckle noise sufficiently. This is described in the next section.

In this paper, we propose two new approaches to solve the problems described above. The first is "Moving diffuser method". This method was tried in the HS with horizontal parallax only (HPO) [20], but details of experimental result have not been reported. We first tried this method to Holographic 3D printer that generates full-parallax HS. In the result, we found through experiments that this method can reduce the speckle-pattern noise localized behind the reconstructed image, while alternative high-frequency noise appeared on the hologram plane. This is described in section 3.1.

To solve the problem of first method, we propose the second one "Multiple exposures method". In this approach, by introducing a multiple exposures and one of digital-diffusers, effective speckle reduction is possible. This will be described in section 3.2.

2. Conventional methods

2.1. Using typical diffuser

If HS is recorded with a typical diffuser, a speckle pattern affects LCD image and is recorded in the elementary hologram. When using the static diffuser, since a same speckle pattern is recorded in all elementary holograms, this pattern becomes localized at infinity of the reconstructed image. An average size of the speckle spot depends on the aperture size, and the smaller the aperture size is made for high-density recording, this pattern becomes more apparent because speckle size becomes large.

Also high spatial coherence of an illuminated light reduces the blur of both the reconstructed image and the speckle-pattern noise. This is critical problem for high quality HS recording.

2.2. Limitation of band-limited diffusers

Band-limited diffusers optimized to Fourier transform holograms has been analyzed for holographic data storage applications and also have been applied to HS recording [15-17]. However, the problem of this method has not been described in detail. In this paper, we describe the characteristics of one of the band-limited diffusers called digital diffuser, which is designed by digital process that is for example pseudorandom phase sequences (PRPS) [11] or iterative Fourier transform (IFTA) [14]. We evaluate smoothness inside the aperture on Fourier plane and a speckle contrast [22] on LCD plane by a numerical simulation. The smoothness and the speckle contrast are written as:

$$Smoothness = \frac{\langle I_F \rangle}{I_{Fmax}}, \quad Speckle\ Contrast = \frac{\sigma_{LCD}}{\langle I_{LCD} \rangle}$$

Where I_F is the intensity of Fourier plane, I_{LCD} is the intensity of LCD plane and σ_{LCD} is the standard deviation of I_{LCD}. The results of 50 times iterations in IFTA are shown in figure 2 with the case of 6L-CPRPS [11], which is one of the PRPS. The result shows that there is a trade-off between the smoothness and the speckle contrast. We can understand that this is a limitation of single exposure with digital diffuser.

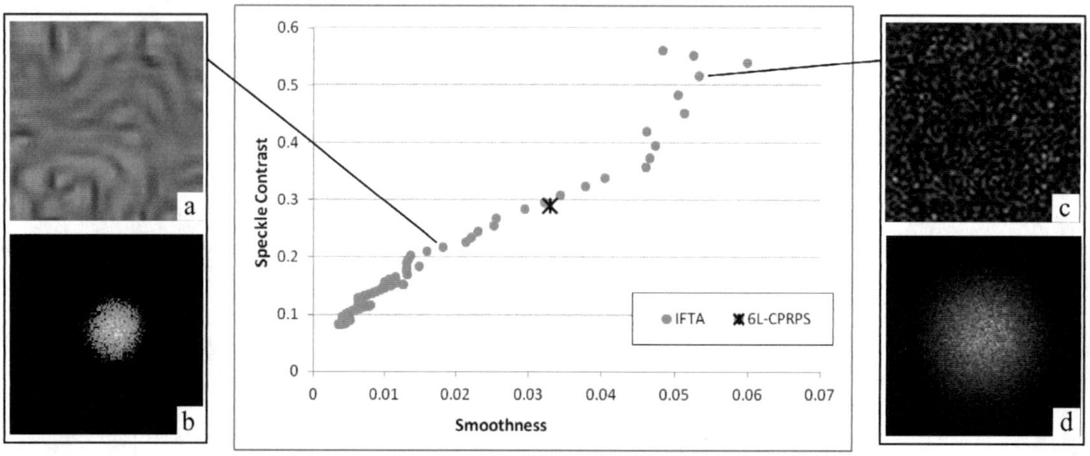

Figure 2. Speckle contrast vs. smoothness of digital diffusers designed by IFTA and 6L-PRPS in numerical simulations. One of the intensity distributions of (a)(c) LCD plane and (b)(d) Fourier plane.

3. New Approach for speckle reduction

3.1. Approach 1. Moving diffuser

Moving the diffuser after each exposure, different speckle patterns are recorded in each elementary hologram. These patterns are superposed at observer's retina in reconstructed step, and then the speckle contrast at infinity seems to be reduced. We evaluate this effect quantitatively through experiments. Figure 3 shows the reconstructed images, and Figure 4 shows the relation between speckle contrast at infinity and the number of phase pattern M, which corresponds to the number of illuminated positions on the diffuser.

If the moving distance is greater than the average size of the speckle on LCD plane, uncorrelated speckle patterns are superposed in observer's retina since HS is reconstructed with incoherent light. Therefore the speckle contrast at infinity reduces in proportion to $1/\sqrt{M}$ theoretically. The number of rays toward the same direction entering the observer's pupil determines limitation of speckle reduction by this approach. When the diameter of the pupil is 3.5mm and the pitch of elementary hologram is 0.2mm, about 250 rays superposed on the retina, so that the speckle contrast reduces in proportion to $1/\sqrt{250} \approx 0.063$.

On the other hand, it was found that high-frequency noise appeared on the surface of the reconstructed HS. This noise is shown in (c) of figure 3. Also the relation between the number of phase pattern M and the energy of this noise are shown in figure 5. When recording by a static diffuser, a very low frequency noise appears at the HS surface, and this noise does not bother the reconstructed image while the infinity noise is more noticeable. When using moving diffuser, since each elementary

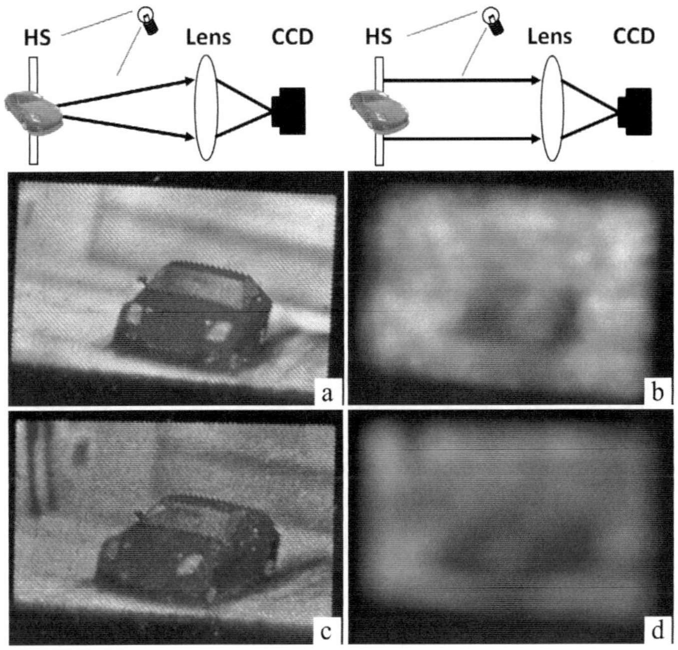

Figure 3. Reconstructed image with typical diffuser (M=1) focus on (a) HS surface and (b) infinity, and with moving diffuser (M=250) focus on (c) HS surface and (d) infinity.

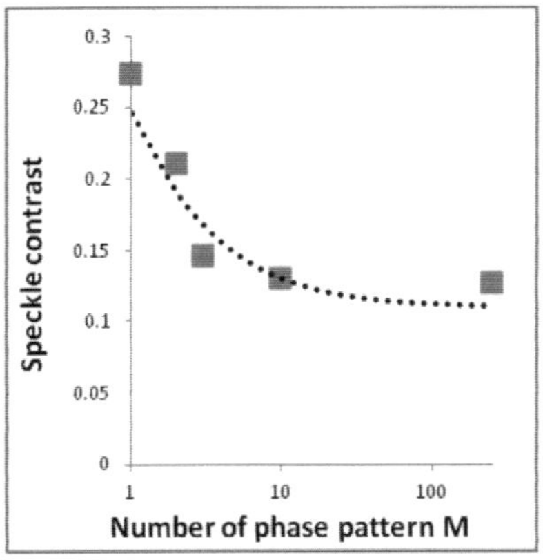

Figure 4. Speckle contrast at infinity with moving diffuser. Dot-line is the theoretical value.

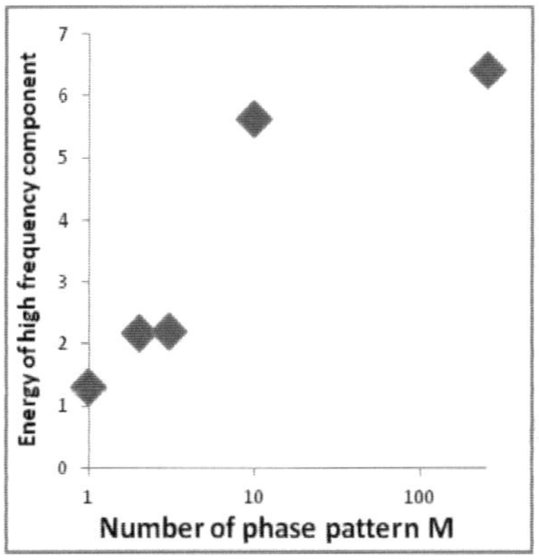

Figure 5. High frequency noise at HS surface with moving diffuser.

hologram reconstructs speckle pattern that is different from each other, observer perceives various brightness at the elementary holograms when he or she focuses on the HS surface. This noise is perceived when the value M become large to some extent, and the maximum value of the frequency depends on the pitch of elementary hologram. In addition, when the point of view moves, the high frequency noise is perceived as a flicker. This is the problem of moving diffuser.

3.2. Approach 2. Multiple exposures with 4L-PRPS

3.2.1 Principle. "Moving diffuser method" is effective to reduce the speckle-pattern noise behind the reconstructed image, but has the problem of the high frequency noise at the HS surface. We propose a new method to solve this problem. In photo-polymers for holographic data storage applications, multiple exposures are employed. Therefore, we consider applying the multiple exposures to reduce

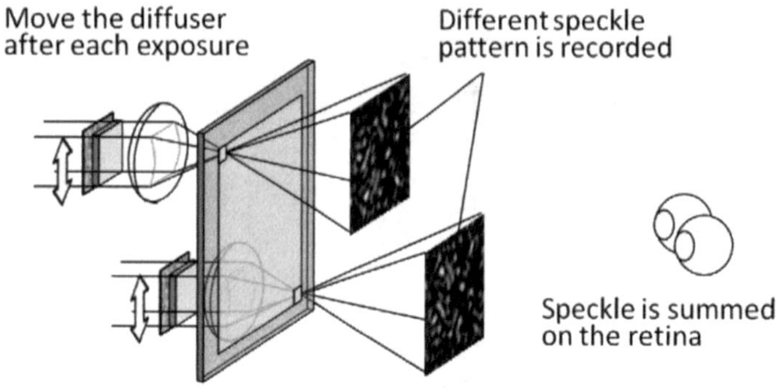

Figure 6. Different speckle patterns are reconstructed and superposed in retina by recording with moving diffuser.

the speckle noise in HS. For example, an elementary hologram is recorded by multiple exposures to several times (M times) while the diffuser is shifted after each exposure. Then, if each speckle pattern recorded in the single elementary hologram is uncorrelated, the speckle noise reconstructed from the single elementary hologram reduces in proportion to $1/\sqrt{M}$, so that speckle noise on retina also reduced, and then high frequency noise at the HS surface does not appear in case of the multiple exposures using same sets of positions on the diffuser. In this way, the speckle noise behind the reconstructed image can be reduced without increasing high-frequency noise on the hologram plane.

However, the number of multiple exposures cannot be increase so much because it takes a large recording time to shift the diffuser and the diffraction efficiency is affected by multiple exposures. Therefore the factor of $1/\sqrt{\mathrm{M}}$ is inefficient. If it is possible that each speckle pattern recorded by multiple exposures have a negative correlation, we can reduce the speckle contrast effectively smaller than $1/\sqrt{\mathrm{M}}$ [19].

According to this concept, we propose a new method. Let us consider the phase modulation applied to the digital diffuser designed by 4Level-PRPS (4L-PRPS). It has the random phase sequences in which the phase step at each boundary is either $+\pi/2$ or $-\pi/2$. When the size of the elementary hologram is $2\lambda f/p$, where λ is the wavelength, f is the focal length of Fourier lens and p is the pitch of the phase cell in digital diffuser, the reconstructed image recorded by 4L-PRPS becomes dot-like pattern. Exposing an elementary hologram four times with shifting the diffuser half-length of the phase cell pitch, dot-like pattern reconstructed from the elementary hologram is superposed on observer's retina. Then, since dot-like pattern and the shifted pattern have negative correlation, the speckle contrast is effectively reduced as shown in figure 7.

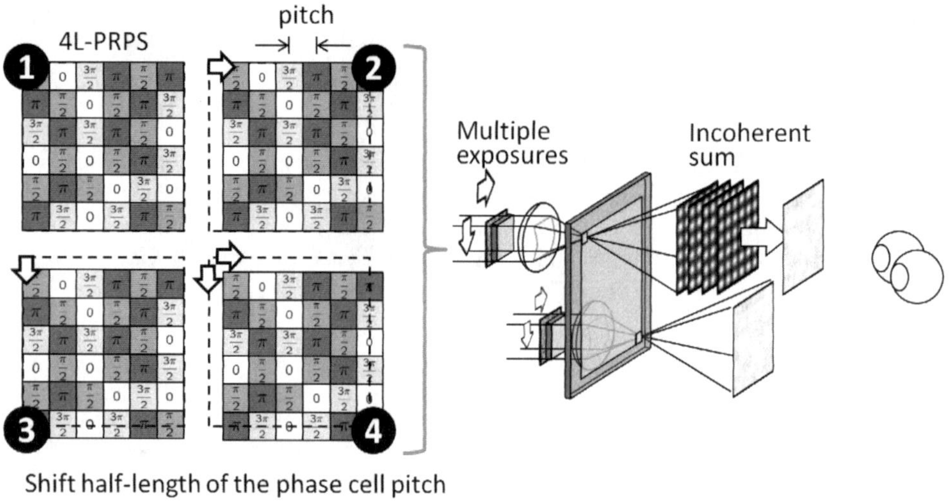

Figure 7. Speckle reduction by multiple exposures with 4L-PRPS.

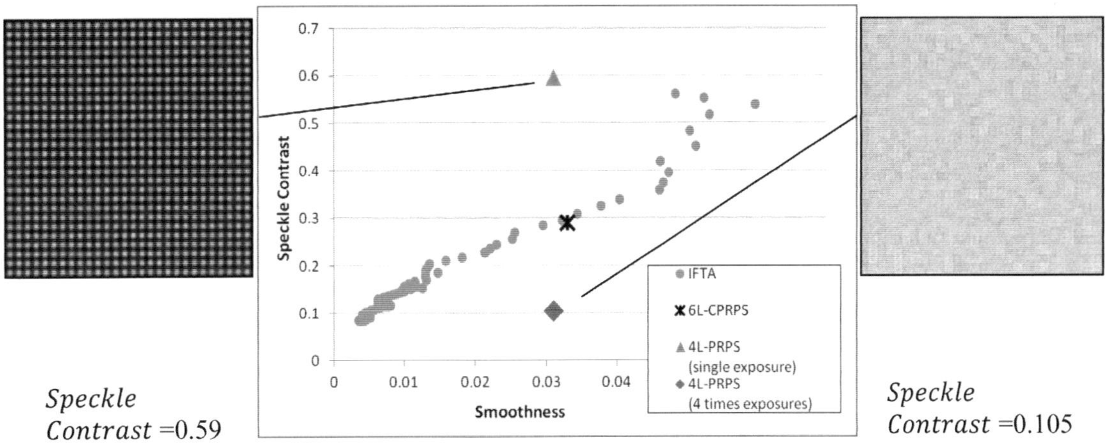

Figure 8. Simulation results of proposed method.

3.2.2 Numerical simulations. We show the effect of proposed method by numerical simulation. Recording of one elementary hologram is divided into four exposures with shifting the digital diffuser after each exposure. Shifting directions are vertical, horizontal and both of them, and the each shift amount is half-length of the phase cell pitch. The result is shown in figure 8. Both the smoothness on the Fourier plane and the speckle contrast of reconstructed image in proposed method show good results beyond the trade-off line of single exposure with optimized digital diffuser. The speckle contrast is reduced almost 0.1 even as $1/\sqrt{4} \approx 0.5$ in case of M=4, avoiding the high frequency noise at the HS surface. This means that it is effective approximately five times compared to the approach with a typical moving diffuser and multiple exposures.

4. Discussions

In the second proposed method, there are some issues for the optical implementation. The digital diffuser, that is a diffractive optical element (DOE), is implemented by an optical element such as fabricated glass or a phase only SLM. If the former case, it is necessary to move the DOE mechanically. Then, it requires longer moving time and introduces vibration to the recording system. When producing the full-color HS, three DOE and complicated optical system are also needed. In contrast, the latter case, there are not problems described above. But, in case of the phase only SLM such as PAL-SLM [23], a blur of phase modulation is concerned.

The evaluations of human speckle perception are studied for laser projection systems [24]. In future, considering a human perception is also necessary for the speckle noise evaluations of a holographic display in addition to the speckle contrast.

5. Conclusions

We propose two methods for the speckle reduction in holographic 3D printer, which produces full-color full-parallax HS with high-density light ray information. Conventional methods have the limits of speckle reduction such as trade-off between smoothness of Fourier plane and the speckle contrast of reconstructed ray distributions. Then, we evaluated "Moving diffuser method" and found that this method is effective to reduce the speckle-pattern noise behind the reconstructed image, but the problem of the high frequency noise at the HS surface appears. "Multiple exposures method" is also proposed to solve this problem. This approach is shown to be effective by numerical simulation. We will show the experimental result and high quality HS in future works.

References

[1] Yamaguchi M, Ohyama N and Honda T 1992 *Appl. Opt.* **31** 217-22
[2] Klug M A, Halle M W, Lucente M and Plesniak W J 1993 Proc. of SPIE Vol. 1914
[3] Maruyama S, Ono Y and Yamaguchi M 2008 Proc. of SPIE Vol. 6912
[4] Lin L H and Beauchamp H L 1970 *Appl. Opt.* **9** 2088-92
[5] Burckhardt C B 1970 *Appl. Opt.* **9** 695-700
[6] Takeda Y, Oshida Y and Miyamura Y 1972 *Appl. Opt.* **11** 818-22
[7] Kurtz C N, Hoadley H O and DePalma J J 1973 *JOSA* **63** 1080-92
[8] Dallas W J 1973 *Appl. Opt.* **12** 1179-87
[9] Yonezawa S 1976 *Opt. Commun.***19** 370-3
[10] Torii Y 1978 *Opt. Commun.* **24** 175-80
[11] Nakayama Y and Kato M 1979 *JOSA* **69** 1367-72
[12] Iwamoto A 1980 *Appl. Opt.* **19** 215-21
[13] Wrowski F and Bryngdahl O 1988 *JOSA A* **5** 1058-65
[14] Bräuer R, Wyrowski F and Bryngdahl O 1991 *JOSA A* **8** 572-8
[15] Yang J, Bernardo L M and You-Seok Bae 1999 *Appl. Opt* .**38** 5641-5
[16] Yamaguchi M, Endoh H, Honda T and Ohyama N 1994 *Optics Letters* **19** 135-7
[17] Hilaire P S 1995 *Optical Engineering* **34** 83-89
[18] Helseth L E and Singstad I 2001 *Opt. Commun.* **193** 81-86
[19] Hsu W F and Yeh C F 2011 *Appl. Opt.* **50** 50-55
[20] Kihara N *et al.* Japan Patent Kokai H10−20747(H10.1.23)
[21] Klug M A *et al.* USPatent 7,847,992 B2(Dec.7, 2010)
[22] Goodman J W 2007 *Speckle Phenomena in Optics: Theory and Applications* (Roberts and Company, Englewood)
[23] Igasaki Y, Li F, Yoshida N, Toyoda H, Inoue T, Mukohzaka N, Kobayashi Y and Hara T 1999 *optical review* **6** No.4 339-344
[24] Roelandt S, Meuret Y, Craggs G, Verschaffelt G, Janssens P and Thienpont H 2012 *Optics Express* **20** 8770-83

Optical matching system for eliminating lateral position displacement of holographic images under varying laser wavelength

Shou Liu, Xiangsu Zhang[1] and Xuechang Ren

Department of Physics, Xiamen University, Xiamen 361005, China

E-mail: xszhang@xmu.edu.cn

Abstract. The diffraction angle of a hologram is proportional to the wavelength so that a lateral displacement of the holographic image will occur when there is a change in wavelength. Such displacement is more serious when the image is generated at far distance. A matching system including an HOE and a hologram is proposed to eliminate the image lateral displacement under varying wavelength. Theoretical analysis demonstrates that the two elements must have the conditions: (a) They must have the same fringe periods; (b) They must be placed in parallel; (c) The diffracted beam from HOE must be collimated. The two elements were fabricated in the experiment and applied into a holographic sight. A test was carried out on the sight through watching the image lateral movement at 50-meter distance. The experimental result shows that under varying wavelength there is no image lateral displacement observed, verifying the effectiveness of the matching system.

1. Introduction

According to theory, the diffraction angle of a diffractive element is determined by the incident angle and wavelength of the light illuminating the element. Holograms are diffractive elements, so the angular positions of the holographic images are affected by laser wavelength. If the image distance is very long, slight change of diffraction angle caused by wavelength variation will generate large lateral displacement of the image.

A holographic sight uses a laser beam emitted from a laser diode (LD) to illuminate a hologram. The hologram generates a holographic virtual reticle image in the direction perpendicular to the hologram plane at long distance. Through the hologram the shooter can clearly see the reticle image on the target plane therefore can fire immediately, which greatly simplifies the aiming process thus increases the shooting speed. However, when there is a temperature drift on LD, the wavelength of the beam emitted by LD will change. This will cause lateral deviation of the reticle image from the designed position, thus largely reduces the aiming accuracy.

We have designed an optical matching system that can completely eliminate image lateral displacement under varying wavelength. The system consists of a transmission hologram and a transmission holographic optical element (HOE). The HOE is used to compensate the change of diffraction angle of the hologram caused by wavelength change. The matching system has been

[1] To whom any correspondence should be addressed.

applied to holographic sights. Experimental result shows that for the image reconstructed at 50-meter distance, there is no lateral displacement at all under varying laser wavelength.

2. Theory

2.1. Lateral displacement of the holographic image caused by the change of wavelength

Based on grating equation, the angular position of a holographic image reconstructed from a hologram can be expressed as

$$\sin\phi = \frac{\lambda}{d} - \sin\theta\,, \tag{1}$$

where ϕ is the image angle, d is a constant representing the average fringe period of the hologram, θ and λ are incident angle and wavelength of the laser beam on hologram respectively. It is clear from equation (1) that when λ changes, if the light incident angle θ keeps the same, image angle ϕ is only affected by λ. If the image is far from the hologram, small change of ϕ may cause large lateral displacement as shown in Fig. 1. For the requirement of holographic sights, holographic images must be fixed in the direction of the hologram normal. So equation (1) becomes

$$0 = \frac{\lambda}{d} - \sin\theta\,. \tag{2}$$

If the wavelength changes from λ to λ' while the incident beam angle is maintained, the angular position of the image has the expression

$$\sin\phi' = \frac{\lambda'}{d} - \sin\theta\,. \tag{3}$$

Let (3) – (2), we have

$$\sin\phi' = \frac{\lambda' - \lambda}{d}\,. \tag{4}$$

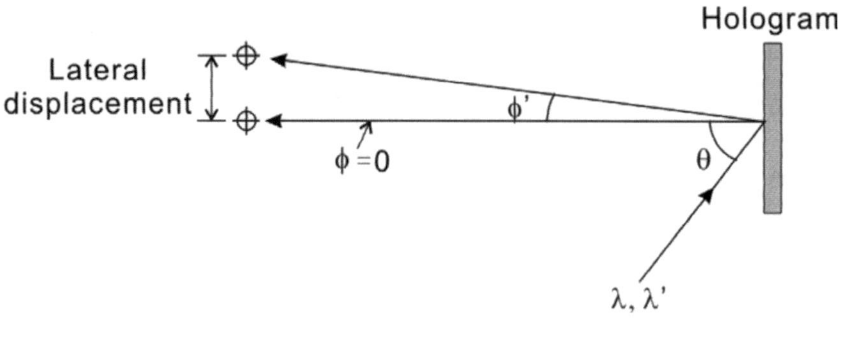

Fig 1. Lateral displacement of the image caused by wavelength change

The wavelength of LDs used in holographic sights is 535 nm. For making the holographic sight compact, the illuminating beam on hologram usually has large incidence. Suppose the beam incident angle θ is 60°, to make $\phi = 0$ under 535 nm illumination, according to equation (2) the period d of the hologram must be 618 nm. Usually one degree temperature change on LD will cause about 0.3 nm wavelength change. Suppose temperature has increased 10 degrees, the wavelength of LD will become 538 nm. Inserting λ = 535 nm, λ' = 538 nm and d = 618 nm into equation (4) we have $\sin\phi'$ = 4.9× 10^{-3}. If the image distance is 50 m, the lateral displacement of the image will be 24 cm. Holographic sights are usually used for shooting the targets in the distance around tens to one hundred meters, so the lateral displacement will largely affect the aiming accuracy.

2.2. Matching system

From equation (2) we can see that to keep $\phi = 0$ while λ is varying, the only way is to change θ accordingly so that the impact on ϕ by varying λ can be canceled out. For the convenience of analyzing, equation (2) is rewritten as

$$d \sin\theta = \lambda \,. \tag{5}$$

To make angle θ change accordingly with varying wavelength, an HOE is introduced as the compensation element to form a matching system with the hologram. The HOE is placed before the hologram in the light path, hence the diffracted beam from HOE serves as the illuminating beam to the hologram. The HOE must be able to generate a $\Delta\theta$ while there is a $\Delta\lambda$ so that the following equation is satisfied

$$d \sin(\theta + \Delta\theta) = \lambda + \Delta\lambda \,. \tag{6}$$

The angular relation of HOE can be expressed as

$$d'(\sin\alpha + \sin\beta) = \lambda \,, \tag{7}$$

where α and β are angles of incident and diffracted beams of HOE respectively, d' is the average period of HOE. Let laser beam fall on HOE in normal direction, so α=0. Equation (7) becomes

$$d' \sin\beta = \lambda \,. \tag{8}$$

When there is a wavelength change $\Delta\lambda$, the diffraction angle of HOE will have a change $\Delta\beta$. Equation (8) can be rewritten as

$$d' \sin(\beta + \Delta\beta) = \lambda + \Delta\lambda \,. \tag{9}$$

Substituting (6) into (9) we have the condition for keeping the holographic image in the direction of hologram normal under varying wavelength

$$d' \sin(\beta + \Delta\beta) = d \sin(\theta + \Delta\theta) \,. \tag{10}$$

Comparing left and right sides in equation (10), we have

$$d' = d \tag{11}$$

and

$$\begin{gathered} \sin(\beta + \Delta\beta) = \sin(\theta + \Delta\theta) \\ \downarrow \\ \beta + \Delta\beta = \theta + \Delta\theta \end{gathered} \,. \tag{12}$$

Equation (11) demonstrates that the periods of HOE and hologram must be equal. This can be fulfilled through the fabrication processes of the two elements. Equation (12) demonstrates that the diffraction angle of HOE and the incident angle of hologram must always be equal. This condition can only be fulfilled if the two elements are positioned in parallel and the diffracted beam of HOE is collimated.

3. Experiment and result

The above analysis indicates that to achieve accurate aiming of holographic sights, the matching system including an HOE and the hologram can be used to maintain the holographic image in the direction of the hologram normal in spite of the wavelength change. In the matching system, the HOE and the hologram must have the following conditions: (a) They must have the same fringe periods; (b) They must be placed in parallel; (c) The diffracted beam from HOE must be collimated.

3.1. Fabrications of HOE and the hologram

In fabricating holographic elements, the periods of the elements are determined by the angles between the object and the reference beams. In order to obtain equal fringe period of the HOE and the hologram, in their recording geometries the angles between the two recording beams were set to be equal. Collimated beam was used as the reference beam in both recordings, and the recording material used in the experiment was silver-halide.

For generating collimated diffraction beam, HOE must be designed as a lens. For making the optical setup in the sight compact, in the experiment the HOE was fabricated as an off-axis transmission holographic lens[1]. Fig. 2 shows the geometry for recording HOE. A laser beam is splitted by a splitter BS into two beams: one is expanded by expander BE_1 then falls on the holographic recording plate H, with the central line of the beam along the plate normal, the other beam is expanded by expander BE_2 and collimated by lens L then falls on the plate H with large incident angle.

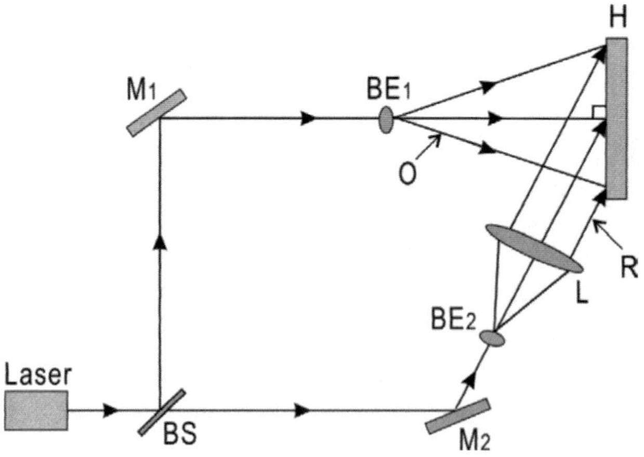

Fig. 2. Recording geometry of HOE
(BS – beam splitter, M – mirror, BE – beam expander, L – lens,
H – holographic plate, O – object beam, R – reference beam)

The optical geometry for fabricating hologram was different from that normally used for fabricating 2-dimensional (2D) transmission holograms. For a holographic sight, ideally the distance of the holographic image should be infinite so that clear reticle image can be seen on target plane at any distance. This requires that in recording process the 2D object should be placed at infinite distance. To realize this, a lens was used in the geometry to transform finite object distance into infinite. The lens is marked L_1 in Fig. 3, placed in between of the object and the recording plate. According to Gaussian lens formula[2], if an object is placed at the distance equal to the focal length of the imaging lens, the image distance will be infinite. So in the geometry, the distance between L_1 and the object is chosen to be the focal length of L_1. The light path of the reference beam was the same as that used in recording HOE. In image reconstruction, the original reference beam must be used to illuminate the hologram to obtain a virtual image[3].

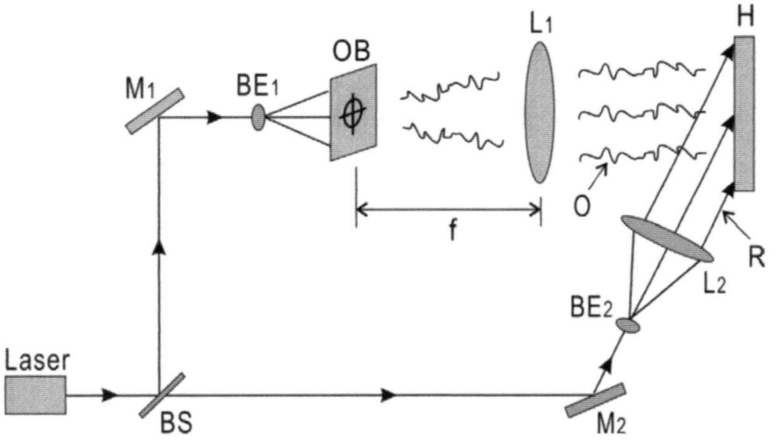

Fig. 3. Recording geometry of the hologram
(BS – beam splitter, M – mirror, BE – beam expander, OB – 2D object, L – lens, H – holographic plate, O – object beam, R – reference beam)

3.2. Optical setup with the matching system for holographic sights

Fig. 4 shows the optical setup with the matching system designed for holographic sights. The HOE and the hologram are both placed vertically but in different height, with the diffracted beam from HOE traveling upward to the hologram An LD with wavelength 535 nm is placed at the focal point of the HOE with the central line of the beam propagating along HOE normal. The diffracted beam of HOE is collimated with diffraction angle β and illuminates the hologram with incident angle θ. It is obvious that $\beta = \theta$. By placing the eye close to the hologram, a virtual reticle image is generated at far distance in hologram normal.

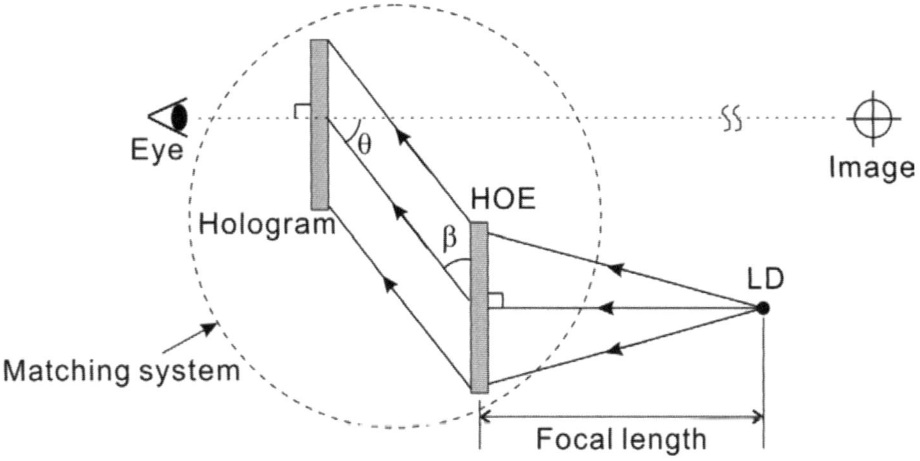

Fig. 4. Optical setup of the holographic sight with matching system

3.3. Experimental result

The optical setup with the matching system has been assembled into holographic sights. Fig. 5 is the cross-sectional view of the sight, illustrating the layout of optical components and the light path.

The test of the effect of the matching system was carried out by watching the lateral movement of the holographic image under varying wavelength. The sight was pointed at an object 50 m away through a window. After switching on the sight, through the hologram a virtual reticle image could be seen clearly on the object. The room temperature was increased to cause wavelength variation. By watching the image position carefully while the temperature was going up, no image lateral movement was observed.

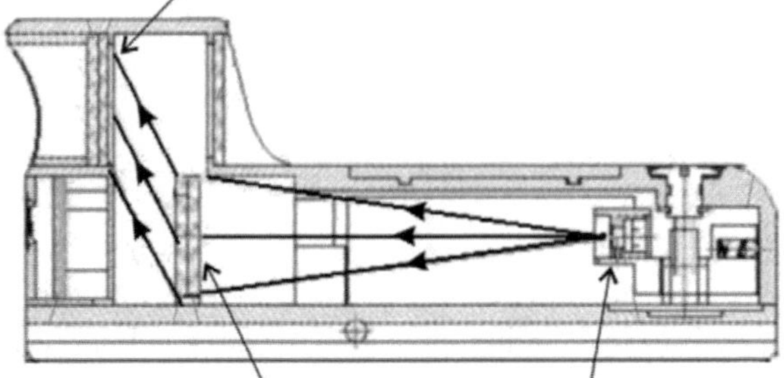

Fig. 5. Cross-sectional view of the holographic sight, illustrating the layout of the optical components and the light path

4. Conclusion

An optical matching system has been proposed to eliminate lateral displacement of the holographic images under varying wavelength. The system includes an HOE and a hologram. The two elements must have the following conditions: (a) They must have the same fringe periods; (b) They must be placed in parallel; (c) The diffracted beam from the HOE must be collimated. The matching system is especially useful for far distance aiming,. The experimental result obtained from a holographic sight verifies that the proposed matching system can effectively maintain the holographic image in the fixed direction despite the change of wavelength, so accurate aiming can be guaranteed.

Acknowledgments

The work was jointly supported by National Science and Technology Ministry under project TERS (No. 2011YQ03012406) and the Natural Science Foundation of Fujian Province (No. 2012J01285).

References

[1] Yu M W 1984 *Optical Holography and Information Processing* (Beijing: National Defence Publishing House)
[2] Hecht E and Zajac A 1974 *Optics* (London: Addison-Wesley Publishing Company) p 108
[3] Collier R J, Burckhardt C B and Lin L H 1971 *Optical Holography* (New York: Academic Press) chapter 1 pp 14–17

AUTHOR INDEX

AUTHOR INDEX

Institute of Physics
Dirac House, Temple Back
Bristol BS1 6BE UK

CURRAN ASSOCIATES INC
proceedings
.com